NOTATION

RELATIONS		
	$A \times B$	the cross, or Cartesian, product of sets A, B
	$\mathcal{R} \subseteq A \times B$	$\mathcal{R}$ is a relation from A to B
	$a \mathcal{R} b$; $(a, b) \in \mathcal{R}$	a is related to b
	$a \not\mathcal{R} b$; $(a, b) \notin \mathcal{R}$	a is not related to b
	$\mathcal{R}^c$	the converse of relation $\mathcal{R}$: $(a, b) \in \mathcal{R}$ iff $(b, a) \in \mathcal{R}^c$
	$\mathcal{R} \circ \mathcal{S}$	the composite relation for $\mathcal{R} \subseteq A \times B$, $\mathcal{S} \subseteq B \times C$: $(a, c) \in \mathcal{R} \circ \mathcal{S}$ if $(a, b) \in \mathcal{R}$, $(b, c) \in \mathcal{S}$ for some $b \in B$
	lub $\{a, b\}$	the least upper bound of a and b
	glb $\{a, b\}$	the greatest lower bound of a and b
	$[a]$	the equivalence class of element a (relative to an equivalence relation $\mathcal{R}$ on a set A): $\{x \in A \,|\, x \mathcal{R} a\}$

FUNCTIONS		
	$f: A \rightarrow B$	f is a function from A to B
	$f(A_1)$	for $f: A \rightarrow B$ and $A_1 \subseteq A$, $f(A_1)$ is the image of A_1 under f, i.e., $\{f(a) \,|\, a \in A_1\}$
	$f(A)$	for $f: A \rightarrow B$, $f(A)$ is the range of f.
	$f: A \times A \rightarrow A$	f is a binary operation on A
	$1_A: A \rightarrow A$	the identity function on A: $1_A(a) = a$ for each $a \in A$
	$f|_{A_1}$	the restriction of $f: A \rightarrow B$ to $A_1 \subseteq A$
	$g \circ f$	the composite function for $f: A \rightarrow B$, $g: B \rightarrow C$
	f^{-1}	the inverse of function f
	$f^{-1}(B_1)$	the inverse image of $B_1 \subseteq B$ for $f: A \rightarrow B$

THE ALGEBRA OF STRINGS		
	Σ	a finite set of symbols called an alphabet
	λ	the empty string
	$\|x\|$	the length of string x
	Σ^n	$\{x_1 x_2 \ldots x_n \,|\, x_i \in \Sigma\}$, $n \in \mathbf{Z}^+$
	Σ^0	$\{\lambda\}$
	Σ^+	$\bigcup_{n \in \mathbf{Z}^+} \Sigma^n$: the set of all strings of positive length
	Σ^*	$\bigcup_{n \geq 0} \Sigma^n$: the set of all finite strings
	$A \subseteq \Sigma^*$	A is a language
	AB	the concatenation of languages $A, B \subseteq \Sigma^*$: $\{ab \,|\, a \in A, b \in B\}$
	A^n	$\{a_1 a_2 \ldots a_n \,|\, a_i \in A \subseteq \Sigma^*\}$, $n \in \mathbf{Z}^+$
	A^0	$\{\lambda\}$
	A^+	$\bigcup_{n \in \mathbf{Z}^+} A^n$
	A^*	$\bigcup_{n \geq 0} A^n$: the Kleene closure of language A
	$M = (S, \mathcal{I}, \mathbb{O}, \nu, \omega)$	A finite state machine M with internal states S, input alphabet $\mathcal{I}$, output alphabet $\mathbb{O}$, next state function $\nu: S \times \mathcal{I} \rightarrow S$ and output function $\omega: S \times \mathcal{I} \rightarrow \mathbb{O}$

Discrete and Combinatorial Mathematics

Discrete
and
Combinatorial
Mathematics

An Applied Introduction

Ralph P. Grimaldi

ROSE-HULMAN INSTITUTE OF
TECHNOLOGY

▲▼ ADDISON-WESLEY
PUBLISHING COMPANY

READING, MASSACHUSETTS ▪ MENLO PARK, CALIFORNIA ▪ DON MILLS, ONTARIO ▪ WOKINGHAM, ENGLAND

AMSTERDAM ▪ SYDNEY ▪ SINGAPORE ▪ TOKYO ▪ MEXICO CITY ▪ BOGOTÁ ▪ SANTIAGO ▪ SAN JUAN

Jeffrey Pepper, Sponsoring Editor

Loretta Bailey, Art Editor
Sylvia Clark, Copy Editor
Ann Delacey, Manufacturing Supervisor
Illustrated Arts, Illustrator
Herbert Merritt, Production Editor
Martha H. Stearns, Managing Editor
Maria Szmauz, Text and Cover Designer
Cheryl Wurzbacher, Production Manager

Library of Congress Cataloging in Publication Data

Grimaldi, Ralph P.
    Discrete and combinatorial mathematics.

    Includes index.
    1. Mathematics — 1961— . 2. Electronic data
processing — Mathematics. 3. Combinatorial analysis.
I. Title.
QA39.2.G748   1985   510   84-9359
ISBN 0-201-12590-0

Reprinted with corrections, June 1985

DEDICATED TO
MY MOTHER AND FATHER

P R E F A C E

Resulting from the technological advances of the last two decades, changes in the undergraduate curriculum have included the introduction of courses in (1) combinatorics (or counting) and (2) discrete methods (that appeal to the *finite* nature of certain problems). One reason for this is the abundance of applications of these mathematical disciplines in computer science and operations research. Discrete methods and combinatorial reasoning underly the areas of data structures, as well as computational complexity and the analysis of algorithms. Consequently, many majors in computer science are also required to take courses in these disciplines. In addition to applications in computer science, one also finds applications in engineering and the physical and life sciences, and in statistics and the social sciences. Therefore the area of discrete mathematics provides valuable training for students in areas besides mathematics and computer science.

The objective of this book is to provide an introductory survey in both combinatorial and discrete mathematics. Since it is intended for the beginning student, there are a great number of examples with detailed explanations. (Each example is separately numbered and an open square is used to denote the end of each example.) In addition, wherever proofs are given, they too are presented with sufficient detail.

The text strives to accomplish the following goals:

1. To introduce the student at the sophomore-junior level, if not earlier, to the topics and techniques of combinatorial reasoning and discrete methods. Problems in counting, or enumeration, require a careful analysis of structure (e.g., whether or not order is important) and logical possibilities. There may even be a question of existence for some situations. Following such a careful analysis, we shall often find that the solution of a problem requires simple techniques for counting the possible outcomes that evolve from the breakdown of the given problem into smaller subproblems.

2. To introduce a wide variety of applications. In this regard, where structures from abstract algebra are required, only the basic theory needed for the application is developed. Furthermore, the solutions of some applications lend themselves to iterative procedures that lead to specific algorithms. The algorithmic approach to the solution of problems is fundamental in discrete mathematics. This approach reinforces the close ties between this discipline and the area of computer science.

3. To develop the mathematical maturity of the student through the study of an area that is so different from the calculus. Here, for example, there is the opportunity to establish results by counting a certain collection of objects in more than one way. This provides what are called combinatorial identities and introduces a novel proof technique. Proofs by mathematical induction are also used throughout the text, following their development in Chapter 6. Prior to Chapter 6 the ideas behind induction appear in certain recursive definitions and examples.

With regard to proofs in general, an attempt has been made to motivate theorems from observations on specific examples. In addition, whenever a finite situation provides a result that is not true for the infinite case, this situation is singled out for attention. Closed squares are used in the text to indicate the end of a proof. Proofs that are extremely long and/or rather special in nature are omitted. However, for the small number of proofs that are omitted, references are supplied for the reader interested in seeing the validation of these results. (The amount of emphasis placed on proofs will depend on the goals of the individual instructor and his or her student audience.)

4. To present an adequate survey of topics for the computer science student who will be taking more advanced courses in data structures and algorithm analysis. The coverage here on groups, rings, and fields will also provide an applied introduction for mathematics majors who wish to continue their study of abstract algebra.

The prerequisites needed for this book are primarily a sound background in high school algebra and an interest in attacking and solving a variety of problems. No particular programming ability is assumed. There are a few programs that appear in the text, but these are designed to reinforce particular examples. Such results may be skipped without any loss of continuity. With regard to calculus, we shall mention later in this preface its extent in Chapters 10 and 11.

My major motivation for writing this book is the encouragement I've received over the past seven years from my students and colleagues. This text reflects both my interests and those of my students, as well as the current recommendations of the Committee on the Undergraduate Program in Mathematics and the Association of Computing Machinery.

Since the areas of discrete and combinatorial mathematics are fairly new to the undergraduate curriculum there are many opinions as to which topics should be included in such courses. As each instructor and student will have different interests and needs, the coverage here is rather broad, as a survey course mandates. Yet there are many topics that some readers may feel should also be covered. Furthermore, there may be some differences of opinion with regard to the order in which the topics included here are presented. The order here rests upon my conviction that enumeration can reinforce the study of structure and vice versa.

Despite the interweaving of structure and enumeration, the chapters following Chapter 6 have been developed as independently as possible. The first six chapters form the underlying core of the text and provide enough material for a one-quarter or one-semester course. A second course that emphasizes combinatorics should include Chapters 7, 10, 11, and sections 1, 2, 3, 9, 10, 11 of Chapter 12. (In Chapter 10 some results from calculus are used; namely, the differentiation of algebraic functions, and partial fraction decompositions. For those who wish to skip this chapter, the first three sections of Chapter 11 can still be covered.) For a course in the role of discrete structures in computer science, the material in Chapters 9, 14, 15, 16, and sections 1–8 of Chapter 12 provides applications on switching functions and coding theory, and an introduction to graph theory and trees, and their role in optimization. Finally, a course in applied algebra can be developed after Chapter 6, and this should include coverage of Chapters 8, 9, 12, and 13. Other possible courses can be developed by consideration of the following dependency table.

Chapter	Dependence on Prior Chapters
1	No dependence
2	1
3	1, 2
4	1, 2, 3
5	1, 2, 3, 4
6	Minor dependence on 1, 2, 3
7	1, 2
8	3, 5, 6 (The Euler ϕ function is used here. This function is established in Section 7.1 but the result can be used in Chapter 8 without doing Chapter 7.)
9	2, 3, 5
10	1, 2
11	1, 2, 10
12	1, 2, 3, 5, 6
13	2, 3, 5, 6, 8
14	1, 2, 3, 6 (Although some graph theoretic ideas are mentioned in Chapters 4, 5, 7, and 9, this chapter is developed with no dependence on these earlier results.)
15	1, 2, 3, 6, 14
16	2, 14, 15.

In regard to the dependence of one section of a chapter on earlier sections, one should anticipate some dependence in the section exercises. Also, at the end of each chapter is a set of miscellaneous exercises where ideas from several chapters may be needed for the solutions. The overall role of the exercises is a key one. The exercises at the end of each section are designed to: (1) review the material in that section; (2) tie together ideas from earlier sections of the chapter; and (3) develop further concepts related to the material in the section. A few exercises call for computer programs to implement a given example or algorithm. These are designed for students with a minimal amount of programming experience. Answers are provided at the back of the text for almost all of the odd-numbered exercises.

In addition to the miscellaneous exercises, each chapter is concluded with a summary and historical review of the major ideas covered in that chapter. This should provide an overview of the development of the concepts in the chapter and provide information on further applications. A list of references for further reading also appears at the end of each chapter.

If space permitted, I would mention each of the students who took courses in discrete mathematics and combinatorics from me and suggested putting my class notes into a book. To those students who worked from the mimeographed version of this book I owe many thanks for finding mistakes and suggesting ways to improve the exposition. Most helpful in this category were Paul Griffith, Meredith Vannauker, Paul Barloon,

Byron Bishop, Lee Beckham, Brett Hunsaker, Tom Vanderlaan, Michael Bryan, Charles Wilson, and Richard Nichols. I thank Lawrence Alldredge and Martin Rivers for reviewing several chapters of the text, and Lawrence Alldredge, Barry Farbrother, Paul Hogan, Dennis Lewis, and Charles Kyker for their enlightening comments on some of the programs and applications mentioned in the text. I gratefully acknowledge the persistent enthusiasm and optimism of my editors, Wayne Yuhasz and Jeff Pepper, as well as Mary Crittenden, Herb Merritt, and Maria Szmauz, among other members of the Addison-Wesley staff who assisted in the fulfillment of this project. The overall reviewers — Robert Crawford of Western Kentucky University, Carl Eckberg of San Diego State University, and especially Douglas Shier of Clemson University — deserve a special note of thanks for their very thorough work. I am also indebted to my colleagues, John Kinney, Gary Sherman, and especially Alfred Schmidt, for their encouragement throughout the two years spent on writing this book. I believe they are somewhat responsible for a great deal of what is of value here. However, if there is one person to whom I owe the greatest note of thanks, it is definitely the ever-patient and encouraging Mary Lou McCullough who typed and retyped and . . . to bring out the best in the manuscript. Alas, any remaining errors, ambiguities, or misleading results rest upon my shoulders alone.

Terre Haute, Indiana   R.P.G.
December 1984

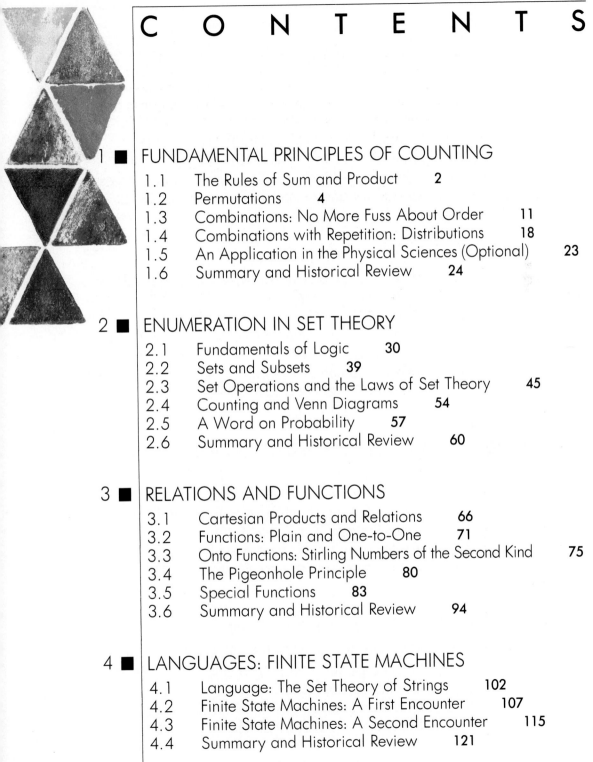

C O N T E N T S

Fundamental
Principles
of Counting

Enumeration, or counting, may strike one as an obvious process that a student learns when first studying arithmetic. But then, it seems, that very little attention is given to any further development in counting as the student turns to "more difficult" areas in mathematics, such as algebra, geometry, trigonometry, and calculus. Consequently, this first chapter should provide some warning as to the seriousness and difficulty of "mere" counting. Enumeration does not end with arithmetic; there is quite a bit more to it. And as we enter this fascinating field of mathematics, we shall come upon many problems that are very simple to state but somewhat "sticky" to solve.

Beware of formulas! Without an analysis of each problem, a mere knowledge of formulas will prove next to useless. Instead, welcome the challenge to solve problems different from routine problems or past experiences. Seek solutions based on your own scrutiny, regardless of whether it follows what the author provides. There are often several ways to solve a given problem.

1.1 ■ THE RULES OF SUM AND PRODUCT

Our study of discrete and combinatorial mathematics begins with two basic principles of counting: the rules of sum and product. These appear quite simple in statement and initial application. In analyzing more complicated problems, one often is able to decompose such problems into parts that can be handled using these basic principles. We want to develop the ability to decompose such problems and piece together the partial solutions to arrive at the final answer. This will be done by analyzing and solving many diverse enumeration problems, with recognition of the principles being used in the solutions. Our first principle of counting can be stated as follows.

The Rule of Sum: If a first task can be performed in m ways, while a second task can be performed in n ways, and the two tasks cannot be performed simultaneously, then performing either task can be accomplished in any of $m + n$ ways.

Before demonstrating its use, we state here an observation that will be true throughout the entire text: When we say that a particular occurrence, such as a first task, can come about in m ways, these m ways are assumed to be distinct, unless a statement is made to the contrary.

EXAMPLE 1.1 A college library has 40 textbooks on sociology and 50 textbooks dealing with anthropology. By the rule of sum a student at this college can select among $40 + 50 = 90$ textbooks in order to learn more about one of these two subjects. □

EXAMPLE 1.2 The rule can be extended beyond two tasks as long as no pair of them can occur simultaneously. For instance, a computer science instructor who has, say, five introductory books each on APL, BASIC, FORTRAN and Pascal can select any one of these 20 to give to a student interested in learning a first programming language. □

EXAMPLE 1.3

The computer science instructor of Example 1.2 has two colleagues. One of these colleagues has three textbooks on algorithm analysis, and the other has five such textbooks. If n denotes the number of books this instructor can borrow on this topic, then $5 \leq n \leq 8$, for here the colleagues may be in possession of copies of the same textbook. □

With the rule of sum taken care of we now consider the following example, which will introduce us to the rule of product.

EXAMPLE 1.4

In trying to reach a decision on plant expansion, an administrator assigns 12 of her employees to two committees. Committee A consists of five members and is to investigate possible favorable results from such an expansion. The other seven employees, committee B, will scrutinize possible unfavorable repercussions. Should the administrator decide to speak to just one committee member before making her decision, then by the rule of sum there are 12 employees she can call upon for input. However, in order to be a bit more unbiased, she decides to speak with a member of committee A on Monday and then a member of committee B on Tuesday, before reaching a decision. Using the following principle we find that she can speak with two such employees in $5 \times 7 = 35$ ways. □

> **The Rule of Product:** If a procedure can be broken down into first and second stages, and if there are m possible outcomes for the first stage and n for the second stage, then the total procedure can be carried out, in the designated order, in mn ways.

This rule is sometimes referred to as the *principle of choice*.

EXAMPLE 1.5

The drama club of Central University is having tryouts for a spring play. With six men and eight women auditioning for the leading male and female roles, by the rule of product the director can cast his leading couple in $6 \times 8 = 48$ ways. □

EXAMPLE 1.6

Here various extensions of the rule are illustrated by considering the manufacture of license plates consisting of two letters followed by four digits.

a) If no letter or digit can be repeated, there are $26 \times 25 \times 10 \times 9 \times 8 \times 7 = 3,276,000$ different possible plates.

b) With repetitions of letters and digits allowed, we find $26 \times 26 \times 10 \times 10 \times 10 \times 10 = 6,760,000$ license plates possible.

c) If repetitions are allowed, how many of the plates in part (b) have both letters vowels (a, e, i, o, u) and all digits even? (0 is an even integer.) □

EXAMPLE 1.7

At times it is necessary to combine several different counting principles in the solution of one problem. Here we find that both the rules of sum and product are needed to attain the answer.

In certain versions of the programming language BASIC, a variable name consists of a single letter (A, B, C, ...) or a single letter followed by a single digit. Since the computer does not distinguish between capitals and lowercase letters, a and A are considered the same variable name, as are E7 and e7. By the rule of product there are $26 \times 10 = 260$ variable names consisting of a letter followed by a digit; and since there are 26 variable names consisting of a single letter, by the rule of sum there are a total of $26 + 260 = 286$ variable names in this programming language.   □

1.2 ■ PERMUTATIONS

Using the rule of product, we turn now to counting arrangements of objects in a specified order or design. These arrangements are often called *permutations,* and we shall develop some systematic methods for dealing with them. We start with a typical example.

EXAMPLE 1.8 In a class of ten students, five are to be chosen and seated in a row for a picture. How many such linear arrangements are possible?

The key word here is *arrangement,* which designates the importance of order. If A, B, C, ..., I, J denote the ten students, then BCEFI, CEFIB, and ABCFG are three such different arrangements, even though the first two involve the same five students.

To answer the question, we consider the positions and possible numbers of students we can choose from in order to fill each position. The filling of a position is a stage of our procedure.

$$10 \quad \times \quad 9 \quad \times \quad 8 \quad \times \quad 7 \quad \times \quad 6$$

| 1st position | 2nd position | 3rd position | 4th position | 5th position |

Although any of the ten students can occupy the first position in the row, since repetitions are not possible here, we can select only one of the nine remaining students to fill the second position; continuing in this way, we find only six students to select from to fill the fifth and final position. This yields a total of 30,240 possible arrangements of five students selected from the class of ten.

We should realize that we get the same answer if we fill in the positions in the opposite order ($6 \times 7 \times 8 \times 9 \times 10$) or if we fill in the positions in the order 2nd, 1st, 4th, 5th, and 3rd ($9 \times 10 \times 7 \times 6 \times 8$).   □

Definition 1.1 ▶ For an integer $n \geq 0$, n *factorial,* denoted $n!$, is defined by

$$0! = 1,$$

$$n! = (n)(n-1)(n-2)\cdots(3)(2)(1), \quad \text{for} \quad n \geq 1.$$

One finds that $1! = 1$, $2! = 2$, $3! = 6$, $4! = 24$, $5! = 120$. In addition, for any $n \geq 0$, $(n+1)! = (n+1)(n)!$.

Utilizing the factorial notation, we find that the answer to Example 1.8 can be expressed in the following more compact form:

$$10 \times 9 \times 8 \times 7 \times 6 = 10 \times 9 \times 8 \times 7 \times 6 \times \frac{5 \times 4 \times 3 \times 2 \times 1}{5 \times 4 \times 3 \times 2 \times 1} = \frac{10!}{5!}$$

Definition 1.2

▶ Given a collection of n objects, any arrangement of these objects is called a *permutation* of the collection.

Starting with the letters a, b, c, we find that there are six ways to arrange all of the letters: abc, acb, bac, bca, cab, cba. If we are interested in arranging only two of the letters at a time, there are six such size-two permutations from the collection: ab, ba, ac, ca, bc, cb.

In general, if there are n objects, denoted $a_1, a_2, \ldots, a_n$, and r is an integer, with $1 \le r \le n$, then by the rule of product, the number of arrangements or permutations of size r for the n objects is

$$\underset{\substack{\text{1st} \\ \text{position}}}{n} \times \underset{\substack{\text{2nd} \\ \text{position}}}{(n-1)} \times \underset{\substack{\text{3rd} \\ \text{position}}}{(n-2)} \times \cdots \times \underset{\substack{r\text{th} \\ \text{position}}}{(n-r+1)} =$$

$$(n)(n-1)(n-2)\cdots(n-r+1)$$
$$\times \frac{(n-r)(n-r-1)\cdots(3)(2)(1)}{(n-r)(n-r-1)\cdots(3)(2)(1)} = \frac{n!}{(n-r)!}.$$

We denote this number by $P(n,r)$. For $r = 0$, $P(n,0) = 1 = n!/(n-0)!$, so $P(n,r) = n!/(n-r)!, 0 \le r \le n$. A special case of this result is Example 1.8, where $n = 10$, $r = 5$, and $P(10,5) = 30{,}240$. When permuting all of the n objects in the collection, we have $r = n$ and find that $P(n,n) = n!/0! = n!$.

The number of permutations of size r, where $0 \le r \le n$, from a collection of n objects, is $P(n,r) = n!/(n-r)!$. However, if repetitions are allowed, then by the rule of product there are n^r possible arrangements, with $r \ge 0$.

EXAMPLE 1.9

The number of permutations of the letters in the word COMPUTER is 8!. If only four of the letters are used, the number of permutations (of size four) is $P(8,4) = 8!/(8-4)! = 8!/4! = 1680$. If repetitions of letters are allowed, the number of possible 12-letter sequences is $8^{12} \doteq 6.872 \times 10^{10}$.   □

EXAMPLE 1.10

Unlike Example 1.9, the number of permutations of the letters in the word BALL is 12, not 4!, or 24. The reason for this is that we do not have four distinct letters to permute. To get the 12 arrangements, we can list them as shown in Table 1.1(a).

If the two L's are distinguished as L_1, L_2, then we can use our previous ideas on permutations of distinct objects; with the four distinct symbols B, A, L_1, L_2, we have $4! = 24$ permutations. These are listed in Table 1.1(b). Examining Table 1.1 we see

Table 1.1

A B L L	A B L₁ L₂	A B L₂ L₁
A L B L	A L₁ B L₂	A L₂ B L₁
A L L B	A L₁ L₂ B	A L₂ L₁ B
B A L L	B A L₁ L₂	B A L₂ L₁
B L A L	B L₁ A L₂	B L₂ A L₁
B L L A	B L₁ L₂ A	B L₂ L₁ A
L A B L	L₁ A B L₂	L₂ A B L₁
L A L B	L₁ A L₂ B	L₂ A L₁ B
L B A L	L₁ B A L₂	L₂ B A L₁
L B L A	L₁ B L₂ A	L₂ B L₁ A
L L A B	L₁ L₂ A B	L₂ L₁ A B
L L B A	L₁ L₂ B A	L₂ L₁ B A

(a)   (b)

that to each permutation for which the L's are indistinguishable there corresponds a *pair* of permutations with distinct L's. Consequently,

2 × (Number of permutations of the symbols B, A, L, L)

= (Number of permutations of the symbols B, A, L₁, L₂),

and the answer to the original problem of finding all permutations of the letters in BALL is $4!/2 = 12$. □

EXAMPLE 1.11

Using the idea developed in Example 1.10, we now consider the permutations of all letters in PEPPER.

There are $3! = 6$ permutations with the P's distinguished for each permutation in which the P's are not distinguished. For example, $P_1 E P_2 P_3 E R$, $P_1 E P_3 P_2 E R$, $P_2 E P_1 P_3 E R$, $P_2 E P_3 P_1 E R$, $P_3 E P_1 P_2 E R$, and $P_3 E P_2 P_1 E R$ all correspond to PEPPER when we remove the subscripts on the P's. In addition, to the permutation $P_1 E P_2 P_3 E R$ there corresponds the pair $P_1 E_1 P_2 P_3 E_2 R$, $P_1 E_2 P_2 P_3 E_1 R$, when the E's are distinguished. Consequently,

$(2!)(3!)$ (Number of permutations of the unsubscripted letters in PEPPER)

= (Number of permutations of the subscripted letters in $P_1 E_1 P_2 P_3 E_2 R$),

so the number of permutations of the letters in PEPPER is $6!/(2!\,3!) = 60$. □

Before stating a general principle for permutations with repetitions, we want to point out that in our prior two examples we solved a new type of problem by relating it to previous enumeration principles. This is a common practice in mathematics in general and will often occur in discrete and combinatorial formula derivations.

In general, if there are n objects with n_1 of a first type, n_2 of a second type, ..., and n_r of an rth type, where $n_1 + n_2 + \cdots + n_r = n$, then there are $\dfrac{n!}{n_1!\,n_2!\cdots n_r!}$ permutations of the given n objects.

EXAMPLE 1.12   The MASSASAUGA is a brown and white venomous snake indigenous to North America. Permuting the letters MASSASAUGA, we find that there are

$$\frac{10!}{4!\,3!\,1!\,1!\,1!} = 25{,}200$$

possible arrangements. Among these there are

$$\frac{7!}{3!\,1!\,1!\,1!\,1!} = 840$$

in which all four A's are together. To get this last result we considered all permutations of the seven symbols AAAA (one symbol), S, S, S, M, U, G.   □

EXAMPLE 1.13   We now do something a bit more abstract and prove that if n and k are positive integers with $n = 2k$, then $n!/2^k$ is an integer. Since our argument relies on counting, it is an example of a combinatorial proof.

Consider the n symbols $x_1, x_1, x_2, x_2, \ldots, x_k, x_k$. The number of ways we can arrange all of these $n = 2k$ symbols is an integer that is

$$\underbrace{\frac{n!}{2!\,2!\cdots 2!}}_{k\text{ factors of }2!} = \frac{n!}{2^k}.$$ □

Before closing the door on permutations for the first time, we welcome an opportunity to apply what has been developed so far to a situation in which the arrangements are no longer linear.

EXAMPLE 1.14   If six people, designated as A, B, ..., F, are seated about a round table, how many different circular arrangements are possible, if arrangements are considered the same when one can be obtained from the other by rotation? (In Fig. 1.1, arrangements (a) and (b) are considered identical, while (b), (c), (d) are three distinct arrangements.)

As in so many new circumstances we try to relate to past ideas. Considering Figs. 1.1(a) and 1.1(b), starting at the top of the circle and moving clockwise, we list the distinct linear arrangements ABEFCD and CDABEF, which correspond to the same circular arrangement. In addition to these two, four other linear arrangements—

Figure 1.1

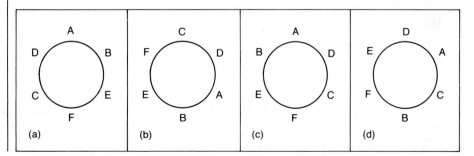

BEFCDA, DABEFC, EFCDAB, and FCDABE—are found to correspond to the same circular arrangement as in (a) or (b). So as each circular arrangement corresponds with six linear arrangements, we have 6 × (Number of circular arrangements of A, B, ..., F) = (Number of linear arrangements of A, B, ..., F) = 6!.

Consequently, there are 6!/6 = 5! = 120 arrangements of A, B, ..., F around the circular table.   □

EXAMPLE 1.15   Suppose now that the six people of Example 1.14 are three married couples, with A, B, and C female. We want to arrange the six people around the table so that the sexes alternate. (Once again arrangements are considered identical if one can be obtained from the other by rotation.)

Before solving this problem we solve Example 1.14 by an alternative method, which will assist us in solving our present problem.

Figure 1.2

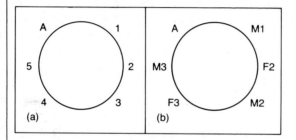

If we place A at the table, as shown in Fig. 1.2(a), we then have five locations (clockwise from A) to fill. The filling of these locations with B, C, ..., F is the problem of permuting B, C, ..., F in a linear manner, and this can be done in 5! = 120 ways.

To solve the new problem of alternating the sexes, consider placing A as in Fig. 1.2(b). The next position, clockwise from A, is marked M1 (Male 1) and can be filled in three ways. Continuing clockwise from A, position F2 (Female 2) can be filled in two ways. Proceeding in this manner, by the rule of product there are 3 × 2 × 2 × 1 × 1 = 12 ways that these six people can be arranged with no two men or women seated next to each other.   □

EXERCISES

1. Buick automobiles come in 4 models, 12 colors, 3 engine sizes, and 2 transmission types.

   a)   How many distinct Buicks can be manufactured?

   b)   If one of the available colors is blue, how many different blue Buicks can be manufactured?

   c)   If one engine size is V-8, how many distinct blue Buicks have a V-8 engine?

2. In how many ways can the symbols a, b, c, d, e, e, e, e, e be arranged so that no e is adjacent to another e?

3. Determine the number of six-digit integers (no leading zeros) in which (a) no digit may be repeated, or (b) digits may be repeated. Answer parts (a) and (b) with the extra condition that the six-digit integer is (i) even, (ii) divisible by 5, or (iii) divisible by 4.

4. Find the value(s) of n in each of the following: (a) $P(n, 2) = 90$, (b) $P(n, 3) = 3P(n, 2)$, and (c) $2P(n, 2) + 50 = P(2n, 2)$.

5. A certain "Burger Joint" advertises that a customer can have his or her hamburger with or without any or all of the following: catsup, mustard, mayonnaise, lettuce, tomato, onion, pickle, cheese, or mushrooms. How many different kinds of hamburger orders are possible?

6. An alphabet of 40 symbols is used for transmitting messages in a communication system. How many distinct messages (lists of symbols) of 25 symbols can the transmitter generate if symbols can be repeated in the message? How many if 10 of the 40 symbols can appear only as the first and/or last symbols of the message, the other 30 symbols can appear anywhere, and repetitions of all symbols are allowed?

7. In a certain implementation of the programming language Pascal, an identifier consists of a letter or a letter followed by up to seven symbols, which may be letters or digits. (We assume that the computer does not distinguish between capital and lowercase letters; there are 26 letters, 10 digits.) How many distinct identifiers are possible for this version of Pascal?

8. The production of a machine part consists of four stages. There are six assembly lines for the first stage, four assembly lines for the second stage, five for the third stage, and five for the last. Determine the number of different ways a machine part can be totally assembled in this production process.

9. Write a computer program to compute $n!$ for any integer $n \geq 0$.

10. Write a computer program to compute $P(n, r)$ for any integers $n, r \geq 0$.

11. Show that for integers $n, r \geq 0$, with $n + 1 > r$,

$$P(n + 1, r) = \left(\frac{n + 1}{n + 1 - r} \right) P(n, r).$$

12. A computer science professor has seven different programming books on a bookshelf. Three of the books deal with FORTRAN, while the other four are concerned with BASIC. In how many ways can the professor arrange the books on the shelf if (a) there are no restrictions? (b) the languages should alternate? (c) all the FORTRAN books must be next to each other? (d) all FORTRAN books must be next to each other and all BASIC books must be next to each other? (e) the three FORTRAN books are placed on the shelf with two BASIC books on either side?

13. a) In how many ways can the letters in VISITING be arranged?

    b) For the arrangements of part (a), how many have all three I's together?

14. Consider the following BASIC program:

```
10 FOR I = 1 TO 12
20      FOR J = 5 TO 10
30          FOR K = 15 TO 8 STEP -1
40              PRINT (I - J) * K
50          NEXT K
60      NEXT J
70 NEXT I
80 END
```

a) How many times is the PRINT statement of line 40 executed?

b) What counting principle is being used here?

15. A sequence of letters of the form abcba, where the expression is unchanged upon reversing order, is an example of a *palindrome* (of five letters).

a) If a letter may appear more than twice, how many palindromes of five letters are there? of six letters?

b) Repeat part (a) under the condition that no letter appears more than twice.

16. a) In how many ways can a student answer a ten-question True–False examination if the student is merely guessing? 2^{10}

b) In how many ways can the student answer the test in part (a) if it is possible to leave a question unanswered in order to avoid an extra penalty for a wrong answer?

c) Answer parts (a) and (b) if the examination consists of ten multiple-choice questions, each question followed by four choices.

17. How many different paths in the *xy*-plane are there from $(0, 0)$ to $(7, 7)$ if a path proceeds one step at a time by going either one space to the right (R) or one space upward (U)? How many such paths are there from $(2, 7)$ to $(9, 14)$? Is there any general statement that can be made that incorporates these two results?

18. a) How many permutations are there of all the letters in SOCIOLOGICAL?

b) In how many of these permutations are A and G adjacent?

c) In how many of the permutations in part (a) are all the vowels adjacent?

d) How many of the permutations in part (c) have the vowels in alphabetical order? How many in part (a)?

19. a) In how many ways can seven people be arranged about a circular table?

b) If two of the people insist on sitting next to each other, how many arrangements are possible? $6! = 5!$

20. a) In how many ways can eight people, denoted A, B, ..., H, be seated about the square table shown in Fig. 1.3, where Figs. 1.3(a) and 1.3(b) are considered the same, but distinct from Fig. 1.3(c)?

b) If two of the eight people, say A and B, do not get along well, how many different seatings are possible with A and B not sitting next to each other?

$2(6!) + 2(6!)$

Figure 1.3

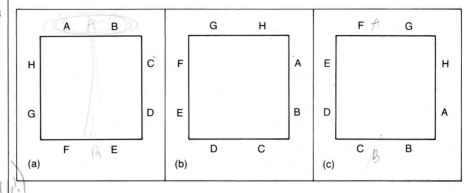

(a)   (b)   (c)

c) For the seatings in part (b), how many avoid having A and B sitting across the table from each other? $2 6!$

21. Provide a combinatorial argument to show that if n, k are positive integers with $n = 3k$, then $n!/(3!)^k$ is an integer.

1.3 ■ COMBINATIONS: NO MORE FUSS ABOUT ORDER

The standard deck of playing cards consists of 52 cards comprising four suits: clubs, diamonds, hearts, and spades. Each suit has 13 cards: ace, 2, 3, ..., 9, 10, jack, queen, king.

If we are asked to draw three cards from a standard deck, in succession and without replacement, then by the rule of product there are

$$52 \times 51 \times 50 = \frac{52!}{49!} = P(52, 3)$$

possibilities, one of which is AH (ace of hearts), 9C (nine of clubs), KD (king of diamonds). If instead we simply select three cards at one time from the deck so that the order of selection of the cards is no longer important, then the six permutations AH–9C–KD, AH–KD–9C, 9C–AH–KD, 9C–KD–AH, KD–9C–AH, and KD–AH–9C all correspond to just one (unordered) selection. Consequently, each selection, or combination, of three cards, *with no reference to order,* corresponds with 3! permutations, or arrangements, of three cards. In equation form this translates into

(3!) × (Number of selections of size three from a deck of 52)

= Number of permutations of the 52 cards taken three at a time

$$= P(52, 3) = \frac{52!}{49!}.$$

Consequently, three cards can be drawn, without replacement, from a standard deck in $52!/(3!\,49!) = 22{,}100$ ways.

In general, if we start with n distinct objects, each *selection* or *combination* of r of these objects, with no reference to order, corresponds to $r!$ permutations of size r from the n objects. Thus the number of combinations of size r from a collection of size n, denoted $C(n, r)$, $0 \le r \le n$, satisfies $(r!) \times C(n, r) = P(n, r)$ and

$$C(n, r) = \frac{P(n, r)}{r!} = \frac{n!}{r!(n - r)!}, \qquad 0 \le r \le n.$$

In addition to $C(n, r)$ the symbol $\binom{n}{r}$ is also frequently used.

A word to the wise! When dealing with any counting problem, we should ask ourselves about the importance of order in the problem. If order is relevant, we think in terms of permutations and the rule of product. When order is not relevant, combinations could play a key role in solving the problem.

EXAMPLE 1.16 A hostess is having a dinner party for some members of her charity committee. Because of the size of her home, she can invite only 11 of the 20 committee members. Since order is not important, she can invite "the lucky 11" in $\binom{20}{11} = 20!/(11!\,9!) = 167{,}960$ ways. However, once the 11 arrive, how she arranges them around her rectangular dining table is a permutation problem. Unfortunately, neither combinations nor permutations can help our hostess in dealing with "the offended nine" who did not get invited. □

EXAMPLE 1.17 a) A student taking a history examination is directed to answer any seven of ten essay questions. As there is no concern about order here, the student can answer the examination in

$$\binom{10}{7} = \frac{10!}{7!\,3!} = \frac{10 \times 9 \times 8}{3 \times 2 \times 1} = 120 \text{ ways.}$$

b) If the student must answer three questions from the first five and four questions from the last five, three questions can be selected from the first five in $\binom{5}{3} = 10$ ways, and the other four questions can be selected in $\binom{5}{4} = 5$ ways. So by the rule of product the student can complete the examination in $\binom{5}{3}\binom{5}{4} = 10 \times 5 = 50$ ways. □

EXAMPLE 1.18 a) At Rydell High School, the gym teacher must select nine girls from the junior and senior classes for a volleyball team. If there are 28 juniors and 25 seniors, the selection can be made in $\binom{53}{9} = 4{,}431{,}613{,}550$ ways.

b) If two juniors and one senior are the best spikers and must be on the team, then the rest of the team can be chosen in $\binom{50}{6} = 15{,}890{,}700$ ways.

c) For a certain tournament the team must comprise four juniors and five seniors. Under these constraints the teacher can select her team in $\binom{28}{4}\binom{25}{5} =$ 1,087,836,750 ways.   □

Some problems can be treated from the viewpoint of either permutations or combinations, depending on how one analyzes the situation. The following example demonstrates this.

EXAMPLE 1.19 The gym teacher of example 1.18 must make up four volleyball teams of nine girls each from the 36 freshman girls in her P.E. class. In how many ways can she select these four teams? Call the teams A, B, C, D.

a) To form team A, she can select any nine girls from the 36 enrolled in $\binom{36}{9}$ ways. For team B the selection process yields $\binom{27}{9}$ possibilities. This then leaves $\binom{18}{9}$ and $\binom{9}{9}$ possible ways to select teams C and D, respectively. So by the rule of product the four teams can be chosen in

$$\binom{36}{9}\binom{27}{9}\binom{18}{9}\binom{9}{9} = \left(\frac{36!}{27!\,9!}\right)\left(\frac{27!}{18!\,9!}\right)\left(\frac{18!}{9!\,9!}\right)\left(\frac{9!}{9!\,0!}\right)$$

$$= \frac{36!}{9!\,9!\,9!\,9!} \doteq 2.145 \times 10^{19} \text{ ways.}$$

b) For an alternative solution, consider the 36 students lined up as follows:

1st	2nd	3rd		35th	36th
student	student	student	...	student	student

In order to select the four teams, we must distribute nine A's, nine B's, nine C's and nine D's in the 36 spaces. The number of ways in which this can be done is the number of permutations of 36 letters comprising nine each of A, B, C, D. This is now the familiar problem of permutations of nondistinct objects, and the answer is

$$\frac{36!}{9!\,9!\,9!\,9!}, \quad \text{as in part (a).} \qquad \square$$

Our next example points out how some problems require the concepts of both permutations and combinations for their solutions.

EXAMPLE 1.20 The number of permutations of the letters in TALLAHASSEE is

$$\frac{11!}{3!\,2!\,2!\,2!\,1!\,1!} = 831,600.$$

How many of these have no adjacent A's?   *8 non- A*
    Disregarding the A's, there are
                                                          3 A

$$\frac{8!}{2!\,2!\,2!\,1!\,1!} = 5040$$

ways to arrange the remaining letters. One of these 5040 ways is shown in the following figure, where the upward arrows indicate nine possible locations for the three A's.

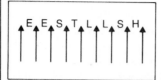

Three of these locations can be selected in $\binom{9}{3} = 84$ ways; and since this is also possible for all of the other 5039 arrangements of E, E, S, T, L, L, S, H, by the rule of product, there are $5040 \times 84 = 423,360$ permutations of the letters in TALLAHASSEE with no consecutive A's.   □

We shall close this section with three results related to the concept of combinations.

First we note that for integers n, r, with $n \geq r \geq 0$, $\binom{n}{r} = \binom{n}{n-r}$. This can be established algebraically from the formula for $\binom{n}{r}$, but we prefer to observe that when dealing with a selection of size r from a collection of n distinct objects, the selection process leaves behind $n - r$ objects. Consequently, $\binom{n}{r} = \binom{n}{n-r}$ affirms the existence of a correspondence between the selections of size r (objects chosen) and the selections of size $n - r$ (objects left behind). An example of this correspondence is shown in Table 1.2 where $n = 5$, $r = 2$, and the distinct objects are 1, 2, 3, 4, 5. This type of correspondence will be more formally defined in Chapter 3 and used in other counting situations.

Table 1.2

Selections of size $r = 2$ (objects chosen)		Selections of size $n - r = 3$ (objects left behind)	
1. 1, 2	6. 2, 4	1. 3, 4, 5	6. 1, 3, 5
2. 1, 3	7. 2, 5	2. 2, 4, 5	7. 1, 3, 4
3. 1, 4	8. 3, 4	3. 2, 3, 5	8. 1, 2, 5
4. 1, 5	9. 3, 5	4. 2, 3, 4	9. 1, 2, 4
5. 2, 3	10. 4, 5	5. 1, 4, 5	10. 1, 2, 3

Our second result is a theorem from our past experience in algebra.

Theorem 1.1 ▶ (*The Binomial Theorem*) If x and y are variables and n is a positive integer, then

$$(x + y)^n = \binom{n}{0} x^0 y^n + \binom{n}{1} x^1 y^{n-1} + \binom{n}{2} x^2 y^{n-2} + \cdots$$

$$+ \binom{n}{n-1} x^{n-1} y^1 + \binom{n}{n} x^n y^0 = \sum_{k=0}^{n} \binom{n}{k} x^k y^{n-k}.$$

Proof | In the product

$$(x + y)(x + y)(x + y)\cdots(x + y)$$

1st 2nd 3rd   nth
factor factor factor   factor

the coefficient of $x^k y^{n-k}$, $0 \le k \le n$, is the number of different ways we can select k x's (and consequently $(n - k)$ y's) from the n x's available in the n factors. (One way, for example, is to choose x from the first k factors and y from the last $n - k$ factors.) The total number of such selections of size k from a collection of size n is $C(n, k) = \binom{n}{k}$, and from this the theorem follows. ∎

From this result we note that

$$(x + y)^n = \sum_{k=0}^{n} \binom{n}{n - k} x^k y^{n-k}.$$

Also, $\binom{n}{k}$ is often referred to as a *binomial coefficient* because of this theorem.

EXAMPLE 1.21 | a) From the binomial theorem it follows that the coefficient of $x^5 y^2$ in $(x + y)^7$ is $\binom{7}{5} = \binom{7}{2} = 21$.

b) The coefficient of $a^5 b^2$ in $(2a - 3b)^7$ is $\binom{7}{5}(2)^5(-3)^2$. This is obtained from the binomial theorem by setting $x = 2a$ and $y = -3b$. □

Corollary 1.1 | ▶ For any integer $n > 0$,

a) $\binom{n}{0} + \binom{n}{1} + \binom{n}{2} + \cdots + \binom{n}{n} = 2^n$, and

b) $\binom{n}{0} - \binom{n}{1} + \binom{n}{2} - \cdots + (-1)^n \binom{n}{n} = 0$.

Proof | Part (a) follows from the binomial theorem by setting $x = y = 1$. When $x = -1$ and $y = 1$, part (b) results. ∎

Our third and final result generalizes the binomial theorem and is called the *multinomial theorem*.

Theorem 1.2 | ▶ For positive integers n, t, the coefficient of $x_1^{n_1} x_2^{n_2} x_3^{n_3} \cdots x_t^{n_t}$ in $(x_1 + x_2 + x_3 + \cdots + x_t)^n$ is

$$\frac{n!}{n_1! \, n_2! \, n_3! \cdots n_t!},$$

where each n_i is an integer with $0 \le n_i \le n$, for all $1 \le i \le t$, and $n_1 + n_2 + n_3 + \cdots + n_t = n$.

Proof | As in the proof of the binomial theorem, the coefficient of $x_1^{n_1} x_2^{n_2} x_3^{n_3} \cdots x_t^{n_t}$ is the number of ways we can select x_1 from n_1 of the n factors, x_2 from n_2 of the $n - n_1$ remaining factors, x_3 from n_3 of the $n - n_1 - n_2$ now remaining factors, ..., and x_t from n_t of the last $n - n_1 - n_2 - n_3 - \cdots - n_{t-1} = n_t$ remaining factors. This can be carried out, as in part (a) of Example 1.19, in

$$\binom{n}{n_1}\binom{n - n_1}{n_2}\binom{n - n_1 - n_2}{n_3}\cdots\binom{n - n_1 - n_2 - \cdots - n_{t-1}}{n_t}$$

ways. We leave to the reader the details showing that this is equal to

$$\frac{n!}{n_1! \, n_2! \, n_3! \cdots n_t!},$$

which is also written as

$$\binom{n}{n_1, n_2, n_3, \ldots, n_t}$$

and is called a *multinomial coefficient*. ∎

EXERCISES

1. A committee of 12 is to be selected from ten men and ten women. In how many ways can the selection be carried out if a) there are no restrictions? b) there must be six men and six women? c) there must be an even number of women? d) there must be more women than men? e) there must be at least eight men?

2. In how many ways can a gambler draw five cards from a standard deck and get (a) a flush (five cards of the same suit)? (b) four aces? (c) four of a kind? (d) three aces and two jacks? (e) a full house (three of a kind and a pair)? (f) three of a kind? (g) two pair?

3. Show that if n is an integer, with $n \geq 1$, then

$$\binom{2n}{n} + \binom{2n}{n - 1} = \frac{1}{2}\binom{2n + 2}{n + 1}.$$

4. How many ways are there to pick a five-person basketball team from 12 possible players? How many selections include the weakest and strongest players?

5. A student is to answer seven out of ten questions on an examination. In how many ways can he make his selection if (a) there are no restrictions? (b) he must answer the first two questions? (c) he must answer at least three of the first five questions?

6. In ordering the daily special at a diner a customer has a choice of three entrees and may select any two of six available vegetables.

   a) How many different dinners can she select if (i) she must select two different vegetables? (ii) she is permitted to have two helpings of the same vegetable?

   b) Answer parts (i) and (ii) of (a) if she also has a choice of tomato juice, orange juice, or bean soup as an appetizer.

7. In how many ways can 12 different books be distributed among four children so that (a) each child gets three books? (b) the two oldest children get four books each while the two youngest get two books each?

8. How many permutations of the letters in MISSISSIPPI have no consecutive S's?

9. A gym coach must select 11 seniors to play on a football team. If he can make his selection in 12,376 ways, how many seniors are eligible to play?

10. a) Fifteen points, no three collinear, are given on a plane. How many lines do they determine?

   b) Twenty-five points, no four coplanar, are given in space. How many triangles do they determine? How many planes? How many tetrahedra (pyramidlike solids with four triangular faces)?

11. How many triangles are determined by the vertices of a regular polygon of n sides? How many if no side of the polygon is to be a side of any triangle?

12. Determine the coefficient of x^9y^3 in (a) $(x + y)^{12}$, (b) $(x + 2y)^{12}$, and (c) $(2x - 3y)^{12}$.

13. Complete the details in the proof of the multinomial theorem.

14. Determine the coefficient of

   a) xyz^2 in $(x + y + z)^4$ b) xyz^2 in $(w + x + y + z)^4$

   c) xyz^2 in $(2x - y - z)^4$ d) xyz^{-2} in $(x - 2y + 3z^{-1})^4$

   e) $w^3x^2yz^2$ in $(2w - x + 3y - 2z)^8$.

15. Determine the sum of all the coefficients in

   a) $(x + y)^3$ b) $(x + y)^{10}$ c) $(x + y + z)^{10}$

   d) $(w + x + y + z)^5$ e) $(2s - 3t + 5u + 6v - 11w + 3x + 2y)^{10}$.

16. With n a positive integer, evaluate the sum

$$\binom{n}{0} + 2\binom{n}{1} + 2^2\binom{n}{2} + \cdots + 2^k\binom{n}{k} + \cdots + 2^n\binom{n}{n}.$$

17. For x a real number and n a positive integer, show that

   a) $1 = (1 + x)^n - \binom{n}{1}x^1(1 + x)^{n-1} + \binom{n}{2}x^2(1 + x)^{n-2} - \cdots$

   $\quad + (-1)^n\binom{n}{n}x^n.$

   b) $1 = (2 + x)^n - \binom{n}{1}(x + 1)(2 + x)^{n-1}$

   $\quad + \binom{n}{2}(x + 1)^2(2 + x)^{n-2} - \cdots + (-1)^n\binom{n}{n}(x + 1)^n.$

   c) $2^n = (2 + x)^n - \binom{n}{1}x^1(2 + x)^{n-1} + \binom{n}{2}x^2(2 + x)^{n-2} - \cdots$

   $\quad + (-1)^n\binom{n}{n}x^n.$

1.4 ■ COMBINATIONS WITH REPETITION: DISTRIBUTIONS

When repetitions are allowed, we have seen that for n distinct objects an arrangement of size r of these objects can be obtained in n^r ways, for an integer $r \geq 0$. We now turn to the comparable problem for combinations and once again obtain a related problem whose solution follows from our previous enumeration principles.

EXAMPLE 1.22 On their way home from track practice, seven high-school freshmen stop at a fast food restaurant, where each of them has one of the following: a cheeseburger, a hot dog, a taco, or a fish sandwich. How many different purchases are possible?

Let c, h, t, f represent cheeseburger, hot dog, taco, and fish sandwich, respectively. Here we are concerned with how many of each item are purchased, and not the order in which they are purchased, so the problem is one of selections or combinations with repetition.

In Table 1.3 we have listed some possible purchases in column (a) and another means of representing each purchase in column (b).

Table 1.3

(a)	(b)
1. c, c, h, h, t, t, f	1. xx | xx | xx | x
2. c, c, c, c, h, t, f	2. xxxx | x | x | x
3. c, c, c, c, c, c, f	3. xxxxxx | | | x
4. h, t, t, f, f, f, f	4. | x | xx | xxxx
5. t, t, t, t, t, f, f	5. | | xxxxx | xx
6. t, t, t, t, t, t, t	6. | | xxxxxxx |
7. f, f, f, f, f, f, f	7. | | | xxxxxxx

For a purchase in column (b) of Table 1.3 we realize that each x to the left of the first bar (\|) represents a c, each x between the first and second bars represents an h, the x's between the second and third bars are for t's, and any x to the right of the third bar stands for an f. The third purchase, for example, has three consecutive bars, since no one bought a hot dog or taco; the bar at the start of the fourth purchase indicates no cheeseburgers in that purchase.

Once again a correspondence has been established between two collections of objects, where we know how to count the number in one collection. For column (b) of Table 1.3, we are enumerating all permutations of ten symbols consisting of seven x's and three \|'s, so by our correspondence the number of different orders for column (a) is

$$\frac{10!}{7!\,3!} = \binom{10}{7}.$$

In this example we note that the seven x's (one for each freshman) correspond to the size of the selection and that the three bars are needed to separate the $3 + 1 = 4$ possible food items that can be chosen.   □

In general, given n distinct objects, where we wish to select, *with repetition*, r of these objects, we find, as in Table 1.3, that we are considering all permutations of r x's and $n - 1$ |'s and their number is

$$\frac{(n + r - 1)!}{r!(n - 1)!} = \binom{n + r - 1}{r}.$$

Consequently, the number of combinations of n objects taken r at a time, *with repetition*, is $C(n + r - 1, r)$.

(In Example 1.22, $n = 4$, $r = 7$, so it is possible for r to exceed n when repetitions are allowed.)

EXAMPLE 1.23 A donut shop offers 20 different kinds of donuts. Assuming that there are at least a dozen of each kind when we enter the shop, we can select a dozen donuts in $C(20 + 12 - 1, 12) = C(31, 12) = 141,120,525$ ways. (Here $n = 20$, $r = 12$.) □

EXAMPLE 1.24 President Helen has four secretaries: (1) Betty, (2) Goldie, (3) Mary Lou, and (4) Mona, and wishes to distribute \$100, in ten-dollar bills, as a Christmas bonus to them.

a) Allowing the situation in which one or more of the secretaries gets nothing, President Helen is making a selection of size 10 (one for each ten-dollar bill) from a collection of size 4 (four secretaries), with repetition. This can be done in $C(4 + 10 - 1, 10) = C(13, 10) = 286$ ways.

b) If there are to be no hard feelings, each secretary must get at least \$10. With this restriction President Helen is now faced with making a selection of size 6 (the remaining six ten-dollar bills) from the same collection of size 4, and the choices now number $C(4 + 6 - 1, 6) = C(9, 6) = 84$. (For example, here the selection 2, 3, 3, 4, 4, 4 means that Betty does not get anything extra, while Goldie gets an additional \$10, Mary Lou an additional \$20, and Mona ends up with \$40 total.)

c) If each secretary must get at least \$10 and Mona, as head secretary, gets at least \$50, the number of ways President Helen can distribute the bonus money is

$$\underbrace{C(3 + 2 - 1, 2)}_{\substack{\text{Mona gets} \\ \text{exactly \$50}}} + \underbrace{C(3 + 1 - 1, 1)}_{\substack{\text{Mona gets} \\ \text{exactly \$60}}}$$

$$+ \underbrace{C(3 + 0 - 1, 0)}_{\substack{\text{Mona gets} \\ \text{exactly \$70}}} = 10 = \underbrace{C(4 + 2 - 1, 2)}_{\substack{\text{Using the} \\ \text{technique in part (b)}}} \quad □$$

Having seen examples utilizing combinations with repetition, we now consider two examples involving other counting principles as well.

EXAMPLE 1.25

In how many ways can we distribute seven apples and six oranges among four children so that each child receives at least one apple?

Giving each child an apple, we have $C(4 + 3 - 1, 3) = 20$ ways to distribute the other three apples and $C(4 + 6 - 1, 6) = 84$ ways to distribute the six oranges. So by the rule of product there are $20 \times 84 = 1680$ ways to distribute the fruit under the stated conditions.   □

EXAMPLE 1.26

A message made up of 12 different symbols is to be transmitted through a communication channel. In addition to the 12 symbols, the transmitter will also send a total of 45 (blank) spaces between the symbols, with at least three spaces between each pair of consecutive symbols. In how many ways can the transmitter send the message?

There are 12! ways to arrange the twelve different symbols, and for any of these arrangements there are 11 positions between the 12 symbols. Since there must be at least three spaces between successive symbols, we use up 33 of the 45 spaces and must now locate the remaining 12 spaces. This is now a selection, with repetition, of size 12 (the spaces) from a collection of size 11 (the locations), and this can be accomplished in $C(11 + 12 - 1, 12) = 646,646$ ways.

Consequently, by the rule of product the transmitter can send the message with the required spacing in $(12!)\binom{22}{12} \doteq 3.097 \times 10^{14}$ ways.   □

In the next example an idea is introduced that appears to have more to do with number theory than combinations or permutations. Nonetheless, the solution of this example will turn out to be equivalent to counting combinations with repetitions.

EXAMPLE 1.27

Determine all integer solutions to the equation

$$x_1 + x_2 + x_3 + x_4 = 7, \qquad \text{where } x_i \geq 0 \quad \text{for all } 1 \leq i \leq 4.$$

One solution of the equation is $x_1 = 3$, $x_2 = 3$, $x_3 = 0$, $x_4 = 1$; a possible interpretation for this is that we are distributing seven pennies (identical objects) among four children (distinct containers), and here we have given the first two children three pennies each, nothing to the third child, and the last penny to the fourth child. Continuing with this interpretation we see that each nonnegative integer solution of the equation corresponds with a selection, with repetition, of size 7 (the *identical* pennies) from a collection of size 4 (the *distinct* children), so there are $C(4 + 7 - 1, 7) = 120$ solutions.   □

At this point it is crucial that we recognize the equivalence of the following:

a) The number of integer solutions of the equation

$$x_1 + x_2 + \cdots + x_n = r, \qquad x_i \geq 0, \qquad 1 \leq i \leq n.$$

b) The number of selections, with repetition, of size r from a collection of size n.

c) The number of ways r identical objects can be distributed among n distinct containers.

In terms of distributions, (c) is valid only when the r objects being distributed are identical and the n containers are distinct. When both the r objects and the n containers are distinct, we can select any of the n containers for each one of the objects and get n^r distributions by the rule of product.

When the objects are distinct but the containers are identical, we shall solve the problem using the Stirling numbers of the second kind (Chapter 3). For the final case, in which both objects and containers are identical, the theory of partitions of integers (Chapter 10) will provide some necessary results.

EXAMPLE 1.28

In how many ways can one distribute ten (identical) white marbles among six distinct containers?

The solution to this problem is equivalent to finding the number of nonnegative integer solutions to the equation $x_1 + x_2 + \cdots + x_6 = 10$. That number is the number of selections of size 10, with repetition, from a collection of size 6. Hence the answer is $C(6 + 10 - 1, 10) = 3003$. □

We close this section with two final examples related to the theme of the section.

EXAMPLE 1.29

From Example 1.28 we know that there are 3003 nonnegative integer solutions to the equation $x_1 + x_2 + \cdots + x_6 = 10$. How many such solutions are there to the inequality $x_1 + x_2 + \cdots + x_6 < 10$?

One approach that may seem feasible in solving this inequality is to determine the number of such solutions to $x_1 + x_2 + \cdots + x_6 = k$, where k is an integer, $0 \le k \le 9$. Although feasible now, the technique becomes unrealistic if 10 is replaced by a somewhat larger number, say 100. In Chapter 2, however, we shall establish a combinatorial identity that will help us obtain an alternative solution to the problem using this approach.

For the present we transform the problem by noting the correspondence between the nonnegative integer solutions of

$$x_1 + x_2 + \cdots + x_6 < 10 \tag{1}$$

and the integer solutions of

$$x_1 + x_2 + \cdots + x_6 + x_7 = 10, \qquad 0 \le x_i, \qquad 1 \le i \le 6, \qquad 0 < x_7. \tag{2}$$

The number of solutions of (2) is the same as the number of nonnegative integer solutions of $y_1 + y_2 + \cdots + y_6 + y_7 = 9$, where $y_i = x_i$, $1 \le i \le 6$, $y_7 = x_7 - 1$. This is $C(7 + 9 - 1, 9) = 5005$. □

Our final result takes us back to the binomial and multinomial expansions.

EXAMPLE 1.30

In the binomial expansion for $(x + y)^n$, each term is of the form $\binom{n}{k}x^k y^{n-k}$, and so the total number of terms in the expansion is the number of nonnegative integer solutions of $x_1 + x_2 = n$ (x_1 is the exponent for x, x_2 the exponent for y). This number is $C(2 + n - 1, n) = n + 1$.

One may feel that we have used a rather long-winded argument to get the above result. Many of us would probably be willing to believe the result on the basis of our experiences in computing $(x + y)^n$ for various small values of n.

Although experience is worthwhile in pattern recognition, it is not always enough to find a general principle. Here it would prove of little value if we want to know how many terms there are in the expansion of $(w + x + y + z)^{10}$.

Each distinct term here is of the form $\binom{10}{n_1, n_2, n_3, n_4} w^{n_1} x^{n_2} y^{n_3} z^{n_4}$, where $0 \le n_i$, $1 \le i \le 4$, and $n_1 + n_2 + n_3 + n_4 = 10$. This last equation can be solved in $C(4 + 10 - 1, 10) = 286$ ways, and so there are 286 terms in $(w + x + y + z)^{10}$.   □

EXERCISES

1. In how many ways can ten (identical) dimes be distributed among five children if (a) there are no restrictions? (b) each child gets at least one dime? (c) the oldest child gets at least two dimes?

2. In how many ways can 15 (identical) candy bars be distributed among five children so that the youngest only gets one or two of them?

3. Determine the number of ways to select 20 coins from four large containers filled with pennies, nickels, dimes, and quarters.

4. A certain ice-cream store has 31 flavors of ice cream available. In how many ways can we order a dozen ice-cream cones if (a) we do not want the same flavor more than once? (b) a flavor may be ordered as many times as 12? (c) a flavor may be ordered no more than 11 times? (d) more than half of them must be chocolate?

5. a) In how many ways can we select five coins from a collection of ten consisting of one penny, one nickel, one dime, one quarter, one half dollar, and five (identical) Susan B. Anthony dollars?

   b) In how many ways can we select n objects from a collection of size $2n$, consisting of n distinct and n identical objects?

6. Answer Example 1.26 where the 12 symbols being transmitted are four A's, four B's, and four C's.

7. Determine the number of integer solutions of $x_1 + x_2 + x_3 + x_4 = 32$ where

   a) $x_i \ge 0, \quad 1 \le i \le 4.$ b) $x_i > 0, \quad 1 \le i \le 4.$

   c) $x_1, x_2 \ge 5, \quad x_3, x_4 \ge 7.$ d) $x_i \ge 8, \quad 1 \le i \le 4.$

   e) $x_i \ge -2, \quad 1 \le i \le 4.$ f) $x_1, x_2, x_3 > 0, \quad 0 < x_4 \le 25.$

8. In how many ways can a teacher distribute eight chocolate donuts and seven jelly donuts among three student helpers if each helper wants at least one donut of each kind?

9. Two n-digit integers (leading zeros allowed) are considered equivalent if one is a permutation of the other. (For example, 12033, 20331, and 01332 are considered equivalent five-digit integers.)

   a) How many five-digit integers are not equivalent?

   b) If the digits 1, 3, 7 can appear at most once, how many nonequivalent five-digit integers are there?

10. Determine the number of integer solutions for $x_1 + x_2 + x_3 + x_4 + x_5 < 40,$

where

a) $x_i \geq 0$, $1 \leq i \leq 5$. b) $x_i \geq -3$, $1 \leq i \leq 5$.

11. In how many ways can we distribute eight identical white balls into four distinct containers so that (a) no container is left empty? (b) the fourth container has an odd number of balls in it?

12. a) Find the coefficient of $v^2 w^4 xz$ in $(3v + 2w + x + y + z)^8$.

    b) How many distinct terms arise in the expansion in part (a)?

13. How many ways are there to place 12 marbles, of the same size, in five distinct jars if (a) the marbles are all black? (b) each marble is a different color?

14. Find the number of nonnegative integer solutions for

    a) $2x_1 + x_2 + x_3 + x_4 = 10$;

    b) $2x_1 + 2x_2 + x_3 + x_4 = 10$.

15. In how many ways can one quarter, one dime, one nickel, and 25 1982 pennies be distributed among five children (a) with no restrictions? (b) so that the oldest child gets either 20¢ or 25¢?

16. A chemistry teacher has seven cartons each containing 36 test tubes of "unknowns" for a laboratory experiment. The first carton's 36 unknowns comprise four different compounds occurring 5, 12, 7, 12 times. In how many ways can the contents of this carton be distributed among five different chemistry labs?

17. a) How many nonnegative integer solutions are there to the pair of equations $x_1 + x_2 + x_3 + \cdots + x_7 = 37$, $x_1 + x_2 + x_3 = 6$?

    b) How many solutions in part (a) have x_1, x_2, $x_3 > 0$?

18. In how many ways can 24 pieces of chalk be distributed among four classrooms so that the largest classroom has at least as many pieces as the other three classrooms combined?

19. Show that the number of ways to place n distinct objects into r different containers, with the objects in each container ordered, is $P(r + n - 1, r - 1)$.

20. a) Given positive integers m, n with $m \geq n$, show that the number of ways to distribute m identical objects into n distinct containers with no container left empty is $C(m - 1, m - n) = C(m - 1, n - 1)$.

    b) Show that the number of distributions in part (a) where each container holds at least r objects ($m \geq nr$) is $C(m - 1 + (1 - r)n, n - 1)$.

1.5 ■ AN APPLICATION IN THE PHYSICAL SCIENCES (OPTIONAL)

In this section we discuss an application of the counting techniques developed earlier in this chapter. This application arises in statistical mechanics and statistical thermodynamics. In these areas, one is interested in the number of ways n subatomic particles can be distributed among r distinct energy states, where different energy states can occupy the same energy level.

For the Maxwell-Boltzmann model the particles are assumed to be distinct, and any number of them can be in any energy state. Here we get r^n possible distributions, since there are r possible energy states for each of the n particles. (Modern theory of quantum mechanics has shown that this model is inappropriate for those subatomic particles known at present.)

Two more successful models are the Bose-Einstein and Fermi-Dirac models, which were originally based on the intrinsic angular momenta of the particles.

The Bose-Einstein model requires the n particles to be identical, with any number of them allowed in any energy state. The number of different distributions here is the number of nonnegative integer solutions to $x_1 + x_2 + \cdots + x_r = n$, which is $C(r + n - 1, n)$. Particles with integer spin (in units of $(h/2\pi)$, where $h = $ Planck's constant) follow this model. Such particles, called *bosons*, include photons, pi mesons, and the conduction electrons in superconductive material, like lead or tin.

For particles with half-integer spin, such as protons, electrons, and neutrons, the Fermi-Dirac model is more appropriate, and the particles are called *fermions*. In this model the n particles are again identical, but an energy state can contain at most one particle. Consequently, here $n \leq r$, and the number of possible distributions is $\binom{r}{n}$. (This model is quite useful in the study of semiconductor band theory.)

1.6 ■ SUMMARY AND HISTORICAL REVIEW

In this first chapter we introduced the fundamentals for counting combinations and permutations in a large variety of problems. The breakdown of problems into components requiring the same or different formulas for their solutions provided a key insight into the areas of discrete and combinatorial mathematics. As we continue to investigate further principles of enumeration, as well as discrete mathematical structures for applications in coding theory, enumeration, optimization, and sorting schemes in computer science, we shall still fall back on the fundamental ideas of this chapter.

The notion of permutation can be found in the Hebrew work *Sefer Yetzirah* (The Book of Creation), a mystic manuscript written sometime between 200 and 600 A.D. The first textbook dealing with some of the material of this chapter was *Ars Conjectandi* by Jakob Bernoulli (1654–1705). The text was published posthumously in 1731 and contained a reprint of the first formal treatise on probability, written by Christiaan Huygens in 1657.

The binomial theorem for $n = 2$ appears in the work of Euclid (300 B.C.), but it was not until the sixteenth century that the term "binomial coefficient" was actually introduced by Michel Stifel (1486–1567). In his *Arithmetica Integra* (1544) he gives the binomial coefficients up to the order of $n = 17$. Blaise Pascal (1623–1662), in his research on probability, published in the 1650's a treatise dealing with the relationships among binomial coefficients, combinations, and polynomials. These results were used by Jakob Bernoulli in proving the general form of the binomial theorem in a manner like that presented in this chapter. Actual use of the symbol $\binom{n}{r}$ did not occur until the nineteenth century, when it was used by Andreas von Ettingshausen (1796–1878).

It was not until the twentieth century, however, that the advent of the computer provided the necessary means for a systematic analysis of processes and algorithms to generate permutations and combinations. We shall examine one such algorithm in section 11.1.

The first comprehensive textbook dealing with topics in combinations and permutations was written by W. Whitworth [6]. Chapter 2 of D. Cohen [1], Chapter 1 of C. L. Liu [2], Chapter 1 of H. Ryser [4], and Chapter 2 of A. Tucker [5] also deal with the material of this chapter. For more on the ideas of statistical mechanics and statistical thermodynamics, the textbook by R. Reed and R. Roy [3] should be scrutinized.

REFERENCES

1. Cohen, Daniel I. A., *Basic Techniques of Combinatorial Theory*, John Wiley & Sons, New York, 1978.

2. Liu, C. L., *Introduction to Combinatorial Mathematics*, McGraw-Hill, New York, 1968.

3. Reed, Robert D., and Roy, R. R., *Statistical Physics for Students of Science and Engineering*, Intext Educational Publishers, Scranton, Pennsylvania, 1971.

4. Ryser, H. J., *Combinatorial Mathematics*, Published by the Mathematical Association of America, distributed by John Wiley & Sons, New York, 1963.

5. Tucker, Alan, *Applied Combinatorics*, John Wiley & Sons, New York, 1980.

6. Whitworth, W. A., *Choice and Chance*, reprint of the 5th edition (1901), Hafner, New York, 1965.

◼ MISCELLANEOUS EXERCISES

1. In the manufacture of a certain type of automobile there are four kinds of major defects and seven kinds of minor defects. For those situations in which defects do occur, in how many ways can there be twice as many minor defects as there are major ones?

2. There are nine different dials on a machine and each dial has five settings labeled 0, 1, 2, 3, 4.

   a) In how many ways can all the dials on the machine be set?

   b) If the nine dials are arranged in a line at the top of the machine, how many of the machine settings have no two adjacent dials with the same setting?

   c) How many machine settings in part (b) use only 0, 2, 4 as dial settings?

   d) How many machine settings in part (a) use only three different dial settings, where each dial setting appears three times?

3. a) How many distinct paths are there from $(-1, 2, 0)$ to $(1, 3, 7)$ in Euclidean three-space if each move is one of the following types:

   (H), $(x, y, z) \rightarrow (x + 1, y, z)$;   (V), $(x, y, z) \rightarrow (x, y + 1, z)$;

   (A), $(x, y, z) \rightarrow (x, y, z + 1)$.

    b)   How many such paths are there from $(1, 0, 5)$ to $(8, 1, 7)$?

    c)   Generalize the results in parts (a) and (b).

    d)   How many of the paths in part (a) start with an even number of A's?

4. A choir director must select six hymns for a Sunday church service. She has three hymn books and each contains 25 hymns (there are 75 different hymns in all). In how many ways can she select the hymns if she wishes to select (a) two hymns from each book? (b) at least one hymn from each book?

5. How many ways are there to place 25 different flags on ten numbered flagpoles if the order of the flags on a flagpole is (a) not relevant? (b) relevant? (c) relevant and every flagpole flies at least one flag?

6. How many distinct four-digit integers can one make from the digits 1, 3, 3, 7, 7, 8?

7. There are 12 men at a dance. (a) In how many ways can eight of them be selected to form a cleanup crew? (b) How many ways are there to pair off eight women at the dance with eight of these 12 men?

8. How many n-digit quaternary $(0, 1, 2, 3)$ sequences have exactly r 1's?

9. In how many ways can the letters of WONDERING be arranged with exactly two consecutive vowels?

10. An organic solvent is made by mixing six different liquid compounds. After a first compound is poured into a vat, the other compounds are added in a prescribed order. All possible orders are tested to find which produces the best yield. How many tests are needed?

11. In how many ways can the ten identical horses on a carousel be painted so that three are brown, three are white, and four are black?

12. In how many ways can a teacher distribute 12 different science books among 16 students if (a) no student gets more than one book? (b) the oldest student gets two books but no other student gets more than one book?

13. Given the following list of numbers: $-5, -4, -3, -2, -1, 1, 2, 3, 4$, four numbers are selected. (a) In how many ways can the selections be made so that the product of the four numbers is positive and (i) the numbers are distinct? (ii) each number may be selected as many as four times? (iii) each number may be selected at most three times? (b) Answer part (a) with the product of the four numbers negative.

14. a)   Find the coefficient of x^2yz^2 in $[(x/2) + y - 3z]^5$.

    b)   How many distinct terms are there in the complete expansion of

$$\left(\frac{x}{2} + y - 3z\right)^5?$$

    c)   What is the sum of all coefficients in the complete expansion?

15. a)   In how many ways can ten people, denoted $A, B, \ldots, I, J$, be seated about the rectangular table shown in Fig. 1.4, where Figs. 1.4(a) and 1.4(b) are considered the same, but different from Fig. 1.4(c)?

Figure 1.4

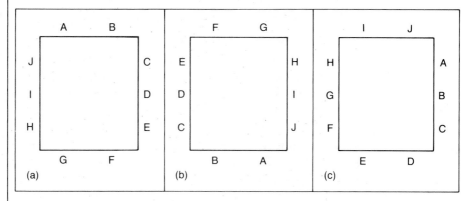

b)   In how many of the arrangements of part (a) are A and B seated on longer sides of the table across from each other?

16. a)   Determine the number of nonnegative integer solutions to the pair of equations $x_1 + x_2 + x_3 = 6$, $x_1 + x_2 + \cdots + x_5 = 15$, $x_i \geq 0$, $1 \leq i \leq 5$.

   b)   Answer part (a) with the pair of equations replaced by the pair of inequalities $x_1 + x_2 + x_3 \leq 6$, $x_1 + x_2 + \cdots + x_5 \leq 15$, $x_i \geq 0$, $1 \leq i \leq 5$.

17. A closed system consists of four photons and satisfies the Bose-Einstein model for statistical mechanics. The total energy of the system is $4E$, while each photon can have an energy level of kE, with k an integer, $0 \leq k \leq 4$, and E a positive constant. A photon of energy kE can occupy any one of $k^2 + 1$ distinct energy states at that energy level. In terms of energy states occupied by the photons, how many different configurations can this closed system assume?

18. A closed system consists of four electrons and satisfies the Fermi-Dirac model for statistical mechanics. The total energy of the system is $4E$, while each electron can have an energy level of kE, with k an integer, $0 \leq k \leq 4$, and E a positive constant. An electron of energy kE can occupy any one of $2(k^2 + 1)$ distinct energy states at that energy level. In terms of energy states occupied by electrons, how many different configurations can this closed system assume?

19. In how many ways can a dozen apples be distributed among five children so that no one gets more than seven apples?

20. For any given set in a tennis tournament, opponent A can beat opponent B in seven different ways. (At 6–6 they play a tie breaker.) The first opponent to win three sets wins the tournament. (a) In how many ways can scores be recorded with A winning in five sets? (b) In how many ways can scores be recorded with the tournament requiring at least four sets?

21. Let n be odd. In how many ways can we arrange n 1's and r 0's with a run (list of consecutive identical symbols) of exactly k 1's with $k \leq n < 2k$?

22. Given n distinct objects, in how many ways can r of these objects be arranged in a circle, where arrangements are considered the same if one can be obtained from the other by rotation?

23. For any positive integer n, show that

$$\binom{n}{0} + \binom{n}{2} + \binom{n}{4} + \cdots = \binom{n}{1} + \binom{n}{3} + \binom{n}{5} + \cdots .$$

24. a)   In how many ways can the letters in UNUSUAL be arranged?

b)   For the arrangements in part (a), how many have all three U's together?

c)   How many of the arrangements in part (a) have no consecutive U's?

d)   How many of the arrangements in part (a) have all four vowels together?

2

Enumeration In
Set Theory

Underlying the mathematics we study in algebra, geometry, combinatorics, and almost every other area of contemporary mathematics is the notion of a set. Very often this concept provides an underlying structure for a concise formulation of the mathematical topic being investigated. Consequently, many books on mathematics have an introductory chapter on set theory or mention in an appendix those parts of the theory that are needed in the text. Here it may appear that we have neglected set theory. We started with a first chapter on fundamentals of counting. In actuality we have relied upon intuition, since each time the word "collection" was mentioned in Chapter 1, we were dealing with a set.

Trying to define a set is rather difficult and often results in the circular use of such synonyms as class, collection, and aggregate. When we first began the study of geometry, we used our intuition for the ideas of point, line, and incidence. Then we started to define new terms and prove theorems, using these intuitive notions along with certain axioms and postulates. In our study of set theory, intuition is invoked once again, this time for the comparable ideas of element, set, and membership.

Before we start to study set theory, however, we introduce some material on logic. This material is closely tied to set theory and is the basis for many of the proofs that appear in the text.

2.1 ■ FUNDAMENTALS OF LOGIC

In the development of any mathematical theory, assertions are made in the form of sentences. Such verbal or written assertions, called *statements* or *propositions,* are declarative sentences that are either true or false. (Statements that require an individual's opinion, and hence may not be either true or false, will be avoided throughout this discussion.) For example, the following are propositions:

a) p: Combinatorics is a required course,

b) q: I am a computer science major, and

c) r: The plotter is out of order today.

These propositions can be thought of as *primitive* statements, for there is really no way to break them down into anything simpler. Primitive statements are used with *logical connectives* to form *compound* statements.

We can negate a proposition or combine two propositions as follows.

1. Negation: The *negation* of a statement p is denoted by $\bar{p}$, which is read "Not p." For p as above, $\bar{p}$ is the statement "Combinatorics is not a required course."

2. Conjunction: The *conjunction* of p, q is denoted by $p \wedge q$, which is read "p and q." In our example the compound statement $p \wedge q$ is read "Combinatorics is a required course, **and** I am a computer science major."

3. Disjunction: The expression $p \vee q$ denotes the *disjunction* of p, q, which is read "p or q." Hence, "Combinatorics is a required course, **or** I am a computer science major" is the verbal translation for $p \vee q$. We use the word "or" in the *inclusive* sense here. Consequently, $p \vee q$ is true if one or the other *or both* of the statements p, q are true. In English we sometimes write "and/or" to point this out. The

exclusive "or" is denoted by $p \veebar q$. The compound statement $p \veebar q$ is true if one or the other but *not both* of the statements p, q is true. One way to express $p \veebar q$ for the example here is, "Combinatroics is a required course, or I am a computer science major, but not both."

4. Implication: We say that "p implies q" and write $p \to q$ to designate the *implication* of q by p. Alternatively, we can say (a) If p, then q; (b) p is sufficient for q; (c) p only if q; and (d) q is necessary for p. A verbal translation of $p \to q$ for our example is "If combinatorics is a required course, then I am a computer science major." The statement p is called the *hypothesis* of the implication; q is called the *conclusion*. When statements are combined in this manner, there need not be any causal relation between the statements. We simply write $p \to q$ with the above definition for implication, whether p, q are related in some way or not.

5. Equivalence: Last, the *equivalence* of two statements p, q, is denoted by $p \leftrightarrow q$, which is read as "p is equivalent to q," "p if and only if q," or "p is necessary and sufficient for q." For p, q above, "Combinatorics is a required course if and only if I am majoring in computer science" conveys the meaning of $p \leftrightarrow q$.

In part 3 of the discussion above, we mentioned the circumstances under which the statements $p \vee q$, $p \veebar q$ are considered to be true, based on the truth of their primitive components p, q. This idea of determining the truth or falsity of a compound statement as a function of the truth values of its primitive components is worth further investigation. The *truth tables*, Tables 2.1 and 2.2, summarize the truth and falsity of the negation and other compound statements, based on the truth values of their primitive components. In constructing truth tables we write "0" for false and "1" for true.

Table 2.1   Table 2.2

p	$\bar{p}$
0	1
1	0

p	q	$p \wedge q$	$p \vee q$	$p \veebar q$	$p \to q$	$p \leftrightarrow q$
0	0	0	0	0	1	1
0	1	0	1	1	1	0
1	0	0	1	1	0	0
1	1	1	1	0	1	1

The listings of the four possible truth assignments for p, q can be carried out in any order. For later work the listing presented here will prove to be useful.

We see that the columns of truth values for p, $\bar{p}$ are the opposite of each other. The statement $p \wedge q$ is true only when both p, q are true, while $p \vee q$ is false only when both primitives are false. As stated above, $p \veebar q$ is true when exactly one of p, q is true.

For the implication $p \to q$, the result is true in all cases except where p is true and q is false. We do not want a true statement to lead us into believing something that is false. However, we regard as true a statement such as "If $2 + 3 = 6$, then $2 + 4 = 7$," even though the statements "$2 + 3 = 6$" and "$2 + 4 = 7$" are both false.

(In computer science the If-Then statement and If-Then-Else statement arise in such languages as BASIC and Pascal. The hypothesis p is often a relational expression such as $x > 2$. This expression then becomes a (logical) statement that has truth value 0 or 1, depending on the value of the variable x at that point in the program. The conclusion q is a statement that may be an executable statement directing the program to another line or causing some result to be printed. When dealing with "If p Then q," the computer executes q only on the condition that p is true. For p false the computer goes to the next (numbered) line in the program sequence. For a statement of the form "If p Then q Else r," q is executed when p is true and r is executed when p is false.)

The statement $p \leftrightarrow q$ is true exactly when the primitive statements have the same truth value.

Finally, we note that the results in Tables 2.1 and 2.2 also apply when compound statements are substituted for the symbols p, q. Example 2.1 demonstrates this.

EXAMPLE 2.1   Let us examine the truth table for the compound statement "I am a computer science major, and if the plotter is not out of order today, then combinatorics is a required course." In symbolic notation this is $q \wedge (\bar{r} \rightarrow p)$. The last column of Table 2.3 contains the truth values for this result. The preceding columns show how the truth table is built up by considering smaller parts of the compound statement and using the results from Tables 2.1 and 2.2.   □

Table 2.3

p	q	r	$\bar{r}$	$\bar{r} \rightarrow p$	$q \wedge (\bar{r} \rightarrow p)$
0	0	0	1	0	0
0	0	1	0	1	0
0	1	0	1	0	0
0	1	1	0	1	1
1	0	0	1	1	0
1	0	1	0	1	0
1	1	0	1	1	1
1	1	1	0	1	1

Table 2.4 shows a truth table for the statements $\bar{p} \vee q$ and $p \rightarrow q$. We see that the truth values for $\bar{p} \vee q$ and $p \rightarrow q$ are exactly the same. This situation leads us to the following idea.

Definition 2.1   ▶   Two statements s_1, s_2 are said to be *logically equivalent*, written $s_1 \Leftrightarrow s_2$, when the truth tables for s_1, s_2 are exactly the same.

Table 2.4

p	q	$\bar{p}$	$\bar{p} \vee q$	$p \rightarrow q$
0	0	1	1	1
0	1	1	1	1
1	0	0	0	0
1	1	0	1	1

As a result of this concept, we see that we do not need the connective for implication. The same result can be achieved by using negation and disjunction. In the same manner, from the result in Table 2.5 we have $(p \leftrightarrow q) \Leftrightarrow (p \rightarrow q) \wedge (q \rightarrow p)$. Using the logical equivalence from Table 2.4, we can also write $(p \leftrightarrow q) \Leftrightarrow (\bar{p} \vee q) \wedge (\bar{q} \vee p)$. This reduces by one, again, the number of logical connectives we need to study. Examining Table 2.6, we find that negation, along with the connectives $\wedge$ and $\vee$, are all we really need. In fact, we may even eliminate either $\wedge$ or $\vee$. However, for the applications we want to study, we need both $\wedge$ and $\vee$ as well as negation.

Table 2.5

p	q	$p \rightarrow q$	$q \rightarrow p$	$(p \rightarrow q) \wedge (q \rightarrow p)$	$p \leftrightarrow q$
0	0	1	1	1	1
0	1	1	0	0	0
1	0	0	1	0	0
1	1	1	1	1	1

Table 2.6

p	q	$p \veebar q$	$(p \vee q)$	$\overline{(p \wedge q)}$	$(p \vee q) \wedge \overline{(p \wedge q)}$
0	0	0	0	1	0
0	1	1	1	1	1
1	0	1	1	1	1
1	1	0	1	0	0

Consider the truth table in Table 2.7. The proposition $p \rightarrow (p \vee q)$ is always true; the proposition $p \wedge (\bar{p} \wedge q)$ is always false. These examples illustrate the following idea.

Table 2.7

p	q	$p \vee q$	$p \rightarrow (p \vee q)$	$(\bar{p} \wedge q)$	$p \wedge (\bar{p} \wedge q)$
0	0	0	1	0	0
0	1	1	1	1	0
1	0	1	1	0	0
1	1	1	1	0	0

Definition 2.2

▶ A proposition that is always true is called a *tautology;* one that is always false is labeled a *contradiction*.

We use the symbol T_0 to denote any tautology, F_0 to designate any contradiction.

For statements s_1, s_2, if $s_1 \leftrightarrow s_2$ is a tautology, then s_1 and s_2 must have the same truth tables, so $s_1 \Leftrightarrow s_2$. Whenever $s_1 \rightarrow s_2$ is a tautology, we say that s_1 *logically implies* s_2 and write $s_1 \Rightarrow s_2$. For example, for any propositions p, q, $p \Rightarrow p \vee q$.

We will use the idea of logical implication to prove theorems in the text. For the most part we deal with $[p \wedge (p \rightarrow q)] \Rightarrow q$, a *rule of inference* often referred to as "*modus ponens*." In addition, $[(p \rightarrow q) \wedge \bar{q}] \Rightarrow \bar{p}$ (*modus tollens*) and $[(p \rightarrow q) \wedge (q \rightarrow r)] \Rightarrow (p \rightarrow r)$ (*law of the syllogism*) are also used to some degree. Modus tollens is often referred to as the "proof by contradiction." There are many other such rules, but we choose not to develop this topic any further. Nonetheless, the reader who is interested in pursuing this topic should consult the chapter references.

Using the concepts of logical equivalence, tautology, and contradiction, we state the following list of laws for the algebra of propositions.

The Laws of Logic

For any propositions p, q, r,

1. $\bar{\bar{p}} \Leftrightarrow p$ Law of *Double Negation*

2. $\overline{p \vee q} \Leftrightarrow \bar{p} \wedge \bar{q}$ *DeMorgan's* Laws
   $\overline{p \wedge q} \Leftrightarrow \bar{p} \vee \bar{q}$

3. $p \vee q \Leftrightarrow q \vee p$ *Commutative* Laws
   $p \wedge q \Leftrightarrow q \wedge p$

4. $p \vee (q \vee r) \Leftrightarrow (p \vee q) \vee r$ *Associative* Laws
   $p \wedge (q \wedge r) \Leftrightarrow (p \wedge q) \wedge r$

5. $p \vee (q \wedge r) \Leftrightarrow (p \vee q) \wedge (p \vee r)$ *Distributive* Laws
   $p \wedge (q \vee r) \Leftrightarrow (p \wedge q) \vee (p \wedge r)$

6. $p \vee p \Leftrightarrow p$ *Idempotent* Laws
   $p \wedge p \Leftrightarrow p$

7. $p \vee F_0 \Leftrightarrow p$ *Identity* Laws
   $p \wedge T_0 \Leftrightarrow p$

8. $p \vee \bar{p} \Leftrightarrow T_0$ *Inverse* Laws
   $p \wedge \bar{p} \Leftrightarrow F_0$

9. $p \vee T_0 \Leftrightarrow T_0$ *Domination* Laws
   $p \wedge F_0 \Leftrightarrow F_0$

10. $p \vee (p \wedge q) \Leftrightarrow p$ *Absorption* Laws
    $p \wedge (p \vee q) \Leftrightarrow p$

To prove all these results, we could write down the truth tables and compare the corresponding truth values in each case. However, aside from the Law of Double Negation, these laws seem to fall naturally into pairs.

Definition 2.3 ▶ If s is a proposition, the *dual* of s, denoted s^d, is the proposition obtained by replacing each occurrence of $\wedge$ ($\vee$) in s by $\vee$ ($\wedge$), and each occurrence of T_0 (F_0) by F_0 (T_0).

The statements $p \vee \bar{p} \Leftrightarrow T_0$ and $p \wedge \bar{p} \Leftrightarrow F_0$ are duals of each other.

We now state and use a theorem without proving it. However, in Chapter 9 we shall justify the result that appears here.

Theorem
2.1

▶ (*The Principle of Duality*) If s is a theorem in the algebra of propositions, then its dual, s^d, is likewise a theorem.

As a result, laws 2–10 in our list can be established by proving one of the laws in each pair and then invoking this principle.

We close this section now with an application in simplifying switching networks. Here the laws of logic prove quite helpful.

A switching network is made up of wires and switches connecting two terminals T_1 and T_2. In such a network, any switch is either open (0), so that no current flows through it, or closed (1), so that current does flow through it.

In Fig. 2.1 we have in (a) a network with one switch. Each of (b) and (c) contains two (independent) switches.

For the network in (b), current flows from T_1 to T_2 if either of the switches p, q is closed. We call this a *parallel* network and represent it by $p \vee q$. The network in (c) requires that each of the switches p, q be closed in order for current to flow from T_1 to T_2. Here the switches are in *series,* and this network is represented by $p \wedge q$.

Figure 2.1

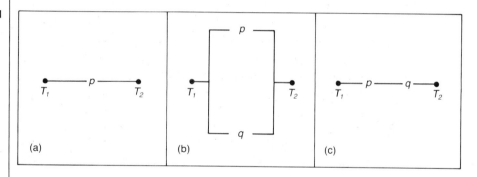

(a)   (b)   (c)

EXAMPLE 2.2

The switches in a network need not act independently of each other. Consider the network in Fig. 2.2(a). Here the switches marked by t and $\bar{t}$ are not independent. We have coupled these two switches so that t is open (closed) if and only if $\bar{t}$ is simulta-

Figure 2.2

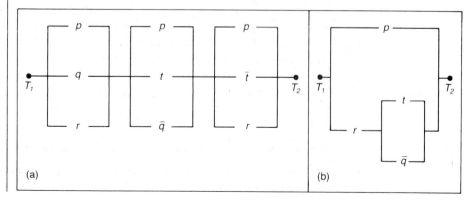

(a)   (b)

neously closed (open). The same is true of the switches at q, $\bar{q}$. (Also, for example, the three switches labeled with p are not independent.)

This network is represented by $(p \vee q \vee r) \wedge (p \vee t \vee \bar{q}) \wedge (p \vee \bar{t} \vee r)$. Using the laws of logic, we simplify this proposition, which represents the network, as follows.

	Reason
$(p \vee q \vee r) \wedge (p \vee t \vee \bar{q}) \wedge (p \vee \bar{t} \vee r)$	**Reason**
$\Leftrightarrow p \vee [(q \vee r) \wedge (t \vee \bar{q}) \wedge (\bar{t} \vee r)]$	Distributive Law of $\vee$ over $\wedge$
$\Leftrightarrow p \vee [([(q \vee r) \wedge t] \vee [(q \vee r) \wedge \bar{q}]) \wedge (\bar{t} \vee r)]$	Distributive Law of $\wedge$ over $\vee$
$\Leftrightarrow p \vee [([(q \vee r) \wedge t] \vee [(q \wedge \bar{q}) \vee (r \wedge \bar{q})]) \wedge (\bar{t} \vee r)]$	Distributive Law of $\wedge$ over $\vee$
$\Leftrightarrow p \vee [([(q \vee r) \wedge t] \vee (r \wedge \bar{q})) \wedge (\bar{t} \vee r)]$	$q \wedge \bar{q} \Leftrightarrow F_0$, the identity for $\vee$
$\Leftrightarrow p \vee [([(q \vee r) \wedge t] \wedge (\bar{t} \vee r)) \vee ((r \wedge \bar{q}) \wedge (\bar{t} \vee r))]$	Why?
$\Leftrightarrow p \vee [[(q \vee r) \wedge (t \wedge (\bar{t} \vee r))] \vee [(r \wedge \bar{q} \wedge \bar{t}) \vee (r \wedge \bar{q} \wedge r)]]$	Why?
$\Leftrightarrow p \vee [[(q \vee r) \wedge (t \wedge r)] \vee [(r \wedge \bar{q} \wedge \bar{t}) \vee (r \wedge \bar{q})]]$	Why?
$\Leftrightarrow p \vee [[(q \vee r) \wedge (t \wedge r)] \vee (r \wedge \bar{q})]$	Absorption Law
$\Leftrightarrow p \vee [(q \wedge t \wedge r) \vee (r \wedge t \wedge r) \vee (r \wedge \bar{q})]$	Distributive Law of $\wedge$ over $\vee$
$\Leftrightarrow p \vee [((q \wedge t \wedge r) \vee (r \wedge t)) \vee (r \wedge \bar{q})]$	Idempotent Law
$\Leftrightarrow p \vee [(r \wedge t) \vee (r \wedge \bar{q})]$	Absorption Law
$\Leftrightarrow p \vee [r \wedge (t \vee \bar{q})]$	Distributive Law of $\wedge$ over $\vee$

Hence $(p \vee q \vee r) \wedge (p \vee t \vee \bar{q}) \wedge (p \vee \bar{t} \vee r) \Leftrightarrow p \vee [r \wedge (t \vee \bar{q})]$, and the network in Fig. 2.2(b) is equivalent to the original network in the sense that current flows from T_1 to T_2 in network (a) exactly when it does in network (b). But in (b) the network has only four switches, five fewer than in network (a). □

EXERCISES

1. Let p, q, r, s denote the statements: p: I finish writing my computer program before lunch; q: I play tennis in the afternoon; r: The sun is shining; s: The humidity is low. Write the following in symbolic form.

a) If the sun is shining, I shall play tennis this afternoon.

b) Finishing the writing of my computer program before lunch is necessary for my playing tennis this afternoon.

c) The sun is shining and I shall play tennis this afternoon.

d) Low humidity and sunshine are sufficient for me to play tennis this afternoon.

2. Let p, q, r denote the following statements: p: Triangle ABC is isosceles; q: Triangle ABC is equilateral; r: Triangle ABC is equiangular. Translate each of the following into an English sentence.

   a) $q \rightarrow p$;   b) $\bar{p} \rightarrow \bar{q}$;   c) $q \leftrightarrow r$;   d) $p \wedge \bar{q}$;   e) $r \rightarrow p$

3. Construct a truth table for each of the following propositions.

   a) $p \rightarrow (p \vee q)$ b) $p \rightarrow (q \rightarrow r)$ c) $(p \rightarrow q) \rightarrow r$

   d) $(p \rightarrow q) \rightarrow (q \rightarrow p)$ e) $[p \wedge (p \rightarrow q)] \rightarrow q$ f) $(p \wedge q) \rightarrow p$

   g) $q \leftrightarrow (\bar{p} \vee \bar{q})$ h) $[(p \rightarrow q) \wedge (q \rightarrow r)] \rightarrow (p \rightarrow r)$

4. Are any of the propositions in Exercise 3 tautologies?

5. a) If p, q are propositions, show that $(p \rightarrow q) \Leftrightarrow (\bar{q} \rightarrow \bar{p})$ and $(q \rightarrow p) \Leftrightarrow (\bar{p} \rightarrow \bar{q})$.

   b) The statement $q \rightarrow p$ is called the *converse* of $p \rightarrow q$; $\bar{p} \rightarrow \bar{q}$ is its *inverse*, while $\bar{q} \rightarrow \bar{p}$ is its *contrapositive*.

   Write the converse, inverse, and contrapositive of the statement: If quadrilateral $ABCD$ is a square, then quadrilateral $ABCD$ is a rectangle.

6. Negate each of the following, and simplify the resulting proposition.

   a) $p \wedge (q \vee r) \wedge (\bar{p} \vee \bar{q} \vee r)$ b) $(p \wedge q) \rightarrow r$

   c) $p \rightarrow (\bar{q} \wedge r)$ d) $p \vee q \vee (\bar{p} \wedge \bar{q} \wedge r)$

7. a) How many rows are needed for the truth table of the proposition $(p \vee \bar{q}) \leftrightarrow [(\bar{r} \wedge s) \rightarrow t]$?

   b) If $p_1, p_2, \ldots, p_n$ are primitive propositions and the compound statement p contains at least one occurrence of each p_i, $1 \leq i \leq n$, how many rows are needed to construct the truth table for p?

8. Determine all truth value assignments, if any, for p, q, r, s, t, that make each of the following propositions false.

   a) $(p \wedge q \wedge r) \rightarrow (s \vee t)$

   b) $(p \wedge q \wedge r) \rightarrow (s \veebar t)$

9. a) Prove that $(\bar{p} \vee q) \wedge (p \wedge (p \wedge q)) \Leftrightarrow (p \wedge q)$.

   b) Write the dual of the result in part (a).

10. Write the duals for (a) $p \rightarrow q$, (b) $p \leftrightarrow q$, and (c) $p \veebar q$.

11. Define the connective "Nand," or "Not . . . and," by $(p \uparrow q) \Leftrightarrow (\overline{p \wedge q})$, for propositions p, q. Represent the following using only this connective:

    a) $\bar{p}$ b) $p \vee q$ c) $p \wedge q$ d) $p \rightarrow q$ e) $p \leftrightarrow q$

12. The connective "Nor," or "Not...or," is defined by $(p \downarrow q) \Leftrightarrow (\overline{p \vee q})$, for propositions p, q. Represent the propositions in parts (a)–(e) of Exercise 11, using only this connective.

13. If p, q are propositions, prove that
    a) $(\overline{p \downarrow q}) \Leftrightarrow (\overline{p} \uparrow \overline{q})$
    b) $(\overline{p \uparrow q}) \Leftrightarrow (\overline{p} \downarrow \overline{q})$

14. Simplify each of the networks in Fig. 2.3.

Figure 2.3

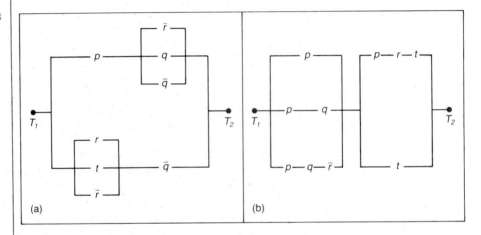

(a)

(b)

15. For the following BASIC program, how many times is the PRINT statement of line 40 executed?

```
10 X = 10
20 FOR I = 1 TO 7
30      FOR J = 1 TO I + 3
40              IF  ((X > 8) OR (I > 5 AND J < 10)) THEN   PRINT X
50      NEXT J
60      X = X - 1
70 NEXT I
80 END
```

16. A segment of a Pascal program contains a Repeat-Until loop structured as:

    Repeat
    ....

    ....

    Until (x <> 0 And y > 0) Or (Not (w > 0 And t = 3));

    For each of the following assignments for the variables x, y, w, t, determine whether or not the loop terminates.

    a) $x = 7, y = 2, w = 5, t = 3$.
    b) $x = 0, y = 2, w = -3, t = 3$.
    c) $x = 0, y = -1, w = 1, t = 3$.
    d) $x = 1, y = -1, w = 1, t = 3$.
    e) $x = 0, y = 3, w = 7, t = 4$.

2.2 ■ SETS AND SUBSETS

We have a "gut feeling" that a set should be a well-defined collection of elements. These elements are often called *objects* and are said to be members of the set. The adjective "well-defined" implies that for any object we care to consider, we are able to determine whether or not it is in the set under scrutiny. Consequently, we avoid dealing with sets that depend upon opinion, such as the set of outstanding major league pitchers for the 1970's.

We use capital letters, such as A, B, C, ..., for representing sets and the lowercase letters for elements. For a set A we write $x \in A$ if x is an element of A; $y \notin A$ indicates that y is not a member of A.

EXAMPLE 2.3

Using set braces a set A can be designated $A = \{1, 2, 3, 4, 5\}$. Here $2 \in A$ but $6 \notin A$. Alternatively, this set can be described verbally as the set of positive integers from 1 to 5, inclusive. In addition we can also write A as $\{x \mid x$ is an integer, $1 \le x \le 5\}$. Here the properties following "$\mid$" help us determine the elements of A.

Beware! $\{x \mid 1 \le x \le 5\}$ does not adequately describe set A since $x = 3/2$, for example, satisfies the property $1 \le x \le 5$, but is not a member of A. One may feel that since $x = 3/2$ is not an integer it should be eliminated from consideration. The real dilemma here is ignorance as to which elements (in this case, numbers) we should be considering. When dealing with a particular problem there is a *universe*, or *universal set*, stated or implied, from which we select the elements to form our sets. In this problem if we knew that the universe, denoted $\mathcal{U}$, was the set of all integers or the set of all positive integers, then $\{x \mid 1 \le x \le 5\}$ would adequately describe A. □

A change in the universe can change a set. If $\mathcal{U}$ is the set of positive integers, then $S = \{x \mid -2 \le x \le 3\}$ contains the three elements 1, 2, 3. However, for $\mathcal{U}$ the set of all integers, S becomes $\{-2, -1, 0, 1, 2, 3\}$.

EXAMPLE 2.4

For the universe $\mathcal{U} = \{1, 2, 3, 4, 5\}$, consider a set $A = \{1, 2\}$. If $B = \{x \mid x^2 \in \mathcal{U}\}$ the members of B are 1, 2. Since A and B have the same members they are considered to be the same set. This leads to the following definition. □

Definition 2.4

▶ For a universe $\mathcal{U}$, sets A and B (taken from $\mathcal{U}$) are said to be *equal*, and we write $A = B$, if A and B contain the same elements.

From this definition we find that neither order nor repetition is relevant for a set, so $\{1, 2, 3\} = \{3, 1, 2\} = \{2, 2, 1, 3\} = \{1, 2, 1, 3, 1\}$.

EXAMPLE 2.5

For $\mathcal{U} = \{1, 2, 3, \ldots\}$, the set of positive integers, let

a) $A = \{1, 4, 9, \ldots, 64, 81\} = \{x^2 \mid x \in \mathcal{U}, x^2 < 100\}$

b) $B = \{1, 4, 9, 16\} = \{y^2 \mid y \in \mathcal{U}, y^2 < 20\} = \{y^2 \mid y \in \mathcal{U}, y^2 < 23\}$

c) $C = \{2, 4, 6, 8, \ldots\} = \{2k \mid k \in \mathcal{U}\}$.

Sets A and B are examples of *finite* sets, while C is called an *infinite* set. When dealing with sets like A or C, we can either describe the sets in terms of properties the elements must satisfy or list enough elements to indicate what is, we hope, an obvious pattern. For any finite set A, $|A|$ denotes the number of elements in A and is referred to as the *cardinality*, or *size*, of A.

Here the sets B and A are such that every element of B is also an element of A. This important relationship occurs throughout set theory and its applications, and leads to the following definition.   □

Definition 2.5

▶ If C, D are sets from a universe $\mathcal{U}$, we say that C is a *subset* of D, and write $C \subseteq D$ or $D \supseteq C$, if every element of C is likewise an element of D. If, in addition, there is an element of D that is not in C, C is called a *proper subset* of D, denoted $C \subset D$, or $D \supset C$.

EXAMPLE 2.6

A variable name in ANSI (American National Standards Institute) FORTRAN consists of a single letter followed by at most five characters (letters or digits). If $\mathcal{U}$ is the set of all such variable names, then by the rules of sum and product, $|\mathcal{U}| = 26 + 26(36) + 26(36)^2 + \cdots + 26(36)^5 = 26\sum_{i=0}^{5} 36^i = 1,617,038,306$, so $\mathcal{U}$ is a large, but still finite, set. An integer variable in this programming language must start with I, J, K, L, M, N. So if A is the subset of all integer variables in ANSI FORTRAN, $|A| = 6 + 6(36) + 6(36)^2 + \cdots + 6(36)^5 = 6\sum_{i=0}^{5} 36^i = 373,162,686$.   □

Using the concept of subset, the following theorem characterizes set equality.

Theorem 2.2

▶ Let A, $B \subseteq \mathcal{U}$. Then $A = B$ if and only if $A \subseteq B$ and $B \subseteq A$.

Proof

The proof of this result follows from Definitions 2.4 and 2.5. The details are left for the reader. (Often "iff" will be used to abbreviate "if and only if".)   ■

EXAMPLE 2.7

Let $\mathcal{U} = \{1, 2, 3, 4, 5\}$ with $A = \{1, 2, 3\}$, $B = \{3, 4\}$, $C = \{1, 2, 3, 4\}$. Then the following subset relations hold:

a) $A \subseteq C$ b) $A \subset C$ c) $B \subset C$ d) $A \subseteq A$

e) $B \nsubseteq A$ (i.e., B is not a subset of A)   f) $A \not\subset A$.

In addition to A, B, we are interested in determining how many subsets C has in total. Before answering this, however, we need to introduce the set with no members.   □

Definition 2.6

▶ The *null*, or *empty*, *set* is a set containing no elements, and is denoted by $\emptyset$ or $\{\ \}$.

We note that $|\emptyset| = 0$ but $\{0\} \neq \emptyset$. Also, $\emptyset \neq \{\emptyset\}$ as $\{\emptyset\}$ is a set with one element, namely the null set.

We are primarily interested in the situation where two sets satisfy a subset relation. However, for sets A, B from universe $\mathcal{U}$, it can happen that A is not a subset of B. This is denoted by $A \nsubseteq B$ and occurs when there is an element $x \in A$ such that $x \notin B$.

Theorem 2.3

▶ For any set A, $\emptyset \subseteq A$; $\emptyset \subset A$ if $A \neq \emptyset$.

Proof

If the first result is not true, then $\emptyset \not\subseteq A$, so there is an element x from the universe with $x \in \emptyset$ but $x \notin A$. But $x \in \emptyset$ is impossible. In addition, if $A \neq \emptyset$, then there is an element $a \in A$ (and $a \notin \emptyset$) so $\emptyset \subset A$. ∎

EXAMPLE 2.8

Returning now to Example 2.7 we determine the number of subsets of the set $C = \{1, 2, 3, 4\}$. In constructing a subset of C, we have, for each member x of C, two distinct choices, namely either include it in the subset or exclude it. Consequently there are $2 \times 2 \times 2 \times 2$ total choices, resulting in $2^4 = 16$ subsets of C. These include the empty set $\emptyset$ and the set C itself. Should we need the number of subsets of two elements from C, the result is the number of ways two objects can be selected from a set of four objects, namely $C(4, 2)$ or $\binom{4}{2}$. As a result the total number, 2^4, of subsets of C can also be realized as the sum $\binom{4}{0} + \binom{4}{1} + \binom{4}{2} + \binom{4}{3} + \binom{4}{4}$, where the first summand is for the empty set, the second summand for the four *singleton* subsets, the third summand for the six subsets of size two, etc. So $2^4 = \sum_{k=0}^{4} \binom{4}{k}$. □

Definition 2.7

▶ If A is a set from universe $\mathcal{U}$, the *power set* of A, denoted $\mathcal{P}(A)$, is the collection of all subsets of A.

EXAMPLE 2.9

For the set C of Example 2.8, $\mathcal{P}(C) = \{\emptyset, \{1\}, \{2\}, \{3\}, \{4\}, \{1, 2\}, \{1, 3\}, \{1, 4\}, \{2, 3\}, \{2, 4\}, \{3, 4\}, \{1, 2, 3\}, \{1, 2, 4\}, \{1, 3, 4\}, \{2, 3, 4\}, C\}$.

In general, for any finite set A with $|A| = n \geq 0$, A has 2^n subsets, so $|\mathcal{P}(A)| = 2^n$. For any $0 \leq k \leq n$, there are $\binom{n}{k}$ subsets of size k. Counting the subsets of A according to the number, k, of elements in a subset, we have the combinatorial identity $\binom{n}{0} + \binom{n}{1} + \binom{n}{2} + \cdots + \binom{n}{n} = 2^n$, for $n \geq 0$.

This identity was established earlier in Corollary 1.1(a). The presentation here is an example of a combinatorial proof since the identity is established by counting the same collection of objects (subsets of A) in two different ways.

With the derivation of one combinatorial identity behind us, let us tackle another interesting result.

EXAMPLE 2.10

For integers n, r with $n \geq r \geq 1$,

$$\binom{n+1}{r} = \binom{n}{r} + \binom{n}{r-1}.$$

Although this can be established algebraically from the definition of $\binom{n}{r}$ as $n!/(r!(n-r)!)$, we use a combinatorial approach. Let $A = \{x, a_1, a_2, \ldots, a_n\}$ and consider all subsets of A that contain r elements. There are $\binom{n+1}{r}$ such subsets. These fall into exactly one of the following two cases: those that contain the element x and those that do not. To get a subset C of A, where $x \in C$ and $|C| = r$, place x in C and then select $r - 1$ of the elements $a_1, a_2, \ldots, a_n$. This can be done in $\binom{n}{r-1}$ ways. For the other case we want a subset B of A with $|B| = r$ and $x \notin B$. So we select r elements from among $a_1, a_2, \ldots, a_n$, which we can do in $\binom{n}{r}$ ways. It then follows by the rule of sum that $\binom{n+1}{r} = \binom{n}{r} + \binom{n}{r-1}$.   □

EXAMPLE 2.11   We now investigate how the identity of Example 2.10 can help us solve Example 1.29, where we sought the number of nonnegative integer solutions of the inequality $x_1 + x_2 + \cdots + x_6 < 10$.

For each integer k, $0 \le k \le 9$, the number of solutions to $x_1 + x_2 + \cdots + x_6 = k$ is $\binom{6+k-1}{k} = \binom{5+k}{k}$. So the number of nonnegative integer solutions to $x_1 + x_2 + \cdots + x_6 < 10$ is

$$\binom{5}{0} + \binom{6}{1} + \binom{7}{2} + \binom{8}{3} + \cdots + \binom{14}{9}$$

$$= \left[\binom{6}{0} + \binom{6}{1}\right] + \binom{7}{2} + \binom{8}{3} + \cdots + \binom{14}{9}$$

$$= \left[\binom{7}{1} + \binom{7}{2}\right] + \binom{8}{3} + \cdots + \binom{14}{9} = \left[\binom{8}{2} + \binom{8}{3}\right] + \binom{9}{4} + \cdots + \binom{14}{9}$$

$$= \left[\binom{9}{3} + \binom{9}{4}\right] + \cdots + \binom{14}{9} = \cdots = \binom{14}{8} + \binom{14}{9} = \binom{15}{9} = 5005.   □$$

EXAMPLE 2.12   In Fig. 2.4 we find a part of the rather useful and interesting array of numbers called *Pascal's triangle*.

Figure 2.4

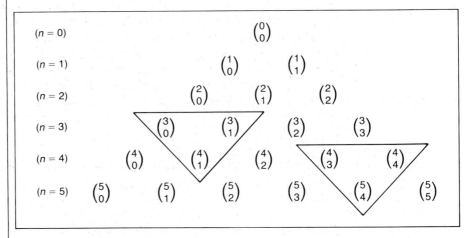

($n = 0$) $\binom{0}{0}$

($n = 1$) $\binom{1}{0}$ $\binom{1}{1}$

($n = 2$) $\binom{2}{0}$ $\binom{2}{1}$ $\binom{2}{2}$

($n = 3$) $\binom{3}{0}$ $\binom{3}{1}$ $\binom{3}{2}$ $\binom{3}{3}$

($n = 4$) $\binom{4}{0}$ $\binom{4}{1}$ $\binom{4}{2}$ $\binom{4}{3}$ $\binom{4}{4}$

($n = 5$) $\binom{5}{0}$ $\binom{5}{1}$ $\binom{5}{2}$ $\binom{5}{3}$ $\binom{5}{4}$ $\binom{5}{5}$

Notice that in this partial listing the two triangles shown satisfy the condition that the binomial coefficient at the bottom of the inverted triangle is the sum of the other two terms in the triangle. This follows from the identity in Example 2.10.

Replacing each of the binomial coefficients by its numerical value, the Pascal triangle appears as in Fig. 2.5.   □

Figure 2.5

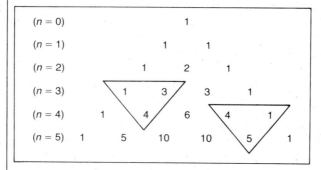

There are certain sets of numbers that appear frequently throughout the text. Consequently, we close this section by giving them the following designations.

a) $\mathbf{Z}$ = the set of *integers* = $\{0, 1, -1, 2, -2, 3, -3, \ldots\}$
b) $\mathbf{N}$ = the set of *nonnegative integers* or *natural numbers* = $\{0, 1, 2, 3, \ldots\}$
c) $\mathbf{Z}^+$ = the set of *positive integers* = $\{1, 2, 3, \ldots\}$ = $\{x \in \mathbf{Z} \mid x > 0\}$
d) $\mathbf{Q}$ = the set of *rational numbers* = $\{a/b \mid a, b \in \mathbf{Z}, b \neq 0\}$
e) $\mathbf{Q}^+$ = the set of *positive rational numbers* = $\{a/b \mid a/b \in \mathbf{Q}, a/b > 0\}$
f) $\mathbf{Q}^*$ = the set of *nonzero rational numbers*
g) $\mathbf{R}$ = the set of *real numbers*
h) $\mathbf{R}^+$ = the set of *positive real numbers*
i) $\mathbf{R}^*$ = the set of *nonzero real numbers*
j) $\mathbf{C}$ = the set of *complex numbers* = $\{x + yi \mid x, y \in \mathbf{R}, i^2 = -1\}$
k) $\mathbf{C}^*$ = the set of *nonzero complex numbers*
ℓ) For any $n \in \mathbf{Z}^+$, $\mathbf{Z}_n = \{0, 1, 2, \ldots, n - 1\}$
m) For real numbers a, b with $a < b$, $[a, b] = \{x \mid x \in \mathbf{R}, a \leq x \leq b\}$, $(a, b) = \{x \mid x \in \mathbf{R}, a < x < b\}$, $[a, b) = \{x \mid x \in \mathbf{R}, a \leq x < b\}$, $(a, b] = \{x \mid x \in \mathbf{R}, a < x \leq b\}$. The first set is called a *closed interval,* the second set an *open interval,* and the other two sets are called *half-open intervals.*

EXERCISES

1. Which of the following sets are equal?
  a) $\{1, 2, 3\}$ b) $\{3, 2, 1, 3\}$ c) $\{3, 1, 2, 3\}$ d) $\{1, 2, 2, 3\}$
2. Let $A = \{1, \{1\}, \{2\}\}$. Which of the following statements are true?
  a) $1 \in A$ b) $\{1\} \in A$ c) $\{1\} \subseteq A$ d) $\{\{1\}\} \subseteq A$
  e) $\{2\} \in A$ f) $\{2\} \subseteq A$ g) $\{\{2\}\} \subseteq A$ h) $\{\{2\}\} \subset A$
3. For $A = \{1, 2, \{2\}\}$, which of the eight statements in Exercise 2 are true?
4. Which of the following statements are true?
  a) $\emptyset \in \emptyset$ b) $\emptyset \subset \emptyset$ c) $\emptyset \subseteq \emptyset$ d) $\emptyset \in \{\emptyset\}$ e) $\emptyset \subset \{\emptyset\}$ f) $\emptyset \subseteq \{\emptyset\}$

5. If a set A has 63 proper subsets, what is $|A|$?

6. For $A = \{1, 2, 3, 4, 5, 6, 7\}$, determine the number of

   a) subsets of A.

   b) nonempty subsets of A.

   c) proper subsets of A.

   d) nonempty proper subsets of A.

   e) subsets of A containing three elements.

   f) subsets of A containing $1, 2$.

   g) subsets of A containing five elements, including $1, 2$.

   h) proper subsets of A containing $1, 2$.

   i) subsets of A with an even number of elements.

   j) subsets of A with an odd number of elements.

   k) subsets of A with an odd number of elements, including the element 3.

7. a) If a set A has 64 subsets of odd cardinality, what is $|A|$?

   b) Generalize the result of part (a).

8. Which of the following sets are nonempty?

   a) $\{x \mid x \in \mathbf{N},\ 2x + 7 = 3\}$ b) $\{x \mid x \in \mathbf{Z},\ 3x + 5 = 9\}$

   c) $\{x \mid x \in \mathbf{Q},\ x^2 + 4 = 6\}$ d) $\{x \mid x \in \mathbf{R},\ x^2 + 4 = 6\}$

   e) $\{x \mid x \in \mathbf{R},\ x^2 + 5 = 4\}$ f) $\{x \mid x \in \mathbf{R},\ x^2 + 3x + 3 = 0\}$

   g) $\{x \mid x \in \mathbf{C},\ x^2 + 3x + 3 = 0\}$

9. About to leave a restaurant counter a man sees that he has one penny, one nickel, one dime, one quarter, and one half-dollar. In how many ways can he leave some (at least one) of his coins for a tip if

   a) there are no restrictions?

   b) he wants to have some change left?

   c) he wants to leave at least 10 cents?

10. Prove Theorem 2.2.

11. Give an example of three sets W, X, Y such that $W \in X$, $X \in Y$ but $W \notin Y$.

12. Write the next three rows for the Pascal triangle in Fig. 2.5.

13. For sets A, B, C prove each of the following.

    a) If $A \subseteq B$, $B \subseteq C$, then $A \subseteq C$.   b) If $A \subset B$, $B \subseteq C$, then $A \subset C$.

    c) If $A \subseteq B$, $B \subset C$, then $A \subset C$.   d) If $A \subset B$, $B \subset C$, then $A \subset C$.

14. For sets $A, B, C \subseteq \mathcal{U}$, prove or disprove (with a counterexample), the statement: If $A \subseteq B$, $B \nsubseteq C$, then $A \nsubseteq C$.

15. One quarter of the five-element subsets of $\{1, 2, 3, \ldots, n\}$ contain the element 7. Determine n.

16. Establish the identity in Example 2.10 algebraically.

17. Give a combinatorial argument to show that for integers n, r with $n \geq r \geq 2$,

$$\binom{n+2}{r} = \binom{n}{r} + 2\binom{n}{r-1} + \binom{n}{r-2}.$$

18. For positive integers n, r show that

$$\binom{n+r+1}{r} = \binom{n+r}{r} + \binom{n+r-1}{r-1} + \cdots$$

$$+ \binom{n+2}{2} + \binom{n+1}{1} + \binom{n}{0}.$$

19. Let $A = \{1, 2, 3, \ldots, 39, 40\}$.
    a) Write a computer program to generate six-element subsets of A.
    b) For $B = \{2, 3, 5, 7, 11, 13, 17, 19, 23, 29, 31, 37\}$, write a computer program to generate a six-element subset of A and then determine if it is a subset of B.

2.3 ■ SET OPERATIONS AND THE LAWS OF SET THEORY

After learning how to count, a student usually faces methods for combining counting numbers. Most often this is accomplished through addition. Usually the student's world of arithmetic revolves about the set $\mathbf{Z}^+$ (or a subset of $\mathbf{Z}^+$ that can be spoken and written about, as well as punched out on a hand-held calculator) where the addition of two elements from $\mathbf{Z}^+$ results in a third element of $\mathbf{Z}^+$, called the sum. Hence the student can concentrate on addition without having to enlarge his or her arithmetic world beyond $\mathbf{Z}^+$. This is also true for the operation of multiplication. Thus addition and multiplication are said to be *binary* (acting on two elements) *operations* on $\mathbf{Z}^+$.
    In a comparable way we introduce the following binary operations for sets.

Definition 2.8 ▶ For A, $B \subseteq \mathcal{U}$ we define the following:

a) $A \cup B$ (the *union* of A and B) $= \{x \mid x \in A \text{ or } x \in B\}$. (Here the word "or" is used in the inclusive sense so that an element of $\mathcal{U}$ that happens to be in both A and B is listed as an element of their union.)

b) $A \cap B$ (the *intersection* of A and B) $= \{x \mid x \in A \text{ and } x \in B\}$.

c) $A \triangle B$ (the *symmetric difference* of A and B) $= \{x \mid x \in A \text{ or } x \in B \text{ but } x \notin A \cap B\}$.   □

Note that if A, $B \subseteq \mathcal{U}$, then $A \cap B$, $A \cup B$, $A \triangle B \subseteq \mathcal{U}$. Consequently, $\cap$, $\cup$, $\triangle$ are binary operations on $\mathcal{P}(\mathcal{U})$, or we can say that $\mathcal{P}(\mathcal{U})$ is *closed* under these operations.

EXAMPLE 2.13

With $\mathcal{U} = \{1, 2, 3, \ldots, 9, 10\}$, $A = \{1, 2, 3, 4, 5\}$, $B = \{3, 4, 5, 6, 7\}$, $C = \{7, 8, 9\}$ we have:

a) $A \cap B = \{3, 4, 5\}$ b) $A \cup B = \{1, 2, 3, 4, 5, 6, 7\}$

c) $B \cap C = \{7\}$ d) $A \cap C = \emptyset$

e) $A \bigtriangleup B = \{1, 2, 6, 7\}$ f) $A \cup C = \{1, 2, 3, 4, 5, 7, 8, 9\}$

g) $A \bigtriangleup C = \{1, 2, 3, 4, 5, 7, 8, 9\}$ □

From parts (d), (f), and (g) of Example 2.13 we introduce the following general ideas.

Definition 2.9

▶ If $S, T \subseteq \mathcal{U}$, then S and T are called *disjoint,* or *mutually disjoint,* when $S \cap T = \emptyset$.

Theorem 2.4

▶ If $S, T \subseteq \mathcal{U}$, then S and T are disjoint iff $S \cup T = S \bigtriangleup T$.

Proof

We start with S, T disjoint. (To prove that $S \cup T = S \bigtriangleup T$ we use Theorem 2.2. As this is the first instance of its use we shall be somewhat cautious and very detailed.) If $x \in S \cup T$, then $x \in S$ or $x \in T$ (or perhaps both). But with S and T disjoint, $x \notin S \cap T$ so $x \in S \bigtriangleup T$. Consequently, as $x \in S \cup T$ implies $x \in S \bigtriangleup T$, we have $S \cup T \subseteq S \bigtriangleup T$. For the opposite inclusion, if $y \in S \bigtriangleup T$, then $y \in S$ or $y \in T$. (But $y \notin S \cap T$; we don't actually use this here.) So $y \in S \cup T$. Therefore $S \bigtriangleup T \subseteq S \cup T$ and by Theorem 2.2 it follows that $S \bigtriangleup T = S \cup T$.

Conversely, if $S \cup T = S \bigtriangleup T$ and $S \cap T \neq \emptyset$, let $x \in S \cap T$. Then $x \in S$ and $x \in T$, so $x \in S \cup T$. However, $x \notin S \bigtriangleup T$, contradicting the given set equality. Consequently, S and T are disjoint.   ∎

In proving the first part of Theorem 2.4 we showed that if S, T are any sets, then $S \bigtriangleup T \subseteq S \cup T$. The disjointedness of S and T was needed only for the opposite inclusion.

After mastering the skill of addition, one usually comes next to subtraction. Here the set **N** causes some difficulty. For example, **N** contains 2 and 5 but $2 - 5 = -3$, and $-3 \notin$ **N**. Hence **N** is not closed under subtraction, and subtraction is not a binary operation on **N**. However it is a binary operation on the *superset* **Z** of **N**. So with **Z** we can introduce the *unary,* or *monary, operation* of negation where we "take the minus" of a number such as 3, getting -3.
— We now introduce a comparable unary operation for sets.

Definition 2.10

▶ For a set $A \subseteq \mathcal{U}$, the *complement* of A, denoted $\mathcal{U} - A$, or $\overline{A}$, is given by $\{x \mid x \in \mathcal{U}, x \notin A\}$.

EXAMPLE 2.14

For the sets of Example 2.13, $\overline{A} = \{6, 7, 8, 9, 10\}$, $\overline{B} = \{1, 2, 8, 9, 10\}$ $\overline{C} = \{1, 2, 3, 4, 5, 6, 10\}$.   □

Related to the concept of the complement we have the following:

Definition
2.11

▶ For $A, B \subseteq \mathcal{U}$, the *(relative) complement* of A in B, denoted $B - A$, is given by $\{x \mid x \in B, x \notin A\}$.

EXAMPLE 2.15

For the sets of Example 2.13 we have:

a) $B - A = \{6, 7\}$ b) $A - B = \{1, 2\}$ c) $A - C = A$

d) $C - A = C$ e) $A - A = \emptyset$ f) $\mathcal{U} - A = \overline{A}$ □

Our next result will again make use of Theorem 2.2. It provides links among the notions of subset, union, intersection, and complement.

As in Section 2.1, for statements s_1 and s_2, if s_1 logically implies s_2 we write $s_1 \Rightarrow s_2$. If, in addition, we have $s_2 \Rightarrow s_1$, then s_1 and s_2 are called *(logically) equivalent* and we write $s_1 \Leftrightarrow s_2$.

Theorem
2.5

▶ For sets $A, B \subseteq \mathcal{U}$, the following statements are equivalent:

a) $A \subseteq B$;   b) $A \cup B = B$;   c) $A \cap B = A$ d) $\overline{B} \subseteq \overline{A}$.

Proof

In order to prove the theorem we prove that (a) $\Rightarrow$ (b), (b) $\Rightarrow$ (c), (c) $\Rightarrow$ (d) and (d) $\Rightarrow$ (a).

i) (a) $\Rightarrow$ (b) If A, B are any sets, then $B \subseteq A \cup B$. For the opposite inclusion, if $x \in A \cup B$, then $x \in A$ or $x \in B$, but since $A \subseteq B$, in either case we have $x \in B$. So $A \cup B \subseteq B$ and the equality follows.

ii) (b) $\Rightarrow$ (c) Given sets A, B, we always have $A \supseteq A \cap B$. For the opposite inclusion, let $y \in A$. With $A \cup B = B$, $y \in A \Rightarrow y \in A \cup B \Rightarrow y \in B \Rightarrow y \in A \cap B$, so $A \subseteq A \cap B$ and we conclude that $A = A \cap B$.

iii) (c) $\Rightarrow$ (d) $z \in \overline{B} \Rightarrow z \notin B \Rightarrow z \notin A \cap B$ since $A \cap B \subseteq B$. With $A \cap B = A$, $z \notin A \cap B \Rightarrow z \notin A \Rightarrow z \in \overline{A}$, so $\overline{B} \subseteq \overline{A}$.

iv) (d) $\Rightarrow$ (a) Last, $w \in A \Rightarrow w \notin \overline{A}$, and since $\overline{B} \subseteq \overline{A}$, $w \notin \overline{A} \Rightarrow w \notin \overline{B}$. Then $w \notin \overline{B} \Rightarrow w \in B$, so $A \subseteq B$.   ■

With a bit of theorem proving under our belts we now introduce the major laws that govern set theory. These bear a marked similarity to the Laws of Logic given in Section 2.1. In many instances these set theoretic principles bear similarities to the arithmetic properties of the real numbers, where "$\cup$" plays the role of "+" and "$\cap$" the role of "$\times$". However, there are several differences.

The Laws of Set Theory

For any sets A, B, C taken from a universe $\mathcal{U}$

1. $\overline{\overline{A}} = A$ Law of *Double Complement*

2. $\overline{A \cup B} = \overline{A} \cap \overline{B}$ *DeMorgan's* Laws
    $\overline{A \cap B} = \overline{A} \cup \overline{B}$

3. $A \cup B = B \cup A$ *Commutative* Laws
    $A \cap B = B \cap A$

4. $A \cup (B \cup C) = (A \cup B) \cup C$ *Associative* Laws
   $A \cap (B \cap C) = (A \cap B) \cap C$

5. $A \cup (B \cap C) = (A \cup B) \cap (A \cup C)$ *Distributive* Laws
   $A \cap (B \cup C) = (A \cap B) \cup (A \cap C)$

6. $A \cup A = A$ *Idempotent* Laws
   $A \cap A = A$

7. $A \cup \emptyset = A$ *Identity* Laws
   $A \cap \mathcal{U} = A$

8. $A \cup \overline{A} = \mathcal{U}$ *Inverse* Laws
   $A \cap \overline{A} = \emptyset$

9. $A \cup \mathcal{U} = \mathcal{U}$ *Domination* Laws
   $A \cap \emptyset = \emptyset$

10. $A \cup (A \cap B) = A$ *Absorption* Laws
    $A \cap (A \cup B) = A$

All the laws above can be established as in the first part of the proof of Theorem 2.4. We demonstrate this by establishing the second distributive law, that of intersection over union.

Proof   For each $x \in \mathcal{U}$, $x \in A \cap (B \cup C) \Leftrightarrow (x \in A)$ and $(x \in B \cup C) \Leftrightarrow (x \in A)$ and $(x \in B$ or $x \in C) \Leftrightarrow (x \in A$ and $x \in B)$ or $(x \in A$ and $x \in C) \Leftrightarrow (x \in A \cap B)$ or $(x \in A \cap C) \Leftrightarrow x \in (A \cap B) \cup (A \cap C)$. As we have equivalent statements throughout, we have established both subset relations simultaneously, so $A \cap (B \cup C) = (A \cap B) \cup (A \cap C)$. (The equivalence of the third and fourth statements follows from the comparable principle in the Laws of Logic, namely the distributive law of conjunction over disjunction.)   ∎

The reader undoubtedly expects the pairing of the laws in items 2–10 to have some importance. As with the Laws of Logic, these pairs of statements are called *duals*. One statement can be obtained from the other by interchanging all occurrences of $\cup$ by $\cap$ and vice versa, and all occurrences of $\mathcal{U}$ by $\emptyset$ and vice versa. As in Section 2.1, we will state and use the following theorem. We shall prove a more general result in Chapter 9.

Theorem 2.6   ▶   *(The Principle of Duality)* If s is a theorem dealing with set theory, then the dual of s, denoted s^d, is also a theorem in the theory of sets.

Using this principle cuts our work down considerably. For each pair of laws in 2–10 one need only prove one of the statements, and then invoke this principle to obtain the other statement in the pair.

When at most four sets are involved in a set-equality or subset statement, we can investigate the problems graphically. This does not constitute a rigorous proof but is useful nonetheless.

Named in honor of the English logician John Venn (1834–1883), a *Venn diagram* is constructed as follows: $\mathcal{U}$ is depicted as the interior of a rectangle, while subsets of

$\mathcal{U}$ are represented by the interiors of circles and other closed curves. Fig. 2.6 shows two Venn diagrams. The shaded region in Fig. 2.6(a) represents the set A, while $\overline{A}$ is represented by the unshaded area. The shaded region in Fig. 2.6(b) represents $A \cup B$. The set $A \cap B$ is the cross-hatched region in this figure.

Figure 2.6

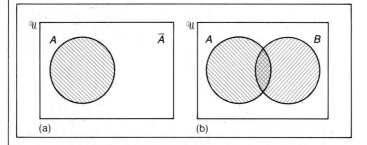

(a)   (b)

In Fig. 2.7 Venn diagrams are used to establish one of the DeMorgan laws. Fig. 2.7(a) has everything except $A \cap B$ shaded. In Fig. 2.7(b), $\overline{A}$ is the region shaded by the lines going from the lower left to the upper right; the lines going from the upper left to the lower right shade the region representing $\overline{B}$. For these Venn diagrams, the shaded area in part (b) is the same as that in part (a). Consequently, $\overline{A \cap B} = \overline{A} \cup \overline{B}$.

Figure 2.7

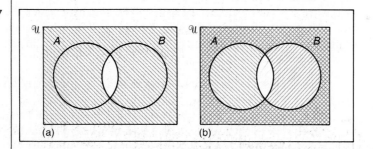

(a)   (b)

We further illustrate the use of these diagrams by showing that for sets A, B, $C \subseteq \mathcal{U}$, $(\overline{A \cup B}) \cup C = (\overline{A} \cap \overline{B}) \cup C$. Instead of shading regions, another approach, also using Venn diagrams, numbers the regions as in Fig. 2.8 where, for example, region 3 is $\overline{A} \cap B \cap \overline{C}$ and region 7 is $A \cap \overline{B} \cap C$. Each region is a set of the form $S_1 \cap S_2 \cap S_3$, where S_1 is replaced by A or $\overline{A}$, S_2 by B or $\overline{B}$, and S_3 by C or $\overline{C}$. Consequently by the rule of product there are eight possible regions.

Consulting Fig. 2.8 we see that $A \cup B$ is made up of regions 2, 3, 5, 6, 7, 8, so $\overline{A \cup B}$ comprises regions 1 and 4. In forming $(\overline{A \cup B}) \cup C$ regions 6, 7, 8 are picked up, so $(\overline{A \cup B}) \cup C$ is made up of regions 1, 4, 6, 7, 8. The set $\overline{A}$ consists of regions 1, 3, 4, 6, while regions 1, 2, 4, 7 make up $\overline{B}$, so $\overline{A} \cap \overline{B}$ comprises regions 1 and 4. Taking the union of $\overline{A} \cap \overline{B}$ with C, we then finish with regions 1, 4, 6, 7, 8, as we did for $(\overline{A \cup B}) \cup C$.

Figure 2.8

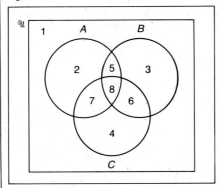

Table 2.8

A	B	A ∩ B	A ∪ B
0	0	0	0
0	1	0	1
1	0	0	1
1	1	1	1

(a)

A	$\overline{A}$
0	1
1	0

(b)

One more technique for establishing set equalities is the *membership table*. (This method is akin to the truth table of Section 2.1.)

We observe that for sets $A, B \subseteq \mathcal{U}$, an element $x \in \mathcal{U}$ satisfies exactly one of the following four situations:

a) $x \notin A, x \notin B$ b) $x \notin A, x \in B$

c) $x \in A, x \notin B$ d) $x \in A, x \in B$.

When x is an element of a given set we write a 1 in the column representing that set; when x is not in the set, we write a 0. Using this notation the membership tables for $A \cap B$, $A \cup B$, $\overline{A}$ are given in Table 2.8. Here, for example, the third row in part (a) of the table tells us that when an element $x \in \mathcal{U}$ is in set A but not in B then it is not in $A \cap B$ but it is in $A \cup B$.

These operations on 0 and 1 are the same as in ordinary arithmetic except that $1 \cup 1 = 1$.

Using membership tables, the equality of two sets can be established by comparing their respective columns in the table. Table 2.9 demonstrates this for the distributive law of union over intersection.

Table 2.9

A	B	C	B ∩ C	A ∪ (B ∩ C)	A ∪ B	A ∪ C	(A ∪ B) ∩ (A ∪ C)
0	0	0	0	0	0	0	0
0	0	1	0	0	0	1	0
0	1	0	0	0	1	0	0
0	1	1	1	1	1	1	1
1	0	0	0	1	1	1	1
1	0	1	0	1	1	1	1
1	1	0	0	1	1	1	1
1	1	1	1	1	1	1	1

Since these columns are identical we conclude that $A \cup (B \cap C) = (A \cup B) \cap (A \cup C)$.

Now that we have the laws of set theory, what can we do with them? The following examples will, hopefully, demonstrate how the laws can be used to simplify a complicated set expression or derive new set equalities.

EXAMPLE 2.16 Simplify the expression $\overline{\overline{(A \cup B) \cap C} \cup \overline{B}}$.

$\overline{\overline{(A \cup B) \cap C} \cup \overline{B}}$	**Reasons**
$= \overline{\overline{((A \cup B) \cap C)}} \cap \overline{\overline{B}}$	DeMorgan's Law
$= ((A \cup B) \cap C) \cap B$	Law of Double Complement
$= (A \cup B) \cap (C \cap B)$	Associative Law of Intersection
$= (A \cup B) \cap (B \cap C)$	Commutative Law of Intersection
$= [(A \cup B) \cap B] \cap C$	Associative Law of Intersection
$= B \cap C$	Absorption Law

EXAMPLE 2.17 Express $\overline{A - B}$ in terms of $\cup$ and $^{-}$.

From the definition of relative complement, $A - B = \{x \mid x \in A, x \notin B\} = A \cap \overline{B}$. Therefore,

$\overline{A - B} = \overline{A \cap \overline{B}}$	**Reasons**
$= \overline{A} \cup \overline{\overline{B}}$	DeMorgan's Law
$= \overline{A} \cup B$	Law of Double Complement

EXAMPLE 2.18 From the observation made in Example 2.17 we have $A \triangle B = \{x \mid x \in A \cup B, x \notin A \cap B\} = (A \cup B) - (A \cap B) = (A \cup B) \cap \overline{(A \cap B)}$, so

$\overline{A \triangle B} = \overline{(A \cup B) \cap \overline{(A \cap B)}}$	**Reasons**
$= \overline{(A \cup B)} \cup \overline{\overline{(A \cap B)}}$	DeMorgan's Law
$= \overline{(A \cup B)} \cup (A \cap B)$	Law of Double Complement
$= (A \cap B) \cup \overline{(A \cup B)}$	Commutative Law of $\cup$
$= (A \cap B) \cup (\overline{A} \cap \overline{B})$	DeMorgan's Law
$= [(A \cap B) \cup \overline{A}] \cap [(A \cap B) \cup \overline{B}]$	Distributive Law of $\cup$ over $\cap$
$= [(A \cup \overline{A}) \cap (B \cup \overline{A})] \cap$	Distributive Law of $\cup$ over $\cap$
$\quad [(A \cup \overline{B}) \cap (B \cup \overline{B})]$	
$= [\mathcal{U} \cap (B \cup \overline{A})] \cap [(A \cup \overline{B}) \cap \mathcal{U}]$	Inverse Law
$= (B \cup \overline{A}) \cap (A \cup \overline{B})$	Identity Law
$= (\overline{A} \cup B) \cap (A \cup \overline{B})$	Commutative Law of $\cup$
$= (\overline{A} \cup B) \cap \overline{(\overline{A} \cap B)}$	DeMorgan's Law
$= \overline{A} \triangle B$	
$= (A \cup \overline{B}) \cap (\overline{A} \cup B)$	Commutative Law of $\cap$
$= (A \cup \overline{B}) \cap \overline{(A \cap \overline{B})}$	DeMorgan's Law
$= A \triangle \overline{B}$	

In closing this section we extend the set operations of $\cup$ and $\cap$ beyond three sets.

Definition
2.12

▶ Let I denote a set of *indices*. If for each *index* $i \in I$ there is a set $A_i \subseteq \mathcal{U}$, then

$$\bigcup_{i \in I} A_i = \{x \mid x \in A_i \text{ for at least one } i \in I\}, \text{ and}$$

$$\bigcap_{i \in I} A_i = \{x \mid x \in A_i \text{ for every } i \in I\}.$$

Note that $x \notin \bigcup_{i \in I} A_i$ if $x \notin A_i$ for *every* index $i \in I$. If $x \notin A_i$ for *at least one* index $i \in I$, then $x \notin \bigcap_{i \in I} A_i$.

If the index set I is the set $\mathbf{Z}^+$, we can write

$$\bigcup_{i \in \mathbf{Z}^+} A_i = A_1 \cup A_2 \cup \ldots = \bigcup_{i=1}^{\infty} A_i, \qquad \bigcap_{i \in \mathbf{Z}^+} A_i = A_1 \cap A_2 \cap \ldots = \bigcap_{i=1}^{\infty} A_i.$$

EXAMPLE 2.19   Let $I = \{3, 4, 5, 6, 7\}$, and for $i \in I$ let $A_i = \{1, 2, 3, \ldots, i\} \subseteq \mathcal{U} = \mathbf{Z}^+$. Then $\bigcup_{i \in I} A_i = \bigcup_{i=3}^{7} A_i = A_3 \cup A_4 \cup \ldots \cup A_7 = \{1, 2, 3, \ldots, 7\} = A_7$, while $\bigcap_{i \in I} A_i = \{1, 2, 3\} = A_3$. □

EXAMPLE 2.20   Let $\mathcal{U} = \mathbf{R}$ and $I = \mathbf{R}^+$. If for each $r \in \mathbf{R}^+$, $A_r = [-r, r]$, then $\bigcup_{r \in I} A_r = \mathbf{R}$, $\bigcap_{r \in I} A_r = \{0\}$. □

When dealing with generalized unions and intersections, membership tables and Venn diagrams are unfortunately next to useless, but the rigorous approach, as demonstrated in the proof of Theorem 2.4, is still available.

Theorem
2.7

▶ (*Generalized DeMorgan Laws*) Let I be an index set where for each $i \in I$, $A_i \subseteq \mathcal{U}$. Then

a) $\overline{\bigcup_{i \in I} A_i} = \bigcap_{i \in I} \overline{A_i}$;   b) $\overline{\bigcap_{i \in I} A_i} = \bigcup_{i \in I} \overline{A_i}$

Proof   We shall prove Theorem 2.7(a) and leave the proof of part (b) for the reader. For each $x \in \mathcal{U}$, $x \in \overline{\bigcup_{i \in I} A_i} \Leftrightarrow x \notin \bigcup_{i \in I} A_i \Leftrightarrow x \notin A_i$, for all $i \in I \Leftrightarrow x \in \overline{A_i}$, for all $i \in I \Leftrightarrow x \in \bigcap_{i \in I} \overline{A_i}$. ∎

EXERCISES

1. For $\mathcal{U} = \{1, 2, 3, \ldots, 9, 10\}$ let $A = \{1, 2, 3, 4, 5\}$, $B = \{1, 2, 4, 8\}$, $C = \{1, 2, 3, 5, 7\}$, $D = \{2, 4, 6, 8\}$. Determine each of the following.

a) $(A \cup B) \cap C$ b) $A \cup (B \cap C)$ c) $\overline{C} \cup \overline{D}$

d) $\overline{C \cap D}$ e) $(A \cup B) - C$ f) $A \cup (B - C)$

g) $(B - C) - D$ h) $B - (C - D)$ i) $(A \cup B) - (C \cap D)$

2. If $A = [0, 3]$, $B = [2, 7)$, with $\mathcal{U} = \mathbf{R}$, determine each of the following:

a) $A \cap B$ b) $A \cup B$ c) $\overline{A}$

d) $A \triangle B$ e) $A - B$ f) $B - A$

3. For each pair of statements s_1, s_2, determine which of the following is true:

i) $s_1 \Leftrightarrow s_2$;   ii) $s_1 \Rightarrow s_2, s_2 \not\Rightarrow s_1$;

iii) $s_2 \Rightarrow s_1, s_1 \not\Rightarrow s_2$;   iv) $s_1 \not\Rightarrow s_2, s_2 \not\Rightarrow s_1$.

   a) s_1: Triangle ABC is equilateral.
        s_2: Triangle ABC is equiangular.

   b) s_1: A, B are sets with $A \subseteq B$.
        s_2: A, B are sets with $A = B$.

   c) s_1: Triangle ABC is equilateral.
        s_2: Triangle ABC is isosceles.

   d) s_1: Quadrilateral $ABCD$ is equilateral.
        s_2: Quadrilateral $ABCD$ is equiangular.

4. Prove each of the following results without using Venn diagrams or membership tables. (Assume a universe $\mathcal{U}$.)

   a)   If $A \subseteq B$, $C \subseteq D$, then $A \cap C \subseteq B \cap D$ and $A \cup C \subseteq B \cup D$.

   b) $A \subseteq B$ iff $A \cap \overline{B} = \emptyset$ c) $A \subseteq B$ iff $\overline{A} \cup B = \mathcal{U}$.

5. Prove or disprove each of the following:

   a)   For sets A, B, $C \subseteq \mathcal{U}$, $A \cap C = B \cap C \Rightarrow A = B$.

   b)   For sets A, B, $C \subseteq \mathcal{U}$, $A \cup C = B \cup C \Rightarrow A = B$.

   c)   For sets A, B, $C \subseteq \mathcal{U}$, $A \cap C = B \cap C$, $A \cup C = B \cup C \Rightarrow A = B$.

   d)   For sets A, B, $C \subseteq \mathcal{U}$, $A \triangle C = B \triangle C \Rightarrow A = B$.

6. Using Venn diagrams, investigate the truth or falsity of each of the following, for sets A, B, $C \subseteq \mathcal{U}$.

   a) $A \triangle (B \cap C) = (A \triangle B) \cap (A \triangle C)$

   b) $A \cap (B \triangle C) = (A \cap B) \triangle (A \cap C)$

   c) $A \triangle (B \cup C) = (A \triangle B) \cup (A \triangle C)$

   d) $A \cup (B \triangle C) = (A \cup B) \triangle (A \cup C)$

   e) $A - (B \cup C) = (A - B) \cap (A - C)$

   f) $A - (B \cap C) = (A - B) \cup (A - C)$

   g) $A \triangle (B \triangle C) = (A \triangle B) \triangle C$

7. If $A = \{a, b, d\}$, $B = \{d, x, y\}$, $C = \{x, z\}$, how many proper subsets are there for the set $(A \cap B) \cup C$? How many for the set $A \cap (B \cup C)$?

8. For a given universal set $\mathcal{U}$, each subset A of $\mathcal{U}$ satisfies the idempotent laws of union and intersection. (a) Are there any real numbers which satisfy an idempotent property for addition? (I.e., can we find any real number(s) x so that $x + x = x$?) (b) Answer part (a) replacing addition by multiplication.

9. Write the dual of each of the following set theoretic results.

   a) $\mathcal{U} = (A \cap B) \cup (A \cap \overline{B}) \cup (\overline{A} \cap B) \cup (\overline{A} \cap \overline{B})$

   b) $A = A \cap (A \cup B)$

   c) $A \cup B = (A \cap B) \cup (A \cap \overline{B}) \cup (\overline{A} \cap B)$

   d) $A = (A \cup B) \cap (A \cup \emptyset)$

10. Let $\mathcal{U} = \{a, b, c, \ldots, x, y, z\}$, with $A = \{a, b, c\}$, $C = \{a, b, d, e\}$. If $|A \cap B| = 2$ and $(A \cap B) \subset B \subset C$, determine B.

11. Prove or disprove each of the following for sets $A, B \subseteq \mathcal{U}$.

   a) $\mathcal{P}(A \cup B) = \mathcal{P}(A) \cup \mathcal{P}(B)$ b) $\mathcal{P}(A \cap B) = \mathcal{P}(A) \cap \mathcal{P}(B)$.

12. Use membership tables to establish each of the following:

   a) $\overline{A \cap B} = \overline{A} \cup \overline{B}$ b) $A \cup A = A$ c) $A \cup (A \cap B) = A$

   d) $(A \cap B) \cup (\overline{A} \cap C) = (A \cap \overline{B}) \cup (\overline{A} \cap \overline{C})$

13. a) How many rows are needed to construct the membership table for $A \cap (B \cup C) \cap (D \cup \overline{E} \cup \overline{F})$?

   b) How many rows are needed to construct the membership table for a set made up from the sets $A_1, A_2, \ldots, A_n$, using $\cap$, $\cup$ and $\bar{}$?

   c) Given the membership tables for two sets A, B, how can the relation $A \subseteq B$ be recognized?

   d) Use membership tables to determine whether or not $(A \cap B) \cup \overline{(B \cap C)} \supseteq A \cup \overline{B}$.

14. Using the laws of set theory, simplify each of the following:

   a) $A \cap (B - A)$ b) $(A \cap B) \cup (A \cap B \cap \overline{C} \cap D) \cup (\overline{A} \cap B)$

   c) $(A - B) \cup (A \cap B)$ d) $\overline{A} \cup \overline{B} \cup (A \cap B \cap \overline{C})$

   e) $\overline{A} \cup (A \cap \overline{B}) \cup (A \cap B \cap \overline{C}) \cup (A \cap B \cap C \cap \overline{D}) \cup \ldots$

15. Let $\mathcal{U} = \mathbf{R}$ and let $I = \mathbf{Z}^+$. For each $n \in \mathbf{Z}^+$ let $A_n = [-2n, 3n]$. Determine each of the following:

   a) A_3 b) A_4 c) $A_3 - A_4$ d) $A_3 \triangle A_4$

   e) $\bigcup_{n=1}^{7} A_n$ f) $\bigcap_{n=1}^{7} A_n$ g) $\bigcup_{n \in \mathbf{Z}^+} A_n$ h) $\bigcap_{n=1}^{\infty} A_n$

16. Provide the details for the proof of Theorem 2.7(b).

17. Given a universe $\mathcal{U}$ and an index set I, for each $i \in I$, let $B_i \subseteq \mathcal{U}$. Prove that for $A \subseteq \mathcal{U}$, $A \cap (\bigcup_{i \in I} B_i) = \bigcup_{i \in I} (A \cap B_i)$, and $A \cup (\bigcap_{i \in I} B_i) = \bigcap_{i \in I} (A \cup B_i)$. [Generalized Distributive Laws]

2.4 ■ COUNTING AND VENN DIAGRAMS

With all of the theoretical work and theorem proving of the last section, now is a good time to return to the theme of the chapter and examine some additional counting problems.

For sets A, B from a finite universe $\mathcal{U}$, the following Venn diagrams will help us obtain counting formulas for $|\overline{A}|$, $|A \cup B|$ in terms of $|A|$, $|B|$, $|A \cap B|$.

As Fig. 2.9 demonstrates, $A \cup \overline{A} = \mathcal{U}$ and $A \cap \overline{A} = \emptyset$, so by the rule of sum, $|A| + |\overline{A}| = |\mathcal{U}|$ or $|\overline{A}| = |\mathcal{U}| - |A|$. The sets A, B, in Fig. 2.10, have no intersection, so here the rule of sum leads us to $|A \cup B| = |A| + |B|$, and necessitates that A, B be finite but does not require any condition on the cardinality of $\mathcal{U}$.

Turning to the case where A, B are not disjoint, we motivate the formula for $|A \cup B|$ with the following example.

Figure 2.9

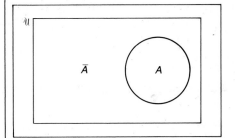

Figure 2.10

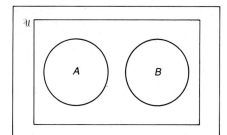

Figure 2.11

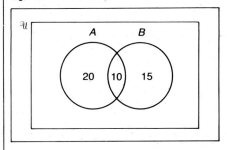

EXAMPLE 2.21   In a class of 50 college freshmen, 30 are studying BASIC, 25 are studying Pascal, and 10 are studying both languages. How many freshmen are studying either computer language?

We let $\mathcal{U}$ be the class of 50 freshmen, A the subset of those studying BASIC, and B the subset of students studying Pascal. In order to answer the question, we need $|A \cup B|$. In Fig. 2.11 the numbers in the regions are obtained from the given information: $|A| = 30$, $|B| = 25$, $|A \cap B| = 10$. Consequently, $|A \cup B| = 45 \neq |A| + |B|$, since $|A| + |B|$ counts the students in $A \cap B$ twice. To remedy this overcount we subtract $|A \cap B|$ from $|A| + |B|$ to obtain the correct formula: $|A \cup B| = |A| + |B| - |A \cap B|$.   □

For the general situation, given finite sets A, B, $|A \cup B| = |A| + |B| - |A \cap B|$. Consequently finite sets A, B are mutually disjoint iff $|A \cup B| = |A| + |B|$.

This situation extends to three sets as seen in the following example.

EXAMPLE 2.22   A 14-pin logic chip has four AND gates, each with two inputs and one output. (See Fig. 2.12.) The first AND gate (i.e., the gate at pins 1, 2, 3) can have any or all of the following defects:

D_1: the first input (pin 1) is stuck at 0.
D_2: the second input (pin 2) is stuck at 0.
D_3: the output (pin 3) is stuck at 1.

For a sample of 100 such logic chips we let A, B, C be the subsets having defects D_1, D_2, D_3 respectively. With $|A| = 23$, $|B| = 26$, $|C| = 30$, $|A \cap B| = 7$, $|A \cap C| = 8$, $|B \cap C| = 10$, $|A \cap B \cap C| = 3$, how many chips in the sample are defective?

Figure 2.12

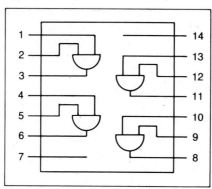

Figure 2.13

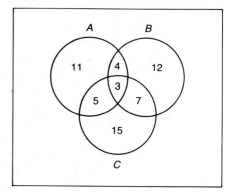

Working backwards from $|A \cap B \cap C| = 3$ to $|A| = 23$, we label the regions as shown in Fig. 2.13 and find that $|A \cup B \cup C| = 57 = |A| + |B| + |C| - |A \cap B| - |A \cap C| - |B \cap C| + |A \cap B \cap C|$.   □

From the formula for $|A \cup B \cup C|$ and DeMorgan's law we find that if A, B, C are sets from a finite universe $\mathcal{U}$, then $|\overline{A} \cap \overline{B} \cap \overline{C}| = |\overline{A \cup B \cup C}| = |\mathcal{U}| - |A \cup B \cup C| = |\mathcal{U}| - |A| - |B| - |C| + |A \cap B| + |A \cap C| + |B \cap C| - |A \cap B \cap C|$.

We close this section with a problem that uses this result.

EXAMPLE 2.23 A student visits an arcade each day after school and plays one game of either Laser-Man, Millipede, or Space Conquerors. In how many ways can he play one game each day so that he plays each of the three types at least once during a given school week?

Here there is a slight twist. The set $\mathcal{U}$ consists of all permutations of size five taken from the set of three games, with repetitions allowed. The set A represents the subset of all sequences of five games played during the week *without playing Laser Man*. The sets B, C are defined similarly, leaving out Millipede and Space Conquerors, respectively. The enumeration techniques of Chapter 1 give $|\mathcal{U}| = 3^5$, $|A| = |B| = |C| = 2^5$, $|A \cap B| = |A \cap C| = |B \cap C| = 1^5 = 1$, $|A \cap B \cap C| = 0$, so by the formula above there are $|\overline{A} \cap \overline{B} \cap \overline{C}| = 3^5 - 3 \cdot 2^5 + 3 \cdot 1^5 - 0 = 150$ ways the student can select his daily games during a school week, playing each type at least once.

This example can be expressed in an equivalent distribution form, since we are seeking the number of ways to distribute five distinct objects (Monday, Tuesday, . . . , Friday) among three distinct containers (the computer games) with no container left empty. More will be said about this in the next chapter.   □

2.5 ■ A WORD ON PROBABILITY

When one performs an *experiment* such as tossing a single die or selecting two students from a class of 20 to work on a project, a listing of all possible outcomes for the situation is called a *sample space*. Consequently, $\{1, 2, 3, 4, 5, 6\}$ serves as a sample space for the first experiment mentioned, while $\{\{a_i, a_j\} \mid 1 \leq i, j \leq 20, i \neq j\}$ could be used for the latter experiment, with a_i denoting the ith student, $1 \leq i \leq 20$.

Unfortunately, for a given experiment there can be more than one sample space. If we toss a coin three times and consider the outcomes, a possible sample space for our experiment is $\{0, 1, 2, 3\}$, where the number i, $0 \leq i \leq 3$, refers to the number of heads that come up in the three tosses. The outcomes can also be given by the sample space $\mathcal{S} = \{HHH, THH, HTH, HHT, TTH, THT, HTT, TTT\}$, where we feel that each of the eight possible outcomes has the *same likelihood* of occurrence. This is not the case with our first sample space $\{0, 1, 2, 3\}$, where we feel that there is a better chance of getting one head in the three tosses than there is of getting no heads.

In this text we shall always use a sample space where each element has the same likelihood of occurrence. Under this assumption of equal likelihood we shall use a definition for probability that was first given by the French mathematician Pierre-Simon de Laplace (1749–1827) in his *Analytic Theory of Probability*.

For a sample space $\mathcal{S}$ of an experiment $\mathcal{E}$ any subset A of $\mathcal{S}$ is called an *event*. Each element of $\mathcal{S}$ is called an *elementary event*, so if $|\mathcal{S}| = n$ and $a \in \mathcal{S}, A \subseteq \mathcal{S}$,

$$Pr(a) = \textit{The probability that a occurs} = \frac{1}{n} = \frac{|\{a\}|}{|\mathcal{S}|}, \text{ and}$$

$$Pr(A) = \textit{The probability that A occurs} = \frac{|A|}{|\mathcal{S}|} = \frac{|A|}{n} = \frac{1}{n}\left(\sum_{a \in A} |\{a\}|\right),$$

since A is the disjoint union of its singletons.

We demonstrate these ideas in the following examples.

EXAMPLE 2.24   In tossing a die once, what is the probability of getting a 5 or a 6?

Here $\mathcal{S} = \{1, 2, 3, 4, 5, 6\}$ and the event A we want to consider is $\{5, 6\}$. Hence

$$Pr(A) = \frac{|A|}{|\mathcal{S}|} = \frac{2}{6} = \frac{1}{3}. \qquad \square$$

EXAMPLE 2.25   If a coin is tossed four times, what is the probability of getting two heads and two tails?

The sample space here consists of all sequences of the form x, x, x, x where each x is replaced by a T or an H, so $|\mathcal{S}| = 2^4 = 16$. The event A we are concerned about contains all permutations of the four symbols H, H, T, T, so $|A| = 4!/(2! \, 2!) = 6$. Consequently, $Pr(A) = 6/16 = 3/8$. $\qquad \square$

For Example 2.25, each toss is *independent* of the outcomes of any previous toss. Such an occurrence is called a *Bernoulli trial*. These trials are investigated in a first course in probability in conjunction with the Bernoulli distribution, an important example of a discrete probability distribution. We shall come upon this again in Chapter 12 when we study the application of abelian groups in coding theory.

In our last example we use Venn diagrams with probability.

EXAMPLE 2.26

In a survey of 120 passengers, an airline found that 48 preferred wine with their meals, 78 preferred mixed drinks, and 66, iced tea. In addition, 36 enjoyed any given pair of these beverages and 24 passengers enjoyed them all. If two passengers are selected at random from the survey sample of 120, what is the probability that

(Event A) they both want only iced tea with their meals?

(Event B) they both enjoy exactly two of the three beverage offerings?

From the information provided we construct the Venn diagram in Fig. 2.14. The sample space $\mathscr{S}$ consists of the pairs of passengers we can select from the sample of 120, so $|\mathscr{S}| = \binom{120}{2} = 7140$. The Venn diagram indicates that there are 18 passengers who drink only iced tea, so $|A| = \binom{18}{2}$ and $Pr(A) = 51/2380$. The reader should verify that $Pr(B) = 3/34$.   □

Figure 2.14

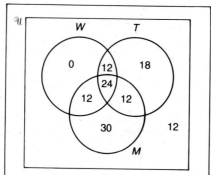

EXERCISES

1. How many linear arrangements of the 26 different letters of the alphabet are there that contain (a) either the pattern "OUT" or the pattern "DIG"? (b) neither the pattern "MAN" nor the pattern "ANT"?

2. A six-character variable name in ANSI FORTRAN starts with a letter of the alphabet. Each of the other five characters can be either a letter or a digit. (Repetitions are allowed.) How many six-character variable names contain the pattern "FUN" or the pattern "TIP"?

3. How many linear arrangements of the digits 0, 1, 2, ..., 9 either start with a 3 or end with a 7, or both?

4. A professor has two dozen introductory textbooks on computer science and is concerned about their coverage of the topics (A) compilers; (B) data structures; and (C) interpreters. The following data lists the number of books which contain material on these topics:

$$|A| = 8 \qquad |C| = 13 \qquad |A \cap C| = 3 \qquad |A \cap B \cap C| = 2$$
$$|B| = 13 \qquad |A \cap B| = 5 \qquad |B \cap C| = 6$$

(a) How many of the textbooks include material on exactly one of these topics? (b) How many do not deal with any of the topics? (c) How many have no material on compilers?

5. A letter is selected at random from the word "chemistry." (a) What is the probability that it is either e or i? (b) What is the probability that it does not occur before q in the alphabet?

6. In selecting a new computer for its computing center, a college examines 15 different models with some attention given to the following considerations: (A) magnetic tape drive; (B) terminal for graphics display; (C) semiconductor memory (in addition to core memory). The numbers of computers with any or all of these features are given as follows: $|A| = |B| = |C| = 6$, $|A \cap B| = |B \cap C| = 1, |A \cap C| = 2, |A \cap B \cap C| = 0$. (a) How many of the models have exactly one of the features being considered? (b) How many have none of the features? (c) If a model is selected at random, what is the probability that it has exactly two of these features?

7. a) How many permutations of the letters in MISCELLANEOUS have no pair of consecutive letters the same?

   b) If a permutation of these letters is randomly generated, what is the probability that no pair of identical consecutive letters occurs?

8. How many arrangements of the letters in CHEMIST are there with either H before E, or E before T, or T before M? (Here "before" means anywhere before, not just immediately before.)

9. An integer is selected at random from 3–17 inclusive. If A is the event where a number divisible by three is chosen and B is the event where the number exceeds 10, determine $Pr(A)$, $Pr(B)$, $Pr(A \cap B)$, $Pr(A \cup B)$. How is $Pr(A \cup B)$ related to $Pr(A)$, $Pr(B)$, and $Pr(A \cap B)$?

10. The freshman class of a private engineering college has 300 students. It is known that 180 can program in Pascal, 120 in FORTRAN, 30 in APL, 12 in Pascal and APL, 18 in FORTRAN and APL, 12 in Pascal and FORTRAN, and 6 in all three languages.

   a) If a student is selected at random, what is the probability that she can program in exactly two languages?

   b) If two students are selected at random, what is the probability they can

   i) both program in Pascal?

   ii) both program only in Pascal?

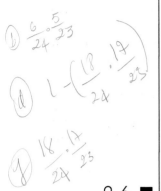

11. Eight different books, three in physics and five in electrical engineering, are placed at random on a library shelf. Find the probability that the three physics books are all together.

12. A shipment of 24 new cars contains 15 in excellent condition, six with minor defects and three with major defects. If two cars are selected from the shipment, what is the probability that (a) both are in excellent condition? (b) both have minor defects? (c) at most one has a minor defect? (d) at least one has a minor defect? (e) exactly one has a minor defect? (f) neither has a minor defect?

How are the results in parts (b), (e), and (f) related?

2.6 ■ SUMMARY AND HISTORICAL REVIEW

In this chapter we introduced some of the fundamentals of logic and set theory with consideration to enumeration and elementary probability.

The first systematic study of logical reasoning is found in the work of the Greek philosopher Aristotle (384–322 B.C.). In a modified form, this type of logic was taught up to and throughout the Middle Ages. Little change took place until the nineteenth century when the English mathematician George Boole (1815–1864) created a system of mathematical logic which he published in 1848 in the pamphlet *The Mathematical Analysis of Logic*. Following Boole's death, interest in mathematical logic was minimal until the publication of Russell and Whitehead's *Principia Mathematica* (1910–1913). Here the foundations set down by Boole were finally appreciated. From this and the work of other twentieth century mathematicians and logicians, the more polished techniques of contemporary mathematical logic are now available.

The algebra of set theory evolved during the nineteenth and early twentieth centuries. In England, George Peacock (1791–1858) was a pioneer in mathematical reforms and among the first, in his *Treatise on Algebra,* to revolutionize the entire conception of algebra and arithmetic. His ideas were further developed by Duncan Gregory (1813–1844), William Rowan Hamilton (1805–1865), and Augustus DeMorgan (1806–1871), who attempted to remove ambiguity from elementary algebra and place its study in the strict postulational form. It was not until 1854, however, when Boole published his *Investigation of the Laws of Thought,* that an algebra dealing with sets and logic was formalized, extending the work of Peacock and his contemporaries.

The presentation here is primarily concerned with finite sets. However, the investigation of infinite sets and their cardinalities has occupied the minds of many mathematicians and philosophers. The intuitive approach to set theory was taken until the time of Georg Cantor (1845–1918) who defined a set, in 1895, in a way comparable to the "gut feeling" we mentioned at the start of Section 2.2. His definition, however, was one of the obstacles Cantor was never able to remove entirely from his theory of sets.

In the 1870's when Cantor was researching trigonometric series and series of real numbers, he needed a device to compare the sizes of infinite sets of numbers. His treatment of the infinite as an actuality, on the same level as the finite, was quite revolutionary. Consequently some of his work was rejected, but it won enough acceptance so that by 1890 the theory of sets, finite and infinite, was considered a branch of mathematics by itself.

By the turn of the century the theory was widely accepted, but in 1901 the paradox now known as Russell's paradox showed that set theory, as originally proposed, was inconsistent. The difficulty seemed to be in the unrestricted way in which sets could be defined; the idea of a set being a member of itself was considered particularly suspect. In their work *Principia Mathematica,* Bertrand Russell (1872–1970) and Alfred North Whitehead (1861–1947) developed a hierarchy in the theory of sets known as the *theory of types*. This axiomatic set theory, among other twentieth century formulations, avoided the Russell paradox.

In Section 2.2 we mentioned the array of numbers known as Pascal's triangle. We could have done this in Chapter 1 with the binomial theorem, but waited until we had some needed combinatorial identities to verify how the triangle is constructed. The array appears in the work of the Chinese algebraist Chu Shi-kie (1303), but its first appearance in Europe was not until the sixteenth century, on the title page of a book by Petrus Apianus (1495–1552). Niccolo Tartaglia (1499–1559) used the triangle in computing powers of $(x + y)$. Because of his work on the properties and applications of this triangle, the array has come to honor the French mathematician Blaise Pascal (1623–1662).

Last, although probability originated with games of chance and enumeration problems, we mention it here because set theory has evolved as the exact medium needed to state and solve problems in this important contemporary area of applied mathematics.

The text by E. Mendelson [4] provides an interesting introduction for those readers who wish to pursue the topic of mathematical logic further.

More on the history and development of set theory can be found in Chapter 26 of C. Boyer [1]. Formal developments of set theory, including results on infinite sets, can be found in P. Halmos [3] and P. Suppes [6].

An interesting history of the origins of probability and statistical ideas up to the Newtonian era can be found in F. N. David [2]. Chapters 1 and 2 of P. Meyer [5] are an excellent source for those interested in learning more about discrete probability.

REFERENCES

1. Boyer, Carl B., *History of Mathematics,* John Wiley & Sons, New York, 1968.

2. David, Florence Nightingale, *Games, Gods, and Gambling,* Hafner, New York, 1962.

3. Halmos, Paul R., *Naive Set Theory,* Van Nostrand, New York, 1960.

4. Mendelson, Elliott, *Introduction to Mathematical Logic,* Van Nostrand, Princeton, New Jersey, 1964.

5. Meyer, Paul L., *Introductory Probability and Statistical Applications,* 2nd ed., Addison-Wesley, Reading, Massachusetts, 1970.

6. Suppes, Patrick C., *Axiomatic Set Theory,* Van Nostrand, New York, 1960.

■ MISCELLANEOUS EXERCISES

1. A set of propositions $p_1, p_2, \ldots, p_n$ is said to logically imply a conclusion q if $(p_1 \wedge p_2 \wedge \ldots \wedge p_n) \Rightarrow q$. In this situation the propositions $p_1, p_2, \ldots, p_n$ are said to present a *valid argument* for q.

   Test the validity of each of the following arguments.

   a) p_1: If Roger studies then he will pass discrete mathematics.

      p_2: If Roger doesn't play tennis then he'll study.

      p_3: Roger failed discrete mathematics.

      q: Therefore, Roger played tennis.

      Let p: Roger studies; r: Roger passes discrete mathematics; s: Roger plays tennis.

      Does $[(p \rightarrow r) \wedge (\bar{s} \rightarrow p) \wedge \bar{r}] \Rightarrow s$?

   b) p_1: If it is warm this Friday, I'll wear my suede jacket if the pockets are mended.

      p_2: The forecast for this Friday is for warm weather but those pockets won't be mended.

      q: Therefore, I won't be wearing my suede jacket this Friday.

   c) p_1: The contract is fulfilled if and only if the new windows are installed in June.

      p_2: If the new windows are installed in June, we can move in on the first of July.

      p_3: If we can't move in on July 1, we must pay the rent on our apartment for July.

      p_4: If the windows are not installed, we must pay the rent on our apartment for July.

      q: Therefore, we don't pay the July rent on our apartment.

2. Prove that for any propositions $p, q, r, p \leftrightarrow (q \leftrightarrow r) \Leftrightarrow (p \leftrightarrow q) \leftrightarrow r$.

3. A set A has 128 subsets of even cardinality. (a) How many subsets of A have odd cardinality? (b) What is $|A|$?

4. For positive integers m, n, r with $r \leq \min\{m, n\}$, show that

$$\binom{m+n}{r} = \binom{m}{0}\binom{n}{r} + \binom{m}{1}\binom{n}{r-1} + \binom{m}{2}\binom{n}{r-2} + \cdots$$

$$+ \binom{m}{r}\binom{n}{0} = \sum_{k=0}^{r} \binom{m}{k}\binom{n}{r-k}.$$

5. a) In how many ways can a teacher divide up a group of seven students into two teams each containing at least one student? two students?

   b) Answer (a) replacing seven by a positive integer $n \geq 4$.

6. Determine whether each of the following statements is true or false. For each false statement give a counterexample.

   a) If A and B are infinite sets, then $A \cap B$ is infinite.

   b) If B is infinite and $A \subseteq B$, then A is infinite.

   c) If $A \subseteq B$ with B finite, then A is finite.

   d) If $A \subseteq B$ with A finite, then B is finite.

7. a) Determine the number of linear arrangements of m 1's and r 0's with no adjacent 1's. (State any needed condition(s) for m, r.)

   b) If $\mathcal{U} = \{1, 2, 3, \ldots, n\}$, how many sets $A \subseteq \mathcal{U}$ are such that $|A| = k$ with A containing no consecutive integers? (State any needed condition(s) for n, k.)

8. Let $A = \{1, 2, 3, \ldots, 15\}$.

   a) How many subsets of A contain all the odd integers in A?

   b) How many subsets of A contain exactly three odd integers?

   c) How many eight-element subsets of A contain exactly three odd integers?

   d) Write a computer program to generate an eight-element subset of A and have it print out how many of the eight elements are odd.

9. Let $\mathcal{U} = \mathbf{R}$ and the index set $I = \mathbf{Q}^{+}$. For each $q \in \mathbf{Q}^{+}$, let $A_q = [0, 2q]$, $B_q = (0, 3q]$. Determine each of the following:

   a) $A_{7/3}$ b) $B_{3/5}$ c) $A_3 - B_4$ d) $A_3 \triangle B_4$

   e) $\bigcup_{q \in I} A_q$ f) $\bigcup_{q \in I} B_q$ g) $\bigcap_{q \in I} A_q$ h) $\bigcap_{q \in I} B_q$

10. For a universe $\mathcal{U}$ and sets $A, B \subseteq \mathcal{U}$, prove each of the following.

   a) $A \triangle B = B \triangle A$ b) $A \triangle \overline{A} = \mathcal{U}$ c) $A \triangle \mathcal{U} = \overline{A}$

   d) $A \triangle \emptyset = A$, so $\emptyset$ is the identity for $\triangle$, as well as for $\cup$.

11. Consider the membership table (Table 2.10). If we are given the condition that $A \subseteq B$, then we need only consider those rows of the table for which this is true, namely rows 1, 2, and 4 as indicated by the arrows. For these rows, the columns for B and $A \cup B$ are exactly the same, and so the membership table here shows that $A \subseteq B \Rightarrow A \cup B = B$.

Table 2.10

	A	B	A ∪ B
→	0	0	0
→	0	1	1
	1	0	1
→	1	1	1

Use membership tables to verify each of the following:

   a) $A \subseteq B \Rightarrow A \cap B = A$

   b) $(A \cap B = A)$ and $(B \cup C = C) \Rightarrow A \cup B \cup C = C$.

    c) $C \subseteq B \subseteq A \Rightarrow (A \cap \overline{B}) \cup (B \cap \overline{C}) = A \cap \overline{C}$

    d) $A \triangle B = C \Rightarrow A \triangle C = B$ and $B \triangle C = A$.

12. a) Use the equivalence $A \subseteq B \Leftrightarrow A \cup B = B$ to show that the dual of $A \subseteq B$ is $B \subseteq A$.

    b) State the dual of each theorem in Exercise 11.

13. In how many ways can 15 laboratory assistants be assigned to work on one, two, or three different experiments so that each experiment has at least one person spending some time on it?

14. At a high school science fair 34 students received awards for scientific projects. Fourteen awards were given for projects in biology, 13 in chemistry and 21 in physics. If three students received awards in all three subject areas, how many received awards for exactly (a) one subject area? (b) two subject areas?

15. If eight people are seated around a circular table, what is the probability that a given pair are sitting next to each other?

16. If the letters in the word BOOLEAN are arranged at random, what is the probability that the two O's remain together in the arrangement?

17. If 16 chocolate-chip cookies are distributed among four children, what is the probability that every child gets (a) at least one cookie? (b) at least two cookies? (c) four cookies?

18. Fifty students, each with 75¢, visited the arcade of Example 2.23. Seventeen of the students played each of the three computer games and 37 of them played at least two of them. No student played any other game at the arcade nor did any student play a given game more than once. Each game costs 25¢ to play and the total proceeds from the student visit were $24.25. How many of these students preferred to watch and played none of the games?

3

Relations
and Functions

In this chapter we extend the set theory of Chapter 2 to include the concepts of relation and function. Algebra, trigonometry, and calculus all involve functions. Here, however, we shall study functions from a set theoretic approach that includes finite functions, and introduce new counting ideas in the study.

We take a path along which we shall find the answers to the following (closely related) six problems:

1. The Defense Department has seven different contracts that deal with a high security project. Four companies can manufacture the distinct parts called for in each contract, and to keep the project as well protected as possible it is best to have all four companies working on some part. In how many ways can the contracts be awarded so that every company is involved?

2. How many seven-digit quaternary $(0, 1, 2, 3)$ sequences have at least one occurrence of each of the digits 0, 1, 2, 3?

3. An $m \times n$ *zero-one matrix* is a matrix A with m rows and n columns, such that in row i, $1 \le i \le m$, and column j, $1 \le j \le n$, the entry a_{ij} that appears is either 0 or 1. How many 7×4 zero-one matrices have exactly one 1 in each row and at least one 1 in each column? (The zero-one matrix is a data structure that arises in computer science. We'll see more about it in later chapters.)

4. Seven (unrelated) people enter the lobby of a building which has four additional floors, and get on an elevator. What is the probability that the elevator must stop at every floor to let passengers off?

5. For positive integers m, n with $m < n$,

$$\sum_{k=0}^{n} (-1)^k \binom{n}{n-k} (n-k)^m = 0.$$

6. For every positive integer n,

$$n! = \sum_{k=0}^{n} (-1)^k \binom{n}{n-k} (n-k)^n.$$

Do you recognize the connection among the first four problems? The first three are the same problem in different settings. However, it is not obvious that the last two problems are related or that there is a connection between them and the first four. These identities, however, will be established using the same counting techniques developed to solve the first four problems.

3.1 ■ CARTESIAN PRODUCTS AND RELATIONS

Definition 3.1 ▶ For sets $A, B \subseteq \mathcal{U}$, the *Cartesian*, or *cross, product* of A, B is denoted by $A \times B$ and equals $\{(a, b) \mid a \in A, b \in B\}$.

We say that the elements of $A \times B$ are *ordered pairs*. For $(a, b), (c, d) \in A \times B$, we have $(a, b) = (c, d)$ if and only if $a = c$ and $b = d$.

If A, B are finite, it follows from the rule of product that $|A \times B| = |A| \cdot |B|$. Although we generally will not have $A \times B = B \times A$, we will have $|A \times B| = |B \times A|$. Also, although A, $B \subseteq \mathcal{U}$ it is not necessary that $A \times B \subseteq \mathcal{U}$, so $\mathcal{U}$ is not necessarily closed under this operation.

EXAMPLE 3.1   Let $\mathcal{U} = \{1, 2, 3, \ldots, 7\}$, $A = \{2, 3, 4\}$, $B = \{4, 5\}$. Then

a) $A \times B = \{(2, 4), (2, 5), (3, 4), (3, 5), (4, 4), (4, 5)\}$.

b) $B \times A = \{(4, 2), (4, 3), (4, 4), (5, 2), (5, 3), (5, 4)\}$.

c) $B^2 = B \times B = \{(4, 4), (4, 5), (5, 4), (5, 5)\}$.

d) $B^3 = B \times B \times B = \{(a, b, c) \mid a, b, c \in B\}$; $(4, 5, 5) \in B^3$.   □

EXAMPLE 3.2   With $\mathcal{U} = \mathbf{R}$, $\mathbf{R} \times \mathbf{R} = \{(x, y) \mid x, y \in \mathbf{R}\}$ is recognized as the real plane of coordinate geometry and two-dimensional calculus. The subset $\mathbf{R}^+ \times \mathbf{R}^+$ is the interior of the first quadrant of this plane. Likewise $\mathbf{R}^3$ represents Euclidean three-space, where such three-dimensional surfaces as spheres and planes are subsets of importance.   □

EXAMPLE 3.3   An experiment $\mathcal{E}$ is conducted as follows: A single die is rolled and its outcome noted and then a coin is flipped and its outcome noted. Determine a sample space $\mathcal{S}$ for $\mathcal{E}$.

Let $\mathcal{E}_1$ denote the first part of experiment $\mathcal{E}$, and let $\mathcal{S}_1 = \{1, 2, 3, 4, 5, 6\}$ be a sample space for $\mathcal{E}_1$. Likewise let $\mathcal{S}_2 = \{\mathrm{H}, \mathrm{T}\}$ be a sample space for $\mathcal{E}_2$, the second part of the experiment. Then $\mathcal{S} = \mathcal{S}_1 \times \mathcal{S}_2$ is a sample space for $\mathcal{E}$.

This sample space can be represented pictorially with a *tree diagram* which exhibits all the possible outcomes of experiment $\mathcal{E}$. In Fig. 3.1 we have such a tree diagram, which proceeds from left to right. From the leftmost endpoint six branches

Figure 3.1

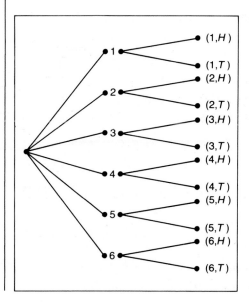

originate for the six outcomes of the first stage of the experiment $\mathscr{E}$. From each point, numbered $1, 2, \ldots, 6$, two branches indicate the outcomes in tossing the coin. The 12 ordered pairs at the right endpoints constitute the sample space $\mathscr{S}$.   □

In addition to their tie-in with Cartesian products, tree diagrams also arise in other situations.

E X A M P L E   3 . 4   At the Wimbledon Tennis Championships, women play at most three sets in a match. The winner is the first to win two sets. If we let N and E denote the two players, the tree diagram in Fig. 3.2 indicates the six ways in which this match can be won. For example, the starred line segment (edge) indicates that player E won the first set. The double-starred edge indicates that player N has won the match by winning the first and third sets.   □

Figure 3.2

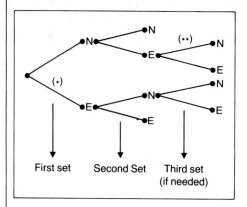

First set   Second Set   Third set
(if needed)

Tree diagrams are examples of a general structure called a *tree*. Trees and graphs are important structures that arise in computer science and optimization theory. These will be investigated in later chapters.

Returning to the cross product of two sets, we shall find the subsets of this structure of great interest.

Definition 3.2   ▶   For sets $A, B \subseteq \mathcal{U}$, any subset of $A \times B$ is called a *relation* from A to B. Any subset of $A \times A$ is called a *binary relation* on A.

E X A M P L E   3 . 5   With $A, B, \mathcal{U}$ as in Example 3.1, the following are relations from A to B.

a) $\emptyset$ b) $\{(2, 4)\}$ c) $\{(2, 4), (2, 5)\}$

d) $\{(2, 4), (3, 4), (4, 4)\}$ e) $\{(2, 4), (3, 4), (4, 5)\}$ f) $A \times B$

Since $|A \times B| = 6$, it follows from Definition 3.2 that there are 2^6 possible relations from A to B.   □

> In general, for finite sets A, B with $|A| = m$, $|B| = n$, there are 2^{mn} relations from A to B, including the empty relation as well as the relation $A \times B$ itself.

EXAMPLE 3.6

Let $B = \{1, 2\} \subseteq \mathbf{N}$, $\mathcal{U} = \mathcal{P}(B)$ and $A = \mathcal{U} = \{\emptyset, \{1\}, \{2\}, \{1, 2\}\}$. The following is an example of a *binary relation* on A: $\mathcal{R} = \{(\emptyset, \emptyset), (\emptyset, \{1\}), (\emptyset, \{2\}), (\emptyset, \{1, 2\}), (\{1\}, \{1\}), (\{1\}, \{1, 2\}), (\{2\}, \{2\}), (\{2\}, \{1, 2\}), (\{1, 2\}, \{1, 2\})\}$. We can say that the relation $\mathcal{R}$ is the *subset relation* where $(C, D) \in \mathcal{R}$ iff $C, D \subseteq B$ and $C \subseteq D$. □

EXAMPLE 3.7

With $A = \mathcal{U} = \mathbf{Z}^+$, we define a binary relation $\mathcal{R}$ on set A as $\{(x, y) \mid x \le y\}$. This is the familiar "is less than or equal to" relation for the set of positive integers. It can be represented graphically as the set of points with positive integer components located on or above the line $y = x$ in the Euclidean plane, as partially shown in Fig. 3.3. Here we cannot list the entire relation as we did in Example 3.6 but we note, for example, that $(7, 7), (7, 11) \in \mathcal{R}$, but $(8, 2) \notin \mathcal{R}$. $(7, 11) \in \mathcal{R}$ can also be denoted by $7 \, \mathcal{R} \, 11$; $(8, 2) \notin \mathcal{R}$ becomes $8 \, \mathcal{\not R} \, 2$. $7 \, \mathcal{R} \, 11$ and $8 \, \mathcal{\not R} \, 2$ are examples of the *infix* notation in a relation. □

Figure 3.3

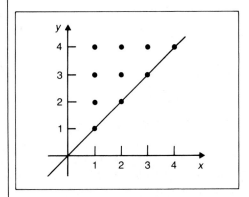

EXAMPLE 3.8

When the Pascal compiler translates a source program into the machine language of the object program, a *symbol table* which contains the following sets is constructed by the compiler:

1. S: the set of symbolic names such as variables, constants, and types.

2. A: the set of possible attributes for elements of S, such as integer, real, boolean, character.

3. L: the set of locations, or addresses, in memory where the elements of S are stored.

The information in the table provides relations from S to A and S to L. □

We close this section with these final observations.

a) For any set $A \subseteq \mathcal{U}$, $A \times \emptyset = \emptyset$. (If $A \times \emptyset \neq \emptyset$, let $(a, b) \in A \times \emptyset$. Then $a \in A$ and $b \in \emptyset$. Impossible!) Likewise, $\emptyset \times A = \emptyset$.

b) The Cartesian product and the binary operations of union and intersection are interrelated in the following theorem.

Theorem 3.1

▶ For any sets A, B, $C \subseteq \mathcal{U}$.

a) $A \times (B \cap C) = (A \times B) \cap (A \times C)$

b) $A \times (B \cup C) = (A \times B) \cup (A \times C)$

c) $(A \cap B) \times C = (A \times C) \cap (B \times C)$

d) $(A \cup B) \times C = (A \times C) \cup (B \times C)$

Proof

We prove Theorem 3.1(a) and leave the other parts for the reader. For any a, $b \in \mathcal{U}$, $(a, b) \in A \times (B \cap C) \Leftrightarrow a \in A$ and $b \in B \cap C \Leftrightarrow a \in A$ and $b \in B$, $C \Leftrightarrow a \in A$, $b \in B$, and $a \in A$, $b \in C \Leftrightarrow (a, b) \in A \times B$ and $(a, b) \in A \times C \Leftrightarrow (a, b) \in (A \times B) \cap (A \times C)$. ■

EXERCISES

1. If $\mathcal{U} = \mathbf{N}$, $A = \{1, 2, 3, 4\}$, $B = \{2, 5\}$, $C = \{3, 4, 7\}$, determine $A \times B$; $B \times A$; $A \cup (B \times C)$; $(A \cup B) \times C$; $(A \times C) \cup (B \times C)$.

2. If $\mathcal{U} = \{1, 2, 3, 4, 5\}$, $A = \{1, 2, 3\}$, $B = \{2, 4, 5\}$, give examples of:
    a) three nonempty relations from A to B.
    b) three nonempty binary relations on A.

+3. For A, B, $\mathcal{U}$ as in Exercise 2, determine the following: (a) $|A \times B|$ (b) The number of relations from A to B. (c) The number of binary relations on A. (d) The number of relations from A to B that contain $(1, 2)$ and $(1, 5)$. (e) The number of relations from A to B that contain exactly five ordered pairs. (f) The number of binary relations on A that contain at least seven elements.

+4. For sets A, $B \subseteq \mathcal{U}$, when is it true that $A \times B = B \times A$?

5. For $\mathcal{U} = \mathbf{N}$, $A = \{2, 3, 4, 5, 6, 7\}$, $B = \{10, 11, 12, 13, 14\}$, list the elements of the relation $\mathcal{R} \subseteq A \times B$ where $a \, \mathcal{R} \, b$ if a (evenly) divides b.

6. The men's final at Wimbledon is won by the first player to win three sets of the five-set match. Let C and M denote the players. Draw a tree diagram to show all of the ways the match can be decided.

7. If $\mathcal{U} = \mathbf{R}$, sketch the relation $\{(x, y) \mid x^2 + y^2 = 4\}$. What happens if $\mathcal{U}$ is $\mathbf{R}^+$?

8. Logic chips are taken from a container, tested individually, and labeled defective or good. The testing process is continued until either two defective chips are found or five chips are tested in total. Using a tree diagram, exhibit a sample space for this process.

9. Complete the proof of Theorem 3.1.

10. A rumor is spread as follows. The originator calls two people. Each of these people phones three friends, who in turn each call five associates. If no one

receives more than one call, how many people now know the rumor? How many phone calls were made?

11. For A, B, $C \subseteq \mathcal{U}$, prove that $A \times (B - C) = (A \times B) - (A \times C)$.

12. Let A, B be sets with $|B| = 3$. If there are 4096 relations from A to B, what is $|A|$?

3.2 ■ FUNCTIONS: PLAIN AND ONE-TO-ONE

In this section we concentrate on a special kind of relation called a *function*. General relations will appear again in Chapter 5.

Definition 3.3

▶ For nonempty sets A, B, a *function*, or *mapping, f from A to B*, denoted $f: A \to B$, is a relation from A to B in which every element of A appears exactly once as the first component of an ordered pair in the relation.

We often write $f(a) = b$ when (a, b) is an ordered pair in the function f. For $(a, b) \in f$, b is called *the image* of a under f, while a is *an antecedent* of b. In addition, the definition suggests that f is a method for *associating* with each $a \in A$ a *unique* $b \in B$, denoting this process by $f(a) = b$. Consequently, (a, b), $(a, c) \in f$ implies $b = c$.

EXAMPLE 3.9

For $A = \{1, 2, 3\}$, $B = \{w, x, y, z\}$, $f = \{(1, w), (2, x), (3, x)\}$ is a function, and consequently a relation, from A to B. $\mathcal{R}_1 = \{(1, w), (2, x)\}$, $\mathcal{R}_2 = \{(1, w), (2, w), (2, x), (3, z)\}$ are relations but not functions, from A to B. (Why not?) □

Definition 3.4

▶ For the function $f: A \to B$, A is called the *domain* of f, and B the *codomain* of f. The subset of B consisting of those elements that appear as second components in the ordered pairs of f is called the *range* of f, and is also denoted by $f(A)$.

In Example 3.9, domain of $f = \{1, 2, 3\}$, codomain of $f = \{w, x, y, z\}$ and the range of $f = f(A) = \{w, x\}$.

A pictorial representation of these ideas can be given by a Venn diagram, as in Fig. 3.4. This diagram suggests that a be regarded as *input* that is *transformed* by f

Figure 3.4

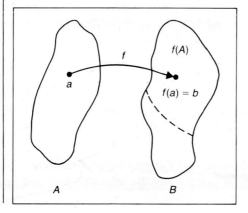

into the corresponding *output, f(a)*. In this context a FORTRAN compiler can be thought of as a function that transforms a source program (the input) into its corresponding object program (the output).

EXAMPLE 3.10   Many interesting functions arise in computer science.

a)   In the programming language BASIC, the greatest integer function, denoted INT, is an integer-valued function defined for the domain of all real numbers. Hence we write INT: $\mathbf{R} \to \mathbf{Z}$ where INT(X) = X, if X is an integer; INT(X) = the integer to the immediate left of X on the real number line, if X is not an integer. For example, INT(5.1) = 5 = INT(5), and INT(−7.8) = −8 = INT(−8). Here the codomain and range are the set $\mathbf{Z}$.

b)   The function trunc (for truncation) arises in Pascal and is an integer-valued function defined on $\mathbf{R}$. The function deletes the fractional part of a real number. For example, trunc (3.78) = 3, trunc (5) = 5, trunc (−7.22) = −7.

c)   In storing a matrix in a one-dimensional array, many computer languages use the *row major* implementation. Here, if $A = (a_{ij})_{n \times n}$ is an $n \times n$ matrix, the first row of A is stored in locations 1, 2, 3, ... , n of the array if we start with a_{11} in location 1. The entry a_{21} is then found in position $n + 1$, while entry a_{34} occupies position $2n + 4$ in the array. In order to determine the location of any entry a_{ij}, $1 \le i, j \le n$, from A, one defines the *access function f* from the entries of A to the positions 1, 2, 3, ... , n^2 of the array. A formula for the access function here is $f(a_{ij}) = (i - 1)n + j$.   □

In Example 3.9 there are 2^{12} = 4096 relations from A to B. We have examined one function among these relations, and now we wish to count the total number of functions from A to B.

For the general case, let A, B be sets with $|A| = m$, $|B| = n$. Consequently, if $A = \{a_1, a_2, \ldots, a_m\}$, $B = \{b_1, b_2, \ldots, b_n\}$, a typical function $f: A \to B$ can be described by $\{(a_1, x_1), (a_2, x_2), \ldots, (a_m, x_m)\}$. We can select any of the n elements of B for x_1, and then do the same for x_2. (We can select any element of B for x_2 so that the same element of B may be selected for both x_1 and x_2.) We continue this selection process until one of the n elements of B is finally selected for x_m. In this way, using the rule of product, there are $n^m = |B|^{|A|}$ functions from A to B.

Therefore for A, B in Example 3.9, there are $4^3 = |B|^{|A|}$ = 64 functions from A to B, and $3^4 = |A|^{|B|}$ = 81 functions from B to A.

Now that we have the concept of a function as a special type of relation, we turn our attention to a special type of function.

Definition 3.5   ▶   A function $f: A \to B$ is called *one-to-one*, or *injective*, if each element of B appears at most once as the second component of an ordered pair in f.

If $f: A \rightarrow B$ is one-to-one, with A, B finite, we must have $|A| \leq |B|$. For general sets A, B, if $f: A \rightarrow B$ is one-to-one, then for a_1, $a_2 \in A$, $f(a_1) = f(a_2) \Rightarrow a_1 = a_2$. Consequently, a function of this type can be characterized as one for which each element of the range is the image of exactly one element from the domain.

EXAMPLE 3.11 | Let $A = \{1, 2, 3\}$, $B = \{1, 2, 3, 4, 5\}$. The function

$$f = \{(1, 1), (2, 3), (3, 4)\}$$

is an injective function from A to B;

$$g = \{(1, 1), (2, 3), (3, 3)\}$$

is a function from A to B but it fails to be injective since $g(2) = g(3)$ but $2 \neq 3$. ☐

For A, B in Example 3.11 there are 2^{15} relations from A to B and 5^3 of these are functions. The next question we naturally want to answer is how many functions $f: A \rightarrow B$ are one-to-one. Again we argue for general finite sets.

With $A = \{a_1, a_2, a_3, \ldots, a_m\}$, $B = \{b_1, b_2, b_3, \ldots, b_n\}$, $m \leq n$, a one-to-one function $f: A \rightarrow B$ has the form $\{(a_1, x_1), (a_2, x_2), (a_3, x_3), \ldots, (a_m, x_m)\}$, where there are n choices for x_1 (i.e., any element of B), $n - 1$ choices for x_2 (i.e., any element of B except the one chosen for x_1), $n - 2$ choices for x_3, and so on, finishing with $n - (m - 1) = n - m + 1$ choices for x_m. By the rule of product the number of one-to-one functions from A to B is

$$n(n - 1)(n - 2) \cdots (n - m + 1) = n!/(n - m)! = P(n, m) = P(|B|, |A|).$$

Consequently, for A, B in Example 3.11 there are $5 \cdot 4 \cdot 3 = 60$ one-to-one functions $f: A \rightarrow B$.

Definition 3.6 | ▶ If $f: A \rightarrow B$ and $A_1 \subseteq A$, then

$$f(A_1) = \{b \in B \mid b = f(a) \quad \text{for some } a \in A_1\},$$

and $f(A_1)$ is called the *image of A_1 under f*.

EXAMPLE 3.12 | Let $f: \mathbf{R} \rightarrow \mathbf{R}$ be given by $f(x) = x^2$. Then $f(\mathbf{R}) =$ the range of $f = [0, +\infty)$. The image of $\mathbf{Z}$ under f is $f(\mathbf{Z}) = \{0, 1, 4, 9, 16, \ldots\}$. ☐

Theorem 3.2 | ▶ Let $f: A \rightarrow B$, with A_1, $A_2 \subseteq A$. Then

a) $f(A_1 \cup A_2) = f(A_1) \cup f(A_2)$.   b) $f(A_1 \cap A_2) \subseteq f(A_1) \cap f(A_2)$.

c) $f(A_1 \cap A_2) = f(A_1) \cap f(A_2)$ when f is injective.

Proof | We prove (b) and leave the remaining parts for the reader.

For any $b \in B$, $b \in f(A_1 \cap A_2) \Leftrightarrow b = f(a)$, for some $a \in A_1 \cap A_2 \Leftrightarrow (b = f(a), a \in A_1)$ and $(b = f(a), a \in A_2) \Rightarrow b \in f(A_1)$ and $b \in f(A_2) \Rightarrow b \in f(A_1) \cap f(A_2)$, so $f(A_1 \cap A_2) \subseteq f(A_1) \cap f(A_2)$. ∎

Definition
3.7

▶ If $f: A \to B$ and $A_1 \subseteq A$, $f|_{A_1}: A_1 \to B$ is called the *restriction of f to A_1* if $f|_{A_1}(a) = f(a)$ for all $a \in A_1$.

Definition
3.8

▶ Let $A_1 \subseteq A$ and $f: A_1 \to B$. If $g: A \to B$ and $g(a) = f(a)$ for all $a \in A_1$, we call g an *extension of f to A*.

EXAMPLE 3.13

Let $A = \{w, x, y, z\}$, $B = \{1, 2, 3, 4, 5\}$, $A_1 = \{w, y, z\}$. Let $f: A \to B$, $g: A_1 \to B$ be represented by the diagrams in Fig. 3.5. Then $g = f|_{A_1}$ and f is an extension of g from A_1 to A. We note that for the given function $g: A_1 \to B$ there are five ways to extend g from A_1 to A.   □

Figure 3.5

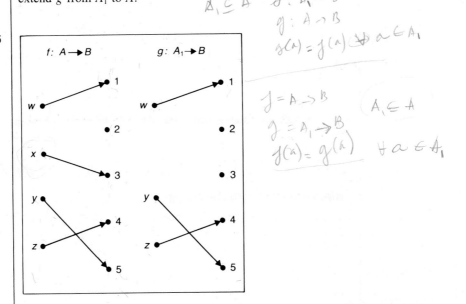

EXERCISES

1. Determine whether or not each of the following relations is a function. If a relation is a function, find its range.

a) $\{(x, y) \mid x, y \in \mathbf{Z}, y = x^2 + 7\}$.   b) $\{(x, y) \mid x, y \in \mathbf{R}, y^2 = x\}$.

c) $\{(x, y) \mid x, y \in \mathbf{R}, y = 3x + 1\}$.   d) $\{(x, y) \mid x, y \in \mathbf{Q}, x^2 + y^2 = 1\}$.

e) $\mathcal{R}$ is a relation from A to B where $|A| = 5$, $|B| = 6$, $|\mathcal{R}| = 6$.

2. Can the formula $f(x) = 1/(x^2 - 2)$ be used to define a function $f: \mathbf{R} \to \mathbf{R}$? A function $f: \mathbf{Z} \to \mathbf{R}$?

3. Let $A = \{1, 2, 3, 4\}$, $B = \{x, y, z\}$. (a) List five functions from A to B. (b) How many functions $f: A \to B$ are there? (c) How many functions $f: A \to B$ are one-to-one? (d) How many functions $g: B \to A$ are there? (e) How many functions $g: B \to A$ are one-to-one? (f) How many functions $f: A \to B$ satisfy $f(1) = x$? (g) How many functions $f: A \to B$ satisfy $f(1) = f(2) = x$? (h) How many functions $f: A \to B$ satisfy $f(1) = x, f(2) = y$?

4. If there are 2187 functions $f: A \to B$ and $|B| = 3$, what is $|A|$?

5. For each of the following functions, determine whether the function is one-to-one and determine its range.

   a) $f: \mathbf{Z} \to \mathbf{Z}, \ f(x) = 2x + 1$ b) $f: \mathbf{Q} \to \mathbf{Q}, \ f(x) = 2x + 1$

   c) $f: \mathbf{Z} \to \mathbf{Z}, \ f(x) = x^3 - x$ d) $f: \mathbf{R} \to \mathbf{R}, \ f(x) = e^x$

   e) $f: [-\pi/2, \pi/2] \to \mathbf{R}, \ f(x) = \sin x$ f) $f: [0, \pi] \to \mathbf{R}, \ f(x) = \sin x$

6. Let $f: \mathbf{R} \to \mathbf{R}, \ f(x) = x^2$. Determine $f(A)$ for the following subsets $A \subseteq \mathbf{R}$.

   a) $A = \{2, 3\}$ b) $A = \{-3, -2, 2, 3\}$ c) $A = (-3, 3)$

   d) $A = (-3, 2]$ e) $A = [-7, 2]$ f) $A = (-4, -3] \cup [5, 6]$

7. Let $A = \{1, 2, 3, 4, 5\}$, $B = \{w, x, y, z\}$, $A_1 = \{2, 3, 5\} \subseteq A$, $g: A_1 \to B$. In how many ways can g be extended to a function $f: A \to B$?

8. Give an example of a function $f: A \to B$, with $A_1, A_2 \subseteq A$ and $f(A_1 \cap A_2) \neq f(A_1) \cap f(A_2)$.

9. Prove parts (a) and (c) of Theorem 3.2.

10. If $A = \{1, 2, 3, 4, 5\}$ and there are 6720 injective functions $f: A \to B$, what is $|B|$?

3.3 ■ ONTO FUNCTIONS: STIRLING NUMBERS OF THE SECOND KIND

The results we develop in this section will provide the answers to the first five problems stated at the beginning of this chapter. We find that the *onto* function is the key to all of the answers.

Definition 3.9 ▶ If $f: A \to B$, f is called *onto*, or *surjective*, if $f(A) = B$ (i.e., if for all $b \in B$ there is at least one $a \in A$ with $f(a) = b$).

EXAMPLE 3.14 The function $f: \mathbf{R} \to \mathbf{R}$ defined by $f(x) = x^3$ is an onto function but the function $g: \mathbf{R} \to \mathbf{R}$ given by $g(x) = x^2$ is not, since $g(\mathbf{R}) = [0, +\infty) \subset \mathbf{R}$.   □

EXAMPLE 3.15 With $A = \{1, 2, 3, 4\}$, $B = \{x, y, z\}$,

$$f_1 = \{(1, z), (2, y), (3, x), (4, y)\} \quad \text{and} \quad f_2 = \{(1, x), (2, x), (3, y), (4, z)\}$$

are both functions from A onto B. The function $g = \{(1, x), (2, x), (3, y), (4, y)\}$ is not onto, since $g(A) = \{x, y\} \subset B$.   □

  If A, B are finite sets, for any onto function $f: A \to B$ to possibly exist we must have $|A| \geq |B|$. Considering the development in the first two sections of this chapter, the reader undoubtedly feels it is time once again to use the rule of product and count the number of onto functions $f: A \to B$ where $|A| = m \geq n = |B|$. Unfortunately the rule of product proves to be insufficient here. We shall obtain the needed result for some specific examples and then conjecture a general formula. In Chapter 7 the conjecture will be established using the Principle of Inclusion and Exclusion.

EXAMPLE 3.16   If $A = \{x, y, z\}$ and $B = \{1, 2\}$, then all functions $f: A \rightarrow B$ are onto except for $f_1 = \{(x, 1), (y, 1), (z, 1)\}$ and $f_2 = \{(x, 2), (y, 2), (z, 2)\}$, the *constant* functions. So there are $|B|^{|A|} - 2 = 2^3 - 2 = 6$ onto functions from A to B.   □

In general, if $|A| = m \geq 2$ and $|B| = 2$, there are $2^m - 2$ onto functions from A to B. (Does this formula tell us anything when $m = 1$?)

EXAMPLE 3.17   For $A = \{w, x, y, z\}$ and $B = \{1, 2, 3\}$, there are 3^4 functions from A to B. Considering subsets of B of size 2, there are 2^4 functions from A to $\{1, 2\}$, 2^4 functions from A to $\{2, 3\}$, and 2^4 functions from A to $\{1, 3\}$. So we have $3(2^4) = \binom{3}{2}2^4$ functions from A to B that are definitely not onto. However, before we swallow $3^4 - \binom{3}{2}2^4$ as the final answer, we must realize that not all of these $\binom{3}{2}2^4$ functions are distinct. For when we consider all the functions from A to $\{1, 2\}$ we are removing, among these, the function $\{(w, 2), (x, 2), (y, 2), (z, 2)\}$. Then considering the functions from A to $\{2, 3\}$ we remove the same function: $\{(w, 2), (x, 2), (y, 2), (z, 2)\}$. Consequently, in the result $3^4 - \binom{3}{2}2^4$ we have twice removed each of the constant functions $f: A \rightarrow B$, where $f(A)$ is one of the sets $\{1\}$, $\{2\}$, or $\{3\}$. Adjusting our present result for this we find that there are $3^4 - \binom{3}{2}2^4 + 3 = \binom{3}{3}3^4 - \binom{3}{2}2^4 + \binom{3}{1}1^4$ onto functions from A to B.

Keeping $B = \{1, 2, 3\}$, for any set A, with $|A| = m \geq 3$, there are $\binom{3}{3}3^m - \binom{3}{2}2^m + \binom{3}{1}1^m$ functions from A onto B. (What result does this formula yield when $m = 1, 2$?)   □

The prior two examples suggest a pattern which we now state, without proof, as our general formula.

For finite sets A, B with $|A| = m \geq n = |B|$, there are

$$\binom{n}{n}n^m - \binom{n}{n-1}(n-1)^m + \binom{n}{n-2}(n-2)^m - \cdots$$

$$+ (-1)^{n-2}\binom{n}{2}2^m + (-1)^{n-1}\binom{n}{1}1^m = \sum_{k=0}^{n-1}(-1)^k\binom{n}{n-k}(n-k)^m$$

$$= \sum_{k=0}^{n}(-1)^k\binom{n}{n-k}(n-k)^m$$

onto functions from A to B.

EXAMPLE 3.18   Let $A = \{1, 2, 3, 4, 5, 6, 7\}$ and $B = \{w, x, y, z\}$. Applying the general formula with $m = 7$, $n = 4$ there are

$$\binom{4}{4}4^7 - \binom{4}{3}3^7 + \binom{4}{2}2^7 - \binom{4}{1}1^7 = \sum_{k=0}^{3}(-1)^k\binom{4}{4-k}(4-k)^7$$

$$= \sum_{k=0}^{4}(-1)^k\binom{4}{4-k}(4-k)^7 = 8400 \text{ functions from } A \text{ onto } B.   □$$

The result in Example 3.18 is also the answer to the first three questions proposed at the start of this chapter. Once we remove the unnecessary vocabulary we recognize that in all three cases we want to distribute seven different objects into four distinct containers with no container left empty. This can be solved in terms of onto functions.

For Problem 4 we have a sample space $\mathscr{S}$ consisting of the 4^7 ways seven people can each select one of the four floors. (Note that 4^7 is also the total number of functions $f: A \rightarrow B$ where $|A| = 7$, $|B| = 4$.) The event we are concerned with contains 8400 of those selections, so the probability that the elevator must stop at every floor is $8400/16384 \doteq 0.5127$, slightly more than half of the time.

Finally, for Problem 5, since $\sum_{k=0}^{n}(-1)^k\binom{n}{n-k}(n-k)^m$ is the number of onto functions $f: A \rightarrow B$ for $|A| = m$, $|B| = n$, when $m < n$ there are no such functions and the summation is 0.

Problem 6 will be settled in Section 3.5.

Before going on to anything new, we consider one more problem.

EXAMPLE 3.19   At the CH Company, Joan, the supervisor, has a secretary, Teresa, and three other administrative assistants. If there are seven accounts to be processed, in how many ways can Joan assign the accounts so that each assistant works on at least one account and Teresa's work includes, if nothing else, the most expensive account?

First and foremost, the answer is not 8400 as in Example 3.18. Here we must consider two disjoint subcases and then apply the rule of sum.

a) If Teresa, the secretary, works only on the most expensive account, then the other six accounts can be distributed among the three administrative assistants in $\sum_{k=0}^{3}(-1)^k\binom{3}{3-k}(3-k)^6 = 540$ ways. ($540 =$ the number of onto functions $f: A \rightarrow B$ with $|A| = 6$, $|B| = 3$.)

b) If Teresa does more than just the most expensive account, the assignments can be made in $\sum_{k=0}^{4}(-1)^k\binom{4}{4-k}(4-k)^6 = 1560$ ways. ($1560 =$ the number of onto functions $g: C \rightarrow D$ with $|C| = 6$, $|D| = 4$.)

Consequently, the assignments can be given under the prescribed conditions in $540 + 1560 = 2100$ ways.   □

We now continue our discussion with the distribution of distinct objects into containers with none left empty, but now the containers become identical.

EXAMPLE 3.20   If $A = \{a, b, c, d\}$, $B = \{1, 2, 3\}$ there are 36 onto functions from A to B, or, equivalently, 36 ways to distribute four distinct objects into three distinguishable containers, with no container empty (and no regard for the location of objects in a given container). Among these 36 distributions we find the following collection of six (one of six such possible collections of six),

(1) $\{a, b\}_1$ $\{c\}_2$ $\{d\}_3$ (2) $\{a, b\}_1$ $\{d\}_2$ $\{c\}_3$

(3) $\{c\}_1$ $\{a, b\}_2$ $\{d\}_3$ (4) $\{c\}_1$ $\{d\}_2$ $\{a, b\}_3$

(5) $\{d\}_1$ $\{a, b\}_2$ $\{c\}_3$ (6) $\{d\}_1$ $\{c\}_2$ $\{a, b\}_3$,

where, for example, the notation $\{c\}_2$ means that c is in the second container. Now if we no longer distinguish the containers, these $6 = 3!$ distributions become identical, so there are $36/(3!) = 6$ ways to distribute the distinct objects a, b, c, d among three identical containers, leaving no container empty.   □

Generally, for $m \geq n$ there are $\sum_{k=0}^{n}(-1)^k\binom{n}{n-k}(n-k)^m$ ways to distribute m different objects into n numbered (but otherwise identical) containers with no container left empty. Removing the numbers on the containers, so that they are now identical in appearance, we find that one distribution into these n (nonempty) identical containers corresponds with $n!$ such distributions into the numbered containers. So the number of ways to distribute the m distinct objects into n identical containers, with no container left empty, is

$$\frac{1}{n!}\sum_{k=0}^{n}(-1)^k\binom{n}{n-k}(n-k)^m.$$

This will be denoted by $S(m, n)$, and is called a *Stirling number of the second kind.*

We note that for $|A| = m \geq n = |B|$, there are $n! \cdot S(m, n)$ onto functions from A to B.

Table 3.1 lists some Stirling numbers of the second kind.

Table 3.1

m \ n	1	2	3	4	5	6	7	8
1	1							
2	1	1						
3	1	3	1					
4	1	7	6	1				
5	1	15	25	10	1			
6	1	31	90	65	15	1		
7	1	63	301	350	140	21	1	
8	1	127	966	1701	1050	266	28	1

$S(m, n)$

EXAMPLE 3.21   For $m \geq n$, $\sum_{i=1}^{n} S(m, i)$ is the number of ways to distribute m distinct objects into n identical containers with empty containers possible. From the fourth row of Table 3.1 we see that there are $1 + 7 + 6 = 14$ ways to distribute the objects a, b, c, d among three identical containers, with some container(s) possibly empty.   □

We close this section with the derivation of an identity involving Stirling numbers. The proof is combinatorial in nature.

Theorem
3.3

▶ Let n be a positive integer with $1 < n \leq m$. Then

$$S(m + 1, n) = S(m, n - 1) + nS(m, n).$$

Proof   Let $A = \{a_1, a_2, \ldots, a_m, a_{m+1}\}$. Then $S(m + 1, n)$ counts the number of ways the objects of A can be distributed among n identical containers, with no container left empty.

To distribute $a_1, a_2, \ldots, a_m$ among $n - 1$ identical containers, with none left empty, there are $S(m, n - 1)$ ways. Then placing a_{m+1} in the remaining empty container results in $S(m, n - 1)$ of the distributions counted in $S(m + 1, n)$, namely those distributions where a_{m+1} is in a container by itself. Alternatively, distributing $a_1, a_2, \ldots, a_m$ among the n identical containers with none left empty, we have $S(m, n)$ distributions. Now, however, for each of these $S(m, n)$ distributions the n containers become distinguished by their contents. Selecting one of the n distinct containers for a_{m+1}, we have $nS(m, n)$ distributions of the total $S(m + 1, n)$, namely those where a_{m+1} is in the same container as another object from A. The result then follows by the rule of sum.   ∎

This identity can now be used to extend Table 3.1 if necessary. For the triangle shown in the table, the largest number corresponds with $S(m + 1, n)$, for $m = 7$, $n = 3$ and we see that $S(7 + 1, 3) = 966 = 63 + 3(301) = S(7, 2) + 3S(7, 3)$.

EXERCISES

1. Give an example of finite sets A, B with $|A|, |B| \geq 4$ and a function $f: A \to B$ such that (a) f is neither one-to-one nor onto; (b) f is one-to-one but not onto; (c) f is onto but not one-to-one; (d) f is onto and one-to-one.

2. For each of the following functions $f: \mathbf{Z} \to \mathbf{Z}$ determine whether the function is one-to-one and whether it is onto. If the function is not onto, determine the range $f(\mathbf{Z})$.

   a) $f(x) = x + 7$ b) $f(x) = 2x - 3$ c) $f(x) = -x + 5$
   d) $f(x) = x^2$ e) $f(x) = x^2 + x$ f) $f(x) = x^3$

3. For each of the following functions $g: \mathbf{R} \to \mathbf{R}$ determine whether the function is one-to-one and whether it is onto. If the function is not onto, determine the range $g(\mathbf{R})$.

   a) $g(x) = x + 7$ b) $g(x) = 2x - 3$ c) $g(x) = -x + 5$
   d) $g(x) = x^2$ e) $g(x) = x^2 + x$ f) $g(x) = x^3$

4. Let $A = \{1, 2, 3, 4\}$, $B = \{1, 2, 3, 4, 5, 6\}$.

   a) How many functions are there from A to B? How many of these are one-to-one? How many are onto?

   b) How many functions are there from B to A? How many of these are onto? How many are one-to-one?

5. Verify that $\sum_{k=0}^{n} (-1)^k \binom{n}{n-k} (n - k)^m = 0$ for $n = 5$, and $m = 2, 3, 4$.

6. A research chemist who has five laboratory assistants is engaged in a research project that calls for nine compounds that must be synthesized. In how many ways can the chemist assign these syntheses to the five assistants so that each of them is working on at least one synthesis?

7. Use the fact that every polynomial equation having real number coefficients and odd degree has a real root to show that the function $f: \mathbf{R} \to \mathbf{R}$, defined by $f(x) = x^5 - 2x^2 + x$, is an onto function. Is f one-to-one?

8. Suppose we have seven different colored balls and four containers numbered I, II, III, and IV. (a) In how many ways can we distribute the balls so that no container is left empty? (b) In this collection of seven colored balls one of them is blue. In how many ways can we distribute the balls so that no container is empty and the blue ball is in container II? (c) If we remove the numbers from the containers so that we can no longer distinguish them, in how many ways can we distribute the seven colored balls among the four identical containers, with some container(s) possibly empty?

9. Determine the next two rows ($m = 9, 10$) of Table 3.1 for the Stirling numbers $S(m, n)$, $1 \leq n \leq m$.

10. Write a computer program to compute the Stirling numbers $S(m, n)$, $1 \leq m \leq 12$, $1 \leq n \leq m$.

11. Let $A = \{1, 2, 3, 4, 5, 6\}$ and $B = \{1, 2, 3, 4\}$. Write a computer program to generate a function from A to B and have it print out whether or not the generated function is onto.

3.4 ■ THE PIGEONHOLE PRINCIPLE

It is time now for a change of pace as we introduce an interesting distribution principle. This principle may seem to have nothing in common with what we have been doing so far but it will prove to be helpful nonetheless.

In mathematics one sometimes finds that an almost obvious idea, when applied in a rather subtle manner, is the key needed to solve a troublesome problem. On the list of such obvious ideas many would undoubtedly place the following rule, known as the *Pigeonhole Principle*.

The Pigeonhole Principle: If m pigeons occupy n pigeonholes, then $m > n$ iff there is at least one pigeonhole with two or more pigeons roosting in it.

Now what can pigeons roosting in pigeonholes have to do with mathematics — discrete, combinatorial, or otherwise? It turns out that this principle can be applied in various problems where we seek to establish whether or not a certain situation can actually occur. We illustrate this in the following examples. We will find the principle useful in Section 3.5 and at other points in the text.

EXAMPLE 3.22   In an office there are 13 file clerks, so at least two of them must have birthdays during the same month. (Here we have 13 pigeons (the file clerks) and 12 pigeonholes (the months of the year).)   □

Here is a second rather immediate application of our principle.

EXAMPLE 3.23   Larry returns from the laundromat with 12 pairs of socks (each pair a different color) in a laundry bag. Drawing the socks from the bag randomly, he'll have to draw at most 13 of them in order to get a matched pair.   □

From this point on application of the pigeonhole principle will be more subtle.

EXAMPLE 3.24   Any subset of size six from the set $S = \{1, 2, 3, \ldots, 9\}$ must contain two elements whose sum is 10.

Here the numbers $1, 2, 3, \ldots, 9$ are the pigeons while the pigeonholes are the subsets $\{1, 9\}$, $\{2, 8\}$, $\{3, 7\}$, $\{4, 6\}$, $\{5\}$. When six pigeons go to their respective pigeonholes they must fill at least one of the two-element subsets whose members sum to 10.   □

EXAMPLE 3.25   Triangle ACE is equilateral with $AC = 1$. If five points are selected from the interior of the triangle there are at least two whose distance apart is less than $1/2$.

Figure 3.6

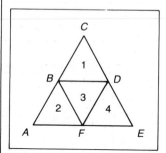

For the triangle in Fig. 3.6, the four smaller triangles are congruent equilateral triangles and $AB = 1/2$. We break up the interior of triangle ACE into the following four regions which are mutually disjoint in pairs:

R_1: the interior of triangle BCD together with the points on the segment BD, excluding B, D.

R_2: the interior of triangle ABF.

R_3: the interior of triangle BDF together with the points on the segments BF, DF, excluding B, D, F.

R_4: the interior of triangle FDE.

Now applying the pigeonhole principle, five points in the interior of triangle ACE must be such that at least two of them are in one of the four regions R_i, $1 \leq i \leq 4$, where any two points are separated by a distance less than $1/2$.   □

EXAMPLE 3.26

Let S be a set of six positive integers whose maximum is at most 14. Show that the sums of the elements in all the nonempty subsets of S cannot be distinct.

For any nonempty subset A of S, the sum of the elements in A, denoted s_A, satisfies $1 \leq s_A \leq 9 + 10 + \cdots + 14 = 69$, and there are $2^6 - 1 = 63$ nonempty subsets of S. We should like to draw the conclusion from the pigeonhole principle by letting the possible sums, from 1 to 69, be the pigeonholes, with the 63 nonempty subsets of S as the pigeons, but then we have too few pigeons.

So instead of considering all nonempty subsets of S we cut back to those nonempty subsets A where $|A| \leq 5$. Then $1 \leq s_A \leq 10 + 11 + \cdots + 14 = 60$. Since there are 62 nonempty subsets of S with $|A| \leq 5$, the elements of at least two of these subsets must yield the same sum. ☐

EXAMPLE 3.27

While on a four-week vacation Herbert will play at least one set of tennis each day, but won't play more than 40 sets total during this time. Prove that no matter how he distributes his sets during the four weeks, there is a consecutive span of days during which he plays exactly 15 sets.

For $1 \leq i \leq 28$, let x_i be the total number of sets Herbert has played from the start of the vacation to the end of the ith day. Then $1 \leq x_1 < x_2 < \ldots < x_{28} \leq 40$, and $x_1 + 15 < \ldots < x_{28} + 15 \leq 55$. We now have the 28 distinct numbers $x_1, x_2, \ldots, x_{28}$ and the 28 distinct numbers $x_1 + 15, x_2 + 15, \ldots, x_{28} + 15$. These 56 numbers can take on only 55 different values, so at least two of them must be equal and we conclude that there exist $1 \leq j < i \leq 28$ with $x_i = x_j + 15$. Hence, from the start of day $j + 1$ to the end of day i, Herbert plays exactly 15 sets of tennis. ☐

EXERCISES

1. In Example 3.23, what plays the roles of the pigeons and pigeonholes in our application of the pigeonhole principle?

2. Show that if there are eight people in a room, at least two of them have birthdays that occur on the same day of the week.

3. How many times must we roll a single die in order to get the same score at least twice?

4. a) Show that if any 14 integers are selected from the set $S = \{1, 2, 3, \ldots, 25\}$, there are at least two whose sum is 26.

   b) State a result that generalizes part (a) and Example 3.24.

5. Let triangle ABC be equilateral, with $AB = 1$. Show that if we select 10 points in the interior of this triangle, there must be at least two whose distance apart is less than $1/3$.

6. Let $ABCD$ be a square with $AB = 1$. Show that if we select five points in the interior of this square, there are at least two whose distance apart is less than $1/\sqrt{2}$.

7. Let S be a set of five positive integers the maximum of which is at most 9. Prove that the sums of the elements in all the nonempty subsets of S cannot be distinct.

8. At a certain high school there are 45 students who are training for eight different track and field events. If each student trains for only one event, and no event has more than 10 students training for it, prove that at least three events have five or more students in training for each.

9. During the first six weeks of his senior year in college, Brace sends his resumé out to different companies. If he sends out at least one resumé each and every day, but no more than 60 resumés in total, show that there is a period of consecutive days during which he sends out exactly 23 resumés.

3.5 ■ SPECIAL FUNCTIONS

In Chapter 2 we mentioned that addition is a binary operation on the set $\mathbf{Z}^+$, while $\cap$ is a binary operation on $\mathcal{P}(\mathcal{U})$ for any given universe $\mathcal{U}$. We now make this notion of a binary operation more precise in terms of a function.

Definition 3.10

▶ For any set A, any function $f: A \times A \to A$ is called a *binary operation* on A.

EXAMPLE 3.28

The function $f: \mathbf{Z} \times \mathbf{Z} \to \mathbf{Z}$, defined by $f(a, b) = a - b$, is a binary operation on $\mathbf{Z}$. □

EXAMPLE 3.29

Let $\mathcal{U} = \{1, 2, 3, 4\}$. For any sets $A, B \subseteq \mathcal{U}$, $g: \mathcal{P}(\mathcal{U}) \times \mathcal{P}(\mathcal{U}) \to \mathcal{P}(\mathcal{U})$, defined by $g(A, B) = A \cup B$, is a binary operation on $\mathcal{P}(\mathcal{U})$. □

Definition 3.11

▶ Let $f: A \times A \to A$.

a) f is said to be *commutative* if $f(a, b) = f(b, a)$ for all $(a, b) \in A \times A$;

b) f is said to be *associative* if for $a, b, c \in A$, $f(f(a, b), c) = f(a, f(b, c))$.

EXAMPLE 3.30

The binary operation of Example 3.29 is commutative and associative, while that in Example 3.28 is neither. □

Having seen one example of a function whose domain involves a cross product of sets, we now investigate other functions where the domain is a subset of such a cross product.

Definition 3.12

▶ For sets A, B, if $D \subseteq A \times B$, then $\pi_A: D \to A$ defined by $\pi_A(a, b) = a$ is called a *projection* on the first coordinate.

The function π_B is defined similarly and we note that if $D = A \times B$, both π_A and π_B are onto.

We now extend the notion of projection as follows: if $A_1, A_2, \ldots, A_n$ are sets, and $\{i_1, i_2, \ldots, i_m\} \subseteq \{1, 2, \ldots, n\}$ with $i_1 < i_2 < \ldots < i_m$, $m \le n$, for $D \subseteq A_1 \times A_2 \times \cdots \times A_n = \times_{i=1}^{n} A_i$, the function $\pi: D \to A_{i_1} \times A_{i_2} \times \cdots \times A_{i_m}$ defined by $\pi(a_1, a_2, \ldots, a_n) = (a_{i_1}, a_{i_2}, \ldots, a_{i_m})$ is a projection of D on the i_1th, i_2th, $\ldots, i_m$th coordinates. The elements of D are called *n-tuples*, while an element in $\pi(D)$ is an *m-tuple*.

These projections arise in a natural way in the study of *relational data bases,* a standard technique for organizing and describing large quantities of data by modern large-scale computing systems. In situations like credit card transactions, not only must existing data be organized but new data must be inserted, as when credit cards are processed for new cardholders. When bills on existing accounts are paid, or when new purchases are made on these accounts, data must be updated. Another example arises when records are searched for special considerations, as when a college admissions office searches educational records seeking, for their mailing lists, students with certain levels of mathematical achievement because of the majors the college offers.

The following example demonstrates the use of projections in a method for organizing and describing data on a somewhat smaller scale.

EXAMPLE 3.31   At a certain Eastern university the following sets are related for purposes of registration:

$A_1 =$ the set of course numbers for courses offered in mathematics.

$A_2 =$ the set of course titles offered in mathematics.

$A_3 =$ the set of mathematics faculty.

$A_4 =$ the set of letters of the alphabet.

Consider the *table,* or relation, $D \subseteq A_1 \times A_2 \times A_3 \times A_4$ given in Table 3.2.

Table 3.2

Course Number	Course Title	Professor	Section Letter
MA 111	Calculus I	G. Sherman	A
MA 111	Calculus I	V. Larney	B
MA 112	Calculus II	J. Kinney	A
MA 112	Calculus II	A. Schmidt	B
MA 112	Calculus II	R. Mines	C
MA 113	Calculus III	J. Kinney	A
MA 113	Calculus III	A. Schmidt	B

The sets A_1, A_2, A_3, A_4 are called the *domains of the relational data base* and *table D* is said to have *degree* 4. Each element of D is often called a *list*.

The projection of D on $A_1 \times A_3 \times A_4$ is shown in Table 3.3. Table 3.4 shows the results for the projection of D on $A_1 \times A_2$.

Tables 3.3 and 3.4 are another way of representing the same data that appear in Table 3.2. Given Tables 3.3 and 3.4, one can recapture Table 3.2.   □

The theory of relational data bases is concerned with representing data in different ways and the operations, such as projections, needed for such representations. Special consideration for the computer implementation of such techniques is also considered.

Table 3.3

Course Number	Professor	Section Letter
MA 111	G. Sherman	A
MA 111	V. Larney	B
MA 112	J. Kinney	A
MA 112	A. Schmidt	B
MA 112	R. Mines	C
MA 113	J. Kinney	A
MA 113	A. Schmidt	B

Table 3.4

Course Number	Course Title
MA 111	Calculus I
MA 112	Calculus II
MA 113	Calculus III

Much more can be said on this topic and additional ideas are mentioned in the exercises and chapter references.

Having examined functions that are one-to-one and those that are onto, we turn now to functions with both of these properties.

Definition 3.13 ► If $f: A \to B$, f is said to be *bijective*, or to be a *one-to-one correspondence*, if f is both one-to-one and onto.

EXAMPLE 3.32 If $A = \{1, 2, 3, 4\}$, $B = \{w, x, y, z\}$, then $f = \{(1, w), (2, x), (3, y), (4, z)\}$ is a one-to-one correspondence from A (on)to B. □

For any set A there is always a very simple but important one-to-one correspondence, as seen in the following.

Definition 3.14 ► The function $1_A: A \to A$, defined by $1_A(a) = a$ for all $a \in A$, is called the *identity function* for A.

If $f: A \to A$ is bijective, we have $f(A) = A$ and can think of $f = \{(a, f(a)) \mid a \in A\}$ as a *permutation* of A.

It should be pointed out that whenever the term "correspondence" was used in Chapter 1, the adjective "one-to-one" was implied though never stated.

Definition 3.15 ► If $f, g: A \to B$ we say that f and g are *equal*, and write $f = g$, if $f(a) = g(a)$ for all $a \in A$.

A common pitfall in dealing with the equality of functions occurs when f, g are functions with a common domain A and $f(a) = g(a)$ for all $a \in A$. It may *not* be the case that $f = g$. The pitfall results from not paying attention to the codomains of the functions.

EXAMPLE 3.33 Let $f: \mathbf{Z} \to \mathbf{Z}$, $g: \mathbf{Z} \to \mathbf{Q}$ where $f(x) = x = g(x)$, for all $x \in \mathbf{Z}$. Then f, g share the common domain $\mathbf{Z}$, have the same range $\mathbf{Z}$, and act the same on every element of $\mathbf{Z}$. Yet $f \neq g$! For f is a one-to-one correspondence, while g is one-to-one but not onto, so the codomains do make a difference. □

Having seen operations that combine integers as well as some that operate on sets, we now introduce an operation to combine two appropriate functions.

Definition 3.16

▶ If $f: A \to B$, $g: B \to C$ we define the *composite function*, which is denoted $g \circ f: A \to C$, by $(g \circ f)(a) = g(f(a))$, for each $a \in A$.

EXAMPLE 3.34

Let $A = \{1, 2, 3, 4\}$, $B = \{a, b, c\}$, $C = \{w, x, y, z\}$ with $f: A \to B$, $g: B \to C$ given by $f = \{(1, a), (2, a), (3, b), (4, c)\}$ and $g = \{(a, x), (b, y), (c, z)\}$. For each element of A we find:

$$(g \circ f)(1) = g(f(1)) = g(a) = x \qquad (g \circ f)(3) = g(f(3)) = g(b) = y$$

$$(g \circ f)(2) = g(f(2)) = g(a) = x \qquad (g \circ f)(4) = g(f(4)) = g(c) = z$$

So $$g \circ f = \{(1, x), (2, x), (3, y), (4, z)\}.$$ ☐

EXAMPLE 3.35

Let $f: \mathbf{R} \to \mathbf{R}$, $g: \mathbf{R} \to \mathbf{R}$ be defined by $f(x) = x^2$, $g(x) = x + 5$. Then

$$(g \circ f)(x) = g(f(x)) = g(x^2) = x^2 + 5,$$

while

$$(f \circ g)(x) = f(g(x)) = f(x + 5) = (x + 5)^2 = x^2 + 10x + 25.$$

Here $g \circ f: \mathbf{R} \to \mathbf{R}$ and $f \circ g: \mathbf{R} \to \mathbf{R}$ but $(g \circ f)(1) = 6 \neq 36 = (f \circ g)(1)$, so even though both composites $f \circ g$ and $g \circ f$ can be formed, we do not have $f \circ g = g \circ f$. Consequently, the composition of functions is not, in general, a commutative operation. ☐

In the definition and examples for composite functions it was required that co-domain of f = domain of g. If range of $f \subseteq$ domain of g, this will actually be enough to yield the composite function $g \circ f: A \to C$. Also, for any $f: A \to B$, we observe that $f \circ 1_A = f = 1_B \circ f$.

An important recurring idea in mathematics is the investigation of whether the combining of two entities with a common property yields a result with this property. For example, if A and B are finite sets, then $A \cap B$, $A \cup B$ are also finite. However, for infinite sets A, B, $A \cup B$ is infinite but $A \cap B$ could be finite. (Give an example.) For the composition of functions we have the following result.

Theorem 3.4

▶ Let $f: A \to B$, $g: B \to C$.

a) If f, g are one-to-one, then $g \circ f$ is one-to-one.

b) If f, g are onto, then $g \circ f$ is onto.

Proof

a) To prove that $g \circ f: A \to C$ is one-to-one, let a_1, $a_2 \in A$ with $(g \circ f)(a_1) = (g \circ f)(a_2)$. Then $(g \circ f)(a_1) = (g \circ f)(a_2) \Leftrightarrow g(f(a_1)) = g(f(a_2)) \Rightarrow f(a_1) = f(a_2)$, since g is one-to-one. Also, $f(a_1) = f(a_2) \Rightarrow a_1 = a_2$, since f is one-to-one. Consequently, $g \circ f$ is one-to-one.

b) For $g \circ f: A \to C$, let $z \in C$. Since g is onto, there exists $y \in B$ with $g(y) = z$. With f onto, there exists $x \in A$ with $f(x) = y$. Hence $z = g(y) = g(f(x)) = (g \circ f)(x)$, so the range of $(g \circ f) = C =$ codomain of $(g \circ f)$, and $g \circ f$ is onto. ∎

Although function composition is not commutative, if $f: A \to B$, $g: B \to C$, $h: C \to D$, what can we say about the functions $h \circ (g \circ f)$ and $(h \circ g) \circ f$, i.e., is function composition associative?

Theorem 3.5

▶ Given $f: A \to B$, $g: B \to C$, $h: C \to D$, then $(h \circ g) \circ f = h \circ (g \circ f)$.

Proof

Since the two functions have the same domain, A, and codomain, D, the result will follow by showing that for every $x \in A$, $((h \circ g) \circ f)(x) = (h \circ (g \circ f))(x)$. (See the Venn diagram in Fig. 3.7.)

Figure 3.7

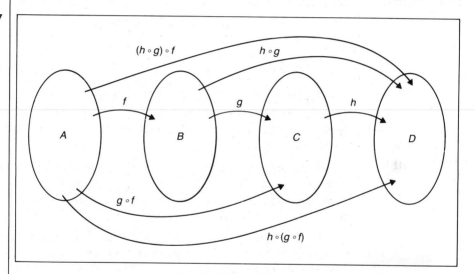

Using the definition of composite function we find that

$$((h \circ g) \circ f)(x) = (h \circ g)(f(x)) = h(g(f(x))),$$

while

$$(h \circ (g \circ f))(x) = h((g \circ f)(x)) = h(g(f(x))).$$

Consequently, the composition of functions is an associative operation. ∎

By virtue of the associative property for function composition we can write $h \circ g \circ f$, $(h \circ g) \circ f$ or $h \circ (g \circ f)$ without any problem of ambiguity. In addition this property allows us to define powers of functions, where appropriate.

Definition 3.17

▶ If $f: A \to A$ we define $f^1 = f$ and for $n \in \mathbf{Z}^+$, $f^{n+1} = f \circ (f^n)$.

This definition is the first example we have introduced where the result is defined *recursively*. With $f^{n+1} = f \circ (f^n)$, we see the dependence of f^{n+1} on a previous power, namely f^n. A great deal more will be done with recursion throughout the text, especially in Chapter 11.

EXAMPLE 3.36

With $A = \{1, 2, 3, 4\}$ and $f: A \to A$ defined by $f = \{(1, 2), (2, 2), (3, 1), (4, 3)\}$ we have $f^2 = f \circ f = \{(1, 2), (2, 2), (3, 2), (4, 1)\}$, and $f^3 = f \circ f^2 = f \circ f \circ f = \{(1, 2), (2, 2), (3, 2), (4, 2)\}$. (What are f^4, f^5?)   □

We now come to the last new idea for this chapter: the existence of the invertible function and some of its properties.

Definition 3.18

▶ For sets $A, B \subseteq \mathcal{U}$ if $\mathcal{R}$ is a relation from A to B, the *converse* of $\mathcal{R}$, denoted $\mathcal{R}^c$, is a relation from B to A defined by $\mathcal{R}^c = \{(b, a) \mid (a, b) \in \mathcal{R}\}$.

To get $\mathcal{R}^c$ from $\mathcal{R}$ we simply interchange the components of each ordered pair in $\mathcal{R}$. So if $A = \{1, 2, 3, 4\}$, $B = \{w, x, y\}$ and $\mathcal{R} = \{(1, w), (2, w), (3, x)\}$ then $\mathcal{R}^c = \{(w, 1), (w, 2), (x, 3)\}$, a relation from B to A.

For the same sets A, B as above, let $f: A \to B$ where $f = \{(1, w), (2, x), (3, y), (4, x)\}$. Then $f^c = \{(w, 1), (x, 2), (y, 3), (x, 4)\}$, a relation, but not a function, from B to A. We wish to investigate the condition(s) under which the converse of a function yields a function, but before getting too abstract let us consider the following example.

EXAMPLE 3.37

For $A = \{1, 2, 3\}$, $B = \{w, x, y\}$, let $f: A \to B$ be given by $f = \{(1, w), (2, x), (3, y)\}$. Then $f^c = \{(w, 1), (x, 2), (y, 3)\}$, a function from B to A. The function f^c is called an *inverse function* for f and we find that $f^c \circ f = 1_A$ and $f^c = 1_B$.   □

What has happened in this finite example leads us to the following.

Definition 3.19

▶ If $f: A \to B$, f is said to be *invertible* if there is a function $g: B \to A$ such that $g \circ f = 1_A$ and $f \circ g = 1_B$.

EXAMPLE 3.38

Let $f, g: \mathbf{R} \to \mathbf{R}$ be defined by $f(x) = 2x + 5$, $g(x) = (1/2)(x - 5)$. Then $(g \circ f)(x) = g(f(x)) = g(2x + 5) = (1/2)[(2x + 5) - 5] = x$, and $(f \circ g)(x) = f(g(x)) = f((1/2)(x - 5)) = 2[(1/2)(x - 5)] + 5 = x$, so $f \circ g = 1_{\mathbf{R}}$ and $g \circ f = 1_{\mathbf{R}}$. Consequently f and g are invertible functions.   □

Having seen some examples of invertible functions, we now wish to show that the function g of Definition 3.19 is unique, and then we'll find the means to identify an invertible function.

Theorem 3.6

▶ If a function $f: A \to B$ is invertible, then its inverse function $g: B \to A$ is unique.

Proof | With g an inverse of f we have $g \circ f = 1_A$, $f \circ g = 1_B$. If g is not unique, there is another function $h: B \to A$ with $h \circ f = 1_A$, $f \circ h = 1_B$. Consequently, $h = h \circ 1_B = h \circ (f \circ g) = (h \circ f) \circ g = 1_A \circ g = g$.   ∎

As a result of this theorem we can now denote the inverse of an invertible function f by f^{-1}. But when is a function invertible?

Theorem 3.7

▶ A function $f: A \to B$ is invertible if and only if it is one-to-one and onto.

Proof | Before actually starting the proof we note that the invertible functions of Examples 3.37 and 3.38 are bijective.

Assuming $f: A \to B$ is invertible, we have a unique function $g: B \to A$ with $g \circ f = 1_A$, $f \circ g = 1_B$. If $a_1, a_2 \in A$ with $f(a_1) = f(a_2)$, then $g(f(a_1)) = g(f(a_2))$, or $(g \circ f)(a_1) = (g \circ f)(a_2)$. With $g \circ f = 1_A$ it follows that $a_1 = a_2$, so f is one-to-one. For the onto property, let $b \in B$. Then $g(b) \in A$ so we can talk about $f(g(b))$. As $f \circ g = 1_B$ we have $b = 1_B(b) = (f \circ g)(b) = f(g(b))$, so f is onto.

Conversely, suppose $f: A \to B$ is bijective. Since f is onto, for each $b \in B$ there is an $a \in A$ with $f(a) = b$. Consequently we define a function $g: B \to A$ by $g(b) = a$, where $f(a) = b$. This definition yields a unique function. The only problem that could arise is if $g(b) = a_1 \neq a_2 = g(b)$ because $f(a_1) = b = f(a_2)$. However, this situation cannot arise because f is one-to-one. Since our definition of g is such that $g \circ f = 1_A$, $f \circ g = 1_B$, we find that f is invertible, with $g = f^{-1}$.   ∎

EXAMPLE 3.39 | From Theorem 3.7 the function $f_1: \mathbf{R} \to \mathbf{R}$ defined by $f_1(x) = x^2$ is not invertible (since it is not one-to-one), but $f_2: [0, +\infty) \to [0, +\infty)$ defined by $f_2(x) = x^2$ is invertible with $f_2^{-1}(x) = \sqrt{x}$.   □

The next result combines the ideas of function composition and inverse functions. The proof is left to the reader.

Theorem 3.8

▶ If $f: A \to B$, $g: B \to C$ are invertible functions, then $g \circ f: A \to C$ is invertible and $(g \circ f)^{-1} = f^{-1} \circ g^{-1}$.

Having seen some examples of functions and their inverses, one might wonder if there is an algebraic method to determine the inverse of an invertible function. If the function is finite we simply interchange the components of the given ordered pairs. But what if the function is defined by a formula, as in Example 3.39? Fortunately, the algebraic manipulations prove to be little more than a careful analysis of "interchanging the components of the ordered pairs." This is demonstrated in the following examples.

EXAMPLE 3.40 | For $m, b \in \mathbf{R}$, $m \neq 0$, $f: \mathbf{R} \to \mathbf{R}$ defined by $f = \{(x, y) \mid y = mx + b\}$ is an invertible function.

To get f^{-1} we note that

$$f^{-1} = \{(x, y) \mid y = mx + b\}^{-1} = \{(y, x) \mid y = mx + b\}$$
$$= \underbrace{\{(x, y) \mid x = my + b\}}_{} = \{(x, y) \mid y = (1/m)(x - b)\}.$$

This is where we wish to change the
components of the ordered pairs of f.

So $f: \mathbf{R} \to \mathbf{R}$ is defined by $f(x) = mx + b$, while $f^{-1}: \mathbf{R} \to \mathbf{R}$ is defined by $f^{-1}(x) = (1/m)(x - b)$. □

EXAMPLE 3.41   Let $f: \mathbf{R} \to \mathbf{R}^+$ be defined by $f(x) = e^x$, where $e = 2.7183$, the base for the natural logarithm. From the graph in Fig. 3.8 we see that f is one-to-one and onto, so $f^{-1}: \mathbf{R}^+ \to \mathbf{R}$ does exist. Then $f^{-1} = \{(x, y) \mid y = e^x\}^{-1} = \{(x, y) \mid x = e^y\} = \{(x, y) \mid y = \ln x\}$, so $f^{-1}(x) = \ln x$.

Figure 3.8

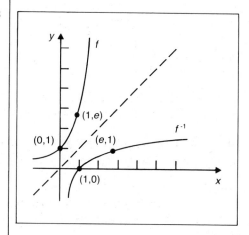

We should note that what happens in Fig. 3.8 happens in general, i.e., the graphs of f and f^{-1} are symmetric about the line $y = x$. For example, the line segment connecting the points $(1, e)$ and $(e, 1)$ is bisected by the line $y = x$. This is true for any corresponding pair of points $(x, f(x))$ and $(f(x), f^{-1}(f(x)))$.

This example yields the following formulas:

$$x = 1_{\mathbf{R}}(x) = (f^{-1} \circ f)(x) = \ln(e^x), \quad \text{for all } x \in \mathbf{R},$$
$$x = 1_{\mathbf{R}^+}(x) = (f \circ f^{-1})(x) = e^{\ln x}, \quad \text{for all } x > 0. \quad □$$

The result $x = e^{\ln x}$, for $x > 0$, is quite useful. In the standard implementation of Pascal, there is no exponentiation. To determine 2^3 one can resort to repeated multiplication but this is futile when dealing with the evaluation of a number like $(5.73)^{4.32}$. As exp and ln are defined functions in Pascal we can determine $(5.73)^{4.32}$ by rewriting it as $e^{4.32 \ln(5.73)}$, since $5.73 = e^{\ln(5.73)}$ from the formula above. This becomes exp$(4.32 * \ln(5.73))$ in Pascal.

Even when a function $f: A \rightarrow B$ is not invertible, we find use for the symbol f^{-1} in the following sense.

Definition 3.20 ▶ If $f: A \rightarrow B$ and $B_1 \subseteq B$, then $f^{-1}(B_1) = \{x \in A \,|\, f(x) \in B_1\}$. $f^{-1}(B_1)$ is called the *inverse image of B_1 under f*.

Be careful! Although we have the concept of an inverse image for any function, not every function has an inverse function. Consequently, we cannot assume the existence of an inverse for a function f just because we find the symbol f^{-1} being used. A little caution is needed here.

EXAMPLE 3.42 Let $f: \mathbf{Z} \rightarrow \mathbf{R}$ be defined by $f(x) = x^2 + 5$. Table 3.5 lists $f^{-1}(B)$ for various subsets B of the codomain $\mathbf{R}$. □

Table 3.5

B	$f^{-1}(B)$	B	$f^{-1}(B)$
$\{6\}$	$\{-1, 1\}$	$[-4, 5)$	$\emptyset$
$[6, 7]$	$\{-1, 1\}$	$[-4, 5]$	$\{0\}$
$[6, 10]$	$\{-2, -1, 1, 2\}$	$[5, +\infty)$	$\mathbf{Z}$

The concept of an inverse image appears in conjunction with the set operations of intersection, union, and complementation in the following result.

Theorem 3.9 ▶ If $f: A \rightarrow B$ and $B_1, B_2 \subseteq B$, then (a) $f^{-1}(B_1 \cap B_2) = f^{-1}(B_1) \cap f^{-1}(B_2)$, (b) $f^{-1}(B_1 \cup B_2) = f^{-1}(B_1) \cup f^{-1}(B_2)$, and (c) $f^{-1}(\overline{B_1}) = \overline{f^{-1}(B_1)}$.

Proof We prove part (b) and leave parts (a) and (c) for the reader.
For $a \in A$, $a \in f^{-1}(B_1 \cup B_2) \Leftrightarrow f(a) \in B_1 \cup B_2 \Leftrightarrow f(a) \in B_1$ or $f(a) \in B_2 \Leftrightarrow a \in f^{-1}(B_1)$ or $a \in f^{-1}(B_2) \Leftrightarrow a \in f^{-1}(B_1) \cup f^{-1}(B_2)$. ■

Using the notion of the inverse image, we see that a function $f: A \rightarrow B$ is one-to-one iff $|f^{-1}(b)| \leq 1$ for each $b \in B$. (Note: $f^{-1}(b)$ is a shorter way to write $f^{-1}(\{b\})$.)

Since discrete mathematics is primarily concerned with finite sets, the last result of this section demonstrates how the property of finiteness can yield results that fail to be true in general. In addition we see an application of the pigeonhole principle.

Theorem 3.10 ▶ Let $f: A \rightarrow B$ for finite sets A, B, where $|A| = |B|$. Then the following statements are equivalent: (a) f is one-to-one; (b) f is onto; and (c) f is invertible.

Proof We have already shown in Theorem 3.7 that (c) $\Rightarrow$ (a), (b) and that together (a), (b) $\Rightarrow$ (c). Consequently, this theorem will follow when we show that for these conditions on A, B, (a) $\Leftrightarrow$ (b). Assuming (b), if f is not one-to-one then there are elements $a_1, a_2 \in A$, with $a_1 \neq a_2$, but $f(a_1) = f(a_2)$. Then by the pigeonhole principle, $|A| > |f(A)| = |B|$, contradicting $|A| = |B|$. Conversely, let $f: A \rightarrow B$ be one-to-one, so $|A| = |f(A)|$. If f is not onto, then $|f(A)| < |B|$, which contradicts $|A| = |B|$. ■

Using this theorem we now verify the combinatorial identity introduced in Problem 6 at the start of this chapter. For if $n \in \mathbf{Z}^+$ and $|A| = |B| = n$, there are $n!$ one-to-one functions from A to B and $\sum_{k=0}^{n} (-1)^k \binom{n}{n-k} (n-k)^n$ onto functions from A to B. The equality $n! = \sum_{k=0}^{n} (-1)^k \binom{n}{n-k} (n-k)^n$ is then the numerical equivalent of parts (a) and (b) of Theorem 3.10.

EXERCISES

1. Let $|A| = 5$. (a) What is $|A \times A|$? (b) How many functions $f : A \times A \to A$ are there? (c) How many binary operations are there on A? (d) How many of these binary operations are commutative?

2. Let $A = \{a_1, a_2, a_3, \ldots, a_n\}$. (a) How many binary operations $f : A \times A \to A$ are there with $f(a_1, a_2) = a_3$? (b) How many of the binary operations in part (a) are commutative?

3. Each of the following functions $f : \mathbf{Z} \times \mathbf{Z} \to \mathbf{Z}$ is a binary operation on $\mathbf{Z}$. Determine in each case whether f is commutative and/or associative.

   a) $f(x, y) = x + y - xy$ b) $f(x, y) = \max\{x, y\}$
   c) $f(x, y) = x^y$ d) $f(x, y) = x + y - 3$

4. For a set A and a binary operation $f : A \times A \to A$, an element $z \in A$ is called an *identity* for f if $f(a, z) = a = f(z, a)$, for all $a \in A$. Which binary operations in Exercise 3 have an identity?

5. If $f : A \times A \to A$ has an identity, it is unique. For suppose that $z_1, z_2 \in A$ are both identities. Then

$$z_1 = f(z_1, z_2) = z_2$$

(Since z_2 is an identity   Why?
element for f.)

6. Let $A = \{z, a, b, c, d\}$. (a) How many binary operations $f : A \times A \to A$ have z as the identity element? (b) How many of these binary operations are commutative?

7. Let A_i, $1 \leq i \leq 5$, be the domains for a table $D \subseteq \times_{i=1}^{5} A_i$, where $A_1 = \{U, V, W, X, Y, Z\}$ (used as code names for different cereals in a test); and $A_2 = A_3 = A_4 = A_5 = \mathbf{Z}^+$. The table D is given as follows:

Table 3.6

Code name of cereal	Grams of sugar per 1 oz serving	% of RDA of vitamin A per 1 oz serving	% of RDA of vitamin C per 1 oz serving	% of RDA of protein per 1 oz serving
U	1	25	25	6
V	7	25	2	4
W	12	25	2	4
X	0	60	40	20
Y	3	25	40	10
Z	2	25	40	10

(RDA = Recommended Daily Allowance)

a) What is the degree of the table?

b) Find the projection of D on $A_3 \times A_4 \times A_5$.

c) A domain of a table is called a *primary key* for the table if its value uniquely identifies each list of D. Determine the primary key(s) for this table.

8. Let A_i, $1 \le i \le 5$, be the domains for a table $D \subseteq \times_{i=1}^{5} A_i$, where $A_1 = \{1, 2\}$ (used to identify the daily vitamin capsule produced by two pharmaceutical companies); $A_2 = \{A, D, E\}$; and $A_3 = A_4 = A_5 = \mathbf{Z}^+$. The table D is given as follows:

Table 3.7

Vitamin capsule	Vitamin present in capsule	Amount of vitamin in capsule in IU	Dosage: capsules/day	No. of capsules per bottle
1	A	10,000	1	100
1	D	400	1	100
1	E	30	1	100
2	A	4,000	1	250
2	D	400	1	250
2	E	15	1	250

(IU = International units)

a) What is the degree of the table?

b) What is the projection of D on $A_1 \times A_2$? on $A_3 \times A_4 \times A_5$?

c) In this table there is no primary key. (See Exercise 7.) We can, however, define a *composite primary key* as the cross product of a *minimal* number of domains of the table, whose components, taken collectively, uniquely identify each list of D. Determine some composite primary keys for this table.

9. Let $f: A \to B$, $g: C \to D$. Define $h: A \times C \to B \times D$ by

$$h(a, c) = (f(a), g(c)).$$

Prove that h is bijective iff f, g are bijective.

10. Let $f, g, h: \mathbf{Z} \to \mathbf{Z}$ be defined by $f(x) = x - 1$, $g(x) = 3x$,

$$h(x) = \begin{cases} 0, & x \text{ even} \\ 1, & x \text{ odd}. \end{cases}$$

Determine (a) $f \circ g$, $g \circ f$, $g \circ h$, $h \circ g$, $f \circ (g \circ h)$, $(f \circ g) \circ h$; (b) f^2, f^3, g^2, g^3, h^2, h^3, h^{500}.

11. If $\mathcal{U}$ is a given universe with $S, T \subseteq \mathcal{U}$, define $g: \mathcal{P}(\mathcal{U}) \to \mathcal{P}(\mathcal{U})$ by $g(A) = T \cap (S \cup A)$ for $A \subseteq \mathcal{U}$. Prove that $g^2 = g$.

12. Let $g: \mathbf{N} \to \mathbf{N}$ be defined by $g(n) = 2n$. If $A = \{1, 2, 3, 4\}$ and $f: A \to \mathbf{N}$ is given by $f = \{(1, 2), (2, 3), (3, 5), (4, 7)\}$, find $g \circ f$.

13. Let $f, g: \mathbf{R} \to \mathbf{R}$, where $g(x) = 1 - x + x^2$, $f(x) = ax + b$. If $(g \circ f)(x) = 9x^2 - 9x + 3$, determine a, b.

14. Let $f: A \to B$, $g: B \to C$. Then prove that (a) $g \circ f: A \to C$ onto $\Rightarrow g$ onto; (b) $g \circ f: A \to C$ one-to-one $\Rightarrow f$ one-to-one.

15. If $A, B \subseteq \mathcal{U}$ and $\mathcal{R}_1, \mathcal{R}_2 \subseteq A \times B$, prove that (a) $(\mathcal{R}_1 \cup \mathcal{R}_2)^c = \mathcal{R}_1^c \cup \mathcal{R}_2^c$; (b) $(\mathcal{R}_1 \cap \mathcal{R}_2)^c = \mathcal{R}_1^c \cap \mathcal{R}_2^c$; and (c) $(\mathcal{R}_1^c)^c = \mathcal{R}_1$.

16. For each of the following functions $f: \mathbf{R} \to \mathbf{R}$ determine whether or not f is invertible and, if so, determine f^{-1}.

   a) $f = \{(x, y) \mid 2x + 3y = 7\}$ b) $f = \{(x, y) \mid ax + by = c, b \neq 0\}$
   c) $f = \{(x, y) \mid y = x^3\}$ d) $f = \{(x, y) \mid y = x^4 - x\}$

17. a) Find the inverse of the function $f: \mathbf{R} \to \mathbf{R}^+$ defined by $f(x) = e^{2x+5}$.
   b) Show that $f \circ f^{-1} = 1_{\mathbf{R}^+}$, $f^{-1} \circ f = 1_{\mathbf{R}}$.
   c) Graph f, f^{-1} on the same set of axes.

18. Determine f^{-1} for (a) $f: \mathbf{R} \to \mathbf{R}$, $f(x) = -x$; (b) $f: \mathbf{R}^2 \to \mathbf{R}^2$, $f(x, y) = (y, x)$; (c) $f: \mathbf{R}^2 \to (\mathbf{R} \times \mathbf{R}^+)$, $f(x, y) = (5x, e^y)$.

19. Prove Theorem 3.8.

20. Let $f: \mathbf{R} \to \mathbf{R}$ be defined by $f(x) = x^2$. For each of the following subsets B of $\mathbf{R}$, find $f^{-1}(B)$.

   a) $B = \{0, 1\}$ b) $B = \{-1, 0, 1\}$ c) $B = [0, 1]$
   d) $B = [0, 1)$ e) $B = [-1, 1]$ f) $B = [0, 4]$
   g) $B = [0, 1] \cup [4, 9]$ h) $B = (0, 1] \cup (4, 9)$.

   Determine three infinite subsets B of $\mathbf{R}$ for which $f^{-1}(B) = \emptyset$.

21. Let $f: \mathbf{R} \to \mathbf{R}$ be defined by $f(x) = $ the greatest integer in x. This is generally denoted by $f(x) = \lfloor x \rfloor$. Hence, for any $x \in \mathbf{R}$, $\lfloor x \rfloor = x$ if $x \in \mathbf{Z}$, and for $x \notin \mathbf{Z}$, $\lfloor x \rfloor$ is the integer directly to the left of x on the real number line. Find $f^{-1}(B)$ for each of the following subsets B of $\mathbf{R}$.

   a) $B = \{0, 1\}$ b) $B = \{-1, 0, 1\}$ c) $B = [0, 1)$
   d) $B = [0, 2)$ e) $B = [-1, 2)$ f) $B = [0, 1]$
   g) $B = [-1, 2]$ h) $B = [-1, 0) \cup (1, 3]$.

22. If $f: \mathbf{R} \to \mathbf{R}$ is given by $f(x) = \sin x$, find $f^{-1}(B)$ for (a) $B = \{0\}$; (b) $B = \{0, 1\}$; and (c) $B = [0, 1/2]$.

23. Prove parts (a) and (c) of Theorem 3.9.

24. Prove that $f: A \to B$ is one-to-one iff $|f^{-1}(b)| \leq 1$ for all $b \in B$.

25. a) Give an example of a function $f: \mathbf{Z} \to \mathbf{Z}$ where f is (i) one-to-one but not onto; (ii) onto but not one-to-one.
   b) Do the examples in part (a) contradict Theorem 3.10?

26. If $|A| = |B| = 5$, how many functions $f: A \to B$ are invertible?

3.6 ■ SUMMARY AND HISTORICAL REVIEW

In this chapter we developed the function concept, which is of great importance in all areas of mathematics. Although we were primarily concerned with finite, or discrete, functions, the definition applies equally well to infinite sets and includes the functions of trigonometry and calculus. However, we did emphasize the role of a discrete function as one that transforms a finite set into a finite set. In this setting computer output can be thought of as a function of computer input, and a compiler can be regarded as a function that transforms a (source) program into a set of machine language instructions (object program).

The actual word "function," in its Latin form, was introduced in 1694 by Gottfried Wilhelm Leibniz (1646–1716) to denote a quantity associated with a curve (such as the slope of the curve, or the coordinates of a point of the curve). By 1718, under the direction of Johann Bernoulli (1667–1748), a function was regarded as an algebraic expression made up of constants and a variable. Equations or formulas involving constants and variables came later with Leonhard Euler (1707–1783). His idea is generally the definition of function found in high school mathematics. Also, in about 1734, we find in the work of Euler and Alexis Clairaut (1713–1765) the notation $f(x)$, which is still in use today.

Euler's idea remained intact until the time of Joseph B. J. Fourier (1768–1830), who found the need for a more general type of function in his investigation of trigonometric series. In 1837, Peter Gustav Lejeune Dirichlet (1805–1859) set down a more rigorous formulation of the concepts of variable, function, and the correspondence between the independent variable x and the dependent variable y, when $y = f(x)$. Dirichlet's work emphasized the relationship between two sets of numbers and did not call for the existence of a formula or expression connecting the two sets. With the developments in set theory during the nineteenth and twentieth centuries came the generalization of the function as a particular type of relation.

In addition to his fundamental work on the definition of a function, Dirichlet was also quite active in applied mathematics and in number theory, where he found need for the pigeonhole principle, which is sometimes referred to as the Dirichlet drawer principle.

The nineteenth and twentieth centuries saw the use of the special function, one-to-one correspondence, in the study of the infinite. In about 1888, Richard Dedekind (1831–1916) defined an infinite set as one that can be placed into a one-to-one correspondence with a proper subset of itself. (Galileo (1564–1642) had observed this for the set $\mathbf{Z}^+$.) Two infinite sets that could be placed in a one-to-one correspondence with each other were said to have the same *transfinite cardinal number*. In a series of articles Georg Cantor (1845–1918) developed the idea of levels of infinity and showed that $|\mathbf{Z}| = |\mathbf{Q}|$ but $|\mathbf{Z}| < |\mathbf{R}|$. A set A with $|A| = |\mathbf{Z}|$ is called *countable*, or *denumerable*, and we write $|\mathbf{Z}| = \aleph_0$, as Cantor did, using the Hebrew letter aleph with the subscripted 0, to denote the first level of infinity. To show that $|\mathbf{Z}| < |\mathbf{R}|$, or that the real numbers were *uncountable*, Cantor devised a technique now referred to as the Cantor diagonal method.

Finally, the Stirling numbers of Section 3.3 are named in honor of James Stirling (1692–1770), a pioneer in the development of generating functions, a topic we will investigate later in the text. He was an associate of Sir Isaac Newton (1642–1727) and was using the Maclaurin series in his work 25 years before Colin Maclaurin (1698–1746). However, although his name is not attached to this infinite series, it does appear in the approximation known as Stirling's formula: $n! \doteq (2\pi n)^{1/2} e^{-n} n^n$, which, as justice would have it, was actually developed by Abraham DeMoivre (1667–1754).

For more on infinite sets and the work of Cantor, consult Chapter 8 of H. Eves and C. V. Newsom [3], or Chapter IV of R. L. Wilder [5]. Generalizations of the pigeonhole principle have been a great area of research during this century, culminating in the subject of Ramsey theory, named for Frank Plumpton Ramsey (1903–1930). An interesting introduction to Ramsey theory can be found in Chapter 5 of D. I. A. Cohen [1]. For even more, the text by R. L. Graham, B. L. Rothschild, and J. H. Spencer [4] will provide worthwhile reading. Lastly, extensive coverage on the topic of relational data bases can be found in the work of C. J. Date [2].

REFERENCES

1. Cohen, Daniel I. A., *Basic Techniques of Combinatorial Theory,* John Wiley & Sons, New York, 1978.

2. Date, C. J., *An Introduction to Database Systems,* 3rd ed., Addison-Wesley, Reading, Massachusetts, 1982.

3. Eves, Howard, and Newsom, Carroll V., *An Introduction to the Foundations and Fundamental Concepts of Mathematics,* rev. ed., Holt, Rinehart and Winston, New York, 1965.

4. Graham, Ronald L., Rothschild, Bruce L., and Spencer, Joel H., *Ramsey Theory,* John Wiley & Sons, New York, 1980.

5. Wilder, Raymond L., *Introduction to the Foundations of Mathematics,* 2nd ed., John Wiley & Sons, New York, 1965.

■ MISCELLANEOUS EXERCISES

1. a) For sets A, B, C, D, prove that $A \subseteq C, B \subseteq D \Rightarrow A \times B \subseteq C \times D$.

   b) Does $A \times B \subseteq C \times D$ necessarily imply that $A \subseteq C$ and $B \subseteq D$?

2. Determine whether each of the following statements is true or false. For each false statement give a counterexample.

   a) If $f: A \rightarrow B$ is a function and $(a, b), (a, c) \in f$, then $b = c$.

   b) If $f: A \rightarrow B$ is a one-to-one correspondence and A, B are finite, then $A = B$.

   c) If $f: A \rightarrow B$ is one-to-one, then f is invertible.

   d) If $f: A \rightarrow B$ is invertible, then f is one-to-one.

   e) If $f: A \rightarrow B$ is one-to-one and $g, h: B \rightarrow C$ with $g \circ f = h \circ f$, then $g = h$.

   f) If $f: A \rightarrow B$, $A_1, A_2 \subseteq A$, then $f(A_1 \cap A_2) = f(A_1) \cap f(A_2)$.

   g) If $f: A \rightarrow B$, $B_1, B_2 \subseteq B$, then $f^{-1}(B_1 \cap B_2) = f^{-1}(B_1) \cap f^{-1}(B_2)$.

3. With $\mathcal{U} = \mathbf{Z}$, let $A, B \subseteq \mathcal{U}$ with $A = \{2, 3, 5\}$, $B = \{1, 2, 5, 6\}$.

   a) Diagram $A \times B$ as a subset of the Euclidean plane.

   b) If $\mathcal{R}$ is a relation from A to B defined as $\{(a, b) \mid a + b \text{ is odd}\}$, diagram $\mathcal{R}$ as a subset of the Euclidean plane.

   c) How many relations from A to B are not functions from A to B?

4. Let $\mathcal{U} = \mathbf{N}$ and let $A, B \subseteq \mathcal{U}$ with $1 < |A| < |B|$. If there are 262,144 relations from A to B, determine all possibilities for $|A|$ and $|B|$.

5. If $\mathcal{U}_1$, $\mathcal{U}_2$ are universal sets with $A, B \subseteq \mathcal{U}_1$, $C, D \subseteq \mathcal{U}_2$, prove that
   a) $(A \cap B) \times (C \cap D) = (A \times C) \cap (B \times D)$; and
   b) $(A \cup B) \times (C \cup D) = (A \times C) \cup (B \times D) \cup (A \times D) \cup (B \times C)$. (So in general, $(A \cup B) \times (C \cup D) \supseteq (A \times C) \cup (B \times D)$.)

6. Let $A = \{1, 2, 3, 4, 5\}$, $B = \{1, 2, 3, 4, 5, 6\}$. How many injective functions $f: A \to B$ satisfy (a) $f(1) = 3$? (b) $f(1) = 3$, $f(2) = 6$?

7. Ackermann's function, $A(m, n)$, is defined *recursively* (i.e., for $m > 0$, the value of $A(m, n)$ depends on the value(s) of A for one or more nonnegative integers smaller than m or n) for integers m, $n > 0$ by:

$$A(0, n) = n + 1, \quad n \geq 0; \qquad A(m, 0) = A(m - 1, 1), \quad m > 0;$$

$$A(m, n) = A(m - 1, A(m, n - 1)), \quad m > 0, \quad n > 0.$$

Calculate $A(2, 3)$.

8. Let A_1, A, B be sets with $\{1, 2, 3, 4, 5\} = A_1 \subset A$, $B = \{s, t, u, v, w, x\}$, and $f: A_1 \to B$. If f can be extended to A in 216 ways, what is $|A|$?

9. Let $A = \{1, 2, 3, 4, 5\}$, $B = \{t, u, v, w, x, y, z\}$. (a) If a function $f: A \to B$ is randomly generated, what is the probability it is one-to-one? (b) Write a computer program to generate functions $f: A \to B$ and have it print out how many functions it generates until it generates one that is one-to-one.

10. Let S be a set of seven positive integers the maximum of which is at most 24. Prove that the sums of the elements in all the nonempty subsets of S cannot be distinct.

11. In a ten-day period a secretary typed 84 letters to different clients. She typed 12 of these letters on the first day, 7 on the second day, 3 on the ninth day, and finished the last 8 on the tenth day. Show that for a period of three consecutive days she typed at least 25 letters.

12. With both of their parents working, Thomas, Stuart, and Craig must handle ten weekly chores among themselves. (a) In how many ways can they divide up the work so that everyone is responsible for at least one chore? (b) In how many ways can the chores be assigned if Thomas, as the eldest, must mow the lawn (one of the ten weekly chores) and no one is allowed to be idle?

13. Let $n \in \mathbf{N}$, $n \geq 2$. Show that $S(n, 2) = 2^{n-1} - 1$.

14. Let $f: X \to Y$, and for each $i \in I$, let $A_i \subseteq X$. Prove that
   a) $f(\bigcup_{i \in I} A_i) = \bigcup_{i \in I} f(A_i)$;
   b) $f(\bigcap_{i \in I} A_i) \subseteq \bigcap_{i \in I} f(A_i)$;   and
   c) $f(\bigcap_{i \in I} A_i) = \bigcap_{i \in I} f(A_i)$,   for f injective.

15. Given 8 books on Pascal, 17 FORTRAN books, 6 APL books, 12 COBOL books, and 20 BASIC books, how many of these books must we select to insure that we have 10 books dealing with the same computer language?

16. If $\{x_1, x_2, \ldots, x_7\} \subseteq \mathbf{Z}^+$, show that there exist $i, j, i \neq j$, where either $x_i + x_j$ or $x_i - x_j$ is divisible by 10.

17. If $f: \mathbf{R} \rightarrow \mathbf{R}$ with $f(x) = x^n$, for which $n \in \mathbf{Z}^+$ is f invertible?

18. Let A be a set with $|A| = n$.

    a) How many binary operations are there on A?

    b) A ternary (3-ary) operation on A is a function $f: A \times A \times A \rightarrow A$. How many ternary operations are there on A?

    c) A k-ary operation on A is a function $f: A_1 \times A_2 \times \cdots \times A_k \rightarrow A$, where $A_i = A$, for all $1 \leq i \leq k$. How many k-ary operations are there on A?

    d) A k-ary operation for A is called *commutative* if

    $$f(a_1, a_2, \ldots, a_k) = f(\pi(a_1), \pi(a_2), \ldots, \pi(a_k)),$$

    where $a_1, a_2, \ldots, a_k \in A$ (repetitions allowed), and $\pi(a_1), \pi(a_2), \ldots, \pi(a_k)$ is any rearrangement of $a_1, a_2, \ldots, a_k$. How many of the k-ary operations on A are commutative?

19. A function $f: \mathbf{R} \rightarrow \mathbf{R}$ is said to be *increasing* if for real numbers $x, y, x < y \Rightarrow f(x) < f(y)$. Prove that if $f, g: \mathbf{R} \rightarrow \mathbf{R}$ are increasing functions, then $g \circ f: \mathbf{R} \rightarrow \mathbf{R}$ is increasing.

20. If $\mathcal{U}$ is a universe and $A \subseteq \mathcal{U}$, we define the *characteristic function* of A by $\chi_A: \mathcal{U} \rightarrow \{0, 1\}$, where

    $$\chi_A(x) = \begin{cases} 1, & x \in A \\ 0, & x \notin A \end{cases}$$

    For sets $A, B \subseteq \mathcal{U}$, prove each of the following:

    a) $\chi_{A \cap B} = \chi_A \cdot \chi_B$ where $(\chi_A \cdot \chi_B)(x) = \chi_A(x) \cdot \chi_B(x)$;

    b) $\chi_{A \cup B} = \chi_A + \chi_B - \chi_{A \cap B}$; and

    c) $\chi_{\bar{A}} = 1 - \chi_A$, where $(1 - \chi_A)(x) = 1(x) - \chi_A(x) = 1 - \chi_A(x)$.

    (For $\mathcal{U}$ finite, placing the elements of $\mathcal{U}$ in a fixed order results in a one-to-one correspondence between subsets A of $\mathcal{U}$ and the arrays of 0's and 1's obtained as the images of $\mathcal{U}$ under χ_A. These arrays can then be used for the computer storage and manipulation of certain subsets of $\mathcal{U}$.)

21. With $A = \{x, y, z\}$, let $f, g: A \rightarrow A$ be given by $f = \{(x, y), (y, z), (z, x)\}$, $g = \{(x, y), (y, x), (z, z)\}$. Determine each of the following: $f \circ g, g \circ f, f^{-1}, g^{-1}, (g \circ f)^{-1}, f^{-1} \circ g^{-1}, g^{-1} \circ f^{-1}$.

22. Let $f: \mathbf{R} \rightarrow \mathbf{R}$ be defined by $f(x) = (2 + \sin(x + 1))^3, x \in \mathbf{R}$. Find four functions $f_i: \mathbf{R} \rightarrow \mathbf{R}, 1 \leq i \leq 4$, so that $f = f_4 \circ f_3 \circ f_2 \circ f_1$.

23. For $A = \{1, 2, 3\}$ find the number of functions $f: A \rightarrow A$ where (a) $f^2 = 1_A$; (b) $f^2 = f$.

24. a) If $f: \mathbf{R} \to \mathbf{R}$ is defined by $f(x) = 5x + 3$, find $f^{-1}(8)$.

   b) If $g: \mathbf{R} \to \mathbf{R}$, where $g(x) = |x^2 + 3x + 1|$, find $g^{-1}(1)$.

   c) For $h: \mathbf{R} \to \mathbf{R}$, given by

$$h(x) = \left| \frac{x}{x + 2} \right|,$$

   find $h^{-1}(4)$.

25. If $f: X \to Y$, show that (a) for $A \subseteq X$, $A \subseteq f^{-1}(f(A))$; and (b) for $B \subseteq Y$, $B \supseteq f(f^{-1}(B))$. State any needed condition(s) that f must satisfy in order to have equality in (a); in (b).

26. Let $f: X \to Y$ and let I be an index set, where for each $i \in I$, $B_i \subseteq Y$. Prove that

   a) $f^{-1}(\bigcup_{i \in I} B_i) = \bigcup_{i \in I} f^{-1}(B_i)$;   b) $f^{-1}(\bigcap_{i \in I} B_i) = \bigcap_{i \in I} f^{-1}(B_i)$.

27. In the programming language Pascal the functions pred and succ (for predecessor and successor, respectively) are functions from $\mathbf{Z}$ to $\mathbf{Z}$ where $\text{pred}(x) = \pi(x) = x - 1$ and $\text{succ}(x) = \sigma(x) = x + 1$.

   a) Determine $(\pi \circ \sigma)(x)$, $(\sigma \circ \pi)(x)$.

   b) Determine π^2, π^3, π^n ($n \geq 2$), σ^2, σ^3, σ^n ($n \geq 2$).

   c) Determine π^{-2}, π^{-3}, π^{-n} ($n \geq 2$), σ^{-2}, σ^{-3}, σ^{-n} ($n \geq 2$), where, for example, $\sigma^{-2} = \sigma^{-1} \circ \sigma^{-1} = (\sigma \circ \sigma)^{-1} = (\sigma^2)^{-1}$.

28. In the study of the analysis of algorithms, the *asymptotic complexity* of an algorithm involves the following concept: if $f, g: \mathbf{Z}^+ \to \mathbf{R}$, g is said to *asymptotically dominate* f (or f is *asymptotically dominated* by g) if there exist constants $m \in \mathbf{R}^+$, $k \in \mathbf{Z}^+$ such that $|f(n)| \leq m|g(n)|$ for all $n \in \mathbf{Z}^+$, $n \geq k$.

   a) Prove that if $f, g: \mathbf{Z}^+ \to \mathbf{R}$ are defined by $f(n) = 5$, $g(n) = n$, $n \in \mathbf{Z}^+$, then f is asymptotically dominated by g.

   b) Prove that if $f, g, h: \mathbf{Z}^+ \to \mathbf{R}$ are defined by $f(n) = n$, $g(n) = an + b$, $a, b \in \mathbf{R}$, $a > 0$, $h(n) = n + (1/n)$, then each function asymptotically dominates the other two.

   (When f is asymptotically dominated by g we find the notation $f \in \mathcal{O}(g)$, where $\mathcal{O}(g)$ is read "order g" or "big-Oh of g".)

   c) Let $f, g: \mathbf{Z}^+ \to \mathbf{R}$ be defined by $f(n) = n + 100$, $g(n) = n^2$. Show that $f \in \mathcal{O}(g)$.

   d) For $f, g: \mathbf{Z}^+ \to \mathbf{R}$ we have $g \notin \mathcal{O}(f)$ if for all $m \in \mathbf{R}^+$ and all $k \in \mathbf{Z}^+$ there exists $n \in \mathbf{Z}^+$ such that $n \geq k$ and $|g(n)| > m|f(n)|$. Show that for f, g in part (c), $g \notin \mathcal{O}(f)$.

29. In the programming language BASIC a function that is not built into the language can be defined in a given program using the DEF command, followed by FN and a variable name, with the function variables in parentheses separated by commas. For example, the real functions $f(x) = x^3 - 5x$, $g(x, y) = x^2 - xy$ are given by

```
    DEF FNF(X) = X ** 3 - 5 * X
    DEF FNG(X,Y) = X ** 2 - X * Y.
```

Consider the following BASIC program:

```
10 DEF FNH(X,Y) = X ** 2 - 7 * X + 3 * Y + 4
20     FOR I = 1 TO 100 STEP 0.01
30         FOR J = 50 TO -10 STEP -0.5
40             PRINT FNH(I, J)
50         NEXT J
60     NEXT I
70 END
```

If it takes approximately 200 nanoseconds (200×10^{-9} seconds) for each execution of the command on line 40, how much time is spent in printing all of the results in this program?

30. Mrs. Blasi has five sons (Michael, Rick, David, Kenneth, and Donald) who enjoy reading books about sports. With Christmas approaching she visits a bookstore where she finds 12 different books on sports.

    a) In how many ways can she select nine of these books?

    b) Having made her purchase, in how many ways can she distribute the books among her sons so that each of them gets at least one book?

    c) State a problem for the situation presented here, where the answer is the product of the answers in parts (a) and (b).

    d) In her purchase of nine books, two of the books deal with basketball, Donald's favorite sport. In how many ways can Mrs. Blasi distribute the books among her sons so that Donald gets at least the two books on basketball?

31. Let $m, n \in \mathbf{Z}^{+}$ with $n \geq m$. (a) In how many ways can one distribute n distinct objects among m different containers with no container left empty? (b) In the expansion of $(x_1 + x_2 + \cdots + x_m)^n$, what is the sum of all the multinomial coefficients $\binom{n}{n_1, n_2, \ldots, n_m}$ where $n_1 + n_2 + \cdots + n_m = n$ and $n_i > 0$, $1 \leq i \leq m$.

32. If $n \in \mathbf{Z}^{+}$, $n \geq 4$, verify that $S(n, n - 2) = \binom{n}{3} + 3\binom{n}{4}$.

33. Let $n \in \mathbf{Z}^{+}$, n odd. If $i_1, i_2, \ldots, i_n$ is a permutation of the integers $1, 2, \ldots, n$, prove that $(1 - i_1)(2 - i_2) \cdots (n - i_n)$ is an even integer. (Which counting principle is at work here?)

4

Languages: Finite State Machines

In this era of computers and telecommunications, we find ourselves confronted every day with input-output situations. In purchasing a soft drink from a vending machine, we *input* some coins and then press a button to get our expected *output*, the soft drink desired. The first coin that we input sets the machine in motion. Although we usually don't care about what happens inside the machine (unless there is some kind of breakdown and we suffer a loss), we should realize that somehow the machine is keeping track of the coins we insert, until the correct total has been inserted. It is only then, and not before, that the vending machine should output the desired soft drink. Consequently, for the vendor to make his expected profit per soft drink, the machine must *internally remember*, as each coin is inserted, what sum of money has been deposited.

A computer is another example of an input-output device. Here the input is generally some type of information and the output is the result obtained after processing this information. How the input is processed depends upon the internal workings of the computer where it must have the ability to remember past information as it works on the information it is presently processing.

Using the concepts developed on sets and functions, in this chapter we investigate an abstract model called a *finite state machine*, or *sequential circuit*. Such circuits are one of two basic types of control circuits found in digital computers. (The other type is a *combinational circuit* or *gating network*, which is examined in Chapter 9.) They are also found in other systems such as our vending machine, as well as in the controls for elevators, and in traffic-light systems and compiler construction.

As the name indicates, a finite state machine has a finite number of internal states where the machine remembers certain information, when it is in that particular state. However, before getting into this concept we need some set theoretic material in order to talk about what constitutes valid input for such a machine.

4.1 ■ LANGUAGE: THE SET THEORY OF STRINGS

Strings of symbols, or characters, play a key role in the processing of information by a computer. As computer programs are representable in terms of finite sequences of characters, some algebraic way is needed for handling such finite sequences, or *strings*.

Throughout this section we use Σ to denote a nonempty *finite* set of symbols, collectively called an *alphabet*. We do not list in Σ elements that can be formed from the other elements of Σ by juxtaposition. Consequently, we never have situations like $\Sigma = \{0, 1, 2, 11, 12\}$ or $\Sigma = \{x, y, z, yx, xx\}$. (In this way no confusion arises in Definition 4.1 when we deal with the length of a string.)

If Σ is an alphabet, we define $\Sigma^2 = \{xy \mid x, y \in \Sigma\}$, and call the elements of Σ^2 strings, or *words*, of length two. For $n \geq 1$, Σ^{n+1} is defined recursively by $\Sigma^{n+1} = \{xy \mid x \in \Sigma, y \in \Sigma^n\}$, where $\Sigma^1 = \Sigma$. Also, $\Sigma^0 = \{\lambda\}$, where λ denotes the *empty string*, that is, the string consisting of no symbols taken from Σ. (This symbol, λ, should not be incorrectly identified with the blank (space) which is found in many alphabets.) Since $\emptyset \subseteq \Sigma$ we need to be cautious. Observe that $\{\lambda\} \not\subseteq \Sigma$, since $\lambda \notin \Sigma$. Also, $\{\lambda\} \neq \emptyset$, since $|\{\lambda\}| = 1$ while $|\emptyset| = 0$.

We define $\Sigma^* = \bigcup_{n=0}^{\infty} \Sigma^n$, $\Sigma^+ = \bigcup_{n=1}^{\infty} \Sigma^n = \bigcup_{n \in \mathbf{Z}^+} \Sigma^n$, so $\Sigma^* = \Sigma^+ \cup \{\lambda\}$. Elements of Σ^*, Σ^+ are referred to as strings, words, or *sentences*. Σ^*, Σ^+ are *infinite* sets but their elements are *finite* strings of characters.

Finally, for all $n \in \mathbf{Z}^+$, $|\Sigma^n| = |\Sigma|^n$ since we are dealing with arrangements where repetitions are allowed. For a fixed $n \in \mathbf{Z}^+$, if w_1, $w_2 \in \Sigma^n$, then $w_1 = x_1 x_2 \ldots x_n$, $w_2 = y_1 y_2 \ldots y_n$, where x_i, $y_i \in \Sigma$, $1 \le i \le n$. Here $w_1 = w_2$ if $x_i = y_i$, $1 \le i \le n$. Thus two strings in Σ^* are *equal* only when corresponding symbols from Σ match identically.

Definition 4.1

▶ For $w \in \Sigma^*$, if $w \ne \lambda$, $w = x_1 x_2 \ldots x_n$, where $x_i \in \Sigma$, $1 \le i \le n$. We define the *length* of the string w as n, and denote this by $\|w\|$. Also $\|\lambda\| = 0$.

Consequently, Σ^+ is the set of all strings of positive length constructed from Σ. If $\beta \in \Sigma$ and β denotes a blank, then $\|\beta\| = \|\beta\lambda\| = 1$, $\|\beta\beta\| = 2$, but $\|\lambda\| = 0$.

EXAMPLE 4.1

If $\Sigma = \{0, 1\}$, $\Sigma^2 = \{00, 01, 10, 11\}$. Also, $010 \in \Sigma^3$ and $01010 \in \Sigma^*$. We see that $\|01010\| = 5 = 3 + 2 = \|010\| + \|10\|$. □

Definition 4.2

▶ If x, $y \in \Sigma^*$ then $x = x_1 x_2 \ldots x_m$, $y = y_1 y_2 \ldots y_n$, where each x_i, $1 \le i \le m$, and y_j, $1 \le j \le n$, is in Σ. We form the *concatenation* of x, y—denoted xy—by $xy = x_1 x_2 \ldots x_m y_1 y_2 \ldots y_n$.

Concatenation is a binary operation on Σ^* (and Σ^+). It is associative, but generally not commutative, and the element λ is the identity for the operation. Also, for any x, $y \in \Sigma^*$, $\|xy\| = \|x\| + \|y\|$.

Definition 4.3

▶ If x, $y \in \Sigma^*$ and $w = xy$, then x is called a *prefix* of w, and if $y \ne \lambda$, x is said to be a *proper prefix*. y is called a *suffix* of w, and is a *proper suffix* if $x \ne \lambda$.

EXAMPLE 4.2

If $\|x\| = 5$, $\|y\| = 4$ and $w = xy$, then w has x as a proper prefix and y as a proper suffix. In all, w has nine proper prefixes and nine proper suffixes, since λ is both a proper prefix and proper suffix for every string in Σ^+. □

EXAMPLE 4.3

If w, a, b, c, $d \in \Sigma^*$, then $ab = w = cd \Leftrightarrow b$ is a suffix of d, or d is a suffix of $b \Leftrightarrow a$ is a prefix of c or c is a prefix of a. □

Definition 4.4

▶ If x, y, $z \in \Sigma^*$ and $w = xyz$, then y is called a *substring* of w. If at least one of x, z is different from λ, then y is called a *proper substring*.

Definition 4.5

▶ For $x \in \Sigma^*$, we define $x^0 = \lambda$, $x^1 = x$, $x^2 = xx$, $x^3 = xx^2, \ldots, x^{n+1} = xx^n, \ldots, n \in \mathbf{Z}^+$. (Again we have a recursive or inductive definition where a mathematical entity we presently seek falls back upon previous comparable entities.)

This definition provides a way of dealing with the n-fold concatenation of a string with itself, and includes the special case where the string is just one symbol.

EXAMPLE 4.4   If $\Sigma = \{0, 1\}$ and $x = 01$, then $x^0 = \lambda$, $x^1 = 01$, $x^2 = 0101$, and $x^3 = 010101$. For any $n > 0$, x^n consists of a string of n 0's and n 1's where the first symbol is 0 and the symbols alternate. Here $\|x^2\| = 4 = 2\|x\|$, $\|x^3\| = 6 = 3\|x\|$, and for $n \in \mathbf{N}$, $\|x^n\| = n\|x\|$.   □

EXAMPLE 4.5   For $\Sigma = \{0, 1\}$ the set Σ^* consists of all finite strings of 0's and 1's together with the empty string. For n reasonably small we could actually list all strings in Σ^n.

If $\Sigma = \{\beta, 0, 1, 2, \ldots, 9, +, -, \times, /, (,)\}$, where β denotes the blank (or space), it is harder to describe Σ^* and for $n > 2$ there are many strings to list in Σ^n. Here in Σ^* we find familiar arithmetic expressions like $(7 + 5)/(2 \times (3 - 10))$ as well as jibberish like $+)((7/\times +3/($.   □

When we consider the standard alphabet, including the blank, there are many strings such as *qxio, the wxxy red atzl,* and *aeytl* that do not represent words or parts of sentences that appear in the English language, even though they are elements of Σ^*. Consequently, in order to consider only those words and expressions that have meaning in English we concentrate on a subset of Σ^*.

Definition 4.6   ▶ For a given alphabet Σ, a *language* over Σ is a subset of Σ^*. The subset $\emptyset$ is called the *empty language*.

EXAMPLE 4.6   With $\Sigma = \{0, 1\}$, the sets $A = \{0, 01, 001\}$ and $B = \{0, 01, 001, 0001, \ldots\}$ are languages over Σ.   □

EXAMPLE 4.7   With Σ the alphabet of 26 letters, 10 digits, and special symbols used in a given implementation of Pascal, the collection of executable programs for that implementation constitutes a language. In the same situation each executable program could be considered a language, as could a particular finite set of such programs.   □

Since languages are sets we can form the union, intersection, and symmetric difference of two languages. However, for the work here, *concatenation* of languages proves more useful.

Definition 4.7   ▶ For an alphabet Σ and languages $A, B \subseteq \Sigma^*$, the *concatenation* of A, B, denoted AB, is $\{ab \mid a \in A, b \in B\}$.

We can think of concatenation as the cross product where we remove the comma and parentheses in each ordered pair, so that (a, b) becomes ab. We will see that just as $A \times B \neq B \times A$ in general, we also have $AB \neq BA$, in general. However, for A, B finite we did have $|A \times B| = |B \times A|$, but here $|AB| \neq |BA|$ is possible for finite languages.

EXAMPLE 4.8   Let $\Sigma = \{x, y, z\}$ and let A, B be the finite languages $A = \{x, xy, z\}$, $B = \{\lambda, y\}$. Then $AB = \{x, xy, z, xyy, zy\}$, so $|AB| = 5 < 6 = 3 \cdot 2 = |A||B|$.

The inequality arises because there are two ways to represent xy: (1) xy, $x \in A$, $y \in B$; and, (2) $xy\lambda$, $xy \in A$, $\lambda \in B$. (The concept of uniqueness of representation is a key to the success of many mathematical ideas. We will come across it again in Chapters 6 and 9.) □

The example above indicates that for finite languages A, B, $|AB| \leq |A||B|$, in general.

EXAMPLE 4.9 With $\Sigma = \{x, y\}$, $A = \{x, xy\}$, $B = \{x, yx\}$ we find that $AB = \{xx, xyx, xyyx\}$, $BA = \{xx, xxy, yxx, yxxy\}$. Here, not only is $AB \neq BA$, but $|AB| \neq |BA|$. □

The following deals with some of the properties satisfied by the concatenation of languages.

Theorem 4.1

▶ For an alphabet Σ, let A, B, $C \subseteq \Sigma^*$. Then

a) $A\{\lambda\} = \{\lambda\}A = A$

b) $(AB)C = A(BC)$

c) $A(B \cup C) = AB \cup AC$

d) $(B \cup C)A = BA \cup CA$

e) $A(B \cap C) \subseteq AB \cap AC$

f) $(B \cap C)A \subseteq BA \cap CA$

Proof We prove (d) and (f) and leave the other parts for the reader.

(d) With $x \in \Sigma^*$, $x \in (B \cup C)A \Rightarrow x = yz$, $y \in B \cup C$, $z \in A \Rightarrow (x = yz$, $y \in B$, $z \in A)$ or $(x = yz$, $y \in C$, $z \in A) \Rightarrow x \in BA$ or $x \in CA$, so $(B \cup C)A \subseteq BA \cup CA$. Conversely, $x \in BA \cup CA \Rightarrow x \in BA$ or $x \in CA \Rightarrow x = ba_1$, $b \in B$, $a_1 \in A$ or $x = ca_2$, $c \in C$, $a_2 \in A$. Assume $x = ba_1$, $b \in B$, $a_1 \in A$. Since $B \subseteq B \cup C$, we have $x = ba_1$, where $b \in B \cup C$, $a_1 \in A$. Then $x \in (B \cup C)A$, so $BA \cup CA \subseteq (B \cup C)A$. (The argument is similar if $x = ca_2$.) With both inclusions established we have $(B \cup C)A = BA \cup CA$.

(f) For $x \in \Sigma^*$, $x \in (B \cap C)A \Rightarrow x = yz$, $y \in B \cap C$, $z \in A \Rightarrow (x = yz$, $y \in B$, $z \in A)$ and $(x = yz$, $y \in C$, $z \in A) \Rightarrow x \in BA \cap CA$, so $(B \cap C)A \subseteq BA \cap CA$.

With $\Sigma = \{x, y, z\}$, let $B = \{x, xx, y\}$, $C = \{y, xy\}$, $A = \{y, yy\}$. Then $xyy \in BA \cap CA$ but $xyy \notin (B \cap C)A$. Consequently $(B \cap C)A \subset BA \cap CA$. ■

Comparable to the concepts of Σ^n, Σ^*, Σ^+, the following are given for an arbitrary language $A \subseteq \Sigma^*$.

Definition 4.8

▶ For a language $A \subseteq \Sigma^*$,

a) $A^0 = \{\lambda\}$, $A^1 = A$, and for any $n \in \mathbb{Z}^+$, $A^{n+1} = \{ab \,|\, a \in A, b \in A^n\}$;

b) $A^+ = \bigcup_{n \in \mathbb{Z}^+} A^n$, the *positive closure* of A;

c) $A^* = A^+ \cup \{\lambda\}$. A^* is called the *Kleene closure* of A, in honor of the American logician Stephen Cole Kleene (1909-).

EXAMPLE 4.10 If $\Sigma = \{x, y, z\}$ and $A = \{x\}$, then $A^0 = \{\lambda\}$, $A^n = \{x^n\}$, $n \in \mathbb{N}$, $A^+ = \{x^n \,|\, n \geq 1\}$, $A^* = \{x^n \,|\, n \geq 0\}$. □

EXAMPLE 4.11 With $\Sigma = \{x, y\}$ and $A = \{xx, xy, yx, yy\}$, A^* is the set of all strings in Σ^* which have even length. If $B = \{x, y\}$, then BA^* would yield all strings in Σ^* of odd length, so $BA^* = A^*B$, $\Sigma^* = A^* \cup BA^*$. □

EXAMPLE 4.12

In the algebra of real numbers, if $a, b \in \mathbf{R}$, $a, b > 0$, then $a^2 = b^2 \Rightarrow a = b$. However, in the case of languages, if $\mathbf{\Sigma} = \{x, y\}$, $A = \{\lambda, x, x^3, x^4, \ldots\} = \{x^n \mid n \geq 0\} - \{x^2\}$, $B = \{x^n \mid n \geq 0\}$, then $A^2 = B^2 \ (= B)$, but $A \neq B$. (Note: We never have $\lambda \in \mathbf{\Sigma}$, but it is possible to have $\lambda \in A$.)   □

We close this section with a theorem on the closure of a language.

Theorem 4.2

▶ For an alphabet $\mathbf{\Sigma}$ and languages $A, B \subseteq \mathbf{\Sigma}^*$

a) $A \subseteq AB^*$ b) $A \subseteq B^*A$

c) $A \subseteq B \Rightarrow A^+ \subseteq B^+$ d) $A \subseteq B \Rightarrow A^* \subseteq B^*$

e) $AA^* = A^*A = A^+$ f) $A^*A^* = A^* = (A^*)^* = (A^*)^+ = (A^+)^*$

g) $(A \cup B)^* = (A^* \cup B^*)^* = (A^*B^*)^*$

Proof

We provide the proofs for Theorem 4.2 (c) and (g).

(c) Let $A \subseteq B$, and $x \in A^+$. $x \in A^+ \Rightarrow x \in A^n$, for some $n \in \mathbf{Z}^+ \Rightarrow x = a_1 a_2 \ldots a_n$, where each $a_i \in A$, $1 \leq i \leq n$. Then each $a_i \in B$, $1 \leq i \leq n$, so $x \in B^n \subseteq B^+$ and the result follows.

(g) $[(A \cup B)^* = (A^* \cup B^*)^*]$. $A \subseteq A^*$, $B \subseteq B^* \Rightarrow (A \cup B) \subseteq (A^* \cup B^*) \Rightarrow (A \cup B)^* \subseteq (A^* \cup B^*)^*$ (by part d). Conversely, $A, B \subseteq A \cup B \Rightarrow A^*, B^* \subseteq (A \cup B)^*$ (by part d) $\Rightarrow (A^* \cup B^*) \subseteq (A \cup B)^* \Rightarrow (A^* \cup B^*)^* \subseteq (A \cup B)^*$ (by parts d and f). From both inclusions we have $(A \cup B)^* = (A^* \cup B^*)^*$.

$[(A^* \cup B^*)^* = (A^*B^*)^*]$. $A^*, B^* \subseteq A^*B^*$ (by parts a and b) $\Rightarrow (A^* \cup B^*) \subseteq A^*B^* \Rightarrow (A^* \cup B^*)^* \subseteq (A^*B^*)^*$ (by part d). Conversely, if $xy \in A^*B^*$, $x \in A^*$, $y \in B^*$, then $x, y \in A^* \cup B^*$, so $xy \in (A^* \cup B^*)^*$, and $A^*B^* \subseteq (A^* \cup B^*)^*$. Using part (f) again, $(A^*B^*)^* \subseteq (A^* \cup B^*)^*$, and the result follows.   ■

EXERCISES

1. Let $\mathbf{\Sigma} = \{a, b, c, d, e\}$. (a) What is $|\mathbf{\Sigma}^2|$? $|\mathbf{\Sigma}^3|$? (b) How many strings in $\mathbf{\Sigma}^*$ have length at most five?

2. For $\mathbf{\Sigma} = \{w, x, y, z\}$ determine the number of strings in $\mathbf{\Sigma}^*$ of length five (a) that start with w; (b) with precisely two w's; (c) with no w's; (d) with an even number of w's.

3. If $x \in \mathbf{\Sigma}^*$ and $\|x^3\| = 36$, what is $\|x\|$?

4. Let $\mathbf{\Sigma} = \{\beta, x, y, z\}$ where β denotes a blank, so $x\beta \neq x$, $\beta\beta \neq \beta$, and $x\beta y \neq xy$ but $x\lambda y = xy$. Compute each of the following.

   a) $\|\lambda\|$ b) $\|\lambda\lambda\|$ c) $\|\beta\|$ d) $\|\beta\beta\|$

   e) $\|\beta^3\|$ f) $\|x\beta\beta y\|$ g) $\|\beta\lambda\|$ h) $\|\lambda^{10}\|$

5. Let $\mathbf{\Sigma} = \{v, w, x, y, z\}$ and $A = \bigcup_{n=1}^6 \mathbf{\Sigma}^n$. How many strings in A have xy as a proper prefix?

6. If $A \ (\neq \emptyset)$ is a language and $A^2 = A$, prove that $\lambda \in A$.

7. If A, B, C, D are languages over $\mathbf{\Sigma}$ prove that (a) $A \subseteq B$, $C \subseteq D \Rightarrow AC \subseteq BD$; and (b), $A\emptyset = \emptyset A = \emptyset$.

8. For $\Sigma = \{x, y, z\}$, let $A, B \subseteq \Sigma^*$ be given by $A = \{xy\}$, $B = \{\lambda, x\}$. Determine (a) AB; (b) BA; (c) B^3; (d) B^+; (e) A^*.

9. Provide the proofs for the remaining parts of Theorems 4.1 and 4.2.

10. If A, B are languages over Σ, and $A \subseteq B^*$, prove that $A^* \subseteq B^*$.

11. For a given alphabet Σ, let I denote an index set, where for each $i \in I$, $B_i \subseteq \Sigma^*$. If $A \subseteq \Sigma^*$, prove that (a) $A(\bigcup_{i \in I} B_i) = \bigcup_{i \in I} AB_i$; and (b) $(\bigcup_{i \in I} B_i)A = \bigcup_{i \in I} B_i A$. (These results generalize parts (c) and (d) of Theorem 4.1.)

12. For $\Sigma = \{x, y\}$, use finite languages from Σ^*, together with set operations, as in Example 4.11, to describe the set of strings in Σ^* that (a) contain exactly one occurrence of x; (b) contain exactly two occurrences of x; (c) begin with x; (d) end in yxy; (e) begin with x or end in yxy or both; (f) begin with x or end in yxy but not both.

4.2 ■ FINITE STATE MACHINES: A FIRST ENCOUNTER

We return now to the vending machine mentioned at the start of this chapter, and analyze it in the following circumstance.

At a metropolitan office, a vending machine dispenses two kinds of soft drinks in cans: cola (C) and root beer (RB). The cost of a can of either soft drink is 20¢. The machine accepts nickels, dimes and quarters, and returns the necessary change. One day Mary Jo decides to have a can of root beer. She goes to the vending machine and inserts two nickels and a dime, in that order, presses the white button, denoted W, and out comes her can of root beer. (To get a can of cola one presses the black button, denoted B.)

What Mary Jo has done, in making her purchase, can be represented as in Table 4.1 where t_0 is the initial time, when she inserts her first nickel, and t_1, t_2, t_3, t_4 are later moments in time, with $t_1 < t_2 < t_3 < t_4$.

Table 4.1

	t_0	t_1	t_2	t_3	t_4
State	(1) s_0	(4) s_1 (5¢)	(7) s_2 (10¢)	(10) s_3 (20¢)	(13) s_0
Input	(2) 5¢	(5) 5¢	(8) 10¢	(11) W	
Output	(3) Nothing	(6) Nothing	(9) Nothing	(12) RB	

The numbers (1), (2), ..., (12), (13) in this table indicate the order of events in the purchase of Mary Jo's root beer. For each input at time t_i, $0 \le i \le 3$, there is at that time a corresponding output, and then a change in state. The new state at time t_{i+1} depends upon both the input and (present) state at time t_i.

The machine is in a state of readiness at state s_0. It waits for a customer to start inserting coins that will total 20¢ or more and then press a button to get a soft drink. If at any time the total of the coins inserted exceeds 20¢ the machine provides the needed change (before the customer presses the button to get the soft drink).

At time t_0 Mary Jo provides the machine with her first input, 5¢. She receives nothing at this time but at the later time t_1 the machine is in state s_1, where it *remembers* her total of 5¢ and waits for her second input (of 5¢ at time t_1). The machine again (at time t_1) provides no output but at the next time, t_2, it is in state s_2, remembering a total of 10¢ = 5¢ (remembered at state s_1) + 5¢ (inserted at time t_1). Providing her dime (at time t_2) as the next input to the machine, Mary Jo receives no soft drink yet, since the machine doesn't *know* which type Mary Jo prefers, but it does *know* now (t_3) that she has inserted the necessary total of 20¢ = 10¢ (remembered at state s_2) + 10¢ (inserted at time t_2). At last Mary Jo presses the white button and at time t_3 the machine dispenses the output (her can of root beer), and then returns, at time t_4, to the starting state s_0, just in time for Mary Jo's friend Rizzo to deposit a quarter, receive her nickel change, press the black button and obtain the can of cola she desires. The purchase made by Rizzo is analyzed in Table 4.2.

Table 4.2

	t_0		t_1		t_2	
State	(1) s_0		(4) s_3 (20¢)		(7) s_0	
Input	(2) 25¢		(5) B			
Output	(3) 5¢ change		(6) C			

What has happened in the case of this vending machine can be abstracted to help in the analysis of certain aspects of digital computers and telephone communication systems.

The major features of such a machine are as follows:

1. The machine can be in only one of *finitely many states* at a given time. These states are called the *internal states* of the machine, and at a given time the total memory available to the machine is the knowledge of which internal state it is in at that moment.

2. The machine will accept as *input* only a finite number of symbols, and collectively these are referred to as the *input alphabet* $\mathcal{I}$. In the vending machine example the input alphabet is {nickel, dime, quarter, W, B}, each item of which is recognized by each internal state.

3. An *output* and *next state* are determined by each combination of inputs and internal states. The finite set of all possible outputs constitutes the *output alphabet* $\mathcal{O}$ for the machine.

4. We assume that the sequential processings of the machine are *synchronized* by separate and distinct clock pulses, and that the machine operates in a *deterministic* manner, where the output is completely determined by the total input provided and the starting state of the machine.

With these observations we are led to the following definition.

Definition 4.9

▶ A *finite state machine* is a five-tuple $M = (S, \mathcal{I}, \mathcal{O}, \nu, \omega)$ where $S =$ the set of internal states for M; $\mathcal{I} =$ the input alphabet for M; $\mathcal{O} =$ the output alphabet for M; $\nu: S \times \mathcal{I} \to S$, the *next state function*; and $\omega: S \times \mathcal{I} \to \mathcal{O}$, the *output function*.

Using the notation of this definition, if the machine is in state s at time t_i and we input x at this time, then the output at time t_i is $\omega(s, x)$. This output is followed by a transition of the machine at time t_{i+1} to the next internal state given by $\nu(s, x)$.

We assume that when a finite state machine receives its first input, we are at time $t_0 = 0$ and the machine is in a designated starting state denoted by s_0. Our development will concentrate primarily on the output and state transitions that take place sequentially, with little or no reference to the sequence of clock pulses at times $t_0, t_1, t_2, \ldots$.

Since the sets S, $\mathcal{I}$, $\mathcal{O}$ are finite, it is possible to represent ν and ω, for a given finite state machine, by means of a table that lists $\nu(s, x)$ and $\omega(s, x)$ for all $s \in S$, $x \in \mathcal{I}$. Such a table is referred to as the *state table* or *transition table* for the given machine. A second representation of the machine is made by means of a *state diagram*.

We demonstrate the state table and state diagram in the following examples.

EXAMPLE 4.13

Consider the finite state machine $M = (S, \mathcal{I}, \mathcal{O}, \nu, \omega)$ where $S = \{s_0, s_1, s_2\}$, $\mathcal{I} = \mathcal{O} = \{0, 1\}$, and ν, ω are given by the *state table* in Table 4.3. The first column of the table lists the (*present*) *states* for the machine. The entries in the second row are the elements of the input alphabet $\mathcal{I}$, listed once under ν and then again under ω. The numbers in the last two columns are elements of the output alphabet $\mathcal{O}$.

Table 4.3

	ν		ω	
	0	1	0	1
s_0	s_0	s_1	0	0
s_1	s_2	s_1	0	0
s_2	s_0	s_1	0	1

To calculate $\nu(s_1, 1)$, for example, we find s_1 in the column of present states and proceed horizontally over from s_1 until we are below the entry 1 in the section of the table for ν. This entry gives $\nu(s_1, 1) = s_1$. In the same way we find $\omega(s_1, 1) = 0$.

Designating s_0 as the starting state, if the input provided to M is the string 1010 then the output is 0010 as demonstrated in Table 4.4. Here the machine is left in state s_2, so that if we had another input string we would provide the first character of that string, here 0, at state s_2 unless the machine is *reset* to start once again at s_0.   □

Table 4.4

State	s_0	$\nu(s_0, 1) = s_1$	$\nu(s_1, 0) = s_2$	$\nu(s_2, 1) = s_1$	$\nu(s_1, 0) = s_2$
Input	1	0	1	0	0
Output	$\omega(s_0, 1) = 0$	$\omega(s_1, 0) = 0$	$\omega(s_2, 1) = 1$	$\omega(s_1, 0) = 0$	

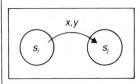

Figure 4.1

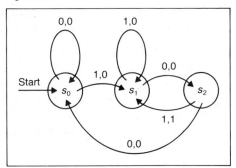

Figure 4.2

Since we are primarily interested in the output, not in the sequence of transition states, the same machine can be represented by means of a *state diagram*. Here we can obtain the output string without actually listing the transition states. In such a diagram each internal state s is represented by a circle with s inside of it. For states s_i, s_j if $\nu(s_i, x) = s_j$ for $x \in \mathcal{I}$ and $\omega(s_i, x) = y$, $y \in \mathcal{O}$, we represent this in the state diagram by drawing a directed arc from the circle for s_i to the circle for s_j and labeling the arc with the input x and output y as shown in Fig. 4.1.

With these conventions, the state diagram for the machine M of Table 4.3 is shown in Fig. 4.2. Although the table is more compact, the diagram allows us to follow an input string through each transition state it determines, picking up each of the corresponding output symbols before each transition. Here if the input string is 00110101, starting at state s_0, the first input of 0 yields an output of 0 and returns us to s_0. The next input of 0 yields the same result, but for the third input, 1, the output is 0 and we are now in state s_1. Continuing in this manner we arrive at the output string 00000101 and finish in state s_1. (We note that the input string 00110101 is an element of $\mathcal{I}^*$, the Kleene closure of $\mathcal{I}$ and the output string is in $\mathcal{O}^*$, the Kleene closure of $\mathcal{O}$.)

Starting at s_0, what is the output for the input string 1100101101?

EXAMPLE 4.14   For the vending machine described earlier in this section we have the state table, Table 4.5, with

1. $S = \{s_0, s_1, s_2, s_3, s_4\}$, where at state s_k, the machine remembers the insertion of a total of $5k$ cents;

Table 4.5

	ν					ω				
	5¢	10¢	25¢	B	W	5¢	10¢	25¢	B	W
s_0	s_1	s_2	s_4	s_0	s_0	n	n	5¢	n	n
s_1	s_2	s_3	s_4	s_1	s_1	n	n	10¢	n	n
s_2	s_3	s_4	s_4	s_2	s_2	n	n	15¢	n	n
s_3	s_4	s_4	s_4	s_3	s_3	n	5¢	20¢	n	n
s_4	s_4	s_4	s_4	s_0	s_0	5¢	10¢	25¢	C	RB

2. $\mathcal{I} = \{5\cent, 10\cent, 25\cent, B, W\}$, where B denotes the black button one presses for a cola, and W the white button for root beer.

3. $\mathcal{O} = \{n(\text{nothing}), RB(\text{root beer}), C(\text{cola}), 5\cent, 10\cent, 15\cent, 20\cent, 25\cent\}$. □

As observed at the end of Example 4.13, for a general finite state machine $M = (S, \mathcal{I}, \mathcal{O}, \nu, \omega)$, the input can be realized as an element of $\mathcal{I}^*$, with the output from $\mathcal{O}^*$. Consequently, it is to our advantage to extend the domains of ν and ω from $S \times \mathcal{I}$ to $S \times \mathcal{I}^*$. For ω we enlarge the codomain to $\mathcal{O}^*$, recalling, should the need arise, that both $\mathcal{I}^*$, $\mathcal{O}^*$ contain an empty string, λ. With these extensions, if $x_1x_2 \ldots x_k \in \mathcal{I}^*$, $k \in \mathbf{Z}^+$, then starting at any state $s_1 \in S$, we have

$$\nu(s_1, x_1) = s_2$$

$$\nu(s_1, x_1x_2) = \nu(\nu(s_1, x_1), x_2) = \nu(s_2, x_2) = s_3$$

$$\nu(s_1, x_1x_2x_3) = \nu(\nu(\underbrace{\nu(s_1, x_1)}_{s_2}, x_2), x_3) = s_4$$

$$\nu(s_2, x_2) = s_3$$

. . . .

$$\nu(s_1, x_1x_2 \ldots x_k) = \nu(s_k, x_k) = s_{k+1}, \quad \text{and}$$

$$\omega(s_1, x_1) = y_1$$

$$\omega(s_1, x_1x_2) = \omega(s_1, x_1)\omega(\nu(s_1, x_1), x_2) = \omega(s_1, x_1)\omega(s_2, x_2) = y_1y_2$$

$$\omega(s_1, x_1x_2x_3) = \omega(s_1, x_1)\omega(s_2, x_2)\omega(s_3, x_3) = y_1y_2y_3$$

. . . .

$$\omega(s_1, x_1x_2 \ldots x_k) = \omega(s_1, x_1)\omega(s_2, x_2) \ldots \omega(s_k, x_k) = y_1y_2 \ldots y_k \in \mathcal{O}^*$$

Also, $\nu(s_1, \lambda) = s_1$ for any $s_1 \in S$.
(We shall use these extensions again in Chapter 5.)

We close this section with an example that is relevant in computer science.

Figure 4.3

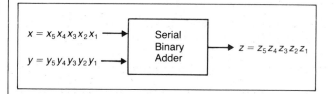

EXAMPLE 4.15 Let $x = x_5x_4x_3x_2x_1 = 00111$, $y = y_5y_4y_3y_2y_1 = 01101$ be binary numbers where x_1, y_1 are the least significant bits. The leading 0's in x, y are there to make the strings for x, y of equal length and to guarantee enough places to complete the sum. A *serial binary adder* is a finite state machine that we can use to obtain $x + y$. The diagram in Fig. 4.3 demonstrates this, where $z = z_5z_4z_3z_2z_1$ has least significant bit z_1.

In the addition $z = x + y$ we have

$$
\begin{array}{rccccc}
x = & 0 & 0 & 1 & 1 & 1 \\
+\, y = & +\,0 & 1 & 1 & 0 & 1 \\
\hline
z = & 1 & 0 & 1 & 0 & 0
\end{array}
$$

third
addition

first
addition

We note that for the first addition $x_1 = y_1 = 1$ and $z_1 = 0$, while for the third addition we have $x_3 = y_3 = 1$ and $z_3 = 1$, because of a *carry* from the addition of x_2 and y_2 (and the *carry* from $x_1 + y_1$). Consequently each output depends on the sum of two inputs and the ability to *remember* a carry of 0 or 1, which is crucial when it is 1.

The serial binary adder is modeled by a finite state machine $M = (S, \mathcal{I}, \mathcal{O}, \nu, \omega)$ as follows. The set $S = \{s_0, s_1\}$, where s_i indicates a carry of i; $\mathcal{I} = \{00, 01, 10, 11\}$ so there is a pair of inputs depending on whether we are seeking $0 + 0, 0 + 1, 1 + 0$, or $1 + 1$, respectively; and $\mathcal{O} = \{0, 1\}$. The functions ν, ω are given in the state table (Table 4.6) and state diagram (Fig. 4.4).

Table 4.6

	ν				ω			
	00	01	10	11	00	01	10	11
s_0	s_0	s_0	s_0	s_1	0	1	1	0
s_1	s_0	s_1	s_1	s_1	1	0	0	1

In Table 4.6 we find, for example, that $\nu(s_1, 01) = s_1$, and $\omega(s_1, 01) = 0$ because s_1 indicates a carry of 1 from the addition of the previous bits. The 01 input indicates that we are adding 0 and 1 (and carrying a 1). Hence the sum is 10 and $\omega(s_1, 01) = 0$ for the 0 in 10. The carry is again remembered in $s_1 = \nu(s_1, 01)$.

Figure 4.4

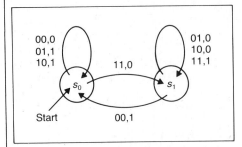

From the state diagram (Fig. 4.4) we see that the starting state must be s_0 since there is no carry prior to the addition of the least significant bits. □

The state diagrams in Figs. 4.2 and 4.4 are examples of *labeled directed graphs*. Much more will be done with graph theory throughout the text, for it has applications not only in computer science and electrical engineering but also in coding theory (prefix codes) and optimization (transport networks).

EXERCISES

1. Using the finite state machine of Example 4.13, find the output for each of the following input strings $x \in \mathcal{I}^*$, and determine the last internal state in the transition process. (Assume that we always start at s_0.)

   a) $x = 1010101$ (b) $x = 1001001$ c) $x = 101001000$

2. For the finite state machine of Example 4.13 an input string x, starting at state s_0, produces the output string 00101. Determine x.

3. Let $M = (S, \mathcal{I}, \mathbb{O}, \nu, \omega)$ be a finite state machine where $S = \{s_0, s_1, s_2, s_3\}$, $\mathcal{I} = \{a, b, c\}$, $\mathbb{O} = \{0, 1\}$, and ν, ω are determined by Table 4.7.

Table 4.7

	ν			ω		
	a	b	c	a	b	c
s_0	s_0	s_3	s_2	0	1	1
s_1	s_1	s_1	s_3	0	0	1
s_2	s_1	s_1	s_3	1	1	0
s_3	s_2	s_3	s_0	1	0	1

   a)   Starting at s_0, what is the output for the input string $abbccc$?

   b)   Draw the state diagram for this finite state machine.

4. Give the state table for the vending machine of Example 4.14 if the cost of a can of cola or root beer is increased to 25¢.

Figure 4.5

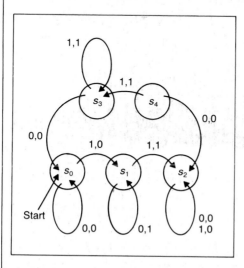

5. A finite state machine $M = (S, \mathcal{I}, \mathbb{O}, \nu, \omega)$ has $\mathcal{I} = \mathbb{O} = \{0, 1\}$ and is determined by the state diagram, Fig. 4.5.

   a)   Determine the output string for the input string 110111, starting at s_0. What is the last transition state?

b) Answer part (a) for the same string but with s_1 as the starting state. What about s_2, s_3 as starting states?

c) Find the state table for this machine.

d) In which state should we start so that the input string 10010 produces the output 10000?

e) Determine an input string $x \in \mathcal{I}^*$ of minimal length, such that $\nu(s_4, x) = s_1$. Is x unique?

6. Machine M has $\mathcal{I} = \{0, 1\} = \mathcal{O}$, and is determined by the state diagram in Fig. 4.6.

Figure 4.6

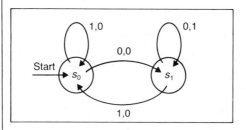

a) Describe in words what this finite state machine does.

b) What must state s_1 remember?

c) Find two languages $A, B \subseteq \mathcal{I}^*$ such that for every $x \in AB$, $\omega(s_0, x)$ ends in 1.

7. a) If S, $\mathcal{I}$, $\mathcal{O}$ are finite sets, with $|S| = 3$, $|\mathcal{I}| = 5$, $|\mathcal{O}| = 2$, determine

    i) $|S \times \mathcal{I}|$;

    ii) the number of functions $\nu: S \times \mathcal{I} \to S$; and

    iii) the number of functions $\omega: S \times \mathcal{I} \to \mathcal{O}$.

b) For S, $\mathcal{I}$, $\mathcal{O}$ in part (a), how many finite state machines do they determine?

Figure 4.7

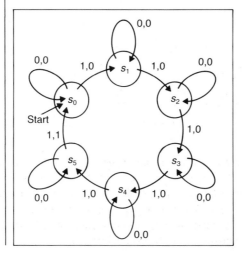

8. Let $M = (S, \mathcal{I}, \mathcal{O}, \nu, \omega)$ be a finite state machine with $\mathcal{I} = \mathcal{O} = \{0, 1\}$ and S, ν, ω determined by the state diagram in Fig. 4.7.

a) Find the output for the input string $x = 0110111011$.

b) Give the transition table for this finite state machine.

c) Starting in state s_0, if the output for an input string x is 0000001, determine all possibilities for x.

d) Describe in words what this finite state machine does.

4.3 ■ FINITE STATE MACHINES: A SECOND ENCOUNTER

Having seen some examples of finite state machines we turn to the study of some additional machines that arise in computer hardware. One important type of machine is the *sequence recognizer*.

EXAMPLE 4.16 Here $\mathcal{I} = \mathcal{O} = \{0, 1\}$ and we want to construct a machine that recognizes each occurrence of the sequence 111 as it is encountered in any input string $x \in \mathcal{I}^*$. For example, if $x = 1110101111$, then the corresponding output should be 0010000011, where a 1 in the ith position of the output indicates that a 1 can be found in positions i, $i - 1$, and $i - 2$ of x. Here overlapping of sequences of 111 can occur, so some characters in the input string can be thought of as characters in more than one triple of 1's.

Letting s_0 denote the starting state, we realize that we must have a state to remember 1 (the possible start of 111) and a state to remember 11. In addition, anytime our input symbol is 0, we go back to s_0 and start the search for three successive 1's over again.

Figure 4.8

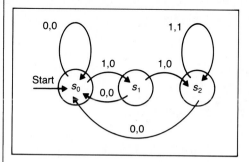

In Fig. 4.8, s_1 remembers a single 1, and s_2 remembers the string 11. If s_2 is reached, then a third "1" indicates the occurrence of the triple in the input string, and the output 1 *recognizes* this occurrence. But this third "1" also means that we have the first two 1's of another possible triple coming up in the string (as happens in 11101011"1"1). So after recognizing the occurrence of 111 with an output of 1, we return to state s_2 to remember the two inputs of 1"1".

If we are concerned with recognizing all strings that end in 111, then for any $x \in \mathcal{I}*$, the machine will recognize such a sequence with final output 1. This machine is then a recognizer of the language $A = \{0, 1\}*\{111\}$.   ☐

Figure 4.9

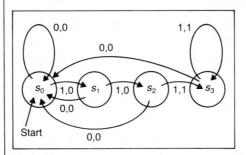

Another finite state machine that recognizes the same triple 111 is shown in Fig. 4.9. The finite state machines represented by the state diagrams in Figs. 4.8 and 4.9 perform the same task and are said to be *equivalent*. The state diagram in Fig. 4.9 has one more state than that in Fig. 4.8, but at this stage we are not overly concerned with getting a finite state machine with a minimal number of states. In Chapter 5 we shall develop a technique to take a given finite state machine M and find one that is equivalent to it and has the smallest number of internal states needed.

The next example is a bit more selective.

EXAMPLE 4.17   Now we want not only to recognize the occurrence of 111 but we want to recognize only those occurrences that end in a position that is a multiple of three. Consequently, with $\mathcal{I} = \mathcal{O} = \{0, 1\}$, if $x \in \mathcal{I}*$, $x = 1110111$, then we want $\omega(s_0, x) = 0010000$, not 0010001. In addition, for $x \in \mathcal{I}*$, $x = 111100111$, the output $\omega(s_0, x)$ is to be 001000001, not 001100001, for here, because of length considerations, overlapping of sequences of 111 is not allowed.

Figure 4.10

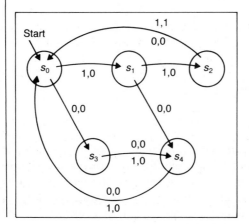

Again we start at s_0 (Fig. 4.10), but now s_1 must remember a first 1 only if it occurs in x in position 1, 4, 7, ... If the input at s_0 is 0, we cannot simply return to s_0 as in Example 4.16. We must remember that this 0 is the *first* of three symbols of no interest. Hence from s_0 we go to s_3 and then to s_4, processing any triple of the form $0yz$ where 0 occurs in x in position $3k + 1$, $k \geq 0$. The same type of situation happens at s_1 if the input is 0. Finally at s_2 the sequence 111 is recognized with an output of 1, if it occurs. The machine then returns to s_0 to input the next symbol of the input string. □

Figure 4.11

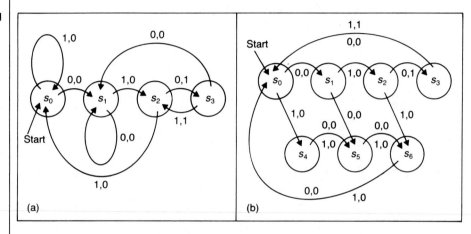

(a)

(b)

EXAMPLE 4.18 | Figure 4.11 shows the state diagrams for finite state machines that will recognize the occurrence of the sequence 0101 in an input string $x \in \mathscr{I}^*$, where $\mathscr{I} = \mathbb{O} = \{0, 1\}$. The machine in Fig. 4.11(a) recognizes with an output of 1 each occurrence of 0101 in an input string, regardless of where it occurs. In Fig. 4.11(b) the machine recognizes with an output of 1 only those prefixes of x whose length is a multiple of four and end in 0101. (So no overlapping is allowed here.) Consequently, for $x = 01010100101$, $\omega(s_0, x) = 00010100001$ for (a), while for (b), $\omega(s_0, x) = 00010000000$. □

A class of finite state machines that is important in the design of digital devices consists of the *k-unit delay machines*, where $k \in \mathbf{Z}^+$. For $k = 1$, we want to construct a machine M such that if $x = x_1 x_2 \ldots x_{m-1} x_m$, then for starting state s_0, $\omega(s_0, x) = 0x_1 x_2 \ldots x_{m-1}$, so that the output is the input delayed one time unit (clock pulse). (The use of 0 as the first symbol in $\omega(s_0, x)$ is conventional.)

EXAMPLE 4.19 | Let $\mathscr{I} = \mathbb{O} = \{0, 1\}$. With starting state s_0, $\omega(s_0, x) = 0$ for $x = 0$ or 1, since the first output is 0; the states s_1, s_2 remember a prior input of 0, 1, respectively. In Fig. 4.12, we label, for example, the arc from s_1 to s_2 with 1, 0 because with an input of 1 we need to go to s_2 where inputs of 1 at time t_i are remembered so that they can become outputs of 1 at time t_{i+1}. The 0 in the label 1, 0 is the output, since starting in s_1 indicates that the prior input was 0, which becomes the present output. The labels on the other arcs are obtained by the same type of reasoning. □

Figure 4.12

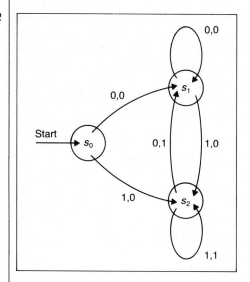

EXAMPLE 4.20

Observing the structure for a one-unit delay, we extend our ideas to the two-unit delay shown in Fig. 4.13. If $x \in \mathscr{I}^*$, $x = x_1 x_2 \ldots x_m$, $m > 2$, and s_0 is the starting state, then $\omega(s_0, x) = 00x_1 \ldots x_{m-2}$. For states s_0, s_1, s_2 the output is 0 for all possible inputs. States s_3, s_4, s_5, s_6 must remember the two prior inputs 00, 01, 10, 11, respectively. To get the other directed arcs in the diagram we shall consider one such arc and then use similar reasoning for the others. For the arc from s_5 to s_3 in Fig. 4.13 (a), let the input

Figure 4.13

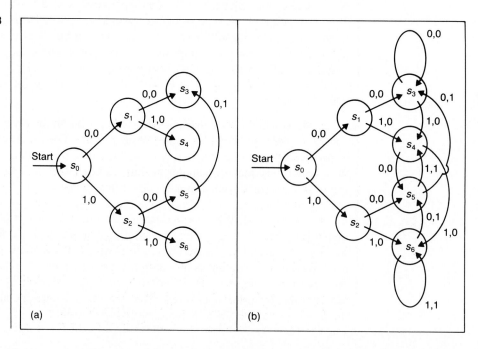

be 0. Since the prior input to s_5 from s_2 is 0, we must go to the state that remembers the two prior inputs 00. This is state s_3. Going back two states from s_5 to s_2 to s_0 we see the input is 1 (from s_0 to s_2). This then becomes the output (delayed two places) for the arc from s_5 to s_3. (The complete machine is shown in part (b).) ☐

We turn now to a study of some additional properties that arise for these machines. The machine in Fig. 4.14 will be used for examples of the terms defined.

Definition 4.10

▶ Let $M = (S, \mathcal{I}, \mathcal{O}, \nu, \omega)$ be a finite state machine.

a) For s_i, $s_j \in S$, s_j is said to be *reachable* from s_i if $s_i = s_j$ or if there is an input string $x \in \mathcal{I}^+$, such that $\nu(s_i, x) = s_j$. (In Fig. 4.14, state s_3 is reachable from s_0, s_1, s_2 but not from s_4, s_5, s_6 or s_7. No state is reachable from s_3 except s_3 itself.)

Figure 4.14

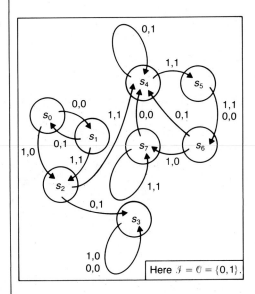

Here $\mathcal{I} = \mathcal{O} = \{0, 1\}$.

b) A state $s \in S$ is said to be *transient* if $\nu(s, x) = s$ for $x \in \mathcal{I}^*$ implies $x = \lambda$, i.e., there is no $x \in \mathcal{I}^+$ with $\nu(s, x) = s$. (For the machine in Fig. 4.14, s_2 is the only transient state.)

c) For s_i, $s_j \in S$, s_i is *conditionally transient* with respect to s_j if there exists $x \in \mathcal{I}^+$, such that $\nu(s_i, x) = s_j$, but for all $x \in \mathcal{I}^+$, $\nu(s_j, x) \neq s_i$. (For the given machine, each of s_0, s_1, s_2 is conditionally transient with respect to each of s_3, s_4, s_5, s_6, s_7.)

d) A state $s \in S$ is called a *sink*, or *sink state*, if $\nu(s, x) = s$, for all $x \in \mathcal{I}^*$. (s_3 is the only sink in Fig. 4.14.)

e) Let $S_1 \subseteq S$, $\mathcal{I}_1 \subseteq \mathcal{I}$. If $\nu_1 = \nu|_{S_1 \times \mathcal{I}_1}$: $S_1 \times \mathcal{I}_1 \rightarrow S_1$ (that is, the restriction of ν to $S_1 \times \mathcal{I}_1 \subseteq S \times \mathcal{I}$) has its range within S_1, then with $\omega_1 = \omega|_{S_1 \times \mathcal{I}_1}$, $M_1 = (S_1, \mathcal{I}_1, \mathcal{O}, \nu_1, \omega_1)$ is called a *submachine* of M. (With $S_1 = \{s_4, s_5, s_6, s_7\}$, and $\mathcal{I}_1 = \{0, 1\}$, we get a submachine M_1 of the machine M in Fig. 4.14.)

f)   A machine is said to be *strongly connected* if for any states s_i, $s_j \in S$, s_j is reachable from s_i. (The machine in Fig. 4.14 is not strongly connected but the submachine M_1 in (e) has this property.)

We close this section with a result that uses a tree diagram.

Definition 4.11

▶   For a finite state machine M, if s_i, $s_j \in S$, $i \neq j$, a shortest input string $x \in \mathcal{I}^+$ is called a *transfer* (or *transition*) *sequence* from s_i to s_j, if (a) $\nu(s_i, x) = s_j$, and (b) if $y \in \mathcal{I}^+$ with $\nu(s_i, y) = s_j$, then $\|y\| \geq \|x\|$.

There can be more than one such sequence for two states s_i, s_j.

EXAMPLE 4.21

For the finite state machine M given by the transition table, Table 4.8, where $\mathcal{I} = \mathbb{O} = \{0, 1\}$, find a transfer sequence from state s_0 to state s_2.

Table 4.8

	ν		ω	
	0	1	0	1
s_0	s_6	s_1	0	1
s_1	s_5	s_0	0	1
s_2	s_1	s_2	0	1
s_3	s_4	s_0	0	1
s_4	s_2	s_1	0	1
s_5	s_3	s_5	1	1
s_6	s_3	s_6	1	1

Figure 4.15

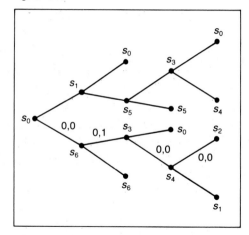

In constructing the tree diagram of Fig. 4.15, we start at state s_0 and find those states that can be reached from s_0 by strings of length one. Here we find s_1 and s_6. Then we do the same thing with s_1 and s_6, finding, as a result, those states reachable from s_0 with input strings of length two. Continuing to expand the tree from left to right, we get to a vertex labeled with the desired state, s_2. Each time we reach a vertex labeled with a state used previously, we terminate that part of the expansion since we cannot get any new states. After we arrive at the state we want, we backtrack to s_0 and use the state table to label the branches, as shown in Fig. 4.15. Hence, for $x = 0000$, $\nu(s_0, x) = s_2$ with $\omega(s_0, x) = 0100$. (Here x is unique.)   □

EXERCISES

1. Let $\mathcal{I} = \mathbb{O} = \{0, 1\}$. (a) Construct a state diagram for a finite state machine that recognizes each occurrence of 0000 in a string $x \in \mathcal{I}^*$. (Here overlapping is allowed.) (b) Construct a state diagram for a finite state machine that recognizes

each string $x \in \mathcal{I}*$ that ends in 0000 and has length $4k$, $k \in \mathbf{Z}^+$. (Here overlapping is not permitted.)

2. Answer Exercise 1 for the sequences 0110 and 1010.

3. For $k \in \mathbf{Z}^+$, k fixed, draw a state diagram for a finite state machine that recognizes a string of k consecutive 0's. (Overlapping is permitted here.)

4. Table 4.9 defines ν, ω for a finite state machine M with $\mathcal{I} = \mathbb{O} = \{0, 1\}$.

Table 4.9

	ν		ω	
	0	1	0	1
s_0	s_0	s_1	0	0
s_1	s_0	s_1	1	1

   a) Draw the state diagram for M.

   b) Determine the output for input sequences (i) $x = 111$; (ii) $x = 1010$; (iii) $x = 00011$, starting at s_0 in each case.

   c) Describe in words what machine M does.

   d) How is this machine related to that of Fig. 4.12?

5. Verify the labels on all of the directed arcs in the state diagram for the two-unit delay machine shown in Fig. 4.13(b).

6. For each of the machines in Table 4.10, determine the transient states, conditionally transient states, sink states, submachines, and strongly connected submachines.

Table 4.10

(a)

	ν		ω	
	0	1	0	1
s_0	s_4	s_1	0	0
s_1	s_4	s_2	0	1
s_2	s_3	s_5	0	0
s_3	s_2	s_5	1	0
s_4	s_4	s_4	1	1
s_5	s_2	s_3	0	1

(b)

	ν		ω	
	0	1	0	1
s_0	s_0	s_1	1	0
s_1	s_0	s_1	0	1
s_2	s_1	s_3	0	0
s_3	s_0	s_4	0	0
s_4	s_4	s_4	1	1

(c)

	ν		ω	
	0	1	0	1
s_0	s_1	s_2	0	1
s_1	s_0	s_2	1	1
s_2	s_2	s_3	1	1
s_3	s_6	s_4	0	0
s_4	s_5	s_5	1	0
s_5	s_3	s_4	1	0
s_6	s_6	s_6	0	0

7. Determine a transfer sequence from state s_2 to state s_5 in the finite state machine (c) of Exercise 6. Is your sequence unique?

4.4 ■ SUMMARY AND HISTORICAL REVIEW

In this chapter we have been introduced to the theory of languages and a discrete structure called a *finite state machine*. Using our prior development of elementary set theory and finite functions we were able to combine some abstract notions and to model digital devices such as sequence recognizers and delays. Comparable coverage of this is given in Chapter 1 of L. Dornhoff and F. Hohn [1], and Chapter 2 of D. Stanat and D. McAllister [9].

The finite state machine we developed is based on the model put forth in 1955 by G. H. Mealy in [6], and is consequently referred to as the "Mealy machine." The model is based on earlier concepts found in the work of D. A. Huffman [4] and E. F. Moore [7]. For further reading on the pioneering work dealing with various aspects and applications of the finite state machine, consult the material edited by E. F. Moore [8]. Additional information on the actual synthesis of such machines, and related hardware considerations, along with an extensive coverage of many related ideas, can be found in Chapters 9–15 of Z. Kohavi [5].

For more on languages and their relation to finite state machines, examine Chapters 7–10 of J. Gersting [2] and Chapters 7 and 8 of A. Gill [3].

REFERENCES

1. Dornhoff, Larry L., and Hohn, Franz E., *Applied Modern Algebra,* Macmillan, New York, 1978.

2. Gersting, Judith L., *Mathematical Structures for Computer Science,* W. H. Freeman, San Francisco, 1982.

3. Gill, Arthur, *Applied Algebra for the Computer Sciences,* Prentice-Hall Series in Automatic Computation, Prentice-Hall, Englewood Cliffs, New Jersey, 1976.

4. Huffman, D. A., "The Synthesis of Sequential Switching Circuits," *Journal of the Franklin Institute,* vol. 257, pp. 161–190, March, 1954; pp. 275–303, April, 1954. Reprinted in Moore [8].

5. Kohavi, Zvi, *Switching and Finite Automata Theory,* 2nd ed., McGraw-Hill, New York, 1978.

6. Mealy, G. H., "A Method for Synthesizing Sequential Circuits," *Bell System Technical Journal,* vol. 34, pp. 1045–1079, September, 1955.

7. Moore, E. F., "Gedanken-experiments on Sequential Machines," *Automata Studies, Annals of Mathematical Studies,* no. 34, pp. 129–153, Princeton University Press, Princeton, New Jersey, 1956.

8. Moore, E. F. (editor), *Sequential Machines: Selected Papers,* Addison-Wesley, Reading, Massachusetts, 1964.

9. Stanat, Donald F., and McAllister, David F., *Discrete Mathematics in Computer Science,* Prentice-Hall, Englewood Cliffs, New Jersey, 1977.

■ MISCELLANEOUS EXERCISES

1. Give an example of a language A over an alphabet Σ, where $(A^2)^* \neq (A^*)^2$.

2. For languages $A, B \subseteq \Sigma^*$, does $A^* \subseteq B^* \Rightarrow A \subseteq B$?

3. For $\Sigma = \{0, 1\}$, let A, $B \subseteq \Sigma^*$ with $A = \{0, 1\} \cup (\{0, 1\}\{1\}^*\{1\})$, $B = \{0, 1\}\{1\}^*$. Prove that $A = B$.

4. For a given alphabet Σ and index set I, let $B_i \subseteq \Sigma^*$, for each $i \in I$. If $A \subseteq \Sigma^*$, prove that (a) $A(\cap_{i \in I} B_i) \subseteq \cap_{i \in I} AB_i$; and (b) $(\cap_{i \in I} B_i)A \subseteq \cap_{i \in I} B_i A$.

5. Let M be the finite state machine in Fig. 4.16. For states s_i, s_j, $0 \leq i, j \leq 2$, let $\mathbb{O}_{ij}$ denote the set of all output strings that M can produce as it goes from state s_i to state s_j. If $i = 2$, $j = 0$, for example, $\mathbb{O}_{20} = \{0\}(\{00\}^*\{1\}^*)^*$.
   Find $\mathbb{O}_{02}$, $\mathbb{O}_{22}$, $\mathbb{O}_{11}$, $\mathbb{O}_{00}$, $\mathbb{O}_{10}$.

Figure 4.16   Figure 4.17

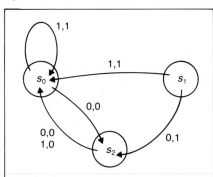   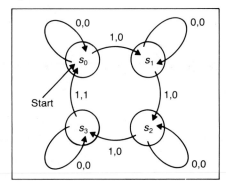

6. Let M be the finite state machine in Fig. 4.17.

   a) Find the state table for this machine.

   b) Explain what this machine does.

   c) How many distinct input strings x are there such that $\|x\| = 8$ and $\nu(s_0, x) = s_0$? How many are there with $\|x\| = 12$?

7. A finite state machine M is called *simply minimal* if, in its state table, no two output rows are identical. For $|S| = n$, $|\mathcal{I}| = m$, $|\mathbb{O}| = p$, show that there are

$$(n^{nm})[(p^m)(p^m - 1)(p^m - 2) \cdots (p^m - (n - 1))] = \frac{(n^{nm})(p^m)!}{(p^m - n)!}$$

   simply minimal machines possible.

8. Let $M = (S, \mathcal{I}, \mathbb{O}, \nu, \omega)$ be a finite state machine with $|S| = n$, and let $0 \in \mathcal{I}$.

   a) Show that for the input string $0000 \ldots$ the output is eventually periodic.

   b) What is the maximum number of 0's we can input before the periodic output starts?

   c) What is the length of the maximum period that can occur?

9. For the finite state machine in part (c) of Table 4.10, determine a transfer sequence from state s_0 to state s_5. Is your sequence unique?

10. Draw the state diagram for a finite state machine $M = (S, \mathcal{I}, \mathcal{O}, \nu, \omega)$, where $\mathcal{I} = \mathcal{O} = \{0, 1\}$, if for any $x \in \mathcal{I}^+$, M puts out its first 1 when it recognizes the substring 1111 and then puts out its second 1 when it recognizes the substring 0000, after which its output is constantly 0.

11. Although the state diagram seems more convenient than the state table when dealing with a finite state machine $M = (S, \mathcal{I}, \mathcal{O}, \nu, \omega)$, as the input strings get longer and the sizes of S, $\mathcal{I}$, $\mathcal{O}$ increase, the state table proves useful when simulating the machine on a computer. The block form of the table suggests the use of a matrix or two-dimensional array for storing ν, ω. Use this observation to develop a program that will simulate the machine in Table 4.11.

Table 4.11

	ν		ω	
	0	1	0	1
s_1	s_2	s_1	0	0
s_2	s_3	s_1	0	0
s_3	s_3	s_1	1	1

5

Relations:
The Second
Time Around

In Chapter 3 we introduced the concept of relation and then concentrated on special kinds of relations called functions. Returning to relations in this chapter, we will emphasize relations on a set A, i.e., subsets of $A \times A$. With the theory of languages and finite state machines from Chapter 4, we find many examples of relations on a set A, where A represents a set of strings from a given alphabet or a set of internal states from a finite state machine. Various properties of relations are developed, along with ways to represent finite relations for computer manipulation. Directed graphs reappear as a second way to represent such relations. Finally, among the many relations possible on a set A, there are two types of particular importance: equivalence relations and partial orders. Equivalence relations arise in many areas in mathematics. For the present we use an equivalence relation on the set of internal states in a finite state machine M to find a machine M_1, with as few internal states as possible, that performs whatever tasks M is capable of performing. The procedure is known as the minimization process.

5.1 ■ RELATIONS REVISITED: PROPERTIES OF RELATIONS

We start by recalling some previous ideas.

Definition 5.1

▶ If A, B are sets, a *relation from A to B* is any subset of $A \times B$. Subsets of $A \times A$ are called *relations on A*.

EXAMPLE 5.1

a) Define the relation $\mathcal{R}$ on the set $\mathbf{Z}$ by $a \mathcal{R} b$, or $(a, b) \in \mathcal{R}$, if $a \leq b$. This is the ordinary "is less than or equal to" relation on the set $\mathbf{Z}$ and can also be given on $\mathbf{Q}$ or $\mathbf{R}$, but not on $\mathbf{C}$.

b) Let $n \in \mathbf{Z}^+$. For $x, y \in \mathbf{Z}$, the *modulo n relation* $\mathcal{R}$ is defined by $x \mathcal{R} y$ iff $x - y$ is a multiple of n. With $n = 7$, we find $9 \mathcal{R} 2$, $-3 \mathcal{R} 11$, $(14, 0) \in \mathcal{R}$, but $3 \not\mathcal{R} 7$.   □

EXAMPLE 5.2

Let $\mathbf{\Sigma}$ be an alphabet, with language $A \subseteq \mathbf{\Sigma}^*$. For $x, y \in A$, define $x \mathcal{R} y$ if x is a prefix of y. Other relations can be defined on A by replacing "prefix" with either "suffix" or "substring."   □

EXAMPLE 5.3

Consider a finite state machine $M = (S, \mathscr{I}, \mathbb{O}, \nu, \omega)$.

a) For $s_1, s_2 \in S$, define $s_1 \mathcal{R} s_2$ if $\nu(s_1, x) = s_2$, for some $x \in \mathscr{I}$. Relation $\mathcal{R}$ establishes the *first level of reachability*.

b) The relation for the *second level of reachability* can also be given for S. Here $s_1 \mathcal{R} s_2$ if $\nu(s_1, x_1 x_2) = s_2$, for $x_1 x_2 \in \mathscr{I}^2$. This can be extended to higher levels, if the need arises. For the general *reachability* relation we have $\nu(s_1, y) = s_2$, for $y \in \mathscr{I}^*$.

c) Given $s_1, s_2 \in S$ the relation of *1-equivalence*, denoted by $s_1 \mathrm{E}_1 s_2$, and read "s_1 is 1-equivalent to s_2", is defined when $\omega(s_1, x) = \omega(s_2, x)$ for all $x \in \mathscr{I}$. $s_1 \mathrm{E}_1 s_2$

indicates that if machine M starts in either state s_1 or s_2 the output is the same for each element of $\mathcal{J}$. This idea can be extended to states being *k-equivalent,* where we write $s_1 \, E_k \, s_2$ if $\omega(s_1, y) = \omega(s_2, y)$, for all $y \in \mathcal{J}^k$. Here the same output is obtained for any string in $\mathcal{J}^k$, if we start at either s_1 or s_2.

If two states are *k*-equivalent for all $k \in \mathbf{Z}^+$, then they are called *equivalent.* We will look further into this idea later in the chapter.   □

We now proceed to an investigation of some of the properties a relation can satisfy.

Definition 5.2

▶ A relation $\mathcal{R}$ on a set A is called *reflexive* if for all $x \in A$, $(x, x) \in \mathcal{R}$.

We say that we want each element x of A to be related to itself, for $\mathcal{R}$ to be reflexive. All the relations in Examples 5.1 and 5.2 are reflexive, but only part (c) of Example 5.3 is reflexive. (What goes wrong in parts (a) and (b) of Example 5.3?)

EXAMPLE 5.4

For $A = \{1, 2, 3, 4\}$, a relation $\mathcal{R} \subseteq A \times A$ will be reflexive if $\mathcal{R} \supseteq \{(1, 1), (2, 2),$ $(3, 3), (4, 4)\}$. Consequently, $\mathcal{R}_1 = \{(1, 1), (2, 2), (3, 3)\}$ is not a reflexive relation on A, whereas $\mathcal{R}_2 = \{(x, y) \mid x, y \in A, x \leq y\}$ is reflexive on A.   □

EXAMPLE 5.5

Given a finite set A with $|A| = n$, we have $|A \times A| = n^2$, so there are 2^{n^2} relations on A. How many of these are reflexive?

If $A = \{a_1, a_2, \ldots, a_n\}$, a relation $\mathcal{R}$ on A is reflexive if $\{(a_i, a_i) \mid 1 \leq i \leq n\} \subseteq \mathcal{R}$. Considering the other $n^2 - n$ ordered pairs in $A \times A$ (those of the form (a_i, a_j), $1 \leq i, j \leq n, i \neq j$) as we construct a reflexive relation $\mathcal{R}$ on A, we either include or exclude each of these ordered pairs, so by the rule of product there are $2^{(n^2-n)}$ reflexive relations on A.   □

Definition 5.3

▶ Relation $\mathcal{R}$ on set A is called *symmetric* if $(x, y) \in \mathcal{R} \Rightarrow (y, x) \in \mathcal{R}$, for $x, y \in A$.

EXAMPLE 5.6

With $A = \{1, 2, 3\}$, we have:

a) $\mathcal{R}_1 = \{(1, 2), (2, 1), (1, 3), (3, 1)\}$, a symmetric, but not reflexive, relation on A;

b) $\mathcal{R}_2 = \{(1, 1), (2, 2), (3, 3), (2, 3)\}$, a reflexive, but not symmetric, relation on A;

c) $\mathcal{R}_3 = \{(1, 1), (2, 2), (3, 3)\}$ and $\mathcal{R}_4 = \{(1, 1), (2, 2), (3, 3), (2, 3), (3, 2)\}$, reflexive and symmetric relations on A;

d) $\mathcal{R}_5 = \{(1, 1), (2, 3), (3, 3)\}$, a relation on A that is neither reflexive nor symmetric.   □

To count the symmetric relations on $A = \{a_1, a_2, \ldots, a_n\}$, we write $A \times A$ as $A_1 \cup A_2$, where $A_1 = \{(a_i, a_i) \mid 1 \leq i \leq n\}$ and $A_2 = \{(a_i, a_j) \mid 1 \leq i, j \leq n, i \neq j\}$, so that every ordered pair in $A \times A$ is in exactly one of A_1, A_2. For A_2, $|A_2| = |A \times A| - |A_1| = n^2 - n = n(n - 1)$, an even integer. The set A_2 contains $(1/2)(n^2 - n)$ subsets of the form $\{(a_i, a_j), (a_j, a_i)\}$, $1 \leq i < j \leq n$. In constructing a symmetric relation $\mathcal{R}$ on A, for each ordered pair in A_1 we have our usual choice of exclusion or inclusion. For each of the $(1/2)(n^2 - n)$ subsets of ordered pairs in A_2 we have the same two choices. So by the rule of product there are $2^n \cdot 2^{(1/2)(n^2-n)} = 2^{(1/2)(n^2+n)}$ symmetric relations on A.

In counting those relations on A that are both reflexive and symmetric, we have only one choice for each ordered pair in A_1. So we have $2^{(1/2)(n^2-n)}$ relations on A that are both reflexive and symmetric.

Definition 5.4

▶ For a set A, a relation $\mathcal{R}$ on A is called *transitive* if $(x, y), (y, z) \in \mathcal{R} \Rightarrow (x, z) \in \mathcal{R}$. (So if x "is related to" y, and y "is related to" z, we want x "related to" z, with y playing the role of "intermediary".)

EXAMPLE 5.7

All the relations in Examples 5.1 and 5.2 are transitive, as are the relations in Example 5.3(c). □

EXAMPLE 5.8

Define the relation $\mathcal{R}$ on the set $\mathbf{Z}^+$ by $a \mathcal{R} b$ if a divides b, i.e., for some $c \in \mathbf{Z}^+$, $b = ca$. Now if $x \mathcal{R} y$ and $y \mathcal{R} z$, do we have $x \mathcal{R} z$? $x \mathcal{R} y \Rightarrow y = sx, s \in \mathbf{Z}^+$; $y \mathcal{R} z \Rightarrow z = ty, t \in \mathbf{Z}^+$. Consequently, $z = ty = t(sx) = (ts)x$, $ts \in \mathbf{Z}^+$, so $x \mathcal{R} z$ and $\mathcal{R}$ is transitive. In addition, $\mathcal{R}$ is reflexive, but not symmetric, since $2 \mathcal{R} 6$ but $6 \not{\mathcal{R}} 2$. □

EXAMPLE 5.9

If $A = \{1, 2, 3, 4\}$, then $\mathcal{R}_1 = \{(1, 1), (2, 3), (3, 4), (2, 4)\}$ is a transitive relation on A, while $\mathcal{R}_2 = \{(1, 3), (3, 2)\}$ is not transitive since $(1, 2) \notin \mathcal{R}_2$. □

At this point the reader is probably ready to start counting the number of transitive relations on a finite set. This is harder than the previous counting arguments, but by the end of this chapter we shall have the necessary ideas to count the relations $\mathcal{R}$ on a finite set, where $\mathcal{R}$ is reflexive, symmetric, and transitive.

We consider one last property for relations.

Definition 5.5

▶ Given a relation $\mathcal{R}$ on a set A, $\mathcal{R}$ is called *antisymmetric* if $a \mathcal{R} b$, $b \mathcal{R} a \Rightarrow a = b$. (Here the only way we can have a "related to" b and b "related to" a is if a and b are one and the same element from A.)

EXAMPLE 5.10

For a given universe $\mathcal{U}$, define the relation $\mathcal{R}$ on $\mathcal{P}(\mathcal{U})$ by $(A, B) \in \mathcal{R}$ if $A \subseteq B$, for $A, B \subseteq \mathcal{U}$. So $\mathcal{R}$ is the subset relation of Chapter 2 and if $A \mathcal{R} B$ and $B \mathcal{R} A$, then we have $A \subseteq B$, $B \subseteq A$, which is equivalent to $A = B$. Consequently, this relation is antisymmetric, as well as reflexive and transitive, but it is not symmetric. □

Before we are led astray into thinking that "not symmetric" is synonymous with "antisymmetric", let us consider the following.

EXAMPLE 5.11

For $A = \{1, 2, 3\}$, relation $\mathcal{R}$ on A given by $\mathcal{R} = \{(1, 2), (2, 1), (2, 3)\}$ is not symmetric because $(3, 2) \notin \mathcal{R}$, and it is not antisymmetric, either, since $(1, 2), (2, 1) \in \mathcal{R}$, but $1 \neq 2$. The relation $\mathcal{R}_1 = \{(1, 1), (2, 2)\}$ is both symmetric and antisymmetric.
How many relations on A are antisymmetric?
Writing

$$A \times A = \{(1, 1), (2, 2), (3, 3)\} \cup \{(1, 2), (2, 1), (1, 3), (3, 1), (2, 3), (3, 2)\},$$

we make two observations as we try to construct an antisymmetric relation $\mathcal{R}$ on A.

1. Any element $(x, x) \in A \times A$ can either be included or excluded with no concern about whether or not $\mathcal{R}$ is antisymmetric.

2. For an element of the form (x, y), $x \neq y$, we must consider both (x, y) and (y, x) and note that for $\mathcal{R}$ to remain antisymmetric we have three alternatives: (a) place (x, y) in $\mathcal{R}$: (b) place (y, x) in $\mathcal{R}$; (c) place neither (x, y) nor (y, x) in $\mathcal{R}$. (What happens if we place both (x, y) and (y, x) in $\mathcal{R}$?)

So by the rule of product the number of antisymmetric relations on A is $(2^3)(3^3) = (2^3)(3^{(3^2-3)/2})$. If $|A| = n > 0$, then there are $(2^n)(3^{(n^2-n)/2})$ antisymmetric relations on A. □

At this point we have seen the four major properties that arise in the study of relations. Before closing this section we define two more notions, each of which involves three of these four properties.

Definition 5.6 ▶ A relation $\mathcal{R}$ on a set A is called a *partial order*, or *partial ordering relation*, if $\mathcal{R}$ is reflexive, antisymmetric, and transitive.

EXAMPLE 5.12 The relation in Example 5.1(a) is a partial order, but the relation in part (b) is not. All the relations of Example 5.2 are partial orders, as is the subset relation of Example 5.10. □

Definition 5.7 ▶ An *equivalence relation* $\mathcal{R}$ on a set A is a relation that is reflexive, symmetric, and transitive.

EXAMPLE 5.13 The relation in Example 5.1(b) and all the relations in Example 5.3(c) are equivalence relations.

For any set A, $A \times A$ is an equivalence relation on A, and if $A = \{a_1, a_2, \ldots, a_n\}$, $\mathcal{R} = \{(a_i, a_i) \,|\, 1 \leq i \leq n\}$ is the smallest equivalence relation on A.

If $\mathcal{R}$ is a relation on a set A, then $\mathcal{R}$ is both an equivalence relation and a partial order on A iff $\mathcal{R}$ is the equality relation on A. □

EXERCISES

1. If $A = \{1, 2, 3, 4\}$, give an example of a relation $\mathcal{R}$ on A that is

   a) reflexive and symmetric, but not transitive.

   b) reflexive and transitive, but not symmetric.

   c) symmetric and transitive, but not reflexive.

2. For relation (b) in Example 5.1, determine five values of x for which $(x, 5) \in \mathcal{R}$.

3. For each of the following relations, determine whether the relation is reflexive, symmetric, antisymmetric, or transitive.

   a) $\mathcal{R} \subseteq \mathbf{Z}^+ \times \mathbf{Z}^+$ where $a \,\mathcal{R}\, b$ if $a \,|\, b$ (read "a divides b").

   b) $\mathcal{R}$ is the relation on $\mathbf{Z}$ where $a \,\mathcal{R}\, b$ if $a \,|\, b$.

c) For a given universe $\mathcal{U}$ and $C \subseteq \mathcal{U}$, define $\mathcal{R}$ on $\mathcal{P}(\mathcal{U})$ as follows: for $A, B \subseteq \mathcal{U}$, define $A \mathcal{R} B$ if $A \cap C = B \cap C$.

d) On the set A of all lines in $\mathbf{R}^2$, define the relation $\mathcal{R}$ for two lines ℓ_1, ℓ_2 by $\ell_1 \mathcal{R} \ell_2$ if ℓ_1 is perpendicular to ℓ_2.

e) $\mathcal{R}$ is the relation on $\mathbf{Z}$ where $x \mathcal{R} y$ if $x + y$ is even (odd).

f) $\mathcal{R}$ is the relation on $\mathbf{Z}$ where $x \mathcal{R} y$ if $x - y$ is even (odd).

g) Let T be the set of all triangles in $\mathbf{R}^2$. Define $\mathcal{R}$ on T by $t_1 \mathcal{R} t_2$ if t_1 and t_2 have an angle of the same measure.

h) $\mathcal{R}$ is the relation on $\mathbf{Z} \times \mathbf{Z}$ where $(a, b) \mathcal{R} (c, d)$ if $a \leq c$. (*Note:* $\mathcal{R} \subseteq (\mathbf{Z} \times \mathbf{Z}) \times (\mathbf{Z} \times \mathbf{Z})$.)

4. Which relations in Exercise 3 are partial orders? Which are equivalence relations?

5. a) Let $\mathcal{R}_1$, $\mathcal{R}_2$ be relations on a set A. Prove or disprove that $\mathcal{R}_1$, $\mathcal{R}_2$ reflexive $\Rightarrow$ $\mathcal{R}_1 \cap \mathcal{R}_2$ reflexive.

   b) Answer part (a) when each occurrence of reflexive is replaced by (i) symmetric; (ii) antisymmetric; (iii) transitive.

6. Answer Exercise 5, replacing each occurrence of $\cap$ by $\cup$.

7. For each of the following statements about relations on a set A, where $|A| = n$, determine whether the statement is true or false. If false, give a counterexample.

   a) If $\mathcal{R}$ is a reflexive relation on A, then $|\mathcal{R}| \geq n$.

   b) If $\mathcal{R}$ is a relation on A and $|\mathcal{R}| \geq n$, then $\mathcal{R}$ is reflexive.

   c) If $\mathcal{R}_1$, $\mathcal{R}_2$ are relations on A and $\mathcal{R}_2 \supseteq \mathcal{R}_1$, then $\mathcal{R}_1$ reflexive (symmetric, antisymmetric, transitive) $\Rightarrow$ $\mathcal{R}_2$ reflexive (symmetric, antisymmetric, transitive).

   d) If $\mathcal{R}_1$, $\mathcal{R}_2$ are relations on A and $\mathcal{R}_2 \supseteq \mathcal{R}_1$, then $\mathcal{R}_2$ reflexive (symmetric, antisymmetric, transitive) $\Rightarrow$ $\mathcal{R}_1$ reflexive (symmetric, antisymmetric, transitive).

   e) If $\mathcal{R}$ is an equivalence relation on A, then $n \leq |\mathcal{R}| \leq n^2$.

8. If $A = \{w, x, y, z\}$, determine the number of relations on A that are (a) reflexive; (b) symmetric; (c) reflexive and symmetric; (d) reflexive and contain (x, y); (e) symmetric and contain (x, y); (f) antisymmetric; (g) antisymmetric and contain (x, y); (h) symmetric and antisymmetric; (i) reflexive, symmetric and antisymmetric.

9. What is wrong with the following argument?

   If A is a set with $\mathcal{R}$ a relation on A that is symmetric and transitive, then $\mathcal{R}$ is reflexive.

   Proof: Let $(x, y) \in \mathcal{R}$. By the symmetric property, $(y, x) \in \mathcal{R}$. Then with (x, y), $(y, x) \in \mathcal{R}$, it follows by the transitive property that $(x, x) \in \mathcal{R}$. Consequently, $\mathcal{R}$ is reflexive.

10. Let A be a set with $|A| = n$, and let $\mathcal{R}$ be a relation on A that is antisymmetric. What is the maximum value for $|\mathcal{R}|$? How many antisymmetric relations can have this size?

11. Let A be a set with $|A| = n$, and let $\mathcal{R}$ be an equivalence relation on A with $|\mathcal{R}| = r$. Why is $r - n$ always even?

5.2 ■ COMPUTER RECOGNITION: ZERO-ONE MATRICES AND DIRECTED GRAPHS

As our interest in relations is centered on those for finite sets, we are concerned with ways of representing such relations so that the properties of Section 5.1 can be identified. For this reason we now develop the necessary tools: relation composition, zero-one matrices, and directed graphs.

Analogous to the composition of functions, relations can be combined in the following circumstances.

Definition 5.8 ▶ If A, B, C are sets with $\mathcal{R}_1 \subseteq A \times B$ and $\mathcal{R}_2 \subseteq B \times C$, the *composite relation* $\mathcal{R}_1 \circ \mathcal{R}_2$ is a relation from A to C defined by $\mathcal{R}_1 \circ \mathcal{R}_2 = \{(x, z) \mid x \in A, z \in C,$ and there exists $y \in B$ with $(x, y) \in \mathcal{R}_1$, $(y, z) \in \mathcal{R}_2\}$.

Beware! The composition of two relations is written in an order opposite to that for function composition. We will see why shortly.

EXAMPLE 5.14 Let $A = \{1, 2, 3, 4\}$, $B = \{w, x, y, z\}$, $C = \{5, 6, 7\}$. If $\mathcal{R}_1 = \{(1, x), (2, x), (3, y), (3, z)\}$ is a relation from A to B and $\mathcal{R}_2 = \{(w, 5), (x, 6)\}$ is a relation from B to C, then $\mathcal{R}_1 \circ \mathcal{R}_2 = \{(1, 6), (2, 6)\}$, a relation from A to C. If $\mathcal{R}_3 = \{(w, 5), (w, 6)\}$, then $\mathcal{R}_1 \circ \mathcal{R}_3 = \emptyset$. □

EXAMPLE 5.15 Let A be the set of employees at a computing center, while B denotes a set of high-level programming languages and C is a list of projects $\{p_1, p_2, \ldots, p_8\}$ for which work assignments must be made using the people in A. Consider $\mathcal{R}_1 \subseteq A \times B$, where an ordered pair of the form (L. Alldredge, Pascal) indicates that employee L. Alldredge is proficient in Pascal (and perhaps other programming languages). The relation $\mathcal{R}_2 \subseteq B \times C$ consists of ordered pairs of the form (Pascal, p_2) where Pascal is considered an essential language needed to work on project p_2. In the composite relation $\mathcal{R}_1 \circ \mathcal{R}_2$ we find (L. Alldredge, p_2). If no other ordered pair in $\mathcal{R}_2$ has p_2 as its second component, we would know that if L. Alldredge was assigned to p_2 it was solely on the basis of his proficiency in Pascal. (Here $\mathcal{R}_1 \circ \mathcal{R}_2$ has been used to set up a matching process between employees and projects, based on knowledge of specific programming languages.) □

Comparable to the associative law for function composition, we find the following result.

Theorem 5.1 ▶ Let A, B, C, D be sets with $\mathcal{R}_1 \subseteq A \times B$, $\mathcal{R}_2 \subseteq B \times C$, $\mathcal{R}_3 \subseteq C \times D$. Then $\mathcal{R}_1 \circ (\mathcal{R}_2 \circ \mathcal{R}_3) = (\mathcal{R}_1 \circ \mathcal{R}_2) \circ \mathcal{R}_3$.

Proof Since both $\mathcal{R}_1 \circ (\mathcal{R}_2 \circ \mathcal{R}_3)$ and $(\mathcal{R}_1 \circ \mathcal{R}_2) \circ \mathcal{R}_3$ are relations from A to D, there is some reason to believe they are equal. If $(a, d) \in \mathcal{R}_1 \circ (\mathcal{R}_2 \circ \mathcal{R}_3)$, then there is an element

$b \in B$ with $(a, b) \in \mathcal{R}_1$ and $(b, d) \in (\mathcal{R}_2 \circ \mathcal{R}_3)$. Also, $(b, d) \in (\mathcal{R}_2 \circ \mathcal{R}_3) \Rightarrow$ $(b, c) \in \mathcal{R}_2$, $(c, d) \in \mathcal{R}_3$ for some $c \in C$. Then $(a, b) \in \mathcal{R}_1$, $(b, c) \in \mathcal{R}_2 \Rightarrow$ $(a, c) \in \mathcal{R}_1 \circ \mathcal{R}_2$. Finally, $(a, c) \in \mathcal{R}_1 \circ \mathcal{R}_2$, $(c, d) \in \mathcal{R}_3 \Rightarrow (a, d) \in (\mathcal{R}_1 \circ \mathcal{R}_2) \circ \mathcal{R}_3$, and $\mathcal{R}_1 \circ (\mathcal{R}_2 \circ \mathcal{R}_3) \subseteq (\mathcal{R}_1 \circ \mathcal{R}_2) \circ \mathcal{R}_3$. The opposite inclusion follows by similar reasoning. ∎

As a result of this theorem no ambiguity arises if we write $\mathcal{R}_1 \circ \mathcal{R}_2 \circ \mathcal{R}_3$ for either of the relations in Theorem 5.1. In addition, we can now define the powers of a relation $\mathcal{R}$ on a set.

Definition 5.9

▶ Given a set A and a relation $\mathcal{R}$ on A we define the *powers of* $\mathcal{R}$ recursively by (a) $\mathcal{R}^1 = \mathcal{R}$; (b) for $n \in \mathbf{Z}^+$, $\mathcal{R}^{n+1} = \mathcal{R} \circ \mathcal{R}^n$.

Note that for $n \in \mathbf{Z}^+$, $\mathcal{R}^n$ is a relation on A.

EXAMPLE 5.16

If $A = \{1, 2, 3, 4\}$ and $\mathcal{R} = \{(1, 2), (1, 3), (2, 4), (3, 2)\}$, then $\mathcal{R}^2 = \{(1, 4), (1, 2), (3, 4)\}$, $\mathcal{R}^3 = \{(1, 4)\}$, and for $n \geq 4$, $\mathcal{R}^n = \emptyset$. □

As the set A and relation $\mathcal{R}$ on A grow larger, calculations such as those in Example 5.16 become tedious. We seek a way to avoid this tedium. The needed tool is the computer, once a way can be found to tell the machine about the set A and the relation $\mathcal{R}$ on A.

Definition 5.10

▶ An $m \times n$ *zero-one matrix* $E = (e_{ij})_{m \times n}$ is a rectangular array of numbers in m rows and n columns, where e_{ij}, $1 \leq i \leq m$, $1 \leq j \leq n$, denotes the entry in the ith row and jth column of E, and each such entry is 0 or 1. (We can also write $(0, 1)$-matrix for this type of matrix.)

EXAMPLE 5.17

$$\text{The matrix } E = \begin{bmatrix} 1 & 0 & 0 & 1 \\ 0 & 1 & 0 & 1 \\ 1 & 0 & 0 & 0 \end{bmatrix}$$

is a 3×4 $(0, 1)$-matrix where, for example, $e_{11} = 1$, $e_{23} = 0$, $e_{31} = 1$. □

In working with these matrices we use the standard operations of matrix addition and multiplication, but *with the stipulation that* $1 + 1 = 1$. (Hence the addition is called Boolean.)

EXAMPLE 5.18

Let $A = \{1, 2, 3, 4\}$ and $\mathcal{R} = \{(1, 2), (1, 3), (2, 4), (3, 2)\}$, as in Example 5.16. Keeping the order of the elements in A fixed, we define the *relation matrix* for $\mathcal{R}$, denoted $M(\mathcal{R})$, as follows: $M(\mathcal{R})$ is a 4×4 $(0, 1)$-matrix whose entries m_{ij}, $1 \leq i, j \leq 4$, are given by:

$$m_{ij} = \begin{cases} 1, & \text{if } (i, j) \in \mathcal{R}, \\ 0, & \text{otherwise.} \end{cases}$$

In this case we find that

$$M(\mathcal{R}) = \begin{bmatrix} 0 & 1 & 1 & 0 \\ 0 & 0 & 0 & 1 \\ 0 & 1 & 0 & 0 \\ 0 & 0 & 0 & 0 \end{bmatrix}.$$

Now how can this be of any use? If we compute $(M(\mathcal{R}))^2$ using the conventions above, we find that

$$(M(\mathcal{R}))^2 = \begin{bmatrix} 0 & 1 & 0 & 1 \\ 0 & 0 & 0 & 0 \\ 0 & 0 & 0 & 1 \\ 0 & 0 & 0 & 0 \end{bmatrix},$$

which happens to be the relation matrix for $\mathcal{R} \circ \mathcal{R} = \mathcal{R}^2$. (Check Example 5.16). Furthermore,

$$(M(\mathcal{R}))^4 = \begin{bmatrix} 0 & 0 & 0 & 0 \\ 0 & 0 & 0 & 0 \\ 0 & 0 & 0 & 0 \\ 0 & 0 & 0 & 0 \end{bmatrix},$$

the relation matrix for $\mathcal{R}^4 = \emptyset$.   □

What has happened here carries over to the general situation. We now state, without proof, some results about relation matrices and their use in studying relations. (The proof (for part c) requires the principle of mathematical induction, which is developed in Section 6.1.)

Let A be a set with $|A| = n$ and $\mathcal{R}$ a relation on A. If $M(\mathcal{R})$ is the relation matrix for $\mathcal{R}$, then

a) $M(\mathcal{R}) = \mathbf{0}$ (the matrix of all 0's) iff $\mathcal{R} = \emptyset$;

b) $M(\mathcal{R}) = \mathbf{1}$ (the matrix of all 1's) iff $\mathcal{R} = A \times A$; and

c) $M(\mathcal{R}^m) = [M(\mathcal{R})]^m$, for $m \in \mathbf{Z}^+$.

Another property of relation matrices is demonstrated in the following.

EXAMPLE 5.19 Consider the sets A, B, C and the relations $\mathcal{R}_1$, $\mathcal{R}_2$ of Example 5.14. With the orders of A, B, C fixed as in that example, the relation matrices for $\mathcal{R}_1$, $\mathcal{R}_2$ are

$$M(\mathcal{R}_1) = \begin{array}{c} \\ (1) \\ (2) \\ (3) \\ (4) \end{array}\begin{array}{cccc} (w) & (x) & (y) & (z) \\ \begin{bmatrix} 0 & 1 & 0 & 0 \\ 0 & 1 & 0 & 0 \\ 0 & 0 & 1 & 1 \\ 0 & 0 & 0 & 0 \end{bmatrix} \end{array}, \qquad M(\mathcal{R}_2) = \begin{array}{c} \\ (w) \\ (x) \\ (y) \\ (z) \end{array}\begin{array}{ccc} (5) & (6) & (7) \\ \begin{bmatrix} 1 & 0 & 0 \\ 0 & 1 & 0 \\ 0 & 0 & 0 \\ 0 & 0 & 0 \end{bmatrix} \end{array}$$

In finding $M(\mathcal{R}_1)$, we are dealing with a relation from A to B, so the elements of A are used to mark the rows of $M(\mathcal{R}_1)$ while the elements of B designate the columns. Then to denote, for example, that $(2, x) \in \mathcal{R}_1$, we place a 1 in the row marked (2) and the column marked (x). The same idea is used to obtain $M(\mathcal{R}_2)$.

Multiplying these matrices:

$$M(\mathcal{R}_1) \cdot M(\mathcal{R}_2) = \begin{bmatrix} 0 & 1 & 0 & 0 \\ 0 & 1 & 0 & 0 \\ 0 & 0 & 1 & 1 \\ 0 & 0 & 0 & 0 \end{bmatrix}\begin{bmatrix} 1 & 0 & 0 \\ 0 & 1 & 0 \\ 0 & 0 & 0 \\ 0 & 0 & 0 \end{bmatrix} = \begin{bmatrix} 0 & 1 & 0 \\ 0 & 1 & 0 \\ 0 & 0 & 0 \\ 0 & 0 & 0 \end{bmatrix} = M(\mathcal{R}_1 \circ \mathcal{R}_2),$$

and in general we have: if $\mathcal{R}_1$ is a relation from A to B and $\mathcal{R}_2$ is a relation from B to C, then $M(\mathcal{R}_1) \cdot M(\mathcal{R}_2) = M(\mathcal{R}_1 \circ \mathcal{R}_2)$. That is, the product of the relation matrices for $\mathcal{R}_1$, $\mathcal{R}_2$, in that order, equals the relation matrix of the composite relation $\mathcal{R}_1 \circ \mathcal{R}_2$. (It was because of this that the composition of two relations was written in the order specified in Definition 5.8.) □

Using the $(0, 1)$-matrix for a relation, we turn to the recognition of the reflexive, symmetric, antisymmetric, and transitive properties.

Definition 5.11 ▶ Let $E = (e_{ij})_{m \times n}$, $F = (f_{ij})_{m \times n}$ be two $(0, 1)$-matrices. We say that E *precedes*, or *is less than*, F, and write $E \le F$, if $e_{ij} \le f_{ij}$, $1 \le i \le m$, $1 \le j \le n$.

EXAMPLE 5.20 With $E = \begin{bmatrix} 1 & 0 & 1 \\ 0 & 0 & 1 \end{bmatrix}$ and $F = \begin{bmatrix} 1 & 0 & 1 \\ 0 & 1 & 1 \end{bmatrix}$, we have $E \le F$. In fact there are eight $(0, 1)$-matrices F for which $E \le F$. □

Definition 5.12 ▶ For $n \in \mathbf{Z}^+$, $I_n = (\delta_{ij})_{n \times n}$ is the $n \times n$ $(0, 1)$-matrix where

$$\delta_{ij} = \begin{cases} 1, & \text{if } i = j \\ 0, & \text{if } i \ne j. \end{cases}$$

Definition 5.13 ▶ Let $A = (a_{ij})_{m \times n}$ be a $(0, 1)$-matrix. The *transpose* of A, written A^{tr}, is the matrix $(a^*_{ji})_{n \times m}$ where for $1 \le j \le n$, $1 \le i \le m$, $a^*_{ji} = a_{ij}$.

EXAMPLE 5.21 For $A = \begin{bmatrix} 0 & 1 \\ 0 & 0 \\ 1 & 1 \end{bmatrix}$, $A^{tr} = \begin{bmatrix} 0 & 0 & 1 \\ 1 & 0 & 1 \end{bmatrix}$.

As this example demonstrates, the ith row (column) of A equals the ith column (row) of A^{tr}. This indicates a way to obtain A^{tr} from A.   □

Theorem 5.2

▶   Given a set A with $|A| = n$, and a relation $\mathcal{R}$ on A, let M denote the relation matrix for $\mathcal{R}$. Then

a) $\mathcal{R}$ is reflexive iff $I_n \leq M$.   b) $\mathcal{R}$ is symmetric iff $M = M^{tr}$.

c) $\mathcal{R}$ is transitive iff $M \cdot M = M^2 \leq M$.

d) $\mathcal{R}$ is antisymmetric iff $M \cap M^{tr} \leq I_n$. (The matrix $M \cap M^{tr}$ is formed by operating on corresponding entries in M and M^{tr} according to the rules: $0 \cap 0 = 0 \cap 1 = 1 \cap 0 = 0, 1 \cap 1 = 1$.)

Proof

The results follow from the definitions of the relation properties and the $(0, 1)$-matrix. We demonstrate this for part (c), using the elements of A to designate the rows and columns in M, as in Example 5.19.

   Let $M^2 \leq M$. If $(x, y), (y, z) \in \mathcal{R}$, then there are 1's in row (x), column (y) and in row (y), column (z) of M. Consequently, in row (x), column (z) of M^2 there is a 1. This 1 must also occur in row (x), column (z) of M, since $M^2 \leq M$. Hence $(x, z) \in \mathcal{R}$ and $\mathcal{R}$ is transitive.

   Conversely, if $\mathcal{R}$ is transitive and M is the relation matrix for $\mathcal{R}$, let s_{xz} be the entry in row (x) and column (z) of M^2, with $s_{xz} = 1$. For s_{xz} to equal 1 in M^2, there must exist at least one $y \in A$ where $m_{xy} = m_{yz} = 1$ in M. This happens only if $x \mathcal{R} y$ and $y \mathcal{R} z$. With $\mathcal{R}$ transitive it follows that $x \mathcal{R} z$. So $m_{xz} = 1$ and $M^2 \leq M$.

   The proofs of the remaining parts are left to the reader.   ■

   The relation matrix is a useful tool for the computer recognition of certain properties of relations. Storing information as described here, this matrix is an example of a *data structure*. Of interest is how the relation matrix is used in the study of graph theory,[†] and how graph theory is used in the recognition of certain properties of relations.

Definition 5.14

▶   Let V be a finite set. A *directed graph* (or *digraph*) G on V is made up of the elements of V, called the *vertices* or *nodes* of G, and a subset E of $V \times V$, called the *edges*, or *arcs*, of G. If $a, b \in V$ and $(a, b) \in E^{‡}$, then there is an edge *from a to b*. Vertex a is called the *origin* or *source* of the edge, with b the *terminus*, or *terminating vertex*, and we say that b is *adjacent from a*, and a is *adjacent to b*. In addition, if $a \neq b$ then $(a, b) \neq (b, a)$. Also, an edge of the form (a, a) is called a *loop* (at a).

EXAMPLE 5.22

For $V = \{1, 2, 3, 4, 5\}$, the diagram in Fig. 5.1 is a directed graph G on V with edge set $\{(1, 1), (1, 2), (1, 4), (3, 2)\}$. Vertex 5 is a part of this graph even though it is not the origin or terminus of an edge. It is referred to as an *isolated* vertex. As we see here, edges need not be straight line segments and there is no concern about the length of an edge.   □

[†]Since the terminology of graph theory is not standard, the reader may find some differences between definitions here and in other texts.

[‡]In this chapter we allow only one edge from a to b. Situations where multiple edges occur are called *multigraphs*. These are studied in Chapter 14.

Fig. 5.1

Figure 5.2

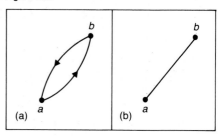

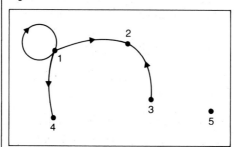

(a)   (b)

When we develop a *flowchart* to study a computer program or algorithm, we deal with a special type of directed graph where the shapes of the vertices are of importance in the algorithm analysis. Road maps, too, are directed graphs, where the cities and towns are represented by vertices, and the highways linking any two localities are given by edges. In road maps, an edge is often directed in both directions. Consequently, if *G* is a directed graph and *a*, *b* ∈ *V*, *a* ≠ *b*, with both (a, b), $(b, a) \in E$, the single undirected edge, $\{a, b\} = \{b, a\}$, in Fig. 5.2(b) is used to represent the two directed edges shown in Fig. 5.2(a). In this case, *a* and *b* are called *adjacent* vertices. (Directions may also be disregarded for loops.)

How are directed graphs used in the study of relations? For a set *A*, with relation $\mathcal{R}$ on *A*, we construct a directed graph *G* with vertex set *A* and edge set $E \subseteq A \times A$, where $(a, b) \in E$ if *a*, *b* ∈ *A* and *a* $\mathcal{R}$ *b*.

EXAMPLE 5.23   For $A = \{1, 2, 3, 4\}$, let $\mathcal{R} = \{(1, 1), (1, 2), (2, 3), (3, 2), (3, 3), (3, 4), (4, 2)\}$. The directed graph associated with $\mathcal{R}$ is shown in Fig. 5.3(a). If the directions are ignored we get the *associated undirected graph* in (b) of the figure. Here we see that the graph is *connected* in the sense that for any two vertices *x*, *y*, with *x* ≠ *y*, there is a *path* starting at *x* and ending at *y*. Such a path consists of a *finite sequence of undirected edges*, so that the edges $\{1, 2\}$, $\{2, 4\}$ provide a path from 1 to 4, while $\{3, 4\}$, $\{4, 2\}$, $\{2, 1\}$ provide a path from 3 to 1. The sequence of edges $\{3, 4\}$, $\{4, 2\}$, $\{2, 3\}$ provides a path from 3 to 3. Such a path, although not a loop, is called a *cycle*. This is an example of an undirected cycle of *length* three, since it has three edges in it. For paths

Figure 5.3

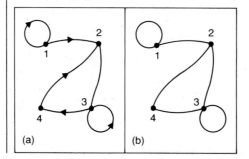

(a)   (b)

and cycles in both undirected and directed graphs, the adjective "simple" is used if the path passes through a vertex at most once; for a simple cycle we are allowed to terminate at the vertex we started from.

Note that loops are cycles of length one and have no bearing on graph connectivity.

□

Definition 5.15

▶ If G is a directed graph on V, then G is said to be *strongly connected* if for all $x, y \in V$, $x \neq y$, there is a path of *directed* edges from x to y.

It is in this sense that we talked about strongly connected machines in Chapter 4. The graph in Fig. 5.3(a) is connected but not strongly connected. For example, there is no directed path from 3 to 1. In Fig. 5.4 the directed graph on $V = \{1, 2, 3, 4\}$ is strongly connected and *loop-free*.

Figure 5.4

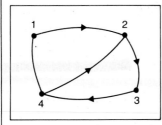

Figure 5.5

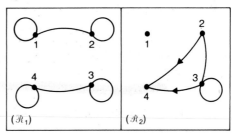

$(\mathcal{R}_1)$ $(\mathcal{R}_2)$

EXAMPLE 5.24

For $A = \{1, 2, 3, 4\}$, consider the relations $\mathcal{R}_1 = \{(1, 1), (1, 2), (2, 1), (2, 2), (3, 3), (3, 4), (4, 3), (4, 4)\}$ and $\mathcal{R}_2 = \{(2, 4), (2, 3), (3, 2), (3, 3), (3, 4)\}$. As seen in Fig. 5.5, the graphs of these relations are *disconnected*. However, each graph is the union of two connected pieces called the *components* of the graph. For $\mathcal{R}_1$ the graph is made up of two strongly connected components. For $\mathcal{R}_2$, one component consists of an isolated vertex and the other component is connected but not strongly connected.

□

EXAMPLE 5.25

The graphs in Fig. 5.6 are examples of undirected graphs that are loop-free and have an edge for every pair of distinct vertices. These graphs are called the *complete graphs* on n vertices, and are denoted by K_n. In Fig. 5.6 we have examples of the complete

Figure 5.6

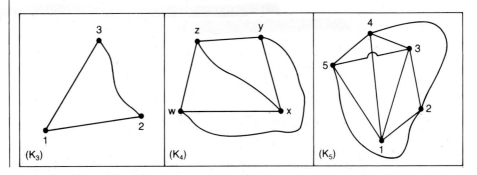

(K_3) (K_4) (K_5)

graphs on three, four, and five vertices, respectively. The complete graph K_2 consists of two points x, y and an edge connecting them, while K_1 consists of one point and no edges, since no loops are allowed.

In K_5, two edges cross, namely $\{3, 5\}$ and $\{1, 4\}$. However, there is no point of intersection creating a new vertex. If we try to avoid the crossing of edges by drawing the graph differently, we will run into the same problem over again. This difficulty will be examined in Chapter 14 when we deal with the planarity of graphs.   □

For a graph G on a vertex set V, the graph gives rise to a relation $\mathcal{R}$ on V where $x \mathcal{R} y$, if (x, y) is an edge in G. Consequently, there is a $(0, 1)$-matrix for G and since this relation matrix comes about from the adjacencies of pairs of vertices, it is referred to as the *adjacency matrix* for G as well as the relation matrix for $\mathcal{R}$.

At this point we tie together the properties of relations and the structure of directed graphs.

EXAMPLE 5.26   If $A = \{1, 2, 3\}$, and $\mathcal{R} = \{(1, 1), (1, 2), (2, 2), (3, 3), (3, 1)\}$, then $\mathcal{R}$ is a reflexive relation on A, but it is neither symmetric nor transitive. The directed graph associated with $\mathcal{R}$ consists of five edges. Three of these edges are loops which result from the reflexive property of $\mathcal{R}$. (See Fig. 5.7.) In general, if $\mathcal{R}$ is a relation on a finite set A, then $\mathcal{R}$ is reflexive iff its directed graph contains a loop at each vertex (element of A).   □

Figure 5.7   Figure 5.8

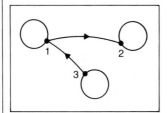

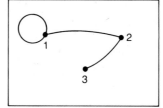

EXAMPLE 5.27   The relation $\mathcal{R} = \{(1, 1), (1, 2), (2, 1), (2, 3), (3, 2)\}$ is symmetric on $A = \{1, 2, 3\}$ but it is neither reflexive nor transitive. The directed graph for $\mathcal{R}$ is found in Fig. 5.8. In general, a relation $\mathcal{R}$ on a finite set A is symmetric iff its directed graph contains only loops and undirected edges.   □

Figure 5.9

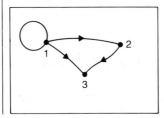

EXAMPLE 5.28

For $A = \{1, 2, 3\}$, consider $\mathcal{R} = \{(1, 1), (1, 2), (2, 3), (1, 3)\}$. The directed graph for $\mathcal{R}$ is shown in Fig. 5.9. Here $\mathcal{R}$ is transitive but not reflexive or symmetric. The directed graph indicates that a relation on a set A will be transitive iff its directed graph satisfies the following: For any x, $y \in A$, if there is a (directed) path from x to y in the associated graph, then there is an edge (x, y) also. (Here $(1, 2)$, $(2, 3)$ is a (directed) path from 1 to 3, and we also have the edge $(1, 3)$ for transitivity.) □

Our final example deals with equivalence relations.

EXAMPLE 5.29

For $A = \{1, 2, 3, 4, 5\}$, the following are equivalence relations on A:

$$\mathcal{R}_1 = \{(1, 1), (1, 2), (2, 1), (2, 2), (3, 3), (3, 4), (4, 3), (4, 4), (5, 5)\},$$

$$\mathcal{R}_2 = \{(1, 1), (1, 2), (1, 3), (2, 1), (2, 2), (2, 3), (3, 1), (3, 2), (3, 3), (4, 4), (4, 5),$$
$$(5, 4), (5, 5)\}.$$

Their associated graphs are shown in Fig. 5.10. If we ignore the loops in each graph we find the graph decomposed into components such as K_1, K_2, K_3. In general, a relation on a finite set A is an equivalence relation iff its associated graph consists of the disjoint union of complete graphs augmented by loops at every vertex. □

Figure 5.10

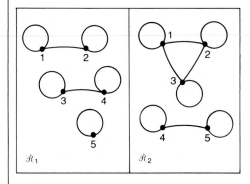

$\mathcal{R}_1$ $\mathcal{R}_2$

EXERCISES

1. For $A = \{1, 2, 3, 4\}$, let $\mathcal{R}$, $\mathcal{S}$ be relations on A defined by $\mathcal{R} = \{(1, 2), (1, 3), (2, 4), (4, 4)\}$, $\mathcal{S} = \{(1, 1), (1, 2), (1, 3), (2, 3), (2, 4)\}$. Find $\mathcal{R} \circ \mathcal{S}$, $\mathcal{S} \circ \mathcal{R}$, $\mathcal{R}^2$, $\mathcal{R}^3$, $\mathcal{S}^2$, $\mathcal{S}^3$.

2. If $\mathcal{R}$ is a reflexive relation on a set A, prove that $\mathcal{R}^2$ is reflexive.

3. Provide the proof for the opposite inclusion in Theorem 5.1.

4. For sets A, B, and C, consider relations $\mathcal{R}_1 \subseteq A \times B$, $\mathcal{R}_2 \subseteq B \times C$, $\mathcal{R}_3 \subseteq B \times C$. Prove that (a) $\mathcal{R}_1 \circ (\mathcal{R}_2 \cup \mathcal{R}_3) = (\mathcal{R}_1 \circ \mathcal{R}_2) \cup (\mathcal{R}_1 \circ \mathcal{R}_3)$; and (b) $\mathcal{R}_1 \circ (\mathcal{R}_2 \cap \mathcal{R}_3) \subseteq (\mathcal{R}_1 \circ \mathcal{R}_2) \cap (\mathcal{R}_1 \circ \mathcal{R}_3)$. (Provide an example to show that a proper inclusion can occur in (b).)

5. For a relation $\mathcal{R}$ on a set A, define $\mathcal{R}^0 = \{(a, a) \mid a \in A\}$. If $|A| = n$, prove that there exist s, $t \in \mathbf{N}$ with $0 \leq s < t \leq 2^{n^2}$ such that $\mathcal{R}^s = \mathcal{R}^t$.

6. With $A = \{1, 2, 3, 4\}$ let $\mathcal{R} = \{(1, 1), (1, 2), (2, 3), (3, 3), (3, 4), (4, 4)\}$ be a relation on A. Find two relations $\mathcal{S}, \mathcal{T}$ on A where $\mathcal{S} \neq \mathcal{T}$ but $\mathcal{R} \circ \mathcal{S} = \mathcal{R} \circ \mathcal{T} = \{(1, 1), (1, 2), (1, 4)\}$.

7. How many 6×6 $(0, 1)$-matrices A are there with $A = A^{tr}$?

8. If $E = \begin{bmatrix} 1 & 0 & 1 & 1 \\ 0 & 1 & 0 & 1 \\ 1 & 0 & 0 & 0 \end{bmatrix}$, how many $(0, 1)$-matrices F satisfy $E \leq F$?

9. Provide the proofs for Theorem 5.2 (a), (b), and (d).

10. Use Theorem 5.2 to develop a computer program for the recognition of equivalence relations on a finite set.

11. For $A = \{1, 2, 3, 4\}$, let $\mathcal{R} = \{(1, 1), (1, 2), (2, 3), (3, 3), (3, 4)\}$ be a relation on A. Draw the directed graph G on A that is associated with $\mathcal{R}$. Do likewise for $\mathcal{R}^2$, $\mathcal{R}^3$, $\mathcal{R}^4$.

12. For $|A| = 5$, how many relations $\mathcal{R}$ on A are there? How many of these relations are symmetric?

13. Let $|A| = 5$. (a) How many directed graphs can one construct on A? (b) How many of the graphs in (a) are actually undirected?

14. How many (undirected) edges are there in the complete graphs K_6, K_7, K_n, $n \in \mathbf{Z}^+$?

15. a) Keeping the order of the elements fixed as 1, 2, 3, 4, 5, determine the $(0, 1)$ relation matrices for the equivalence relations in Example 5.29.

    b) Do the results of part (a) lead to any generalization?

16. a) Let $\mathcal{R}$ be the relation on $A = \{1, 2, 3, 4, 5, 6, 7\}$, where the directed graph associated with $\mathcal{R}$ consists of the components, each a cycle, shown in Fig. 5.11. Find the smallest integer $n > 1$, such that $\mathcal{R}^n = \mathcal{R}$. What is the smallest value of $n > 1$ for which the graph of $\mathcal{R}^n$ contains some loops? Does it ever happen that the graph of $\mathcal{R}^n$ consists of only loops?

Figure 5.11

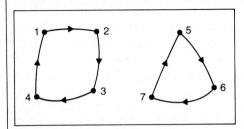

    b) Answer the same questions for the relation $\mathcal{R}$ on $A = \{1, 2, 3, \ldots, 9, 10\}$, whose directed graph is as shown in Fig. 5.12.

    c) Do the results of (a) and (b) indicate anything in general?

Figure 5.12

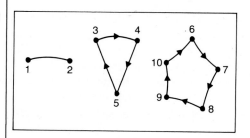

5.3 ■ PARTIAL ORDERS: HASSE DIAGRAMS

If you ask children to recite the numbers they know, you'll hear a uniform response of "1, 2, 3, ..." Without paying attention to it, they list these numbers in increasing order. In this section we take a closer look at this idea of order, something we may have taken for granted. We start by observing something about the sets **N**, **Z**, **Q**, **R**, and **C**.

The set **N** is closed under the binary operations of (ordinary) addition and multiplication, but if we seek an answer to the equation $x + 5 = 2$, we find that no element of **N** provides a solution. So we enlarge **N** to **Z**, where we can perform subtraction as well as addition and multiplication. However, we soon run into trouble trying to solve $2x + 3 = 4$. Enlarging to **Q**, we can perform nonzero division in addition to the other operations. Yet this soon proves to be inadequate, as $x^2 - 2 = 0$ necessitates the introduction of the real but irrational numbers $\pm\sqrt{2}$. Even expanding from **Q** to **R**, more trouble arises in trying to solve $x^2 + 1 = 0$. Finally we arrive at **C**, the complex number system, where any polynomial equation of the form $c_n x^n + c_{n-1} x^{n-1} + \cdots + c_2 x^2 + c_1 x + c_0 = 0$, $c_i \in \mathbf{C}$, $0 \leq i \leq n$, can be solved. (This result is known as the Fundamental Theorem of Algebra. Its proof requires material on functions of a complex variable, so no proof is given here.) As we kept building up from **N** to **C**, gaining more ability to solve polynomial equations, something was lost as we went from **R** to **C**. In **R**, given numbers r_1, r_2, with $r_1 \neq r_2$, we can always decide whether $r_1 < r_2$ or $r_2 < r_1$. However, in **C**, $(2 + i) \neq (1 + 2i)$, but what meaning can we attach to such a statement as "$(2 + i) < (1 + 2i)$"? We have lost the ability to "order" the elements of our number system.

As in Section 5.1, let A be a set and $\mathcal{R}$ a relation on A. The pair $(A, \mathcal{R})$ is called a *partially ordered set,* or *poset,* if relation $\mathcal{R}$ on A is a partial order, or partial ordering relation. If A is called a poset we understand that there is a partial order $\mathcal{R}$ on A that makes A into this poset. Examples 5.1(a), 5.2, and 5.10 are posets.

EXAMPLE 5.30 Let A be the set of courses offered at a college. Define relation $\mathcal{R}$ on A by $x \mathcal{R} y$, if x, y are the same course, or if x is a prerequisite for y. Then $\mathcal{R}$ makes A into a poset. □

EXAMPLE 5.31 Define $\mathcal{R}$ on $A = \{1, 2, 3, 4\}$ by $x \mathcal{R} y$ if $x | y$, i.e., x divides y. Then $\mathcal{R} = \{(1, 1), (2, 2), (3, 3), (4, 4), (1, 2), (1, 3), (1, 4), (2, 4)\}$ is a partial order, and $(A, \mathcal{R})$ is a poset. □

EXAMPLE 5.32   In the construction of a house there are certain jobs, such as digging the foundation, that must be performed before other phases of the construction can be undertaken. If A is a set of tasks that must be performed in building a house or completing a particular manufacturing process, we can define a relation $\mathcal{R}$ on A by $x \, \mathcal{R} \, y$ if x, y denote the same task or if task x must be performed before the start of task y. In this way we place an order on the elements of A, making it into a poset that is sometimes referred to as a PERT (Program Evaluation and Review Technique) network.   □

Figure 5.13

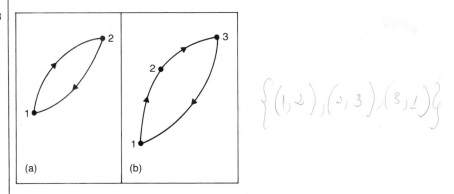

$\{(1,2),(2,3),(3,1)\}$

(a)   (b)

Turning to directed graphs, consider the diagrams in Fig. 5.13. If part (a) were part of the directed graph associated with a relation $\mathcal{R}$, then since $(1, 2), (2, 1) \in \mathcal{R}$ with $1 \neq 2$, $\mathcal{R}$ could not be antisymmetric. For part (b), if the diagram were part of the graph of a transitive relation $\mathcal{R}$, then $(1, 2), (2, 3) \in \mathcal{R} \Rightarrow (1, 3) \in \mathcal{R}$. Since $(3, 1) \in \mathcal{R}$ and $1 \neq 3$, $\mathcal{R}$ is not antisymmetric, so it cannot be a partial order.

From these observations, given a transitive relation $\mathcal{R}$, if the graph G associated with $\mathcal{R}$ contains any directed cycles, other than loops, $\mathcal{R}$ cannot be antisymmetric, and so it fails to be a partial order.

EXAMPLE 5.33   Consider the directed graph for the partial order in Example 5.31. Fig. 5.14 (a) is the graphical representation of $\mathcal{R}$. In part (b) of the figure we have a somewhat easier diagram, called the *Hasse diagram* for $\mathcal{R}$.

Figure 5.14

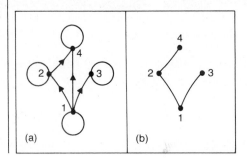

(a)   (b)

When we know that a relation $\mathcal{R}$ is a partial order on a set A, we can eliminate the loops at the vertices of its directed graph. Since $\mathcal{R}$ is also transitive, having the edges $(1, 2)$ and $(2, 4)$ is enough to insure the existence of edge $(1, 4)$, so we need not include that edge. In this way the diagram in Fig. 5.14 (b) is obtained. □

In general, if $\mathcal{R}$ is a partial order on a set A, we construct a Hasse diagram for $\mathcal{R}$ on A by drawing a line segment (undirected edge) from x *up to* y, if $x, y \in A$ with $x \mathcal{R} y$ and, most important, if there is no other element $z \in A$ such that $x \mathcal{R} z$ and $z \mathcal{R} y$. (So there is nothing "in between" x and y.) If we adopt the convention of reading the diagram from bottom to top, then it is not necessary to direct any edges.

Figure 5.15

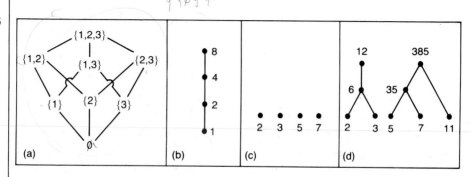

EXAMPLE 5.34

In Fig. 5.15 we have the Hasse diagrams for the following four posets. (a) With $\mathcal{U} = \{1, 2, 3\}$ and $A = \mathcal{P}(\mathcal{U})$, $\mathcal{R}$ is the subset relation on A; (b) Here $\mathcal{R}$ is the "divides" relation applied to $A = \{1, 2, 4, 8\}$; (c) and (d) Same relation as in part (b) but applied to $\{2, 3, 5, 7\}$ in part (c), and $\{2, 3, 5, 6, 7, 11, 12, 35, 385\}$ in part (d). In part (c) we note that a Hasse diagram can have all isolated points; it can also have two (or more) connected pieces, as in part (d). □

EXAMPLE 5.35

On the set $A = \{1, 2, 3, 4, 5\}$, the relation $\mathcal{R}$ on A, defined by $x \mathcal{R} y$ if $x \leq y$, is a partial order, making A into a poset we can denote by $(A, \leq)$. If $B = \{1, 2, 4\} \subset A$, the set $(B \times B) \cap \mathcal{R} = \{(1, 1), (2, 2), (4, 4), (1, 2), (1, 4), (2, 4)\}$ is a partial order on B.

In general if $\mathcal{R}$ is a partial order on A, then for any subset B of A, $(B \times B) \cap \mathcal{R}$ makes B into a poset where the partial order on B is induced from $\mathcal{R}$. □

We turn now to a special type of partial order.

Definition 5.16

▶ If $(A, \mathcal{R})$ is a poset, we say that A is *totally ordered* if for all $x, y \in A$ either $x \mathcal{R} y$ or $y \mathcal{R} x$.

In this case $\mathcal{R}$ is called a *total order*.

EXAMPLE 5.36

(a) On the set **N**, the relation $\mathcal{R}$ defined by $x \mathcal{R} y$ if $x \le y$ is a total order. (b) The subset relation applied to $A = \mathcal{P}(\mathcal{U})$, $\mathcal{U} = \{1, 2, 3\}$, is a partial, but not total, order: $\{1, 2\}$, $\{1, 3\} \in A$ but we have neither $\{1, 2\} \subseteq \{1, 3\}$ nor $\{1, 3\} \subseteq \{1, 2\}$. (c) The Hasse diagram in (b) of Fig. 5.15 shows a total order.   □

Could these notions of partial and total order ever arise in an industrial problem?

We consider the following. A toy manufacturer is about to market a new product and must include a set of instructions for its assembly. He knows that in order to assemble the new toy, there are seven tasks, denoted A, B, C, ..., G, that must be performed in the partial order given by the Hasse diagram of Fig. 5.16. Here we see, for example, that B, A, E must be completed before we can work on task C. Since the set of instructions is to consist of a listing, numbered 1, 2, 3, ..., 7, of these tasks, how can he write the listing and make sure that the partial order of the Hasse diagram is maintained?

Figure 5.16

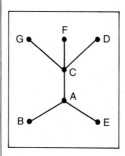

Forgetting the toy and the assembly problem, what we are really asking for here is if we can take the partial order $\mathcal{R}$, given by the Hasse diagram, and find a total order $\mathcal{T}$ on these tasks for which $\mathcal{R} \subseteq \mathcal{T}$. The answer is yes, and the technique is known as *topological sorting*.

Topological Sorting Algorithm

(for a partial order $\mathcal{R}$ on a set A with $|A| = n$)

Step 1: Set $k = 1$. Let H_1 be the Hasse diagram of the partial order.

Step 2: Select a vertex v_k in H_k such that no edge in H_k starts at v_k.

Step 3: If $k = n$, the process is completed and we have a total order

$$\mathcal{T}: v_n < v_{n-1} < \ldots < v_2 < v_1,$$

which contains $\mathcal{R}$.

If $k < n$, remove from H_k the vertex v_k and all edges of H_k that terminate at v_k. Call the result H_{k+1}. Increase k by 1 and return to step 2.

Before we apply this algorithm to the problem at hand we should observe the "a" before the word "vertex" in step 2. This implies that the selection need not be unique and we can get several different total orders $\mathcal{T}$ containing $\mathcal{R}$. Also, in step 3, for vertices v_{i-1}, $2 \leq i \leq n$, the notation $v_i < v_{i-1}$ is used because it is more suggestive of "v_i before v_{i-1}" than is the notation $v_i \mathcal{T} v_{i-1}$.

Figure 5.17

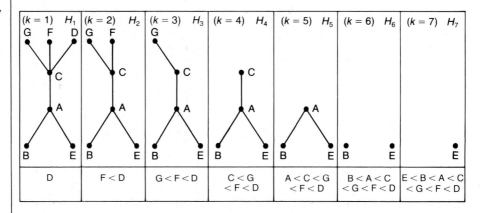

In Fig. 5.17, we show the Hasse diagrams that evolve as we apply the algorithm to the partial order in Fig. 5.16. Below each diagram the total order is listed as it evolves.

If the toy manufacturer writes his instructions in a list as 1-E, 2-B, 3-A, 4-C, 5-G, 6-F, 7-D, he will have a total order which preserves the partial order needed for correct assembly. This total order is one of 12 possible answers.

As is typical in discrete and combinatorial mathematics, we note here how this algorithm provides a procedure which reduces the size of our problem with each successive application.

We turn now to some further properties of posets.

Definition 5.17

▶ If $(A, \mathcal{R})$ is a poset, an element $x \in A$ is called a *maximal* element of A if for all $a \in A$, $a \neq x \Rightarrow x \not{\mathcal{R}} a$. An element $y \in A$ is called a *minimal* element of A if whenever $b \in A$ and $b \neq y$ then $b \not{\mathcal{R}} y$.

EXAMPLE 5.37

Let $\mathcal{U} = \{1, 2, 3\}$, and $A = \mathcal{P}(\mathcal{U})$.

a) Let $\mathcal{R}$ be the subset relation on A. Then $\mathcal{U}$ is maximal while $\emptyset$ is minimal for this poset $(A, \subseteq)$.

b) For B, the collection of proper subsets of $\{1, 2, 3\}$, let $\mathcal{R}$ be the subset relation on B. In the poset $(B, \subseteq)$, $\{1, 2\}, \{1, 3\}, \{2, 3\}$ are all maximal elements, while $\emptyset$ is still the only minimal element.   □

EXAMPLE 5.38   With $\mathcal{R}$ as the "is less than or equal to" relation on the set $\mathbf{Z}$, $(\mathbf{Z}, \leq)$ is a poset with neither a maximal nor a minimal element. The poset $(\mathbf{N}, \leq)$, however, has minimal element 0 but no maximal element.   □

Are there any conditions for which a poset must have a maximal or minimal element?

Theorem 5.3

▶   If $(A, \mathcal{R})$ is a poset and A is finite, then A has both a maximal and a minimal element.

Proof   Let $a_1 \in A$. If there is no element $a \in A$, $a \neq a_1$, with $a_1 \mathcal{R} a$, then a_1 is maximal. Otherwise there is an element $a_2 \in A$, $a_2 \neq a_1$, with $a_1 \mathcal{R} a_2$. If no element $a \in A$, $a \neq a_2$, satisfies $a_2 \mathcal{R} a$, then a_2 is maximal. Otherwise we can find $a_3 \in A$, $a_3 \neq a_2$, $a_3 \neq a_1$ (Why?) with $a_1 \mathcal{R} a_2$ and $a_2 \mathcal{R} a_3$. Continuing in this manner, since A is finite we get to an element $a_n \in A$ with $a_n \cancel{\mathcal{R}} a$ for any $a \neq a_n \in A$, so a_n is maximal.

The proof for a minimal element follows in a similar way.   ∎

Among maximal and minimal elements for a poset, the following, when they exist, are of special concern.

Definition 5.18

▶   If $(A, \mathcal{R})$ is a poset, an element $x \in A$ is called a *least* element if $x \mathcal{R} a$ for all $a \in A$. Element $y \in A$ is called a *greatest* element if $a \mathcal{R} y$ for all $a \in A$.

EXAMPLE 5.39   Let $\mathcal{U} = \{1, 2, 3\}$ and $\mathcal{R}$ be the subset relation.

a)   With $A = \mathcal{P}(\mathcal{U})$, $(A, \subseteq)$ has $\emptyset$ as a least element and $\mathcal{U}$ as a greatest element.

b)   For B = the collection of nonempty subsets of $\mathcal{U}$, $(B, \subseteq)$ has $\mathcal{U}$ as a greatest element. There is no least element, but there are three minimal elements.   □

For a poset $(A, \mathcal{R})$ it is possible to have several maximal and minimal elements. What about least and greatest elements?

Theorem 5.4

▶   If the poset $(A, \mathcal{R})$ has a greatest (least) element, that element is unique.

Proof   Suppose that x, $y \in A$ and that both are greatest elements. Since x is a greatest element, $y \mathcal{R} x$. Likewise, $x \mathcal{R} y$ since y is a greatest element. As $\mathcal{R}$ is antisymmetric, $x = y$.

The proof for the least element is similar.   ∎

Definition 5.19

▶   Let $(A, \mathcal{R})$ be a poset with $B \subseteq A$. An element $x \in A$ is called a *lower bound* of B if $x \mathcal{R} b$ for all $b \in B$. If $y \in A$ and $b \mathcal{R} y$ for all $b \in B$, y is called an *upper bound* of B.

An element $x' \in A$ is called a *greatest lower bound* (glb) of B, if it is a lower bound of B and if for all other lower bounds x'' of B, $x'' \mathcal{R} x'$. Similarly $y' \in A$ is a *least upper bound* (lub) of B if it is an upper bound of B, and if for all other upper bounds y'' of B, $y' \mathcal{R} y''$.

EXAMPLE 5.40   Let $\mathcal{R}$ be the "is less than or equal to" relation for the poset $(A, \mathcal{R})$ in each of the following.

a)   If $A = \mathbf{R}$, and $B = [0, 1]$, then B has glb 0 and lub 1. Note that 0, 1 $\in B$. For $C = (0, 1]$, C has glb 0 and lub 1, and $1 \in C$, but $0 \notin C$.

b)   Keeping $A = \mathbf{R}$, let $B = \{q \in \mathbf{Q} \,|\, q^2 < 2\}$. Then B has $\sqrt{2}$ as a lub and $-\sqrt{2}$ as a glb, and neither of these real numbers is in B.

c)   Now let $A = \mathbf{Q}$, with B as in part (b). Here B has no lub or glb.   □

These examples lead us to the following result.

Theorem 5.5   ▶   If $(A, \mathcal{R})$ is a poset and $B \subseteq A$, then if B has a lub (glb), it is unique.

Proof   We leave the proof to the reader.   ■

We close this section with one last ordered structure.

Definition 5.20   ▶   The poset $(A, \mathcal{R})$ is called a *lattice*, if for any $x, y \in A$, lub$\{x, y\}$ and glb$\{x, y\}$ both exist.

EXAMPLE 5.41   For $A = \mathbf{N}$ and $x, y \in \mathbf{N}$, define $x \mathcal{R} y$ by $x \le y$. Then lub$\{x, y\} = \max\{x, y\}$, glb$\{x, y\} = \min\{x, y\}$, and $(\mathbf{N}, \le)$ is a lattice.   □

EXAMPLE 5.42   For the poset in Example 5.39 (a), if S, $T \subseteq \mathcal{U}$, with lub$\{S, T\} = S \cup T$, glb$\{S, T\} = S \cap T$, then $(\mathcal{P}(\mathcal{U}), \subseteq)$ is a lattice.   □

EXERCISES

1. Draw the Hasse diagram for the poset $(\mathcal{P}(\mathcal{U}), \subseteq)$, where $\mathcal{U} = \{1, 2, 3, 4\}$.

2. Let $A = \{1, 2, 3, 6, 9, 18\}$ and define $\mathcal{R}$ on A by $x \mathcal{R} y$ if $x \,|\, y$. Draw the Hasse diagram for the poset $(A, \mathcal{R})$.

3. Let $(A, \mathcal{R}_1)$, $(B, \mathcal{R}_2)$ be two posets. On $A \times B$ define relation $\mathcal{R}$ by $(a, b) \mathcal{R} (x, y)$ if $a \mathcal{R}_1 x$ and $b \mathcal{R}_2 y$. Prove that $\mathcal{R}$ is a partial order.

4. If $\mathcal{R}_1$, $\mathcal{R}_2$ in Exercise 3 are total orders, is $\mathcal{R}$ a total order?

5. Topologically sort the Hasse diagram in Example 5.34 (a).

6. For $A = \{a, b, c, d, e\}$, the Hasse diagram for the poset $(A, \mathcal{R})$ is shown in Fig. 5.18.

a)   Determine the relation matrix for $\mathcal{R}$.

b)   Construct the directed graph G (on A) that is associated with $\mathcal{R}$.

c)   Topologically sort the poset $(A, \mathcal{R})$.

7. The directed graph G for a relation $\mathcal{R}$ on set $A = \{1, 2, 3, 4\}$ is shown in Fig. 5.19.

a)   Verify that $(A, \mathcal{R})$ is a poset and find its Hasse diagram.

Figure 5.18

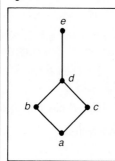

Figure 5.19

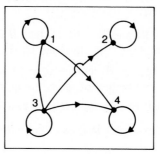

b) Topologically sort $(A, \mathcal{R})$.

c) How many more directed edges are needed in Fig. 5.19 to extend $(A, \mathcal{R})$ to a total order?

8. Let $\mathcal{R}$ be a transitive relation on a set A. Prove that $\mathcal{R}$ is a partial order on A iff $\mathcal{R} \cap \mathcal{R}^c = \{(a, a) \mid a \in A\}$.

9. Prove that a finite poset $(A, \mathcal{R})$ has a minimal element.

10. Prove that if a poset $(A, \mathcal{R})$ has a least element, it is unique.

11. Prove Theorem 5.5.

12. Give an example of a poset with four maximal elements but no greatest element.

13. If $(A, \mathcal{R})$ is a poset, but not a total order, and $\emptyset \neq B \subset A$, does it follow that $(B \times B) \cap \mathcal{R}$ makes B into a poset but not a total order?

14. If $\mathcal{R}$ is a relation on A and G is the associated directed graph, how can one recognize from G that $(A, \mathcal{R})$ is a total order?

15. If G is the directed graph for a relation $\mathcal{R}$ on A, with $|A| = n$, and $(A, \mathcal{R})$ is a total order, how many edges (including loops) are there in G?

16. Let $M(\mathcal{R})$ be the relation matrix for relation $\mathcal{R}$ on A, with $|A| = n$. If $(A, \mathcal{R})$ is a total order, how many 1's appear in $M(\mathcal{R})$?

17. Let $(A, \mathcal{R})$ be a poset. Prove or disprove each of the following.

a) If $(A, \mathcal{R})$ is a lattice, then it is a total order.

b) If $(A, \mathcal{R})$ is a total order, then it is a lattice.

18. If $(A, \mathcal{R})$ is a lattice, with A finite, prove that $(A, \mathcal{R})$ has a greatest element and a least element.

5.4 ■ EQUIVALENCE RELATIONS AND PARTITIONS

As defined earlier, a relation $\mathcal{R}$ on a set A is an equivalence relation if it is reflexive, symmetric, and transitive. For any set $A \neq \emptyset$, the relation of equality is always an equivalence relation on A, where two elements of A are related if they are identical; equality thus establishes the property of "sameness" among the elements of A.

If we consider the relation $\mathscr{R}$ on $\mathbf{Z}$ defined by $x \mathscr{R} y$ if $x - y$ is a multiple of 2, then $\mathscr{R}$ is an equivalence relation on $\mathbf{Z}$ where all even integers are related, as are all odd integers. Here, for example, we do not have $4 = 8$, but we do have $4 \mathscr{R} 8$, for we no longer care about the size of a number but are only concerned with two properties: "evenness" and "oddness". This relation splits $\mathbf{Z}$ into two subsets consisting of the odd and even integers: $\mathbf{Z} = \{\ldots, -3, -1, 1, 3, \ldots\} \cup \{\ldots, -4, -2, 0, 2, 4, \ldots\}$. This splitting up of $\mathbf{Z}$ is an example of a partition, a concept closely related to the equivalence relation. In this section we investigate this relationship and see how it helps us to count the number of equivalence relations on a finite set.

Definition 5.21

▶   Given a set A and an index set I, let $\emptyset \neq A_i \subseteq A$, for each $i \in I$. Then $\{A_i\}_{i \in I}$ is a *partition* of A if

$$\text{(a)} \quad A = \bigcup_{i \in I} A_i; \quad \text{and} \quad \text{(b)} \quad A_i \cap A_j = \emptyset, \quad \text{for } i, j \in I, i \neq j.$$

Each set A_i is called a *cell* or *block* of the partition.

EXAMPLE 5.43   If $A = \{1, 2, 3, \ldots, 10\}$, then each of the following is a partition of A:

a) $A_1 = \{1, 2, 3, 4, 5\}$, $A_2 = \{6, 7, 8, 9, 10\}$;

b) $A_1 = \{1, 3, 5, 7, 9\}$, $A_2 = \{2, 4, 6, 8, 10\}$;

c) $A_1 = \{1, 2, 3\}$, $A_2 = \{4, 6, 7, 9\}$, $A_3 = \{5, 8, 10\}$;

d) $A_i = \{i, i + 5\}$, $1 \leq i \leq 5$.   □

EXAMPLE 5.44   Let $A = \mathbf{R}$ and for each $i \in \mathbf{Z}$, let $A_i = [i, i + 1)$. Then $\{A_i\}_{i \in \mathbf{Z}}$ forms a partition of $\mathbf{R}$.   □

Now just how do partitions come into play with equivalence relations?

Definition 5.22

▶   Let $\mathscr{R}$ be an equivalence relation on a set A. For any $x \in A$, the *equivalence class* of x, denoted $[x]$, is defined by $[x] = \{y \in A \,|\, y \mathscr{R} x\}$.

EXAMPLE 5.45   Define the relation $\mathscr{R}$ on $\mathbf{Z}$ by $x \mathscr{R} y$ if $4 | (x - y)$. For this equivalence relation we find that

$$[0] = \{\ldots, -8, -4, 0, 4, 8, 12, \ldots\} = \{4k \,|\, k \in \mathbf{Z}\}$$

$$[1] = \{\ldots, -7, -3, 1, 5, 9, 13, \ldots\} = \{4k + 1 \,|\, k \in \mathbf{Z}\}$$

$$[2] = \{\ldots, -6, -2, 2, 6, 10, 14, \ldots\} = \{4k + 2 \,|\, k \in \mathbf{Z}\}$$

$$[3] = \{\ldots, -5, -1, 3, 7, 11, 15, \ldots\} = \{4k + 3 \,|\, k \in \mathbf{Z}\}.$$

In addition, we also have $[6] = [2] = [-2]$, $[51] = [3]$, $[17] = [1]$. Most important, $\{[0], [1], [2], [3]\}$ provides a partition of $\mathbf{Z}$.   □

This example leads us to the following general situation.

Theorem
5.6

▶ If $\mathcal{R}$ is an equivalence relation on a set A, and $x, y \in A$, then (a) $x \in [x]$;
(b) $x \, \mathcal{R} \, y$ iff $[x] = [y]$; and (c) $[x] = [y]$ or $[x] \cap [y] = \emptyset$.

Proof

a) This result follows from the reflexive property of $\mathcal{R}$.

b) If $x \, \mathcal{R} \, y$, let $w \in [x]$. Then $w \, \mathcal{R} \, x$ and since $\mathcal{R}$ is transitive, $w \, \mathcal{R} \, y$. So $w \in [y]$,
and $[x] \subseteq [y]$. With $\mathcal{R}$ symmetric, $x \, \mathcal{R} \, y \Rightarrow y \, \mathcal{R} \, x$. So if $t \in [y]$, then $t \, \mathcal{R} \, y$ and
by the transitive property $t \, \mathcal{R} \, x$. Hence $t \in [x]$ and $[y] \subseteq [x]$. Consequently,
$[x] = [y]$.
    Conversely, let $[x] = [y]$. Since $x \in [x]$ by part (a), then $x \in [y]$ or $x \, \mathcal{R} \, y$.

c) This property tells us that equivalence classes can be related in only two ways.
They are either identical or disjoint.
    We assume that $[x] \neq [y]$ and show how it then follows that $[x] \cap [y] = \emptyset$.
If $[x] \cap [y] \neq \emptyset$, then let $v \in A$ with $v \in [x]$ and $v \in [y]$. Then $v \, \mathcal{R} \, x$, $v \, \mathcal{R} \, y$,
and since $\mathcal{R}$ is symmetric, $x \, \mathcal{R} \, v$. Now $x \, \mathcal{R} \, v$, $v \, \mathcal{R} \, y \Rightarrow x \, \mathcal{R} \, y$, by the transitive property. Also $x \, \mathcal{R} \, y \Rightarrow [x] = [y]$ by part (b). This contradicts the
assumption that $[x] \neq [y]$, so we reject the supposition that $[x] \cap [y] \neq \emptyset$, and
the result follows.   ■

Note that if $\mathcal{R}$ is an equivalence relation on A, then by (a), (c) of Theorem 5.6 the
distinct equivalence classes determined by $\mathcal{R}$ form a partition of A.

EXAMPLE 5.46   If $A = \{1, 2, 3, 4, 5\}$ and $\mathcal{R} = \{(1, 1), (2, 2), (2, 3), (3, 2), (3, 3), (4, 4), (4, 5), (5, 4),$
$(5, 5)\}$, then $\mathcal{R}$ is an equivalence relation on A, $[1] = \{1\}$, $[2] = \{2, 3\} = [3]$,
$[4] = \{4, 5\} = [5]$, and $A = [1] \cup [2] \cup [4]$.   □

EXAMPLE 5.47   In ANSI FORTRAN there is a nonexecutable specification statement called the
EQUIVALENCE statement which allows two or more variables in a given program to
refer to the same memory location.
    For example, within a program the statement

$$\text{EQUIVALENCE (A, C, P), (UP, DOWN)}$$

informs the compiler that variables A, C, P will share one memory location and UP
and DOWN will share another. Here the set of all program variables is partitioned by
the equivalence relation $\mathcal{R}$, where $V_1 \, \mathcal{R} \, V_2$ if V_1, V_2 are program variables that share
the same memory location.   □

EXAMPLE 5.48   Having seen examples of how an equivalence relation induces a partition of a set, we
now go backward. If an equivalence relation $\mathcal{R}$ on $A = \{1, 2, 3, 4, 5, 6, 7\}$ induces the
partition $A = \{1, 2\} \cup \{3\} \cup \{4, 5, 7\} \cup \{6\}$, what is $\mathcal{R}$?
    Consider the subset $\{1, 2\}$ of the partition. This subset implies that $[1] = \{1, 2\} =$
$[2]$, and so $(1, 1), (2, 2), (1, 2), (2, 1) \in \mathcal{R}$. (The first two ordered pairs are necessary
for the reflexive property of $\mathcal{R}$, while the others preserve symmetry.)
    In like manner, the subset $\{4, 5, 7\}$ implies that under $\mathcal{R}$ $[4] = [5] = [7] =$
$\{4, 5, 7\}$ and that, as an equivalence relation, $\mathcal{R}$ must contain $\{4, 5, 7\} \times \{4, 5, 7\}$.
In fact, $\mathcal{R} = (\{1, 2\} \times \{1, 2\}) \cup (\{3\} \times \{3\}) \cup (\{4, 5, 7\} \times \{4, 5, 7\}) \cup (\{6\} \times \{6\})$,
and $|\mathcal{R}| = 2^2 + 1^2 + 3^2 + 1^2 = 15$.   □

These examples lead us to the following.

Theorem 5.7   ▶   If A is a set, then

a)   any equivalence relation $\mathcal{R}$ on A induces a partition of A; and,

b)   any partition of A gives rise to an equivalence relation $\mathcal{R}$ on A.

Proof   Part (a) follows from parts (a) and (c) of Theorem 5.6. For part (b), given a partition $\{A_i\}_{i \in I}$ of A, define relation $\mathcal{R}$ on A by $x \mathcal{R} y$, if x, y are in the same cell of the partition. We leave to the reader the details of verifying that $\mathcal{R}$ is an equivalence relation.   ■

On the basis of this theorem and the examples we have examined, we state without proof the following result. (A proof is given on pp. 65-66 of reference [2] by L. Dornhoff and F. Hohn.)

Theorem 5.8   ▶   For any set A, there is a one-to-one correspondence between the equivalence relations on A and the partitions of A.

We are primarily concerned with using this result for finite sets.

EXAMPLE 5.49   a)   If $A = \{1, 2, 3, 4, 5, 6\}$, how many relations on A are equivalence relations?

We solve this problem by counting the partitions of A, realizing that a partition of A is a distribution of the (distinct) elements of A into identical containers, with no container left empty. From Section 3.3 we know, for example, that there are $S(6, 2)$ partitions of A into two identical nonempty containers. Using the Stirling numbers of the second kind, as the number of containers varies from 1 to 6, we have $\sum_{i=1}^{6} S(6, i) = 203$ different partitions of A. Consequently, there are 203 equivalence relations on A.

b)   How many of the equivalence relations in part (a) satisfy $1, 2 \in [4]$?

Identifying 1, 2, 4 as the "same" element under these equivalence relations, we do as in (a) for the set $B = \{1, 3, 5, 6\}$ and find that there are $\sum_{i=1}^{4} S(4, i) = 15$ equivalence relations on A for which $[1] = [2] = [4]$.   □

We close by noting that if A is a finite set with $|A| = n$, then for any $n \le r \le n^2$, there is an equivalence relation $\mathcal{R}$ on A with $|\mathcal{R}| = r$ iff there exist $n_1, n_2, \ldots, n_k \in \mathbf{Z}^+$ with $\sum_{i=1}^{k} n_i = n$ and $\sum_{i=1}^{k} n_i^2 = r$.

5.5 ■ FINITE STATE MACHINES: THE MINIMIZATION PROCESS

In Section 4.3 we encountered two finite state machines that performed the same task but had different numbers of internal states. (See Figs. 4.8, 4.9.) The machine with the larger number of internal states contains *redundant* states, i.e., states that can be eliminated since other states will perform their functions. As the minimization of the number of states in a machine will reduce its complexity and cost, we seek a process for transforming a given machine into one that has no redundant internal states. This process is known as the *minimization process* and its development relies upon the concepts of equivalence relation and partition.

With $M = (S, \mathcal{I}, \mathbb{O}, \nu, \omega)$, we define the relation E_1 on S, by $s_1 \ E_1 \ s_2$ if $\omega(s_1, x) = \omega(s_2, x)$, for all $x \in \mathcal{I}$. This relation E_1 is an equivalence relation on S, and it partitions S into subsets such that two states are in the same subset if they produce the same output for each $x \in \mathcal{I}$. Here the states s_1, s_2 are called *1-equivalent*. For any $k \in \mathbf{Z}^+$, we say that s_1, s_2 are *k-equivalent* if $\omega(s_1, x) = \omega(s_2, x)$ for all $x \in \mathcal{I}^k$. Here ω is the extension of the given output function to $S \times \mathcal{I}^*$. The relation of *k*-equivalence is also an equivalence relation on S; it partitions S into subsets of *k*-equivalent states. We write $s_1 \ E_k \ s_2$ to denote that s_1 and s_2 are *k*-equivalent.

Finally, if $s_1, s_2 \in S$ and s_1, s_2 are *k*-equivalent for all $k \geq 1$, then we call s_1 and s_2 *equivalent*, and write $s_1 \ E \ s_2$. When this happens we find that if we keep s_1 in our machine, s_2 will be redundant and can be removed. Hence our objective is to determine the partition of S induced by E, and select one state for each equivalence class. Then we shall have the minimal realization of the given machine.

Let us start with the following observations.

a) If two states in a machine are not 2-equivalent, could they possibly be 3-equivalent? (or *k*-equivalent, for $k \geq 4$?)

The answer is no. If $s_1, s_2 \in S$ and $s_1 \not{E}_2 s_2$ (i.e., s_1 and s_2 are not 2-equivalent), then there is at least one string $xy \in \mathcal{I}^2$ such that $\omega(s_1, xy) = v_1 v_2 \neq w_1 w_2 = \omega(s_2, xy)$, where $v_1, v_2, w_1, w_2 \in \mathbb{O}$. So with regard to E_3, $s_1 \not{E}_3 s_2$ since for any $z \in \mathcal{I}$, $\omega(s_1, xyz) = v_1 v_2 v_3 \neq w_1 w_2 w_3 = \omega(s_2, xyz)$.

In general, to find states that are $(k + 1)$-equivalent, we look at states that are *k*-equivalent.

b) Now suppose that $s_1, s_2 \in S$ and $s_1 \ E_2 \ s_2$. We wish to determine if $s_1 \ E_3 \ s_2$, i.e., does $\omega(s_1, x_1 x_2 x_3) = \omega(s_2, x_1 x_2 x_3)$ for all strings $x_1 x_2 x_3 \in \mathcal{I}^3$? Consider what happens. First we get $\omega(s_1, x_1) = \omega(s_2, x_1)$, since $s_1 \ E_2 \ s_2 \Rightarrow s_1 \ E_1 \ s_2$. Then there is a transition to the states $\nu(s_1, x_1)$ and $\nu(s_2, x_1)$. Consequently, $\omega(s_1, x_1 x_2 x_3) = \omega(s_2, x_1 x_2 x_3)$, if $\omega(\nu(s_1, x_1), x_2 x_3) = \omega(\nu(s_2, x_1), x_2 x_3)$, (i.e., if $\nu(s_1, x_1) \ E_2 \ \nu(s_2, x_1)$).

In general, for $s_1, s_2 \in S$, $s_1 \ E_{k+1} \ s_2$ if (a) $s_1 \ E_k \ s_2$ and (b) $\nu(s_1, x) \ E_k \ \nu(s_2, x)$ for all $x \in \mathcal{I}$.

With these observations to guide us, we now present an algorithm for the minimization of a finite state machine M.

Step 1: Set $k = 1$. We determine the states that are 1-equivalent by examining the rows in the state table for M. For $s_1, s_2 \in S$, $s_1 \ E_1 \ s_2$ when s_1, s_2 have the same output rows. Let P_1 be the partition of S induced by E_1.

Step 2: Having determined P_k, we obtain P_{k+1} by noting that $s_1 \ E_{k+1} \ s_2$ if $s_1 \ E_k \ s_2$ and $\nu(s_1, x) \ E_k \ \nu(s_2, x)$ for all $x \in \mathcal{I}$. We have $s_1 \ E_k \ s_2$ if s_1, s_2 are in the same subset of the partition P_k. Likewise, $\nu(s_1, x) \ E_k \ \nu(s_2, x)$ if for any $x \in \mathcal{I}$, $\nu(s_1, x), \nu(s_2, x)$ are in the same cell of the partition P_k. In this way P_{k+1} is obtained from P_k.

> **Step 3:** If $P_{k+1} = P_k$, the process is complete. We select one state from each equivalence class and these yield a minimal realization of M.
> If $P_{k+1} \neq P_k$, we increase k by 1 and return to Step 2.

We illustrate the algorithm in the following.

EXAMPLE 5.50   With $\mathcal{I} = \mathcal{O} = \{0, 1\}$, let M be given by the state table, Table 5.1. Looking at the output rows we see that s_3 and s_4 are 1-equivalent as are s_2, s_5, s_6, and E_1 partitions S as

$$P_1: \{s_1\}, \{s_2, s_5, s_6\}, \{s_3, s_4\}.$$

For any $s \in S$, $s \, E_k \, s$, so as we continue this process to determine P_2 we do not concern ourselves with equivalence classes of only one state.

Table 5.1

	ν		ω	
	0	1	0	1
s_1	s_4	s_3	0	1
s_2	s_5	s_2	1	0
s_3	s_2	s_4	0	0
s_4	s_5	s_3	0	0
s_5	s_2	s_5	1	0
s_6	s_1	s_6	1	0

Since $s_3 \, E_1 \, s_4$, there is a chance that we could have $s_3 \, E_2 \, s_4$. Here $\nu(s_3, 0) = s_2$, $\nu(s_4, 0) = s_5$ with $s_2 \, E_1 \, s_5$, and $\nu(s_3, 1) = s_4$, $\nu(s_4, 1) = s_3$ with $s_4 \, E_1 \, s_3$. Hence $\nu(s_3, x) \, E_1 \, \nu(s_4, x)$, for all $x \in \mathcal{I}$, and $s_3 \, E_2 \, s_4$. Similarly, $\nu(s_2, 0) = s_5$, $\nu(s_5, 0) = s_2$ with $s_5 \, E_1 \, s_2$, and $\nu(s_2, 1) = s_2$, $\nu(s_5, 1) = s_5$ with $s_2 \, E_1 \, s_5$. Thus $s_2 \, E_2 \, s_5$. Finally, $\nu(s_5, 0) = s_2$, $\nu(s_6, 0) = s_1$, but $s_2 \, \not\!E_1 \, s_1$, so $s_5 \, \not\!E_2 \, s_6$. (Why don't we investigate the possibility of $s_2 \, E_2 \, s_6$?) Equivalence relation E_2 partitions S as

$$P_2: \{s_1\}, \{s_2, s_5\}, \{s_3, s_4\}, \{s_6\}.$$

Since $P_2 \neq P_1$, we continue the process to get P_3. In determining whether or not $s_2 \, E_3 \, s_5$, we see that $\nu(s_2, 0) = s_5$, $\nu(s_5, 0) = s_2$ and $s_5 \, E_2 \, s_2$. Also, $\nu(s_2, 1) = s_2$, $\nu(s_5, 1) = s_5$ and $s_2 \, E_2 \, s_5$. With $s_2 \, E_2 \, s_5$ and $\nu(s_2, x) \, E_2 \, \nu(s_5, x)$ for all $x \in \mathcal{I}$, we have $s_2 \, E_3 \, s_5$. For s_3, s_4, $(\nu(s_3, 0) = s_2) \, E_2 \, (s_5 = \nu(s_4, 0))$ and $(\nu(s_3, 1) = s_4) \, E_2 \, (s_3 = \nu(s_4, 1))$, so $s_3 \, E_3 \, s_4$ and E_3 induces the partition $P_3: \{s_1\}, \{s_2, s_5\}, \{s_3, s_4\}, \{s_6\}$.

Since $P_2 = P_3$, the process is completed, as indicated in Step 3 of the algorithm. We find that s_5 and s_4 are redundant states. Removing them from the table, and replacing any further occurrence of them by s_2 and s_3, respectively, we arrive at Table 5.2. This is the minimal machine that performs the same tasks as the machine given in Table 5.1.

If we do not want states that skip a subscript, we can always relabel the states in the minimal machine. Here we would have $s_1, s_2, s_3, s_4(=s_6)$, but this s_4 is not the same s_4 we started with in Table 5.1. □

Table 5.2

	ν		ω	
	0	1	0	1
s_1	s_3	s_3	0	1
s_2	s_2	s_2	1	0
s_3	s_2	s_3	0	0
s_6	s_1	s_6	1	0

You may be wondering how we knew that we could stop the process when $P_2 = P_3$. For after all, couldn't it happen that perhaps $P_4 \neq P_3$, or that $P_4 = P_3$ but $P_5 \neq P_4$? To prove that this never occurs we define the following idea.

Definition 5.23

▶ If P_1, P_2 are any partitions of a set A, P_2 is called a *refinement* of P_1, and we write $P_2 \leq P_1$, if every cell of P_2 is contained in a cell of P_1.

In the minimization process of Example 5.50 we had $P_3 = P_2 \leq P_1$. Whenever we apply the algorithm, as we get P_{k+1} from P_k, we always find that $P_{k+1} = P_k$ or $P_{k+1} \leq P_k$, since $(k + 1)$-equivalence implies k-equivalence. So each successive partition is equal to, or refines, the preceding partition.

Theorem 5.9

▶ In applying the minimization process, if $k \geq 1$ and P_k, P_{k+1} are partitions with $P_{k+1} = P_k$, then $P_r = P_{r+1}$ for any $r \geq k + 1$.

Proof

If not, let $r (\geq k + 1)$ be the smallest subscript such that $P_r \neq P_{r+1}$. Then $P_{r+1} < P_r$, so there exist $s_1, s_2 \in S$ with $s_1 \, E_r \, s_2$ but $s_1 \, \not\!\!E_{r+1} \, s_2$. Since $P_r = P_{r-1}$, we have $s_1 \, E_{r-1} \, s_2$. Also, $s_1 \, E_r \, s_2 \Rightarrow \nu(s_1, x) \, E_{r-1} \, \nu(s_2, x)$, for all $x \in \mathcal{I}$. With $P_r = P_{r-1}$, we then find that $\nu(s_1, x) \, E_r \, \nu(s_2, x)$, for all $x \in \mathcal{I}$, so $s_1 \, E_{r+1} \, s_2$. Consequently, $P_r = P_{r+1}$. ∎

We close this section with the following related idea. Let M be a finite state machine with $s_1, s_2 \in S$, and s_1, s_2 not equivalent. Then there must be a smallest integer $k \geq 0$ such that $s_1 \, E_k \, s_2$ but $s_1 \, \not\!\!E_{k+1} \, s_2$. For $k = 0$, we have s_1, s_2 producing different output rows in the state table for M. In this case it is easy to find an $x \in \mathcal{I}$ such that $\omega(s_1, x) \neq \omega(s_2, x)$, and this x distinguishes these inequivalent states. For $k \geq 1$, s_1, s_2 produce the same output rows in the table. Now if we are to distinguish these states we need to find a string $x = x_1x_2 \ldots x_kx_{k+1} \in \mathcal{I}^{k+1}$ such that $\omega(s_1, x) \neq \omega(s_2, x)$, even though $\omega(s_1, x_1x_2 \ldots x_k) = \omega(s_2, x_1x_2 \ldots x_k)$. Such a string x is called a *distinguishing string* for the states s_1, s_2. There may be more than one such string, but each has the (minimal) length $k + 1$.

To find such a string we apply the following algorithm which uses the partitions $P_1, P_2, \ldots, P_k, P_{k+1}$ of the minimization process.

Step 1: Since $s_1 \ E_k \ s_2$ but $s_1 \ \not E_{k+1} \ s_2$, select $x_1 \in \mathscr{I}$ such that $\nu(s_1, x_1) \ \not E_k \ \nu(s_2, x_1)$. (If no such x_1 can be found, then $s_1 \ E_k \ s_2$ and $\nu(s_1, x) \ E_k \ \nu(s_2, x)$ for all $x \in \mathscr{I}$, so $s_1 \ E_{k+1} \ s_2$.)

Step 2: If $k = 1$, $\nu(s_1, x_1) \ \not E_1 \ \nu(s_2, x_1) \Rightarrow$ the states $\nu(s_1, x_1)$, $\nu(s_2, x_1)$ are in different cells of P_1, so they determine different output rows. Find $x_2 \in \mathscr{I}$ with $\omega(\nu(s_1, x_1), x_2) \neq \omega(\nu(s_2, x_1), x_2)$, i.e., $\omega(s_1, x_1x_2) \neq \omega(s_2, x_1x_2)$, and $x = x_1x_2$ is a distinguishing string.

If $k > 1$, $\nu(s_1, x_1) \ \not E_k \ \nu(s_2, x_1)$ but $\nu(s_1, x_1) \ E_{k-1} \ \nu(s_2, x_1)$. (Why?) So the states $\nu(s_1, x_1)$, $\nu(s_2, x_1)$ are in different cells of P_k, and we can find $x_2 \in \mathscr{I}$ such that $\omega(\nu(s_1, x_1), x_2) = \omega(s_1, x_1x_2) = \omega(s_2, x_1x_2) = \omega(\nu(s_2, x_1), x_2)$, but $\nu(s_1, x_1x_2)$, $\nu(s_2, x_1x_2)$ are in different cells of P_{k-1}. (If not, we have $\nu(s_1, x_1) \ E_k \ \nu(s_2, x_1)$.)

Step 3: Using the second part of Step 2 over again, as needed, we construct a string $x_1x_2 \ldots x_k$, and finally arrive at the situation where we have two states $\nu(s_1, x_1x_2 \ldots x_k)$, $\nu(s_2, x_1x_2 \ldots x_k)$ which are in different cells of P_1. At this point we select $x_{k+1} \in \mathscr{I}$ so that $\omega(s_1, x_1x_2 \ldots x_kx_{k+1}) \neq \omega(s_2, x_1x_2 \ldots x_kx_{k+1})$, and $x = x_1x_2 \ldots x_kx_{k+1}$ is a distinguishing string for s_1, s_2.

The algorithm is demonstrated in the following examples.

EXAMPLE 5.51

From Example 5.50 we have the partitions shown below. Here $s_2 \ E_1 \ s_6$ but $s_2 \ \not E_2 \ s_6$. So we seek an input string x of length two such that $\omega(s_2, x) \neq \omega(s_6, x)$. Using Step 1 of the algorithm, we start at P_2, where for s_2, s_6, we find that $\omega(s_2, 0) = 1 = \omega(s_6, 0)$, but $\nu(s_2, 0) = s_5$, $\nu(s_6, 0) = s_1$ are in different cells of P_1. (The input 0 and output 1 provide the labels for the arrows going from the cells of P_2 to those of P_1.) By Step 2 of the algorithm, $\omega(\nu(s_2, 0), 0) = 1 \neq 0 = \omega(\nu(s_6, 0), 0)$. Hence $x = 00$ is a distinguishing string for s_2 and s_6, as $\omega(s_2, 00) = 11 \neq 10 = \omega(s_6, 00)$. □

$$P_2: \{s_1\}, \{s_2, s_5\}, \{s_3, s_4\}, \{s_6\}$$

$$0,1 \qquad \qquad 0,1$$

$$P_1: \{s_1\}, [\{s_2, s_5, s_6\}], \{s_3, s_4\}$$

$$0,0 \quad \mid \quad 0,1 \quad \mid$$

EXAMPLE 5.52

Applying the minimization process to the machine given by the state table in part (a) of Table 5.3, we get the partitions in part (b) of the figure. (Here $P_4 = P_3$.) Applying the algorithm to states s_1, s_4, which are 2-equivalent but not 3-equivalent, we find, as shown in part (b) of Table 5.3, that $x = 111$ is a distinguishing string for these states. (Also note that 11 distinguishes s_2, s_5.) □

Table 5.3

	ν		ω	
	0	1	0	1
s_1	s_4	s_2	0	1
s_2	s_5	s_2	0	0
s_3	s_4	s_2	0	1
s_4	s_3	s_5	0	1
s_5	s_2	s_3	0	0

(a)

P_3: $\{s_1, s_3\}, \{s_2\}, \{s_4\}, \{s_5\}$

$\quad$ 1, 1 $\qquad\qquad$ 1, 1

P_2: $\{s_1, s_3, s_4\}, \{s_2\}, \{s_5\}$

$\quad$ 1, 0 $\qquad$ 1, 0

P_1: $\{s_1, s_3, s_4\}, \{s_2, s_5\}$

$\quad$ 1, 1 $\qquad\quad$ 1, 0

(b)

There is still a great deal more that can be done with finite state machines. Among other omissions, we have carefully avoided any rigorous explanation or proof of why the minimization process works. The interested reader should consult the chapter references for more on this topic.

EXERCISES

1. If $A = \{1, 2, 3, 4, 5\}$ and $\mathcal{R}$ is an equivalence relation on A that induces the partition $A = \{1, 2\} \cup \{3, 4\} \cup \{5\}$, what is $\mathcal{R}$?

2. For $A = \{1, 2, 3, 4, 5, 6\}$, $\mathcal{R} = \{(1, 1), (1, 2), (2, 1), (2, 2), (3, 3), (4, 4), (4, 5), (5, 4), (5, 5), (6, 6)\}$ is an equivalence relation on A. (a) What are [1], [2], [3] under this equivalence relation? (b) What partition of A does $\mathcal{R}$ induce?

3. If $A = A_1 \cup A_2 \cup A_3$, where $A_1 = \{1, 2\}$, $A_2 = \{2, 3, 4\}$, $A_3 = \{5\}$, define relation $\mathcal{R}$ on A by $x \mathcal{R} y$ if x, y are in the same subset A_i, $1 \le i \le 3$. Is $\mathcal{R}$ an equivalence relation?

4. For $A = \mathbf{R}^2$, define $\mathcal{R}$ on A by $(x_1, y_1) \mathcal{R} (x_2, y_2)$ if $x_1 = x_2$.

   a) Verify that $\mathcal{R}$ is an equivalence relation on A.

   b) Describe geometrically the equivalence classes and partition of A induced by $\mathcal{R}$.

5. Let $A = \{1, 2, 3, 4, 5\} \times \{1, 2, 3, 4, 5\}$, and define $\mathcal{R}$ on A by $(x_1, y_1) \mathcal{R} (x_2, y_2)$ if $x_1 + y_1 = x_2 + y_2$.

   a) Verify that $\mathcal{R}$ is an equivalence relation on A.

   b) Determine the equivalence classes [(1, 3)], [(2, 4)], [(1, 1)].

   c) Determine the partition of A induced by $\mathcal{R}$.

6. If $A = \{1, 2, 3, 4, 5, 6, 7\}$, define $\mathcal{R}$ on A by $(x, y) \in \mathcal{R}$ if $x - y$ is a multiple of 3. (a) Show that $\mathcal{R}$ is an equivalence relation on A. (b) Determine the equivalence classes and partition of A induced by $\mathcal{R}$.

7. If $|A| = 30$ and the equivalence relation $\mathcal{R}$ on A partitions A into equivalence classes A_1, A_2, A_3, where $|A_1| = |A_2| = |A_3|$, what is $|\mathcal{R}|$?

8. Let $A = \{v, w, x, y, z\}$. Determine the number of relations on A that are (a) reflexive and symmetric; (b) equivalence relations; (c) reflexive and symmetric but not transitive; (d) equivalence relations which determine exactly two equivalence

classes; (e) equivalence relations where $w \in [x]$; (f) equivalence relations where v, $w \in [x]$; (g) equivalence relations where $w \in [x]$, $y \in [z]$; and (h) equivalence relations where $w \in [x]$, $y \in [z]$, $[x] \neq [z]$.

9. Let $A = \{1, 2, 3, 4, 5, 6, 7\}$. For each of the following values of r, determine an equivalence relation $\mathcal{R}$ on A with $|\mathcal{R}| = r$, or explain why no such relation exists: (a) $r = 6$; (b) $r = 7$; (c) $r = 8$; (d) $r = 9$; (e) $r = 11$; (f) $r = 22$; (g) $r = 23$; (h) $r = 30$; (i) $r = 31$.

10. Provide the details for the proof of Theorem 5.7 (b).

11. Apply the minimization process to each machine in Table 5.4.

Table 5.4

(a)

	ν		ω	
	0	1	0	1
s_1	s_4	s_1	0	1
s_2	s_3	s_3	1	0
s_3	s_1	s_4	1	0
s_4	s_1	s_3	0	1
s_5	s_3	s_3	1	0

(b)

	ν		ω	
	0	1	0	1
s_1	s_6	s_3	0	0
s_2	s_5	s_4	0	1
s_3	s_6	s_2	1	1
s_4	s_4	s_3	1	0
s_5	s_2	s_4	0	1
s_6	s_4	s_6	0	0

(c)

	ν		ω	
	0	1	0	1
s_1	s_6	s_3	0	0
s_2	s_3	s_1	0	0
s_3	s_2	s_4	0	0
s_4	s_7	s_4	0	0
s_5	s_6	s_7	0	0
s_6	s_5	s_2	1	0
s_7	s_4	s_1	0	0

12. For the machine in Exercise 11(c), find a distinguishing string for each given pair of states: (a) s_1, s_5; (b) s_2, s_3; and (c) s_5, s_7.

13. Let M be the finite state machine given in the state diagram shown in Fig. 5.20.

a) Minimize machine M.

b) Find a distinguishing string for each given pair of states: (i) s_3, s_6; (ii) s_3, s_4; and (iii) s_1, s_2.

Figure 5.20

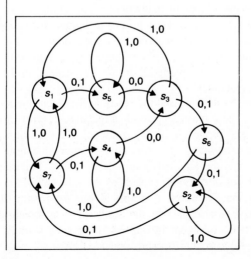

5.6 ■ SUMMARY AND HISTORICAL REVIEW

Once again the relation concept surfaces. In Chapter 3 this idea was introduced as a generalization of the function. Here in Chapter 5 we concentrated on relations and the special properties: reflexive, symmetric, antisymmetric, and transitive. As a result we found ourselves focusing our attention on two special kinds of relations: partial orders and equivalence relations.

A relation $\mathcal{R}$ on a set A is a partial order, making A into a poset, if $\mathcal{R}$ is reflexive, antisymmetric, and transitive. Such a relation generalizes the familiar "is less than or equal to" relation on the real numbers. Try to imagine calculus, or even elementary algebra, without it! For those lacking an appreciation of order, just take a simple computer program and see what happens if the program is entered into the computer haphazardly, permuting the order of the statements. Order is with us wherever we turn. We have grown so accustomed to it that we take it for granted. For a finite poset, the Hasse diagram, a special type of directed graph, provides a pictorial representation of the order in the poset; it also proves useful when a total order, including the given partial order, is needed. The method employed to get a total order is the topological sorting algorithm that is appropriate in the solution of PERT (Program Evaluation and Review Technique) networks.

Although the equivalence relation differs from the partial order in only one property, it is quite different in structure and application. We make no attempt to trace the origin of the equivalence relation, but the ideas behind the reflexive, symmetric, and transitive properties can be found early in *I Principii di Geometrica* (1889), the work of the Italian mathematician Giuseppe Peano (1858–1932). The work of Carl Friedrich Gauss (1777–1855) on *congruence*, which he developed in the 1790's, also utilizes these ideas in spirit, if not in name.

Basically, an equivalence relation $\mathcal{R}$ on a set A generalizes equality; it induces a characteristic of "sameness" among the elements of A. This "sameness" notion then causes the set A to be partitioned into subsets called *equivalence classes*. Conversely, we find that a partition of a set A induces an equivalence relation on A. The partition of a set arises in many places in mathematics and computer science. In computer science many searching algorithms rely upon a technique that successively reduces the size of a given set A that is being searched. By partitioning A into smaller and smaller subsets, the searching procedure is then applied in a more efficient manner. Each successive partition refines its predecessor, the key needed in the minimization process for finite state machines.

Throughout the chapter we used the interplay between relations, directed graphs, and (0, 1)-matrices. These matrices provide a rectangular array of information about a relation, or graph, and prove useful in certain calculations. Storing information like this, in rectangular arrays and in consecutive memory locations, has been used in computer science since the late 1940s and early 1950s. For more on the historical background of such considerations, consult pages 456–462 of D. Knuth [4]. Another way to store information about a graph is the *adjacency list representation*. (See Miscellaneous Exercise 17.) In the study of data structures, *linked lists* and *doubly linked lists* are prominent in implementing such storage. For more on this, consult the text by A. Aho, J. Hopcroft, and J. Ullman [1].

With regard to graph theory, we are in an area of mathematics that dates back to 1736 when the Swiss mathematician Leonhard Euler (1707–1783) solved the problem of the seven bridges of Königsberg. Since then a great deal more has been done in this area, especially in conjunction with data structures in computer science.

For similar coverage of some of the topics in this chapter, the reader should see Chapter 3 of D. F. Stanat and D. F. McAllister [7]. In Chapter 2 (pp. 65–66) of L. Dornhoff and F. Hohn [2], there is a proof of the one-to-one correspondence between the equivalence relations on a set A and the partitions of A. An interesting presentation of the "Equivalence Problem" can be found on pages 353–355 of D. Knuth [4] for those wanting more on the role of the computer with the concept of the equivalence relation.

The early work on the development of the minimization process can be found in the paper by E. F. Moore [6], which builds upon prior ideas of D. A. Huffman [3]. Chapter 10 of Z. Kohavi [5] covers the minimization process for different types of finite state machines and includes some hardware considerations in their design.

REFERENCES

1. Aho, Alfred V., Hopcroft, John E., and Ullman, Jeffrey D., *Data Structures and Algorithms,* Addison-Wesley, Reading, Massachusetts, 1983.

2. Dornhoff, Larry L., and Hohn, Franz E., *Applied Modern Algebra,* Macmillan, New York, 1978.

3. Huffman, D. A., "The Synthesis of Sequential Switching Circuits," *Journal of Franklin Institute,* vol. 257, no. 3. pp. 161–190, 1954; no. 4, pp. 275–303, 1954.

4. Knuth, Donald E., *The Art of Computer Programming,* 2nd ed., Volume 1, *Fundamental Algorithms,* Addison-Wesley, Reading, Massachusetts, 1973.

5. Kohavi, Zvi, *Switching and Finite Automata Theory,* 2nd ed., McGraw-Hill, New York, 1978.

6. Moore, E. F., "Gedanken-experiments on Sequential Machines," in *Automata Studies, Annals of Mathematical Studies,* no. 34, pp. 129–153, Princeton University Press, Princeton, New Jersey, 1956.

7. Stanat, Donald F., and McAllister, David F., *Discrete Mathematics in Computer Science,* Prentice-Hall, Englewood Cliffs, New Jersey, 1977.

■ MISCELLANEOUS EXERCISES

1. Let A be a set and I an index set, where for each $i \in I$, $\mathcal{R}_i$ is a relation on A. Prove or disprove each of the following.

    a) $\bigcup_{i \in I} \mathcal{R}_i$ is reflexive on A iff each $\mathcal{R}_i$ is reflexive on A.

    b) $\bigcap_{i \in I} \mathcal{R}_i$ is reflexive on A iff each $\mathcal{R}_i$ is reflexive on A.

2. Repeat Exercise 1 with reflexive replaced by (i) symmetric; (ii) antisymmetric; and (iii) transitive.

3. For a set A, let $\mathcal{R}_1$, $\mathcal{R}_2$ be symmetric relations on A. If $\mathcal{R}_1 \circ \mathcal{R}_2 \subseteq \mathcal{R}_2 \circ \mathcal{R}_1$, prove that $\mathcal{R}_1 \circ \mathcal{R}_2 = \mathcal{R}_2 \circ \mathcal{R}_1$.

4. If $\mathcal{R}$ is a relation on a set A, prove or disprove: $\mathcal{R}^2$ reflexive $\Rightarrow \mathcal{R}$ reflexive.

5. For sets A, B, C with relations $\mathcal{R}_1 \subseteq A \times B$, $\mathcal{R}_2 \subseteq B \times C$, prove or disprove: $(\mathcal{R}_1 \circ \mathcal{R}_2)^c = \mathcal{R}_2^c \circ \mathcal{R}_1^c$.

6. For each of the following relations on the set specified, determine whether the relation is reflexive, symmetric, antisymmetric, or transitive. Also determine whether it is a partial order or an equivalence relation, and if the latter, describe the partition induced by the relation.

   a) $\mathcal{R}$ is the relation on $\mathbf{Q}$ where $a \mathcal{R} b$ if $|a - b| < 1$.

   b) Let T be the set of all triangles in the plane. For $t_1, t_2 \in T$ define $t_1 \mathcal{R} t_2$ if t_1, t_2 have the same area.

   c) For T as in (b), define $\mathcal{R}$ by $t_1 \mathcal{R} t_2$ if at least two sides of t_1 are contained within the perimeter of t_2.

   d) Let $A = \{1, 2, 3, 4, 5, 6, 7\}$. Define $\mathcal{R}$ on A by $x \mathcal{R} y$ if $xy \geq 10$.

   e) Define $\mathcal{R}$ on $\mathbf{Z}$ by $a \mathcal{R} b$ if $7 \mid (a - b)$.

   f) For $A = \{1, 2, 3, 4\} \times \{1, 2, 3, 4\}$, define $\mathcal{R}$ on A by $(x_1, y_1) \mathcal{R} (x_2, y_2)$ if $(y_1 - x_1) = \pm(y_2 - x_2)$.

   g) Define $\mathcal{R}$ on $\mathbf{R}$ by $x \mathcal{R} y$ if $\cos x = \cos y$.

   h) $\mathcal{R}$ is the relation on $\mathbf{Q}$ where $a \mathcal{R} b$ if $a - b \in \mathbf{Z}$.

7. Give an example of a poset with 5 minimal (maximal) elements but no least (greatest) element.

8. For a set A, let $C = \{P_i \mid P_i \text{ is a partition of } A\}$. Define relation $\mathcal{R}$ on C by $P_i \mathcal{R} P_j$ if $P_i \leq P_j$.

   a) Verify that $\mathcal{R}$ is a partial order on C.

   b) For $A = \{1, 2, 3, 4, 5\}$, let P_i, $1 \leq i \leq 4$, be the following partitions: P_1: $\{1, 2\}$, $\{3, 4, 5\}$; P_2: $\{1, 2\}$, $\{3, 4\}, \{5\}$; P_3: $\{1\}$, $\{2\}$, $\{3, 4, 5\}$; P_4: $\{1, 2\}$, $\{3\}$, $\{4\}$, $\{5\}$.

   Draw the Hasse diagram for $C = \{P_i \mid 1 \leq i \leq 4\}$, where C is partially ordered by refinement.

9. If the complete graph K_n has 45 edges, what is n?

10. Let $A = \{1, 2, 3, 4, 5, 6\} \times \{1, 2, 3, 4, 5, 6\}$; define $\mathcal{R}$ on A by $(x_1, y_1) \mathcal{R} (x_2, y_2)$, if $x_1 y_1 = x_2 y_2$.

    a) Verify that $\mathcal{R}$ is an equivalence relation on A.

    b) Determine the equivalence classes $[(1, 1)]$, $[(2, 2)]$, $[(3, 2)]$, $[(4, 3)]$.

11. For functions f, $g: \mathbf{Z}^+ \to \mathbf{R}$, we say that g *asymptotically dominates* f, if we can find constants $m \in \mathbf{R}^+$, $k \in \mathbf{Z}^+$ such that $|f(n)| \leq m|g(n)|$ for all $n \geq k$, $n \in \mathbf{Z}^+$. (See Exercise 28 in Miscellaneous Exercises, Chapter 3.) Let F be the set of all functions f from $\mathbf{Z}^+$ to $\mathbf{R}$.

a) Define $\mathcal{R}$ on F be $f \mathcal{R} g$ if g asymptotically dominates f. Which relational properties does $\mathcal{R}$ satisfy?

b) Define $\mathcal{R}$ on F by $f \mathcal{R} g$ if f, g asymptotically dominate each other. Verify that $\mathcal{R}$ is an equivalence relation.

c) For $\mathcal{R}$ in part (b), let f, f_1, f_2: $\mathbf{Z}^+ \to \mathbf{R}$ with f_1, $f_2 \in [f]$. If $f_1 + f_2$: $\mathbf{Z}^+ \to \mathbf{R}$ is defined by $(f_1 + f_2)(n) = f_1(n) + f_2(n)$, $n \in \mathbf{Z}^+$, prove that $f_1 + f_2 \in [f]$.

12. If $A = \{2, 4, 8, 20, 28, 56, 112, 224, 336, 772, 1544, 3, 9, 15, 45, 135, 405, 675\}$, the relation $\mathcal{R}$ on A given by $x \mathcal{R} y$ if $x \mid y$ is a partial order whose Hasse diagram is structured in Fig. 5.21. Label the vertices in the diagram to demonstrate the partial order. In how many different ways can we topologically sort $\mathcal{R}$ to get a total order $\mathcal{T}$ where $\mathcal{R} \subseteq \mathcal{T}$? (We sometimes say that $\mathcal{R}$ is *embedded* in $\mathcal{T}$.)

Figure 5.21

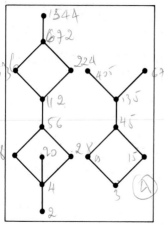

13. A lattice $(A, \mathcal{R})$ is said to be a *complete lattice* if for every nonempty subset $S \subseteq A$, lub S and glb S exist. For $A = \mathbf{Q}$, let $\mathcal{R}$ be the "is less than or equal to" relation on $\mathbf{Q}$. Then $(\mathbf{Q}, \leq)$ is a lattice. Is it a complete lattice?

14. Let M be the $(0, 1)$ adjacency matrix associated with a directed graph G. (We also have M as the $(0, 1)$-matrix for the relation associated with G.) In computing M^2, if we use regular matrix addition and multiplication so that $1 + 1 = 2$ (not Boolean, where $1 + 1 = 1$), what does "2" mean if it occurs in M^2? What about "3" in M^2?

15. For $A = \{a, b, c, d, e, v, w, x, y, z\}$, consider the poset $(A, \mathcal{R})$ whose Hasse diagram is shown in Fig. 5.22. Find

a) glb $\{b, c\}$; b) glb $\{b, w\}$; c) glb $\{e, x\}$; d) lub $\{c, b\}$;

e) lub $\{d, x\}$; f) lub $\{c, e\}$; g) lub $\{a, v\}$.

Is $(A, \mathcal{R})$ a lattice? Is there a maximal element? minimal element? greatest element? least element?

Figure 5.22

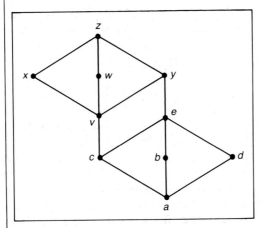

16. For $A = \{1, 2, 3, 4, 5\}$, relation $\mathcal{R}$ on A determines the following $(0, 1)$-matrix.

$$M = \begin{bmatrix} 1 & 0 & 1 & 1 & 0 \\ 0 & 1 & 0 & 0 & 1 \\ 1 & 0 & 1 & 0 & 0 \\ 1 & 0 & 0 & 1 & 1 \\ 0 & 1 & 0 & 1 & 1 \end{bmatrix}$$

a) What is $\mathcal{R}$? What is $\mathcal{R}^2$?

b) Draw the directed graphs associated with $\mathcal{R}$ and $\mathcal{R}^2$.

c) What relational properties does $\mathcal{R}$ satisfy?

17. We have seen that the $(0, 1)$-matrix can be used to represent a graph by means of the adjacency matrix. This proves to be a rather inefficient way to represent the graph when there are many 0's present. A better method for storage of such a graph can be accomplished using the *adjacency list representation*, which is made up of an *adjacency list* for each vertex v and an *index list*. For the graph shown in Fig. 5.23, the representation is given by the two lists in Table 5.5.

Figure 5.23

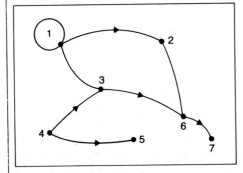

Table 5.5

Adjacency List		Index List	
1	1	1	1
2	2	2	4
3	3	3	5
4	6	4	7
5	1	5	0
6	6	6	9
7	3	7	0
8	5	8	11
9	2		
10	7		

For each vertex v in the graph, we list, preferably in order, each vertex w that is adjacent from v. Hence for 1, we list 1, 2, 3 as the first three adjacencies in our adjacency list. Next to 2 in the index list we place a 4, which tells us where to start looking in the adjacency list for the adjacencies from 2. Since there is a 5 to the right of 3 in the index list, we know that the only adjacency from 2 is 6. Likewise, the 7 to the right of 4 in the index list directs us to the seventh entry in the adjacency list, namely 3, and we find that vertex 4 is adjacent to vertices 3 (seventh vertex in the adjacency list) and 5 (eighth vertex in the adjacency list). We stop at vertex 5 because of the 9 to the right of vertex 6 in the index list. The 0's in the index list next to 5 and 7 indicate that no vertex is adjacent from either 5 or 7. Finally, the last pair of entries in the index list, namely 8 and 11, is a "phantom" that indicates where the adjacency list would pick up from, if there were an eighth vertex in the graph.

Represent each of the graphs in Fig. 5.24 in this manner.

Figure 5.24

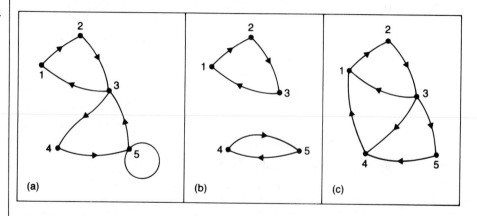

(a)   (b)   (c)

18. The adjacency list representation of a directed graph G is given by the lists in Table 5.6. Construct G from this representation.

Table 5.6

Adjacency List		Index List	
1	2	1	1
2	3	2	4
3	6	3	0
4	3	4	5
5	3	5	8
6	4	6	0
7	5	7	0
8	3	8	10
9	6		

19. Let G be an undirected graph with vertex set V. Define relation $\mathcal{R}$ on V by $v\, \mathcal{R}\, w$ if $v = w$ or if there is a path (of positive length) from v to w (or from w to v, since

G is undirected). (a) Prove that $\mathcal{R}$ is an equivalence relation on V. (b) What can we say about the associated partition?

20. a) For the finite state machine given in Table 5.7, determine a minimal machine that is equivalent to it.

    b) Find a string that distinguishes states s_4 and s_6.

Table 5.7

	ν		ω	
	0	1	0	1
s_1	s_7	s_6	1	0
s_2	s_7	s_7	0	0
s_3	s_7	s_2	1	0
s_4	s_2	s_3	0	0
s_5	s_3	s_7	0	0
s_6	s_4	s_1	0	0
s_7	s_3	s_5	1	0
s_8	s_7	s_3	0	0

21. At the computer center Maria is faced with running 10 computer programs, where, because of priorities, they are restricted by the following conditions: (a) $10 > 8, 3$; (b) $8 > 7$; (c) $7 > 5$; (d) $3 > 9, 6$; (e) $6 > 4, 1$; (f) $9 > 4, 5$; (g) $4, 5, 1 > 2$; where, for example, $10 > 8, 3$ means that program number 10 must be run before programs 8, 3. Determine an order for running these programs so that the priorities are satisfied.

6

The
System
of Integers

Having known about the integers since our first encounters with arithmetic, in this chapter we examine a unique ordering property of this number system. This property will enable us to establish certain mathematical formulas and theorems by a technique called mathematical induction. This principle will place on firm ground any earlier situations in the text where an idea was obtained or defined in a recursive manner.

Although **Z** is not closed under nonzero division, we can still introduce the operation of division for **Z**, and concentrate on special integers called *primes*. These primes turn out to be the "building blocks" of all integers and provide us with a first example of a representation theorem that has come to be called the Fundamental Theorem of Arithmetic.

6.1 ■ THE WELL-ORDERING PRINCIPLE: MATHEMATICAL INDUCTION

At the start of Chapter 5 we saw that if $x, y \in \mathbf{Z}$, then $x + y, xy, x - y \in \mathbf{Z}$. Hence **Z** is closed under the binary operations of addition, multiplication, and subtraction. But so are the sets **Q**, **R**, and **C**!

With regard to order, **Z** is totally ordered under "≤", as are **Q** and **R**. Perhaps then we should forget about **Z** as a separate entity and turn our attention to **R**. We can solve many more equations in **R** than we can in **Z**, and, after all, **Z** is a subset of **R**. Furthermore, where would the continuous notions of calculus be if only **Z** were available?

The world is far from continuous. In the areas of mathematics and computer science we are confronted with discrete situations where the set **Z** (or a subset of **Q**) is the only set of numbers we can (partially) represent on a machine. However, before we abandon **Z**, perhaps in favor of **Q**, let us take a closer look at how the relations "≤" and "<" make **Z** special. In **Z** we can define the set of *positive* elements, or integers, as $\mathbf{Z}^+ = \{x \in \mathbf{Z} \,|\, x > 0\} = \{x \in \mathbf{Z} \,|\, x \geq 1\}$. When we try to do likewise for **Q** and **R** we find that $\mathbf{Q}^+ = \{x \in \mathbf{Q} \,|\, x > 0\}$ and $\mathbf{R}^+ = \{x \in \mathbf{R} \,|\, x > 0\}$, but we cannot represent $\mathbf{Q}^+$ or $\mathbf{R}^+$ using "≥" as for $\mathbf{Z}^+$.

The set $\mathbf{Z}^+$ is different from the sets $\mathbf{Q}^+$ and $\mathbf{R}^+$ in that *every* nonempty subset X of $\mathbf{Z}^+$ contains an integer $a \in X$ such that $a \leq x$, for all $x \in X$ (i.e., X contains a *least* element). This is not so for either $\mathbf{Q}^+$ or $\mathbf{R}^+$. The sets themselves do not contain least elements. There is no smallest positive rational number or smallest positive real number. If we could find, for example, a smallest positive rational number q, then since $0 < q/2 < q$, we would have the smaller positive rational number $q/2$.

These observations lead us to the following property of the set $\mathbf{Z}^+ \subset \mathbf{Z}$.

The Well-Ordering Principle. Any *nonempty* subset of $\mathbf{Z}^+$ contains a smallest element. (We often express this by saying that $\mathbf{Z}^+$ is *well-ordered*.)

This principle serves to distinguish $\mathbf{Z}^+$ from $\mathbf{Q}^+$ and $\mathbf{R}^+$. But does it lead anywhere that is mathematically interesting or useful? The answer is a resounding "Yes!".

It is the basis of a proof technique known as mathematical induction. This technique helps us in proving a general mathematical statement involving a positive integer, when certain instances of that statement suggest a general pattern.

We now establish the basis for this induction technique.

Theorem 6.1

▶ (*Finite Induction Principle*, or *Principle of Mathematical Induction*) Let $S(n)$ denote a mathematical statement (or set of statements) that involves one or more occurrences of the symbol n, which represents a positive integer.

a)   If $S(1)$ is true; and

b)   If whenever $S(k)$ is true for some $k \in \mathbf{Z}^+$, the truth of $S(k + 1)$ is implied by the truth of $S(k)$;

then $S(n)$ is true for all $n \in \mathbf{Z}^+$.

Proof

Let $S(n)$ be such a statement satisfying conditions (a) and (b), and let $F = \{t \in \mathbf{Z}^+ \,|\, S(t) \text{ is false}\}$. We wish to prove that $F = \emptyset$.

If $F \neq \emptyset$, then by the well-ordering principle, F has a least element s. Since $S(1)$ is true, $s \neq 1$, so $s > 1$, and consequently $s - 1 \in \mathbf{Z}^+$. With $s - 1 \notin F$, we have $S(s - 1)$ true. So by condition (b) it follows that $S((s - 1) + 1) = S(s)$ is true, contradicting $s \in F$. This contradiction arose from the assumption that $F \neq \emptyset$. Consequently, $F = \emptyset$.   ■

The reader may have observed a similarity between the two conditions of this theorem and the two parts that came up with our earlier recursive, or inductive, definitions. This principle underlies all such recursive definitions and results.

The choice of 1 in the first condition of Theorem 6.1 is not mandatory. All that is needed is for the statement $S(n)$ to be true for some $n_0 \in \mathbf{Z}$ so that the induction process has a starting place. The integer n_0 could be 5 just as well as 1. It can even be zero or negative, for the set $\mathbf{Z}^+$ in union with $\{0\}$ or any *finite* set of negative integers is well-ordered.

We demonstrate Theorem 6.1 in some of the following results.

EXAMPLE 6.1

For any $n \in \mathbf{Z}^+$, $\sum_{i=1}^{n} i = 1 + 2 + 3 + \cdots + n = (n)(n + 1)/2$.

Proof: For $n = 1$ the statement $S(1)$ is $\sum_{i=1}^{1} i = 1 = (1)(1 + 1)/2$. So $S(1)$ is true and we have a starting point to induct from. Assuming the result true for $n = k$, for some $k \in \mathbf{Z}^+$, we want to show how the truth of $S(k)$ "forces" us to accept the truth of $S(k + 1)$. (The assumption of the truth of $S(k)$ is our *induction hypothesis*.) To establish $S(k + 1)$ we need to show that

$$\sum_{i=1}^{k+1} i = \frac{(k + 1)(k + 2)}{2}.$$

We proceed as follows.

$$\sum_{i=1}^{k+1} i = 1 + 2 + \cdots + k + (k + 1) = \left(\sum_{i=1}^{k} i\right) + (k + 1) = \frac{k(k + 1)}{2} + (k + 1),$$

for we are assuming the truth of $S(k)$. But

$$\frac{k(k + 1)}{2} + (k + 1) = \frac{k(k + 1)}{2} + \frac{2(k + 1)}{2} = \frac{(k + 1)(k + 2)}{2},$$

establishing condition (b) of the theorem.

Consequently, by the principle of finite induction $S(n)$ is true for all $n \in \mathbf{Z}^+$. □

Where would the need for such a summation formula arise?

EXAMPLE 6.2   A wheel of fortune has the numbers from 1 to 36 painted on it, in a random manner. Show that, regardless of how the numbers are situated, there are three in succession which total 55 or more.

Let x_1 be any number on the wheel. Counting clockwise from x_1, label the other numbers $x_2, x_3, \ldots, x_{36}$. For the result to be false, we must have $x_1 + x_2 + x_3 < 55$, $\ldots, x_{34} + x_{35} + x_{36} < 55$, $x_{35} + x_{36} + x_1 < 55$, and $x_{36} + x_1 + x_2 < 55$. In these 36 inequalities, each of the terms $x_1, x_2, \ldots, x_{36}$ appears exactly three times, so each of the integers 1, 2, $\ldots$, 36 appears three times. Adding all 36 inequalities we find that $3 \sum_{i=1}^{36} x_i = 3 \sum_{i=1}^{36} i < 36(55) = 1980$. But

$$\sum_{i=1}^{36} i = \frac{(36)(37)}{2} = 666 \quad \text{so} \quad 1998 = 3(666) < 1980.$$ □

The next summation formula takes us from first powers to squares.

EXAMPLE 6.3   Prove that for any $n \in \mathbf{Z}^+$, $\sum_{i=1}^{n} i^2 = (n)(n + 1)(2n + 1)/6$.

Proof: We start with $S(1)$. Here $\sum_{i=1}^{1} i^2 = 1^2 = 1 = (1)(1 + 1)(2 + 1)/6$, so $S(1)$ is true. Assuming $S(k)$: $\sum_{i=1}^{k} i^2 = (k)(k + 1)(2k + 1)/6$, for $k \in \mathbf{Z}^+$, we want to deduce the truth of

$$S(k + 1): \sum_{i=1}^{k+1} i^2 = (k + 1)((k + 1) + 1)(2(k + 1) + 1)/6$$

$$= (k + 1)(k + 2)(2k + 3)/6.$$

Using the induction hypothesis $S(k)$, we find that

$$\sum_{i=1}^{k+1} i^2 = 1^2 + 2^2 + \cdots + k^2 + (k + 1)^2 = \sum_{i=1}^{k} i^2 + (k + 1)^2$$

$$= [(k)(k + 1)(2k + 1)/6] + (k + 1)^2,$$

From this we have

$$\sum_{i=1}^{k+1} i^2 = (k + 1)[(k)(2k + 1)/6 + (k + 1)] = (k + 1)[(2k^2 + 7k + 6)/6]$$

$$= (k + 1)(k + 2)(2k + 3)/6,$$

and the general result follows by mathematical induction.   □

Consider the following Pascal programs. The program in Fig. 6.1 uses a repeat-until loop to accumulate the sum of the squares. The second program (Fig. 6.2) demonstrates how the result of Example 6.3 can be used in place of such a loop. Each program is used to evaluate the sum of the squares of the first 17 positive integers.

Figure 6.1

```
Program SumOfSquares1 (input,output);
Var
        i,n,s: integer;
Begin
        Writeln('We wish to find the sum of the squares of the first');
        Write('n positive integers where n = ');
        Read(n);
        s := 0;
        i := 0;
        Repeat
              i := i + 1;
              s := s + sqr(i)
        Until i = n;
        Writeln('The sum of the squares of the first ', n:0);
        Write('positive integers is ',s:0)
End.


We wish to find the sum of the squares of the first
n positive integers where n = 17
The sum of the squares of the first 17
positive integers is 1785
```

Figure 6.2

```
Program SumOfSquares2 (input, output);
Var
        n,s: integer;
Begin
        Writeln('We wish to find the sum of the squares of the');
        Write('first n positive integers where n = ');
        Read(n);
        s := (n)*(n + 1)*(2*n + 1) Div 6;
        Writeln('The sum of the squares of the first ', n:0);
        Write('positive integers is ', s:0)
End.


We wish to find the sum of the squares of the
first n positive integers where n = 17
The sum of the squares of the first 17
positive integers is 1785
```

Having seen two applications of mathematical induction, one may wonder if this principle only applies to verifying summation formulas. The following provide two different instances in which induction is the needed key.

EXAMPLE 6.4   Let Σ be an alphabet, with languages $A, B \subseteq \Sigma^*$. If $A \subseteq B$, prove that $A^n \subseteq B^n$, for all $n \in \mathbf{Z}^+$.

Proof: Here $S(1)$ follows, since $A^1 = A \subseteq B = B^1$. Assuming $S(k)$, we have $A^k \subseteq B^k$ for $k \in \mathbf{Z}^+$. Now consider a string $x \in A^{k+1}$. From Definition 4.8, $x = x_1 x_k$, where $x_1 \in A$, $x_k \in A^k$. With $A \subseteq B$ and $A^k \subseteq B^k$, we have $x_1 \in B$, $x_k \in B^k$, so $x = x_1 x_k \in B B^k = B^{k+1}$. So $S(k + 1)$ is true, and by the principle of finite induction, $A^n \subseteq B^n$, $n \in \mathbf{Z}^+$.   □

EXAMPLE 6.5   For $n \geq 0$ let $A_n \subset \mathbf{R}$, where $|A_n| = 2^n$ and the elements of A_n are listed in ascending order. If $r \in \mathbf{R}$, prove that in order to determine if $r \in A_n$, we must compare r with no more than $n + 1$ elements in A_n.

When $n = 0$, $A_0 = \{a\}$ and only one comparison is needed. So the result is true for $n = 0$. For $n = 1$, $A_1 = \{a_1, a_2\}$ with $a_1 < a_2$. To determine if $r \in A_1$, at most two comparisons must be made. Hence the result follows when $n = 1$. Now if $n = 2$, we write $A_2 = \{b_1, b_2, c_1, c_2\} = B_1 \cup C_1$, where $b_1 < b_2 < c_1 < c_2$, $B_1 = \{b_1, b_2\}$, $C_1 = \{c_1, c_2\}$. Comparing r with b_2 we determine which of the possibilities $r \in B_1$ or $r \in C_1$ can occur. Since $|B_1| = |C_1| = 2^1 = 2$, either possibility requires at most two more comparisons (from the prior case where $n = 1$). Consequently, we can determine whether $r \in A_2$ by making no more than $2 + 1 = n + 1$ comparisons.

We now argue in general. Assume the result true for some $k \geq 0$ and consider the case for A_{k+1}, where $|A_{k+1}| = 2^{k+1}$. Let $A_{k+1} = B_k \cup C_k$, where $|B_k| = |C_k| = 2^k$, and the elements of B_k, C_k are in ascending order with the largest element x in B_k smaller than the least element in C_k. Let $r \in \mathbf{R}$. In order to determine if $r \in A_{k+1}$ we consider whether $r \in B_k$ or $r \in C_k$.

a)   First we compare r and x. (One comparison)

b)   If $r \leq x$, then since $|B_k| = 2^k$ it follows by the induction hypothesis that we can determine if $r \in B_k$ by making no more that $k + 1$ additional comparisons. Consequently, at most $(k + 1) + 1$ comparisons are made in all.

c)   If $r > x$ we do likewise with the elements in C_k. We make at most $k + 1$ additional comparisons to see if $r \in C_k$.

The general result now follows by the principle of finite induction.   □

We close this section with an alternative form of Theorem 6.1. This result is sometimes referred to as *complete induction*.

Theorem 6.2   ▶ (*Finite Induction Principle–Alternative Form*) Let $S(n)$ denote a mathematical statement (or set of statements) involving a positive integer n.

a)   If $S(1)$ is true; and

b)   If $S(k + 1)$ is true, for $k \in \mathbf{Z}^+$, whenever $S(1), S(2), \ldots, S(k)$ are true;

then $S(n)$ is true for all $n \in \mathbf{Z}^+$.

Proof   The proofs of this result and its equivalence to Theorem 6.1 are left to the exercises.   ■

Theorem 6.2 will be used to prove a result about a special type of undirected graph called a *tree*. We write $G = (V, E)$ to denote that G is a graph with vertex set V and edge set E.

Definition ▶ An undirected graph $G = (V, E)$ is called a *tree* if G is connected and contains
6.1   no cycles.[†]

Figure 6.3

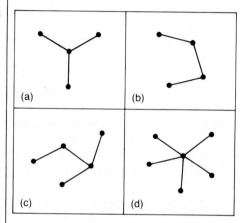

(a)   (b)

(c)   (d)

The graphs in Fig. 6.3 show four examples of trees. The trees in (a) and (b) each have four vertices and three edges. For the tree in (c) there are five vertices and four edges; the tree in (d) has six vertices and five edges. These examples suggest the following result.

Theorem ▶ If $G = (V, E)$ is a tree, then $|V| = |E| + 1$.
6.3

Proof The proof is by induction on $|E|$. If $|E| = 0$, then the tree consists of a single isolated vertex as in Fig. 6.4 (a). Here $|V| = 1 = |E| + 1$. Parts (b) and (c) of the figure verify the result for the cases where $|E| = 1$ or 2.

Figure 6.4

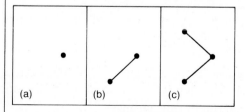

(a)   (b)   (c)

Assume the theorem is true for any tree that contains at most k edges, where $k \geq 0$. Now consider a tree $G = (V, E)$, as in Fig. 6.5, where $|E| = k + 1$. (The dotted edges indicate that some of the tree doesn't appear in the figure.) If the edge

[†]Since a loop is a cycle of length 1, a tree is loop-free.

Figure 6.5

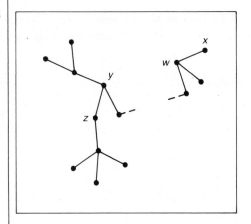

with endpoints y, z is removed from G, we obtain two *subtrees* $G_1 = (V_1, E_1)$ and $G_2 = (V_2, E_2)$ where $|V| = |V_1| + |V_2|$ and $|E_1| + |E_2| + 1 = |E|$. (One of these subtrees could consist of just a single vertex if, for example, the edge with endpoints w, x were removed.) Since $0 \le |E_1|, |E_2| \le k$, by the induction hypothesis, $|E_i| + 1 = |V_i|$, $i = 1, 2$. Consequently, $|V| = |V_1| + |V_2| = (|E_1| + 1) + (|E_2| + 1) = (|E_1| + |E_2| + 1) + 1 = |E| + 1$, and the theorem follows by the alternative form of finite induction. ∎

EXERCISES

1. Prove each of the following by the principle of finite induction.

   a) $1 + 3 + 5 + \cdots + (2n - 1) = n^2$.

   b) $1^2 + 3^2 + 5^2 + \cdots + (2n - 1)^2 = (n)(2n - 1)(2n + 1)/3$.

   c) $1 \cdot 3 + 2 \cdot 4 + 3 \cdot 5 + \cdots + n(n + 2) = (n)(n + 1)(2n + 7)/6$.

   d) $\displaystyle\sum_{i=1}^{n} \frac{1}{i(i + 1)} = \frac{n}{n + 1}$.

   e) $\displaystyle\sum_{i=1}^{n} 2^{i-1} = \sum_{i=0}^{n-1} 2^i = 2^n - 1$

   f) $\displaystyle\sum_{i=1}^{n} i^3 = \frac{n^2(n + 1)^2}{4} = \left(\sum_{i=1}^{n} i\right)^2$.

2. Let Σ be an alphabet. Let $x_i \in \Sigma$ for $1 \le i \le 100$, where $x_i \ne x_j$, $1 \le i < j \le 100$. How many nonempty substrings are there for the string $s = x_1 x_2 \ldots x_{100}$?

3. Consider the following BASIC program:

```
10     FOR I  = 1 TO 123
20          FOR J = 1 TO I
30               PRINT I * J
40          NEXT J
50     NEXT I
60     END
```

    a)   How many times is the PRINT statement at line 30 executed?

    b)   Replace I in line 20 by I**2, and answer the question in part (a).

4. a)   Prove that $(\cos \theta + i \sin \theta)^2 = \cos 2\theta + i \sin 2\theta$, where $i \in \mathbf{C}$, $i^2 = -1$.

    b)   Using induction, prove that for any $n \in \mathbf{Z}^+$, $(\cos \theta + i \sin \theta)^n = \cos n\theta + i \sin n\theta$. (This result is known as *DeMoivre's Theorem*.)

    c)   Verify that $1 + i = \sqrt{2}\,(\cos 45° + i \sin 45°)$, and compute $(1 + i)^{100}$.

5. For $n \in \mathbf{Z}^+$, $n > 10$, prove that

$$n - 2 < \frac{n^2 - n}{12}.$$

6. Given a language $A \subseteq \Sigma^*$, prove that if $A^2 = A$, then $A = A^*$.

7. Prove that for any $n \in \mathbf{Z}^+$, $n > 4$, $2^n > n^2$.

8. For $n \in \mathbf{Z}^+$, let $P(n) =$ the (approximate) number of bacteria in a culture at the end of n hours. If $P(1) = 1000$, $P(2) = 2000$, and $P(n) = P(n - 1) + P(n - 2)$, for all $n > 2$, show that

$$P(n) = \left(\frac{1000}{\sqrt{5}}\right)\left[\left(\frac{1 + \sqrt{5}}{2}\right)^{n+1} - \left(\frac{1 - \sqrt{5}}{2}\right)^{n+1}\right].$$

(In Chapter 11 we will see more on how to obtain such a solution for $P(n)$.)

9. Let $S(n)$ be the statement: for $n \in \mathbf{Z}^+$,

$$\sum_{i=1}^{n} i = \frac{(n + (1/2))^2}{2}.$$

Show that the truth of $S(k)$ implies the truth of $S(k + 1)$, for any $k \in \mathbf{Z}^+$. Is $S(n)$ true for all $n \in \mathbf{Z}^+$?

10. Let S_1 and S_2 be two sets where $|S_1| = m$, $|S_2| = n$, for m, $n \in \mathbf{Z}^+$, and the elements of S_1, S_2 are in ascending order. It can be shown that the elements in S_1, S_2 can be merged into ascending order by making no more than $m + n - 1$ comparisons. (See Lemma 15.1.) Use this result to establish the following.

    For $n \geq 0$, let S be a set with $|S| = 2^n$. Prove that the number of comparisons needed to place the elements of S in ascending order is bounded by $n \cdot 2^n$.

11. Verify Theorem 6.2.

12. a)   Let $n \in \mathbf{Z}^+$, $n \neq 1$, 3. Prove that n can be expressed as a sum of 2's and/or 5's.

    b)   For any $n \in \mathbf{Z}^+$ show that if $n \geq 14$, then n can be written as a sum of 3's and/or 8's.

13. For $n \geq 2$, verify the following by mathematical induction.

    a) $A \cap (B_1 \cup B_2 \cup \ldots \cup B_n) = (A \cap B_1) \cup (A \cap B_2) \cup \ldots \cup (A \cap B_n)$

    (The Generalized Distributive Law of Intersection over Union)

    b) $\overline{A_1 \cap A_2 \cap \ldots \cap A_n} = \overline{A_1} \cup \overline{A_2} \cup \ldots \cup \overline{A_n}$ (Generalized DeMorgan's Law)

14. a) For $n \geq 2$, if $p_1, p_2, p_3, \ldots, p_n, p_{n+1}$ are statements, prove that $[(p_1 \rightarrow p_2) \wedge (p_2 \rightarrow p_3) \wedge \ldots \wedge (p_n \rightarrow p_{n+1})] \Rightarrow [(p_1 \wedge p_2 \wedge p_3 \wedge \ldots \wedge p_n) \rightarrow p_{n+1}]$.

   b) Prove that Theorem 6.2 implies Theorem 6.1.

   c) Use Theorem 6.1 to establish the following: If $\emptyset \neq S \subseteq \mathbf{Z}^+$ and $n \in S$, then S contains a least element.

   d) Show that Theorem 6.1 implies Theorem 6.2.

6.2 ■ THE DIVISION ALGORITHM: PRIME NUMBERS

Although the set $\mathbf{Z}$ is not closed under nonzero division, there are many instances where one integer divides another, for example, 2 divides 6 and 7 divides 21. Here the division is exact and there is no remainder. Thus 2 dividing 6 implies the existence of a quotient, namely 3, such that $6 = 2 \cdot 3$.

Definition 6.2 ▶ If $a, b \in \mathbf{Z}$ and $b \neq 0$, we say that b *divides* a, and write $b \mid a$, if there is an integer c such that $a = bc$. When this occurs we say that b is a *divisor* of a, or a is a *multiple* of b.

With this definition we are able to speak of division inside of $\mathbf{Z}$ without going to $\mathbf{Q}$. Furthermore, when $ab = 0$ for $a, b \in \mathbf{Z}$, then either $a = 0$ or $b = 0$, and we say that $\mathbf{Z}$ has no *proper divisors of 0*. This property allows us to *cancel* as in $2x = 2y \Rightarrow x = y$, for $x, y \in \mathbf{Z}$, because $2x = 2y \Rightarrow 2(x - y) = 0 \Rightarrow 2 = 0$ or $x - y = 0 \Rightarrow x = y$. (Note that at no time did we mention multiplying both sides of the equation $2x = 2y$ by $\frac{1}{2}$. The number $\frac{1}{2}$ is outside the system $\mathbf{Z}$.)

We now summarize some properties of this division operation. Whenever we divide by an integer a, we assume that $a \neq 0$.

Theorem 6.4 ▶ For $a, b, c \in \mathbf{Z}$

a) $1 \mid a$ and $a \mid 0$.

b) $a \mid b, b \mid a \Rightarrow a = \pm b$.

c) $a \mid b, b \mid c \Rightarrow a \mid c$.

d) $a \mid b \Rightarrow a \mid bx$ for all $x \in \mathbf{Z}$.

e) If $x = y + z$, for $x, y, z \in \mathbf{Z}$, and a divides two of the three integers x, y, z, then a divides the remaining integer.

f) $a \mid b, a \mid c \Rightarrow a \mid (bx + cy)$, for all $x, y \in \mathbf{Z}$. (The expression $bx + cy$ is called a *linear combination* of b, c.)

g) For $1 \leq i \leq n$, let $c_i \in \mathbf{Z}$. If a divides each c_i, then $a \mid (c_1 x_1 + c_2 x_2 + \cdots + c_n x_n)$, for any $x_i \in \mathbf{Z}$, $1 \leq i \leq n$.

Proof We prove part (f) and leave the remaining parts for the reader.

If $a \mid b, a \mid c$ then $b = am$, $c = an$, for $m, n \in \mathbf{Z}$. So $bx + cy = (am)x + (an)y = a(mx + ny)$. (By the associative law of multiplication and the distributive law of multiplication over addition, both of which $\mathbf{Z}$ satisfies.) With $bx + cy = a(mx + ny)$ it follows that $a \mid (bx + cy)$. ■

With this operation of integer division we enter the area of *number theory*. As we examine the set $\mathbf{Z}^+$ further we notice that for all $n \in \mathbf{Z}^+$, $n > 1$, n has at least two positive divisors, namely 1 and n itself. Some numbers, like 2, 3, 5, 7, ..., have exactly two positive divisors. These integers are called *primes*. All other positive integers (greater than 1 and not prime) are called *composite*. An immediate connection between prime and composite integers is found in the following.

Lemma 6.1

▶ If $n \in \mathbf{Z}^+$ and n is composite, then there is a prime p such that $p \mid n$.

Proof

If not, let S be the set of all composite numbers that have no prime divisor(s). If $S \neq \emptyset$, then by the Well-Ordering Principle, S has a least element m. m composite $\Rightarrow$ $m = m_1 m_2$, where $1 < m_1 < m$, $1 < m_2 < m$. Since $m_1 \notin S$, m_1 is prime or divisible by a prime. Consequently, there exists a prime p such that $p \mid m$, and $S = \emptyset$. ∎

Now why did we call the above result a *lemma* instead of a theorem? It had to be proved like all other theorems in the book so far. A lemma is still a theorem, but its major role is to establish a result that will be used in proving another theorem.

In listing the primes we are inclined to believe there are infinitely many such numbers. We now verify that this is true.

Theorem 6.5

▶ (Euclid) There are infinitely many primes.

Proof

If not, let $p_1, p_2, \ldots, p_k$ be the finite list of all primes, and let $B = p_1 p_2 \cdots p_k + 1$. Since $B > p_i$, $1 \leq i \leq k$, B cannot be a prime. Hence B is composite, so by Lemma 6.1 there is a prime p_j, $1 \leq j \leq k$, where $p_j \mid B$. Since $p_j \mid B$ and $p_j \mid p_1 p_2 \cdots p_k$, by Theorem 6.4 (e), $p_j \mid 1$. This contradiction arises from the assumption that there are only finitely many primes; the result follows. ∎

Yes, this is the same Euclid from the fourth century B.C. whose *Elements,* written on 13 parchment scrolls, comprised the first organized coverage of the geometry we studied in high school. One finds, however, that these 13 books are also concerned with number theory. In particular, Books VII, VIII, and IX dwell on this topic.

We turn now to the major idea of this section. This result enables us to deal with nonzero division in $\mathbf{Z}$ when that division is not exact.

Theorem 6.6

▶ *(The Division Algorithm)* If $a, b \in \mathbf{Z}$, with $b > 0$, then there exist unique q, $r \in \mathbf{Z}$ with $a = qb + r$, $0 \leq r < b$.

Proof

If $b \mid a$ the result follows, so consider the case where $b \nmid a$ (i.e., b does not divide a). Let $S = \{a - tb \mid t \in \mathbf{Z}, a - tb > 0\}$. If $a > 0$ and $t = 0$, $a \in S$ and $S \neq \emptyset$. For $a \leq 0$, let $t = a - 1$. Then $a - tb = a - (a - 1)b = a(1 - b) + b$, with $(1 - b) \leq 0$, since $b \geq 1$. So $a - tb > 0$ and $S \neq \emptyset$. Hence, for any $a \in \mathbf{Z}$, S is a nonempty subset of $\mathbf{Z}^+$. By the Well-Ordering Principle, S has a least element r, where $0 < r = a - qb$, for some $q \in \mathbf{Z}$. If $r = b$, then $a = (q + 1)b$ and $b \mid a$, contradicting $b \nmid a$. If $r > b$, then $r = b + c$, for some $c \in \mathbf{Z}^+$ and $a - qb = r = b + c \Rightarrow c = a - (q + 1)b \in S$, contradicting r being the least element of S.

This now establishes a quotient q and remainder r, $0 \le r < b$, for the theorem. But are there other q's and r's that also work? If so, let $q_1, q_2, r_1, r_2 \in \mathbf{Z}$ with $a = q_1 b + r_1$, $0 \le r_1 < b$ and $a = q_2 b + r_2$, $0 \le r_2 < b$. Then $q_1 b + r_1 = q_2 b + r_2 \Rightarrow b|q_1 - q_2| = |r_2 - r_1| < b$, since $0 \le r_1, r_2 < b$. If $q_1 \ne q_2$, we have the contradiction $b|q_1 - q_2| < b$. Hence $q_1 = q_2$, $r_1 = r_2$, and the quotient and remainder are unique. ∎

Using the division algorithm we close with a *change of base* problem.

EXAMPLE 6.6   Write 6137 in the octal system (base 8). Here we seek nonnegative integers $r_0, r_1, r_2, \ldots, r_k$, with $r_k > 0$, such that $6137 = (r_k \ldots r_2 r_1 r_0)_8$.

With $6137 = r_0 + r_1 \cdot 8 + r_2 \cdot 8^2 + \cdots + r_k \cdot 8^k = r_0 + 8(r_1 + r_2 \cdot 8 + \cdots + r_k \cdot 8^{k-1})$, r_0 is the remainder obtained in the division algorithm when 6137 is divided by 8.

Consequently, $6137 = 1 + 8 \cdot 767$, so $r_0 = 1$, and $767 = r_1 + r_2 \cdot 8 + \cdots + r_k \cdot 8^{k-1} = r_1 + 8(r_2 + r_3 \cdot 8 + \cdots + r_k \cdot 8^{k-2})$. This yields $r_1 = 7$ and $95 = r_2 + r_3 \cdot 8 + \cdots + r_k \cdot 8^{k-2}$. Continuing in this manner we find $r_2 = 7$, $r_3 = 3$, $r_4 = 1$, and $r_i = 0$ for $i \ge 5$, so $6137 = 1 \cdot 8^4 + 3 \cdot 8^3 + 7 \cdot 8^2 + 7 \cdot 8 + 1 = (13771)_8$.

We can arrange the successive divisions by 8 as follows:

Remainders

$8\,\underline{|6137}$
$8\,\underline{|767}$ 1   (r_0)
$8\,\underline{|95}$ 7   (r_1)
$8\,\underline{|11}$ 7   (r_2)
$8\,\underline{|1}$ 3   (r_3)
0 1   (r_4)   ☐

EXERCISES

1. Verify the remaining parts of Theorem 6.4.

2. Let $a, b, c, d \in \mathbf{Z}^+$. Prove that (a) $a|b, c|d \Rightarrow ac|bd$; (b) $a|b \Rightarrow ac|bc$; and (c) $ac|bc \Rightarrow a|b$.

3. If p, q are primes, prove that $p|q$ iff $p = q$.

4. If $a, b, c \in \mathbf{Z}^+$ and $a|bc$, does it follow that $a|b$ or $a|c$?

5. Let $a, b \in \mathbf{Z}^+$. If $b|a$ and $b|(a + 2)$, prove that $b = 1$ or $b = 2$.

6. If $n \in \mathbf{Z}^+$, and n is odd, prove that $8|(n^2 - 1)$.

7. If $a, b \in \mathbf{Z}^+$, and both are odd, prove that $2|(a^2 + b^2)$ but $4 \nmid (a^2 + b^2)$.

8. Let $n \in \mathbf{Z}^+$. If n is composite, prove that there is a prime p such that $p|n$ and $p \le \sqrt{n}$.

9. Write each of the following base 10 numbers in base 2, base 4, and base 8: (a) 137; (b) 6243; (c) 12,345.

10. Write a computer program to convert a number in base 10 to base b, where $2 \leq b \leq 9$.

11. The Division Algorithm can be generalized as follows: for $a, b \in \mathbf{Z}$, $b \neq 0$, there exist unique $q, r \in \mathbf{Z}$ with $a = qb + r$, $0 \leq r < |b|$. Using Theorem 6.6, verify this generalized form of the algorithm for $b < 0$.

12. Write a computer program to output q and r for the generalized Division Algorithm of the previous exercise.

13. For $n \in \mathbf{Z}^+$, write a computer program that prints out all positive divisors of n.

14. If $a, x, y \in \mathbf{Z}$, and $a \neq 0$, prove that $ax = ay \Rightarrow x = y$.

15. Let $n \in \mathbf{Z}^+$ with $n = r_k \cdot 10^k + \cdots + r_2 \cdot 10^2 + r_1 \cdot 10 + r_0$ (the base 10 representation of n). Prove that (a) $2 \mid n$ iff $2 \mid r_0$; (b) $4 \mid n$ iff $4 \mid (r_1 \cdot 10 + r_0)$; and (c) $8 \mid n$ iff $8 \mid (r_2 \cdot 10^2 + r_1 \cdot 10 + r_0)$.

State a general theorem suggested by these results.

6.3 ■ THE GREATEST COMMON DIVISOR: THE EUCLIDEAN ALGORITHM

Continuing with the division operation developed in Section 6.2, we turn our attention to the divisors of a pair of integers.

Definition 6.3

▶ For $a, b \in \mathbf{Z}$, a positive integer c is said to be a *common divisor* of a, b if $c \mid a$ and $c \mid b$.

EXAMPLE 6.7

The common divisors of 42 and 70 are 1, 2, 7, and 14, with 14 the *greatest* of the common divisors.   □

Definition 6.4

▶ Let $a, b \in \mathbf{Z}$, where at least one of $a, b \neq 0$. Then $c \in \mathbf{Z}^+$ is called a *greatest common divisor* (g.c.d.) of a, b if

a) $c \mid a$, $c \mid b$ (i.e., c is a common divisor of a, b); and

b) for any common divisor d of a, b, we have $d \mid c$.

The result in Example 6.7 satisfies these conditions. However, this example deals with two small integers. What would we do with two integers each having 20 digits? We consider the following questions.

1. Given $a, b \in \mathbf{Z}$, does a greatest common divisor of a, b always exist? If so, how does one find such an integer?

2. How many greatest common divisors can a pair of integers have?

In dealing with these questions, we concentrate on $a, b \in \mathbf{Z}^+$.

Theorem 6.7

▶ For any $a, b \in \mathbf{Z}^+$, there exists a unique $c \in \mathbf{Z}^+$ which is the greatest common divisor of a, b.

Proof Given $a, b \in \mathbf{Z}^+$, let $S = \{as + bt \mid s, t \in \mathbf{Z}, as + bt > 0\}$. Since $S \neq \emptyset$, by the Well-Ordering Principle S has a least element c. We claim that c is the greatest common divisor of a, b.

Since $c \in S$, $c = ax + by$, for some $x, y \in \mathbf{Z}$. Consequently, if $d \in \mathbf{Z}$ and $d \mid a$ and $d \mid b$, then by Theorem 6.4(f) $d \mid (ax + by)$, so $d \mid c$.

If $c \nmid a$, then by the division algorithm $a = qc + r$, with $0 < r < c$. Then $r = a - qc = a - q(ax + by) = (1 - qx)a + (-qy)b$, so $r \in S$, contradicting the choice of c as the least element of S. Consequently, $c \mid a$, and by a similar argument $c \mid b$.

Hence any $a, b \in \mathbf{Z}^+$ have a greatest common divisor. If c_1, c_2 both satisfy the two conditions of Definition 6.4, then with c_1 as a greatest common divisor, and c_2 as a common divisor, $c_2 \mid c_1$. Reversing roles, we find that $c_1 \mid c_2$, and since $c_1, c_2 \in \mathbf{Z}^+$, $c_1 = c_2$. ■

We now know that for $a, b \in \mathbf{Z}^+$, the g.c.d. of a, b exists and is unique. We denote this number by (a, b). (This is, unfortunately, the same notation used for an ordered pair of elements. The context will indicate which idea is being examined.) Here $(a, b) = (b, a)$, and for $a \in \mathbf{Z}$, $a \neq 0$, $(a, 0) = |a|$. Also, for $a, b \in \mathbf{Z}^+$, $(-a, b) = (a, -b) = (-a, -b) = (a, b)$. Finally $(0, 0)$ is not defined and is of no interest to us.

From Theorem 6.7 we see that not only does (a, b) exist but that (a, b) is the *smallest positive integer* we can write as a *linear combination* of a and b. Integers a and b are called *relatively prime* when $(a, b) = 1$, i.e., when there exist $x, y \in \mathbf{Z}$ with $ax + by = 1$.

EXAMPLE 6.8 Since $(42, 70) = 14$, we can find $x, y \in \mathbf{Z}$ with $42x + 70y = 14$, or $3x + 5y = 1$. By inspection $x = 2$, $y = -1$ is a solution; $3(2) + 5(-1) = 1$. But for $k \in \mathbf{Z}$, $1 = 3(2 - 5k) + 5(-1 + 3k)$, so $14 = 42(2 - 5k) + 70(-1 + 3k)$, and the solutions for x, y are not unique.

In general, if $(a, b) = d$, then $((a/d), (b/d)) = 1$. (Verify this!) If $(a/d)x_0 + (b/d)y_0 = 1$, then $1 = (a/d)(x_0 - (b/d)k) + (b/d)(y_0 + (a/d)k)$, for any $k \in \mathbf{Z}$. So $d = a(x_0 - (b/d)k) + b(y_0 + (a/d)k)$, yielding infinitely many solutions to $ax + by = d$. □

This example and the prior observations work well enough when a, b are fairly small. But how does one find (a, b) for some arbitrary $a, b \in \mathbf{Z}^+$? For this we turn to the following result, due to Euclid.

Theorem 6.8. ▶ (Euclidean Algorithm) If $a, b \in \mathbf{Z}^+$, we apply the division algorithm as follows:

$$a = q_1 b + r_1, \qquad 0 < r_1 < b$$
$$b = q_2 r_1 + r_2, \qquad 0 < r_2 < r_1$$
$$r_1 = q_3 r_2 + r_3, \qquad 0 < r_3 < r_2$$

. . . .

$$r_i = q_{i+2}r_{i+1} + r_{i+2}, \qquad 0 < r_{i+2} < r_{i+1}$$

$$\cdot \quad \cdot \quad \cdot \quad \cdot \quad \cdot \quad \cdot \quad \cdot \quad \cdot$$

$$r_{k-3} = q_{k-1}r_{k-2} + r_{k-1}, \qquad 0 < r_{k-1} < r_{k-2}$$

$$r_{k-2} = q_{k}r_{k-1} + r_{k}, \qquad 0 < r_{k} < r_{k-1}$$

$$r_{k-1} = q_{k+1}r_{k}.$$

Then r_k, the last nonzero remainder, equals (a, b).

Proof   To verify that $r_k = (a, b)$ we establish the conditions of Definition 6.4.

Start with the first division process listed above. If $c \mid a$, $c \mid b$, then as $a = q_1 b + r_1$, $c \mid r_1$. Next $c \mid b$, $c \mid r_1 \Rightarrow c \mid r_2$, since $b = q_2 r_1 + r_2$. Continuing down through the division processes we get to where $c \mid r_{k-2}$ and $c \mid r_{k-1}$. From the next to last equation we conclude that $c \mid r_k$ and this verifies condition (b) in Definition 6.4.

To establish condition (a) we go in the reverse order. From the last equation, $r_k \mid r_{k-1}$, and so $r_k \mid r_{k-2}$, since $r_{k-2} = q_k r_{k-1} + r_k$. Continuing up through the equations we get to where $r_k \mid r_2$, $r_k \mid r_3$, so $r_k \mid r_1$. Then $r_k \mid r_2$, $r_1 \Rightarrow r_k \mid b$, and finally $r_k \mid r_1$, $b \Rightarrow r_k \mid a$. Hence $r_k = (a, b)$. ■

EXAMPLE 6.9   Find the g.c.d. of 250 and 111, and express the result as a linear combination of these integers.

$$250 = 2(111) + 28, \qquad 0 < 28 < 111$$

$$111 = 3(28) + 27, \qquad 0 < 27 < 28$$

$$28 = 1(27) + 1, \qquad 0 < 1 < 27$$

$$27 = 27(1) + 0.$$

So 1 is the last nonzero remainder. Therefore $(250, 111) = 1$, and 250 and 111 are relatively prime. Working backward from the third equation we have $1 = 28 - 1(27) = 28 - 1[111 - 3(28)] = (-1)(111) + 4(28) = (-1)(111) + 4[250 - 2(111)] = 4(250) - 9(111)$, a linear combination of 250 and 111.

This expression of 1 as a linear combination of 250 and 111 is not unique since $1 = 250[4 - 111k] + 111[-9 + 250k]$, for any $k \in \mathbf{Z}$.

When need arises we also have $(-250, 111) = (250, -111) = (-250, -111) = (250, 111) = 1$. □

EXAMPLE 6.10   Having determined a g.c.d. in Example 6.9, we now use the Euclidean algorithm to write a computer program that will find $(456, 624)$ and $(116, 641)$.

The Pascal program in Fig. 6.6 determines (a, b) for any $a, b \in \mathbf{Z}^+$. It employs the integer function Mod, where for $x, y \in \mathbf{Z}^+$, x Mod $y =$ the remainder after x is divided by y. For example, 7 Mod 3 is 1, and 18 Mod 5 is 3. (We will deal with this idea in more detail in Chapter 8.) □

Figure 6.6

```
Program    EuclideanAlgorithm (input,output);
Var
            a,b,c,d,r: integer;
Begin
            Writeln ('We wish to determine the greatest common');
            Writeln ('divisor of two positive integers a,b.');
            Write ('a = ');
            Read (a);
            Write ('b = ');
            Read (b);
            r := a Mod b;
            c := a;
            d := b;
            While r > 0 Do
                 Begin
                         c := d;
                         d := r;
                         r := c Mod d
                 End;
            Writeln ('The greatest common divisor of ',a:0,' and ');
            Write (b:0,' is ',d:0)
End.
```

We wish to determine the greatest common
divisor of two positive integers a,b.
a = 456
b = 624
The greatest common divisor of 456 and
624 is 24

We wish to determine the greatest common
divisor of two positive integers a,b.
a = 116
b = 641
The greatest common divisor of 116 and
641 is 1

EXAMPLE 6.11   Assisting students in programming classes, Brian finds that on the average he can help a student debug a Pascal program in six minutes, but it takes 10 minutes to debug a program written in APL. If he works continuously for 104 minutes and doesn't waste any time, how many programs can he debug in each language?

Here we seek integers $x, y \geq 0$, where $6x + 10y = 104$, or $3x + 5y = 52$. As $(3, 5) = 1$, we can write $1 = 3(2) + 5(-1)$, so $52 = 3(104 - 5k) + 5(-52 + 3k)$, $k \in \mathbf{Z}$. In order to obtain $0 \leq x = 104 - 5k$, $0 \leq y = -52 + 3k$, we must have $(52/3) \leq k \leq (104/5)$. So $k = 18, 19, 20$ and there are three possible solutions:

a)   ($k = 18$): $x = 14$, $y = 2$.   b)   ($k = 19$): $x = 9$, $y = 5$.

c)   ($k = 20$): $x = 4$, $y = 8$.   □

The equation in Example 6.11 is an example of a *Diophantine equation,* i.e., a linear equation requiring integer solutions. This type of equation was first investigated by the Greek algebraist Diophantus, who lived in the third century A.D.

Having solved one such equation we seek to discover when a Diophantine equation has a solution. The proof is left for the reader.

Theorem 6.9 ▶ If $a, b, c \in \mathbf{Z}^+$, the Diophantine equation $ax + by = c$ has an integer solution $x = x_0$, $y = y_0$ iff $(a, b) \mid c$.

We close this section with the multiplicative counterpart to the greatest common divisor.

Definition 6.5 ▶ For $a, b, c \in \mathbf{Z}^+$, c is called a *common multiple* of a, b if c is a multiple of both a and b. Furthermore, c is the *least common multiple* (l.c.m.) of a, b if it is the smallest of all positive integers that are common multiples of a, b; we denote c by $[a, b]$.

Theorem 6.10 ▶ Let $a, b, c \in \mathbf{Z}^+$, with $c = [a, b]$. If d is a common multiple of a and b, then $c \mid d$.

Proof If not, by the division algorithm $d = qc + r$, $0 < r < c$. Since $c = [a, b]$, $c = ma$, $m \in \mathbf{Z}^+$. Also, $d = na$, $n \in \mathbf{Z}^+$, since d is a multiple of a. Consequently, $na = qma + r \Rightarrow (n - qm)a = r > 0$, and r is a multiple of a. In a similar way r is seen to be a multiple of b, so r is a common multiple of a, b. But with $0 < r < c$, we contradict c being the least common multiple. Hence $c \mid d$. ∎

Our last result for this section ties together the g.c.d. and l.c.m.

Theorem 6.11 ▶ For $a, b \in \mathbf{Z}^+$, $ab = [a, b] (a, b)$.

Proof The proof is left to the reader. ∎

EXERCISES

1. For each pair $a, b \in \mathbf{Z}^+$ determine (a, b) and express it as a linear combination of a, b.

   a) 231, 1820   b) 1369, 2597   c) 2689, 4001   d) 7982, 7983

2. For each pair a, b in Exercise 1, find $[a, b]$.

3. For any $n \in \mathbf{Z}^+$, what is $(n, n + 1)$? What is $[n, n + 1]$?

4. For $a, b \in \mathbf{Z}^+$, $c, d \in \mathbf{Z}$, what can we say about (a, b) if

   a) $as + bt = 2$?   b) $as + bt = 3$?   c) $as + bt = 4$?

   d) $as + bt = 6$?

5. For $a, b \in \mathbf{Z}^+$ and $d = (a, b)$, prove that $(a/d, b/d) = 1$.

6. For $a, b, n \in \mathbf{Z}^+$, prove that $(na, nb) = n(a, b)$.

7. For $a, b, c, d \in \mathbf{Z}^+$, prove that if $d = a + bc$, then $(b, d) = (a, b)$.

8. Let $a, b, c \in \mathbf{Z}^+$ with $(a, b) = 1$. If $a\,|\,c$, $b\,|\,c$, prove that $ab\,|\,c$. Does the result follow if $(a, b) \neq 1$?

9. Define relation $\mathscr{R}$ on $\mathbf{Z}^+$ by $a\,\mathscr{R}\,b$ if $(a, b) = 1$. Is $\mathscr{R}$ reflexive, symmetric, antisymmetric, or transitive?

10. Verify Theorem 6.9.

11. Determine those values of $c \in \mathbf{Z}^+$, $10 < c < 20$, for which the Diophantine equation $84x + 990y = c$ has no solution. Determine the solutions for the remaining values of c.

12. If a, b are relatively prime, prove that $(a - b, a + b) = 1$ or 2.

13. Let $a, b, c \in \mathbf{Z}^+$ with $(a, b) = 1$. If $a\,|\,bc$, prove that $a\,|\,c$.

14. An executive buys \$249.00 worth of toys for the children of her employees. For each girl she gets a doll costing \$3.30; each boy receives a set of soldiers costing \$2.90. How many toys of each type did she buy?

15. Verify Theorem 6.11.

6.4 ■ THE FUNDAMENTAL THEOREM OF ARITHMETIC

In this section we extend Lemma 6.1 and show that for any $n \in \mathbf{Z}^+$, $n > 1$, either n is prime or can be written as a product of primes, where the representation is unique up to order. This result, known as the *Fundamental Theorem of Arithmetic*, can be found in an equivalent form in Book IX of Euclid's *Elements*.

The following results prove to be useful.

Lemma 6.2 ▶ If $a, b \in \mathbf{Z}^+$ and p is a prime, then $p\,|\,ab \Rightarrow p\,|\,a$ or $p\,|\,b$.

Proof If $p\,|\,a$, we are finished. If not, since p is prime, $(p, a) = 1$, and there are integers x, y with $1 = px + ay$. Then $b = p(bx) + (ab)y$, and since $p\,|\,p$ and $p\,|\,ab$, by Theorem 6.4(e), $p\,|\,b$. ■

Lemma 6.3 ▶ Let $a_i \in \mathbf{Z}^+$, $1 \leq i \leq n$. If p is prime and $p\,|\,a_1 a_2 \cdots a_n$, then $p\,|\,a_i$ for some $1 \leq i \leq n$.

Proof We leave the proof of this result to the reader. ■

EXAMPLE 6.12 Prove that $\sqrt{2}$ is irrational.

If not, we can write $\sqrt{2} = a/b$, where $a, b \in \mathbf{Z}^+$, $(a, b) = 1$. Then $\sqrt{2} = a/b \Rightarrow 2b^2 = a^2 \Rightarrow 2\,|\,a^2 \Rightarrow 2\,|\,a$. (Why?) Also, $2\,|\,a \Rightarrow a = 2a_1$, so $4a_1^2 = a^2 = 2b^2$ and $b^2 = 2a_1^2$. But then $2\,|\,b^2$, so $2\,|\,b$, and a, b are both even. This contradicts $(a, b) = 1$. □

We turn now to the main result of the section.

Theorem
6.12

▶ Any integer $n > 1$ can be written as a product of primes uniquely, up to the order of the primes. (Here a single prime is considered as a product of one factor.)

Proof

The proof consists of two parts: the first covers existence and the second deals with uniqueness.

If not, let $m > 1$ be the smallest integer not expressible as a product of primes. Since m is not a prime, $m = m_1 m_2$, where $1 < m_1 < m$, $1 < m_2 < m$. But then m_1, m_2 can be written as products of primes, since they are less than m. Consequently, with $m = m_1 m_2$ we can obtain a prime factorization of m.

For the integer 2, we have a unique prime factorization, and assuming uniqueness of representation for $3, 4, 5, \ldots, n - 1$, suppose $n = p_1^{s_1} p_2^{s_2} \cdots p_k^{s_k} = q_1^{t_1} q_2^{t_2} \cdots q_r^{t_r}$, where each p_i, $1 \le i \le k$, q_j, $1 \le j \le r$, is a prime, where $p_1 < p_2 < \ldots < p_k$, $q_1 < q_2 < \ldots < q_r$, and where $s_i > 0$, $1 \le i \le k$, $t_j > 0$, $1 \le j \le r$.

Since $p_1 | n$, we have $p_1 | q_1^{t_1} q_2^{t_2} \cdots q_r^{t_r}$. By Lemma 6.3, $p_1 | q_j$ for some $1 \le j \le r$. With p_1, q_j primes we have $p_1 = q_j$. In fact $j = 1$, for otherwise $q_1 | n \Rightarrow q_1 = p_e$ for some $1 < e \le k$ and $p_1 < p_e = q_1 < q_j = p_1$. With $p_1 = q_1$, $n_1 = n/p_1 = p_1^{s_1-1} p_2^{s_2} \cdots p_k^{s_k} = q_1^{t_1-1} q_2^{t_2} \cdots q_r^{t_r}$. Since $n_1 < n$, by the induction hypothesis (Theorem 6.2) it follows that $k = r$, $p_i = q_i$, $1 \le i \le k$, $s_1 - 1 = t_1 - 1$ (so $s_1 = t_1$), and $s_i = t_i$, $2 \le i \le k$. Hence the prime factorization of n is unique. ∎

This result is now used in a variety of situations.

EXAMPLE 6.13

For $n \in \mathbf{Z}^+$, we introduce the function $\tau : \mathbf{Z}^+ \to \mathbf{Z}^+$, where $\tau(n) =$ the number of positive divisors of n. For example, $\tau(2) = 2$, $\tau(3) = 2$, $\tau(4) = 3$. How can we determine $\tau(n)$ for any $n \in \mathbf{Z}^+$?

Using Theorem 6.12, write $n = p_1^{e_1} p_2^{e_2} \cdots p_k^{e_k}$, where for $1 \le i \le k$, p_i is a prime and $e_i > 0$. If $m | n$, then $m = p_1^{f_1} p_2^{f_2} \cdots p_k^{f_k}$, $0 \le f_i \le e_i$, $1 \le i \le k$. So by the rule of product $\tau(n) = (e_1 + 1)(e_2 + 1) \cdots (e_k + 1)$.

Since $2{,}520{,}000 = 2^6 3^2 5^4 7$, $\tau(2{,}520{,}000) = (7)(3)(5)(2) = 210$. □

EXAMPLE 6.14

If $m, n \in \mathbf{Z}^+$, let $m = p_1^{e_1} p_2^{e_2} \cdots p_t^{e_t}$, $n = p_1^{f_1} p_2^{f_2} \cdots p_t^{f_t}$, with each p_i prime, $0 \le e_i$, $0 \le f_i$, $1 \le i \le t$. Then if $a_i = \min\{e_i, f_i\}$, $b_i = \max\{e_i, f_i\}$, $1 \le i \le t$, we have

$$(m, n) = \prod_{i=1}^{t} p_i^{a_i} \quad \text{and} \quad [m, n] = \prod_{i=1}^{t} p_i^{b_i}.$$ □

EXAMPLE 6.15

Prove that if 101 integers are selected from the set $S = \{1, 2, 3, \ldots, 200\}$, there are two integers such that one divides the other.

For each $x \in S$, write $x = 2^k y$, $k \ge 0$, $(2, y) = 1$. Then $y \in T = \{1, 3, 5, \ldots, 199\}$, where $|T| = 100$. Since 101 integers are selected from S, by the pigeonhole principle there are two integers of the form $a = 2^m y$, $b = 2^n y$. If $m \le n$, then $a | b$; otherwise $b | a$. □

EXAMPLE 6.16

Lastly, let $f : \mathbf{Z}^+ \to \mathbf{Z}^+$ be defined by $f(1) = 1$, and $f(n) =$ the number of primes in the prime factorization of n for $n \ge 2$. Then for each $t \in \mathbf{Z}^+$, $f^{-1}(t)$ is infinite. □

EXERCISES

1. Write each of the following numbers as a product of primes

$$p_1^{n_1} p_2^{n_2} \cdots p_k^{n_k}, \quad 0 < n_i, \quad 1 \le i \le k, \quad p_1 < p_2 < \ldots < p_k.$$

   a) 148,500   b) 7,114,800   c) 7,882,875

2. a) Determine the g.c.d. and l.c.m. of each pair of numbers in Exercise 1.

   b) Extending the results of Example 6.14, find the g.c.d. and l.c.m. of the three numbers in Exercise 1.

3. Verify Lemma 6.3.

4. If p is any prime, prove that $\sqrt{p}$ is irrational.

5. a) Prove that $\log_{10} 2$ is irrational.

   b) For any prime p, prove that $\log_{10} p$ is irrational.

6. Determine $\tau(n)$ for each integer of Exercise 1.

7. a) Determine the three smallest values of $n \in \mathbf{Z}^+$ for which $\tau(n) = k$, where $k = 2, 3, 4, 5, 6$.

   b) For any $k \in \mathbf{Z}^+$, $k > 1$, prove that $\tau^{-1}(k)$ is infinite.

8. a) Prove that if 151 integers are selected from $\{1, 2, 3, \ldots, 300\}$, then the selection must include two integers x, y where $x \,|\, y$ or $y \,|\, x$.

   b) State a general result that includes (a) and Example 6.15.

9. Write a computer program to find the prime factorization of an integer $n > 1$.

6.5 ■ SUMMARY AND HISTORICAL REVIEW

According to the Prussian mathematician Leopold Kronecker (1823–1891), "God made the integers, all the rest is the work of man. .... All results of the profoundest mathematical investigation must ultimately be expressible in the simple form of properties of the integers." In the spirit of this quotation we find in this chapter how the handiwork of the Almighty has been further developed by man over the last 24 centuries.

Starting in the fourth century B.C. we find in Euclid's *Elements* not only the geometry of our high school experience, but the fundamental ideas of number theory. Propositions 1 and 2 of Euclid's Book VII include an example of an algorithm to determine the greatest common divisor of two positive integers using an efficient technique to solve, in a *finite* number of steps, a specific problem. (The term *algorithm*, as well as its predecessor *algorism*, was unknown to Euclid. In fact, this term did not enter the vocabulary of most people until the late 1950's when the computer revolution began to make its impact upon society.)

In the century following Euclid's work, we find some number theory in the work of Eratosthenes. However, it is not until five centuries later that the first new major accomplishments in the field were made by Diophantus. In his work *Arithmetica*, his integer solutions of linear (and higher order) equations stood as a mathematical beacon in number theory until the work of the French mathematician Pierre de Fermat (1601–1665).

For more on these mathematicians and others who have worked in the theory of numbers, consult L. Dickson [2]. Chapter 5 in I. Niven and H. Zuckerman [8] deals with the solutions of Diophantine equations and their applications. The articles in [7] deal with interesting contemporary developments in number theory.

In the work *Formulario Mathematico,* Giuseppe Peano (1858–1932) formulated the set of nonnegative integers on the basis of three undefined terms: zero, number, and successor. His formulation is as follows:

a) Zero is a number.

b) For any number n, its successor $\sigma(n) = n + 1$ is a number.

c) There is no number n with $\sigma(n) = 0$.

d) If $\sigma(m) = \sigma(n)$, then $m = n$.

e) If T is a set of numbers with $0 \in T$, and with $\sigma(n) \in T$ for every $n \in T$, then T is the set of all nonnegative integers.

In these postulates the notion of order (successor) and the technique called mathematical induction are seen to be intimately related to the idea of number (i.e., nonnegative integer).

The first European to apply the principle of induction in mathematical proofs was the Venetian scientist Francesco Maurocylus (1491–1575), whose arithmetic book, published in 1575, contains such work. In the next century, Pierre de Fermat made further improvements on the technique in his work involving "the method of infinite descent." Blaise Pascal (c. 1653), in proving such combinatorial results as $C(n, k)/C(n, k + 1) = (k + 1)/(n - k)$, $0 \leq k \leq n - 1$, uses induction and refers to the technique as the work of Maurocylus. The actual term "mathematical induction" was not used, however, until the early nineteenth century when it appeared in the work of Augustus DeMorgan (1806–1871). (An interesting survey on this topic is found in the article by W. H. Bussey [1].)

More on the theory of numbers can be found in the texts by G. Hardy and E. Wright [3], W. J. LeVeque [5], [6], and I. Niven and H. Zuckerman [8]. At a level comparable to the one presented in this chapter, Chapter 3 of V. Larney [4] provides an enjoyable introduction to this material.

REFERENCES

1. Bussey, W. H., "Origins of Mathematical Induction," *American Mathematical Monthly,* Volume 24, 1917, pp. 199–207.

2. Dickson, L., *History of the Theory of Numbers,* Carnegie Institution of Washington, Washington, 1919; reprinted, Chelsea, New York, 1950.

3. Hardy, G. H., and Wright, E. M., *An Introduction to the Theory of Numbers,* 4th ed., Clarendon Press, Oxford, England, 1960.

4. Larney, Violet Hachmeister, *Abstract Algebra: A First Course,* Prindle, Weber & Schmidt, Boston, 1975.

5. LeVeque, William J., *Elementary Theory of Numbers,* Addison-Wesley, Reading, Massachusetts, 1962.

6. LeVeque, William J., *Topics in Number Theory,* Volumes I and II, Addison-Wesley, Reading, Massachusetts, 1956.

7. LeVeque, William J., editor, *Studies in Number Theory,* MAA Studies in Mathematics, Volume 6. Published by the Mathematical Association of America; distributed by Prentice-Hall, Englewood Cliffs, New Jersey, 1969.

8. Niven, Ivan, and Zuckerman, Herbert S., *An Introduction to the Theory of Numbers,* 3rd ed., John Wiley & Sons, New York, 1972.

■ MISCELLANEOUS EXERCISES

1. Let a, d be fixed integers. Determine a summation formula for $a + (a + d) + (a + 2d) + \cdots + (a + (n - 1)d)$, for $n \in \mathbf{Z}^+$. Verify your result by mathematical induction.

2. For $n \in \mathbf{Z}^+$, prove each of the following by mathematical induction:

   a) $5 \mid (n^5 - n)$.   b) $6 \mid (n^3 + 5n)$.

3. Let $S(n)$ be the statement: for all $n \in \mathbf{Z}^+$, $n^2 + n + 41$ is prime.

   a) Verify that $S(n)$ is true for all $1 \le n \le 9$.

   b) Does the truth of $S(k)$ imply that of $S(k + 1)$ for all $k \in \mathbf{Z}^+$?

4. After n months of a certain greenhouse experiment, the number $A(n)$ of plants of a particular type satisfies the equation $A(n) = 3A(n - 1) - 2A(n - 2)$, for $n \ge 2$ with $A(0) = 3$, $A(1) = 7$. Show by mathematical induction that $A(n) = 2^{n+2} - 1$, for all $n \in \mathbf{Z}^+$.

5. If $f: A \to A$ is any function, prove that for all m, $n \in \mathbf{Z}^+$, $f^m \circ f^n = f^n \circ f^m$. (First let $m = 1$ and induct on n. Then induct on m. This technique is known as *double induction*.)

6. Let $f: A \to A$ be an invertible function. For $n \in \mathbf{Z}^+$ prove that $(f^n)^{-1} = (f^{-1})^n$. (This result can be used to define f^{-n} as either $(f^n)^{-1}$ or $(f^{-1})^n$.)

7. If $n \in \mathbf{Z}^+$ prove that 57 divides $7^{n+2} + 8^{2n+1}$.

8. Evaluate each of the following: a) $\sum_{i=11}^{33} i$; b) $\sum_{i=11}^{33} i^2$.

9. Prove that if we select 101 integers from the set $S = \{1, 2, 3, \ldots, 200\}$, there exist m, n in the selection where $(m, n) = 1$.

10. Frances spends \$6.20 on candy for prizes in a contest. If a ten oz. box of this candy costs \$.50 and a three oz. box costs \$.20, how many boxes of each size did she purchase?

11. A wheel of fortune has the integers from 1 to 25 placed on it in a random manner. Show that regardless of how the numbers are positioned on the wheel there are three adjacent numbers whose sum is at least 39.

12. Let $a, b, c, d, m, n, s, t \in \mathbf{Z}^+$ with $ad - bc = 1$, $s = am + bn$, $t = cm + dn$.

    a) Solve for m, n in terms of s, t.   b) Prove that $(s, t) = (m, n)$.

13. Given $r \in \mathbf{Z}^+$, $r = r_0 + r_1 \cdot 10 + r_2 \cdot 10^2 + \cdots + r_n \cdot 10^n$, $0 \le r_i \le 9$, $1 \le i \le n - 1$, $0 < r_n \le 9$.

    a) Prove that $9 \mid r$ iff $9 \mid (r_n + r_{n-1} + \cdots + r_2 + r_1 + r_0)$.

    b) Prove that $3 \mid r$ iff $3 \mid (r_n + r_{n-1} + \cdots + r_2 + r_1 + r_0)$.

    c) If $t = 137486\underline{x}225$, where x is a single digit, determine the value(s) of x so that $3 \mid t$. Which values of x make t divisible by 9?

14. a) Draw the Hasse diagram for the set of positive divisors of the integer n where $n = $ (i) 2; (ii) 4; (iii) 6; (iv) 8; (v) 12; (vi) 16; (vii) 24; (viii) 30; and, (ix) 32.

    b) For any $2 \le n \le 35$, show that the Hasse diagram for the set of positive divisors of n looks like one of the nine diagrams in part (a). (Ignore the numbers at the vertices and concentrate on the structure given by the vertices and edges.) What happens for $n = 36$?

    c) Let $m, n \in \mathbf{Z}^+$ and S, T be the sets of all positive divisors of m, n, respectively. The results of (a) and (b) imply if the Hasse diagrams of S, T are structurally the same, then $\tau(m) = \tau(n)$. But is the converse true?

    d) Show that any Hasse diagram in (a) is a lattice if we define glb$\{x, y\} = (x, y)$ and lub$\{x, y\} = [x, y]$.

15. a) Ten students enter a locker room which contains 10 lockers. The first student opens all the lockers. The second student changes the status (from closed to open, or vice versa) of every other locker, starting with the second locker. The third student then changes the status of every third locker, starting at the third locker. In general, for $1 < k \le 10$, the kth student changes the status of every kth locker, starting with the kth locker. After the tenth student has gone through the lockers, which lockers are left open?

    b) Answer part (a) if 10 is replaced by $n \in \mathbf{Z}^+$, $n \ge 2$.

7

The Principle
of Inclusion
and Exclusion

We now depart from our study of the integers and return to the topic of enumeration as we investigate the *Principle of Inclusion and Exclusion*. Extending the ideas in the counting problems on Venn diagrams in Chapter 2, this principle will assist us in establishing the formula we conjectured in Section 3.3 for the number of onto functions $f: A \to B$, where $|A| = m \geq n = |B|$. Other applications of this principle will demonstrate its versatile nature in combinatorial and discrete mathematics as an *indirect* method for certain situations in enumeration.

7.1 ■ THE PRINCIPLE OF INCLUSION AND EXCLUSION

In this section we develop some notation for stating our new counting principle. Then we establish the principle by a combinatorial argument. Examples will then demonstrate how this principle is applied.

Let S be a set with $N = |S|$, and let $c_1, c_2, \ldots, c_t$ be a collection of conditions or properties satisfied by some, or all, of the elements of S. Some elements of S may satisfy more than one of the conditions, while some may not satisfy any of them. For $1 \leq i \leq t$, $N(c_i)$ will denote the number of elements in S that satisfy condition c_i. (Elements of S are counted here when they satisfy only condition c_i, as well as when they satisfy c_i and other conditions c_j, for $j \neq i$.) For $i, j \in \{1, 2, 3, \ldots, t\}$, $i \neq j$, $N(c_i c_j)$ will denote the number of elements in S that satisfy both of the conditions c_i, c_j, and perhaps some others. ($N(c_i c_j)$ does *not* count the elements of S that satisfy *only* c_i, c_j.) Continuing, if $1 \leq i, j, k \leq t$ are three distinct integers, then $N(c_i c_j c_k)$ denotes the number of elements in S satisfying, perhaps among others, each of the conditions c_i, c_j, and c_k.

For $1 \leq i \leq t$, $N(\bar{c_i}) = N - N(c_i)$ denotes the number of elements in S that do not satisfy condition c_i. If $1 \leq i, j \leq t$, $i \neq j$, $N(\bar{c_i}\bar{c_j}) =$ the number of elements in S that do not satisfy either of the conditions c_i or c_j. (This is not the same as $N(\overline{c_i c_j})$.)

From the Venn diagram in Fig. 7.1, we see that if $N(c_i)$ denotes the number of elements in the left-hand circle and $N(c_j)$ denotes the number of elements in the

Figure 7.1

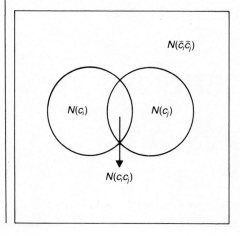

right-hand circle, then $N(c_ic_j)$ is the number of elements in the overlap, while $N(\overline{c}_i\overline{c}_j)$ counts the elements outside the union of these circles.

Consequently, from Fig. 7.1 $N(\overline{c}_i\overline{c}_j) = N - [N(c_i) + N(c_j)] + N(c_ic_j)$, where the last term is added on since it was eliminated twice in the term $[N(c_i) + N(c_j)]$.

Figure 7.2

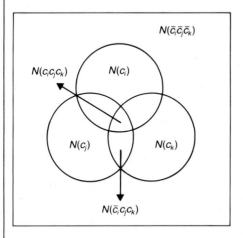

In like manner, from Fig. 7.2

$$N(\overline{c}_i\overline{c}_j\overline{c}_k) = N - [N(c_i) + N(c_j) + N(c_k)] + [N(c_ic_j) + N(c_ic_k) + N(c_jc_k)]$$
$$- N(c_ic_jc_k).$$

From the pattern suggested by these two cases we state the following.

Theorem 7.1

▶ (*The Principle of Inclusion and Exclusion*) Consider a set S, with $|S| = N$, and conditions c_i, $1 \le i \le t$, satisfied by some of the elements of S. The number of elements of S that satisfy *none* of the conditions c_i, $1 \le i \le t$, is denoted by $\overline{N} = N(\overline{c}_1\overline{c}_2\overline{c}_3 \ldots \overline{c}_t)$ where

$$\overline{N} = N - [N(c_1) + N(c_2) + N(c_3) + \cdots + N(c_t)]$$
$$+ [N(c_1c_2) + N(c_1c_3) + \cdots + N(c_1c_t) + N(c_2c_3) + \cdots + N(c_{t-1}c_t)]$$
$$- [N(c_1c_2c_3) + N(c_1c_2c_4) + \cdots + N(c_1c_2c_t) + N(c_1c_3c_4) + \cdots$$
$$+ N(c_1c_3c_t) + \cdots + N(c_{t-2}c_{t-1}c_t)] + \cdots + (-1)^t N(c_1c_2c_3 \ldots c_t), \qquad (1)$$

or

$$\overline{N} = N - \sum_{1 \le i \le t} N(c_i) + \sum_{1 \le i < j \le t} N(c_ic_j) - \sum_{1 \le i < j < k \le t} N(c_ic_jc_k) + \cdots$$
$$+ (-1)^t N(c_1c_2c_3 \ldots c_t). \qquad (2)$$

Proof

Although this result can be established by induction on t, we give a combinatorial argument here.

For each $x \in S$ we show that x contributes the same count, either 0 or 1, to each side of Eq. (2).

If x satisfies none of the conditions, then x is counted once in $\overline{N}$ and once in N, but not in any of the other terms in (2). Consequently, x contributes a count of 1 to each side of the equation.

The other possibility is that x satisfies r of the conditions where $1 \leq r \leq t$. In this case x contributes nothing to $\overline{N}$. But on the right hand side of (2) x is counted:

(1)   One time in N.

(2) r times in $\sum\limits_{1 \leq i \leq t} N(c_i)$. (Once for each of the r conditions.)

(3) $\binom{r}{2}$ times in $\sum\limits_{1 \leq i < j \leq t} N(c_i c_j)$. (Once for each pair of conditions from the r conditions it satisfies.)

(4) $\binom{r}{3}$ times in $\sum\limits_{1 \leq i < j < k \leq t} N(c_i c_j c_k)$. (Why?)

....

$(r + 1)$ $\binom{r}{r} = 1$ time in $\sum N(c_{i_1} c_{i_2} \ldots c_{i_r})$,

where the summation is taken over all selections of size r from the t conditions.

Consequently, on the right-hand side of (2) x is counted

$$1 - r + \binom{r}{2} - \binom{r}{3} + \cdots + (-1)^r \binom{r}{r} = [1 + (-1)]^r = 0^r = 0 \text{ times},$$

by the binomial theorem. Therefore, the two sides of Equation (2) (and Equation (1)) count the same elements of S and the equality is verified.   ∎

An immediate corollary of this principle is given as follows:

Corollary 7.1   ▶   Under the hypotheses of Theorem 7.1, the number of elements in S that satisfy at least one of the conditions c_i, $1 \leq i \leq t$, is given by $N(c_1$ or c_2 or $\ldots$ or $c_t) = N - \overline{N}$.

Before solving some examples we examine some further notation for simplifying the statement of Theorem 7.1.

We write

$S_0 = N$,

$S_1 = [N(c_1) + N(c_2) + \cdots + N(c_t)]$,

$S_2 = [N(c_1 c_2) + N(c_1 c_3) + \cdots + N(c_1 c_t) + N(c_2 c_3) + \cdots + N(c_{t-1} c_t)]$,

and, in general,

$S_k = \sum N(c_{i_1} c_{i_2} \ldots c_{i_k})$, $1 \leq k \leq t$, where the summation is taken over all selections of size k from the list of t conditions. Hence S_k has $\binom{t}{k}$ summands in it.

At this point let us look at how this principle is used to solve certain enumeration problems.

EXAMPLE 7.1   Determine the number of positive integers n where $1 \leq n \leq 100$ and n is *not* divisible by 2, 3, or 5.

Here $S = \{1, 2, 3, \ldots, 100\}$ and $N = 100$. For $n \in S$, n satisfies:

a)   condition c_1 if n is divisible by 2;

b)   condition c_2 if n is divisible by 3; and

c)   condition c_3 if n is divisible by 5.

Then the answer to this problem is $N(\overline{c}_1\overline{c}_2\overline{c}_3)$.
For $r \in \mathbf{R}$, $\lfloor r \rfloor$ denotes the *greatest integer* in r, where

$$\lfloor r \rfloor = \begin{cases} r, & \text{if } r \in \mathbf{Z} \\ \text{the largest integer smaller than } r, & \text{if } r \text{ is not an integer.} \end{cases}$$

So, for example, $\lfloor 5 \rfloor = 5$, $\lfloor \pi \rfloor = 3$ and $\lfloor -\pi \rfloor = -4$.
This function is helpful in this problem, as we find that

$$N(c_1) = \lfloor 100/2 \rfloor = 50, \qquad N(c_2) = \lfloor 100/3 \rfloor = \lfloor 33\ 1/3 \rfloor = 33,$$

$$N(c_3) = \lfloor 100/5 \rfloor = 20, \qquad N(c_1c_2) = \lfloor 100/6 \rfloor = 16,$$

$$N(c_1c_3) = \lfloor 100/10 \rfloor = 10, \qquad N(c_2c_3) = \lfloor 100/15 \rfloor = 6,$$

$$\text{and} \quad N(c_1c_2c_3) = \lfloor 100/30 \rfloor = 3.$$

Applying the inclusion and exclusion principle we find that

$$N(\overline{c}_1\overline{c}_2\overline{c}_3) = S_0 - S_1 + S_2 - S_3 = N - [N(c_1) + N(c_2) + N(c_3)]$$
$$+ [N(c_1c_2) + N(c_1c_3) + N(c_2c_3)] - N(c_1c_2c_3)$$
$$= 100 - [50 + 33 + 20] + [16 + 10 + 6] - 3 = 26.$$

(These 26 numbers are 1, 7, 11, 13, 17, 19, 23, 29, 31, 37, 41, 43, 47, 49, 53, 59, 61, 67, 71, 73, 77, 79, 83, 89, 91, 97.)   □

EXAMPLE 7.2   In Chapter 1 we found the number of nonnegative integer solutions to the equation $x_1 + x_2 + x_3 + x_4 = 18$. We now answer the same question with the extra restriction that $x_i \leq 7$, $1 \leq i \leq 4$.

Here S is the set of solutions of $x_1 + x_2 + x_3 + x_4 = 18$, with $0 \leq x_i$, $1 \leq i \leq 4$. So $|S| = N = \binom{4+18-1}{18} = \binom{21}{18}$.

We say that a solution x_1, x_2, x_3, x_4 satisfies condition c_i, $1 \leq i \leq 4$, if $x_i > 7$ (or $x_i \geq 8$). The answer to the problem is then $N(\overline{c}_1\overline{c}_2\overline{c}_3\overline{c}_4)$.

Here $N(c_1) = N(c_2) = N(c_3) = N(c_4)$. To compute $N(c_1)$ we want the number of integer solutions of $x_1 + x_2 + x_3 + x_4 = 10$, with each $x_i \geq 0$, $1 \leq i \leq 4$. Then we add 8 to the value of x_1 and get the solutions of $x_1 + x_2 + x_3 + x_4 = 18$ that satisfy condition c_1. Hence $N(c_i) = \binom{4+10-1}{10} = \binom{13}{10}$, for each $1 \le i \le 4$, and $S_1 = \binom{4}{1}\binom{13}{10}$.

Likewise, $N(c_1 c_2)$ is the number of integer solutions of $x_1 + x_2 + x_3 + x_4 = 2$, $x_i \ge 0$, $1 \le i \le 4$. So $N(c_1 c_2) = \binom{4+2-1}{2} = \binom{5}{2}$, and $S_2 = \binom{4}{2}\binom{5}{2}$.

Since $N(c_i c_j c_k) = 0$ for any selection of three conditions, and $N(c_1 c_2 c_3 c_4) = 0$, we have

$$N(\bar{c}_1 \bar{c}_2 \bar{c}_3 \bar{c}_4) = N - S_1 + S_2 - S_3 + S_4 = \binom{21}{18} - \binom{4}{1}\binom{13}{10} + \binom{4}{2}\binom{5}{2} - 0 + 0 = 246.$$

So of the 1330 nonnegative integer solutions of $x_1 + x_2 + x_3 + x_4 = 18$, only 246 of them satisfy $x_i \le 7$, $1 \le i \le 4$.   □

Our next example establishes the formula conjectured in Section 3.3 for counting onto functions.

EXAMPLE 7.3   For finite sets A, B, with $|A| = m \ge n = |B|$, let $A = \{a_1, a_2, \ldots, a_m\}$, $B = \{b_1, b_2, \ldots, b_n\}$, and $S = $ the set of all functions $f: A \to B$. Then $N = |S| = n^m$.

For $1 \le i \le n$, let c_i denote the condition on S where a function $f: A \to B$ satisfies c_i if b_i is *not* in the range of f. (Note the difference between c_i here, and c_i in Examples 7.1 and 7.2.) Then $N(\bar{c}_i)$ is the number of functions in S which have b_i in their range, and $N(\bar{c}_1 \bar{c}_2 \ldots \bar{c}_n)$ counts the number of onto functions $f: A \to B$.

For any $1 \le i \le n$, $N(c_i) = (n-1)^m$, since any element of B, except b_i, can be used as the second component of an ordered pair for a function $f: A \to B$, where the range of f does not include b_i. Likewise, for any $1 \le i < j \le n$, there are $(n-2)^m$ functions $f: A \to B$ where the range contains neither b_i nor b_j. From these observations we have $S_1 = [N(c_1) + N(c_2) + \cdots + N(c_n)] = n(n-1)^m = \binom{n}{1}(n-1)^m$, and $S_2 = [N(c_1 c_2) + N(c_1 c_3) + \cdots + N(c_1 c_n) + N(c_2 c_3) + \cdots + N(c_2 c_n) + \cdots + N(c_{n-1} c_n)] = \binom{n}{2}(n-2)^m$. In general, for $1 \le k \le n$,

$$S_k = \sum_{1 \le i_1 < i_2 < \ldots < i_k \le n} N(c_{i_1} c_{i_2} \ldots c_{i_k}) = \binom{n}{k}(n-k)^m$$

It then follows by the principle of inclusion and exclusion that the number of onto functions from A to B is

$$N(\bar{c}_1 \bar{c}_2 \bar{c}_3 \ldots \bar{c}_n) = N - S_1 + S_2 - S_3 + \cdots + (-1)^n S_n$$

$$= n^m - \binom{n}{1}(n-1)^m + \binom{n}{2}(n-2)^m - \binom{n}{3}(n-3)^m$$

$$+ \cdots + (-1)^n (n-n)^m = \sum_{i=0}^{n} (-1)^i \binom{n}{i}(n-i)^m$$

$$= \sum_{i=0}^{n} (-1)^i \binom{n}{n-i}(n-i)^m.$$ □

We now solve a problem similar to those in Chapter 2 that deal with Venn diagrams.

EXAMPLE 7.4

In how many ways can the 26 letters of the alphabet be arranged so that none of the patterns *car, dog, pun,* or *byte* occurs?

Let S denote the set of all arrangements of the 26 letters. Then $|S| = 26!$ For $1 \le i \le 4$, an arrangement in S is said to satisfy condition c_i if it contains car, dog, pun, or byte, respectively.

$$N(c_1) = N(c_2) = N(c_3) = 24!, \qquad N(c_4) = 23!$$

$$N(c_1c_2) = N(c_1c_3) = N(c_2c_3) = 22!, \qquad N(c_ic_4) = 21!, \qquad i \ne 4$$

$$N(c_1c_2c_3) = 20!, \qquad N(c_ic_jc_4) = 19!, \qquad 1 \le i < j \le 3$$

$$N(c_1c_2c_3c_4) = 17!$$

So the number of arrangements in S that contain none of the given patterns is
$N(\bar{c}_1\bar{c}_2\bar{c}_3\bar{c}_4) = 26! - [3(24!) + 23!] + [3(22!) + 3(21!)] - [20! + 3(19!)] + 17!$ □

Our next example deals with a number theory problem.

EXAMPLE 7.5

For $n \in \mathbf{Z}^+$, let $\phi(n)$ be the number of positive integers m, where $1 \le m < n$ and $(m, n) = 1$, i.e., m, n are relatively prime. This function is known as *Euler's phi function* and arises in several situations in abstract algebra involving enumeration. We find that $\phi(2) = 1$, $\phi(3) = 2$, $\phi(4) = 2$, $\phi(5) = 4$, $\phi(6) = 2$. For any prime p, $\phi(p) = p - 1$. We would like to derive a formula for $\phi(n)$ that is related to n so that we need not make a case-by-case comparison for each m, $1 \le m < n$, against the integer n.

The derivation of our formula will use the inclusion-exclusion principle as in Example 7.1. We proceed as follows: For $n \ge 2$, write $n = p_1^{e_1}p_2^{e_2}\cdots p_t^{e_t}$, where $p_1, p_2, \ldots, p_t$ are distinct primes and $e_i \ge 1$, $1 \le i \le t$. We consider the case where $t = 4$. This will be enough to demonstrate the general idea.

With $S = \{1, 2, 3, \ldots, n\}$, $N = |S| = n$, and for $1 \le i \le 4$ we say that $k \in S$ satisfies condition c_i if k is divisible by p_i. For $1 \le k < n$, $(k, n) = 1$ if k is not divisible by any of the primes p_i, $1 \le i \le 4$. Hence $\phi(n) = N(\bar{c}_1\bar{c}_2\bar{c}_3\bar{c}_4)$.

For $1 \le i \le 4$, $N(c_i) = n/p_i$; $N(c_ic_j) = n/(p_ip_j)$, $1 < i < j \le 4$. Also, $N(c_ic_jc_\ell) = n/(p_ip_jp_\ell)$, $1 \le i < j < \ell \le 4$, and $N(c_1c_2c_3c_4) = n/(p_1p_2p_3p_4)$. So

$$\phi(n) = S_0 - S_1 + S_2 - S_3 + S_4$$

$$= n - \left[\frac{n}{p_1} + \cdots + \frac{n}{p_4} \right] + \left[\frac{n}{p_1p_2} + \frac{n}{p_1p_3} + \cdots + \frac{n}{p_3p_4} \right]$$

$$- \left[\frac{n}{p_1p_2p_3} + \cdots + \frac{n}{p_2p_3p_4} \right] + \frac{n}{p_1p_2p_3p_4}$$

$$= n \left[1 - \left(\frac{1}{p_1} + \cdots + \frac{1}{p_4} \right) + \left(\frac{1}{p_1p_2} + \frac{1}{p_1p_3} + \cdots + \frac{1}{p_3p_4} \right) \right.$$

$$\left. - \left(\frac{1}{p_1p_2p_3} + \cdots + \frac{1}{p_2p_3p_4} \right) + \frac{1}{p_1p_2p_3p_4} \right]$$

$$= \frac{n}{p_1 p_2 p_3 p_4} [p_1 p_2 p_3 p_4 - (p_2 p_3 p_4 + p_1 p_3 p_4 + p_1 p_2 p_4 + p_1 p_2 p_3)$$

$$+ (p_3 p_4 + p_2 p_4 + p_2 p_3 + p_1 p_4 + p_1 p_3 + p_1 p_2)$$

$$- (p_4 + p_3 + p_2 + p_1) + 1]$$

$$= \frac{n}{p_1 p_2 p_3 p_4} [(p_1 - 1)(p_2 - 1)(p_3 - 1)(p_4 - 1)]$$

$$= n \left[\frac{p_1 - 1}{p_1} \cdot \frac{p_2 - 1}{p_2} \cdot \frac{p_3 - 1}{p_3} \cdot \frac{p_4 - 1}{p_4} \right] = n \prod_{i=1}^{4} \left(1 - \frac{1}{p_i} \right).$$

In general, $\phi(n) = n \prod_{p \mid n}(1 - (1/p))$, where the product is taken over all primes p dividing n.

Consequently,

$$\phi(23,100) = \phi(2^2 \cdot 3 \cdot 5^2 \cdot 7 \cdot 11)$$

$$= (23,100)(1 - (1/2))(1 - (1/3))(1 - (1/5)) \cdot$$

$$(1 - (1/7))(1 - (1/11)) = 4800. \qquad \square$$

The Pascal program in Fig. 7.3, evaluates $\phi(n)$ for $n \geq 1$. Here we use it to find $\phi(n)$ for $n = 131$; 31,500; and 198,000.

Figure 7.3

```
Program EulerPhiFunction (input,output);
Var
          i,j,k,n,phi,originalvalue: integer;
Begin
          Write ('The value of n is ');
          Read (n);
          phi := n;
          originalvalue := n;
          If n Mod 2 = 0 Then
                    Begin
                              phi := phi Div 2;
                              While n Mod 2 = 0 Do
                                        n := n Div 2
                    End;
          If n Mod 3 = 0 Then
                    Begin
                              phi := (phi * 2) Div 3;
                              While n Mod 3 = 0 Do
                                        n := n Div 3
                    End;
          i := 5;
          While n >= 5 Do
                    Begin
                              j := 1;
```

```
                    Repeat
                         j := j + 1;
                         k := i Mod j
                    Until (k = 0) Or (j = trunc(sqrt(i)));
                    If (k<>0) And (n Mod i = 0) Then
                         Begin
                              phi := (phi * (i − 1)) Div i;
                              While n Mod i = 0 Do
                                        n := n Div i
                         End;
                    i := i + 2
               End;
          Write ('For n = ', originalvalue:0, ' there are ', phi:0, ' numbers');
          Write (' smaller than ', originalvalue:0);
          Writeln (' and relatively prime to it.')
End.
```

The value of n is 131
For n = 131 there are 130 numbers smaller than 131 and relatively prime to it.

The value of n is 31500
For n = 31500 there are 7200 numbers smaller than 31500 and relatively prime to it.

The value of n is 198000
For n = 198000 there are 48000 numbers smaller than 198000 and relatively prime to it.

Our final example recalls some of the graph theory of Chapter 5.

EXAMPLE 7.6 In a certain area of the countryside there are five villages. An engineer is to devise a system of two-way roads so that after the system is completed no village is isolated. In how many ways can he do this?

Calling the villages a, b, c, d, e we seek the number of loop-free undirected graphs on these vertices where no vertex is isolated. Consequently, we want to count situations like (a), (b) in Fig. 7.4, but not situation (c).

Figure 7.4

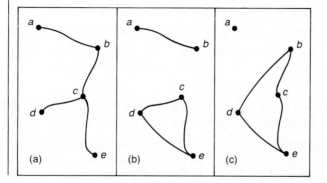

(a)   (b)   (c)

Let S be the set of loop-free undirected graphs G on $V = \{a, b, c, d, e\}$. Then $N = |S| = 2^{10}$, since there are $\binom{5}{2} = 10$ possible two-way roads for these five villages, and each road can be either included or excluded.

For $1 \le i \le 5$, let c_i be the condition that a system of these roads isolates village a, b, c, d, e, respectively. Then the answer to the problem is $N(\bar{c}_1 \bar{c}_2 \bar{c}_3 \bar{c}_4 \bar{c}_5)$. We find that $N(c_1) = 2^6$ and $S_1 = \binom{5}{1}2^6$; $N(c_1 c_2) = 2^3$ and $S_2 = \binom{5}{2}2^3$; $N(c_1 c_2 c_3) = 2^1$ and $S_3 = \binom{5}{3}2^1$; $N(c_1 c_2 c_3 c_4) = 2^0$ and $S_4 = \binom{5}{4}2^0$; and, $N(c_1 c_2 c_3 c_4 c_5) = 2^0$ and $S_5 = \binom{5}{5}2^0$.

Consequently, $N(\bar{c}_1 \bar{c}_2 \bar{c}_3 \bar{c}_4 \bar{c}_5) = 2^{10} - \binom{5}{1}2^6 + \binom{5}{2}2^3 - \binom{5}{3}2^1 + \binom{5}{4}2^0 - \binom{5}{5}2^0 = 768.$ □

EXERCISES

1. Determine the number of positive integers n, $1 \le n \le 2000$, that are
   a) not divisible by 2, 3, or 5;
   b) not divisible by 2, 3, 5, or 7; and
   c) not divisible by 2, 3, or 5, but are divisible by 7.

2. Find all real numbers x such that
   a) $7\lfloor x \rfloor = \lfloor 7x \rfloor$;
   b) $\lfloor 7x \rfloor = 7$;
   c) $\lfloor x + 7 \rfloor = x + 7$;
   d) $\lfloor x + 7 \rfloor = \lfloor x \rfloor + 7$.

3. Determine how many integer solutions there are to $x_1 + x_2 + x_3 + x_4 = 19$, if
   a) $0 \le x_i$, $1 \le i \le 4$
   b) $0 \le x_i < 8$, $1 \le i \le 4$
   c) $0 \le x_1 \le 5$, $0 \le x_2 \le 6$, $3 \le x_3 \le 7$, $3 \le x_4 \le 8$.

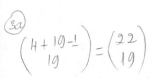

$\left(\begin{array}{c} 4 + 10 - 1 \\ 10 \end{array} \right) = \left(\begin{array}{c} 22 \\ 19 \end{array} \right)$

4. Determine the number of integer solutions there are to $x_1 + x_2 + x_3 + x_4 = 19$ if $-5 \le x_i \le 10$, $1 \le i \le 4$.

5. In how many ways can three x's, three y's, and three z's be arranged so that no consecutive triple of the same letter appears?

6. In how many ways can Troy select nine marbles from a bag of twelve, where three are red, three blue, three white, and three green?

7. Find the number of arrangements of $a, b, c, \ldots, x, y, z$, in which none of the patterns *spin*, *game*, *path*, or *net* occurs.

8. Compute $\phi(n)$ for n equal to (a) 51; (b) 420; and (c) 12300.

9. Let $n \in \mathbf{Z}^+$. (a) Determine $\phi(2^n)$. (b) Determine $\phi(2^n p)$, where p is an odd prime.

10. For $n \in \mathbf{Z}^+$, when is $\phi(n)$ odd?

11. How many positive integers n less than 6000 (a) satisfy $(n, 6000) = 1$? (b) share a common prime divisor with 6000?

12. Answer the question in Example 7.6 for the case of six villages.

13. If eight distinct dice are rolled, what is the probability that all six numbers appear?

14. How many social security numbers (nine digit sequences) have each of the digits 1, 3, and 7 appearing at least once?

15. At a six-week conference in mathematics Sharon met seven of her friends from college. During the conference she met each friend at lunch 15 times, every pair of them 7 times, every trio 5 times, every foursome 4 times, each set of five 3 times, each set of six twice, but never all seven at once. If she had lunch every day during the 42 days of the conference, did she ever have lunch alone?

16. Mrs. Phillips has eight grandchildren all of whom like ice cream. In her freezer she has enough ice cream for 6 vanilla, 3 chocolate, 6 strawberry, and 5 chocolate chip cones. On her birthday all of her grandchildren come to see her, and her oldest grandchild tells her how many requests there are for each flavor. In how many ways can her grandchildren make requests that will embarrass Mrs. Phillips for not having enough of each flavor?

17. For the program in Fig. 7.3, what purpose do the three two-line While loops serve?

    Why is i incremented by 2 in the line $i := i + 2$, instead of by 1? (Would a different result come about if we incremented i by 1?)

7.2 ■ GENERALIZATIONS OF THE PRINCIPLE

Given a set S with $|S| = N$, and conditions $c_1, c_2, \ldots, c_t$ satisfied by some of the elements of S, in Section 7.1 we saw how inclusion and exclusion provides a way to determine $N(\bar{c}_1\bar{c}_2\ldots\bar{c}_t)$, the number of elements in S that satisfy none of the t conditions. If $m \in \mathbf{Z}^+$, $1 \leq m \leq t$, we now want to determine $E_m =$ the number of elements in S that satisfy exactly m of the t conditions. (At present we can obtain E_0.)

We can write such equations as

$$E_1 = N(c_1\bar{c}_2\bar{c}_3\ldots\bar{c}_t) + N(\bar{c}_1 c_2\bar{c}_3\ldots\bar{c}_t) + \cdots + N(\bar{c}_1\bar{c}_2\bar{c}_3\ldots c_t),$$

and

$$E_2 = N(c_1 c_2\bar{c}_3\ldots\bar{c}_t) + N(c_1\bar{c}_2 c_3\ldots\bar{c}_t) + \cdots + N(\bar{c}_1\bar{c}_2\bar{c}_3\ldots\bar{c}_{t-2}c_{t-1}c_t),$$

and although these results do not assist us as much as we should like, we shall find them to be a useful starting place as we examine the Venn diagrams for the cases where $t = 3$ and 4.

For Fig. 7.5 where $t = 3$, we place a numbered condition beside the circle representing those elements of S satisfying that particular condition. Then E_1 equals the number of elements in regions 2, 3, 4. But we can also write

$$E_1 = N(c_1) + N(c_2) + N(c_3) - 2[N(c_1 c_2) + N(c_1 c_3) + N(c_2 c_3)] + 3N(c_1 c_2 c_3).$$

In $N(c_1) + N(c_2) + N(c_3)$ we count the elements in regions 5, 6, and 7 twice and those in region 8 three times. In the next term the elements in regions 5, 6, and 7 are deleted twice. We remove the elements in region 8 six times in $2[N(c_1 c_2) + N(c_1 c_3) +$

Figure 7.5

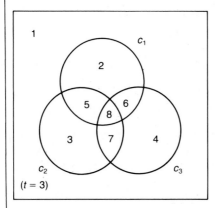

$(t = 3)$

$N(c_2c_3)]$, so we then add on the term $3N(c_1c_2c_3)$ and end up not counting the elements in region 8 at all. Hence we have $E_1 = S_1 - 2S_2 + 3S_3 = S_1 - \binom{2}{1}S_2 + \binom{3}{2}S_3$.

Turning to E_2, our earlier equation indicates that we want to count the elements of S in regions 5, 6, and 7. From the Venn diagram

$$E_2 = N(c_1c_2) + N(c_1c_3) + N(c_2c_3) - 3N(c_1c_2c_3) = S_2 - 3S_3 = S_2 - \binom{3}{1}S_3,$$

and

$$E_3 = N(c_1c_2c_3) = S_3.$$

In Fig. 7.6, the conditions c_1, c_2, c_3 are associated with circular subsets of S while c_4 is paired with the rather irregular shaped area made up of regions 4, 8, 9, 11, 12, 13, 14, and 16. For $1 \le i \le 4$, E_i is determined as follows:

E_1 [regions 2, 3, 4, 5]:

$$
\begin{aligned}
E_1 = {}& [N(c_1) + N(c_2) + N(c_3) + N(c_4)] \\
& - 2[N(c_1c_2) + N(c_1c_3) + N(c_1c_4) + N(c_2c_3) + N(c_2c_4) + N(c_3c_4)] \\
& + 3[N(c_1c_2c_3) + N(c_1c_2c_4) + N(c_1c_3c_4) + N(c_2c_3c_4)] \\
& - 4N(c_1c_2c_3c_4) \\
= {}& S_1 - 2S_2 + 3S_3 - 4S_4 = S_1 - \binom{2}{1}S_2 + \binom{3}{2}S_3 - \binom{4}{3}S_4 .
\end{aligned}
$$

(Note: Taking an element in region 3 we find that it is counted once in E_1 and once in S_1 (in $N(c_3)$). Taking an element in region 6 we find that it is not counted in E_1; it is counted twice in S_1 (in both $N(c_2)$ and $N(c_3)$) but removed twice in $2S_2$ (for it is counted once in S_2 in $N(c_2c_3)$), so overall it is not counted. The reader should now consider an element from region 12 and one from region 16 and show that each contributes a count of 0 to both sides of the formula for E_1.)

E_2 [regions 6–11]:

From Fig. 7.6, $E_2 = S_2 - 3S_3 + 6S_4 = S_2 - \binom{3}{1}S_3 + \binom{4}{2}S_4$. For details on this formula we examine the results in Table 7.1, where next to each summand of S_2, S_3, S_4 we list the regions whose elements are counted in determining that particular

Figure 7.6

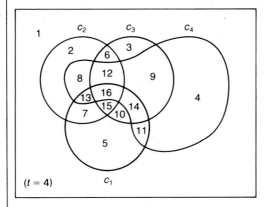

$(t = 4)$

Table 7.1

S_2	S_3	S_4
$N(c_1c_2)$: 7, 13, 15, 16 $N(c_1c_3)$: 10, 14, 15, 16 $N(c_1c_4)$: 11, 13, 14, 16 $N(c_2c_3)$: 6, 12, 15, 16 $N(c_2c_4)$: 8, 12, 13, 16 $N(c_3c_4)$: 9, 12, 14, 16	$N(c_1c_2c_3)$: 15, 16 $N(c_1c_2c_4)$: 13, 16 $N(c_1c_3c_4)$: 14, 16 $N(c_2c_3c_4)$: 12, 16	$N(c_1c_2c_3c_4)$: 16

summand. In calculating $S_2 - 3S_3 + 6S_4$ we find the elements in regions 6–11, which are precisely those that are to be counted in E_2.

Finally, E_3 is made up of regions 12–15 and $E_3 = S_3 - 4S_4 = S_3 - \binom{4}{1}S_4$; E_4 consists of the elements in region 16 and $E_4 = S_4$.

These results suggest the following.

Theorem 7.2

▶ Under the hypotheses of Theorem 7.1, for any $1 \le m \le t$, the number of elements in S that satisfy *exactly* m of the conditions $c_1, c_2, \ldots, c_t$ is given by

$$E_m = S_m - \binom{m+1}{1}S_{m+1} + \binom{m+2}{2}S_{m+2} - \cdots + (-1)^{t-m}\binom{t}{t-m}S_t. \tag{1}$$

(If $m = 0$ we obtain Theorem 7.1.)

Proof

Arguing as in Theorem 7.1, let $x \in S$ and consider the following three cases.

a) If x satisfies fewer than m conditions then it contributes a count of 0 to each of the terms $E_m, S_m, S_{m+1}, \ldots, S_t$, and is not counted on either side of the equation.

b) When x satisfies exactly m of the conditions it is counted once in E_m and once in S_m, but not in $S_{m+1}, \ldots, S_t$. Consequently it is included once in the count for either side of the equation.

c) Suppose x satisfies r of the conditions where $m < r \le t$. Then x contributes nothing to E_m. Yet it is counted $\binom{r}{m}$ times in S_m, $\binom{r}{m+1}$ times in $S_{m+1}, \ldots,$ and $\binom{r}{r}$

times in S_r, but 0 times for any term beyond S_r. So on the right-hand side of the equation x is counted $\binom{r}{m} - \binom{m+1}{1}\binom{r}{m+1} + \binom{m+2}{2}\binom{r}{m+2} - \cdots + (-1)^{r-m}\binom{r}{r-m}\binom{r}{r}$ times.

For $0 \le k \le r - m$,

$$\binom{m+k}{k}\binom{r}{m+k} = \frac{(m+k)!}{k!\,m!} \cdot \frac{r!}{(m+k)!\,(r-m-k)!}$$

$$= \frac{r!}{m!} \cdot \frac{1}{k!\,(r-m-k)!} = \frac{r!}{m!\,(r-m)!} \cdot \frac{(r-m)!}{k!\,(r-m-k)!}$$

$$= \binom{r}{m}\binom{r-m}{k}.$$

Consequently, on the right-hand side of Eq. (1) x is counted

$$\binom{r}{m}\binom{r-m}{0} - \binom{r}{m}\binom{r-m}{1} + \binom{r}{m}\binom{r-m}{2} - \cdots + (-1)^{r-m}\binom{r}{m}\binom{r-m}{r-m}$$

$$= \binom{r}{m}\left[\binom{r-m}{0} - \binom{r-m}{1} + \binom{r-m}{2} - \cdots + (-1)^{r-m}\binom{r-m}{r-m}\right]$$

$$= \binom{r}{m}[1-1]^{r-m} = \binom{r}{m} \cdot 0 = 0 \text{ times},$$

and the formula is verified.   ∎

Based on this result, if L_m denotes the elements of S (under the hypotheses of Theorem 7.1) that satisfy *at least* m of the t conditions, we have the following formula.

Corollary 7.2 ▶ $L_m = S_m - \binom{m}{m-1}S_{m+1} + \binom{m+1}{m-1}S_{m+2} - \cdots + (-1)^{t-m}\binom{t-1}{m-1}S_t$.

Proof A proof is outlined in the exercises at the end of Section 7.3.   ∎

EXAMPLE 7.7 Referring back to Example 7.6 we shall find the number of systems of two-way roads the engineer can devise so that after completion exactly (at least) two of the towns remain isolated.

Using the results previously calculated for this example we have

$$E_2 = S_2 - \binom{3}{1}S_3 + \binom{4}{2}S_4 - \binom{5}{3}S_5 = 80 - 3(20) + 6(5) - 10(1) = 40,$$

$$L_2 = S_2 - \binom{2}{1}S_3 + \binom{3}{1}S_4 - \binom{4}{1}S_5 = 80 - 2(20) + 3(5) - 4(1) = 51.$$ □

7.3 ■ DERANGEMENTS: NOTHING IS IN ITS RIGHT PLACE

From elementary calculus we find that the Maclaurin series for the exponential function is given by

$$e^x = 1 + x + \frac{x^2}{2!} + \frac{x^3}{3!} + \cdots = \sum_{n=0}^{\infty} \frac{x^n}{n!},$$

so that

$$e^{-1} = \sum_{n=0}^{\infty} \frac{(-1)^n}{n!} = 1 - 1 + \frac{1}{2!} - \frac{1}{3!} + \cdots$$

To five places, $e^{-1} = 0.36788$ and $1 - 1 + (1/2!) - (1/3!) + \cdots - (1/7!) = 0.36786$. Consequently, for any $k \in \mathbf{Z}^+$, $k \geq 7$, e^{-1} is a very good approximation to $\sum_{n=0}^{k} ((-1)^n)/n!$.

We find these ideas helpful in the following.

EXAMPLE 7.8 While at the racetrack, Ralph bets on each of the ten horses in a race to come in according to how they are favored. In how many ways can they reach the finish line so that he loses all of his bets?

Removing the words "horses" and "racetrack" from the problem, we really want to know in how many ways we can arrange the numbers 1, 2, 3, ..., 10 so that 1 is not in first place (its natural position), 2 is not in second place (its natural position), ..., and 10 is not in tenth place (its natural position). These arrangements are called the *derangements* of 1, 2, 3, ..., 10.

The principle of inclusion and exclusion provides the key. For $1 \leq i \leq 10$, an arrangement of 1, 2, 3, ..., 10 is said to satisfy condition c_i if integer i is in the ith place. We obtain the number of derangements, denoted by d_{10}, as follows.

$$d_{10} = N(\bar{c}_1\bar{c}_2\bar{c}_3 \ldots \bar{c}_{10}) = 10! - \binom{10}{1}9! + \binom{10}{2}8! - \binom{10}{3}7! + \cdots + \binom{10}{10}0!$$

$$= 10![1 - \binom{10}{1}(9!/10!) + \binom{10}{2}(8!/10!) - \binom{10}{3}(7!/10!) + \cdots + \binom{10}{10}(0!/10!)]$$

$$= 10![1 - 1 + (1/2!) - (1/3!) + \cdots + (1/10!)] \doteq (10!)(e^{-1}).$$

The sample space here consists of the 10! ways the horses can finish. So the *probability* that Ralph will lose every bet is approximately $(10!)(e^{-1})/(10!) = e^{-1}$. This probability remains (more or less) the same if the number of horses in the race is 11, 12, ... On the other hand, for n horses, $n \geq 10$, the probability that our gambler wins at least one bet is $1 - e^{-1} = 0.63212$.   □

EXAMPLE 7.9 At the C–H company Peggy has seven books to be reviewed, so she hires seven people to review them. She wants two reviews per book, so the first week she gives each person one book to read and then redistributes the books at the start of the second week. In how many ways can she make these two distributions so that she gets two reviews (by different people) of each book?

She can distribute the books in 7! ways the first week. Numbering both the books and the reviewers (for the first week) as 1, 2, ..., 7, for the second distribution she must arrange these numbers so that none of them is in its natural position. This she can do in d_7 ways. By the rule of product she can make the two distributions in $(7!)d_7 \doteq (7!)^2(e^{-1})$ ways.   □

EXERCISES

1. For the situation in Examples 7.6 and 7.7, compute E_i, $0 \le i \le 5$ and show that $\sum_{i=0}^{5} E_i = N = |S|$.

2. The following provides an outline for proving Corollary 7.2. Fill in the needed details.

   a) First note that $E_t = L_t = S_t$.

   b) What is E_{t-1}, and how are L_t and L_{t-1} related?

   c) Show that $L_{t-1} = S_{t-1} - \binom{t-1}{t-2}S_t$.

   d) For any $1 \le m \le t - 1$, how are L_m, L_{m+1}, E_m related?

   e) Using the results in (a)–(d), establish the corollary by a backward type of induction.

3. a) Let $A = \{1, 2, 3, \ldots, 7\}$. A function $f: A \to A$ is said to have a *fixed point* if for some $x \in A$, $f(x) = x$. How many one-to-one functions $f: A \to A$ have at least one fixed point?

   b) In how many ways can we devise a secret code by assigning to each letter of the alphabet a different letter to represent it?

4. In how many ways can the integers 1, 2, 3, ..., 10 be arranged in a line so that no even integer is in its natural position?

5. a) In how many ways can the letters in the word ARRANGEMENT be arranged so that there are exactly two pairs of consecutive identical letters? at least two pairs of consecutive identical letters?

   b) Answer part (a), replacing two by three.

6. a) Let $A = \{1, 2, 3, \ldots, 10\}$, $B = \{1, 2, 3, \ldots, 7\}$. How many functions $f: A \to B$ satisfy $|f(A)| = 4$? How many have $|f(A)| \le 4$?

   b) In how many ways can one distribute 10 distinct prizes among four students with exactly two students getting nothing? How many ways have at least two students getting nothing?

7. a) When n balls, numbered 1, 2, 3, ..., n, are taken in succession from a container, a *rencontre* occurs if the mth ball withdrawn is numbered m, $1 \le m \le n$. Find the probability of getting no rencontres; exactly one rencontre; at least one rencontre; r rencontres, $1 \le r \le n$.

   b) Approximate the answers to the questions in part (a).

8. a) In how many ways can the integers 1, 2, 3, ..., n be arranged in a line so that none of the patterns 12, 23, 34, ..., $(n-1)n$ occurs?

   b) Show that the result in part (a) equals $d_{n-1} + d_n$.

9. Answer part (a) of Exercise 8 if the numbers are arranged in a circle, and, as we count clockwise about the circle, none of the patterns 12, 23, 34, ..., $(n-1)n$, $n1$ occurs.

10. Four applicants for a job are to be interviewed for 30 minutes each: 15 minutes with each of supervisors Nancy and Yolanda. (The interviews are in separate

rooms and interviewing starts at 9:00 A.M.) (a) In how many ways can these interviews be scheduled during a one-hour period? (b) One applicant, named Josephine, arrives at 9:00 A.M. What is the probability that she will have her two interviews one after the other? (c) Regina, another applicant, arrives at 9:00 A.M. and hopes to be finished in time to leave by 9:50 A.M. for another appointment. What is the probability that Regina will be able to do this?

11. What is the probability that the gambler in Example 7.8 wins (a) exactly five of his bets? (b) at least five of his bets?

7.4 ■ ROOK POLYNOMIALS

Consider the six-square "chessboard" shown in Fig. 7.7. In chess there is a piece called a *rook* or *castle,* which is allowed at one turn to move horizontally or vertically over as many unoccupied spaces as it wishes. Here a rook in square 3 of the figure can move in one turn to squares 1, 2, or 4. A rook at square 5 could move to square 6 or square 2 (even though there is no square between squares 5 and 2).

For $k \in \mathbf{Z}^+$ we want to determine the number of ways k rooks can be placed on this chessboard so that no two of them can take each other, i.e., no two of them are in the same row or column of the chessboard. This number is denoted by r_k.

For any chessboard, r_1 is the number of squares on the board. Here $r_1 = 6$. Two nontaking rooks can be placed at the following pairs of positions: $(1, 4), (1, 5), (2, 4),$ $(2, 6), (3, 5), (3, 6), (4, 5), (4, 6)$, so $r_2 = 8$. Continuing, we find that $r_3 = 2$, using the locations $(1, 4, 5), (2, 4, 6)$; $r_k = 0$, for $k \geq 4$.

With $r_0 = 1$ the *rook polynomial,* $r(C, x)$, for the chessboard in Fig. 7.7 is defined as $r(C, x) = 1 + 6x + 8x^2 + 2x^3$. For any $k \geq 0$, the coefficient of x^k is the number of ways we can place k nontaking rooks on chessboard C.

Figure 7.7

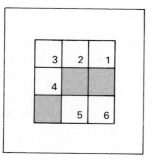

What we have done here using a case-by-case analysis soon proves tedious. As the size of the board increases, we have to consider cases where numbers like r_4 and r_5 are nonzero. Consequently, we make some observations that allow us to make use of small boards, and somehow break up a large board into smaller *subboards.*

The chessboard C in Fig. 7.8 is made up of 11 unshaded squares. We note that C consists of a 2×2 subboard C_1 and a seven-square subboard C_2 located in the

Figure 7.8

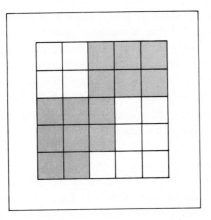

lower right corner. These subboards are *disjoint* as they have no squares in the same row or column of C.

Calculating as we did for our first chessboard, here we find

$$r(C_1, x) = 1 + 4x + 2x^2 \qquad r(C_2, x) = 1 + 7x + 10x^2 + 2x^3$$

$$r(C, x) = 1 + 11x + 40x^2 + 56x^3 + 28x^4 + 4x^5 = r(C_1, x) \cdot r(C_2, x).$$

Hence $r(C, x) = r(C_1, x) \cdot r(C_2, x)$. But did this happen by luck or is something happening here that we should examine more closely? For example, to obtain r_3 for C, we need to know in how many ways three nontaking rooks can be placed on board C. These fall into three cases:

a) All three rooks are on subboard C_2: 2 ways.

b) Two rooks are on subboard C_2 and one is on C_1: $(10)(4) = 40$ ways.

c) One rook is on subboard C_2 and two are on C_1: $(7)(2) = 14$ ways.

Consequently, three nontaking rooks can be placed on board C in $2 + (10)(4) + (7)(2) = 56$ ways. Here we see that 56 arises just as it does as the coefficient of x^3 in the product $r(C_1, x) \cdot r(C_2, x)$.

In general, if C is a chessboard made up of *pairwise* disjoint subboards $C_1, C_2, \ldots, C_n$, then $r(C, x) = r(C_1, x)r(C_2, x) \cdots r(C_n, x)$.

The last result for this section demonstrates the type of principle we have seen in other results in combinatorial and discrete mathematics: Given a large chessboard, break it into smaller subboards whose rook polynomials can be determined by inspection.

Consider the chessboard C in Fig. 7.9. Let $k \geq 1$. For any square of C, such as the one designated by (*), there are two possibilities to examine.

Figure 7.9

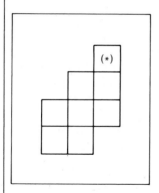

a) Place one rook on the designated square. Then we must remove as possible locations for all of the other $k - 1$ rooks, all other squares of C in the same row or column as the designated square. We use C_s to denote the remaining smaller subboard.

b) We do not use the designated square at all. The k rooks are placed on the subboard C_e (C with the one designated square eliminated).

Since these two cases are all-inclusive and disjoint,

$$r_k(C) = r_{k-1}(C_s) + r_k(C_e).$$

From this we see that

$$r_k(C)x^k = r_{k-1}(C_s)x^k + r_k(C_e)x^k. \tag{1}$$

If $n =$ the number of squares in the chessboard (here $n = 8$), as Eq. (1) is valid for $1 \le k \le n$, we write

$$\sum_{k=1}^{n} r_k(C)x^k = \sum_{k=1}^{n} r_{k-1}(C_s)x^k + \sum_{k=1}^{n} r_k(C_e)x^k. \tag{2}$$

For Eq. (2) we realize that the summations may stop before $k = n$. We have seen cases, as in Fig. 7.7, where r_n and some prior r_k's are 0. The summations start at $k = 1$, for otherwise we could find ourselves with the term $r_{-1}(C_s)x^0$ in the first summand on the right-hand side of Eq. (2).

Equation (2) may be rewritten as

$$\sum_{k=1}^{n} r_k(C)x^k = x \sum_{k=1}^{n} r_{k-1}(C_s)x^{k-1} + \sum_{k=1}^{n} r_k(C_e)x^k \tag{3}$$

or,

$$1 + \sum_{k=1}^{n} r_k(C)x^k = x \cdot r(C_s, x) + \sum_{k=1}^{n} r_k(C_e)x^k + 1$$

from which it follows that

$$r(C, x) = x \cdot r(C_s, x) + r(C_e, x).\tag{4}$$

We now use this final equation to determine the rook polynomial for the chessboard in Fig. 7.9. Each time the idea in Eq. (4) is used, we mark the special square we are using with (∗). Parentheses are placed about each chessboard to denote the rook polynomial of the board.

$$= x^2(1 + 2x) + 2x(1 + 4x + 2x^2) + x(1 + 3x + x^2)$$

$$3x + 12x^2 + 7x^3 + x(1 + 2x) + (1 + 4x + 2x^2) = 1 + 8x + 16x^2 + 7x^3.$$

7.5 ■ ARRANGEMENTS WITH FORBIDDEN POSITIONS

The rook polynomials of the previous section seem interesting on their own. Now we shall find them useful in solving the following problems.

EXAMPLE 7.10   In making seating arrangements for their son's wedding reception, Grace and Nick are down to four relatives, denoted R_i, $1 \leq i \leq 4$, who do not get along with one another. There is a single open seat at each of the five tables T_j, $1 \leq j \leq 5$. Because of family differences

a) R_1 will not sit at T_1 or T_2.   b) R_2 will not sit at T_2.
c) R_3 will not sit at T_3 or T_4.   d) R_4 will not sit at T_4 or T_5.

Figure 7.10

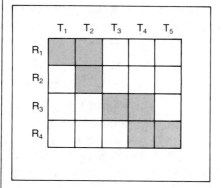

We represent this situation in Fig. 7.10. The number of ways we can seat these four people, and satisfy (a)–(d) is the number of ways four nontaking rooks can be placed on the chessboard made up of the *unshaded* squares. However, since there are only seven shaded squares, as opposed to thirteen unshaded ones, it would be easier to work with the shaded chessboard.

We start with the needed conditions to apply inclusion and exclusion.

For $1 \leq i \leq 4$, let c_i be the condition where a seating assignment of these four people is made with relative R_i in a forbidden (shaded) position. As usual, S denotes the total number of ways we can place the four relatives, one to a table. Then $|S| = N = S_0 = 5!$

To determine S_1 we consider each of the following:

$N(c_1) = 4! + 4!$, for there are 4! ways to seat R_2, R_3, R_4 if R_1 is in forbidden position T_1, and another 4! ways if R_1 is at table T_2, his or her other forbidden position.

$N(c_2) = 4!$, for after placing R_2 at forbidden table T_2, R_1, R_3, and R_4 must be placed at T_1, T_3, T_4, T_5, one person to a table.

$N(c_3) = 4! + 4!$, one summand for R_3 being in forbidden position T_3, and the other summand for the forbidden position T_4.

$N(c_4) = 4! + 4!$, the two summands arising for R_4 being placed at each of the two forbidden positions T_4 and T_5.

Hence $S_1 = 7(4!)$.

Turning to S_2 we have these considerations:

$N(c_1c_2) = 3!$, for after placing R_1 at T_1 and R_2 at T_2, there are three tables (T_3, T_4, T_5) where R_3 and R_4 can be seated.

$N(c_1c_3) = 3! + 3! + 3! + 3!$, since there are four cases where R_1 and R_3 are located at forbidden positions:

i) R_1 at T_1; R_3 at T_3 ii) R_1 at T_2; R_3 at T_3

iii) R_1 at T_1; R_3 at T_4 iv) R_1 at T_2; R_3 at T_4.

In a similar manner we find that $N(c_1c_4) = 4(3!)$, $N(c_2c_3) = 2(3!)$, $N(c_2c_4) = 2(3!)$, $N(c_3c_4) = 3(3!)$. Consequently, $S_2 = 16(3!)$.

Before continuing, we make a few observations about S_1 and S_2. For S_1 we have $7(4!) = 7(5 - 1)!$, where 7 is the number of shaded squares in Fig. 7.10. Also, $S_2 = 16(3!) = 16(5 - 2)!$, where 16 is the number of ways two nontaking rooks can be placed on the shaded chessboard.

In general, for $0 \le i \le 4$, $S_i = r_i(5 - i)!$, where r_i is the number of ways to place i nontaking rooks on the shaded chessboard in Fig. 7.10.

Consequently, to expedite the solution of this problem, we turn to $r(C, x)$, the rook polynomial of the chessboard. Using the decomposition of C into the disjoint subboards in the upper left and lower right corners, we find that

$$r(C, x) = (1 + 3x + x^2)(1 + 4x + 3x^2) = 1 + 7x + 16x^2 + 13x^3 + 3x^4,$$

so

$$N(\bar{c}_1\bar{c}_2\bar{c}_3\bar{c}_4) = S_0 - S_1 + S_2 - S_3 + S_4 = 5! - 7(4!) + 16(3!) - 13(2!) + 3(1!)$$

$$= \sum_{i=0}^{4} (-1)^i r_i(5 - i)! = 25.$$

So Grace and Nick can breathe a sigh of relief. There are 25 ways they can seat these last four relatives at the reception and avoid any squabbling. □

Our last example demonstrates how a bit of rearranging of our chessboard can help in our calculations.

EXAMPLE 7.11   We have a pair of dice with one die red and the other green. Rolling these dice six times, what is the probability that we obtain all six values on both the red die and the green die if we know that the ordered pairs $(1, 2)$, $(2, 1)$, $(2, 5)$, $(3, 4)$, $(4, 1)$, $(4, 5)$ and $(6, 6)$ did not occur? (Here an ordered pair (x, y) indicates x on red and y on green.)

Recognizing this as another problem dealing with permutations and forbidden positions, we construct the chessboard in Fig. 7.11(a), where the shaded squares constitute the forbidden positions. In this figure the shaded squares are scattered. Relabeling the rows and columns, the chessboard can be redrawn as in Fig. 7.11(b), where we have taken shaded squares in the same row (or column) of the board in (a) and made them adjacent. In Fig. 7.11(b), chessboard C is the union of four mutually

Figure 7.11

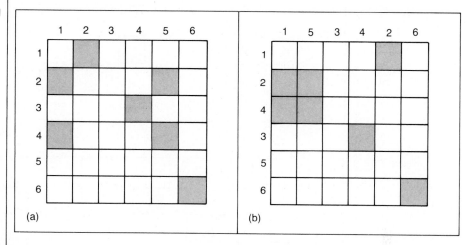

(a)   (b)

disjoint subboards and

$$r(C, x) = (1 + 4x + 2x^2)(1 + x)^3 = 1 + 7x + 17x^2 + 19x^3 + 10x^4 + 2x^5.$$

For $1 \leq i \leq 6$, define c_i as the condition where, having rolled the dice six times, all six values occur on both the red die and green die, but i on the red die is paired with one of the forbidden numbers on the green die. (Note that $N(c_5) = 0$.) Then the number of sequences of the six rolls of the dice for the event we are interested in is

$$(6!)N(\bar{c}_1\bar{c}_2\bar{c}_3\bar{c}_4\bar{c}_5\bar{c}_6) = (6!)\sum_{i=0}^{6}(-1)^i S_i = (6!)\sum_{i=0}^{6}(-1)^i r_i \cdot (6 - i)!$$

$$= 6![6! - 7(5!) + 17(4!) - 19(3!) + 10(2!) - 2(1!) + 0(0!)]$$

$$= 6![192] = 138{,}240.$$

As the sample space consists of all sequences of six ordered pairs selected with repetition from the 29 unshaded squares of the chessboard, the probability of this event is $138{,}240/(29)^6 \doteq 0.00023$.   □

EXERCISES

1. Verify directly the rook polynomials for (a) the unshaded chessboards in Figs. 7.8 and 7.9, and (b) the shaded chessboards in Figs. 7.10 and 7.11(b).

2. Construct or describe a smallest (i.e., least number of squares) chessboard for which $r_{10} \neq 0$.

3. a) Find the rook polynomial for the standard 8×8 chessboard.
   b) Answer (a) with 8 replaced by n, $n \in \mathbf{Z}^+$.

4. Find the rook polynomials for the shaded chessboards in Fig. 7.12.

5. a) Find the rook polynomials for the shaded chessboards in Fig. 7.13.

Figure 7.12

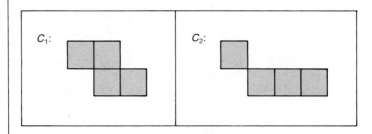

Figure 7.13

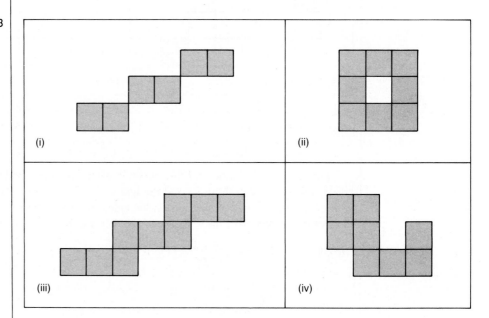

b)   Generalize the chessboard (and rook polynomial) for Fig. 7.13(i).

6. Why do we have 6! in the term $(6!)N(\bar{c}_1\bar{c}_2\ldots\bar{c}_6)$ for the solution of Example 7.11?

7. Professor Cary has five graders to correct programs in his courses in APL, BASIC, FORTRAN, Pascal and PL/I. Jeanne and Charles both dislike FORTRAN. Sandra wants to avoid BASIC and PL/I. Paul detests APL and BASIC, and Todd refuses to work in FORTRAN and Pascal. In how many ways can Professor Cary assign each grader to correct programs in one language, cover all five languages, and keep everybody content?

8. A pair of dice, one red and the other green, is rolled six times. We know that the ordered pairs $(1, 1)$, $(1, 5)$, $(2, 4)$, $(3, 6)$, $(4, 2)$, $(4, 4)$, $(5, 1)$, and $(5, 5)$ did not come up. What is the probability that every value came up on both the red die and the green one?

9. Five professors named Al, Violet, Gary, Jack, and Mary Lou are to be assigned to teach one class each from among calculus I, II, III, statistics, and combinatorics. Al will not teach calculus II or combinatorics, Gary cannot stand statistics, Violet and Mary Lou both refuse to teach calculus I or III, and Jack detests calculus II.

   a) In how many ways can their chairman assign each of these professors one of these five courses and still keep peace in the department?

   b) For the assignments in (a), what is the probability that Violet gets to teach combinatorics?

   c) For the assignments in (a), what is the probability that Gary or Mary Lou gets to teach calculus II?

7.6 ■ SUMMARY AND HISTORICAL REVIEW

In the first two chapters of this text we were concerned with enumeration problems in which we had to be careful of situations where arrangements or selections were overcounted. This became even more involved in Chapter 3 when we tried to count the number of onto functions between two finite sets.

With Venn diagrams to lead the way, in this chapter we obtained a pattern called the Principle of Inclusion and Exclusion. Using this principle, we restated each problem in terms of conditions and subsets. Using enumeration formulas on permutations and combinations developed earlier, we solved some simpler subproblems and let the principle manage our concern about overcounting. As a result, we were able to solve a variety of problems, some dealing with number theory and one with graph theory. We also proved the formula conjectured earlier in Section 3.3 for the number of onto functions between two finite sets.

This principle has an interesting history, being found in different manuscripts under such names as the "Sieve Method" or the "Principle of Cross Classification." A set theoretic version of the principle, which concerned itself with set unions and intersections, is found in *Doctrine of Chances* (1718), a text on probability theory by Abraham DeMoivre (1667–1754). Somewhat earlier, in 1713, Pierre Remond de Montmort (1678–1719) used the idea behind the principle in his solution of the problem generally known as *le probleme des rencontres* (i.e., derangements).

For the way we developed and dealt with the principle, credit must be given to James Joseph Sylvester (1814–1897). The importance of the technique was not generally appreciated, however, until somewhat later, when the publication by W. A. Whitworth [9] made mathematicians more aware of its potential and use.

For more on the use of this principle, examine Chapter 4 of C. L. Liu [3], Chapter 2 of H. J. Ryser [7], or Chapter 5 of A. Tucker [8]. More number theoretic results related to the principle, including the Möbius inversion formula, can be found in Chapter 2 of M. Hall [1], Chapter X of C. L. Liu [4], and Chapter 16 of G. Hardy and E. Wright [2]. An extension of this formula is given in the article by G. C. Rota [6].

The ideas behind the rook polynomials and their applications were developed in the late 1930's and during the 1940's and 1950's. Much more on this topic is found in Chapters 7 and 8 of J. Riordan [5].

REFERENCES

1. Hall, Marshall, Jr., *Combinatorial Theory,* Blaisdell, Waltham, Massachusetts, 1967.

2. Hardy, Godfrey Harold, and Wright, Edward Maitland, *An Introduction to the Theory of Numbers,* 4th ed., Clarendon Press, Oxford, England, 1960.

3. Liu, C. L., *Introduction to Combinatorial Mathematics,* McGraw-Hill, New York, 1968.

4. Liu, C. L., *Topics in Combinatorial Mathematics,* Mathematical Association of America, 1972.

5. Riordan, John, *An Introduction to Combinatorial Analysis,* Princeton University Press, Princeton, New Jersey, 1980. (Originally published in 1958 by John Wiley & Sons.)

6. Rota, Gian Carlo, "On the Foundations of Combinatorial Theory, I. Theory of Möbius Functions," *Zeitschrift für Wahrscheinlichkeits Theorie, 2:* pp. 340–368, 1964.

7. Ryser, Herbert J., *Combinatorial Mathematics,* Carus Mathematical Monograph, No. 14. Published by the Mathematical Association of America, distributed by John Wiley & Sons, New York, 1963.

8. Tucker, Alan, *Applied Combinatorics,* John Wiley & Sons, New York, 1980.

9. Whitworth, William Allen, *Choice and Chance.* Originally published at Cambridge in 1867. Reprint of the 5th ed. (1901), Hafner, New York, 1965.

■ MISCELLANEOUS EXERCISES

1. Determine how many $n \in \mathbf{Z}^+$ satisfy $n \leq 500$ and are not divisible by 2, 3, 5, 6, 8, or 10.

2. Determine all $x \in \mathbf{R}$ such that $\lfloor x \rfloor + \lfloor x + \frac{1}{2} \rfloor = \lfloor 2x \rfloor$.

3. Find three values for $n \in \mathbf{Z}^+$ where $\phi(n) = 16$.

4. Find the number of $n \in \mathbf{Z}^+$ such that $n \leq 1000$ and n is not a perfect square, cube, or fourth power.

5. In how many ways can we arrange the integers 1, 2, 3, ..., 8 in a line so that there are no occurrences of the patterns 12, 23, ..., 78, 81?

6. a) If we have k different colors available, in how many ways can we paint the walls of a pentagonal room if adjacent walls are to be painted with different colors?

b) What is the smallest value of k for which such a coloring is possible?

c) Answer (a) and (b) for a hexagonal room.

7. If 13 cards are dealt from a standard deck of 52, what is the probability that these 13 cards include (a) at least one card from each suit? (b) exactly one void (e.g., no clubs)? (c) exactly two voids?

8. Using the result of Theorem 7.2, prove that the number of ways we can place s different objects in n distinct containers with m containers each containing exactly r of the objects is

$$\frac{(-1)^m n! \, s!}{m!} \sum_{i=m}^{n} \frac{(-1)^i (n-i)^{s-ir}}{(i-m)!\,(n-i)!\,(s-ir)!\,(r!)^i}$$

9. Ten women attend a business luncheon. Each woman checks her coat and attaché case. Upon leaving, each woman is given a coat and case at random. (a) In how many ways can the coats and cases be distributed with no woman getting either of her possessions? (b) In how many ways do we have no woman getting back both of her possessions?

10. Consider the list $x_1, x_1, x_2, x_2, \ldots, x_n, x_n$, consisting of n pairs of distinct symbols. In how many ways can one arrange these $2n$ symbols so that there is no pair of consecutive identical symbols?

11. In how many ways can four w's, four x's, four y's, four z's be arranged so that there is no consecutive quadruple of the same letter?

12. Consider a list of three x's, three y's, three z's.

a) Show that these nine letters can be arranged in $(8!/3!\,3!) - (7!/3!\,3!)$ ways with a pair or triple of adjacent x's.

b) Show that we can arrange these letters in $(7!/3!) - 2(6!/3!) + (5!/3!)$ ways with a pair or triple of adjacent x's and adjacent y's.

c) Arranging these nine letters, show that there are $6! - 3(5!) + 3(4!) - 3!$ arrangements with a pair or triple of every letter occurring.

d) In how many ways can we arrange these nine letters so that no consecutive letters are the same?

e) Why can't we solve (d) by considering the two ways we can arrange the three x's and three y's with no consecutive letters the same, and then determine how to place the three z's so that no two are consecutive?

13. a) Given n distinct objects, in how many ways can we select r of these objects so that each selection includes some particular m of the n objects? ($m \leq r \leq n$).

b) Using the principle of inclusion and exclusion prove that for $m \leq r \leq n$,

$$\binom{n-m}{n-r} = \sum_{i=0}^{m} (-1)^i \binom{m}{i} \binom{n-i}{r}.$$

14. a) Let $\lambda \in \mathbf{Z}^+$. If we have λ different colors available, in how many ways can we color the vertices of the graph in Fig. 7.14(a), so that no adjacent vertices are the same color? This result in λ is called the *chromatic polynomial* of the graph, and the smallest value of λ for which this polynomial is positive is called the *chromatic number* of the graph. What is the chromatic number of this graph? (We deal with this idea further in Chapter 14.)

    b) If there are six colors available, in how many ways can the rooms R_i, $1 \le i \le 5$, in Fig. 7.14(b) be painted so that rooms with a common doorway D_j, $1 \le j \le 5$, are painted with different colors?

Figure 7.14

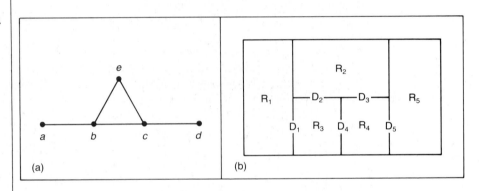

(a)   (b)

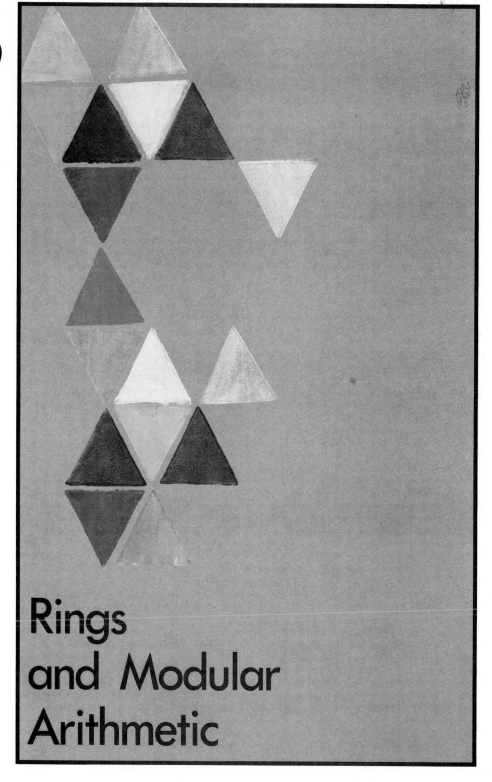

Rings
and Modular
Arithmetic

Our emphasis returns to structure as we begin the investigation of sets of elements in which the structure arises from two binary operations. The concepts of structure and enumeration often reinforce each other. Here we will see this occur as ideas from Chapters 3 and 7 come to the forefront again.

When the set $\mathbf{Z}$ was examined in Chapter 6, it was in conjunction with the binary operations of addition and multiplication. In this chapter we emphasize these operations by writing $(\mathbf{Z}, +, \cdot)$, instead of just $\mathbf{Z}$. From some of the properties of $(\mathbf{Z}, +, \cdot)$ we define the algebraic structure called a *ring*. Without knowing it, we have been dealing with rings in many mathematical settings. Now we will be concerned with finite rings that arise in number theory and computer science applications. Of particular interest we will examine the *hashing function*, which provides a means of identifying records stored in a table.

8.1 ■ THE RING STRUCTURE: DEFINITION AND EXAMPLES

We start by defining the ring structure, realizing as we do that most abstract definitions, like theorems, come about from a study of many examples where one identifies the common idea or ideas present in what may seem to be a collection of unrelated objects.

Definition 8.1

▶ Let R be a nonempty set on which we define two binary operations, denoted by $+$ and $\cdot$ (which may be quite different from the ordinary addition and multiplication we are accustomed to). Then $(R, +, \cdot)$ is a *ring* if for all $a, b, c \in R$ the following conditions are satisfied:

a) $a + b = b + a$;   Commutative Law of $+$

b) $a + (b + c) = (a + b) + c$;   Associative Law of $+$

c) There exists $z \in R$ such that   Existence of an identity for $+$
   $a + z = z + a = a$ for every $a \in R$;

d) For each $a \in R$ there is an element   Existence of inverses under $+$
   $b \in R$ with $a + b = b + a = z$;

e) $a \cdot (b \cdot c) = (a \cdot b) \cdot c$;   Associative Law of $\cdot$

f) $a \cdot (b + c) = a \cdot b + a \cdot c$ Distributive Laws of $\cdot$ over $+$
   $(b + c) \cdot a = b \cdot a + c \cdot a$.

The binary operations here are closed by definition. We are dealing, as in Chapter 3, with two functions, which can be denoted by f, g, where $f, g: R \times R \rightarrow R$ and $f(a, b) = a + b$, $g(a, b) = a \cdot b$. (We often write ab for $a \cdot b$.)

In the next section we will show that the additive identity is unique, as is the additive inverse of each element.

EXAMPLE 8.1

Under ordinary addition and multiplication, $\mathbf{Z}, \mathbf{Q}, \mathbf{R}, \mathbf{C}$ are rings. In all of these rings the additive identity z is the integer 0, and the additive inverse of any number x is $-x$.   □

EXAMPLE 8.2 Let $M_2(\mathbf{Z})$ denote the set of all 2×2 matrices with integer components. ($M_2(\mathbf{Q})$, $M_2(\mathbf{R})$, $M_2(\mathbf{C})$ are defined similarly.) In $M_2(\mathbf{Z})$ two matrices are equal if the corresponding components are equal.

Here we define $+$ and $\cdot$ by

$$\begin{bmatrix} a & b \\ c & d \end{bmatrix} + \begin{bmatrix} e & f \\ g & h \end{bmatrix} = \begin{bmatrix} a+e & b+f \\ c+g & d+h \end{bmatrix}, \begin{bmatrix} a & b \\ c & d \end{bmatrix}\begin{bmatrix} e & f \\ g & h \end{bmatrix} = \begin{bmatrix} ae+bg & af+bh \\ ce+dg & cf+dh \end{bmatrix}$$

Under these operations, $M_2(\mathbf{Z})$ is a ring where $z = \begin{bmatrix} 0 & 0 \\ 0 & 0 \end{bmatrix}$ and the additive inverse of $\begin{bmatrix} a & b \\ c & d \end{bmatrix}$ is $\begin{bmatrix} -a & -b \\ -c & -d \end{bmatrix}$.

A few things happen here that do not occur in the rings of Example 8.1.

$$\begin{bmatrix} 1 & 2 \\ 1 & 1 \end{bmatrix}\begin{bmatrix} 3 & 7 \\ 1 & 0 \end{bmatrix} = \begin{bmatrix} 5 & 7 \\ 4 & 7 \end{bmatrix} \neq \begin{bmatrix} 10 & 13 \\ 1 & 2 \end{bmatrix} = \begin{bmatrix} 3 & 7 \\ 1 & 0 \end{bmatrix}\begin{bmatrix} 1 & 2 \\ 1 & 1 \end{bmatrix}$$

shows that multiplication need not be commutative in a ring. That is why there are two distributive laws. Also,

$$\begin{bmatrix} 1 & -1 \\ -1 & 1 \end{bmatrix}\begin{bmatrix} 2 & 1 \\ 2 & 1 \end{bmatrix} = \begin{bmatrix} 0 & 0 \\ 0 & 0 \end{bmatrix},$$

even though neither $\begin{bmatrix} 1 & -1 \\ -1 & 1 \end{bmatrix}$ nor $\begin{bmatrix} 2 & 1 \\ 2 & 1 \end{bmatrix}$ is the additive identity. Hence a ring may contain *proper divisors of zero*. □

Definition 8.2 ▶ Let $(R, +, \cdot)$ be a ring.

a) If $ab = ba$ for all $a, b \in R$, then R is called a *commutative* ring.

b) R is said to have no *proper divisors of zero* if for any $a, b \in R$, $ab = z \Rightarrow a = z$ or $b = z$.

c) If there is an element $u \in R$ such that $au = ua = a$ for all $a \in R$, we call u the *unity*, or *multiplicative identity*, of R. R is then called a *ring with unity*.

The rings in Example 8.1 are all commutative rings with unity the integer 1. None of them has any proper divisors of zero. Meanwhile, the ring $M_2(\mathbf{Z})$ is a noncommutative ring whose unity is the matrix $\begin{bmatrix} 1 & 0 \\ 0 & 1 \end{bmatrix}$. This ring does contain proper divisors of zero.

We turn now to finite rings.

EXAMPLE 8.3 Let $\mathcal{U} = \{1, 2\}$ and $R = \mathcal{P}(\mathcal{U})$. Define $+$ and $\cdot$ on the elements of R by

$$A + B = A \triangle B = \{x \mid x \in A \text{ or } x \in B, \text{ but not both}\};$$

$$A \cdot B = A \cap B = \text{the intersection of sets } A, B \subseteq \mathcal{U}.$$

We form Tables 8.1(a) and (b) for these operations.

Table 8.1

+ ($\triangle$)	$\emptyset$	{1}	{2}	$\mathcal{U}$
$\emptyset$	$\emptyset$	{1}	{2}	$\mathcal{U}$
{1}	{1}	$\emptyset$	$\mathcal{U}$	{2}
{2}	{2}	$\mathcal{U}$	$\emptyset$	{1}
$\mathcal{U}$	$\mathcal{U}$	{2}	{1}	$\emptyset$

(a)

$\cdot$ ($\cap$)	$\emptyset$	{1}	{2}	$\mathcal{U}$
$\emptyset$	$\emptyset$	$\emptyset$	$\emptyset$	$\emptyset$
{1}	$\emptyset$	{1}	$\emptyset$	{1}
{2}	$\emptyset$	$\emptyset$	{2}	{2}
$\mathcal{U}$	$\emptyset$	{1}	{2}	$\mathcal{U}$

(b)

From results in Chapter 2, R satisfies conditions (a), (b), (e), and (f) for these operations of "addition" and "multiplication." The table for "addition" shows that $\emptyset$ is the additive identity. For any $x \in R$, the additive inverse of x is x itself. The multiplication table is symmetric about the diagonal from the upper left to the lower right, so the operation defined by the table is commutative. The table also indicates that R has *unity* $\mathcal{U}$. So R is a finite commutative ring with unity. The elements {1}, {2} provide an example of proper divisors of zero.   □

EXAMPLE 8.4   For $R = \{a, b, c, d, e\}$ we define $+$ and $\cdot$ by Tables 8.2(a) and (b).

Table 8.2

+	a	b	c	d	e
a	a	b	c	d	e
b	b	c	d	e	a
c	c	d	e	a	b
d	d	e	a	b	c
e	e	a	b	c	d

(a)

$\cdot$	a	b	c	d	e
a	a	a	a	a	a
b	a	b	c	d	e
c	a	c	e	b	d
d	a	d	b	e	c
e	a	e	d	c	b

(b)

Although we do not verify the 125 equalities needed to establish each of the associative laws and the distributive law, $(R, +, \cdot)$ is a finite commutative ring with unity, and has no proper divisors of zero. The element a is the zero of R while b is the unity. Here every nonzero element x has a *multiplicative inverse* y, where $xy = yx = b$, the unity. Elements c and d are multiplicative inverses of each other; b is its own inverse, as is e.   □

Definition 8.3   ▶ Let R be a ring with unity u. If a, $b \in R$ and $ab = ba = u$, then b is called a *multiplicative inverse* of a and a is called a *unit* of R.

Definition 8.4   ▶ Let R be a commutative ring with unity. Then

a) R is called an *integral domain* if R has no proper divisors of zero.

b) R is called a *field* if every nonzero element of R is a unit.

The ring $(\mathbf{Z}, +, \cdot)$ is an integral domain but not a field, while $\mathbf{Q}$, $\mathbf{R}$, $\mathbf{C}$, under ordinary addition and multiplication, are both integral domains and fields. The ring in Example 8.4 is both an integral domain and a field.

EXAMPLE 8.5 For our last finite ring we let $R = \{s, t, v, w, x, y\}$ where $+$ and $\cdot$ are given by Tables 8.3(a) and (b).

Table 8.3

+	s	t	v	w	x	y
s	s	t	v	w	x	y
t	t	v	w	x	y	s
v	v	w	x	y	s	t
w	w	x	y	s	t	v
x	x	y	s	t	v	w
y	y	s	t	v	w	x

·	s	t	v	w	x	y
s	s	s	s	s	s	s
t	s	t	v	w	x	y
v	s	v	x	s	v	x
w	s	w	s	w	s	w
x	s	x	v	s	x	v
y	s	y	x	w	v	t

(a)   (b)

From these tables we see that $(R, +, \cdot)$ is a commutative ring with unity, but it is neither an integral domain nor a field. The element t is the unity, and t and y are the units of R.

We also note that $vv = vy$, and even though v is not the zero element of R, we cannot cancel and say that $v = y$. So a general ring does *not* satisfy the cancellation law of multiplication that we take for granted. We shall look at this idea again in the next section. □

EXERCISES

1. What is the additive inverse of each element in the rings of Examples 8.4 and 8.5?

2. Determine whether or not each of the following sets of numbers is a ring under ordinary addition and multiplication.
   a) $R =$ the set of positive integers and zero.
   b) $R =$ the set of all even integers.
   c) $R = \{kn \mid n \in \mathbf{Z}, k \text{ a fixed positive integer}\}$.
   d) $R = \{a + b\sqrt{2} \mid a, b \in \mathbf{Z}\}$.
   e) $R = \{a + b\sqrt{2} + c\sqrt{3} \mid a \in \mathbf{Z}, b, c \in \mathbf{Q}\}$.

3. Tables 8.4(a) and (b) make $(R, +, \cdot)$ into a ring, where $R = \{s, t, x, y\}$.

   (a) What is the zero for this ring? (b) What is the additive inverse of each element? (c) What is $t(s + xy)$? (d) Is R a commutative ring? (e) Does R have a unity? (f) Find a pair of zero divisors.

Table 8.4

+	s	t	x	y
s	y	x	s	t
t	x	y	t	s
x	s	t	x	y
y	t	s	y	x

·	s	t	x	y
s	y	y	x	x
t	y	y	x	x
x	x	x	x	x
y	x	x	x	x

(a)   (b)

4. For the set R in Example 8.3, keep $A \cdot B = A \cap B$, but define $A + B = A \cup B$. Is $(R, \cup, \cap)$ a ring?

5. Let $R = \{a + bi \mid a, b \in \mathbf{Z}, i^2 = -1\}$, with addition and multiplication defined by $(a + bi) + (c + di) = (a + c) + (b + d)i$; $(a + bi)(c + di) = (ac - bd) + (bc + ad)i$. (a) Verify that R is an integral domain. (b) Determine all units in R.

6. a) · Determine the multiplicative inverse of the matrix $\begin{bmatrix} 1 & 2 \\ 3 & 7 \end{bmatrix}$ in the ring $M_2(\mathbf{Z})$, that is, find a, b, c, d so that

$$\begin{bmatrix} 1 & 2 \\ 3 & 7 \end{bmatrix} \begin{bmatrix} a & b \\ c & d \end{bmatrix} = \begin{bmatrix} a & b \\ c & d \end{bmatrix} \begin{bmatrix} 1 & 2 \\ 3 & 7 \end{bmatrix} = \begin{bmatrix} 1 & 0 \\ 0 & 1 \end{bmatrix}.$$

b) Show that $\begin{bmatrix} 1 & 2 \\ 3 & 8 \end{bmatrix}$ is a unit in the ring $M_2(\mathbf{Q})$ but not a unit in $M_2(\mathbf{Z})$.

7. If $\begin{bmatrix} a & b \\ c & d \end{bmatrix} \in M_2(\mathbf{R})$, prove that $\begin{bmatrix} a & b \\ c & d \end{bmatrix}$ is a unit of this ring if and only if $ad - bc \neq 0$.

8. Give an example of a ring with eight elements. How about one with 16 elements? Generalize.

9. Define a new type of addition and multiplication, denoted by $\oplus$ and $\odot$, on the set $\mathbf{Z}$ as follows. For $x, y \in \mathbf{Z}$, $x \oplus y = x + y - 1$, $x \odot y = x + y - xy$. (a) Prove that $(\mathbf{Z}, \oplus, \odot)$ is a ring. (b) Is this ring commutative? (c) Does the ring have a unity element? What about units? (d) Is the ring an integral domain? a field?

10. For $R = \{s, t, x, y\}$, define $+$ and $\cdot$, making R into a ring, by Table 8.5(a) for $+$ and the partial table for $\cdot$ in Table 8.5(b).

Table 8.5

+	s	t	x	y
s	s	t	x	y
t	t	s	y	x
x	x	y	s	t
y	y	x	t	s

·	s	t	x	y
s	s	s	s	s
t	s	t	?	?
x	s	t	?	y
y	s	?	s	?

(a)   (b)

a)   Using the associative and distributive laws, determine the entries for the missing spaces in the multiplication table.

b)   Is this ring commutative?

c)   Does it have a unity? How about units?

d)   Is the ring an integral domain or field?

8.2 ■ RING PROPERTIES AND SUBSTRUCTURES

In each ring of Section 8.1 we were concerned with *the* zero element of the ring and *the* additive inverse of each ring element. It is time now to show, along with other properties, that these elements are unique.

Theorem 8.1

▶ In any ring $(R, +, \cdot)$,

a)   the zero element z is unique; and,

b)   the additive inverse of each ring element is unique.

Proof   a)   If R has more than one additive identity, let z_1, z_2 denote two such elements. Then

$$z_1 = z_1 + z_2 = z_2.$$

Since z_2 is an additive identity   Since z_1 is an additive identity

b)   For $a \in R$, suppose there are two elements $b, c \in R$ where $a + b = b + a = z$, and $a + c = c + a = z$. Then $b = b + z = b + (a + c) = (b + a) + c = z + c = c$. (The reader should supply the condition that establishes each equality.)   ■

As a result of the uniqueness in (b), from this point on we shall denote *the* additive inverse of $a \in R$ by $-a$. Further, we may now speak of *subtraction* in the ring, where we understand that $a - b = a + (-b)$.

From Theorem 8.1(b) we obtain the following for ring R.

Theorem 8.2

▶ (*The Cancellation Laws of Addition*) For $a, b, c \in R$,

a) $a + b = a + c \Rightarrow b = c$; and

b) $b + a = c + a \Rightarrow b = c$.

Proof   a)   Since $a \in R$, $-a \in R$ and we have

$$a + b = a + c \Rightarrow (-a) + (a + b) = (-a) + (a + c) \Rightarrow [(-a) + a] + b$$

$$= [(-a) + a] + c \Rightarrow z + b = z + c \Rightarrow b = c.$$

b)   We leave this similar proof for the reader.   ■

Theorem 8.3

▶ For any ring R and $a \in R$, $az = za = z$.

Proof | If $a \in R$, $az = a(z + z)$, since $z + z = z$. Hence $z + az = az = az + az$. (Why?) Using the cancellation law of addition, we have $z = az$.

The proof that $za = z$ is done similarly.   ∎

The reader may feel that the result of Theorem 8.3 is obvious. But we are not dealing with just **Z** or **Q** or $M_2(\mathbf{Z})$. Our objective is to show that *any* ring satisfies such a result, and to get the result we may only use the conditions in the definition of a ring and whatever properties we've derived up to this point.

The uniqueness of additive inverses now implies the following.

Theorem 8.4

▶ Given a ring $(R, +, \cdot)$, for any $a, b \in R$,

a) $-(-a) = a$;

b) $a(-b) = (-a)b = -(ab)$; and

c) $(-a)(-b) = ab$.

Proof | a)   By the convention stated after Theorem 8.1, $-(-a)$ denotes *the* additive inverse of $-a$. Since $(-a) + a = z$, a is also an additive inverse for $-a$. Consequently, by the uniqueness of such inverses, $-(-a) = a$.

b)   We shall prove that $a(-b) = -(ab)$, and leave the other part for the reader. We know that $-(ab)$ denotes *the* additive inverse of ab. However, $ab + a(-b) = a[b + (-b)] = az = z$, by Theorem 8.3, so by the uniqueness of additive inverses $a(-b) = -(ab)$.

c)   Here we establish an idea we have used in algebra since our first encounter with *signed numbers*. "Minus times minus does indeed equal plus," and the proof follows from the properties and definition of a ring. From part (b) we have $(-a)(-b) = -[a(-b)] = -[-(ab)]$, and the result then follows from (a).   ∎

Comparable to Theorem 8.1, for the operation of multiplication one finds the following.

Theorem 8.5

▶ Given a ring $(R, +, \cdot)$,

a)   if R has a unity, it is unique; and

b)   if R has a unity, and x is a unit of R, then the multiplicative inverse of x is unique.

Proof | The proofs of these results are left to the reader.   ∎

As a result of this theorem we shall denote *the* multiplicative inverse of any unit x in R by x^{-1}. Also, the definition of a field can now be restated: A field is a commutative ring F with unity, such that for all $x \in F$, $x \neq z \Rightarrow x^{-1} \in F$.

With this notion to assist us we examine some further properties and relations between fields and integral domains.

Theorem 8.6

▶ Let $(R, +, \cdot)$ be a commutative ring with unity. Then R is an integral domain if and only if, for $a, b, c \in R$, $a \neq z$, $ab = ac \Rightarrow b = c$. (Hence, a commutative ring with unity that satisfies the *cancellation law of multiplication* is an integral domain.)

Proof If R is an integral domain and $x, y \in R$, then $xy = z \Rightarrow x = z$ or $y = z$. Now if $ab = ac$, then $ab - ac = a(b - c) = z$, and since $a \neq z$, $b - c = z$ or $b = c$. Conversely, if R is commutative with unity and R satisfies multiplicative cancellation, let $a, b \in R$ with $ab = z$. If $a = z$, we are finished. If not, as $az = z$, we can write $ab = az$, and conclude that $b = z$. So there are no proper divisors of zero and R is an integral domain. ∎

Before going on, let us realize that the cancellation law of multiplication does *not* imply the existence of multiplicative inverses. The integral domain $(\mathbf{Z}, +, \cdot)$ satisfies multiplicative cancellation, but contains only two elements, namely 1 and -1, that are units. Hence, an integral domain need not be a field. But what about a field? Is it an integral domain?

Theorem 8.7

▶ If $(F, +, \cdot)$ is a field, then it is an integral domain.

Proof Let $a, b \in F$ with $ab = z$. If $a = z$, we are finished. If not, a has a multiplicative inverse a^{-1}, since F is a field. Then $ab = z \Rightarrow a^{-1}(ab) = a^{-1}z \Rightarrow (a^{-1}a)b = a^{-1}z \Rightarrow ub = z \Rightarrow b = z$. Hence F has no proper divisors of zero and is an integral domain. ∎

In Chapter 3 we found that functions $f\colon A \to A$ could be one-to-one (onto) without being onto (one-to-one). However, if A were *finite,* such a function f was one-to-one iff it was onto. (See Theorem 3.10.) The same situation occurs with *finite* integral domains. An integral domain need not be a field, but when it is finite we find that this structure is a field.

Theorem 8.8

▶ A *finite* integral domain $(D, +, \cdot)$ is a field.

Proof Since D is finite, we list the elements of D as $\{d_1, d_2, \ldots, d_n\}$. For $d \in D$, $d \neq z$, $dD = \{dd_1, dd_2, \ldots, dd_n\} \subseteq D$, since D is closed under multiplication. As $|D| = n$ and $dD \subseteq D$, if we could show that dD contains n elements, we would have $dD = D$. If $|dD| < n$, then $dd_i = dd_j$, for some $1 \leq i < j \leq n$. But since D is an integral domain and $d \neq z$, we have $d_i = d_j$, when they are supposed to be distinct. So $dD = D$ and for some $1 \leq k \leq n$, $dd_k = u$, the unity of D. Then $dd_k = u \Rightarrow d$ is a unit of D, and as d was chosen arbitrarily, it follows that $(D, +, \cdot)$ is a field. ∎

In the next section we shall find finite fields that are useful in discrete mathematics. Before closing this section, however, let us examine some special subsets of a ring.

When dealing with finite state machines in Chapter 4, we saw instances where subsets of the set of internal states gave rise to machines on their own (when the next state and output functions of the original machines were restricted). These were called submachines. As binary operations are special kinds of functions we examine a similar idea in the following.

Definition
8.5

▶ For a ring $(R, +, \cdot)$, a nonempty subset S of R is called a *subring* of R, if $(S, +, \cdot)$ — i.e., S under the addition and multiplication of R, restricted to S — is a ring.

EXAMPLE 8.6   For any ring R, $\{z\}$ and R itself are always subrings of R.   □

EXAMPLE 8.7
a) The set of all even integers is a subring of $(\mathbf{Z}, +, \cdot)$. In fact, for any $n \in \mathbf{Z}^+$, $n\mathbf{Z} = \{nx \mid x \in \mathbf{Z}\}$ is a subring of $(\mathbf{Z}, +, \cdot)$.

b) $(\mathbf{Z}, +, \cdot)$ is a subring of $(\mathbf{Q}, +, \cdot)$, which is a subring of $(\mathbf{R}, +, \cdot)$, a subring of $(\mathbf{C}, +, \cdot)$.   □

EXAMPLE 8.8   In Example 8.5, $S = \{s, w\}$, and $T = \{s, v, x\}$ are subrings of R.

The next result characterizes those subsets of a ring that are subrings.

Theorem
8.9

▶ Given a ring $(R, +, \cdot)$, a nonempty subset S of R is a subring of R iff

1. for all $a, b \in S$, $a + b$, $ab \in S$ (i.e., S is closed under the binary operations of addition and multiplication defined on R); and

2. for all $a \in S$, $-a \in S$.

Proof   If $(S, +, \cdot)$ is a subring of R, then in its own right it satisfies all the conditions of a ring. Hence it satisfies (1) and (2) of the theorem. Conversely, let S be a nonempty subset of R that satisfies (1) and (2). Conditions (a), (b), (e), and (f) of the definition of a ring are inherited by the elements of S, since they are also elements of R. Thus, all we need to verify here is that S has an additive identity. As $S \neq \emptyset$, there is an element $a \in S$, and by (2), $-a \in S$. Then by (1), $z = a + (-a) \in S$.   ■

Note that $(\mathbf{Z}^+, +, \cdot)$ satisfies (1) above, but not (2), so it is not a subring of $(\mathbf{Z}, +, \cdot)$.

The result in Theorem 8.9 can also be given as follows.

Theorem
8.10

▶ For any ring $(R, +, \cdot)$, if $\emptyset \neq S \subseteq R$,

a) then $(S, +, \cdot)$ is a subring of R iff for $a, b \in S$, $a - b$, $ab \in S$.

b) and if S is finite, then $(S, +, \cdot)$ is a subring of R iff for $a, b \in S$, $a + b$, $ab \in S$. (Once again additional help comes from a finiteness condition.)

Proof   These proofs we leave for the reader.   ■

We close with an important type of subring.

Definition
8.6

▶ A nonempty subset I of a ring R is called an *ideal* of R if for all $a, b \in I$, $r \in R$, (a) $a - b \in I$; and (b) ar, $ra \in I$.

An ideal is a subring, but the converse is not so: $(\mathbf{Z}, +, \cdot)$ is a subring of $(\mathbf{Q}, +, \cdot)$ but not an ideal, since $(1/2)9 \notin \mathbf{Z}$ for $(1/2) \in \mathbf{Q}$, $9 \in \mathbf{Z}$. Meanwhile, all the subrings in Example 8.7(a) are ideals of $(\mathbf{Z}, +, \cdot)$.

EXERCISES

1. Complete the proofs of Theorems 8.2, 8.4, 8.5, and 8.10.

2. If a, b, c are elements in a ring $(R, +, \cdot)$, prove that (a) $a(b - c) = ab - (ac)$; and (b) $(b - c)a = ba - (ca)$.

3. a)   If R is a ring with unity and a, b are units of R, prove that ab is a unit of R and that $(ab)^{-1} = b^{-1}a^{-1}$.

   b)   For the ring $M_2(\mathbf{Z})$, find A^{-1}, B^{-1}, $(AB)^{-1}$, $(BA)^{-1}$, $B^{-1}A^{-1}$ if

$$A = \begin{bmatrix} 4 & 7 \\ 1 & 2 \end{bmatrix}, \qquad B = \begin{bmatrix} 5 & 2 \\ 2 & 1 \end{bmatrix}.$$

4. Prove that a unit in a ring R cannot be a proper divisor of zero.

5. If a is a unit in ring R, prove that $-a$ is also a unit.

6. If a, b are units in a ring R, is $a + b$ necessarily a unit of R?

7. Let S, T be subrings of a ring R. Prove that $S \cap T$ is a subring of R.

8. Answer Exercise 7 replacing "subring" with "ideal."

9. a)   For $R = M_2(\mathbf{Z})$, prove that

$$S = \left\{ \begin{bmatrix} a & 0 \\ 0 & 0 \end{bmatrix} \,\middle|\, a \in \mathbf{Z} \right\}$$

   is a subring of R.

   b)   What is the unity of R?

   c)   Does S have a unity?

   d)   Does S have any properties that R does not have?

10. Let R be a commutative ring with unity u, and let I be an ideal of R. (a) If $u \in I$, prove that $I = R$. (b) If I contains a unit of R, prove that $I = R$.

11. If R is a field, how many ideals does R have?

12. Let $(R, +, \cdot)$ be a commutative ring with unity, given by Tables 8.6(a) and (b).

Table 8.6

+	z	u	a	b
z	z	u	a	b
u	u	z	b	a
a	a	b	z	u
b	b	a	u	z

(a)

$\cdot$	z	u	a	b
z	z	z	z	z
u	z	u	a	b
a	z	a	b	u
b	z	b	u	a

(b)

a)   Verify that R is a field.

b)   Find a subring of R that is not an ideal.

c)   Let x, y be unknowns. Solve the following system of linear equations in R: $bx + y = u$; $x + by = z$.

13. Let R be a commutative ring with unity u, with $u \neq z$.
    a) For any $a \in R$, prove that $aR = \{ar \mid r \in R\}$ is an ideal of R.
    b) If the only ideals of R are $\{z\}$ and R, prove that R is a field.

14. Let $(S, +, \cdot)$ and $(T, +', \cdot')$ be two rings. For $R = S \times T$, define addition "$\oplus$" and multiplication "$\odot$" by

    $$(s_1, t_1) \oplus (s_2, t_2) = (s_1 + s_2, t_1 +' t_2),$$
    $$(s_1, t_1) \odot (s_2, t_2) = (s_1 \cdot s_2, t_1 \cdot' t_2).$$

    a) Prove that under these binary operations, R is a ring.
    b) If both S and T are commutative, prove that R is commutative.
    c) If S has unity u_S and T has unity u_T, what is the unity of R?
    d) If S and T are fields, is R also a field?

15. Let $(R, +, \cdot)$ be a ring with unity u, and $|R| = 8$. On $R^4 = R \times R \times R \times R$, define $+$ and $\cdot$ as suggested by Exercise 14. In the ring R^4, (a) how many elements have exactly two nonzero components? (b) how many elements have all nonzero components? (c) is there a unity? (d) how many units are there, if R has four units?

16. Let $(R, +, \cdot)$ be a ring, with $a \in R$. Define $0a = z$, $1a = a$, and $(n + 1)a = na + a$, for any $n \in \mathbf{Z}^+$. (Here we are multiplying elements of R by elements of $\mathbf{Z}$, so we have an operation different from the multiplication in either $\mathbf{Z}$ or R.) For $n > 0$, we define $(-n)a = n(-a)$, so, for example, $(-3)a = 3(-a) = 2(-a) + (-a) = [(-a) + (-a)] + (-a) = [-(a + a)] + (-a) = -[(a + a) + a] = -[2a + a] = -(3a)$.
    For any $a, b \in R$, $m, n \in \mathbf{Z}$, prove that
    a) $ma + na = (m + n)a$;   b) $m(na) = (mn)a$;
    c) $n(a + b) = na + nb$;   d) $n(ab) = (na)b = a(nb)$; and
    e) $(ma)(nb) = (mn)(ab) = (na)(mb)$.

17. a) For ring $(R, +, \cdot)$ and $a \in R$, we define $a^1 = a$, and for $n \in \mathbf{Z}^+$ $a^{n+1} = a^n a$. Prove that for $m, n \in \mathbf{Z}^+$, $(a^m)(a^n) = a^{m+n}$, and $(a^m)^n = a^{mn}$.
    b) Can you suggest how we might define a^0 or a^{-n}, $n \in \mathbf{Z}^+$, including any necessary conditions R must satisfy for these definitions to make sense?

8.3 ■ THE INTEGERS MODULO n

Enough theory for a while! We shall now concentrate on the construction and use of special finite rings and fields.

Definition 8.7   ▶ Let $n \in \mathbf{Z}^+$, $n > 1$. For $a, b \in \mathbf{Z}$, we say that *a is congruent to b modulo n,* and write $a \equiv b \pmod{n}$, if $n \mid (a - b)$, or $a = b + kn$, $k \in \mathbf{Z}$.

EXAMPLE 8.9   (a) $17 \equiv 2 \pmod 5$; (b) $-7 \equiv -49 \pmod 6$.   □

Theorem
8.11

► Congruence modulo n is an equivalence relation on $\mathbf{Z}$.

Proof

The proof is left for the reader. ■

As an equivalence relation induces a partition, for $n \geq 2$, congruence modulo n partitions $\mathbf{Z}$ into n equivalence classes

$$[0] = \{\ldots, -2n, -n, 0, n, 2n, \ldots\} = \{0 + nx \mid x \in \mathbf{Z}\}$$

$$[1] = \{\ldots, -2n + 1, -n + 1, 1, n + 1, 2n + 1, \ldots\} = \{1 + nx \mid x \in \mathbf{Z}\}$$

$$[2] = \{\ldots, -2n + 2, -n + 2, 2, n + 2, 2n + 2, \ldots\} = \{2 + nx \mid x \in \mathbf{Z}\}$$

$$\begin{matrix} \cdot & & & \cdot \\ \cdot & & & \cdot \\ \cdot & & & \cdot \end{matrix}$$

$$[n - 1] = \{\ldots, -n - 1, -1, n - 1, 2n - 1, 3n - 1, \ldots\}$$

$$= \{(n - 1) + nx \mid x \in \mathbf{Z}\}.$$

For any $t \in \mathbf{Z}$, by the division algorithm we can write $t = qn + r$, $0 \leq r < n$, so $t \in [r]$, or $[t] = [r]$. We use $\mathbf{Z}_n$ to denote $\{[0], [1], [2], \ldots, [n - 1]\}$. (When there is no chance for ambiguity we often replace $[a]$ by a and write $\mathbf{Z}_n = \{0, 1, 2, \ldots, n - 1\}$.) Our objective now is to define addition and multiplication on $\mathbf{Z}_n$ so that we obtain a ring.

For $[a], [b] \in \mathbf{Z}_n$, define $+$ and $\cdot$ by

$$[a] + [b] = [a + b], \quad \text{and} \quad [a] \cdot [b] = [a][b] = [ab].$$

For example, if $n = 7$, then $[2] + [6] = [2 + 6] = [8] = [1]$, and $[2][6] = [12] = [5]$.

Before these definitions are so readily accepted, we must investigate whether or not these operations are *well-defined* in the sense that if $[a] = [c]$, $[b] = [d]$ then $[a] + [b] = [c] + [d]$, $[a][b] = [c][d]$. As $[a] = [c]$ can occur with $a \neq c$, do the results of our addition and multiplication depend on which representatives are chosen from the equivalence classes? We shall prove that the results are independent of such representatives and the operations very definitely well-defined.

First, $[a] = [c] \Rightarrow a = c + sn$; $[b] = [d] \Rightarrow b = d + tn$. Hence $a + b = (c + sn) + (d + tn) = c + d + (s + t)n$, so $(a + b) \equiv (c + d) \pmod{n}$ and $[a + b] = [c + d]$. Also, $ab = (c + sn)(d + tn) = cd + (sd + ct + stn)n$ and $ab \equiv cd \pmod{n}$, or $[ab] = [cd]$.

This result now leads us to the following.

Theorem
8.12

► For $n \in \mathbf{Z}^+$, $n \geq 2$, under the binary operations defined above, $\mathbf{Z}_n$ is a commutative ring with unity [1].

Proof

The proof is left to the reader. Verification of the ring properties follows from the definitions of addition and multiplication in $\mathbf{Z}_n$, and the corresponding properties for the ring $(\mathbf{Z}, +, \cdot)$. ■

Before stating any further results, let us examine two particular examples. In Tables 8.7(a) and (b) and 8.8(a) and (b), we simplify $[a]$ by writing a.

Table 8.7 $\mathbf{Z}_5$

+	0	1	2	3	4
0	0	1	2	3	4
1	1	2	3	4	0
2	2	3	4	0	1
3	3	4	0	1	2
4	4	0	1	2	3

$\cdot$	0	1	2	3	4
0	0	0	0	0	0
1	0	1	2	3	4
2	0	2	4	1	3
3	0	3	1	4	2
4	0	4	3	2	1

(a)   (b)

Table 8.8 $\mathbf{Z}_6$

+	0	1	2	3	4	5
0	0	1	2	3	4	5
1	1	2	3	4	5	0
2	2	3	4	5	0	1
3	3	4	5	0	1	2
4	4	5	0	1	2	3
5	5	0	1	2	3	4

$\cdot$	0	1	2	3	4	5
0	0	0	0	0	0	0
1	0	1	2	3	4	5
2	0	2	4	0	2	4
3	0	3	0	3	0	3
4	0	4	2	0	4	2
5	0	5	4	3	2	1

(a)   (b)

In $\mathbf{Z}_5$ every nonzero element has a multiplicative inverse, so $\mathbf{Z}_5$ is a field. For $\mathbf{Z}_6$, however, 1 and 5 are the only units and 2, 3, 4 are divisors of zero. In $\mathbf{Z}_9$, $3 \cdot 3 = 3 \cdot 6 = 0$, so for $\mathbf{Z}_5$ to be a field we need more than just an odd modulus.

Theorem 8.13 ▶ $\mathbf{Z}_n$ is a field iff n is a prime.

Proof Let n be a prime, and suppose that $0 < a < n$. Then $(a, n) = 1$, so there are integers s, t with $as + tn = 1$. So $as \equiv 1 \pmod{n}$, or $[a][s] = [1]$, and $[a]$ is a unit of $\mathbf{Z}_n$, which is consequently a field.

Conversely, if n is not prime, then $n = n_1 n_2$, where $1 < n_1, n_2 < n$. So $[n_1] \neq [0]$, $[n_2] \neq [0]$ while $[n_1][n_2] = [n_1 n_2] = [0]$, and $\mathbf{Z}_n$ is not even an integral domain, so it cannot be a field. ∎

In $\mathbf{Z}_6$, $[5]$ is a unit, while $[3]$ is a zero divisor. We seek a way to recognize when $[a]$ is a unit in $\mathbf{Z}_n$, for n composite.

Theorem 8.14 ▶ In $\mathbf{Z}_n$, $[a]$ is a unit iff $(a, n) = 1$.

Proof If $(a, n) = 1$, the result follows as in Theorem 8.13. For the converse, let $[a] \in \mathbf{Z}_n$ and $[a]^{-1} = [s]$. Then $[as] = [a][s] = [1]$, so $as \equiv 1 \pmod{n}$ and $as = 1 + tn$, $t \in \mathbf{Z}$. But $1 = as + n(-t) \Rightarrow (a, n) = 1$. ∎

EXAMPLE 8.10

Find $[25]^{-1}$ in $\mathbf{Z}_{72}$.

Since $(25, 72) = 1$, the Euclidean algorithm leads us to

$$72 = 2(25) + 22, \qquad 0 < 22 < 25$$

$$25 = 1(22) + 3, \qquad 0 < 3 < 22$$

$$22 = 7(3) + 1, \qquad 0 < 1 < 3.$$

As 1 is the last nonzero remainder, we have

$$1 = 22 - 7(3) = 22 - 7[25 - 22] = (-7)(25) + (8)(22)$$

$$= (-7)(25) + 8[72 - 2(25)] = 8(72) - 23(25).$$

But

$$1 = 8(72) - 23(25) \Rightarrow 1 \equiv (-23)(25) \equiv (-23 + 72)(25) \pmod{72},$$

so $[1] = [49][25]$ and $[25]^{-1} = [49]$ in $\mathbf{Z}_{72}$. $\qquad \square$

Now $[25]$ is a unit in $\mathbf{Z}_{72}$, but is there any way of knowing how many units this ring has? From Theorem 8.14, if $1 \le a < 72$, then $[a]^{-1}$ exists iff $(a, 72) = 1$. Consequently, the number of units in $\mathbf{Z}_{72}$ is the number of integers a, where $1 \le a < 72$ and $(a, 72) = 1$. This is

$$\phi(72) = \phi(2^3 3^2) = (72)[1 - (1/2)][1 - (1/3)] = (72)(1/2)(2/3) = 24.$$

In general, for any $n \in \mathbf{Z}^+$, $n > 1$, there are $\phi(n)$ units and $n - 1 - \phi(n)$ proper divisors of zero in $\mathbf{Z}_n$.

The last example for this section provides an application in information retrieval.

EXAMPLE 8.11

When searching a table of records stored in a computer, each record is assigned a memory location or *address* in the computer's memory. The record itself is often made up of fields (this has nothing to do with ring structures). For instance, a college registrar keeps a record on each student, with the record containing information on the student's Social Security number, name, and major, for a total of three fields.

In searching for a particular student's record we can use his or her Social Security number as the *key* to the record, since it uniquely identifies that record. As a result we develop a function from the set of keys to the set of addresses in the table.

If the college is small enough we may find that the first four digits of the Social Security number are enough for identification. We develop a *hashing* (or *scattering*) function h from the set of keys (still Social Security numbers) to the set of addresses, determined now by the first four digits of the key. For example, $h(081\text{-}37\text{-}6495)$ identifies the record at the address associated with 0813. In this way we can store the table using at most 10,000 addresses. All is well so long as h is one-to-one. Should a second student have Social Security number 081-39-0207, h would no longer uniquely identify a student's record. When this happens a *collision* is said to occur. Since increasing the size of the stored table often results in more unused storage, we must balance the cost of this storage against the cost of handling such collisions. Techniques for resolving collisions have been devised. These depend on the data structures (such as vectors or linear linked lists) that are used to store the records.

Different kinds of hashing functions that have been developed include the following.

a) The *division* method: Here we restrict the number of addresses we want to use to a fixed integer n. For any key k (a positive integer) we define $h(k) = r \equiv k \pmod{n}$, $0 \le r < n$.

b) The *midsquare* hashing method squares part or all of the key. If we want $0 \le h(\text{key}) \le 10^n - 1$, for some $n \in \mathbf{Z}^+$, we extract n digits from the approximate center of the present value of $h(\text{key})$ to obtain the new value of $h(\text{key})$.

c) Often implemented is the *folding* method where the key is split into parts and the parts added together to give $h(\text{key})$. For example, $h(081\text{-}37\text{-}6495) = 081 + 37 + 6495 = 6613$ utilizes folding, and if we want only three-digit addresses, suppressing the first 6, we can have $h(081\text{-}37\text{-}6495) = 613$.

The importance of choosing a pertinent hashing function cannot be emphasized enough as we try to improve efficiency in terms of greater speed and less unused storage.

Using the modular concept we can develop a hashing function h, using the same keys as above, where

$$h(x_1 x_2 x_3 \text{-} x_4 x_5 \text{-} x_6 x_7 x_8 x_9) = y_1 y_2 y_3,$$

with

$$y_1 \equiv x_1 + x_2 + x_3 \pmod{5}, \qquad 0 \le y_1 < 5$$

$$y_2 \equiv x_4 + x_5 \pmod{3}, \qquad 0 \le y_2 < 3$$

$$y_3 \equiv x_6 + x_7 + x_8 + x_9 \pmod{7}, \qquad 0 \le y_3 < 7.$$

Here, for example, $h(081\text{-}37\text{-}6495) = 413$. □

EXERCISES

1. List four elements in each of the following equivalence classes.

 a) [1] in $\mathbf{Z}_7$; b) [2] in $\mathbf{Z}_{11}$; c) [10] in $\mathbf{Z}_{17}$.

2. Complete the proofs of theorems 8.11 and 8.12.

3. Define relation $\mathcal{R}$ on $\mathbf{Z}^+$ by $a \mathcal{R} b$, if $\tau(a) = \tau(b)$, where $\tau(a) = $ the number of positive divisors of a. For example, $2 \mathcal{R} 3$, $4 \mathcal{R} 25$, but $5 \not\mathcal{R} 9$.

 a) Verify that $\mathcal{R}$ is an equivalence relation on $\mathbf{Z}^+$.

 b) For the equivalence classes $[a]$, $[b]$ induced by $\mathcal{R}$, define operations of addition and multiplication by $[a] + [b] = [a + b]$, $[a][b] = [ab]$. Are these operations well-defined (i.e., does $a \mathcal{R} c$, $b \mathcal{R} d \Rightarrow (a + b) \mathcal{R} (c + d)$, $(ab) \mathcal{R} (cd)$)?

4. Find the multiplicative inverse of each element in $\mathbf{Z}_{11}$, $\mathbf{Z}_{13}$, and $\mathbf{Z}_{17}$.

5. Find $[a]^{-1}$ in $\mathbf{Z}_{1009}$ for (a) $a = 17$; (b) $a = 100$; and (c) $a = 777$.

6. a) Find all subrings of $\mathbf{Z}_{12}$, $\mathbf{Z}_{18}$, and $\mathbf{Z}_{24}$.

   b) Construct the Hasse diagram for each of these collections of subrings, where the partial order arises from set inclusion. Compare these diagrams with those for the set of positive divisors of n ($n = 12$; 18; 24), where the partial order now comes from the divisibility relation.

   c) Find a formula for the number of subrings in $\mathbf{Z}_n$, $n > 1$.

7. How many units and how many (proper) zero divisors are there in (a) $\mathbf{Z}_{17}$? (b) $\mathbf{Z}_{117}$? (c) $\mathbf{Z}_{1117}$?

8. Prove that in any list of n consecutive integers, one of the integers is divisible by n.

9. If three distinct integers are selected from the set $\{1, 2, 3, \ldots, 1000\}$, what is the probability that their sum is divisible by 3?

10. a) For c, d, n, $m \in \mathbf{Z}$, $n > 1$, $m > 0$, prove that if $c \equiv d \pmod{n}$, then $mc \equiv md \pmod{n}$ and $c^m \equiv d^m \pmod{n}$.

    b) If $x_n x_{n-1} \ldots x_1 x_0 = x_n \cdot 10^n + \cdots + x_1 \cdot 10 + x_0$ denotes an $(n + 1)$-digit integer, then prove that

    $$x_n x_{n-1} \ldots x_1 x_0 \equiv x_n + x_{n-1} + \cdots + x_1 + x_0 \pmod{9}.$$

11. For a, b, $n \in \mathbf{Z}^+$, $n > 1$, prove that $a \equiv b \pmod{n} \Rightarrow (a, n) = (b, n)$.

12. a) Show that for any $[a] \in \mathbf{Z}_7$, if $[a] \neq [0]$ then $[a]^6 = [1]$.

    b) Let $n \in \mathbf{Z}^+$ with $(n, 7) = 1$. Prove that $7 \mid (n^6 - 1)$.

13. For the hashing function at the end of Example 8.11, find (a) $h(123\text{-}04\text{-}2275)$; (b) a Social Security number n such that $h(n) = 413$, thus causing a collision with the number 081-37-6495 of the example.

14. Write a computer program that implements the hashing function of Exercise 13.

8.4 ■ RING HOMOMORPHISMS AND ISOMORPHISMS

In this final section we shall examine functions (between rings) that obey special properties which depend upon the binary operations in the rings.

EXAMPLE 8.12   Consider the rings $(\mathbf{Z}, +, \cdot)$, $(\mathbf{Z}_6, +, \cdot)$, where addition and multiplication in $\mathbf{Z}_6$ are as defined in Section 8.3.

Define $f: \mathbf{Z} \to \mathbf{Z}_6$ by $f(x) = [x]$. For example, $f(1) = [1] = [7] = f(7)$, $f(2) = f(8) = f(2 + 6k) = [2]$, $k \in \mathbf{Z}$. (So f is onto though not one-to-one.)

For 2, 3 $\in \mathbf{Z}$, $f(2) = [2]$, $f(3) = [3]$ and we have $f(2 + 3) = f(5) = [5] = [2] + [3] = f(2) + f(3)$, and $f(2 \cdot 3) = f(6) = [0] = [2][3] = f(2) \cdot f(3)$.

In fact for any $x, y \in \mathbf{Z}$,

$$f(x + y) = [x + y] = [x] + [y] = f(x) + f(y),$$

$\uparrow$ Addition in $\mathbf{Z}$ $\uparrow$ Addition in $\mathbf{Z}_6$

and

$$f(x \cdot y) = [xy] = [x][y] = f(x) \cdot f(y).$$ $\square$

$\uparrow$ Multiplication in $\mathbf{Z}$ $\uparrow$ Multiplication in $\mathbf{Z}_6$

This example demonstrates the following definition.

Definition 8.8

▶ Let $(R, +, \cdot)$ and $(S, \oplus, \odot)$ be rings. A function $f: R \to S$ is called a *ring homomorphism* if for all $a, b \in R$,

a) $f(a + b) = f(a) \oplus f(b)$, and

b) $f(a \cdot b) = f(a) \odot f(b)$.

This function is said to *preserve* the ring operations. Consider $f(a + b) = f(a) \oplus f(b)$. *Adding a, b in R* first and then finding the image (under f) in S of this sum, we get the same result as when we first determine the images (under f) in S of a, b, and then *add* these images *in S*. (Hence we have the function operation and the addition operations commuting with each other.) Similar remarks can be made about the multiplicative operations in the rings.

Definition 8.9

▶ Let $f: (R, +, \cdot) \to (S, \oplus, \odot)$ be a ring homomorphism. If f is one-to-one and onto, then f is called a *ring isomorphism* and we say that R and S are *isomorphic rings*.

We can think of isomorphic rings arising when the "same" ring is dealt with in two different languages. The function f then provides a dictionary for translating from one language into the other.

The terms homomorphism and isomorphism come from the Greek, where *morphe* refers to shape or *structure, homo* means similar, and *iso* means identical or same. Hence homomorphic rings may be thought of as similar in structure while isomorphic rings are *abstractly the same structure*.

EXAMPLE 8.13

For the ring R in Example 8.4 and the ring $\mathbf{Z}_5$, the function $f: R \to \mathbf{Z}_5$ given by

$$f(a) = [0], \qquad f(b) = [1], \qquad f(c) = [2], \qquad f(d) = [3], \qquad f(e) = [4]$$

provides us with a ring isomorphism.

For example, $f(c + d) = f(a) = [0] = [2] + [3] = f(c) + f(d)$, while $f(be) = f(e) = [4] = [1][4] = f(b)f(e)$. (There are 25 such equalities that must be verified for the preservation of each of the binary operations.) $\square$

As there are $5! = 120$ one-to-one functions from R onto $\mathbf{Z}_5$, is there any assistance we can call upon in attempting to determine when one of these functions is an isomorphism? Suggested by the example above, the following theorem provides ways of at least starting to determine when functions between rings can be homomorphisms and isomorphisms.

Theorem 8.15

▶ If $f: (R, +, \cdot) \rightarrow (S, \oplus, \odot)$ is a ring homomorphism, then

a) $f(z_R) = z_S$, where z_R, z_S are the zero elements of R, S, respectively;

b) $f(-a) = -f(a)$, for any $a \in R$;   c) $f(na) = nf(a)$, for any $a \in R$, $n \in \mathbf{Z}$;

d) $f(a^n) = [f(a)]^n$, for any $a \in R$, $n \in \mathbf{Z}^+$; and

e) if A is a subring of R, $f(A)$ is a subring of S.

Proof

a) $z_S \oplus f(z_R) = f(z_R) = f(z_R + z_R) = f(z_R) \oplus f(z_R)$. (Why?) So by the cancellation law of addition in S, we have $f(z_R) = z_S$.

b) $z_S = f(z_R) = f(a + (-a)) = f(a) \oplus f(-a)$. Since additive inverses in S are unique and $f(-a)$ is an additive inverse of $f(a)$, $f(-a) = -f(a)$.

c) If $n = 0$, $f(na) = f(z_R) = z_S = nf(a)$. The result is true for $n = 1$, so we assume the truth for $n = k$. Proceeding by mathematical induction, we examine the case where $n = k + 1$. By the results of Exercise 16 of Section 8.2, $f((k + 1)a) = f(ka + a) = f(ka) \oplus f(a) = kf(a) \oplus f(a)$ (Why?) $= (k + 1)(f(a))$ (Why?). (Note: There are three different kinds of addition here.)

  When $n > 0$, $f(-na) = f(n(-a)) = n(f(-a)) = n(-f(a)) = -nf(a)$, from our prior proof by induction, and part (b). Hence the result follows for all $n \in \mathbf{Z}$.

d) We leave this result for the reader to prove.

e) Since $A \neq \emptyset$, $f(A) \neq \emptyset$, and if $x, y \in f(A)$, then $x = f(a)$, $y = f(b)$ for some $a, b \in A$. Then $x \oplus y = f(a) \oplus f(b) = f(a + b)$, and $x \odot y = f(a) \odot f(b) = f(ab)$, with $a + b$, $ab \in A$ (Why?), so $x \oplus y, x \odot y \in f(A)$. Also, for $x \in f(A)$ with $x = f(a)$, $a \in A$, we have $f(-a) = -f(a) = -x$, and as $-a \in A$ (Why?), we have $-x \in f(A)$. Therefore $f(A)$ is a subring of S. ∎

When the homomorphism is onto we obtain the following.

Theorem 8.16

▶ If $f: (R, +, \cdot) \rightarrow (S, \oplus, \odot)$ is a ring homomorphism from R onto S, then

a) if R has unity u_R, $f(u_R)$ is the unity of S;

b) if R has unity u_R and a is a unit in R, then $f(a)$ is a unit in S, and $f(a^{-1}) = [f(a)]^{-1}$;

c) if R is commutative, then S is commutative; and

d) if I is an ideal of R, then $f(I)$ is an ideal of S.

Proof

We shall prove (d) and leave the other parts to the reader. Since I is a subring of R, $f(I)$ is a subring of S, by (e) of Theorem 8.15. To verify that $f(I)$ is an ideal, let $x \in f(I)$, $s \in S$. Then $x = f(a)$, $s = f(r)$, for $a \in I$, $r \in R$. So $s \odot x = f(r) \odot f(a) = f(ra)$, with $ra \in I$, and we have $s \odot x \in f(I)$. Similarly, $x \odot s \in f(I)$, so $f(I)$ is an ideal of S. ∎

These theorems reinforce the way in which homomorphisms and isomorphisms preserve structure. But can we find any use for these functions, aside from using them to prove more theorems? Before calling it quits, we consider one final example.

EXAMPLE 8.14   With **C** the field of complex numbers and S the ring of 2×2 real matrices of the form $\begin{bmatrix} a & b \\ -b & a \end{bmatrix}$, define $f: \mathbf{C} \to S$ by $f(a + bi) = \begin{bmatrix} a & b \\ -b & a \end{bmatrix}$, for $a + bi \in \mathbf{C}$.

Then

$$a + bi = c + di \Leftrightarrow a = c, b = d \Leftrightarrow \begin{bmatrix} a & b \\ -b & a \end{bmatrix} = \begin{bmatrix} c & d \\ -d & c \end{bmatrix},$$

so f is a one-to-one function. It is also onto. (Why?)
Further,

$$f((a + bi) + (x + yi)) = f((a + x) + (b + y)i)$$

$$= \begin{bmatrix} a + x & b + y \\ -(b + y) & a + x \end{bmatrix} = \begin{bmatrix} a & b \\ -b & a \end{bmatrix} + \begin{bmatrix} x & y \\ -y & x \end{bmatrix}$$

$$= f(a + bi) + f(x + yi),$$

and

$$f((a + bi)(x + yi)) = f((ax - by) + (bx + ay)i)$$

$$= \begin{bmatrix} ax - by & bx + ay \\ -(bx + ay) & ax - by \end{bmatrix} = \begin{bmatrix} a & b \\ -b & a \end{bmatrix} \begin{bmatrix} x & y \\ -y & x \end{bmatrix}$$

$$= f(a + bi)f(x + yi),$$

so f is a ring isomorphism.

Figure 8.1

```
10     Dim A(2,2), B(2,2), C(2,2), D(2,2), E(2,2), F(2,2), G(2,2),
       H(2,2), K(2,2), M(2,2)
20     Mat Read A, E, K
30     Data 1.3,3.7,−3.7,1.3,5.2,−7.1,7.1,5.2,−8.3,9.9,−9.9,−8.3
40     Mat B = A
50     For I = 1 To 3
60         Mat C = A*B
70         Mat A = C
80     Next I
90     Mat D = Inv(A)
100    Mat F = E
110    For J = 1 To 6
120        Mat G = E*F
130        Mat E = G
140    Next J
150    Mat H = D*E
160    Mat M = K + H
170    Print "The solution to the problem is"; M(1,1); "+"; M(1,2); "i"
180    End
```

The solution to the problem is 8379.98 + 15122.7 i

But where can we use this isomorphism?

Suppose we need to find the value of the complex number

$$(-8.3 + 9.9i) + \frac{(5.2 - 7.1i)^7}{(1.3 + 3.7i)^4}.$$

If a computer with an implementation of BASIC that included matrix operations were available, then using this isomorphism we could *translate* the problem from complex numbers, which are not implemented in BASIC, into matrices, which the computer can handle. Hence we seek the result

$$\begin{bmatrix} -8.3 & 9.9 \\ -9.9 & -8.3 \end{bmatrix} + \begin{bmatrix} 5.2 & -7.1 \\ 7.1 & 5.2 \end{bmatrix}^7 \left(\begin{bmatrix} 1.3 & 3.7 \\ -3.7 & 1.3 \end{bmatrix}^4 \right)^{-1}.$$

Using the BASIC program in Fig. 8.1, the answer (to two decimal places) to the complex arithmetic problem is found to be $8379.98 + 15122.7i$. □

EXERCISES

1. If R is the ring of Example 8.5, determine an isomorphism $f: R \to \mathbf{Z}_6$.

2. Complete the proofs of Theorems 8.15 and 8.16.

3. a) How many units are there in the ring $\mathbf{Z}_8$?

   b) How many units are there in the ring $\mathbf{Z}_2 \times \mathbf{Z}_2 \times \mathbf{Z}_2$?

   c) Are $\mathbf{Z}_8$ and $\mathbf{Z}_2 \times \mathbf{Z}_2 \times \mathbf{Z}_2$ isomorphic rings?

4. a) How many units are there in $\mathbf{Z}_{15}$? How many in $\mathbf{Z}_3 \times \mathbf{Z}_5$?

   b) Are $\mathbf{Z}_{15}$ and $\mathbf{Z}_3 \times \mathbf{Z}_5$ isomorphic?

5. Are $\mathbf{Z}_4$ and the ring in Example 8.3 isomorphic?

6. If $S = \left\{ \begin{bmatrix} a & 0 \\ 0 & a \end{bmatrix} \middle| a \in \mathbf{R} \right\}$, then S is a ring under matrix addition and multiplication. Prove that $\mathbf{R}$ is isomorphic to S.

7. If R, S, T are rings, and $f: R \to S$, $g: S \to T$ are ring homomorphisms, prove that the composite function $g \circ f: R \to T$ is a ring homomorphism.

8. If $f: R \to S$ is a ring homomorphism onto S, and J is an ideal of S, prove that $f^{-1}(J) = \{a \in R \mid f(a) \in J\}$ is an ideal of R.

9. a) In the solution of Example 8.14, part of the problem required the calculation of

$$\left(\begin{bmatrix} 1.3 & 3.7 \\ -3.7 & 1.3 \end{bmatrix}^4 \right)^{-1}.$$

   Algebraically this equals

$$\left(\begin{bmatrix} 1.3 & 3.7 \\ -3.7 & 1.3 \end{bmatrix}^{-1} \right)^4,$$

but is there any advantage, with respect to computer computation, in using one form instead of the other?

b)   The program in Fig. 8.1 necessitates 10 matrix multiplications. Revise this program so that it is more efficient with respect to the number of matrix multiplications involved.

8.5 ■ SUMMARY AND HISTORICAL REVIEW

Emphasizing structure induced by two binary operations, this chapter has introduced us to the mathematical system called a ring. Throughout the development of mathematics, the ring of integers has played a key role. In the branch of mathematics called number theory we examine the basic properties of $(\mathbf{Z}, +, \cdot)$, as well as the finite rings $(\mathbf{Z}_n, +, \cdot)$. The matrix rings provide familiar examples of noncommutative rings.

This chapter contains the development of an abstract theory. Based on the definition of a ring, we established principles of elementary algebra we have been using since our early encounters with arithmetic, signed numbers, and the manipulation of unknowns. The reader may have found some of the proofs tedious, as we justified all the steps in the derivations. Faced with the challenge of trying to prove a result in abstract mathematics, one should follow the advice given by the Roman rhetorician, Marcus Fabius Quintilianus (first century A.D.), when he said "One should not aim at being possible to understand (or follow), but at being impossible to be misunderstood."

A famous problem in number theory, known as *Fermat's Last Theorem,* claims that the equation $x^n + y^n = z^n$, $n \in \mathbf{Z}^+$, $n > 1$, has no solutions in $\mathbf{Z}^+$ when $n > 2$. The French mathematician, Pierre de Fermat (1601–1665) wrote that he had proved this result but that the proof was too long to include in the margin of his manuscript. Unfortunately, to this day, there are infinitely many unresolved values of n for the theorem. Attempts to solve this problem have resulted in new mathematical ideas and theories.

In trying to prove Fermat's last theorem, the German mathematician Ernst Kummer (1810–1893) developed the foundations for the concept of the *ideal.* This was later discovered and used by his fellow countryman Richard Dedekind (1831–1916) in his research on what are now called Dedekind domains. The term "ring," however, seems to be due to the German mathematician David Hilbert (1862–1943).

Ring homomorphisms and their interplay with ideals were extensively developed by the German mathematician Emmy Noether (1882–1935). This great genius received little remuneration, financial or otherwise, from the governing bodies of her native land, because of the sexual bias that existed in the universities at that time. Emmy Noether was nonetheless recognized for her talents by her colleagues, and was eulogized in the *New York Times* on May 3, 1935 by none other than Albert Einstein (1879–1955). In addition to sexual bias, as a Jew she was forced to flee her homeland in 1933 when the Nazis came to power. She spent the last two years of her life guiding young mathematicians in the United States. For more on the life of this fascinating personality, examine the biography by A. Dick [2].

The special rings called *fields* arise in the rational, real, and complex number systems. But we also found some interesting finite fields. These structures will be examined again in Chapter 13 in connection with combinatorial designs. The field theory developed by the French genius Evariste Galois (1811–1832) established results about the solutions of polynomial equations of degree > 4. These results had baffled mathematicians for centuries, and his ideas, now known as Galois theory, still comprise one of the most elegant mathematical theories ever developed. More on Galois theory is given in the text by O. Zariski and P. Samuel [7].

For supplemental reading on ring theory at the introductory level, the interested reader should examine Chapters 6, 7, and 12 of N. McCoy and T. Berger [4] and Chapter 6 of V. Larney [3].

The development of modular congruence, along with many related ideas, was primarily due to Carl Friedrich Gauss (1777–1855). More on the solution of congruences can be found in the text by I. Niven and H. Zuckerman [5].

Finally, the topic of hashing, or scattering, can be further investigated in Chapter 2 of J. Tremblay and R. Manohar [6]. Chapter 4 of A. Aho, J. Hopcroft, and J. Ullman [1] includes a discussion on efficiency of hashing functions and a probabilistic investigation of the collision problem that arises for these functions.

REFERENCES

1. Aho, Alfred V., Hopcroft, John E., and Ullman, Jeffrey D., *Data Structures and Algorithms,* Addison-Wesley, Reading, Massachusetts, 1983.

2. Dick, Auguste, *Emmy Noether (1882–1935),* translated by Heidi Blocher, Birkhauser-Boston, Boston, 1981.

3. Larney, Violet Hachmeister, *Abstract Algebra: A First Course,* Prindle, Weber & Schmidt, Boston, 1975.

4. McCoy, Neal H., and Berger, Thomas R., *Algebra: Groups, Rings and Other Topics,* Allyn and Bacon, Boston, 1977.

5. Niven, Ivan, and Zuckerman, Herbert S., *An Introduction to the Theory of Numbers,* 3rd ed., John Wiley & Sons, New York, 1972.

6. Tremblay, Jean-Paul, and Manohar, R., *Discrete Mathematical Structures with Applications to Computer Science,* McGraw-Hill, New York, 1975.

7. Zariski, Oscar, and Samuel, Pierre, *Commutative Algebra,* Volume I, D. Van Nostrand, Princeton, New Jersey, 1958.

■ MISCELLANEOUS EXERCISES

1. Determine whether each of the following statements is true or false. For each false statement give a counterexample.

   a) If $(R, +, \cdot)$ is a ring, and $\emptyset \neq S \subseteq R$ with S closed under $+$ and $\cdot$, then S is a subring of R.

b) If $(R, +, \cdot)$ is a ring with unity, and S is a subring of R, then S has a unity.

c) If $(R, +, \cdot)$ is a ring with unity u_R, and S is a subring of R with unity u_S, then $u_R = u_S$.

d) Every field is an integral domain.

e) No subring of an integral domain has any proper divisors of zero.

f) Any subring of a field is a field.

g) A field can have only two subrings.

h) The function $f: \mathbf{Z} \to \mathbf{Z}$ defined by $f(x) = 2x$ is a ring homomorphism.

i) Every finite field has a prime number of elements.

j) $(\mathbf{Q}, +, \cdot)$ has an infinite number of subrings.

k) If S and T are ideals of a ring R, then $S \cup T$ is an ideal of R.

ℓ) If $a, b \in \mathbf{Z}_p$ (p a prime), and $a^2 = b^2$, then $a = b$.

2. Prove that a ring R is commutative if and only if $(a + b)^2 = a^2 + 2ab + b^2$, for all $a, b \in R$.

3. A ring R is called *Boolean* if $a^2 = a$ for all $a \in R$. If R is Boolean, prove that (a) $a + a = 2a = z$, for all $a \in R$; and (b) R is commutative.

4. a) In the field $\mathbf{C}$ of complex numbers, the *conjugate* of a complex number $z = x + iy$ is given by $\bar{z} = x - iy$. Hence $\overline{2 + 3i} = 2 - 3i$, $\overline{5i} = -5i$, while $\bar{7} = 7$.

   If $z, z_1, z_2 \in \mathbf{C}$, prove that

   i) $\overline{z_1 + z_2} = \bar{z_1} + \bar{z_2}$;

   ii) $\overline{z_1 z_2} = \bar{z_1}\,\bar{z_2}$;

   iii) $(\bar{z})^n = (\overline{z^n})$, for all $n \in \mathbf{Z}^+$;

   iv) $(\bar{z})^{-1} = (\overline{z^{-1}})$, $z \neq 0$;

   v) $z + \bar{z} \in \mathbf{R}$; and

   vi) $z\bar{z} \in \mathbf{R}^+$, $z \neq 0$.

   b) Let $f: \mathbf{C} \to \mathbf{C}$ be defined by $f(z) = \bar{z}$. Prove that f is an isomorphism.

5. a) Let $(D, +, \cdot)$ be an integral domain. Define relation $\mathcal{R}$ on D by $a \mathcal{R} b$ if $a = bs$, where s is a unit of D. Prove that $\mathcal{R}$ is an equivalence relation.

   b) Find the partition induced by $\mathcal{R}$ if the integral domain is

      i) $(\mathbf{Z}, +, \cdot)$;   ii) $\{a + bi \,|\, a, b \in \mathbf{Z}, i^2 = -1\}$; and   iii) $\mathbf{Z}_7$.

6. Let $A = \mathbf{R}^+$. Define $\oplus$ and $\odot$ on A by $a \oplus b = ab$, the ordinary product of a, b; and $a \odot b = a^{\log_2 b}$.

   a) Verify that $(A, \oplus, \odot)$ is a commutative ring with unity.

   b) Is it an integral domain or field?

7. Let R be a ring with ideals A and B. Define $A + B = \{a + b \,|\, a \in A, b \in B\}$. Prove that $A + B$ is an ideal of R. (For any ring R, the ideals of R form a poset

under set inclusion. If A and B are ideals of R, with glb $\{A, B\} = A \cap B$, lub $\{A, B\} = A + B$, the poset is a lattice.)

8. a) If p is a prime, prove that p divides $\binom{p}{k}$, $0 < k < p$.

   b) If $a, b \in \mathbf{Z}$, prove that $(a + b)^p \equiv a^p + b^p \pmod{p}$.

9. Given n positive integers $x_1, x_2, \ldots, x_n$, not necessarily distinct, prove that there exist $1 \leq i < j \leq n$ such that $n \mid (x_{i+1} + \cdots + x_{j-1} + x_j)$.

10. Write a computer program that reverses the order of the digits in a given positive integer. For example, the input 1374 results in the output 4731.

9

Boolean
Algebra
and Switching
Functions

Again we confront an algebraic system in which the structure depends primarily upon two binary operations. Unlike the coverage of rings, in dealing with Boolean algebras we shall stress applications more than the abstract nature of the system. Nonetheless, we shall find results here that are quite different from those for rings. Among other things, a finite Boolean algebra cannot have just any number of elements. It must have 2^n elements, where $n \in \mathbf{Z}^+$.

In 1854 the English mathematician George Boole published his monumental work: *An Investigation of the Laws of Thought*. Within this work Boole created a system of mathematical logic that he developed in terms of what is now called a Boolean algebra.

In 1938 Claude Shannon developed the algebra of switching functions and showed how its structure was related to the ideas established by Boole. As a result, an example of abstract mathematics in the nineteenth century became an applied mathematical discipline in the twentieth century.

9.1 ■ SWITCHING FUNCTIONS: DISJUNCTIVE AND CONJUNCTIVE NORMAL FORMS

An electric switch can be turned on (allowing the flow of current) or off (preventing the flow of current). Similarly, in a transistor, current is passing (conducting) or not passing (nonconducting). These are two examples of *two-state devices*. (In Section 2.1 we saw how the electric switch was related to the two-valued logic we introduced there.)

In order to investigate such devices, we abstract these notions of "true" and "false," "on" and "off," as follows.

Let $B = \{0, 1\}$. We define addition, multiplication, and the complements for the elements of B by

---

a) $0 + 0 = 0;\quad 0 + 1 = 1 + 0 = 1 + 1 = 1.$

b) $0 \cdot 0 = 0 = 1 \cdot 0 = 0 \cdot 1;\quad 1 \cdot 1 = 1.$

c) $\bar{0} = 1;\quad \bar{1} = 0.$

A variable x is called a *Boolean variable* if x only takes on values in B. Consequently $x + x = x$, $x^2 = x \cdot x = xx = x$ for any Boolean variable x.

If $n \in \mathbf{Z}^+$, $B^n = \{(b_1, b_2, \ldots, b_n) \mid b_i \in \{0, 1\}, \ 1 \leq i \leq n\}$. A function $f: B^n \to B$ is called a *Boolean*, or *switching, function* of n variables. The n variables are emphasized by writing $f(x_1, x_2, \ldots, x_n)$, where each x_i, $1 \leq i \leq n$, is a Boolean variable.

EXAMPLE 9.1 Let $f: B^3 \to B$, where $f(x, y, z) = xy + z$. (We write xy for $x \cdot y$.) This Boolean function is determined by evaluating f for each of the eight possible assignments of the variables x, y, z. Table 9.1 demonstrates this. □

Table 9.1

x	y	z	xy	$f(x, y, z) = xy + z$
0	0	0	0	0
0	0	1	0	1
0	1	0	0	0
0	1	1	0	1
1	0	0	0	0
1	0	1	0	1
1	1	0	1	1
1	1	1	1	1

Definition 9.1

▶ If $f, g: B^n \to B$, then $f = g$ if their corresponding (function) tables are exactly the same. (The tables show that $f(b_1, b_2, \ldots, b_n) = g(b_1, b_2, \ldots, b_n)$, for all $(b_1, b_2, \ldots, b_n) \in B^n$.)

Definition 9.2

▶ For $f: B^n \to B$, the *complement* of f, denoted $\bar{f}$, is the Boolean function defined on B^n by $\bar{f}(b_1, b_2, \ldots, b_n) = \overline{f(b_1, b_2, \ldots, b_n)}$. If $g: B^n \to B$, we define $f + g$, $f \cdot g: B^n \to B$, the *sum* and *product* of f, g, respectively, by

$$(f + g)(b_1, b_2, \ldots, b_n) = f(b_1, b_2, \ldots, b_n) + g(b_1, b_2, \ldots, b_n),$$

and

$$(f \cdot g)(b_1, b_2, \ldots, b_n) = f(b_1, b_2, \ldots, b_n) \cdot g(b_1, b_2, \ldots, b_n).$$

Table 9.2

1. $\bar{\bar{f}} = f$	$\bar{\bar{x}} = x$	Law of the Double Complement
2. $\overline{f + g} = \bar{f}\,\bar{g}$ $\overline{fg} = \bar{f} + \bar{g}$	$\overline{x + y} = \bar{x}\bar{y}$ $\overline{xy} = \bar{x} + \bar{y}$	DeMorgan's Laws
3. $f + g = g + f$ $fg = gf$	$x + y = y + x$ $xy = yx$	Commutative Laws
4. $f + (g + h) = (f + g) + h$ $f(gh) = (fg)h$	$x + (y + z) = (x + y) + z$ $x(yz) = (xy)z$	Associative Laws
5. $f + gh = (f + g)(f + h)$ $f(g + h) = fg + fh$	$x + yz = (x + y)(x + z)$ $x(y + z) = xy + xz$	Distributive Laws
6. $f + f = f$ $ff = f$	$x + x = x$ $xx = x$	Idempotent Laws
7. $f + 0 = f$ $f \cdot 1 = f$	$x + 0 = x$ $x \cdot 1 = x$	Identity Laws
8. $f + \bar{f} = 1$ $f\bar{f} = 0$	$x + \bar{x} = 1$ $x\bar{x} = 0$	Inverse Laws
9. $f + 1 = 1$ $f \cdot 0 = 0$	$x + 1 = 1$ $x \cdot 0 = 0$	Dominance Laws
10. $f + (fg) = f$ $f(f + g) = f$	$x + (xy) = x$ $x(x + y) = x$	Absorption Laws

As with the laws of set theory and logic in Chapter 2, the properties shown in Table 9.2 are satisfied by any Boolean functions $f, g, h: B^n \to B$, and any Boolean variables x, y, z. (We write fg for $f \cdot g$.)

The function **0** denotes the Boolean function whose value is always 0, and **1** is the function whose only value is 1. [Note: **0**, **1** $\notin B$.]

Once again the idea of duality appears in properties 2–10. If s denotes a theorem about Boolean functions, then s^d, the *dual* of s, is obtained by interchanging in s all occurrences of $+$ $(\cdot)$ by $\cdot$ $(+)$ and all occurrences of **0** (**1**) by **1** (**0**), and s^d is also a theorem. The same is true for a theorem dealing with Boolean variables, except here it is the Boolean values 0 and 1 that are interchanged, not the constant functions **0** and **1**.

We prove property 5 above as follows.

EXAMPLE 9.2   (The Distributive Law of $+$ over $\cdot$) The last two columns of Table 9.3 show that $f + gh = (f + g)(f + h)$. By the principle of duality we obtain $f(g + h) = fg + fh$. We see that $x + (yz) = (x + y)(x + z)$ is a special case of this property for $f, g, h: B^3 \to B$, where $f(x, y, z) = x$, $g(x, y, z) = y$, $h(x, y, z) = z$. Hence, no additional tables are needed to establish these properties for Boolean variables.   □

Table 9.3

f	g	h	gh	$f + g$	$f + h$	$f + gh$	$(f + g)(f + h)$
0	0	0	0	0	0	0	0
0	0	1	0	0	1	0	0
0	1	0	0	1	0	0	0
0	1	1	1	1	1	1	1
1	0	0	0	1	1	1	1
1	0	1	0	1	1	1	1
1	1	0	0	1	1	1	1
1	1	1	1	1	1	1	1

EXAMPLE 9.3   To establish the first absorption property, in place of a table we have the following:

	Reason
$x + xy = x1 + xy$	Identity Law
$\quad = x(1 + y)$	Distributive Law of $\cdot$ over $+$
$\quad = x1$	Dominance Law
$\quad = x$	Identity Law

This result indicates that some of our properties can be derived from others. The question then is which properties we must establish with tables so that we can derive the other properties as we did here. We will consider this later when we study the structure of a Boolean algebra.   □

Up to this point we have repeated for Boolean functions what we did with propositions: namely, given the Boolean function, we construct its table of values. Now we consider the reverse process: given a table of values, find a Boolean function for which it is the correct table.

EXAMPLE 9.4 Given three Boolean variables, x, y, z, find functions f, g, h: $B^3 \rightarrow B$ for the columns specified in Table 9.4.

For the column under f we want a result that has value 1 in the case where $x = y = 0$ and $z = 1$. The function $f(x, y, z) = \bar{x}\bar{y}z$ is one such function. In the same way, $g(x, y, z) = x\bar{y}\bar{z}$ yields the value 1 for $x = 1$, $y = z = 0$, and is 0 in all other cases. As f and g have value 1 in only one case and these cases are distinct from each other, their sum $f + g$ has value 1 in exactly these two cases. So $h(x, y, z) = f(x, y, z) + g(x, y, z) = \bar{x}\bar{y}z + x\bar{y}\bar{z}$ has the column of values given under h. □

Table 9.4

x	y	z	f	g	h
0	0	0	0	0	0
0	0	1	1	0	1
0	1	0	0	0	0
0	1	1	0	0	0
1	0	0	0	1	1
1	0	1	0	0	0
1	1	0	0	0	0
1	1	1	0	0	0

This example leads us to the following.

Definition 9.3 ▶ For any $n \in \mathbf{Z}^+$, if f is a Boolean function on the n variables $x_1, x_2, \ldots, x_n$, we call

a) each term x_i or its complement $\bar{x}_i$, $1 \le i \le n$, a *literal*;

b) a term of the form $y_1 y_2 \cdots y_n$, where each $y_i = x_i$ or $\bar{x}_i$, $1 \le i \le n$, a *fundamental conjunction*; and

c) a representation of f as a sum of fundamental conjunctions a *disjunctive normal form* (d.n.f.) of f.

Although no formal proof is given here, these examples suggest that any $f: B^n \rightarrow B$, $f \ne \mathbf{0}$, has a unique (up to the order of fundamental conjunctions) representation as a d.n.f.

EXAMPLE 9.5 Find the d.n.f. for $f: B^3 \rightarrow B$, where $f(x, y, z) = xy + \bar{x}z$.

From Table 9.5, we see that the column for f contains four 1's. These indicate the four fundamental conjunctions needed in the d.n.f. of f, so $f(x, y, z) = \bar{x}\bar{y}z + \bar{x}yz + xy\bar{z} + xyz$.

Another way to solve this problem is to take each product term and somehow involve whichever variables are missing. Using the properties of these variables we have $xy + \bar{x}z = xy(z + \bar{z}) + \bar{x}(y + \bar{y})z$ (Why?) $= xyz + xy\bar{z} + \bar{x}yz + \bar{x}\bar{y}z$. □

Table 9.5

x	y	z	xy	$\overline{xz}$	f
0	0	0	0	0	0
0	0	1	0	1	1
0	1	0	0	0	0
0	1	1	0	1	1
1	0	0	0	0	0
1	0	1	0	0	0
1	1	0	1	0	1
1	1	1	1	0	1

EXAMPLE 9.6

Find the d.n.f. for $g(w, x, y, z) = wx\overline{y} + wy\overline{z} + xy$.
    We examine each term, as follows:

a) $wx\overline{y} = wx\overline{y}(z + \overline{z}) = wx\overline{y}z + wx\overline{y}\,\overline{z}$

b) $wy\overline{z} = w(x + \overline{x})y\overline{z} = wxy\overline{z} + w\overline{x}y\overline{z}$

c) $xy = (w + \overline{w})xy(z + \overline{z}) = wxyz + wxy\overline{z} + \overline{w}xyz + \overline{w}xy\overline{z}$.

From the idempotent property of $+$, the d.n.f. of g is

$$g(w, x, y, z) = wx\overline{y}z + wx\overline{y}\,\overline{z} + wxy\overline{z} + w\overline{x}y\overline{z} + wxyz + \overline{w}xyz + \overline{w}xy\overline{z}. \qquad \square$$

Consider Table 9.6. If we agree to list the Boolean variables in alphabetical order, we see that the values for x, y, z in any row determine a binary number. These binary numbers for 0, 1, 2, ..., 7 arise for rows 1, 2, ..., 8, respectively. Because of this, the d.n.f. of a nonzero Boolean function can be expressed more compactly. The function f in Example 9.5 can be given by $f = \Sigma m(1, 3, 6, 7)$, where m indicates the *minterms* at rows 2, 4, 7, 8, where the respective binary labels are 1, 3, 6, 7. The word "minterm" arises as follows: $m(1)$, for example, is the minterm for the row where $x = y = 0$, $z = 1$; the prefix "min" refers to the fact that the fundamental conjunction $\overline{x}\,\overline{y}z$, has value 1 a minimal number of times (without being identically 0).
    Lacking a table, the d.n.f. of the function g of Example 9.6 can still be represented as a *sum of minterms*. For each fundamental conjunction $c_1c_2c_3c_4$, where $c_1 = w$ or $\overline{w}$, ..., $c_4 = z$ or $\overline{z}$, we replace each c_i, $1 \le i \le 4$, by 0 if c_i is complemented, and 1, otherwise. In this way the binary number associated with that fundamental conjunction is obtained. As a sum of minterms, $g = \Sigma m(6, 7, 10, 12, 13, 14, 15)$.

Table 9.6

x	y	z
0	0	0
0	0	1
0	1	0
0	1	1
1	0	0
1	0	1
1	1	0
1	1	1

Dual to the disjunctive normal form is the conjunctive normal form, which we discuss before closing this section.

EXAMPLE 9.7 Let $f: B^3 \to B$ be given by Table 9.7. A term of the form $c_1 + c_2 + c_3$, where $c_1 = x$ or $\bar{x}$, $c_2 = y$ or $\bar{y}$, $c_3 = z$ or $\bar{z}$ is called a *fundamental disjunction*. The fundamental disjunction $x + y + z$ has value 1 in all cases except where the values of x, y, z are 0. Similarly, $x + \bar{y} + z$ has value 1 except when $x = z = 0$ and $y = 1$. As each of these fundamental disjunctions has the value 0 in only one case, and these cases do not occur simultaneously, the product $(x + y + z)(x + \bar{y} + z)$ is 0 in precisely the two cases listed above. Continuing in this manner we write the function f as

$$f = (x + y + z)(x + \bar{y} + z)(\bar{x} + \bar{y} + z),$$

and call this the *conjunctive normal form* (c.n.f.) for f.

As $x + y + z$ has value 1 a maximum number of times (without being identically 1), it is called a *maxterm,* especially when we use a binary row label to represent it. Using the binary numbers indexing the rows of the table we write $f = \prod M(0, 2, 6)$, a *product of maxterms.*

Such representations exist for any $f \neq 1$, and are unique up to the order of the fundamental disjunctions or maxterms. □

Table 9.7

x	y	z	f
0	0	0	0
0	0	1	1
0	1	0	0
0	1	1	1
1	0	0	1
1	0	1	1
1	1	0	0
1	1	1	1

EXAMPLE 9.8 Let $g: B^4 \to B$ where $g(w, x, y, z) = (w + x + y)(x + \bar{y} + z)(w + \bar{y})$. To get the c.n.f. for g, each disjunction in the product is rewritten as follows:

a) $w + x + y = w + x + y + 0 = w + x + y + (z\bar{z})$
$$= (w + x + y + z)(w + x + y + \bar{z})$$

b) $x + \bar{y} + z = w\bar{w} + x + \bar{y} + z = (w + x + \bar{y} + z)(\bar{w} + x + \bar{y} + z)$

c) $w + \bar{y} = w + x\bar{x} + \bar{y} = (w + x + \bar{y})(w + \bar{x} + \bar{y})$
$$= (w + x + \bar{y} + z\bar{z})(w + \bar{x} + \bar{y} + z\bar{z})$$
$$= (w + x + \bar{y} + z)(w + x + \bar{y} + \bar{z})(w + \bar{x} + \bar{y} + z) \cdot$$
$$(w + \bar{x} + \bar{y} + \bar{z})$$

Consequently, using the idempotent law of $\cdot$, we have $g(w, x, y, z) =$ $(w + x + y + z)(w + x + y + \bar{z})(w + x + \bar{y} + z)(\bar{w} + x + \bar{y} + z)(w + x + \bar{y} + \bar{z})(w + \bar{x} + \bar{y} + z)(w + \bar{x} + \bar{y} + \bar{z})$.

To obtain g as a product of maxterms, we associate with each fundamental disjunction $d_1 + d_2 + d_3 + d_4$ the binary number $b_1b_2b_3b_4$ where $b_1 = 0$ if $d_1 = w$; $b_1 = 1$ if $d_1 = \overline{w}$; $\ldots$; $b_4 = 0$ if $d_4 = z$; $b_4 = 1$ if $d_4 = \overline{z}$. As a result, $g = \prod M(0, 1, 2, 3, 6, 7, 10)$. □

EXERCISES

1. a) How many rows are needed to construct the (function) table for a Boolean function of n variables?

   b) How many different Boolean functions of n variables are there?

2. If $f: B^3 \to B$ is defined by $f(x, y, z) = \overline{(x + y)} + (\overline{x}z)$,

   a) determine the d.n.f. and c.n.f. for f.

   b) write f as a sum of minterms and a product of maxterms (utilizing binary indices).

3. Let F_6 be the set of all Boolean functions $f: B^6 \to B$.

   a) What is $|F_6|$?

   b) How many fundamental conjunctions (disjunctions) are there in F_6?

   c) How many minterms (maxterms) are there in F_6?

   d) How many functions $f \in F_6$ have the value 1 when exactly (at least) two of its variables have the value 1?

   e) Let u, v, w, x, y, z denote the six Boolean variables for the functions in F_6. How many of these functions are independent of x (i.e., $f(u, v, w, x, y, z) = f(u, v, w, \overline{x}, y, z)$)? How many are independent of x, y, z?

4. a) For $f, g, h: B^n \to B$, prove that $fg + \overline{f}h + gh = fg + \overline{f}h$; and $fg + f\overline{g} + \overline{f}g + \overline{f}\,\overline{g} = 1$.

   b) State the dual of each result in part (a).

5. Let $f: B^n \to B$. If the d.n.f. of f has m fundamental conjunctions and its c.n.f. has k fundamental disjunctions, how are m, n, and k related?

6. If x, y, z are Boolean variables and $x + y + z = xyz$, prove that x, y, z are simultaneously 0 or 1.

7. Simplify the following Boolean expressions.

   a) $xy + (x + y)\overline{z} + y$

   b) $x + y + \overline{(\overline{x} + y + z)}$

   c) $yz + wx + z + [wz(xy + wz)]$

   d) $x_1 + \overline{x}_1x_2 + \overline{x}_1\overline{x}_2x_3 + \overline{x}_1\overline{x}_2\overline{x}_3x_4 + \cdots$

8. Find the values of the Boolean variables w, x, y, z that satisfy the following system of simultaneous (Boolean) equations.

$$x + \overline{x}y = 0 \qquad \overline{x}y = \overline{x}z \qquad \overline{x}y + \overline{x}\,\overline{z} + zw = \overline{z}w$$

9. For $f, g: B^n \to B$, define the binary operation $\oplus$ (Exclusive Or) by $f \oplus g = f\bar{g} + \bar{f}g$.

   a) Determine $f \oplus f$, $f \oplus \bar{f}$, $f \oplus 1$, $f \oplus 0$.

   b) Prove or disprove each of the following:

   i) $f \oplus g = 0 \Rightarrow f = g$ ii) $f \oplus (g \oplus h) = (f \oplus g) \oplus h$

   iii) $f \oplus g = \bar{f} \oplus \bar{g}$ iv) $f \oplus (gh) = (f \oplus g)(f \oplus h)$

   v) $f(g \oplus h) = fg \oplus fh$ vi) $\overline{(f \oplus g)} = \bar{f} \oplus g = f \oplus \bar{g}$

   vii) $f \oplus g = f \oplus h \Rightarrow g = h$

10. Let $f, g: B^n \to B$. Define the relation "$\leq$" on F_n, the set of all Boolean functions of n variables, by $f \leq g$ if the value of g is 1 at least whenever the value of f is 1.

   a) Prove that this relation is a partial order on F_n.

   b) Prove that $fg \leq f$ and $f \leq f + g$.

   c) For $n = 2$, draw the Hasse diagram for the 16 functions in F_2. Where are the minterms and maxterms located in the diagram? Compare this diagram with that for the power set of $\{a, b, c, d\}$ partially ordered under the subset relation.

9.2 ■ GATING NETWORKS: MINIMAL SUMS OF PRODUCTS: KARNAUGH MAPS

The switching functions of Section 9.1 present an interesting mathematical theory. Their importance lies in their implementation by means of *logic gates* (devices in a digital computer that perform specified tasks in the processing of data). The electrical and mechanical components of such gates depend upon the state of the art. Consequently, we shall not concern ourselves here with questions relating to hardware.

   Figure 9.1 contains the logic gates for negation (complement), conjunction, and disjunction in (a), (b), and (c), respectively. Since the Boolean operations of + and · are associative, we may have more than two inputs for an AND gate or an OR gate.

Figure 9.1

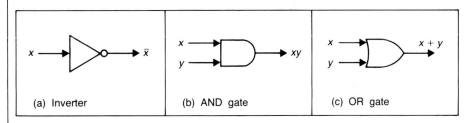

| (a) Inverter | (b) AND gate | (c) OR gate |

   Figure 9.2 shows the *logic*, or *gating*, *network* for the expression $(w + \bar{x}) \cdot (y + xz)$. Symbols on a line to the left of a gate (or inverter) are *inputs*. When they are on line segments to the right of a gate, they are *outputs*. We have *split* the input line for x, so that x may serve as input for both an AND gate and an inverter.

Figure 9.2

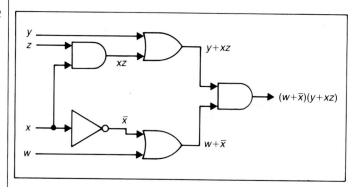

The exercises provide practice in drawing the logic network for a Boolean expression and going from the network to the expression. Meanwhile certain features of these networks need to be emphasized.

1. An input line may be split to provide that input to more than one gate.
2. Input and output lines come together only at gates.
3. There is no doubling back; that is, an output for a gate cannot then be used as an input for the same gate.
4. We assume that the output of a gating network is an instantaneous function of the present inputs. There is no time dependence and we attach no importance to prior inputs as with finite state machines.

With these ideas in mind, the following analyzes the computer addition of binary numbers.

EXAMPLE 9.9

When we add two bits (binary digits), the result consists of a sum s and a carry c. In three of four cases the carry is 0, so we focus on the computation of $1 + 1$. Examining parts (b) and (c) of Table 9.8, we consider the sum, s, and carry, c, as Boolean functions of the variables x and y. Then $c = xy$, and $s = \bar{x}y + x\bar{y} = x \oplus y = (x + y)(\overline{xy})$.

Table 9.8

x	y	Binary sum
0	0	$0 + 0 = 0$
0	1	$0 + 1 = 1$
1	0	$1 + 0 = 1$
1	1	$1 + 1 = 10$

(a)

x	y	sum
0	0	0
0	1	1
1	0	1
1	1	0

(b)

x	y	carry
0	0	0
0	1	0
1	0	0
1	1	1

(c)

Figure 9.3

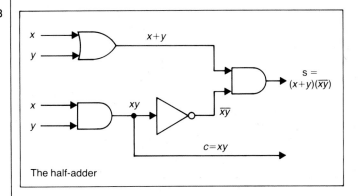

The half-adder

Figure 9.3 is a gating network with two outputs, and is referred to as a *multiple output* network. This device, called a *half-adder,* implements the results in Tables 9.8(b) and (c). Using two half-adders and an OR gate, we construct the *full-adder* shown in Fig. 9.4(a). If $x = x_n x_{n-1} \ldots x_2 x_1 x_0$ and $y = y_n y_{n-1} \ldots y_2 y_1 y_0$, consider the process of adding the binary digits x_i and y_i in finding the sum $x + y$. Here c_{i-1} is the carry from the addition of x_{i-1} and y_{i-1} (and a possible carry c_{i-2}). The input c_{i-1} together with the inputs x_i and y_i produce the sum s_i and carry c_i as shown in the figure. Finally, in Fig. 9.4(b) two full-adders and a half-adder are combined to produce the sum of the two binary numbers $x_2 x_1 x_0$, $y_2 y_1 y_0$ whose sum is $c_2 s_2 s_1 s_0$. $\square$

Figure 9.4

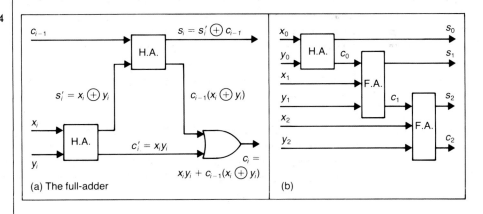

(a) The full-adder

(b)

The next example introduces the main theme of this section—the minimal sum of products representation of a Boolean function.

EXAMPLE 9.10 | Find a gating network for the Boolean function $f(w, x, y, z) = \sum m(4, 5, 7, 8, 9, 11)$.
Consider the order of the variables as w, x, y, z. We can determine f in d.n.f. by writing each minterm number in binary notation and then finding its corresponding fundamental conjunction. For example, (a) $5 = 0101$, indicating the fundamental

conjunction $\overline{w}x\overline{y}z$; and (b) $7 = 0111$ indicates $\overline{w}xyz$. Continuing in this way, we have $f(w, x, y, z) = \overline{w}x\overline{y}z + \overline{w}xyz + w\overline{x}\,\overline{y}\,\overline{z} + w\overline{x}\,\overline{y}z + w\overline{x}yz + \overline{w}x\overline{y}\,\overline{z}$.

Using properties of Boolean variables $f = \overline{w}xz(y + \overline{y}) + w\overline{x}\,\overline{y}(\overline{z} + z) + w\overline{x}yz + \overline{w}x\overline{y}\,\overline{z} = \overline{w}xz + w\overline{x}\,\overline{y} + w\overline{x}yz + \overline{w}x\overline{y}\,\overline{z} = \overline{w}x(z + \overline{y}\,\overline{z}) + w\overline{x}(\overline{y} + yz) = \overline{w}x(z + \overline{y}) + w\overline{x}(\overline{y} + z)$ (Why?) $= \overline{w}x(\overline{y} + z) + w\overline{x}(\overline{y} + z)$, so

a) $f(w, x, y, z) = \overline{w}xz + \overline{w}x\overline{y} + w\overline{x}\,\overline{y} + w\overline{x}z$, a *minimal sum of products;* or

b) $f(w, x, y, z) = \overline{w}x(\overline{y} + z) + w\overline{x}(\overline{y} + z)$.   □

From this point on we shall consider an input of the form $\overline{w}$, which has not passed though any gates, as the exact input, instead of regarding it as the result obtained from inputting w and passing it through an inverter.

In Fig. 9.5(a), we have a gating network implementing the d.n.f. of f. Part (b) of the figure is the gating network for f as a minimal sum of products. This sum is minimal in the sense that it cannot be reduced and still consist of a sum of product terms. Figure 9.5(c) has a gating network for $f = \overline{w}x(\overline{y} + z) + w\overline{x}(\overline{y} + z)$.

The network in (c) has only four logic gates, whereas that in (b) has five such devices. Consequently, we may feel that the network in (c) is better with regard to minimizing cost, since each extra gate increases the cost of production. However, even though there are fewer inputs and fewer gates for the implementation in (c), some of the inputs, namely $\overline{y}$, z, must pass through three *levels of gating* before providing the output f. For the minimal sum of products in (b), there are only two levels of gating. In the study of gating networks, outputs are considered as instantaneous functions of the input. In practice, however, each level of gating adds a delay in the development of the function f. For high-speed digital equipment we want to minimize delay, so we opt for more speed at the price of increasing the manufacturing cost.

As a result of this need to maximize speed we want to represent a Boolean function in a form called a minimal sum of products. In order to accomplish this for functions of not more than six variables we use a pictorial method called the *Karnaugh map,* developed by M. Karnaugh in 1953. The d.n.f. of a Boolean function is a major key behind this technique.

In simplifying the d.n.f. of f in Example 9.10, we combined the two fundamental conjunctions $\overline{w}x\overline{y}z$, $\overline{w}xyz$ into the product term $\overline{w}xz$, since $\overline{w}x\overline{y}z + \overline{w}xyz = \overline{w}xz(\overline{y} + y) = \overline{w}xz(1) = \overline{w}xz$. This indicates that if two fundamental conjunctions differ in exactly one literal, then they can be combined into a product term with that literal missing.

For $g: B^4 \rightarrow B$, where $g(w, x, y, z) = wx\overline{y}\,\overline{z} + wx\overline{y}z + wxyz + wxy\overline{z}$, each fundamental conjunction differs from its predecessor in exactly one literal. Hence we can simplify g as $g = wx\overline{y}(\overline{z} + z) + wxy(z + \overline{z}) = wx\overline{y} + wxy = wx(\overline{y} + y) = wx$. We can also write $g = wx(\overline{y}\,\overline{z} + \overline{y}z + yz + y\overline{z}) = wx(y + \overline{y})(z + \overline{z}) = wx$.

The key to this reduction process is the recognition of pairs (quadruples, . . . , 2^n-tuples) of fundamental conjunctions where any two adjacent terms differ in exactly one literal. If $h: B^4 \rightarrow B$, and the d.n.f. of h has 12 terms, can we move these terms around to recognize the best reductions? The Karnaugh map organizes these terms for us.

Figure 9.5

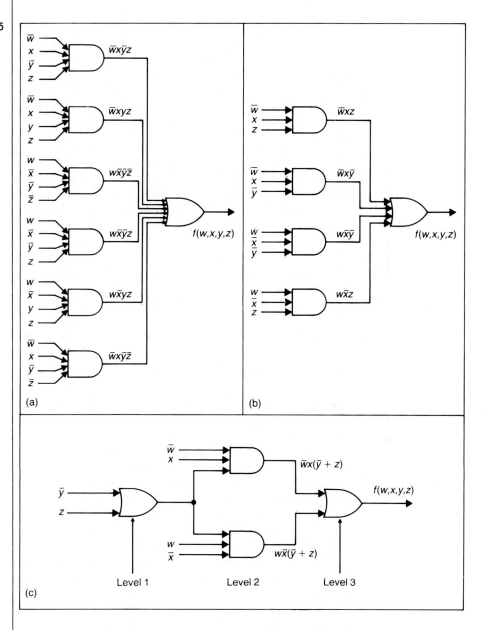

(a)

(b)

(c)

Level 1   Level 2   Level 3

We start with the case of two variables, w and x. Table 9.9 shows the Karnaugh maps for $f(w, x) = wx$ and $g(w, x) = w + x$.

In (a), the 1 interior to the table indicates the fundamental conjunction wx. This occurs in the row for $w = 1$ and the column for $x = 1$, the one case when $wx = 1$. In (b), there are three 1's in the table. The top 1 is for $\overline{w}x$, which has value 1 exactly when $w = 0$, $x = 1$. The bottom two 1's are for $w\overline{x}$ and wx, as we read the bottom row from left to right.

Table 9.9

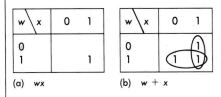

(a)   wx   (b)   w + x

Table 9.9(b) represents the d.n.f. $\overline{w}x + w\overline{x} + wx$. Due to their adjacency in the bottom row, the table indicates that $w\overline{x}$ and wx differ in only one literal and can be combined to yield w. By the idempotent law of addition we can use wx again. In the table the adjacency in the second column indicates the combining of $\overline{w}x$ and wx to get x. (In the x column all possibilities for w, namely w, $\overline{w}$, appear. This is a way to recognize x as the result for that column.)

EXAMPLE 9.11   We now consider three Boolean variables w, x, y. In Table 9.10, the first new idea we encounter is in the column headings for xy. These are not the same as the ones we had for the rows in the function tables. We see here, in going from left to right, that 00 differs from 01 in exactly one place, 01 differs from 11 in exactly one place, etc.

Table 9.10

w \ xy	00	01	11	10
0	1			1
1	1		1	

If $f(w,x,y) = \Sigma m(0,2,4,7)$, then as $0 = 000$ ($\overline{w}\,\overline{x}\,\overline{y}$), $2 = 010$ ($\overline{w}x\overline{y}$), $4 = 100$ ($w\overline{x}\,\overline{y}$), and $7 = 111$ (wxy), we can represent these terms by placing 1's as in Table 9.10. The 1 for wxy is not adjacent to any other 1 so it is *isolated;* we will have wxy in the minimal sum of products representing f. The 1 for $\overline{w}x\overline{y}$ is not isolated, for we consider the table as wrapping around, making this 1 adjacent to the 1 for $\overline{w}\,\overline{x}\,\overline{y}$. These combine to give $\overline{w}\,\overline{y}$. Finally, the 1's in the column for $x = y = 0$ indicate a reduction of $\overline{w}\,\overline{x}\,\overline{y} + w\overline{x}\,\overline{y}$ to $\overline{x}\,\overline{y}$. Hence, as a minimal sum of products $f = wxy + \overline{w}\,\overline{y} + \overline{x}\,\overline{y}$. ☐

Table 9.11

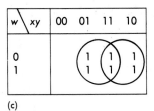

w \ xy	00	01	11	10
0	1			1
1	1			1

(a)

w \ xy	00	01	11	10
0	1	1	1	1
1				

(b)

w \ xy	00	01	11	10
0		1	1	1
1		1	1	1

(c)

EXAMPLE 9.12   From Table 9.11 we have:

a) $f(w, x, y) = \sum m(0, 2, 4, 6) = \bar{y}$, the only variable whose value does not change when the four terms designated by the 1's are considered. (Since the value of y is 0 here, $f(w, x, y) = \bar{y}$.)

b) $f(w, x, y) = \sum m(0, 1, 2, 3) = \bar{w}$.

c) $f(w, x, y) = \sum m(1, 2, 3, 5, 6, 7) = y + x$.   □

Advancing to four variables, we have the following.

EXAMPLE 9.13   Find a minimal sum of products for $f(w, x, y, z) = \sum m(0, 1, 2, 3, 8, 9, 10)$.

Table 9.12

The Karnaugh map for f in Table 9.12 combines the 1's in the four corners to give the term $\bar{x}\bar{z}$. The four 1's in the top row combine to give $\bar{w}\bar{x}$. (Using only the middle two 1's we do not make use of all the available adjacencies and get the term $\bar{w}\bar{x}z$ which has one more literal than $\bar{w}\bar{x}$.) Finally, the 1 in the row ($w = 1$, $x = 0$) and column ($y = 0$, $z = 1$) can be combined with the 1 on its left, *and* these can then be combined with the first two 1's in the top row to give $\bar{x}\bar{y}$. Hence $f(w, x, y, z) = \bar{x}\bar{z} + \bar{w}\bar{x} + \bar{x}\bar{y}$.   □

Table 9.13

EXAMPLE 9.14   The map for $f(w, x, y, z) = \sum m(9, 10, 11, 12, 13)$ appears in Table 9.13. The only 1 in the table that has not been combined with another term is adjacent to a 1 on its right (this combination yields $w\bar{x}z$) and to a 1 above it (this combination yields $w\bar{y}z$). Consequently, we can represent f as a minimal sum of products in two ways: $wx\bar{y} + w\bar{x}y + w\bar{x}z$ and $wx\bar{y} + w\bar{x}y + w\bar{y}z$. So this type of representation is not unique. However, the same number of product terms and the same total number of literals appear in each case.   □

Table 9.14

(a)   (b)

EXAMPLE 9.15
There is a right and a wrong way to use a Karnaugh map.

Let $f(w, x, y, z) = \sum m(3, 4, 5, 7, 9, 13, 14, 15)$. In Table 9.14(a) we combine a block of four 1's into the term xz. But when we account for the other four 1's we do what is shown in (b). So the result in (b) will yield f as a sum of four terms (each with three literals), while the method suggested in (a) adds on the extra (unneeded) term xz. □

The following suggestions are based on what we have done so far. We state them now so that they may be used for larger maps.

1. Start by combining those terms in the table where there is at most one possibility for simplification.
2. Check the four corners of a table. They may contain adjacent 1's even though the 1's appear isolated.
3. In all simplifications try to obtain the largest possible block of adjacent 1's in order to get a minimal product term. (Recall that 1's can be used more than once, if necessary, because of the idempotent law.)
4. If there is a choice in simplifying an entry in the table, try to use adjacent 1's that have not been used in any prior simplification.

EXAMPLE 9.16
If $f(v, w, x, y, z) = \sum m(1, 5, 10, 11, 14, 15, 18, 26, 27, 30, 31)$, we construct two 4×4 tables, one for $v = 0$, the other for $v = 1$. (Table 9.15)

Table 9.15

($v = 0$)   ($v = 1$)

Following the order of the variables, when we write, for example, $5 = 00101$, this indicates the need for a 1 in the second row and second column of the table for $v = 0$. Filling in the other 1's, we see that the 1 in the first row, fourth column of the table for $v = 1$ can only be combined with another term in one way, yielding

$\overline{v}x\overline{y}\overline{z}$. This is true for the two 1's in the second column of the $(v = 0)$ table, also. These give $\overline{v}\,\overline{w}\,\overline{y}z$. The block of eight 1's yields wy, and $f(v, w, x, y, z) = wy + \overline{v}\,\overline{w}\,\overline{y}z + \overline{v}x\overline{y}\overline{z}$. □

A function f of six variables, t, v, w, x, y, z requires four tables for each of the cases (a) $t = 0$, $v = 0$; (b) $t = 0$, $v = 1$; (c) $t = 1$, $v = 1$; (d) $t = 1$, $v = 0$. Beyond six variables, this method becomes overly complicated. Another procedure, the *Quine-McCluskey Method,* can be used. For a large number of variables the method is tedious to perform by hand, but it is a systematic procedure suitable for computer implementation, particularly for computers possessing some type of "binary compare" command. (More about this technique is given in Chapter 7 of Reference 3.)

We close with an example on a *minimal product of sums.*

Table 9.16

wx \ yz	00	01	11	10
00		0		
01		0	0	
11		0	0	0
10		0		0

EXAMPLE 9.17

For $g(w, x, y, z) = \prod M(1, 5, 7, 9, 10, 13, 14, 15)$ we place a 0 in each of the positions for the binary equivalents of the maxterms listed. This yields the results shown in Table 9.16.

The 0 in the lower right-hand corner can be combined only with the 0 above it, and $(\overline{w} + x + \overline{y} + z)(\overline{w} + \overline{x} + \overline{y} + z) = (\overline{w} + \overline{y} + z) + (x\overline{x}) = (\overline{w} + \overline{y} + z) + 0 = \overline{w} + \overline{y} + z$. The block of four 0's simplifies to $\overline{x} + \overline{z}$, while the four 0's in the second column yield $y + \overline{z}$. So $g(w, x, y, z) = (\overline{w} + \overline{y} + z)(\overline{x} + \overline{z})(y + \overline{z})$, a minimal product of sums. □

EXERCISES

1. Using inverters, AND gates, and OR gates, construct the gates shown in Fig. 9.6.

2. Using only NAND gates (see Fig. 9.6), construct the inverter, AND gate, and OR gate.

Figure 9.6

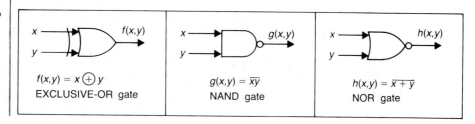

$f(x, y) = x \oplus y$	$g(x, y) = \overline{xy}$	$h(x, y) = \overline{x + y}$
EXCLUSIVE-OR gate	NAND gate	NOR gate

3. Answer Exercise 2 replacing NAND by NOR.

4. Using inverters, AND gates, and OR gates, construct gating networks for

    a) $f(x, y, z) = x\bar{z} + y\bar{z} + x$;   b) $g(x, y, z) = (x + \bar{z})(y + \bar{z})\bar{x}$; and

    c) $h(x, y, z) = \overline{(xy \oplus yz)}$

5. For the network in Fig. 9.7, express f as a function of w, x, y, z.

Figure 9.7

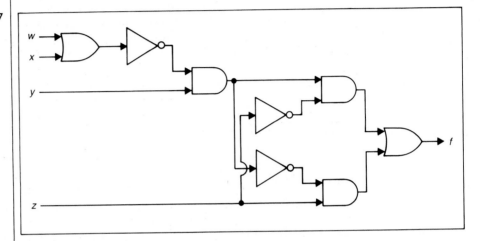

6. Find a minimal sum of products representation for

    a) $f(w, x, y) = \Sigma m(1, 2, 5, 6)$ b) $f(w, x, y) = \prod M(0, 1, 4, 5)$

    c) $f(w, x, y, z) = \Sigma m(0, 2, 5, 7, 8, 10, 13, 15)$

    d) $f(w, x, y, z) = \Sigma m(0, 2, 5, 7, 8, 9, 10, 13, 15)$

    e) $f(w, x, y, z) = \Sigma m(5, 6, 8, 11, 12, 13, 14, 15)$

    f) $f(w, x, y, z) = \Sigma m(7, 9, 10, 11, 14, 15)$

    g) $f(v, w, x, y, z) = \Sigma m(1, 2, 3, 4, 10, 17, 18, 19, 22, 23, 27, 28, 30, 31)$

7. Obtain a minimal product of sums representation for $f(w, x, y, z) = \prod M(0, 1, 2, 4, 5, 10, 12, 13, 14)$.

8. Let $f: B^n \to B$ be a function of the Boolean variables $x_1, x_2, \ldots, x_n$. Determine n if the number of 1's needed to express x_1 in the Karnaugh map for f is (a) 2; (b) 4; (c) 8; (d) 2^k, $k \in \mathbf{Z}^+$.

9. If $g: B^7 \to B$ is a Boolean function of the Boolean variables $x_1, x_2, \ldots, x_7$, how many 1's are needed in the Karnaugh map of g for the product term (a) x_1; (b) $x_1 x_2$; (c) $x_1 \bar{x}_2 x_3$; (d) $x_1 x_3 x_5 x_7$?

9.3 ■ FURTHER APPLICATIONS: DON'T CARE CONDITIONS

Our objective now is to use the ideas we have developed in the first two sections in a variety of applications.

EXAMPLE 9.18 As head of her church bazaar, Paula has volunteered to bake a cake to be sold. Members of the bazaar committee volunteer to donate the needed ingredients as shown in Table 9.17.

Table 9.17

	Flour	Milk	Butter	Pecans	Eggs
Sue	x		x		
Dorothy			x	x	
Bettie	x	x			
Theresa		x			x
Ruthanne		x	x	x	

Paula sends her daughter Amy to pick up the ingredients. Write a Boolean expression to help Paula determine which sets of volunteers she should consider so that Amy can collect all of the necessary ingredients (and nothing extra).

Let s, d, b, t, r denote five Boolean variables corresponding, respectively, to the women listed in the first column of the table. To get the flour Amy must visit Sue or Bettie. In Boolean terminology, we can say that flour determines the sum $s + b$. This term will be part of a product of sums. For the other ingredients the following sums denote the choices.

milk: $b + t + r$ butter: $s + d + r$ pecans: $d + r$ eggs: t

To answer the question posed here, we seek a minimal sum of products for the function $f(s, d, b, t, r) = (s + b)(b + t + r)(s + d + r)(d + r)t$. The answer can be obtained by multiplying everything out and then simplifying the result, or by using a Karnaugh map. This time we'll use the map (in Table 9.18).

Table 9.18

db \ tr	00	01	11	10
00	0	0	0	0
01	0	0	1	0
11	0	0	1	1
10	0	0	0	0

$(s = 0)$

db \ tr	00	01	11	10
00	0	0	1	0
01	0	0	1	0
11	0	0	1	1
10	0	0	1	1

$(s = 1)$

Here $s + b$, for example, is represented by the eight 0's in the first and fourth rows of the table for $s = 0$; t requires the 16 0's in the first two columns of both tables. Filling in the 0's for the other three sums in the product we arrive at the table shown. Now we need a minimal sum of products for the nine 1's in the table. We find the result is $srt + sdt + brt + dbt$. (Verify this.)   □

Definition
9.4

▶ Let $G = (V, E)$ denote a graph (undirected) with vertex set V and edge set E. A subset D of V is called a *dominating set* for G if for every $v \in V$, $v \in D$ or v is adjacent to a vertex in D.

For the graph in Fig. 9.8, $\{a, d\}$, $\{a, c, e\}$, and $\{b, d, e, f\}$ are examples of dominating sets. The set $\{a, c, e\}$ is a *minimal dominating* set, for if any of the three vertices a, c, or e is removed, the remaining two no longer dominate the graph. The set $\{a, d\}$ is also minimal but $\{b, d, e, f\}$ is not, since $\{b, d, e\}$ dominates G.

Figure 9.8

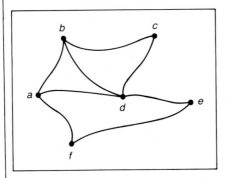

EXAMPLE 9.19

For the graph in Fig. 9.8, let the vertices represent cities and the edges highways. We wish to build hospitals in some of these cities so that each city either has a hospital or is adjacent to a city that does. In how many ways can this be accomplished by building a minimal number of hospitals in each case?

To answer this question we need the minimal dominating sets for G. Consider vertex a. To guarantee that a will satisfy our objective, we must build a hospital in a, or b, or d, or f. Hence we have the sum term $a + b + d + f$. For b to satisfy our objective we *generate* the term $a + b + c + d$. Continuing with the other four locations, the answer is then a minimal sum of products representation for the Boolean function $g(a, b, c, d, e, f) = (a + b + d + f)(a + b + c + d)(b + c + d) \cdot (a + b + c + d + e)(d + e + f)(a + e + f)$. Using the properties of Boolean variables we have

$g = (a + b + d + f)(b + c + d) \cdot$	Absorption Law
$\quad (d + e + f)(a + e + f)$	
$= [(a + f)c + (b + d)][(da) + (e + f)]$	Distributive Law of $+$ over $\cdot$
$= [ac + fc + b + d][da + e + f]$	Distributive Law of $\cdot$ over $+$
$= acda + ace + acf + fcda + fce + fcf$	Distributive Law of $\cdot$ over $+$
$\quad + bda + be + bf + dda + de + df$	
$= ace + cf + ad + be + bf + de + df$	Commutative and Idempotent Laws of $\cdot$, and the Absorption Law

Consequently, in six of the cases the objective can be achieved by building only two hospitals. If a and c have the largest populations and we want to locate hospitals in each of these cities, then we would also have to construct a hospital at e. ☐

The final application introduces the notion of *"don't care" conditions.*

EXAMPLE 9.20   The four input lines for the gating network in Fig. 9.9 provide the binary equivalents of the digits 0, 1, 2, ..., 9, with each number represented as $abcd$ (d is least significant). Construct a gating network with two levels of gating such that the output function f equals 1 for the input digits 0, 3, 6, 9 (i.e., f detects digits divisible by 3).

Figure 9.9

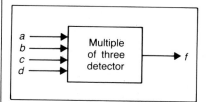

Before concluding that $f = 0$ for the other 12 cases, we examine Table 9.19, where an "x" appears for the value of f in the last six cases. Since these input combinations do not occur (due to certain external constraints), we *don't care* what the value of f is in these situations. For such occurrences the outputs are referred to as *unspecified* and f is called *incompletely specified.* We write $f = \sum m(0, 3, 6, 9) + d(10, 11, 12, 13, 14, 15)$. When seeking a minimal sum of products for f, any or all of these don't care conditions can be used in the simplification process.

From the Karnaugh map in Table 9.20, we write f as a minimal sum of products, obtaining $f = \overline{a}\overline{b}\overline{c}\overline{d} + \overline{b}cd + bc\overline{d} + ad$. The first summand in f is for recognition of 0; $\overline{b}cd$ provides recognition for 3 since it stands for 0011 ($\overline{a}\overline{b}cd$), as 1011 ($a\overline{b}cd$) does not occur. Likewise $bc\overline{d}$ is needed to recognize 6, while ad takes care of 9. Fig. 9.10 provides the interior details (minus the inverters) of Fig. 9.9. (Note that in Table 9.20, there are some don't care conditions that were not used.) ☐

Table 9.19

a	b	c	d	f
0	0	0	0	1
0	0	0	1	0
0	0	1	0	0
0	0	1	1	1
0	1	0	0	0
0	1	0	1	0
0	1	1	0	1
0	1	1	1	0

a	b	c	d	f
1	0	0	0	0
1	0	0	1	1
1	0	1	0	x
1	0	1	1	x
1	1	0	0	x
1	1	0	1	x
1	1	1	0	x
1	1	1	1	x

Table 9.20

ab \ cd	00	01	11	10
00	1			1
01				1
11		x	x	x
10	1		x	x

Figure 9.10

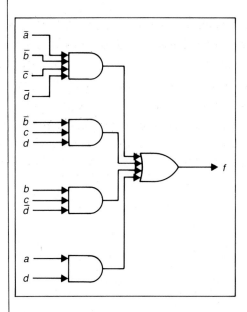

EXERCISES

1. For his tenth birthday, Mona wants to buy her son Jason some stamps for his collection. At the hobby shop she finds six different packages (which we shall call u, v, w, x, y, z). The kinds of stamps in each of these packages are as follows:

Table 9.21

	United States	European	Asian	African
u		✓		✓
v	✓		✓	
w	✓	✓		
x	✓			
y	✓			✓
z			✓	✓

Determine all minimal combinations of packages Mona can buy so that Jason will get some stamps from all four geographical locations.

2. Rework Example 9.19 using a Karnaugh map on six variables.

3. Determine whether each of the following statements is true or false. If the statement is false, provide a counterexample.

Let $G = (V, E)$ be an undirected graph with D_1, $D_2 \subseteq V$.

    a)   If D_1, D_2 are dominating sets of G, $D_1 \cup D_2$ is likewise.

    b)   If D_1, D_2 are dominating sets of G, $D_1 \cap D_2$ is also.

    c)   If D_1 is a dominating set of G and $D_1 \subseteq D_2$, then D_2 dominates G.

    d)   If D_1 is a dominating set of G and $\emptyset \neq D_2 \subseteq D_1$, then D_2 dominates G.

    e)   If $D_1 \cup D_2$ dominates G, then at least one of D_1, D_2 dominates G.

4. a)   Determine all minimal dominating sets for the graph G in Fig. 9.11.

    b)   Let each vertex of G represent a city and each edge denote a communication link between the cities that are its endpoints. What role does a minimal dominating set play in such an interpretation?

Figure 9.11

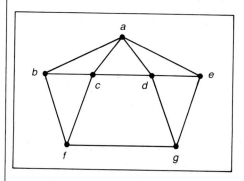

5. Find a minimal sum of products representation for

    a) $f(w, x, y, z) = \Sigma m(1, 3, 5, 7, 9) + d(10, 11, 12, 13, 14, 15)$

    b) $f(w, x, y, z) = \Sigma m(0, 5, 6, 8, 13, 14) + d(4, 9, 11)$

    c) $f(v, w, x, y, z) = \Sigma m(0, 2, 3, 4, 5, 6, 12, 19, 20, 24, 28)$
                                 $+ d(1, 13, 16, 29, 31)$

6. The four input lines for the gating network in Fig. 9.12 provide the binary equivalents of the numbers 0, 1, 2, . . . , 15, where each number is represented as $abcd$, with d the least significant bit.

    a)   Determine the d.n.f. of f, whose value is 1 for $abcd$ prime, and 0 otherwise.

    b)   Draw the two-level gating network for f as a minimal sum of products.

    c)   We are informed that the network is part of a larger network and that as a result, the binary equivalents of 10–15 are never provided as input. Design a two-level gating network for f under these circumstances.

Figure 9.12

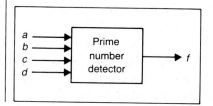

9.4 ■ THE STRUCTURE OF A BOOLEAN ALGEBRA (OPTIONAL)

In this last section we analyze the structure of a Boolean algebra and determine those $n \in \mathbf{Z}^+$ for which there is a Boolean algebra of n elements.

Definition 9.5

▶ Let $\mathcal{B}$ be a nonempty set which contains two special elements 0 and 1 and on which we define binary operations $+, \cdot$ and a monary (or unary) operation $\overline{}$. Then $(\mathcal{B}, +, \cdot, \overline{}, 0, 1)$ is called a *Boolean algebra* if the following conditions are satisfied for all $x, y, z \in \mathcal{B}$.

a) $x + y = y + x$	a)′ $xy = yx$	Commutative Laws
b) $x(y + z) = xy + xz$	b)′ $x + (yz) = (x + y)(x + z)$	Distributive Laws
c) $x + 0 = x$	c)′ $x \cdot 1 = x$	Identity Laws
d) $x + \overline{x} = 1$	d)′ $x \cdot \overline{x} = 0$	Inverse Laws
e) $0 \neq 1$		

As seen above, we often write xy for $x \cdot y$. When the operations and identity elements are known, we write $\mathcal{B}$ instead of $(\mathcal{B}, +, \cdot, \overline{}, 0, 1)$.

From our past experience we have the following examples.

EXAMPLE 9.21

If $\mathcal{U}$ is a (finite) set, then $\mathcal{B} = \mathcal{P}(\mathcal{U})$ is a Boolean algebra where for $A, B \subseteq \mathcal{U}$, $A + B = A \cup B$, $AB = A \cap B$, $\overline{A} =$ the complement of A (in $\mathcal{U}$), and $0 = \emptyset$, $1 = \mathcal{U}$. □

EXAMPLE 9.22

For $n \in \mathbf{Z}^+$, $F_n = \{f : B^n \rightarrow B\}$, the set of Boolean functions on n Boolean variables is a Boolean algebra where $+, \cdot, \overline{}$ are as defined in Definition 9.2, and $0 = \mathbf{0}$, $1 = \mathbf{1}$. □

Let us now examine a new type of Boolean algebra.

EXAMPLE 9.23

Let $\mathcal{B}$ be the set of all positive integer divisors of 30: $\mathcal{B} = \{1, 2, 3, 5, 6, 10, 15, 30\}$. For any $x, y \in \mathcal{B}$, define $x + y = [x, y]$, the l.c.m. of x, y; $xy = (x, y)$, the g.c.d. of x, y; and, $\overline{x} = 30/x$. Then with 1 as the 0 element and 30 as the 1 element, $(\mathcal{B}, +, \cdot, \overline{}, 0, 1)$ is a Boolean algebra.

We shall establish one of the distributive laws for this Boolean algebra and leave the other conditions for the reader to check.

For the first distributive law we want to show that $(x, [y, z]) = [(x, y), (x, z)]$, for any $x, y, z \in \mathcal{B}$. Write $x = 2^{k_1}3^{k_2}5^{k_3}$, $y = 2^{m_1}3^{m_2}5^{m_3}$, $z = 2^{n_1}3^{n_2}5^{n_3}$, where for $1 \leq i \leq 3$, $0 \leq k_i, m_i, n_i \leq 1$.

Then $[y, z] = 2^{s_1}3^{s_2}5^{s_3}$ where $s_i = \max\{m_i, n_i\}$, $1 \leq i \leq 3$, so $(x, [y, z]) = 2^{t_1}3^{t_2}5^{t_3}$ where $t_i = \min\{k_i, \max\{m_i, n_i\}\}$, $1 \leq i \leq 3$.

Also, $(x, y) = 2^{u_1}3^{u_2}5^{u_3}$, where $u_i = \min\{k_i, m_i\}$, $1 \leq i \leq 3$, and $(x, z) = 2^{v_1}3^{v_2}5^{v_3}$, with $v_i = \min\{k_i, n_i\}$. So $[(x, y), (x, z)] = 2^{w_1}3^{w_2}5^{w_3}$, where $w_i = \max\{u_i, v_i\}$, $1 \leq i \leq 3$.

So $w_i = \max\{u_i, v_i\} = \max\{\min\{k_i, m_i\}, \min\{k_i, n_i\}\}$, and $t_i = \min\{k_i, \max\{m_i, n_i\}\}$. In order to verify the result we need $w_i = t_i$, $1 \leq i \leq 3$. If $k_i = 0$, then $w_i = 0 = t_i$. If $k_i = 1$, then $w_i = \max\{m_i, n_i\} = t_i$. Since this exhausts all possibilities, $w_i = t_i$, $1 \leq i \leq 3$ and $(x, [y, z]) = [(x, y), (x, z)]$.

If we analyze this result further we find that 30 can be replaced by any number $m = p_1 p_2 p_3$, where p_1, p_2, p_3 are distinct primes. In fact, the result follows for the set of all divisors of $p_1 p_2 \cdots p_n$, a product of n distinct primes. (Note that such a product is square-free, i.e., there is no $k \in \mathbf{Z}^+$, $k > 1$, with k^2 dividing it.)   □

EXAMPLE 9.24   A word about the propositional calculus. If p, q are two propositions, we may feel that the collection of all statements obtained from p, q, using $\vee, \wedge, {}^{-}$ should be a Boolean algebra. After all, just look at the laws of logic and the way they compare with the comparable results for set theory and Boolean functions. There is one main difference. In logic we have, for example, $p \wedge q \Leftrightarrow q \wedge p$, not $p \wedge q = q \wedge p$. To get around this we define a relation $\mathcal{R}$ on the set S of all propositions so obtained from p, q, where $s_1 \mathcal{R} s_2$ if $s_1 \Leftrightarrow s_2$. Then $\mathcal{R}$ is an equivalence relation on S and partitions S, in this case, into 16 equivalence classes. If we define ${}^{-}, \cdot, +$ on these equivalence classes by $\overline{[s_1]} = [\bar{s}_1]$, $[s_1][s_2] = [s_1 \wedge s_2]$, $[s_1] + [s_2] = [s_1 \vee s_2]$, and recognize $[T_0]$ as the 1 element and $[F_0]$ as the 0 element, we get a Boolean algebra.   □

In the definition of a Boolean algebra, there are nine conditions. Yet in the lists of properties for set theory, logic, and Boolean functions, we listed 19 properties. And there were even more! Undoubtedly, there is a way to get the remaining properties, and others not listed among the 19, from the ones given in the definition.

Theorem 9.1   ▶ (*The Idempotent Laws*) For any $x \in \mathcal{B}$, a Boolean algebra, (i) $x + x = x$; and, (ii) $xx = x$.

Proof   (To the right of each equality appearing in this proof we list the letter of the condition from Definition 9.5 that verifies it.)

(i) $x = x + 0$ c)   (ii) $x = x \cdot 1$ c)'

$\quad = x + (x\bar{x})$ d)' $\quad = x(x + \bar{x})$ d)

$\quad = (x + x)(x + \bar{x})$ b)' $\quad = xx + x\bar{x}$ b)

$\quad = (x + x) \cdot 1$ d) $\quad = xx + 0$ d)'

$\quad = x + x$ c)' $\quad = xx$ c)   ■

In proving this theorem we can obtain the proof of (ii) from that of (i) by changing all occurrences of $+$ to $\cdot$, and vice versa, and all occurrences of 0 to 1, and vice versa. Also, the reasons verifying the corresponding steps constitute a pair of conditions in Definition 9.5. As in the past these pairs are said to be *duals* of each other; condition (e) is called *self-dual*. This now leads us to the following result.

Theorem 9.2   ▶ (*The Principle of Duality*) If s is a theorem about a Boolean algebra, and s can be proved from the conditions in Definition 9.5 and properties derived from these same conditions, then its dual s^d is likewise a theorem.

Proof Let s be such a theorem. Dualizing all the steps and reasons in the proof of s (as in the proof of Theorem 9.1), we obtain a proof for s^d. ∎

We now list some further properties for a Boolean algebra. We shall prove some of these properties and leave the remaining proofs for the reader.

Theorem 9.3

▶ For any Boolean algebra $\mathscr{B}$, if x, y, $z \in \mathscr{B}$, then

a) $x \cdot 0 = 0$ a)′ $x + 1 = 1$

b) $x(x + y) = x$ b)′ $x + xy = x$ Absorption Laws

c) $xy = xz, \bar{x}y = \bar{x}z \Rightarrow y = z$ Cancellation Laws
c)′ $x + y = x + z, \bar{x} + y = \bar{x} + z \Rightarrow y = z$

d) $x(yz) = (xy)z$ d)′ $x + (y + z) = (x + y) + z$ Associative Laws

e) $x + y = 1, xy = 0 \Rightarrow y = \bar{x}$ Uniqueness of Inverses

f) $\bar{\bar{x}} = x$ Law of the Double Complement

g) $\overline{xy} = \bar{x} + \bar{y}$ g)′ $\overline{x + y} = \bar{x}\,\bar{y}$ DeMorgan's Laws

h) $\bar{0} = 1$ h)′ $\bar{1} = 0$

i) $x\bar{y} = 0$ iff $xy = x$ i)′ $x + \bar{y} = 1$ iff $x + y = x$

Proof c) $y = 1 \cdot y = (x + \bar{x})y = xy + \bar{x}y = xz + \bar{x}z = (x + \bar{x})z = 1 \cdot z = z$ (Verify all equalities.)

c)′ This is the dual of (c).

d) To establish this result we use (c)′ and arrive at the conclusion by showing that $x + [x(yz)] = x + [(xy)z]$ and $\bar{x} + [x(yz)] = \bar{x} + [(xy)z]$. Using the idempotent and absorption laws $x + [x(yz)] = (x + x)(x + yz) = x(x + yz) = x$. Likewise $x + [(xy)z] = [x + (xy)](x + z) = x(x + z) = x$. Then $\bar{x} + [x(yz)] = (\bar{x} + x)(\bar{x} + yz) = 1 \cdot (\bar{x} + yz) = \bar{x} + yz$, while $\bar{x} + [(xy)z] = (\bar{x} + xy) \cdot (\bar{x} + z) = ((\bar{x} + x)(\bar{x} + y))(\bar{x} + z) = (1 \cdot (\bar{x} + y))(\bar{x} + z) = (\bar{x} + y) \cdot (\bar{x} + z) = \bar{x} + yz$ (Verify all equalities.)

The result now follows by the cancellation law in (c)′.

d)′ Fortunately, this is the dual of (d).

e) $\bar{x} = \bar{x} + 0 = \bar{x} + xy = (\bar{x} + x)(\bar{x} + y) = 1 \cdot (\bar{x} + y) = (\bar{x} + y) \cdot 1 = (\bar{x} + y)(x + y) = \bar{x}x + y = 0 + y = y$ (Verify all equalities.)

We note that (e) is self-dual. Statement (f) is a corollary of (e) since $\bar{\bar{x}}$ and x are both complements (inverses) of $\bar{x}$.

g) This result will follow from (e) if we can show that $\bar{x} + \bar{y}$ is a complement of xy.
$xy + (\bar{x} + \bar{y}) = (xy + \bar{x}) + \bar{y} = (x + \bar{x})(y + \bar{x}) + \bar{y} = 1 \cdot (y + \bar{x}) + \bar{y} = (y + \bar{y}) + \bar{x} = 1 + \bar{x} = 1.$
Also, $xy(\bar{x} + \bar{y}) = (xy\bar{x}) + (xy\bar{y}) = ((x\bar{x})y) + (x(y\bar{y})) = 0 \cdot y + x \cdot 0 = 0 + 0 = 0.$
Consequently, $\bar{x} + \bar{y}$ is a complement of xy, and by uniqueness of complements, $\overline{xy} = \bar{x} + \bar{y}$. ∎

Enough proving for a while! Let us now consider the Hasse diagrams for the following two Boolean algebras.

a) $(\mathcal{P}(\mathcal{U}), \cup, \cap, \overline{}, \emptyset, \mathcal{U})$, where $\mathcal{U} = \{1, 2, 3\}$, and the partial order is induced by the subset relation.

b) $(\mathcal{S}, +, \cdot, \overline{}, 1, 30)$, where $\mathcal{S} = \{1, 2, 3, 5, 6, 10, 15, 30\}$, $x + y = [x, y]$, $xy = (x, y)$, $\overline{x} = 30/x$. Here the 0 element is the divisor 1 and the 1 element is the divisor 30. The relation $\mathcal{R}$ on $\mathcal{S}$, defined by $x \mathcal{R} y$ if x divides y, makes $\mathcal{S}$ into a poset.

Figure 9.13

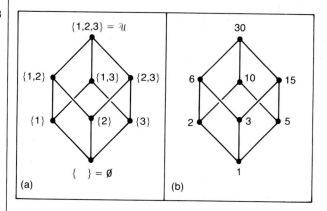

(a)

(b)

Figure 9.13 shows the Hasse diagrams for these two Boolean algebras. Ignoring the labels at the vertices in each diagram, we see that the underlying structures are the same. Hence we have an *isomorphism of Boolean algebras*.

These examples also suggest two other ideas.

1. Can we partially order any finite Boolean algebra?

2. Looking at Fig. 9.13(a), we see that the nonzero elements just above $\emptyset$ are such that every element other than $\emptyset$ can be obtained as a Boolean sum of these three. For example, $\{1, 3\} = \{1\} \cup \{3\}$, $\{1, 2, 3\} = \{1\} \cup \{2\} \cup \{3\}$. For part (b), the numbers 2, 3, and 5 are such that every divisor other than 1 is realized as the Boolean sum of these three. For example, $6 = [2, 3]$, $30 = [2, [3, 5]]$.

We now start to deal formally with these suggestions.

When dealing with sets in Chapter 2 we related the operations of $\overline{}$, $\cap$, $\cup$ to the subset relation by the equivalence of the statements: (a) $A \subseteq B$; (b) $A \cap B = A$; (c) $A \cup B = B$; (d) $\overline{B} \subseteq \overline{A}$, where $A, B \subseteq \mathcal{U}$. We now use (a) and (b) in an attempt to partially order any Boolean algebra $\mathcal{B}$.

Definition 9.6 ▶ If $x, y \in \mathcal{B}$, define $x \leq y$ if $xy = x$.

Hence we define a new concept, namely "$\leq$", in terms of notions we have in $\mathcal{B}$, namely $\cdot$ and the notion of equality. We can make up definitions! But does this one lead us anywhere?

Theorem
9.4

▶ The relation "$\leq$", defined above, is a partial order.

Proof

Since $xx = x$ for any $x \in \mathcal{B}$, $x \leq x$ and the relation is reflexive. To establish anti-symmetry, suppose that $x, y \in \mathcal{B}$ with $x \leq y$ and $y \leq x$. Then $xy = x$ and $yx = y$. By the commutative property, $xy = yx$, so $x = y$. Finally, if $x \leq y$ and $y \leq z$, then $xy = x$ and $yz = y$, so $x = xy = x(yz) = (xy)z = xz$, and with $x = xz$, we have $x \leq z$, so the relation is transitive.   ■

So now we can partially order any Boolean algebra. Before going on, however, let us consider the Boolean algebra consisting of the divisors of 30. How do we apply Theorem 9.4 in this example? Here the partial order is given by $x \leq y$ if $xy = x$. Since xy is (x, y), if $(x, y) = x$, then x divides y. But this was precisely the partial order we had on this Boolean algebra when we started.

With this partial order we return to the observations we made about the elements in the Hasse diagrams of Fig. 9.13.

Definition
9.7

▶ Let 0 denote the zero element of a Boolean algebra $\mathcal{B}$. An element $x \in \mathcal{B}$, $x \neq 0$, is called an *atom* of $\mathcal{B}$ if for all $y \in \mathcal{B}$, $0 \leq y \leq x \Rightarrow y = 0$ or $y = x$.

EXAMPLE 9.25

a) For the Boolean algebra of all subsets of $\mathcal{U} = \{1, 2, 3\}$, the atoms are $\{1\}$, $\{2\}$, $\{3\}$.

b) In dealing with the positive integer divisors of 30, the atoms of this Boolean algebra are 2, 3, 5.

c) The atoms in the Boolean algebra $F_n = \{f : B^n \to B\}$ are the minterms.   □

The atoms of a finite Boolean algebra satisfy the following properties.

Theorem
9.5

▶ a) If x is an atom in a Boolean algebra $\mathcal{B}$, then for all $y \in \mathcal{B}$, $xy = 0$ or $xy = x$.

b) If x_1, x_2 are atoms of $\mathcal{B}$ and $x_1 \neq x_2$, then $x_1 x_2 = 0$.

Proof

a) For any $x, y \in \mathcal{B}$, $xy \leq x$, since $(xy)x = xy$. For x an atom, $xy \leq x \Rightarrow xy = 0$ or $xy = x$.

b) This follows from (a). The reader should supply the details.   ■

Theorem
9.6

▶ If x_1, x_2, $\ldots$, x_n are all the atoms in a Boolean algebra $\mathcal{B}$ and $x \in \mathcal{B}$ with $xx_i = 0$, $1 \leq i \leq n$, then $x = 0$.

Proof

If $x \neq 0$, let $S = \{y \in \mathcal{B} \mid 0 < y \leq x\}$. ($0 < y$ denotes $0 \leq y$, $0 \neq y$.) With $x \in S$, $S \neq \emptyset$. Since S is finite we can find an element z in $\mathcal{B}$ where $0 < z \leq x$ and no element of $\mathcal{B}$ is between 0 and z. Then z is an atom and $0 = xz = z > 0$. Since this possibility leads to a contradiction we cannot have $x \neq 0$, i.e., $x = 0$.   ■

This leads us to the following result on representation.

Theorem 9.7

▶ Given a finite Boolean algebra $\mathcal{B}$ with atoms $x_1, x_2, \ldots, x_n$, any $x \in \mathcal{B}$, $x \neq 0$, can be written as a sum of atoms uniquely, up to order.

Proof Since $x \neq 0$, by Theorem 9.6, $S = \{x_i \mid xx_i \neq 0\} \neq \emptyset$. Let $S = \{x_{i_1}, x_{i_2}, \ldots, x_{i_k}\}$, and $y = x_{i_1} + x_{i_2} + \cdots + x_{i_k}$. Then $xy = x(x_{i_1} + x_{i_2} + \cdots + x_{i_k}) = xx_{i_1} + xx_{i_2} + \cdots + xx_{i_k} = x_{i_1} + x_{i_2} + \cdots + x_{i_k}$, by Theorem 9.5(a). So $xy = y$.

Now consider $(x\bar{y})x_i$, $1 \leq i \leq n$. If $x_i \notin S$, then $xx_i = 0$, and $(x\bar{y})x_i = 0$. For $x_i \in S$, $(x\bar{y})x_i = xx_i(\overline{x_{i_1} + x_{i_2} + \cdots + x_{i_k}}) = xx_i(\bar{x}_{i_1}\bar{x}_{i_2}\cdots\bar{x}_{i_k}) = x(x_i\bar{x}_i)(z)$, where z is the product of the complements of all elements in $S - \{x_i\}$. As $x_i\bar{x}_i = 0$, $(x\bar{y})x_i = 0$. So $(x\bar{y})x_i = 0$ for all x_i, $1 \leq i \leq n$. By Theorem 9.6, $x\bar{y} = 0$.

With $xy = y$ and $x\bar{y} = 0$, $x = x \cdot 1 = x(y + \bar{y}) = xy + x\bar{y} = xy + 0 = y = x_{i_1} + x_{i_2} + \cdots + x_{i_k}$, a sum of atoms.

To show that this representation is unique up to order, suppose $x = x_{j_1} + x_{j_2} + \cdots + x_{j_\ell}$.

If x_{j_1} does not appear in $x_{i_1} + x_{i_2} + \cdots + x_{i_k}$, then $x_{j_1} = x_{j_1} \cdot x_{j_1} = x_{j_1}(x_{j_1} + x_{j_2} + \cdots + x_{j_\ell})$ (By Theorem 9.5(b)) $= x_{j_1}x = x_{j_1}(x_{i_1} + x_{i_2} + \cdots + x_{i_k}) = 0$ (Again by Theorem 9.5(b)). Hence x_{j_1} must appear as a term in $x_{i_1} + x_{i_2} + \cdots + x_{i_k}$, as must x_{j_2}, $\ldots, x_{j_\ell}$. So $\ell \leq k$. By the same reasoning we get $k \leq \ell$, and find the representations identical, except for order. ∎

From this result we see that if $\mathcal{B}$ is a finite Boolean algebra with atoms $x_1, x_2, \ldots, x_n$, then each $x \in \mathcal{B}$ can be uniquely written as $\sum_{i=1}^{n} c_i x_i$, each $c_i \in \{0, 1\}$. If $c_i = 0$, this indicates that x_i is not in the representation of x; $c_i = 1$ indicates that it is. Consequently, each $x \in \mathcal{B}$ is associated with an n-tuple $(c_1, c_2, \ldots, c_n)$ and there are 2^n such n-tuples. Therefore we have proved the following result.

Theorem 9.8

▶ If $\mathcal{B}$ is a finite Boolean algebra with n atoms, then $|\mathcal{B}| = 2^n$.

There is one final question to resolve. If $n \in \mathbf{Z}^+$ how many different Boolean algebras of size 2^n are there? Looking at the Hasse diagrams in Fig. 9.13, we see two different pictures. But if we ignore the labels on the vertices, the underlying structures are exactly the same. Hence these two Boolean algebras are abstractly identical or isomorphic.

Definition 9.8

▶ If $(\mathcal{B}_1, +, \cdot, {}^{-}, 0, 1)$ and $(\mathcal{B}_2, +, \cdot, {}^{-}, 0, 1)$ are Boolean algebras, $\mathcal{B}_1$, $\mathcal{B}_2$ are called *isomorphic* if there is a one-to-one correspondence $f: \mathcal{B}_1 \to \mathcal{B}_2$ so that for all x_1, $y_1 \in \mathcal{B}_1$,

a) $f(\underset{\text{(in } \mathcal{B}_1)}{\uparrow} x_1 + y_1) = f(x_1) \underset{\text{(in } \mathcal{B}_2)}{\uparrow} + f(y_1);$ b) $f(x_1 \underset{\text{(in } \mathcal{B}_1)}{\uparrow} \cdot y_1) = f(x_1) \underset{\text{(in } \mathcal{B}_2)}{\uparrow} \cdot f(y_1);$ and,

c) $f(\bar{x}_1) = \overline{f(x_1)}$. (In $f(\bar{x}_1)$ we take the complement in $\mathcal{B}_1$, while for $\overline{f(x_1)}$ the complement is taken in $\mathcal{B}_2$.)

Such a function f *preserves* the operations of the algebraic structures.

EXAMPLE 9.26

For the two Boolean algebras in Fig. 9.13, define f by

$$f: \emptyset \to 1 \qquad f: \{2\} \to 3 \qquad f: \{1,2\} \to 6 \qquad f: \{2,3\} \to 15$$
$$f: \{1\} \to 2 \qquad f: \{3\} \to 5 \qquad f: \{1,3\} \to 10 \qquad f: \{1,2,3\} \to 30$$

Note the following.

a) The 0 elements correspond under f, as do the 1 elements;
b) $f(\{1\} \cup \{2\}) = f(\{1,2\}) = 6 = $ l.c.m. of 2, 3 $= [f(\{1\}), f(\{2\})]$;
c) $f(\{1,2\} \cap \{2,3\}) = f(\{2\}) = 3 = $ g.c.d. of 6, 15 $= (f(\{1,2\}), f(\{2,3\}))$;
d) $f(\overline{\{2\}}) = f(\{1,3\}) = 10 = 30/3 = \overline{3} = \overline{f(\{2\})}$; and
e) The image of any atom ($\{1\}, \{2\}, \{3\}$) is an atom (2, 3, 5, respectively).

This function is an isomorphism. Once we correspond the 0 elements and the atoms, the remaining correspondences are determined from these by Theorem 9.7 and the preservation of the operations under f. □

From this example we have our final result.

Theorem 9.9

▶ Any finite Boolean algebra $\mathcal{B}$ is isomorphic to a Boolean algebra of sets.

Proof

Since $\mathcal{B}$ is finite, $\mathcal{B}$ has n atoms x_i, $1 \le i \le n$, and $|\mathcal{B}| = 2^n$. Let $\mathcal{U} = \{1, 2, \ldots, n\}$ and $\mathcal{P}(\mathcal{U})$ be the Boolean algebra of subsets of $\mathcal{U}$.

If we define $f: \mathcal{B} \to \mathcal{P}(\mathcal{U})$ by $f(x_i) = \{i\}$, $1 \le i \le n$, and then extend this function to all the other elements of $\mathcal{B}$, preserving the operations of $+$ in $\mathcal{B}$ and $\cup$ in $\mathcal{P}(\mathcal{U})$, (e.g., $f(x_1 + x_2) = \{1,2\}$, $f(x_2 + x_4 + x_7) = \{2,4,7\}$) and set $f(0) = \emptyset$, this function f establishes an isomorphism. ■

EXERCISES

1. Verify the second distributive law and the identity and inverse laws for Example 9.23.

2. Complete the proof of Theorem 9.3.

3. Let $\mathcal{B}$ be the set of positive integer divisors of 210, and define $+$, $\cdot$, $^-$ for $\mathcal{B}$ as in Example 9.23. Determine each of the following:
   a) $30 + 5 \cdot 7$ b) $(30 + 5) \cdot (30 + 7)$ c) $\overline{(14 + 15)}$
   d) $21(2 + \overline{10})$ e) $(2 + 3) + 5$ f) $(6 + 35)(7 + 10)$

4. For a Boolean algebra $\mathcal{B}$ the relation "$\le$" on $\mathcal{B}$, defined by $x \le y$ if $xy = x$, was shown to be a partial order. Prove that: (a) if $x \le y$ then $x + y = y$; and (b) if $x \le y$ then $\overline{y} \le \overline{x}$.

5. If $\mathcal{B}$ is a Boolean algebra, partially ordered by $\le$ and $x, y \in \mathcal{B}$, what is the dual of the statement $x \le y$?

6. Let $F_n = \{f: B^n \to B\}$ be the Boolean algebra of all Boolean functions on n variables. How many atoms does F_n have?

7. Verify Theorem 9.5(b).

8. Given a Boolean algebra $\mathscr{B}$, a nonempty subset $\mathscr{B}_1$ of $\mathscr{B}$ is called a *subalgebra* if for all $x, y \in \mathscr{B}_1$, $x + y$, xy, $\bar{x} \in \mathscr{B}_1$.

    a) Prove that $0, 1 \in \mathscr{B}_1$. (Hence $|\mathscr{B}_1| \geq 2$).

    b) Find two examples of subalgebras in Example 9.23.

    c) For $\mathcal{U} = \{1, 2, 3\}$ and $\mathscr{B} = $ the Boolean algebra of subsets of $\mathcal{U}$, how many different subalgebras can we find?

    d) Show that the definition above can be shortened by requiring only that whenever $x, y \in \mathscr{B}_1$, $x + y$ and $\bar{x} \in \mathscr{B}_1$.

    e) If $\mathscr{B}_1$, $\mathscr{B}_2$ are subalgebras of $\mathscr{B}$, is $\mathscr{B}_1 \cap \mathscr{B}_2$ a subalgebra?

9. Given a Boolean algebra $\mathscr{B}$, let $\emptyset \neq \mathscr{B}_1 \subseteq \mathscr{B}$. If $0, 1 \in \mathscr{B}_1$, and $\mathscr{B}_1$ is closed under $+$ and $\cdot$, is $\mathscr{B}_1$ a subalgebra of $\mathscr{B}$?

10. If $\mathscr{B}$ is a Boolean algebra, prove that the zero element and the one element of $\mathscr{B}$ are unique.

11. Let $f: \mathscr{B}_1 \rightarrow \mathscr{B}_2$ be an isomorphism of Boolean algebras. Prove that

    a) $f(0) = 0$ b) $f(1) = 1$

    c) if $x, y \in \mathscr{B}_1$ with $x \leq y$, then in $\mathscr{B}_2$, $f(x) \leq f(y)$.

    d) if x is an atom of $\mathscr{B}_1$, then $f(x)$ is an atom in $\mathscr{B}_2$.

    e) if $\mathscr{S}_1$ is a subalgebra of $\mathscr{B}_1$, then $f(\mathscr{S}_1)$ is a subalgebra of $\mathscr{B}_2$.

12. Let $\mathscr{B}_1$ be the Boolean algebra of all positive integer divisors of 2310, with $\mathscr{B}_2$ the Boolean algebra of subsets of $\{a, b, c, d, e\}$.

    a) Define $f: \mathscr{B}_1 \rightarrow \mathscr{B}_2$ so that $f(2) = \{a\}$, $f(3) = \{b\}$, $f(5) = \{c\}$, $f(7) = \{d\}$, $f(11) = \{e\}$. For f to be an isomorphism, what must be the images of 35, 110, 210, 330?

    b) How many different isomorphisms can one define between $\mathscr{B}_1$ and $\mathscr{B}_2$?

13. a) If $\mathscr{B}_1$, $\mathscr{B}_2$ are Boolean algebras and $f: \mathscr{B}_1 \rightarrow \mathscr{B}_2$ is one-to-one and onto and such that $f(x + y) = f(x) + f(y)$, $f(\bar{x}) = \overline{f(x)}$, for all $x, y \in \mathscr{B}_1$, prove that f is an isomorphism.

    b) State and prove another result comparable to that in (a). (What principle is used here?)

9.5 ■ SUMMARY AND HISTORICAL REVIEW

The modern concept of abstract algebra was developed by George Boole in his study of general abstract systems, as opposed to particular examples of such systems. In his 1854 publication, *An Investigation of the Laws of Thought,* he formulated the mathematical structure now called a Boolean algebra. Although abstract in nature during the nineteenth century, the study of Boolean algebra was investigated in the twentieth century for its applicative value.

    Starting in 1938, C. E. Shannon [8] made the first major contribution in applied Boolean algebra. He devised the algebra of switching circuits and showed its relation to the algebra of logic. During the 1940's and 1950's additional developments that were made in this area can be found in the paper by C. E. Shannon [9] and in the report of the Harvard University Computation Laboratory [10].

We found that switching functions can be represented by their disjunctive and conjunctive normal forms. These allowed us to write such functions in a compact way using binary indices. The minimization process showed how to obtain a given Boolean function as a minimal sum of products, or minimal product of sums. Based on the map method by E. W. Veitch [11], M. Karnaugh's modification [4] was developed here as a pictorial method for the simplification of Boolean functions. Another technique that we mentioned in the text is the tabulation algorithm known as the Quine-McCluskey method. Originally developed by W. V. Quine, in [6] and [7], this technique was modified by E. J. McCluskey, Jr. [5]. It is very useful for functions with more than six variables and lends itself to computer implementation. The interested reader can find more about Karnaugh maps in Chapter 6 of F. Hill and G. Peterson [3]. Chapter 7 of [3] provides an excellent coverage of the Quine-McCluskey method. A. Friedman and P. Menon, in [2], examine gating networks in the light of contemporary technology, while T. Booth [1] investigates more specific applications of logic design in the study of computers.

Although the major part of this chapter was applied in nature, Section 9.4 found us investigating the structure of a Boolean algebra. Unlike commutative rings with unity, which come in all possible sizes, we found that a Boolean algebra can contain only 2^n elements, $n \in \mathbf{Z}^+$. Uniqueness of representation appeared as we found the atoms of a Boolean algebra used to build the rest of the algebra. The Boolean algebra of sets that we studied in Chapter 2 was found to represent all Boolean algebras in the sense that a finite Boolean algebra with n atoms is isomorphic to the algebra of all subsets of $\{1, 2, 3, \ldots, n\}$.

REFERENCES

1. Booth, Taylor L., *Digital Networks and Computer Systems,* 2nd ed., John Wiley & Sons, New York, 1978.

2. Friedman, Arthur D., and Menon, P. R., *Theory and Design of Switching Circuits,* Computer Science Press, Woodland Hills, California, 1975.

3. Hill, Frederick J., and Peterson, Gerald R., *Introduction to Switching Theory and Logical Design,* 3rd ed., John Wiley & Sons, New York, 1981.

4. Karnaugh, M., "The Map Method for Synthesis of Combinational Logic Circuits," *Transactions of the AIEE,* part I, vol. 72, no. 9, pp. 593–599, 1953.

5. McCluskey, E. J., Jr., "Minimization of Boolean Functions," *Bell System Technical Journal,* vol. 35, no. 6, pp. 1417–1444, November, 1956.

6. Quine, W. V., "The Problem of Simplifying Truth Functions," *American Mathematical Monthly,* vol. 59, no. 8, pp. 521–531, October, 1952.

7. Quine, W. V., "A Way to Simplify Truth Functions," *American Mathematical Monthly,* vol. 62, no. 9, pp. 627–631, November, 1955.

8. Shannon, C. E., "A Symbolic Analysis of Relay and Switching Circuits," *Transactions of the AIEE,* vol. 57, pp. 713–723, 1938.

9. Shannon, C. E., "The Synthesis of Two-terminal Switching Circuits," *Bell System Technical Journal,* vol. 28, pp. 59–98, 1949.

10. Staff of the Computation Laboratory, *Synthesis of Electronic Computing and Control Circuits,* Annals 27, Harvard University Press, Cambridge, Massachusetts, 1951.

11. Veitch, E. W., "A Chart Method for Simplifying Truth Functions," *Proceedings of the ACM,* pp. 127–133, Pittsburgh, Pennsylvania, May, 1952.

MISCELLANEOUS EXERCISES

1. a) Let $n \geq 2$. If x_i is a Boolean variable, $1 \leq i \leq n$, prove that

   i) $\overline{(x_1 + x_2 + \cdots + x_n)} = \bar{x}_1 \bar{x}_2 \cdots \bar{x}_n$

   ii) $\overline{(x_1 x_2 \cdots x_n)} = \bar{x}_1 + \bar{x}_2 + \cdots + \bar{x}_n$

   iii) $\left(\sum_{i=1}^{n} x_i\right)\left(\overline{\prod_{i=1}^{n} x_i}\right) = x_1 \bar{x}_2 + x_2 \bar{x}_3 + x_3 \bar{x}_4 + \cdots + x_{n-1}\bar{x}_n + x_n \bar{x}_1$

   b) State the dual of (iii) in part (a).

2. Let $f, g: B^5 \to B$ be Boolean functions, where $f = \sum m(1, 2, 4, 7, x)$ and $g = \sum m(0, 1, 2, 3, y, z, 16, 25)$. If $f \leq g$, what are x, y, z?

3. Eileen is having a party and finds herself confronted with decisions about inviting five of her friends.

   a) If she invites Margaret, she must also invite Joan.

   b) If Kathleen is invited, Nettie and Margaret must also be invited.

   c) She can invite Cathy or Joan, but definitely not both of them.

   d) Neither Cathy nor Nettie will show up if the other is not invited.

   e) Either Kathleen or Nettie or both must be invited.

   Determine which subsets of these five friends Eileen can invite to her party and still satisfy conditions (a)–(e).

4. Let $f, g: B^4 \to B$, $f = \sum m(2, 4, 6, 8)$, $g = \sum m(1, 2, 3, 4, 5, 6, 7, 8, 9, 11, 13, 15)$. Find a function $h: B^4 \to B$ such that $f = gh$.

5. Use a Karnaugh map to find a minimal sum of products representation for

   a) $f(w, x, y, z) = \sum m(0, 1, 2, 3, 6, 7, 14, 15)$

   b) $g(v, w, x, y, z) = \prod M(1, 2, 4, 6, 9, 10, 11, 14, 17, 18, 19, 20, 22, 25, 26, 27, 30)$.

6. In the feedback network of Fig. 9.14, we have a unit delay (one of the finite state machines of Chapter 4). This device takes the output f at time t and feeds its complement back into the AND gate, along with new values for the inputs x, y, at the next unit of time, or clock pulse, $t + 1$.

Figure 9.14

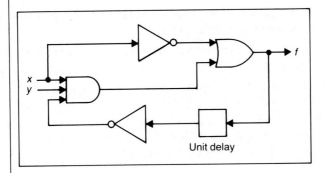

Complete Table 9.22 which provides the output f at each time t for the values of x and y given.

Table 9.22

t	0	1	2	3	4	5	6
x	0	0	1	0	1	1	0
y	0	1	1	0	0	1	1
f	1	1	?	?	?	?	?

7. For n Boolean variables there are 2^{2^n} Boolean functions, each of which can be represented by a function table.

   a) A Boolean function f on n variables $x_1, x_2, \ldots, x_n$ is called *self-dual* if $f(x_1, x_2, \ldots, x_n) = \overline{f(\bar{x}_1, \bar{x}_2, \ldots, \bar{x}_n)}$. How many Boolean functions on n variables are self-dual?

   b) Let $f: B^3 \to B$. Then f is called *symmetric* if $f(x, y, z) = f(x, z, y) = f(y, x, z) = f(y, z, x) = f(z, x, y) = f(z, y, x)$. So the value of f is unchanged when we rearrange the three columns of values listed under x, y, z in the table for f. How many such functions are there on three Boolean variables? How many are there on n Boolean variables?

Figure 9.15

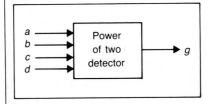

8. The four input lines for the network in Fig. 9.15 provide the binary equivalents of 0, 1, 2, $\ldots$, 15, where each number is represented as $abcd$, with d the least significant bit.

a) Find the d.n.f. of g, whose value is 1 exactly when $abcd$ is the binary equivalent of 1, 2, 4, or 8.

b) Draw the two-level gating network for g as a minimal sum of products.

c) If this network is part of a larger network and, consequently, the binary equivalents of 10–15 never occur as inputs, design a two-level gating network for g in this case.

9. For (a) $n = 60$, and (b) $n = 120$, explain why the positive divisors of n do not yield a Boolean algebra. (Here $x + y = [x, y]$, $xy = (x, y)$, $\bar{x} = n/x$, 1 is the zero element and n is the one element.)

10

Generating
Functions

In this chapter and the next we return to problems of enumeration, introducing at this time the important concept of the *generating function*.

The problem of making selections, with repetitions allowed, was studied in Chapter 1. There we sought the number of integer solutions to the equation $c_1 + c_2 + c_3 + c_4 = 25$, $c_i \geq 0$, $1 \leq i \leq 4$. With the principle of inclusion and exclusion, in Chapter 7 we were able to solve a more restricted version of the problem, such as $c_1 + c_2 + c_3 + c_4 = 25$, $0 \leq c_i < 10$, $1 \leq i \leq 4$. If, in addition, we wanted c_2 to be even, and c_3 a multiple of 3, we could apply the results of Chapters 1 and 7 to several subcases.

The power of the generating function rests upon its ability to not only solve our past problems, but aid us in new situations involving additional restrictions.

10.1 ■ INTRODUCTORY EXAMPLES

Instead of defining a generating function at this point, we shall examine some examples that motivate the idea. We may find that we have dealt with this concept in some past situations.

EXAMPLE 10.1 While shopping one Saturday, Mildred buys 12 oranges for her children, Grace, Mary, and Frank. In how many ways can she distribute the oranges so that Grace gets at least four, and Mary and Frank get at least two, but Frank gets no more than five? Table 10.1 lists all the possible distributions. We see that we have all the integer solutions to the equation $c_1 + c_2 + c_3 = 12$, $4 \leq c_1$, $2 \leq c_2$, $2 \leq c_3 \leq 5$.

$\qquad\qquad\qquad\qquad\qquad\qquad\qquad\qquad G \qquad M \qquad\quad F$

Table 10.1

G	M	F	G	M	F
4	3	5	6	2	4
4	4	4	6	3	3
4	5	3	6	4	2
4	6	2	7	2	3
5	2	5	7	3	2
5	3	4	8	2	2
5	4	3			
5	5	2			

Considering the first two cases in this table, we find the solutions $4 + 3 + 5 = 12$ and $4 + 4 + 4 = 12$. Now where in our prior algebraic experiences did anything like this happen? When multiplying polynomials we add the powers of the variable, and we see here that in multiplying the three polynomials,

$$(x^4 + x^5 + x^6 + x^7 + x^8)(x^2 + x^3 + x^4 + x^5 + x^6)(x^2 + x^3 + x^4 + x^5),$$

two of the ways to obtain x^{12} are from $x^4 x^3 x^5$ and $x^4 x^4 x^4$. Examining this product more closely, we realize that we obtain $x^i x^j x^k$ for every triple (i, j, k) that appears in Table 10.1. Consequently, the coefficient of x^{12} in

$$f(x) = (x^4 + x^5 + \cdots + x^8)(x^2 + x^3 + \cdots + x^6)(x^2 + x^3 + \cdots + x^5)$$

counts the number of distributions, namely 14, that we seek. The function $f(x)$ is called a *generating function* for the distributions.

But where did the factors in this product come from?

The factor $x^4 + x^5 + x^6 + x^7 + x^8$, for example, indicates that we can give Grace 4 *or* 5 *or* 6 *or* 7 *or* 8 of the oranges. Once again we make use of the interplay between the exclusive *or* and ordinary addition. The coefficient of each power of x is 1 because, considering the oranges as identical objects, there is only one way to give Grace four oranges, one way for five oranges, etc. Since each of Mary and Frank must receive at least two oranges, the other terms $(x^2 + x^3 + \cdots + x^6)$ and $(x^2 + x^3 + \cdots + x^5)$ start with x^2, and for Frank we stop at x^5 so that he doesn't receive more than five oranges. (Why does the term for Mary stop at x^6?)

Most of us are reasonably convinced now that the coefficient of x^{12} in $f(x)$ yields the answer. Some, however, may be a bit skeptical about this new idea. It seems that we could list the cases in Table 10.1 faster than we could multiply out $f(x)$ or calculate the coefficient of x^{12} in $f(x)$. At present that may seem true. But as we progress to problems with more unknowns and larger quantities to distribute, the generating function will more than demonstrate its worth. (The reader may realize that the rook polynomials of Chapter 7 are examples of generating functions. In addition, when determining the minimal dominating sets for a graph in Chapter 9 we used a comparable Boolean notion.) For now we consider two more examples. □

EXAMPLE 10.2

If there is an unlimited number (or at least 24 of each color) of red, green, white, and black jelly beans, in how many ways can a child select 24 of these candies so that he has an even number of white beans and at least six black ones?

The polynomials associated with the jelly bean colors are as follows: red (green): $1 + x + x^2 + \cdots + x^{24}$, where the leading 1 is for $1x^0$, since one possibility for the red (and green) jelly beans is that none of that color is selected;

white: $(1 + x^2 + x^4 + x^6 + \cdots + x^{24})$; and

black: $(x^6 + x^7 + x^8 + \cdots + x^{24})$.

So the answer to the problem is the coefficient of x^{24} in the generating function

$$f(x) = (1 + x + x^2 + \cdots + x^{24})^2(1 + x^2 + x^4 + \cdots + x^{24}) \cdot$$
$$(x^6 + x^7 + \cdots + x^{24}).$$

One such selection is five red, three green, eight white, and eight black jelly beans. This arises from x^5 in the first factor, x^3 in the second factor, and x^8 in the last two factors. □

One more example before closing this section!

EXAMPLE 10.3

How many integer solutions are there for the equation $c_1 + c_2 + c_3 + c_4 = 25$, $0 \le c_i$, $1 \le i \le 4$?

We can alternatively ask in how many ways 25 (identical) pennies can be distributed among four children.

For each child the possibilities can be described by the polynomial $1 + x + x^2 + x^3 + \cdots + x^{25}$. Then the answer to this problem is the coefficient of x^{25} in the generating function

$$f(x) = (1 + x + x^2 + \cdots + x^{25})^4.$$

The answer can also be obtained as the coefficient of x^{25} in the generating function

$$g(x) = (1 + x + x^2 + x^3 + \cdots + x^{25} + x^{26} + \cdots)^4,$$

if we rephrase the question in terms of distributing, from a large (or unlimited) number of pennies, 25 pennies among four children. (Where $f(x)$ is a polynomial, $g(x)$ is a *power series* in x.) Note that the terms x^k, $k \geq 26$, are never used. So why bother with them? There will be times when it is easier to compute with a power series than with a polynomial.   □

EXERCISES

1. For each of the following, determine a generating function and indicate the coefficient in the function that is needed to solve the problem. (Give both the polynomial and power series forms of the generating function, wherever appropriate.)
   Find the number of integer solutions for the following equations:

   a) $c_1 + c_2 + c_3 + c_4 = 20$, $0 \leq c_i \leq 7$, $1 \leq i \leq 4$.

   b) $c_1 + c_2 + c_3 + c_4 = 20$, $0 \leq c_i$, $1 \leq i \leq 4$, c_2, c_3 even.

   c) $c_1 + c_2 + c_3 + c_4 + c_5 = 30$, $2 \leq c_1 \leq 4$, $3 \leq c_i \leq 8$, $2 \leq i \leq 5$.

   d) $c_1 + c_2 + c_3 + c_4 + c_5 = 30$, $0 \leq c_i$, $1 \leq i \leq 5$, c_2 even, c_3 odd.

2. Determine the generating function for the number of ways to distribute 35 pennies (from an unlimited supply) among five children if (a) there are no restrictions; (b) each child gets at least 1¢; (c) each child gets at least 2¢; (d) the oldest child gets at least 10¢; (e) the two youngest children must each get at least 10¢.

3. a) Find the generating function for the number of ways to select 10 candy bars from large supplies of six different kinds.

   b) Find the generating function for the number of ways to select, with repetitions allowed, r objects from a collection of n distinct objects.

4. a) Explain why the generating function for the number of ways to have n cents in pennies and nickels is $(1 + x + x^2 + x^3 + \cdots)(1 + x^5 + x^{10} + \cdots)$.

   b) Find the generating function for the number of ways to have n cents in pennies, nickels, and dimes.

5. Find the generating function for the number of integer solutions to the equation $c_1 + c_2 + c_3 + c_4 = 20$, $-3 \leq c_1, c_2$, $-5 \leq c_3 \leq 5$, $0 \leq c_4$.

6. For $S = \{a, b, c\}$, $f(x) = (1 + ax)(1 + bx)(1 + cx)$ is the generating function for the subsets of S. Setting $x = 1$ we get a listing of the subsets of S where 1

represents the empty subset. (Going one step further and setting $a = b = c = 1$, we get 8, the number of subsets of S.)

a)   Give the generating function for the subsets of $S = \{a, b, c, \ldots, r, s, t\}$.

b)   Answer part (a) for selections where each element can be rejected or selected as many as three times.

10.2 ■ DEFINITION AND EXAMPLES: CALCULATIONAL TECHNIQUES

Definition 10.1

▶ Let $a_0, a_1, a_2, \ldots$ be a sequence of real numbers. The function

$$f(x) = a_0 + a_1 x + a_2 x^2 + \cdots = \sum_{i=0}^{\infty} a_i x^i$$

is called the *generating function* for the given sequence.

Where could this idea have come from?

EXAMPLE 10.4   For any $n \in \mathbf{Z}^+$,

$$(1 + x)^n = \binom{n}{0} + \binom{n}{1}x + \binom{n}{2}x^2 + \cdots + \binom{n}{n}x^n,$$

so $(1 + x)^n$ is the generating function for the sequence

$$\binom{n}{0}, \binom{n}{1}, \binom{n}{2}, \ldots, \binom{n}{n}, 0, 0, 0, \ldots.$$ □

EXAMPLE 10.5

a)   For $n \in \mathbf{Z}^+$, $(1 - x^{n+1}) = (1 - x)(1 + x + x^2 + x^3 + \cdots + x^n)$. So $(1 - x^{n+1})/(1 - x) = (1 + x + x^2 + \cdots + x^n)$ and $(1 - x^{n+1})/(1 - x)$ is the generating function for the sequence $1, 1, 1, \ldots, 1, 0, 0, 0, \ldots$, where the first $n + 1$ terms are 1.

b)   Extending the idea in (a), $1 = (1 - x)(1 + x + x^2 + x^3 + x^4 + \cdots)$, so $1/(1 - x)$ is the generating function for the sequence $1, 1, 1, 1, \ldots$. $(1/(1 - x) = 1 + x + x^2 + x^3 + \cdots$ is valid for all real x where $|x| < 1$; it is for this range of values that the *geometric series* $1 + x + x^2 + \cdots$ *converges*. However, in our work with generating functions we will be more concerned with the coefficients of the powers of x than with matters of convergence.)

c)   With $1/(1 - x) = 1 + x + x^2 + x^3 + \cdots = \sum_{i=0}^{\infty} x^i$, taking the derivative of each side we have

$$\frac{d}{dx}\frac{1}{1 - x} = \frac{d}{dx}(1 - x)^{-1} = (-1)(1 - x)^{-2}(-1)$$

$$= \frac{d}{dx}(1 + x + x^2 + x^3 + \cdots) = 1 + 2x + 3x^2 + 4x^3 + \cdots$$

,Consequently, $1/(1 - x)^2$ is the generating function for the sequence 1, 2, 3, 4, . . . , while $x/(1 - x)^2$ generates the sequence 0, 1, 2, 3, ...

d) Continuing from (c), $(d/dx)[x/(1 - x)^2] = (d/dx)[0 + x + 2x^2 + \cdots]$, or $(x + 1)/(1 - x)^3 = 1 + 2^2x + 3^2x^2 + 4^2x^3 + \cdots$. Hence, $(x + 1)/(1 - x)^3$ generates $1^2, 2^2, 3^2, \ldots$, and $x(x + 1)/(1 - x)^3$ generates $0^2, 1^2, 2^2, 3^2, \ldots$. $\square$

For any $n \in \mathbf{Z}^+$, by the binomial theorem $(1 + x)^n = \binom{n}{0} + \binom{n}{1}x + \binom{n}{2}x^2 + \cdots + \binom{n}{n}x^n$. We want to extend this idea to cases where (a) $n < 0$; and (b) n is not necessarily an integer.

With $n, r \in \mathbf{Z}^+$, $n \geq r > 0$, we have

$$\binom{n}{r} = \frac{n!}{[r!(n - r)!]} = \frac{[n(n - 1)(n - 2)\cdots(n - r + 1)]}{r!}$$

If $n \notin \mathbf{Z}^+$ we use $[n(n - 1)(n - 2)\cdots(n - r + 1)]/r!$ as the definition of $\binom{n}{r}$. If $n \in \mathbf{Z}^+$, then

$$\binom{-n}{r} = [(-n)(-n - 1)(-n - 2)\cdots(-n - r + 1)]/r!$$

$$= (-1)^r(n)(n + 1)(n + 2)\cdots(n + r - 1)]/r!$$

$$= (-1)^r(n + r - 1)!/[(n - 1)!r!] = (-1)^r\binom{n + r - 1}{r}.$$

Finally, for any *real* number n, we define $\binom{n}{0} = 1$.

EXAMPLE 10.6   For $n \in \mathbf{Z}^+$, the Maclaurin series expansion for $(1 + x)^{-n}$ is given by

$$(1 + x)^{-n} = 1 + (-n)x + (-n)(-n - 1)x^2/2!$$
$$+ (-n)(-n - 1)(-n - 2)x^3/3! + \cdots$$

$$= 1 + \sum_{r=1}^{\infty} \frac{(-n)(-n - 1)(-n - 2)\cdots(-n - r + 1)}{r!}x^r$$

$$= \sum_{r=0}^{\infty} (-1)^r\binom{n + r - 1}{r}x^r.$$

Hence, $(1 + x)^{-n} = \binom{-n}{0} + \binom{-n}{1}x + \binom{-n}{2}x^2 + \cdots = \sum_{r=0}^{\infty}\binom{-n}{r}x^r$. This generalizes the binomial theorem of Chapter 2, and $(1 + x)^{-n}$ is the generating function for the sequence $\binom{-n}{0}, \binom{-n}{1}, \binom{-n}{2}, \binom{-n}{3}, \ldots$. $\square$

EXAMPLE 10.7   Find the coefficient of x^5 in $(1 - 2x)^{-7}$.

With $y = -2x$, use the result of Example 10.6 to obtain $(1 - 2x)^{-7} = (1 + y)^{-7} = \sum_{r=0}^{\infty}\binom{-7}{r}y^r = \sum_{r=0}^{\infty}\binom{-7}{r}(-2x)^r$. Consequently, the coefficient of x^5 is $\binom{-7}{5}(-2)^5 = (-1)^5\binom{7+5-1}{5}(-32) = (32)\binom{11}{5} = 14{,}784$. $\square$

EXAMPLE 10.8

For any real number n, the Maclaurin series for $(1 + x)^n$ is

$$1 + nx + n(n - 1)x^2/2! + (n)(n - 1)(n - 2)x^3/3! + \cdots$$

$$= 1 + \sum_{r=1}^{\infty} \frac{n(n - 1)(n - 2)\cdots(n - r + 1)}{r!}x^r.$$

Hence

$$(1 + 3x)^{-1/3} = 1 + \sum_{r=1}^{\infty} \frac{(-1/3)(-4/3)(-7/3)\cdots((-3r + 2)/3)}{r!}(3x)^r$$

$$= 1 + \sum_{r=1}^{\infty} \frac{(-1)(-4)(-7)\cdots(-3r + 2)}{r!}x^r,$$

and $(1 + 3x)^{-1/3}$ generates the sequence $1, -1, (-1)(-4)/2!, (-1)(-4)(-7)/3!,$
$\ldots, (-1)(-4)(-7)\cdots(-3r + 2)/r!, \ldots.$ □

Before continuing we collect, for future reference, the following identities.

For any $m, n \in \mathbf{Z}^+, a \in \mathbf{R}$,

1. $(1 + x)^n = \binom{n}{0} + \binom{n}{1}x + \binom{n}{2}x^2 + \cdots + \binom{n}{n}x^n$

2. $(1 + ax)^n = \binom{n}{0} + \binom{n}{1}ax + \binom{n}{2}a^2x^2 + \cdots + \binom{n}{n}a^nx^n$

3. $(1 + x^m)^n = \binom{n}{0} + \binom{n}{1}x^m + \binom{n}{2}x^{2m} + \cdots + \binom{n}{n}x^{nm}$

4. $(1 - x^{n+1})/(1 - x) = 1 + x + x^2 + \cdots + x^n$

5. $1/(1 - x) = 1 + x + x^2 + x^3 + \cdots = \sum_{i=0}^{\infty} x^i$

6. $1/(1 + x)^n = \binom{-n}{0} + \binom{-n}{1}x + \binom{-n}{2}x^2 + \cdots$
   $\qquad = \sum_{i=0}^{\infty} \binom{-n}{i}x^i$
   $\qquad = 1 + (-1)\binom{n+1-1}{1}x + (-1)^2\binom{n+2-1}{2}x^2 + \cdots$
   $\qquad = \sum_{i=0}^{\infty}(-1)^i\binom{n+i-1}{i}x^i$

7. $1/(1 - x)^n = \binom{-n}{0} + \binom{-n}{1}(-x) + \binom{-n}{2}(-x)^2 + \cdots$
   $\qquad = \sum_{i=0}^{\infty}\binom{-n}{i}(-x)^i$
   $\qquad = 1 + (-1)\binom{n+1-1}{1}(-x) + (-1)^2\binom{n+2-1}{2}(-x)^2 + \cdots$
   $\qquad = \sum_{i=0}^{\infty}\binom{n+i-1}{i}x^i$

8. if $f(x) = \sum_{i=0}^{\infty} a_i x^i$, $g(x) = \sum_{i=0}^{\infty} b_i x^i$, and $h(x) = f(x)g(x)$, then $h(x) = \sum_{i=0}^{\infty} c_i x^i$, where for any $k \geq 0$, $c_k = a_0 b_k + a_1 b_{k-1} + \cdots + a_{k-1}b_1 + a_k b_0$.

EXAMPLE 10.9

Determine the coefficient of x^{15} in $f(x) = (x^2 + x^3 + x^4 + \cdots)^4$.
   Since $(x^2 + x^3 + x^4 + \cdots) = x^2(1 + x + x^2 + \cdots) = x^2/(1 - x)$, the coefficient of x^{15} in $f(x)$ is the coefficient of x^{15} in $(x^2/(1 - x))^4 = x^8/(1 - x)^4$. Hence the coefficient sought is that of x^7 in $(1 - x)^{-4}$, namely $\binom{-4}{7}(-1)^7 = (-1)^7\binom{4+7-1}{7} \cdot (-1)^7 = \binom{10}{7} = 120$.
   In general, for $n \in \mathbf{Z}^+$, the coefficient of x^n in $f(x)$ is 0, when $0 \leq n \leq 7$; for $n \geq 8$, the coefficient of x^n in $f(x)$ is the coefficient of x^{n-8} in $(1 - x)^{-4}$, and this is $\binom{-4}{n-8}(-1)^{n-8} = \binom{n-5}{n-8}$. □

EXAMPLE 10.10

In how many ways can we select, with repetitions allowed, r objects from n distinct objects?

For each of the n distinct objects the geometric series $1 + x + x^2 + x^3 + \cdots$ represents the possible choices for that object (namely none, one, two, ...). Considering all of the n distinct objects, the generating function is

$$f(x) = (1 + x + x^2 + x^3 + \cdots)^n,$$

and the answer is the coefficient of x^r in $f(x)$. Now

$$(1 + x + x^2 + x^3 + \cdots)^n = \left(\frac{1}{(1 - x)}\right)^n = \frac{1}{(1 - x)^n} = \sum_{i=0}^{\infty} \binom{n + i - 1}{i} x^i,$$

so the coefficient of x^r is

$$\binom{n + r - 1}{r},$$

the result we found in Chapter 1.   □

EXAMPLE 10.11

In how many ways can a police captain distribute 24 rifle shells to four patrolmen so that each patrolman gets at least three, but not more than eight, shells?

The choices for the number of shells each patrolman receives are given by $x^3 + x^4 + \cdots + x^8$. Since there are four patrolmen, the resulting generating function is $f(x) = (x^3 + x^4 + \cdots + x^8)^4$.

We seek the coefficient of x^{24} in $f(x)$. With $(x^3 + x^4 + \cdots + x^8)^4 = x^{12}(1 + x + x^2 + \cdots + x^5)^4 = x^{12}((1 - x^6)/(1 - x))^4$, the answer is the coefficient of x^{12} in $(1 - x^6)^4(1 - x)^{-4} = [1 - \binom{4}{1}x^6 + \binom{4}{2}x^{12} - \cdots + x^{24}][\binom{-4}{0} + \binom{-4}{1}(-x) + \binom{-4}{2} \cdot (-x)^2 + \cdots]$, which is $[\binom{-4}{12}(-1)^{12} - \binom{4}{1}\binom{-4}{6}(-1)^6 + \binom{4}{2}\binom{-4}{0}] = [\binom{15}{12} - \binom{4}{1}\binom{9}{6} + \binom{4}{2}] = 125$.   □

EXAMPLE 10.12

Verify that for all $n \in \mathbf{Z}^+$, $\binom{2n}{n} = \sum_{i=0}^{n} \binom{n}{i}^2$.

Since $(1 + x)^{2n} = [(1 + x)^n]^2$, by comparison of coefficients (of like powers of x), the coefficient of x^n in $(1 + x)^{2n}$, which is $\binom{2n}{n}$, must equal the coefficient of x^n in $[\binom{n}{0} + \binom{n}{1}x + \binom{n}{2}x^2 + \cdots + \binom{n}{n}x^n]^2$, and this is $\binom{n}{0}\binom{n}{n} + \binom{n}{1}\binom{n}{n-1} + \binom{n}{2}\binom{n}{n-2} + \cdots + \binom{n}{n}\binom{n}{0}$. With $\binom{n}{r} = \binom{n}{n-r}$, for $0 \le r \le n$, the result follows.   □

EXAMPLE 10.13

Determine the coefficient of x^8 in $1/(x - 3)(x - 2)^2$.

Since $1/(x - a) = (-1/a)(1/(1 - (x/a))) = (-1/a)[1 + (x/a) + (x/a)^2 + \cdots]$ for any $a \ne 0$, we could solve this problem by finding the coefficient of x^8 in $1/(x - 3)(x - 2)^2$ expressed as $(-1/3)[1 + (x/3) + (x/3)^2 + \cdots](1/4)[\binom{-2}{0} + \binom{-2}{1}(-x/2) + \binom{-2}{2}(-x/2)^2 + \cdots]$.

An alternative technique uses the partial fraction decomposition:

$$\frac{1}{(x - 3)(x - 2)^2} = \frac{A}{x - 3} + \frac{B}{x - 2} + \frac{C}{(x - 2)^2}.$$

This decomposition implies that for all x,

$$1 = A(x - 2)^2 + B(x - 2)(x - 3) + C(x - 3),$$

or

$$1 = (A + B)x^2 + (-4A - 5B + C)x + (4A + 6B - 3C).$$

By comparison of coefficients, $A + B = 0$, $-4A - 5B + C = 0$, $4A + 6B - 3C = 1$. This yields $A = 1$, $B = -1$, $C = -1$. Hence

$$\frac{1}{(x - 3)(x - 2)^2} = \frac{1}{x - 3} - \frac{1}{x - 2} - \frac{1}{(x - 2)^2}$$

$$= \left(\frac{-1}{3}\right)\frac{1}{1 - (x/3)} + \left(\frac{1}{2}\right)\frac{1}{1 - (x/2)} + \left(\frac{-1}{4}\right)\frac{1}{(1 - (x/2))^2}$$

$$= \left(\frac{-1}{3}\right)\sum_{i=0}^{\infty}\left(\frac{x}{3}\right)^i + \left(\frac{1}{2}\right)\sum_{i=0}^{\infty}\left(\frac{x}{2}\right)^i$$

$$+ \left(\frac{-1}{4}\right)\left[\binom{-2}{0} + \binom{-2}{1}\left(\frac{-x}{2}\right) + \binom{-2}{2}\left(\frac{-x}{2}\right)^2 + \cdots\right].$$

The coefficient of x^8 is $(-1/3)(1/3)^8 + (1/2)(1/2)^8 + (-1/4)\binom{-2}{8}(-1/2)^8 = -[(1/3)^9 + 7(1/2)^{10}]$. □

EXAMPLE 10.14 Use generating functions to determine how many four-element subsets of $S = \{1, 2, 3, \ldots, 15\}$ contain no consecutive integers.

a) Consider one such subset, say 1, 3, 7, 10, and write $1 \leq 1 < 3 < 7 < 10 \leq 15$. We see that this set of inequalities determines the differences $1 - 1 = 0$, $3 - 1 = 2$, $7 - 3 = 4$, $10 - 7 = 3$, $15 - 10 = 5$, and these differences sum to 14. Considering another such subset, say 2, 5, 11, 15, we write $1 \leq 2 < 5 < 11 < 15 \leq 15$; these inequalities yield the differences 1, 3, 6, 4, 0, which also sum to 14.

These examples suggest a one-to-one correspondence between the four-element subsets to be counted and the integer solutions to $c_1 + c_2 + c_3 + c_4 + c_5 = 14$, $0 \leq c_1, c_5$, $2 \leq c_2, c_3, c_4$. The answer is the coefficient of x^{14} in

$$f(x) = (1 + x + x^2 + x^3 + \cdots)(x^2 + x^3 + x^4 + \cdots)^3 \cdot$$
$$(1 + x + x^2 + x^3 + \cdots) = x^6(1 - x)^{-5}.$$

This then is the coefficient of x^8 in $(1 - x)^{-5}$, which is $\binom{-5}{8}(-1)^8 = \binom{5 + 8 - 1}{8} = \binom{12}{8} = 495$.

b) Another way to look at the problem is as follows.

For the subset 1, 3, 7, 10, we examine the strict inequalities $0 < 1 < 3 < 7 < 10 < 16$ and consider how many integers there are between any consecutive pair of them. Here we get 0, 1, 3, 2, 5: 0 since there is no integer between 0 and 1, 1 for the integer 2 between 1 and 3, 3 for the integers 4, 5, 6 between 3 and 7, etc. These 5 integers sum to 11. Doing the same thing to the subset 2, 5, 11, 15, the strict inequalities $0 < 2 < 5 < 11 < 15 < 16$ yield the results 1, 2, 5, 3, 0 which sum to 11.

These results suggest a one-to-one correspondence with the integer solutions to $b_1 + b_2 + b_3 + b_4 + b_5 = 11$, $0 \le b_1$, b_5, $1 \le b_2$, b_3, b_4. The number of these solutions is the coefficient of x^{11} in

$$g(x) = (1 + x + x^2 + \cdots)(x + x^2 + x^3 + \cdots)^3(1 + x + x^2 + \cdots)$$

$$= x^3(1 - x)^{-5}.$$

The answer is $\binom{-5}{8}(-1)^8 = 495$, as above. [The reader may wish to look back at Miscellaneous Exercise 7 in Chapter 2.]   □

EXERCISES

1. Find generating functions for the following sequences:

a) $\binom{8}{0}, \binom{8}{1}, \binom{8}{2}, \ldots, \binom{8}{8}$ b) $\binom{8}{1}, 2\binom{8}{2}, 3\binom{8}{3}, \ldots, 8\binom{8}{8}$

c) $1, -1, 1, -1, 1, -1, \ldots$ d) $0, 0, 0, 1, 1, 1, 1, 1, 1, \ldots$

e) $0, 0, 0, 6, -6, 6, -6, 6, \ldots$ f) $1, 0, 1, 0, 1, 0, 1, \ldots$

g) $1, 2, 4, 8, 16, \ldots$ h) $0, 0, 1, a, a^2, a^3, \ldots, a \ne 0$

2. Determine the constant in $(3x^2 - (2/x))^{15}$.

3. a) Find the coefficient of x^7 in $(1 + x + x^2 + x^3 + \cdots)^{15}$.

   b) Find the coefficient of x^7 in $(1 + x + x^2 + x^3 + \cdots)^n$, $n \in \mathbf{Z}^+$.

4. Find the coefficient of x^{50} in $(x^7 + x^8 + x^9 + \cdots)^6$.

5. Find the coefficient of x^{20} in $(x^2 + x^3 + x^4 + x^5 + x^6)^5$.

6. For $n \in \mathbf{Z}^+$, find in $(1 + x + x^2)(1 + x)^n$ the coefficient of (a) x^7; (b) x^8; and (c) x^r, $0 \le r \le n + 2$, $r \in \mathbf{Z}$.

7. Find the coefficient of x^{15} in each of the following:

a) $x^3(1 - 2x)^{10}$ b) $(x^3 - 5x)/(1 - x)^3$ c) $(1 + x)^4/(1 - x)^4$

8. In how many ways can two dozen identical robots be assigned to four assembly lines with (a) at least three robots assigned to each line? (b) at least three, but no more than nine, robots assigned to each line?

9. In how many ways can 3000 identical envelopes be divided up, in packages of 25, among four student groups so that each group gets at least 150, but not more than 1000, of the envelopes?

10. Two cases of soft drinks, 24 bottles of one type and 24 of another, are distributed among five surveyors who are conducting taste tests. In how many ways can the 48 bottles be distributed so that each surveyor gets (a) at least two bottles of each type? (b) at least two bottles of one particular type, and at least three of the other?

11. If a die is rolled 12 times, what is the probability that the sum of the rolls is 30?

12. Carol is collecting money from her cousins to have a party for her aunt. If eight of the cousins each promise to give $2, $3, $4, or $5 and two others each give $5 or $10, what is the probability she collects exactly $40?

13. In how many ways can Traci select n marbles from a large supply of blue, red, and yellow marbles if the selection must include an even number of blue ones?

14. How can Mary split up 12 hamburgers and 16 hot dogs among her sons Richard, Peter, Christopher, and James so that James gets at least one hamburger and three hot dogs, while his brothers each get at least two hamburgers but at most five hot dogs?

15. Verify that $(1 - x - x^2 - x^3 - x^4 - x^5 - x^6)^{-1}$ is the generating function for the number of ways the sum n, $n \in \mathbf{Z}^+$, can be obtained when a single die is rolled an arbitrary number of times.

16. Show that $(1 - 4x)^{-1/2}$ generates the sequence $\binom{2n}{n}$, $n \in \mathbf{Z}^+$.

17. Find a, b, k so that $(a + bx)^k$ is the generating function for the sequence 1, $2/5$, $12/25$, $88/125$, ...

18. Find the first five terms in the following expansions:
    a) $(1 + 2x)^{1/3}$ b) $(1 - 3x)^{1/2}$ c) $(2x - 1)^{-1/3}$ d) $(2 - x)^{1/4}$

19. In how many ways can we select seven nonconsecutive integers from $\{1, 2, 3, \ldots, 50\}$?

10.3 ■ PARTITIONS OF INTEGERS

In the study of number theory we are confronted with partitioning a positive integer n into positive summands and seeking the number of such partitions. This number is denoted by $p(n)$. For example,

$p(1) = 1$: 1

$p(2) = 2$: $2 = 1 + 1$

$p(3) = 3$: $3 = 2 + 1 = 1 + 1 + 1$ (Order is not important.)

$p(4) = 5$: $4 = 3 + 1 = 2 + 2 = 2 + 1 + 1 = 1 + 1 + 1 + 1$.

We should like to obtain $p(n)$ for a given n without writing all the partitions. A tool is needed to keep track of the number of 1's, 2's, ..., n's that are used as summands for n.

If $n \in \mathbf{Z}^+$, the number of 1's we can use is 0 or 1 or 2 or .... The series $1 + x + x^2 + x^3 + x^4 + \cdots$ keeps account of this for us. In like manner, $1 + x^2 + x^4 + x^6 + \cdots$ keeps track of the number of 2's in the partition of n. For example, to determine $p(10)$ we want the coefficient of x^{10} in $f(x) = (1 + x + x^2 + x^3 + \cdots) \cdot (1 + x^2 + x^4 + x^6 + \cdots) \cdots (1 + x^{10} + x^{20} + \cdots)$, or in $g(x) = (1 + x + x^2 + x^3 + \cdots + x^{10})(1 + x^2 + x^4 + \cdots + x^{10}) \cdots (1 + x^{10})$.

We prefer to work with $f(x)$, since it can be written in the more compact form

$$f(x) = \frac{1}{(1 - x)} \frac{1}{(1 - x^2)} \cdots \frac{1}{(1 - x^{10})} = \prod_{i=1}^{10} \frac{1}{(1 - x^i)}.$$

If this product is extended beyond $i = 10$, we get $P(x) = \prod_{i=1}^{\infty}[1/(1 - x^i)]$, which generates the sequence $p(0), p(1), p(2), p(3), \ldots$, where we define $p(0) = 1$.

Though $P(x)$ is theoretically complete, it is difficult to calculate an infinite product of terms. If we consider only $\prod_{i=1}^{r}[1/(1 - x^i)]$ for some fixed r, then the coefficient of x^n is the number of partitions of n into summands that do not exceed r.

Despite the difficulty in calculating $p(n)$ from $P(x)$ for large values of n, the idea of the generating function will be useful in studying certain kinds of partitions.

EXAMPLE 10.15   Find the generating function for the number of ways an advertising agent can purchase n minutes ($n \in \mathbf{Z}^+$) of air time if time slots come in blocks of 30, 60, or 120 seconds.

Let 30 seconds represent one time unit. Then the answer is the number of integer solutions to the equation $a + 2b + 4c = 2n$, $0 \le a, b, c$.

The associated generating function is

$$f(x) = (1 + x + x^2 + \cdots)(1 + x^2 + x^4 + \cdots)(1 + x^4 + x^8 + \cdots)$$

$$= \frac{1}{1 - x}\frac{1}{1 - x^2}\frac{1}{1 - x^4},$$

and the coefficient of x^{2n} is the number of partitions of $2n$ into 1's, 2's, and 4's, the answer to the problem.   □

EXAMPLE 10.16   Find the generating function for $p_d(n)$, the number of partitions of a positive integer n into *distinct* summands.

Before we start, let us consider the 11 partitions of 6:

1) $1 + 1 + 1 + 1 + 1 + 1$ 2) $1 + 1 + 1 + 1 + 2$ 3) $1 + 1 + 1 + 3$

4) $1 + 1 + 4$ 5) $1 + 1 + 2 + 2$ 6) $1 + 5$

7) $1 + 2 + 3$ 8) $2 + 2 + 2$ 9) $2 + 4$

10) $3 + 3$ 11) 6

Partitions (6), (7), (9), and (11) have distinct summands, so $p_d(6) = 4$.

In calculating $p_d(n)$, for any $k \in \mathbf{Z}^+$ there are two choices: either k is not used as one of the summands of n, or it is. This can be accounted for by the polynomial $1 + x^k$, and consequently, the generating function for these partitions is

$$P_d(x) = (1 + x)(1 + x^2)(1 + x^3)\cdots = \prod_{i=1}^{\infty}(1 + x^i).$$

For any $n \in \mathbf{Z}^+$, $p_d(n)$ is the coefficient of x^n in $(1 + x)(1 + x^2)\cdots(1 + x^n)$. (We define $p_d(0) = 1$.) When $n = 6$, the coefficient of x^6 in $(1 + x)(1 + x^2)\cdots$ $(1 + x^6)$ is 4.   □

EXAMPLE 10.17   Considering the partitions in Example 10.16, we see that the number of partitions of 6 into odd summands is 4, which is also $p_d(6)$. Is this a coincidence?

Let $p_o(n)$ denote the number of partitions of n into odd summands, when $n \ge 1$.

We define $p_o(0) = 1$. The generating function for the sequence $p_o(0), p_o(1), p_o(2), \ldots$ is given by

$$P_o(x) = (1 + x + x^2 + x^3 + \cdots)(1 + x^3 + x^6 + \cdots)(1 + x^5 + x^{10} + \cdots) \cdot$$

$$(1 + x^7 + \cdots) \cdots = \frac{1}{1 - x}\frac{1}{1 - x^3}\frac{1}{1 - x^5}\frac{1}{1 - x^7} \cdots.$$

Now as

$$1 + x = \frac{1 - x^2}{1 - x}, \qquad 1 + x^2 = \frac{1 - x^4}{1 - x^2}, \qquad 1 + x^3 = \frac{1 - x^6}{1 - x^3}, \qquad \cdots,$$

we have

$$P_d(x) = (1 + x)(1 + x^2)(1 + x^3)(1 + x^4) \cdots$$

$$= \frac{1 - x^2}{1 - x}\frac{1 - x^4}{1 - x^2}\frac{1 - x^6}{1 - x^3}\frac{1 - x^8}{1 - x^4} \cdots = \frac{1}{1 - x}\frac{1}{1 - x^3} \cdots = P_o(x).$$

From the equality of the generating functions $p_d(n) = p_o(n)$, for all $n \geq 0$. ☐

We close this section with an idea called the *Ferrer's graph*. This graph uses rows of dots to represent a partition of an integer where the number of dots per row does not increase as we go from any row to the one below it.

In Fig. 10.1 we find the Ferrer's graphs for two partitions of 14: (a) $4 + 3 + 3 + 2 + 1 + 1$, and (b) $6 + 4 + 3 + 1$. The graph in (b) is said to be the *transposition* of the graph in (a), and vice versa, since one graph can be obtained from the other by interchanging rows and columns.

Figure 10.1

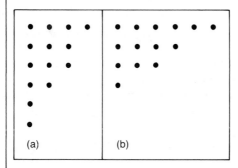

(a)   (b)

These graphs often suggest results about partitions. Here we see a partition of 14 into summands, where 4 is the largest summand, and a second partition of 14 into exactly four summands. As there is a one-to-one correspondence between a Ferrer's graph and its transposition, this example demonstrates a particular instance of the general result: the number of partitions of an integer n into m parts is equal to the number of partitions of n into summands where m is the largest summand.

EXERCISES

1. Find all partitions of 5 and 7.

2. Determine the generating function for the sequence $a_0, a_1, a_2, \ldots$ where a_n is the number of partitions of the integer n into (a) even summands; (b) distinct even summands; and (c) distinct odd summands.

3. In $f(x) = [1/(1 - x)][1/(1 - x^2)][1/(1 - x^3)]$, the coefficient of x^6 is 7. Interpret this result in terms of partitions of 6.

4. Find the generating function for the number of integer solutions of
   a) $2w + 3x + 5y + 7z = n$, $0 \le w, x, y, z.$
   b) $2w + 3x + 5y + 7z = n$, $0 \le w, \ 4 \le x, y, \ 5 \le z.$
   c) $2w + 3x + 5y + 7z = n$, $2 \le w \le 4 \le x \le 7 \le y \le 10 \le z.$

5. Find a generating function for the number of partitions of the integer n into summands where (a) each summand must appear an even number of times; (b) each summand must be even.

6. Find a generating function for the number of partitions of n into summands that (a) cannot occur more than five times; (b) cannot exceed 12 and cannot occur more than five times.

7. Determine the generating function for the number of ways to have n cents, $0 \le n \le 99$, in pennies, nickels, dimes, quarters, and half dollars.

8. Verify the identity

$$\frac{1}{1 - x} = (1 + x + x^2 + \cdots + x^9)(1 + x^{10} + x^{20} + \cdots + x^{90}) \cdot$$
$$(1 + x^{100} + x^{200} + \cdots + x^{900}) \cdots,$$

   and interpret the result in terms of partitions of an integer $n \ge 0$.

9. Using a Ferrer's graph, show that the number of partitions of an integer n into summands not exceeding m is equal to the number of partitions of n into at most m summands.

10. Using a Ferrer's graph, show that the number of partitions of n is equal to the number of partitions of $2n$ into n summands.

10.4 ■ THE EXPONENTIAL GENERATING FUNCTION

The generating function we have been dealing with is often referred to as the *ordinary* generating function of a given sequence. This function arose in selection problems, where order was not relevant. However, turning now to problems of arrangement, where order is crucial, we seek a comparable tool. To find such a tool we return to the binomial theorem.

For any $n \in \mathbf{Z}^+$, $(1 + x)^n = \binom{n}{0} + \binom{n}{1}x + \binom{n}{2}x^2 + \cdots + \binom{n}{n}x^n$, so $(1 + x)^n$ is the (ordinary) generating function for the sequence $\binom{n}{0}, \binom{n}{1}, \binom{n}{2}, \ldots, \binom{n}{n}, 0, 0, \ldots$.

When dealing with this idea in Chapter 1 we also wrote $\binom{n}{r} = C(n, r)$ when we wanted to emphasize that $\binom{n}{r}$ represented the number of combinations of n objects taken r at a time, $0 \le r \le n$. Consequently, $(1 + x)^n$ generates the sequence $C(n, 0)$, $C(n, 1)$, $C(n, 2)$, $\ldots$, $C(n, n)$, 0, 0, $\ldots$. Now for any $0 \le r \le n$,

$$C(n, r) = \frac{n!}{r!\,(n - r)!} = \left(\frac{1}{r!}\right) P(n, r),$$

so

$$(1 + x)^n = C(n, 0) + C(n, 1)x + C(n, 2)x^2 + C(n, 3)x^3 + \cdots + C(n, n)x^n$$

$$= P(n, 0) + P(n, 1)x + P(n, 2)\frac{x^2}{2!} + P(n, 3)\frac{x^3}{3!} + \cdots + P(n, n)\frac{x^n}{n!}.$$

So if in $(1 + x)^n$ we consider the coefficients of $x^r/r!$, $0 \le r \le n$, we obtain $P(n, r)$. Based on this observation we have the following.

Definition 10.2

▶   For a sequence a_0, a_1, a_2, a_3, $\ldots$ of real numbers,

$$f(x) = a_0 + a_1 x + a_2 \frac{x^2}{2!} + a_3 \frac{x^3}{3!} + \cdots = \sum_{i=0}^{\infty} a_i \frac{x^i}{i!}$$

is called the *exponential generating function* for the given sequence.

EXAMPLE 10.18   Examining the expansion for e^x we find

$$e^x = 1 + x + \frac{x^2}{2!} + \frac{x^3}{3!} + \frac{x^4}{4!} + \cdots = \sum_{i=0}^{\infty} \frac{x^i}{i!},$$

so e^x is the exponential generating function for the sequence 1, 1, 1, $\ldots$. (e^x is the ordinary generating function for the sequence 1, 1, $1/2!$, $1/3!$, $1/4!$, $\ldots$.)   □

Our next example shows how this idea can help us count certain types of arrangements.

EXAMPLE 10.19   In how many ways can four of the letters from ENGINE be arranged?

Table 10.2

E E N N	4!/(2!2!)	E G N N	4!/2!
E E G N	4!/2!	E I N N	4!/2!
E E I N	4!/2!	G I N N	4!/2!
E E G I	4!/2!	E I G N	4!

In Table 10.2 we list the possible selections of size 4 from the letters E, N, G, I, N, E, along with the number of permutations those four letters determine.

We now obtain the answer by means of an exponential generating function. For the letter E, we use $[1 + x + (x^2/2!)]$, since there are 0, 1, or 2 E's to arrange. Note that the coefficient of $x^2/2!$ is 1, the number of distinct ways to arrange two E's. In like manner, we have $[1 + x + (x^2/2!)]$ for the arrangements of 0, 1, or 2 N's. The arrangements for each of the letters G and I is represented by $(1 + x)$.

Consequently, the exponential generating function is $f(x) = [1 + x + (x^2/2!)]^2 \cdot (1 + x)^2$, and the answer is the coefficient of $x^4/4!$ in $f(x)$. In computing $f(x)$, the term involving x^4 is

$$\left(\frac{x^4}{2!2!} + \frac{x^4}{2!} + \frac{x^4}{2!} + \frac{x^4}{2!} + \frac{x^4}{2!} + \frac{x^4}{2!} + \frac{x^4}{2!} + x^4 \right)$$

$$= \left[\left(\frac{4!}{2!2!} \right) + \left(\frac{4!}{2!} \right) + \left(\frac{4!}{2!} \right) + \left(\frac{4!}{2!} \right) + \left(\frac{4!}{2!} \right) + \left(\frac{4!}{2!} \right) + \left(\frac{4!}{2!} \right) + 4! \right] \left(\frac{x^4}{4!} \right),$$

and the coefficient of $x^4/4!$ is the answer (102 arrangements) listed in the table.   □

EXAMPLE 10.20   Consider the expansions of e^x and e^{-x}.

$$e^x = 1 + x + \frac{x^2}{2!} + \frac{x^3}{3!} + \frac{x^4}{4!} + \cdots \qquad e^{-x} = 1 - x + \frac{x^2}{2!} - \frac{x^3}{3!} + \frac{x^4}{4!} - \cdots$$

Adding the series

$$e^x + e^{-x} = 2 \left(1 + \frac{x^2}{2!} + \frac{x^4}{4!} + \cdots \right),$$

or

$$\frac{e^x + e^{-x}}{2} = 1 + \frac{x^2}{2!} + \frac{x^4}{4!} + \cdots.$$

Subtracting e^{-x} from e^x yields

$$\frac{e^x - e^{-x}}{2} = x + \frac{x^3}{3!} + \frac{x^5}{5!} + \cdots.$$ □

EXAMPLE 10.21   A ship carries 48 flags, 12 each of red, white, blue, and black. Twelve of these flags are placed on a vertical pole in order to communicate a signal to other ships. How many of these signals use an even number of blue and an odd number of black flags?

The exponential generating function

$$f(x) = \left(1 + x + \frac{x^2}{2!} + \frac{x^3}{3!} + \cdots \right)^2 \left(1 + \frac{x^2}{2!} + \frac{x^4}{4!} + \cdots \right) \left(x + \frac{x^3}{3!} + \frac{x^5}{5!} + \cdots \right)$$

considers all such signals made up of n flags, where $n \geq 1$. The last two factors in $f(x)$ restrict the signals to an even number of blue flags and an odd number of black flags, respectively.

As

$$f(x) = (e^x)^2 \left(\frac{e^x + e^{-x}}{2} \right) \left(\frac{e^x - e^{-x}}{2} \right) = \left(\frac{1}{4} \right) (e^{2x}) (e^{2x} - e^{-2x}) = \frac{1}{4} (e^{4x} - 1)$$

$$= \frac{1}{4} \left(\sum_{i=0}^{\infty} \frac{(4x)^i}{i!} - 1 \right) = \left(\frac{1}{4} \right) \sum_{i=1}^{\infty} \frac{(4x)^i}{i!},$$

the coefficient of $x^{12}/12!$ in $f(x)$ yields $(1/4)(4^{12}) = 4^{11}$ signals made up of 12 flags with an even number of blue and an odd number of black flags.   □

Our final example is reminiscent of past results.

EXAMPLE 10.22 A company hires 11 new employees. Each of these employees is to be assigned to one of four subdivisions with each subdivision getting at least one new employee. In how many ways can these assignments be made?

Calling the subdivisions A, B, C, D, we can equivalently count the number of 11-letter sequences in which there is at least one occurrence of each of the letters A, B, C, D. The exponential generating function for these arrangements is

$$f(x) = \left(x + \frac{x^2}{2!} + \frac{x^3}{3!} + \frac{x^4}{4!} + \cdots\right)^4 = (e^x - 1)^4 = e^{4x} - 4e^{3x} + 6e^{2x} - 4e^x + 1.$$

The answer then is the coefficient of $x^{11}/11!$ in $f(x)$. This is

$$4^{11} - 4(3^{11}) + 6(2^{11}) - 4(1^{11}) = \sum_{i=0}^{4}(-1)^i\binom{4}{i}(4 - i)^{11}.$$

This form of the answer should bring to mind some of the enumeration problems in Chapter 3. Once the vocabulary is set aside, we are counting the number of onto functions $g: X \rightarrow Y$ where $|X| = 11$, $|Y| = 4$.   □

EXERCISES

1. Find the exponential generating function for each of the following.

   a) $1, -1, 1, -1, 1, -1, \ldots$ b) $1, 2, 2^2, 2^3, 2^4, \ldots$
   c) $1, -a, a^2, -a^3, a^4, \ldots, a \in \mathbf{R}.$ d) $1, a^2, a^4, a^6, \ldots, a \in \mathbf{R}.$
   e) $a, a^3, a^5, a^7, \ldots, a \in \mathbf{R}.$ f) $0, 1, 2(2), 3(2^2), 4(2^3), \ldots$

2. a) For the ship in Example 10.21, how many signals use at least one flag of each color? (Solve this with an exponential generating function.)

   b) Restate part (a) in an alternative way that uses the concept of an onto function.

   c) How many signals are there in Example 10.21 where the total number of blue and black flags is even?

3. Find the exponential generating function for the number of ways to arrange n letters, $n \geq 0$, selected from each of the following words:

   a) HAWAII   b) MISSISSIPPI   c) ISOMORPHISM

4. For Exercise 3(b), what is the exponential generating function if the arrangement must contain at least two I's?

5. If the company in Example 10.22 hires 25 new employees, give the exponential generating function for the number of ways to assign these people to the four subdivisions so that each subdivision receives at least 3 but no more than 10 new people.

6. Given the sequences $a_0, a_1, a_2, \ldots,$ and $b_0, b_1, b_2, \ldots,$ with exponential generating functions $f(x), g(x)$, respectively, show that if $h(x) = f(x)g(x)$, then $h(x)$ is the exponential generating function of the sequence $c_0, c_1, c_2, \ldots,$ where $c_n = \sum_{i=0}^{n} \binom{n}{i} a_i b_{n-i}$, for each $n \geq 0$.

7. If a 20-digit ternary (0, 1, 2) sequence is randomly generated, what is the probability that:

   a) It has an even number of 1's?

   b) It has an even number of 1's and an even number of 2's?

   c) It has an odd number of 0's?

   d) The total number of 0's and 1's is odd?

   e) The total number of 0's and 1's is even?

8. How many 20-digit quaternary (0, 1, 2, 3) sequences are there where:

   a) There is at least one 2 and an odd number of 0's?

   b) No symbol occurs exactly twice?

   c) No symbol occurs exactly three times?

   d) There are exactly two 3's or none at all?

9. Twenty-five contracts (for 25 different shuttle components) are to be distributed among five companies, $c_1, c_2, \ldots, c_5$. Find the exponential generating function for the number of ways to award these contracts so that (a) company c_1 gets at least five contracts and all other companies get at least two; (b) every company gets at least one contract, company c_2 gets more contracts than company c_1, and company c_2 gets at most five contracts.

10.5 ■ THE SUMMATION OPERATOR

This final section introduces a technique that helps us go from the (ordinary) generating function for the sequence $a_0, a_1, a_2, \ldots,$ to the generating function for the sequence $a_0, a_0 + a_1, a_0 + a_1 + a_2, \ldots$.

For $f(x) = a_0 + a_1x + a_2x^2 + a_3x^3 + \cdots$, consider the function $f(x)/(1 - x)$.

$$f(x)/(1 - x) = [a_0 + a_1x + a_2x^2 + a_3x^3 + \cdots][1 + x + x^2 + x^3 + \cdots]$$

$$= a_0 + (a_0 + a_1)x + (a_0 + a_1 + a_2)x^2$$
$$+ (a_0 + a_1 + a_2 + a_3)x^3 + \cdots,$$

so $f(x)/(1 - x)$ generates the sequence of sums $a_0, a_0 + a_1, a_0 + a_1 + a_2, a_0 + a_1 + a_2 + a_3, \ldots$.

We find this handy in the following.

EXAMPLE 10.23   Find a formula to express $0^2 + 1^2 + 2^2 + \cdots + n^2$ as a function of n.

As in Section 10.2, we start with $g(x) = 1/(1 - x) = 1 + x + x^2 + \cdots$.

Then

$$(-1)(1 - x)^{-2}(-1) = \frac{1}{(1 - x)^2} = \frac{dg(x)}{dx} = 1 + 2x + 3x^2 + 4x^3 + \cdots,$$

so $x/(1 - x)^2$ is the generating function for 0, 1, 2, 3, 4, ... Repeating this technique we find that

$$x\frac{d}{dx}\left[x\left(\frac{dg(x)}{dx}\right)\right] = \frac{x(1 + x)}{(1 - x)^3} = x + 2^2x^2 + 3^2x^3 + \cdots,$$

so $x(1 + x)/(1 - x)^3$ generates 0^2, 1^2, 2^2, 3^2, ... As a consequence of our earlier observation,

$$\frac{x(1 + x)}{(1 - x)^3}\frac{1}{(1 - x)} = \frac{x(1 + x)}{(1 - x)^4}$$

is the generating function for 0^2, $0^2 + 1^2$, $0^2 + 1^2 + 2^2$, $0^2 + 1^2 + 2^2 + 3^2$, ... Hence the coefficient of x^n in $x(1 + x)/(1 - x)^4$ is $\sum_{i=0}^{n} i^2$. But the coefficient of x^n in $x(1 + x)/(1 - x)^4$ can also be calculated as follows:

$$\frac{x(1 + x)}{(1 - x)^4} = (x + x^2)(1 - x)^{-4}$$

$$= (x + x^2)\left[\binom{-4}{0} + \binom{-4}{1}(-x) + \binom{-4}{2}(-x)^2 + \cdots\right],$$

where the coefficient of x^n is

$$\binom{-4}{n - 1}(-1)^{n-1} + \binom{-4}{n - 2}(-1)^{n-2}$$

$$= (-1)^{n-1}\left(\frac{4 + (n - 1) - 1}{n - 1}\right)(-1)^{n-1}$$

$$+ (-1)^{n-2}\left(\frac{4 + (n - 2) - 1}{n - 2}\right)(-1)^{n-2}$$

$$= \binom{n + 2}{n - 1} + \binom{n + 1}{n - 2} = \frac{(n + 2)!}{3!(n - 1)!} + \frac{(n + 1)!}{3!(n - 2)!}$$

$$= \tfrac{1}{6}[(n + 2)(n + 1)(n) + (n + 1)(n)(n - 1)]$$

$$= \tfrac{1}{6}(n)(n + 1)[(n + 2) + (n - 1)] = (n)(n + 1)(2n + 1)/6.$$

$\square$

EXERCISES

1. Continue the development of the ideas set forth in Example 10.23 and derive the formula $\sum_{i=0}^{n} i^3 = [n(n + 1)/2]^2$.

2. a) Find the generating function for the sequence of products $0 \cdot (-1)$, $1 \cdot 0$, $2 \cdot 1$, $3 \cdot 2$, $4 \cdot 3$, ..., $i \cdot (i - 1)$, ...

   b) Use the result from part (a) to obtain a formula for $\sum_{i=0}^{n} i(i - 1)$.

3. Let $f(x)$ be the generating function for the sequence a_0, a_1, a_2, ... For what sequence is $(1 - x)f(x)$ the generating function?

4. Let $f(x) = \sum_{i=0}^{\infty} a_i x^i$ with $f(1) = \sum_{i=0}^{\infty} a_i$, a finite number. Verify that the quotient $[f(x) - f(1)]/(x - 1)$ is the generating function for the sequence $s_0, s_1, s_2, \ldots,$ where $s_n = \sum_{i=n+1}^{\infty} a_i$.

10.6 ■ SUMMARY AND HISTORICAL REVIEW

In the early thirteenth century the Italian mathematician Leonardo of Pisa (c. 1175–1250), in his *Liber Abaci*, originated the study of the sequence 0, 1, 1, 2, 3, 5, 8, 13, 21, . . . , which can be given recursively by $F_0 = 0, F_1 = 1, F_{n+2} = F_{n+1} + F_n, n \geq 0$. Since Leonardo was the son of Bonaccio, the sequence has come to be called the Fibonacci numbers. (Filius Bonaccii is the Latin form for son of Bonaccio.)

If we consider the formula

$$F_n = \frac{1}{\sqrt{5}}\left[\left(\frac{1 + \sqrt{5}}{2}\right)^n - \left(\frac{1 - \sqrt{5}}{2}\right)^n\right], \qquad n \geq 0,$$

we find $F_0 = 0, F_1 = 1, F_2 = 1, F_3 = 2, F_4 = 3, \ldots$. Yes, this formula determines each Fibonacci number as a function of n. (Here we have the general solution for the recursive Fibonacci relation.) This formula was not derived, however, until 1718, when Abraham DeMoivre (1667–1754) obtained the result from the generating function

$$f(x) = \frac{x}{1 - x - x^2} = \frac{1}{\sqrt{5}}\left[\frac{1}{1 - \left(\dfrac{1 + \sqrt{5}}{2}\right)x} - \frac{1}{1 - \left(\dfrac{1 - \sqrt{5}}{2}\right)x}\right].$$

Extending the existing techniques of the generating function, Leonhard Euler (1707–1783) advanced the study of the partitions of integers in his 1748 two-volume opus, *Introductio in Analysin Infinitorum*. With

$$P(x) = \frac{1}{1 - x}\frac{1}{1 - x^2}\frac{1}{1 - x^3}\cdots = \prod_{i=1}^{\infty}\frac{1}{1 - x^i},$$

we have the generating function for $p(0), p(1), p(2), \ldots$ where $p(n)$ is the number of partitions of n into positive summands.

In the latter part of the eighteenth century, further developments on generating functions arose in conjunction with ideas in probability theory, especially with what is now called the "moment generating function." These related notions were presented in their first complete treatment by the great scholar Pierre-Simon de Laplace (1749–1827) in his 1812 publication, *Théorie Analytique des Probabilités*.

Finally, we mention Norman Macleod Ferrers (1829–1903) after whom the diagram we called the Ferrer's graph is named.

For us the study of the ordinary and exponential generating functions provided a powerful technique that unified ideas found in Chapters 1, 3, 7, and 9. Extending our prior experience with polynomials to power series, and extending the binomial theorem to $(1 + x)^n$ for the cases where n need not be positive or even an integer, we found the necessary tools to compute the coefficients in these generating functions.

This was more than worth the effort, since the algebraic calculations we performed took into account all of the selection processes we were trying to consider. We also found that we had seen some generating functions in prior chapters and saw how they arose in the study of partitions.

For comparable coverage of the material presented in this chapter the interested reader should consult Chapter 2 of C. Liu [3] and Chapter 3 of A. Tucker [6]. An interesting survey article on generating functions, written by Richard P. Stanley, can be found in the text edited by G. Rota [5].

The reader interested in more on the theory of partitions should consult Chapter 10 of I. Niven and H. Zuckerman [4].

Finally, a great deal can be found about the moment generating function and its use in probability theory in Chapter 3 of H. Larson [2] and Chapter XI of the comprehensive work by W. Feller [1].

REFERENCES

1. Feller, William, *An Introduction to Probability Theory and Its Applications,* Volume I, 3rd ed., John Wiley & Sons, New York, 1968.

2. Larson, Harold J., *Introduction to Probability Theory and Statistical Inference,* 2nd ed., John Wiley & Sons, New York, 1969.

3. Liu, C. L., *Introduction to Combinatorial Mathematics,* McGraw-Hill, New York, 1968.

4. Niven, Ivan, and Zuckerman, Herbert, *An Introduction to the Theory of Numbers,* 3rd ed., John Wiley & Sons, New York, 1960.

5. Rota, Gian-Carlo, ed., *Studies in Combinatorics,* Studies in Mathematics, Volume 17, The Mathematical Association of America, Washington, D.C., 1978.

6. Tucker, Alan, *Applied Combinatorics,* John Wiley & Sons, New York, 1980.

■ MISCELLANEOUS EXERCISES

1. Find the generating function for each of the following sequences:

   a) 7, 8, 9, 10, ...   b) 1, a, a^2, a^3, a^4, ..., $a \in \mathbf{R}$.

   c) 1, $(1 + a)$, $(1 + a)^2$, $(1 + a)^3$, ..., $a \in \mathbf{R}$.

   d) 2, $1 + a$, $1 + a^2$, $1 + a^3$, ..., $a \in \mathbf{R}$.

2. Let $S = \{a, b, c, d, e, f, g\}$.

   a) Find the generating function for the subsets of S. How many summands are in the coefficient of x^2 in this function?

   b) Find the generating function for the number of ways to select n objects, $0 \leq n \leq 14$, from S, where each object can be selected at most twice. How many summands are in the coefficient of x^2 in this function?

3. Use a partial fraction decomposition to verify that

$$\frac{x}{1 - x - x^2} = \frac{1}{\sqrt{5}}\left[\frac{1}{1 - \left(\dfrac{1 + \sqrt{5}}{2}\right)x} - \frac{1}{1 - \left(\dfrac{1 - \sqrt{5}}{2}\right)x}\right].$$

4. Find the coefficient of x^{83} in $f(x) = (x^5 + x^8 + x^{11} + x^{14} + x^{17})^{10}$.

5. Sergeant Bueti must distribute 40 bullets (20 for rifles, 20 for hand guns) among four patrolmen so that each patrolman gets at least two, but no more than seven, bullets of each type. In how many ways can he do this?

6. Find a generating function for the number of ways to partition a positive integer n into positive integer summands, where each summand appears an odd number of times or not at all.

7. How many 10-digit telephone numbers use only the digits 1, 3, 5, 7, with each digit appearing at least twice or not at all?

8. Use the first five terms of the binomial expansion of $(1 + x)^{1/3}$ to approximate the cube root of 2.

9. a) For what sequence of numbers is $g(x) = (1 - 2x)^{-5/2}$ the exponential generating function?

   b) Find a and b so that $(1 - ax)^b$ is the exponential generating function for the sequence $1, 7, 7 \cdot 11, 7 \cdot 11 \cdot 15, \ldots$.

10. For integers $n, k \ge 0$ let

    P_1 be the number of partitions of n;

    P_2 be the number of partitions of $2n + k$, where $n + k$ is the greatest summand; and,

    P_3 be the number of partitions of $2n + k$ into precisely $n + k$ summands.

    Using the concept of the Ferrer's graph, prove that $P_1 = P_2$, $P_2 = P_3$, thus concluding that the number of partitions of $2n + k$ into precisely $n + k$ summands is the same for all k.

11. In a rural area there are 12 mailboxes located at a general store.

    a) If a newsboy has 20 identical fliers, in how many ways can he distribute the fliers so that each mailbox gets at least one flier?

    b) If the mailboxes are in two rows of six, what is the probability that a distribution from part (a) will have 10 fliers distributed to the top six boxes and 10 to the bottom six?

12. Let S be a set containing n distinct objects. Verify that $e^x/(1 - x)^k$ is the exponential generating function for the number of ways to choose m of the objects in S, $0 \le m \le n$, and distribute these objects among k distinct containers, with the order of the objects in any container relevant for the distribution.

11

Recurrence
Relations

In various sections of the text we have seen a number of recursive definitions and constructions, where mathematical concepts of size $n + 1$ were obtained from comparable concepts of size n once we were able to establish such a concept for a first value of n, such as 0 or 1. These ideas fell back upon prior results in order to obtain present and future ones. Now we shall find ourselves doing the same thing once again. We shall investigate numeric functions $a(n)$, preferably written as a_n, $n \geq 0$, where a_n depends upon some of the prior terms $a_{n-1}, a_{n-2}, \ldots, a_1, a_0$. This study is called either *recurrence relations* or *difference equations,* and is the discrete counterpart to ideas in ordinary differential equations.

Our development will not employ any ideas from differential equations but will start with the notion of a geometric series. As further ideas are developed we shall see some of the many applications that make this topic so important.

11.1 ■ THE FIRST-ORDER LINEAR RECURRENCE RELATION

A *geometric series* is an infinite sequence of numbers, such as 5, 15, 45, 135, $\ldots$, where the division of any term, other than the first, by its immediate predecessor is a constant, called the *common ratio.* For our sequence this common ratio is 3 since $15 = 3(5)$, $45 = 3(15)$, etc. If $a_0, a_1, a_2, \ldots$ is a geometric series, then $a_1/a_0 = a_2/a_1 = \cdots = a_{n+1}/a_n = \cdots = r$, the common ratio. In this particular geometric series we have $a_{n+1} = 3a_n$, $n \geq 0$.

The *recurrence relation* $a_{n+1} = 3a_n$, $n \geq 0$, does not define a unique geometric series. The sequence 7, 21, 63, 189, $\ldots$ also satisfies the relation. To pinpoint the particular sequence described by $a_{n+1} = 3a_n$, we need to know one of the terms of that sequence. Hence,

$$a_{n+1} = 3a_n, \qquad n \geq 0, \qquad a_0 = 5,$$

uniquely defines the sequence 5, 15, 45, $\ldots$, whereas

$$a_{n+1} = 3a_n, \qquad n \geq 0, \qquad a_1 = 21,$$

identifies 7, 21, 63, $\ldots$ as the geometric series under study.

We call the equation $a_{n+1} = 3a_n$, $n \geq 0$, a recurrence relation because the value of a_{n+1} (the present consideration) is dependent on a_n (a prior consideration). As a_{n+1} depends only upon its immediate predecessor, the relation is said to be of *first order.* In particular this is a *first-order linear homogeneous* recurrence relation *with constant coefficients.* (We'll say more about these ideas later.) The general form of such an equation can be written $a_{n+1} - ca_n = 0$, $n \geq 0$, where c is a constant.

When values such as a_0 or a_1 are given in addition to the recurrence relation, we call these values *boundary conditions.* The expression $a_0 = k$, where k is a constant, is also referred to as an *initial condition.* Our examples show the importance of the boundary condition in determining the unique solution.

Let us return now to the recurrence relation

$$a_{n+1} = 3a_n, \qquad n \geq 0, \qquad a_0 = 5.$$

Writing out the first four terms of this sequence, we find that

$a_0 = 5,$

$a_1 = 3a_0 = 3(5),$

$a_2 = 3a_1 = 3(3a_0) = 3^2(5)$

$a_3 = 3a_2 = 3(3^2(5)) = 3^3(5).$

Consequently, for any $n \geq 0$, $a_n = 5(3^n)$. This is called the *general solution* of the given recurrence relation. In the general solution the value of a_n is a function of n and there is no longer any dependence on prior terms of the sequence. If we wish to compute a_{10}, we simply calculate $5(3^{10}) = 295,245$; there is no need to start at a_0 and build up to a_9 in order to obtain a_{10}.

From this example it follows that the general solution of the recurrence relation $a_{n+1} = ca_n$, $n \geq 0$, c a constant, is unique and is given by $a_n = a_0c^n$, $n \geq 0$.

Thus, the solution $a_n = a_0c^n$, $n \geq 0$, defines a discrete function whose domain is the set **N** of all nonnegative integers.

EXAMPLE 11.1

Solve the recurrence relation $a_n = 7a_{n-1}$, $n \geq 1$, $a_2 = 98$.

This is just an alternative form of the relation $a_{n+1} = 7a_n$, $n \geq 0$, $a_2 = 98$. Hence the general solution has the form $a_n = a_0(7^n)$. Since $a_2 = 98 = a_0(7^2)$, $a_0 = 2$, and $a_n = 2(7^n)$, $n \geq 0$, is the unique solution.   □

EXAMPLE 11.2

A bank pays 6% (annual) interest on savings, compounding the interest monthly. If Bonnie deposits $1000 on the first day of May, how much will this deposit be worth a year later?

Since the annual interest rate is 6%, the monthly rate is $6\%/12 = 0.5\% = 0.005$. For $0 \leq n \leq 12$, let p_n denote the value of Bonnie's deposit at the end of n months. Then $p_{n+1} = p_n + 0.005p_n$ (the interest earned on p_n during the month), $0 \leq n \leq 11$, $p_0 = \$1000$.

The relation $p_{n+1} = (1.005)p_n$, $p_0 = \$1000$, has the solution $p_n = p_0(1.005)^n = \$1000(1.005)^n$. Consequently, at the end of one year, Bonnie's deposit is worth $\$1000(1.005)^{12} = \$1061.68$.   □

The recurrence relation $a_{n+1} - ca_n = 0$ is called *linear* since each subscripted term appears to the first power (as x and y do in the equation of a line in the plane). Sometimes a nonlinear recurrence relation can be made into a linear one by an algebraic substitution.

EXAMPLE 11.3

Find a_{12} if $a_{n+1}^2 = 5a_n^2$, $a_n > 0$, $n \geq 0$, $a_0 = 2$.

Although this recurrence relation is not linear in a_n, if we let $b_n = a_n^2$, then the new relation $b_{n+1} = 5b_n$, $n \geq 0$, $b_0 = 4$ is a linear relation whose solution is $b_n = 4 \cdot 5^n$. Therefore, $a_n = 2(\sqrt{5})^n$, $n \geq 0$, and $a_{12} = 31,250$.   □

The general first-order linear recurrence relation with constant coefficients has the form $a_{n+1} + ca_n = f(n)$, $n \geq 0$, where c is a constant and $f(n)$ is a function on the set **N** of nonnegative integers.

When $f(n) = 0$ for all $n \in \mathbf{N}$, the relation is called *homogeneous;* otherwise the relation is called *nonhomogeneous*. So far we have dealt only with homogeneous relations. Now we shall solve a nonhomogeneous relation. We shall develop specific techniques that work for all linear homogeneous recurrence relations with constant coefficients. However, when dealing with a nonhomogeneous problem, many different techniques prove useful, although none allow us to solve everything than can arise. But if we can recognize some pattern from our past experiences, we are more likely to succeed.

EXAMPLE 11.4   In sorting numeric data, perhaps the most popular, though not the most efficient, method is a technique called the *bubble sort*. Starting with the last item, $A(n)$, in a given list we compare it with its immediate predecessor, $A(n - 1)$, and if $A(n - 1) > A(n)$ we interchange $A(n - 1)$ and $A(n)$. Then we compare $A(n - 1)$ (which may have originally been $A(n)$) against its immediate predecessor, $A(n - 2)$. If the original $A(n - 1)$ and $A(n)$ are such that $A(n - 1) \le A(n)$, we simply compare $A(n - 1)$ against $A(n - 2)$. If $A(n - 2) > A(n - 1)$ we interchange them. Continuing this process, after $n - 1$ comparisons the smallest number in the list is stored in $A(1)$. Then we repeat the process for the numbers now stored in $A(2), A(3), \ldots, A(n)$. In this way, each time this process is carried out, the smallest number in the remaining sublist "bubbles up" to the start of that sublist. (The BASIC program in Fig. 11.1 uses this process to sort in ascending order the 10 numbers supplied at line 50.)

Figure 11.1

```
10   DIM A(10)
20   FOR I = 1 TO 10
30       READ A(I)
40   NEXT I
50   DATA 2, 3, 7, 1, -5, 4, 2, 8, -3, 6
60   FOR J = 2 TO 10
70       FOR K = 10 TO J STEP -1
80           IF A(K) >= A(K - 1) THEN 110
90           B = A(K - 1)
100          A(K - 1) = A(K): A(K) = B
110      NEXT K
120  NEXT J
130  FOR I = 1 TO 10
140      PRINT A(I);
150  NEXT I
160  END

RUNNH
-5 -3 1 2 2 3 4 6 7 8
```

What we want to count here is the total number of comparisons made in order to sort (in ascending order) n given numbers.

If a_n denotes the number of comparisons needed to sort n numbers this way, then we get the following recurrence relation:

$$a_n = a_{n-1} + (n - 1), \qquad n \geq 1, \qquad a_1 = 0.$$

This arises as follows. Given a list of n numbers, we make $n - 1$ comparisons to bubble the smallest number up to the start of the list. The remaining sublist of $n - 1$ numbers then requires a_{n-1} comparisons in order to be completely sorted.

This relation is a linear first-order relation with constant coefficients, but the term $n - 1$ makes it nonhomogeneous. Since we have no technique for attacking such a relation, let us list some terms and see if there is a recognizable pattern.

$a_1 = 0$

$a_2 = a_1 + (2 - 1) = 1$

$a_3 = a_2 + (3 - 1) = 1 + 2$

$a_4 = a_3 + (4 - 1) = 1 + 2 + 3$

$\cdots \quad \cdots \qquad \cdots \quad \cdots$

In general $a_n = 1 + 2 + \cdots + (n - 1) = [(n - 1)n]/2 = (n^2 - n)/2$. (As a result, the bubble sort determines a function $f: \mathbf{Z}^+ \to \mathbf{R}$ given by $f(n) = a_n = (n^2 - n)/2$, which is a measure of the running time for the algorithm. The function f is dominated by $g: \mathbf{Z}^+ \to \mathbf{R}$ where $g(n) = n^2$ and $f \in \mathbb{O}(n^2)$, so the bubble sort is said to require $\mathbb{O}(n^2)$ comparisons. See Miscellaneous Exercises 28 (Chapter 3) and 11 (Chapter 5).) □

At this point we shall examine a recurrence relation with a variable coefficient.

EXAMPLE 11.5
Solve the relation $a_n = n \cdot a_{n-1}, n \geq 1, a_0 = 1$.

Writing the first five terms defined by the relation we have

$a_0 = 1$

$a_1 = 1 \cdot a_0 = 1$

$a_2 = 2 \cdot a_1 = 2 \cdot 1$

$a_3 = 3 \cdot a_2 = 3 \cdot 2 \cdot 1$

$a_4 = 4 \cdot a_3 = 4 \cdot 3 \cdot 2 \cdot 1$

$\cdots \qquad \cdots \qquad \cdots$

Therefore, $a_n = n!$ and the solution is the discrete numeric function $a_n, n \geq 0$, which counts the number of permutations of n objects. □

While on the subject of permutations, we close with a recursive algorithm for generating the permutations of $\{1, 2, 3, \ldots, n - 1, n\}$ from those for $\{1, 2, 3, \ldots, n - 1\}$.

There is only one permutation of $\{1\}$. Examining the permutations of $\{1, 2\}$,

1   2

2   1

we see that after writing the permutation 1 twice we intertwine the number 2 about 1 to get the permutations listed. Writing each of these two permutations three times we intertwine the number 3 and obtain

	1	2	3
	1	3	2
3	1		2
3	2		1
	2	3	1
	2	1	3

We see here that the first permutation is 123 and that each of the next two permutations is obtained from its immediate predecessor by interchanging two numbers: 3 and the integer to its left. When 3 reaches the left side of the permutation we examine the remaining numbers and permute them according to the list of permutations we generated for $\{1, 2\}$. (This makes the procedure recursive.) After that we interchange 3 with the integer on its right until 3 is on the right side of the permutation. (We note that if we interchange 1 and 2 in the last permutation we get 123, the first permutation listed.)

Table 11.1

	c1	c2	c3	c4	c5	c6	c7
(1)		1		2		3	4
(2)		1		2	4	3	
(3)		1	4	2		3	
(4)	4	1		2		3	
(5)	4	1		3		2	
(6)		1	4	3		2	
(7)		1		3	4	2	
(8)		1		3		2	4
(9)		3		1		2	4
(10)		3		1	4	2	
(11)		3	4	1		2	
...		...		...		...	
(15)		3		2	4	1	
(16)		3		2		1	4
(17)		2		3		1	4
...		...		...		...	
(22)		2	4	1		3	
(23)		2		1	4	3	
(24)		2		1		3	4

Continuing for $S = \{1, 2, 3, 4\}$, we first list the six permutations of $\{1, 2, 3\}$ four times. Starting with the permutation 1234 we intertwine the 4 throughout the remaining 23 permutations as indicated in Table 11.1. The only new idea here develops as follows. When progressing from permutation (5) to (6) to (7) to (8) we interchange 4 with the integer to its right. At permutation (8) where 4 has reached the right side, we obtain permutation (9) by keeping the location of 4 fixed and replacing the permutation 132 by 312 from the list of permutations of $\{1, 2, 3\}$. After that we continue as for the first eight permutations until we reach permutation (16) where 4 is again on the right. We then permute 321 to obtain 231 and continue intertwining 4 until all 24 permutations have been generated. Once again, if 1 and 2 are interchanged in the last permutation we obtain the first permutation in our list.

For more on recursive procedures in generating permutations and combinations the reader should consult the chapter references.

EXERCISES

1. Find the recurrence relation, with initial condition, that uniquely determines the geometric series

   a)   2, 10, 50, 250, ...   b)   6, −18, 54, −162, ...

   c)   1, 1/3, 1/9, 1/27, ...   d)   7, 14/5, 28/25, 56/125, ...

2. Find the general solution of the recurrence relations:

   a) $a_{n+1} - 1.5a_n = 0$, $n \geq 0$.   b) $4a_n - 5a_{n-1} = 0$, $n \geq 1$.

   c) $3a_{n+1} - 4a_n = 0$, $n \geq 0$, $a_1 = 5$.

   d) $2a_n - 3a_{n-1} = 0$, $n \geq 1$, $a_4 = 81$.

3. If a_n, $n \geq 0$, is a solution of the recurrence relation $a_{n+1} - ca_n = 0$, and $a_3 = 153/49$, $a_5 = 1377/2401$, what is c?

4. The number of bacteria in a culture is 1000 (approximately) and this number increases 250% every 2 hours. Use a recurrence relation to determine the number of bacteria present after one day.

5. If Laura invests $100 at 6% interest compounded quarterly, how many months must she wait for her money to double? (She cannot withdraw the money before the quarter is up.)

6. Paul invested the stock profits he received 15 years ago in an account that paid 8% interest compounded quarterly. If his account now has $7218.27 in it, what was his initial investment?

7. Find a_{10} if $a_n^3 = 7a_{n-1}^3$, $n \geq 1$, $a_0 = 3$.

8. Solve for a_n, $n \geq 2$, if $5na_n + 2na_{n-1} = 2a_{n-1}$, $n \geq 3$, $a_3 = -30$.

9. Let $x_1, x_2, \ldots, x_{20}$ be a list of distinct real numbers to be sorted by the bubble sort technique of Example 11.4. (a) After how many comparisons will the 10 smallest numbers of the original list be arranged in ascending order? (b) How many more comparisons are needed to finish this sorting job?

10. Complete the list of permutations shown in Table 11.1.

11. In generating the permutations of $\{1, 2, 3, 4, 5\}$ by the procedure developed after Example 11.5, (a) what is the last permutation in the list? (b) what two

permutations precede 25134? (c) what three permutations follow 25134? (d) where do the permutations 21543, 35421, and 31524 occur in the list of 120 permutations?

11.2 ■ THE SECOND-ORDER LINEAR HOMOGENEOUS RECURRENCE RELATION WITH CONSTANT COEFFICIENTS

Let $k \in \mathbf{Z}^+$ and $c_n \ (\neq 0), c_{n-1}, c_{n-2}, \ldots, c_{n-k} \ (\neq 0)$ be real numbers. If $a_n, n \geq 0$, is a discrete numeric function, then

$$c_n a_n + c_{n-1} a_{n-1} + c_{n-2} a_{n-2} + \cdots + c_{n-k} a_{n-k} = f(n), \qquad n \geq k,$$

is a linear recurrence relation (with constant coefficients) of *order k*. When $f(n) = 0$ for all $n \geq 0$, the relation is called *homogeneous,* otherwise it is *nonhomogeneous.*

In this section we shall concentrate on the homogeneous relation of order two: $c_n a_n + c_{n-1} a_{n-1} + c_{n-2} a_{n-2} = 0, n \geq 2$. Based on our work in Section 11.1, we seek a solution of the form $a_n = cr^n$, where $c \neq 0, r \neq 0$.

Substituting into $c_n a_n + c_{n-1} a_{n-1} + c_{n-2} a_{n-2} = 0$ we obtain

$$c_n cr^n + c_{n-1} cr^{n-1} + c_{n-2} cr^{n-2} = 0.$$

With $c, r \neq 0$, this becomes $c_n r^2 + c_{n-1} r + c_{n-2} = 0$, a quadratic equation. Hence the roots r_1, r_2 of this equation, which is called the *characteristic equation,* fall into the following three cases: (a) r_1, r_2 are distinct real numbers; (b) r_1, r_2 form a complex conjugate pair; or, (c) r_1, r_2 are real, but $r_1 = r_2$. In all cases r_1 and r_2 are called the *characteristic roots.*

CASE (A): (DISTINCT REAL ROOTS)

EXAMPLE 11.6   Solve the recurrence relation $a_n + a_{n-1} - 6a_{n-2} = 0, n \geq 2, a_0 = 1, a_1 = 2$.

If $a_n = cr^n, c, r \neq 0$, we obtain $cr^n + cr^{n-1} - 6cr^{n-2} = 0$ from which the characteristic equation $r^2 + r - 6 = 0$ follows.

$$0 = r^2 + r - 6 = (r + 3)(r - 2) \Rightarrow r = 2, -3.$$

Since we have two distinct real roots, $a_n = 2^n$ and $a_n = (-3)^n$ are both solutions. They are *linearly independent* since one is not a multiple of the other. So $a_n = c_1(2)^n + c_2(-3)^n$ denotes the general solution, where c_1, c_2 are arbitrary constants.

With $a_0 = 1, a_1 = 2, c_1$ and c_2 are determined as follows:

$$1 = a_0 = c_1(2)^0 + c_2(-3)^0 = c_1 + c_2$$

$$2 = a_1 = c_1(2)^1 + c_2(-3)^1 = 2c_1 - 3c_2.$$

Solving this system of equations one finds $c_1 = 1, c_2 = 0$. Therefore $a_n = 2^n$ is the *unique* solution of the given relation.   □

An interesting second-order homogeneous recurrence relation is the *Fibonnaci relation.*

EXAMPLE 11.7 Solve the recurrence relation $F_{n+2} = F_{n+1} + F_n$, $n \geq 0$, $F_0 = 0$, $F_1 = 1$.
As in the previous example, let $F_n = cr^n$, $c, r \neq 0$, $n \geq 0$. Upon substitution we get $cr^{n+2} = cr^{n+1} + cr^n$. This gives the characteristic equation $r^2 - r - 1 = 0$. The characteristic roots are $r = (1 \pm \sqrt{5})/2$, so the general solution is $F_n = c_1[(1 + \sqrt{5})/2]^n + c_2[(1 - \sqrt{5})/2]^n$.
To solve for c_1, c_2, we use the given initial values and write $0 = F_0 = c_1 + c_2$, $1 = F_1 = c_1[(1 + \sqrt{5})/2] + c_2[(1 - \sqrt{5})/2]$. Since $-c_1 = c_2$, we have $2 = c_1(1 + \sqrt{5}) - c_1(1 - \sqrt{5})$, and $c_1 = 1/\sqrt{5}$. The general solution is given by

$$F_n = \frac{1}{\sqrt{5}}\left[\left(\frac{1 + \sqrt{5}}{2}\right)^n - \left(\frac{1 - \sqrt{5}}{2}\right)^n\right].$$

□

EXAMPLE 11.8 For $n \geq 0$, let $S = \{1, 2, 3, \ldots, n\}$ (when $n = 0$, $S = \emptyset$), and let a_n denote the number of subsets of S that contain no consecutive integers. Find and solve a recurrence relation for a_n.
For $0 \leq n \leq 4$, we have $a_0 = 1$, $a_1 = 2$, $a_2 = 3$, $a_3 = 5$, $a_4 = 8$. (For example, $a_3 = 5$ since $S = \{1, 2, 3\}$ has $\emptyset$, $\{1\}$, $\{2\}$, $\{3\}$, $\{1, 3\}$ as subsets with no consecutive integers.) These first five terms are reminiscent of the Fibonacci sequence. But do things change as we continue?
Let $n \geq 2$, and $S = \{1, 2, 3, \ldots, n - 2, n - 1, n\}$. If $A \subseteq S$ and A is to be counted in a_n, there are two possibilities:

a) $n \in A$: When this happens $(n - 1) \notin A$, and $A - \{n\}$ would be counted in a_{n-2}.

b) $n \notin A$: For this case A would be counted in a_{n-1}.

Since these two cases are exhaustive and mutually disjoint, $a_n = a_{n-1} + a_{n-2}$, $n \geq 2$, $a_0 = 1$, $a_1 = 2$ is the recurrence relation for the problem. Now we could solve for a_n, but if we notice that $a_n = F_{n+2}$, $n \geq 0$, the result of Example 11.7 implies that

$$a_n = \frac{1}{\sqrt{5}}\left[\left(\frac{1 + \sqrt{5}}{2}\right)^{n+2} - \left(\frac{1 - \sqrt{5}}{2}\right)^{n+2}\right], \qquad n \geq 0.$$

□

Let us examine a comparable relation in a computer science application.

EXAMPLE 11.9 In many programming languages one may consider those legal arithmetic expressions, *without parentheses,* that are made up of the digits 0, 1, 2, $\ldots$, 9 and the binary operation symbols $+$, $*$, $/$. For example, $3 + 4$ and $2 + 3 * 5$ are legal arithmetic expressions; $8 + * 9$ is not. Here $2 + 3 * 5 = 17$, since there is a hierarchy of operations: multiplication and division are performed before addition. Operations at the same level are performed in their order of appearance as the expression is scanned from left to right.
For $n \in \mathbf{Z}^+$, let a_n be the number of these (legal) arithmetic expressions that are made up of n symbols. Then $a_1 = 10$, since the arithmetic expressions of one symbol are the 10 digits. Next $a_2 = 100$. This accounts for the expressions 00, 01, $\ldots$, 09, 10, 11, $\ldots$, 99. (There are no unnecessary leading plus signs.) When $n \geq 3$, we consider two cases for a_n:

1. If x is an arithmetic expression of $n - 1$ symbols, the last symbol must be a digit. Adding one more digit to the right of x we get $10a_{n-1}$ arithmetic expressions of n symbols where the last two symbols are digits.

2. Now let y be an expression of $n - 2$ symbols. To obtain an expression with n symbols (that is not counted in Case 1) we adjoin to the right of y one of the 29 two symbol expressions $+1, \ldots, +9, +0, *1, \ldots, *9, *0, /1, \ldots, /9$.

 From these two cases we have $a_n = 10a_{n-1} + 29a_{n-2}$, $n \geq 2$, $a_1 = 10$, $a_2 = 100$. Here the characteristic roots are $5 \pm 3\sqrt{6}$ and the solution is $a_n = (5/3\sqrt{6})[(5 + 3\sqrt{6})^n - (5 - 3\sqrt{6})^n]$. (Verify this.)   □

Now we turn to an application in the physical sciences.

EXAMPLE 11.10   Consider the linear network in Fig. 11.2, where there are k one-ohm and k three-ohm resistors connected by wires to a generator that supplies a voltage V. Find a formula for v_n, $0 \leq n \leq k$, that gives the voltage at each junction point as a function of n. (This is also the voltage drop across the one-ohm resistor below that junction point.)

 We know that $v_k = V$, the voltage supplied by the generator. To find a recurrence relation we apply the following principles:

1. *Kirchhoff's Law:* At any junction point the sum of the currents flowing into the junction point equals the sum of the currents leaving that point.

2. *Ohm's Law:* If the voltage drop across a resistor of R ohms is $V_2 - V_1$, then the current in that resistor is $(V_2 - V_1)/R$.

Figure 11.2

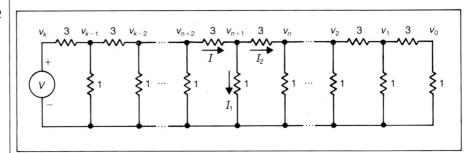

 Consider the current at the junction point labeled v_{n+1}. By Kirchhoff's law $I = I_1 + I_2$. By Ohm's law, $I = (v_{n+2} - v_{n+1})/3$, $I_1 = v_{n+1}/1$, $I_2 = (v_{n+1} - v_n)/3$. Consequently,

$$\frac{v_{n+2} - v_{n+1}}{3} = \frac{v_{n+1}}{1} + \frac{v_{n+1} - v_n}{3}, \qquad 0 \leq n \leq k - 2$$

or, $v_{n+2} - 5v_{n+1} + v_n = 0$.

 Let $v_n = cr^n$, $c, r \neq 0$, $0 \leq n \leq k$. The characteristic equation is $r^2 - 5r + 1 = 0$ with roots $\alpha = (5 + \sqrt{21})/2$, $\beta = (5 - \sqrt{21})/2$. From this, $v_n = c_1\alpha^n + c_2\beta^n$.

Now whatever v_0 is, by the two laws we have $(v_1 - v_0)/3 = v_0/1$, or $v_1 = 4v_0$. So $v_0 = c_1 + c_2$ and $4v_0 = v_1 = c_1\alpha + c_2\beta$. From this it follows that $c_1 = v_0(\beta - 4)/(\beta - \alpha)$, $c_2 = v_0(\alpha - 4)/(\alpha - \beta)$, and

$$v_n = v_0\left[\frac{(\beta - 4)}{(\beta - \alpha)}\alpha^n + \frac{(\alpha - 4)}{(\alpha - \beta)}\beta^n\right].$$

Since

$$V = v_k = v_0\left[\frac{(\beta - 4)\alpha^k - (\alpha - 4)\beta^k}{(\beta - \alpha)}\right],$$

we find $v_0 = (\beta - \alpha)V/[(\beta - 4)\alpha^k - (\alpha - 4)\beta^k]$ and this yields

$$v_n = V\left[\frac{(\beta - 4)\alpha^n - (\alpha - 4)\beta^n}{(\beta - 4)\alpha^k - (\alpha - 4)\beta^k}\right]. \qquad \square$$

The next example demonstrates the use of auxiliary variables.

EXAMPLE 11.11   Find a recurrence relation for the number of binary sequences of length n that have no consecutive 0's.

For $n \geq 0$, let a_n be the number of such sequences of length n. Let $a_n^{(0)}$ count those that end in 0 and $a_n^{(1)}$ those ending in 1. Hence $a_n = a_n^{(0)} + a_n^{(1)}$, and

$$a_n = 2 \cdot a_{n-1}^{(1)} + 1 \cdot a_{n-1}^{(0)}$$

The nth position can be 0 or 1.   The nth position can only be 1.

Consequently, $a_n = a_{n-1}^{(1)} + [a_{n-1}^{(1)} + a_{n-1}^{(0)}] = a_{n-1}^{(1)} + a_{n-1} = a_{n-1} + a_{n-2}$. (Why does $a_{n-1}^{(1)} = a_{n-2}$?) Therefore the recurrence relation for this problem is $a_n = a_{n-1} + a_{n-2}$, $n \geq 2$, $a_0 = 1$, $a_1 = 2$. (We leave the details of the solution for the reader.) $\square$

The last example for case (A) shows how to extend the results for second-order recurrence relations to those of higher order.

EXAMPLE 11.12   Solve the recurrence relation

$$2a_{n+3} = a_{n+2} + 2a_{n+1} - a_n, \qquad n \geq 0, \quad a_0 = 0, \quad a_1 = 1, \quad a_2 = 2.$$

Letting $a_n = cr^n$, $c, r \neq 0$, $n \geq 0$, we obtain the characteristic equation $2r^3 - r^2 - 2r + 1 = 0 = (2r - 1)(r - 1)(r + 1)$. The characteristic roots are $1/2$, 1, -1 and the solution is $a_n = c_1(1)^n + c_2(-1)^n + c_3(1/2)^n = c_1 + c_2(-1)^n + c_3(1/2)^n$. (The solutions 1, $(-1)^n$, $(1/2)^n$ are called linearly independent since it is impossible to get any one of them as a linear combination of the other two.) From $0 = a_0$, $1 = a_1$, $2 = a_2$, we get $c_1 = 5/2$, $c_2 = 1/6$, $c_3 = -8/3$. $\square$

CASE (B): (COMPLEX ROOTS)

Before getting into the complex case we recall DeMoivre's Theorem: $(\cos\theta + i\sin\theta)^n = \cos n\theta + i\sin n\theta$, $n \geq 0$. (This is Exercise 4 of Section 6.1.)

If $z = x + iy \in \mathbf{C}$, we can write $z = r(\cos\theta + i\sin\theta)$, where $r = \sqrt{x^2 + y^2}$ and $(y/x) = \tan\theta$. Then by DeMoivre's Theorem $z^n = r^n(\cos n\theta + i\sin n\theta)$, $n \geq 0$.

Figure 11.3

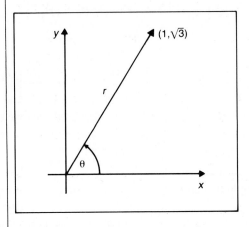

EXAMPLE 11.13

Determine $(1 + \sqrt{3}\,i)^{10}$. Fig. 11.3 shows a geometric way to represent the complex number $1 + \sqrt{3}\,i$ as the point $(1, \sqrt{3})$ in the xy-plane. Here $r = \sqrt{1^2 + (\sqrt{3})^2} = 2$, and $\theta = \pi/3$.

So $1 + \sqrt{3}\,i = 2(\cos(\pi/3) + i\sin(\pi/3))$ and $(1 + \sqrt{3}\,i)^{10} = 2^{10}(\cos(10\pi/3) + i\sin(10\pi/3)) = 2^{10}(\cos(4\pi/3) + i\sin(4\pi/3)) = 2^{10}((-1/2) - (\sqrt{3}/2)i) = (-2^9)(1 + \sqrt{3}\,i)$. $\square$

We'll use such results in the following examples.

EXAMPLE 11.14

Solve the recurrence relation $a_n = 2(a_{n-1} - a_{n-2})$, $n \geq 2$, $a_0 = 1$, $a_1 = 2$.

Letting $a_n = cr^n$, $c, r \neq 0$, we obtain the characteristic equation $r^2 - 2r + 2 = 0$, whose roots are $1 \pm i$. Consequently the general solution has the form $c_1(1 + i)^n + c_2(1 - i)^n$, where c_1 and c_2 presently denote arbitrary *complex* constants. (As in case (A), there are two independent solutions: $(1 + i)^n$, $(1 - i)^n$.)

$$1 + i = \sqrt{2}\,(\cos(\pi/4) + i\sin(\pi/4))$$

and

$$1 - i = \sqrt{2}\,(\cos(-\pi/4) + i\sin(-\pi/4)) = \sqrt{2}\,(\cos(\pi/4) - i\sin(\pi/4)).$$

This yields

$$a_n = c_1(\sqrt{2})^n(\cos(n\pi/4) + i\sin(n\pi/4)) + c_2(\sqrt{2})^n(\cos(n\pi/4) - i\sin(n\pi/4))$$

$$= (\sqrt{2})^n[k_1 \cos(n\pi/4) + k_2 \sin(n\pi/4)],$$

where $k_1 = c_1 + c_2$ and $k_2 = (c_1 - c_2)i$.

$1 = a_0 = [k_1 \cos 0 + k_2 \sin 0] = k_1$
$2 = a_1 = \sqrt{2}[1 \cdot \cos(\pi/4) + k_2 \sin(\pi/4)]$, or $2 = 1 + k_2$, and $k_2 = 1$.

The solution for the given initial value problem is then given by $a_n = (\sqrt{2})^n \cdot [\cos(n\pi/4) + \sin(n\pi/4)]$, $n \geq 0$. (Note that this solution contains no complex numbers. A small point may bother the reader here. Namely, how did we start with c_1, c_2 complex and end up with $k_1 = c_1 + c_2$, $k_2 = (c_1 - c_2)i$ real? This happens if c_1, c_2 are complex conjugates.)   □

Let us now examine an application from linear algebra.

EXAMPLE 11.15   For $b \in \mathbf{R}^+$, consider the $n \times n$ determinant, D_n, given by

$$
\begin{vmatrix}
b & b & 0 & 0 & 0 & 0 & \cdots & 0 & 0 & 0 & 0 & 0 \\
b & b & b & 0 & 0 & 0 & \cdots & 0 & 0 & 0 & 0 & 0 \\
0 & b & b & b & 0 & 0 & \cdots & 0 & 0 & 0 & 0 & 0 \\
0 & 0 & b & b & b & 0 & \cdots & 0 & 0 & 0 & 0 & 0 \\
0 & 0 & 0 & b & b & b & \cdots & 0 & 0 & 0 & 0 & 0 \\
 & & & & & & \ddots & & & & & \\
0 & 0 & 0 & 0 & 0 & 0 & \cdots & b & b & b & 0 & 0 \\
0 & 0 & 0 & 0 & 0 & 0 & \cdots & 0 & b & b & b & 0 \\
0 & 0 & 0 & 0 & 0 & 0 & \cdots & 0 & 0 & b & b & b \\
0 & 0 & 0 & 0 & 0 & 0 & \cdots & 0 & 0 & 0 & b & b
\end{vmatrix}.
$$

Find the value of D_n as a function of n.

Let a_n, $n \geq 1$, denote the value of the $n \times n$ determinant D_n. Then

$$
a_1 = |b| = b, \quad \text{and} \quad a_2 = \begin{vmatrix} b & b \\ b & b \end{vmatrix} = 0.
$$

Expanding D_n by its first row, we have $D_n =$

$$
b \underbrace{\begin{vmatrix}
b & b & 0 & 0 & 0 & \cdots & 0 & 0 & 0 & 0 \\
b & b & b & 0 & 0 & \cdots & 0 & 0 & 0 & 0 \\
0 & b & b & b & 0 & \cdots & 0 & 0 & 0 & 0 \\
0 & 0 & b & b & b & \cdots & 0 & 0 & 0 & 0 \\
 & & & & & \ddots & & & & \\
0 & 0 & 0 & 0 & 0 & \cdots & b & b & b & 0 \\
0 & 0 & 0 & 0 & 0 & \cdots & 0 & b & b & b \\
0 & 0 & 0 & 0 & 0 & \cdots & 0 & 0 & b & b
\end{vmatrix}}_{\text{(This is } D_{n-1}.)}
$$

$$
- b \begin{vmatrix}
b & b & 0 & 0 & 0 & \cdots & 0 & 0 & 0 & 0 \\
0 & b & b & 0 & 0 & \cdots & 0 & 0 & 0 & 0 \\
0 & b & b & b & 0 & \cdots & 0 & 0 & 0 & 0 \\
0 & 0 & b & b & b & \cdots & 0 & 0 & 0 & 0 \\
 & & & & & \ddots & & & & \\
0 & 0 & 0 & 0 & 0 & \cdots & b & b & b & 0 \\
0 & 0 & 0 & 0 & 0 & \cdots & 0 & b & b & b \\
0 & 0 & 0 & 0 & 0 & \cdots & 0 & 0 & b & b
\end{vmatrix}
$$

Expanding the second determinant by its first column, it follows that $D_n = bD_{n-1} - (b)(b)D_{n-2} = bD_{n-1} - b^2 D_{n-2}$. This translates into the relation $a_n = ba_{n-1} - b^2 a_{n-2}$, $n \geq 3$, $a_1 = b$, $a_2 = 0$.

Letting $a_n = cr^n$, $c, r \neq 0$, $n \geq 1$, the characteristic equation produces the roots $b[(1/2) \pm i\sqrt{3}/2]$.

Hence $a_n = c_1[b((1/2) + i\sqrt{3}/2)]^n + c_2[b((1/2) - i\sqrt{3}/2)]^n = b^n \cdot [c_1(\cos(\pi/3) + i\sin(\pi/3))^n + c_2(\cos(\pi/3) - i\sin(\pi/3))^n] = b^n[k_1 \cos(n\pi/3) + k_2 \sin(n\pi/3)]$.

$b = a_1 = b[k_1 \cos(\pi/3) + k_2 \sin(\pi/3)]$, so $1 = k_1(1/2) + k_2(\sqrt{3}/2)$, or $k_1 + \sqrt{3}k_2 = 2$.

$0 = a_2 = b^2[k_1 \cos(2\pi/3) + k_2 \sin(2\pi/3)]$ and $0 = (k_1)(-1/2) + k_2(\sqrt{3}/2)$, or $k_1 = \sqrt{3}k_2$.

Hence $k_1 = 1$, $k_2 = 1/\sqrt{3}$ and the value of D_n is

$$b^n[\cos(n\pi/3) + (1/\sqrt{3})\sin(n\pi/3)].$$ □

And now for the problem of multiple roots.

CASE (C): (REPEATED REAL ROOTS)

EXAMPLE 11.16   Solve the recurrence relation $a_{n+2} = 4a_{n+1} - 4a_n$, $n \geq 0$, $a_0 = 1$, $a_1 = 3$.

As in the other two cases, we let $a_n = cr^n$, where $c, r \neq 0$ and $n \geq 0$. Then the characteristic equation is $r^2 - 4r + 4 = 0$ and the characteristic roots are both $r = 2$. (So $r = 2$ is called "a root of multiplicity 2.") Unfortunately, we now lack two independent solutions, as 2^n and 2^n are definitely multiples of each other. We need one more (independent) solution. Let us try $n(2^n)$.

Substituting $a_n = n(2^n)$ in the given relation we have $4a_{n+1} - 4a_n = 4(n+1) \cdot 2^{n+1} - 4n2^n = 2(n+1)2^{n+2} - n2^{n+2} = [2n+2-n]2^{n+2} = (n+2)2^{n+2} = a_{n+2}$, so $n(2^n)$ is the second independent solution. (It is independent because it is impossible to have $n2^n = k2^n$ for all $n \geq 0$, if k is a constant.)

The general solution is of the form $a_n = c_1(2^n) + c_2 n(2^n)$. With $a_0 = 1$, $a_1 = 3$ we find $a_n = 2^n + (1/2)n(2^n) = 2^n + n(2^{n-1})$, $n \geq 0$. □

In general, if $c_n a_n + c_{n-1} a_{n-1} + \cdots + c_{n-k} a_{n-k} = 0$ with $c_n(\neq 0)$, c_{n-1}, $\ldots$, $c_{n-k}(\neq 0)$ real constants, and r a characteristic root of multiplicity m, $2 \leq m \leq k$, then the part of the general solution that involves the root r has the form

$$A_0 r^n + A_1 n r^n + A_2 n^2 r^n + \cdots + A_{m-1} n^{m-1} r^n$$

$$= (A_0 + A_1 n + A_2 n^2 + \cdots + A_{m-1} n^{m-1})r^n,$$

where $A_0, A_1, A_2, \ldots, A_{m-1}$ are arbitrary constants.

Our last example involves a little probability.

EXAMPLE 11.17   If a first case of measles is recorded in a certain school system, let p_n denote the probability that at least one more case is reported during the nth week after this first recorded case. School records provide evidence that $p_n = p_{n-1} - (0.25)p_{n-2}$, $n \geq 2$.

As $p_0 = 0$, $p_1 = 1$, if the first case (of a new outbreak) is recorded on Monday, March 1, 1982, when did the probability for the occurrence of a new case decrease to less than 0.01 for the first time?

With $p_n = cr^n$, c, $r \neq 0$, the characteristic equation for the recurrence relation is $r^2 - r + (1/4) = 0 = (r - (1/2))^2$. The general solution has the form $p_n = (c_1 + c_2 n)(1/2)^n$, $n \geq 0$. For $p_0 = 0$, $p_1 = 1$, we get $c_1 = 0$, $c_2 = 2$, so $p_n = n2^{-n+1}$, $n \geq 0$.

The first integer n for which $p_n < 0.01$ is 12. Hence, it was not until the week of May 16, 1982, that the probability of another new case occurring was less than 0.01.   □

EXERCISES

1. Solve the following recurrence relations. (No final answer should involve complex numbers.)

   a) $a_n = 5a_{n-1} + 6a_{n-2}$, $n \geq 2$, $a_0 = 1$, $a_1 = 3$.

   b) $2a_{n+2} - 11a_{n+1} + 5a_n = 0$, $n \geq 0$, $a_0 = 2$, $a_1 = -8$.

   c) $3a_{n+1} = 2a_n + a_{n-1}$, $n \geq 1$, $a_0 = 7$, $a_1 = 3$.

   d) $a_{n+2} + a_n = 0$, $n \geq 0$, $a_0 = 0$, $a_1 = 3$.

   e) $a_{n+2} + 4a_n = 0$, $n \geq 0$, $a_0 = a_1 = 1$.

   f) $a_n - 6a_{n-1} + 9a_{n-2} = 0$, $n \geq 2$, $a_0 = 5$, $a_1 = 12$.

   g) $a_n + 2a_{n-1} + 2a_{n-2} = 0$, $n \geq 2$, $a_0 = 1$, $a_1 = 3$.

   h) $a_n = 7a_{n-1} - 10a_{n-2}$, $n \geq 2$, $a_0 = 3$, $a_1 = 15$.

   i) $9a_{n+2} + 12a_{n+1} + 4a_n = 0$, $n \geq 0$, $a_0 = 1$, $a_1 = 4$.

   j) $a_n = 3a_{n-2} + 2a_{n-3}$, $n \geq 3$, $a_0 = 1$, $a_1 = 3$, $a_2 = 7$.

2. a) Verify the final solutions in Examples 11.9 and 11.16.

   b) Solve the recurrence relation in Example 11.11.

3. If $a_0 = 0$, $a_1 = 1$, $a_2 = 4$, $a_3 = 37$ satisfy the recurrence relation $a_{n+2} + ba_{n+1} + ca_n = 0$, $n \geq 0$, where b and c are constants, solve for a_n.

4. Find and solve a recurrence relation for the number of ways to park motorcycles and compact cars in a row of n spaces if each cycle requires one space while each compact needs two spaces. (All cycles are identical in appearance, as are the cars, and we want to use up all the n spaces.)

5. In the Fibonacci sequence, $F_n = F_{n+2} - F_{n+1}$, $n \geq 0$. Hence, upon adding the equations

$$F_0 = F_2 - F_1$$

$$F_1 = F_3 - F_2$$

$$F_2 = F_4 - F_3$$

$$\cdots \quad \cdots \quad \cdots$$

$$F_n = F_{n+2} - F_{n+1},$$

we obtain $\sum_{i=0}^{n} F_i = F_{n+2} - F_1 = F_{n+2} - 1$.

This is one of many such properties discovered by François Lucas (1842–1891). Prove the following additional properties.

a) $F_1 + F_3 + F_5 + \cdots + F_{2n-1} = F_{2n}$, $n \geq 1$.

b) $F_0 + F_2 + F_4 + \cdots + F_{2n} = F_{2n+1} - 1$, $n \geq 0$.

c) $F_0 + F_1 - F_2 + F_3 - F_4 + \cdots + F_{2n-1} - F_{2n} = -F_{2n-1} + 1$,
$n \geq 1$.

6. a)   Prove that

$$\lim_{n \to \infty} \frac{F_{n+1}}{F_n} = \frac{1 + \sqrt{5}}{2}.$$

(This limit has come to be known as the *golden ratio*.)

b)   Consider a regular pentagon $ABCDE$ inscribed in a circle, as in Fig. 11.4.

   i)   Use the law of sines and the double angle formula for the sine to show that $AC/AX = 2 \cos 36°$.

   ii)   As $\cos 18° = \sin 72° = 4 \sin 18° \cos 18°(1 - 2 \sin^2 18°)$ (Why?), show that $\sin 18°$ is a root of the polynomial equation $8x^3 - 4x + 1 = 0$, and deduce that $\sin 18° = (\sqrt{5} - 1)/4$.

c)   Verify that $AC/AX = (1 + \sqrt{5})/2$.

Figure 11.4

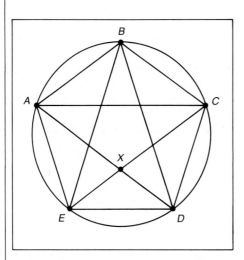

7. For $n \geq 0$, let a_n count the number of ways a sequence of 1's and 2's will sum to n. For example, $a_3 = 3$ since (1) 1, 1, 1; (2) 1, 2; and (3) 2, 1 sum to 3. Find and solve a recurrence relation for a_n.

8. Find and solve a recurrence relation for the number of n-digit ternary $(0, 1, 2)$ sequences with no pair of consecutive 0's.

9. Find and solve a recurrence relation for the number of ways to stack n red, white, green, and blue poker chips so that there are no consecutive blue chips.

10. For $n \geq 1$, let D_n be the $n \times n$ determinant

$$
\begin{vmatrix}
2 & 1 & 0 & 0 & 0 & \cdots & 0 & 0 & 0 & 0 \\
1 & 2 & 1 & 0 & 0 & \cdots & 0 & 0 & 0 & 0 \\
0 & 1 & 2 & 1 & 0 & \cdots & 0 & 0 & 0 & 0 \\
0 & 0 & 1 & 2 & 1 & \cdots & 0 & 0 & 0 & 0 \\
\cdot\cdot & \cdots & \cdots & \cdots & \cdots\cdots & & \cdots & \cdots & \cdot\cdot \\
0 & 0 & 0 & 0 & 0 & \cdots & 1 & 2 & 1 & 0 \\
0 & 0 & 0 & 0 & 0 & \cdots & 0 & 1 & 2 & 1 \\
0 & 0 & 0 & 0 & 0 & \cdots & 0 & 0 & 1 & 2
\end{vmatrix}.
$$

Find and solve a recurrence relation for the value of D_n.

11. Solve the recurrence relation $a_{n+2}^2 - 5a_{n+1}^2 + 4a_n^2 = 0$, $n \geq 0$, $a_0 = 4$, $a_1 = 13$.

12. Determine the constants b and c if $a_n = c_1 + c_2(7^n)$, $n \geq 0$, is the general solution of the relation $a_{n+2} + ba_{n+1} + ca_n = 0$, $n \geq 0$.

11.3 ■ THE NONHOMOGENEOUS RECURRENCE RELATION

We turn now to the recurrence relations

$$a_n + ca_{n-1} = f(n), \qquad n \geq 1 \tag{1}$$

$$a_n + ba_{n-1} + ca_{n-2} = f(n), \qquad n \geq 2, \tag{2}$$

where b and c are constants, $c \neq 0$, and $f(n)$ is not identically 0.

Although there is no general method for solving all nonhomogeneous relations, when the function $f(n)$ has a certain form we shall find a successful technique.

We start with the special case of Eq. (1) when $c = -1$.

For the nonhomogeneous relation $a_n - a_{n-1} = f(n)$, we have

$$a_1 = a_0 + f(1)$$

$$a_2 = a_1 + f(2) = a_0 + f(1) + f(2)$$

$$a_3 = a_2 + f(3) = a_0 + f(1) + f(2) + f(3)$$

$$\vdots$$

$$a_n = a_0 + f(1) + \cdots + f(n) = a_0 + \sum_{i=1}^{n} f(i).$$

This type of relation can be solved in terms of $\sum_{i=1}^{n} f(i)$, if we can find a summation formula from some of our earlier work.

EXAMPLE 11.18   Solve the relation $a_n - a_{n-1} = 3n^2$, $n \geq 1$, $a_0 = 7$.

Here $f(n) = 3n^2$, so the general solution is

$$a_n = a_0 + \sum_{i=1}^{n} f(i) = 7 + 3\sum_{i=1}^{n} i^2 = 7 + \frac{1}{2}(n)(n+1)(2n+1). \qquad \square$$

When a formula for the summation is not known the following will handle Eq. (1), regardless of the value of $c (\neq 0)$, and also work for the second-order nonhomogeneous relation in Eq. (2), for certain functions $f(n)$. Known as the *method of undetermined coefficients*, it relies upon the associated homogeneous relation obtained when $f(n)$ is replaced by 0.

For either of Eq. (1) or Eq. (2) we let $a_n^{(h)}$ denote the general solution of the associated homogeneous relation, and $a_n^{(p)}$ a solution of the given nonhomogeneous relation. The term $a_n^{(p)}$ is called a *particular* solution. Then $a_n = a_n^{(h)} + a_n^{(p)}$ is the general solution of the given relation. To get $a_n^{(p)}$ we use the form of $f(n)$ to suggest a form for $a_n^{(p)}$.

EXAMPLE 11.19

Solve the recurrence relation $a_n - 3a_{n-1} = 5(7^n)$, $n \geq 1$, $a_0 = 2$.

The solution of the associated homogeneous relation is $a_n^{(h)} = c(3^n)$. Since $f(n) = 5(7^n)$, we seek a particular solution $a_n^{(p)}$ of the form $A(7^n)$. As $a_n^{(p)}$ is to be a solution of the given nonhomogeneous relation, we place $a_n^{(p)} = A(7^n)$ into the given relation and find that $A(7^n) - 3A(7^{n-1}) = 5(7^n)$, $n \geq 1$. Dividing by 7^{n-1}, $7A - 3A = 5(7)$, so $A = 35/4$, and $a_n^{(p)} = (35/4)7^n = (5/4)7^{n+1}$, $n \geq 0$. The general solution is $a_n = c(3^n) + (5/4)7^{n+1}$. With $2 = a_0 = c + (5/4)(7)$, $c = -27/4$ and $a_n = (5/4)(7^{n+1}) - (1/4)(3^{n+3})$, $n \geq 0$.   □

EXAMPLE 11.20

Solve the relation $a_n - 3a_{n-1} = 5(3^n)$, $n \geq 1$, $a_0 = 2$.

As in Example 11.19, $a_n^{(h)} = c(3^n)$, but here $a_n^{(h)}$ and $f(n)$ are not linearly independent. As a result we consider a particular solution $a_n^{(p)}$ of the form $Bn(3^n)$. (What happens if we substitute $a_n^{(p)} = B(3^n)$ into the given relation?)

Substituting $a_n^{(p)} = Bn3^n$ into the given relation we have

$$Bn(3^n) - 3B(n-1)(3^{n-1}) = 5(3^n), \quad \text{or} \quad Bn - B(n-1) = 5, \quad \text{so} \quad B = 5.$$

Hence $a_n = a_n^{(h)} + a_n^{(p)} = (c + 5n)3^n$, $n \geq 0$. With $a_0 = 2$, the general solution is $a_n = (2 + 5n)(3^n)$.   □

From these two examples we see that if $f(n) = k(r^n)$, k a constant, $n \geq 0$, in the first-order relation, then $a_n^{(p)} = Ar^n$ if r^n is not a solution of the associated homogeneous relation. When it is, $a_n^{(p)} = Bnr^n$.

For the second-order relation, if $f(n) = k(r^n)$, then

a) $a_n^{(p)} = Ar^n$ if r^n is not a homogeneous solution;

b) $a_n^{(p)} = Bnr^n$, if $a_n^{(h)} = c_1r^n + c_2r_1^n$, where $r_1 \neq r$; and

c) $a_n^{(p)} = Cn^2r^n$, if $a_n^{(h)} = (c_1 + c_2n)r^n$.

EXAMPLE 11.21

(The Towers of Hanoi) Consider n circular disks with holes in their centers. These disks can be stacked on any of the pegs shown in Fig. 11.5. In the figure, $n = 5$ and the disks are stacked on peg 1 with no disk resting upon a smaller one. The objective is to transfer the disks one at a time so that we end up with the original stack on peg 3. Peg 2 may be used as a temporary location for any disk(s) but at no time are we allowed to have a larger disk on top of a smaller one on any peg. How many moves are needed to do this for n disks?

For $n \geq 0$, let a_n be the number of moves it takes to transfer n disks from peg 1 to peg 3 in the manner described. Then for $n + 1$ disks we do the following:

a) Transfer the top n disks from peg 1 to peg 2, using peg 3 as a temporary location. This takes a_n steps.

b) Transfer the largest disk from peg 1 to peg 3. This takes one step.

c) Finally, using peg 1 as a temporary location, transfer the n disks on peg 2 onto the largest disk now on peg 3. This requires another a_n moves.

This results in the relation $a_{n+1} = 2a_n + 1$, $n \geq 0$, $a_0 = 0$.

For $a_{n+1} - 2a_n = 1$, $a_n^{(h)} = c(2^n)$. Since $f(n) = 1$ is not a solution of $a_{n+1} - 2a_n = 0$, we set $a_n^{(p)} = A(1)^n = A$ and find from the given relation that $A = 2A + 1$, so $A = -1$ and $a_n = c(2^n) - 1$. From $a_0 = 0 = c - 1$, $c = 1$ and $a_n = 2^n - 1$, $n \geq 0$. □

Figure 11.5

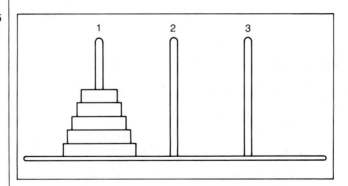

The next example arises from the mathematics of finance.

EXAMPLE 11.22   Pauline takes out a loan of S dollars that is to be paid back in T periods of time. If i is the interest rate per period for the loan, what (constant) payment P must she make at the end of each period?

We let a_n denote the amount still owed on the loan at the end of the nth period (following the nth payment). Then

$$a_{n+1} = a_n + ia_n - P, \qquad 0 \leq n \leq T - 1, \qquad a_0 = S, \qquad a_T = 0.$$

For this relation $a_n^{(h)} = c(1 + i)^n$, while $a_n^{(p)} = A$ since $-P$ is not a solution of the associated homogeneous relation. With $a_n^{(p)} = A$ we find $A - (1 + i)A = -P$, so $A = P/i$. From $a_0 = S$, we obtain $a_n = (S - (P/i))(1 + i)^n + (P/i)$, $0 \leq n \leq T$.

Since $0 = a_T = (S - (P/i))(1 + i)^T + (P/i)$, $(P/i) = ((P/i) - S)(1 + i)^T$ and $P = (Si)[1 - (1 + i)^{-T}]^{-1}$. □

We now consider a problem in the analysis of algorithms.

EXAMPLE 11.23

For $n \geq 1$, let S be a set containing 2^n real numbers. How many comparisons must be made between pairs of numbers in S in order to determine the maximum and minimum elements in S?

If a_n denotes the number of needed comparisons, then $a_1 = 1$. When $n = 2$, $|S| = 2^2 = 4$, so $S = \{x_1, x_2, y_1, y_2\} = S_1 \cup S_2$ where $S_1 = \{x_1, x_2\}$, $S_2 = \{y_1, y_2\}$. Since $a_1 = 1$, it takes one comparison to determine the maximum and minimum elements in each of S_1, S_2. Comparing the minimum elements of S_1 and S_2 and then their maximum elements, we will know the maximum and minimum elements in S, and find that $a_2 = 4 = 2a_1 + 2$. In general, if $|S| = 2^{n+1}$, then write $S = S_1 \cup S_2$ where $|S_1| = |S_2| = 2^n$. To determine the maximum and minimum elements in each of S_1 and S_2 requires a_n comparisons. Comparing the maximum (minimum) elements of S_1 and S_2 requires one more comparison and consequently, $a_{n+1} = 2a_n + 2$, $n \geq 1$.

Here $a_n^{(h)} = c(2^n)$ and $a_n^{(p)} = A$, a constant. Substituting $a_n^{(p)}$ into the relation we find $A = 2A + 2$, or $A = -2$. So $a_n = c2^n - 2$ and with $a_1 = 1 = 2c - 2$, $c = 3/2$. Therefore $a_n = (3/2)(2^n) - 2$.   □

Our last example deals with a second-order relation.

EXAMPLE 11.24

Environmental records show that for a certain lake the population of a specific species of snail increases at a rate three times that of the prior year. Starting with 3000 such snails, and finding 3500 of them the following year, we remove 200 of them from this lake to increase their numbers in other lakes. Continuing to remove 200 of the snails at the end of each year, if a_n represents the snail population in the original lake after n years, find and solve a recurrence relation for a_n, $n \geq 0$.

Here $(a_{n+2} - a_{n+1}) = 3(a_{n+1} - a_n) - 200$, $n \geq 0$. This yields the second-order relation $a_{n+2} - 4a_{n+1} + 3a_n = -200$, for which $a_n^{(h)} = c_1(3^n) + c_2(1^n) = c_1(3^n) + c_2$. Since $f(n) = -200 = -200(1^n)$ is a solution of the associated homogeneous relation, here $a_n^{(p)} = An$ for some constant A. This leads us to

$$A(n + 2) - 4A(n + 1) + 3An = -200, \quad \text{so} \quad -2A = -200, \quad A = 100.$$

Hence $a_n = c_1(3^n) + c_2 + 100n$. With $a_0 = 3000$, and $a_1 = 3500 - 200 = 3300$, we have $a_n = 100(3^n) + 2900 + 100n$, $n \geq 0$.   □

We close this section with a summary that includes the results of Examples 11.19–11.24.

Given a linear nonhomogeneous recurrence relation (with constant coefficients) of the form $c_n a_n + c_{n-1} a_{n-1} + \cdots + c_{n-k} a_{n-k} = f(n)$, where c_n, $c_{n-k} \neq 0$, let $a_n^{(h)}$ denote the homogeneous part of the solution a_n.

1. If $f(n)$ is a constant multiple of one of the following forms and is not a solution of the associated homogeneous relation then $a_n^{(p)}$ has the form shown in Table 11.2. (Here A, B, A_0, A_1, A_2, ..., A_{t-1}, A_t are constants determined by substituting $a_n^{(p)}$ into the given relation; t, r, α are also constants.)

$f(n)$	$a_n^{(p)}$
c, a constant	A, a constant
n	$A_1 n + A_0$
n^2	$A_2 n^2 + A_1 n + A_0$
n^t, $t \in \mathbf{Z}^+$	$A_t n^t + A_{t-1} n^{t-1} + \cdots + A_1 n + A_0$
r^n, $r \in \mathbf{R}$	$A r^n$
$\sin \alpha n$	$A \sin \alpha n + B \cos \alpha n$
$\cos \alpha n$	$A \sin \alpha n + B \cos \alpha n$
$n^t r^n$	$r^n (A_t n^t + A_{t-1} n^{t-1} + \cdots + A_1 n + A_0)$
$r^n \sin \alpha n$	$A r^n \sin \alpha n + B r^n \cos \alpha n$
$r^n \cos \alpha n$	$A r^n \sin \alpha n + B r^n \cos \alpha n$

2. When $f(n)$ comprises a sum of constant multiples of terms like those shown in the table for (1), then $a_n^{(p)}$ is made up of the sum of the corresponding terms in the column headed by $a_n^{(p)}$.

3. Things get trickier if a summand $f_1(n)$ of $f(n)$ contains a factor that is a solution of the associated homogeneous relation. This happens, for example, when $f(n)$ contains summands like cr^n or $(c_1 + c_2 n)r^n$, and r is a characteristic root. If $f_1(n)$ causes this problem, we multiply the particular solution $a_{n_1}^{(p)}$ corresponding to $f_1(n)$ by the smallest power of n, say n^s, for which no summand of $n^s f_1(n)$ is a solution of the associated homogeneous relation. Then $n^s a_{n_1}^{(p)}$ is the corresponding part of $a_n^{(p)}$.

EXERCISES

1. Solve each of the following recurrence relations.

   a) $a_{n+1} - a_n = 2n + 3$, $n \geq 0$, $a_0 = 1$.

   b) $a_{n+1} - a_n = 3n^2 - n$, $n \geq 0$, $a_0 = 3$.

   c) $a_{n+1} - 2a_n = 5$, $n \geq 0$, $a_0 = 1$.

   d) $a_{n+1} - 2a_n = 2^n$, $n \geq 0$, $a_0 = 1$.

2. Use a recurrence relation to derive the formula for $\sum_{i=0}^{n} i^2$.

3. a) Let n lines be drawn in the plane, so that each line intersects every other line, but no three lines are ever coincident. For $n \geq 0$, let a_n count the number of regions into which the plane is separated by the n lines. Find and solve a recurrence relation for a_n.

   b) For the situation in part (a), let b_n count the number of infinite regions that result. Find and solve a recurrence relation for b_n.

4. On the first day of a new year Joseph deposits $1000 in an account that pays 6% interest compounded monthly. At the beginning of each month he adds $200 to his account. If he continues to do this for the next four years (so that he makes 47 additional deposits of $200), how much will his account be worth exactly four years after he opened it?

5. Solve the following recurrence relations.

   a) $a_{n+2} + 3a_{n+1} + 2a_n = 3^n$, $\quad n \geq 0$, $\quad a_0 = 0$, $\quad a_1 = 1$.

   b) $a_{n+2} + 4a_{n+1} + 4a_n = 7$, $\quad n \geq 0$, $\quad a_0 = 1$, $\quad a_1 = 2$.

   c) $a_{n+2} + 4a_{n+1} + 4a_n = n^2$, $\quad n \geq 0$, $\quad a_0 = 0$, $\quad a_1 = 2$.

   d) $a_{n+2} - a_n = \sin(n\pi/2)$, $\quad n \geq 0$, $\quad a_0 = 1$, $\quad a_1 = 1$.

6. Solve the recurrence relation $a_{n+2} - 6a_{n+1} + 9a_n = 3(2^n) + 7(3^n)$, $n \geq 0$, $a_0 = 1$, $a_1 = 4$.

7. Solve the recurrence relation $a_{n+3} - 3a_{n+2} + 3a_{n+1} - a_n = 3 + 5n$, $n \geq 0$.

8. Determine the number of n-digit quaternary $(0, 1, 2, 3)$ sequences where there is never a 3 anywhere to the right of a 0.

9. Meredith borrows $2500, at 12% compounded monthly, to buy a computer. If the loan is to be paid back over two years, what is his monthly payment?

10. The general solution of the recurrence relation $a_{n+2} + b_1 a_{n+1} + b_2 a_n = b_3 n + b_4$, $n \geq 0$, b_i constant, $1 \leq i \leq 4$, is $c_1 2^n + c_2 3^n + n - 7$. Find b_i, $1 \leq i \leq 4$.

11. Solve the following recurrence relations.

   a) $a_{n+2}^2 - 5a_{n+1}^2 + 6a_n^2 = 7n$, $\quad n \geq 0$, $\quad a_0 = a_1 = 1$.

   b) $a_n + na_{n-1} = n!$, $\quad n \geq 1$, $\quad a_0 = 1$.

   c) $a_n^2 - 2a_{n-1} = 0$, $\quad n \geq 1$, $\quad a_0 = 2$. (Let $b_n = \log_2 a_n$, $n \geq 0$.)

11.4 ■ THE METHOD OF GENERATING FUNCTIONS

With all the different cases we had to consider for the nonhomogeneous linear recurrence relation, we now get some assistance from the generating function. This technique will find both the homogeneous and particular solutions for a_n, and incorporate the given initial conditions as well. Furthermore, we'll be able to do even more with this method.

We demonstrate this technique in the following examples.

EXAMPLE 11.25 Solve the relation $a_n - 3a_{n-1} = n$, $n \geq 1$, $a_0 = 1$.

This relation represents an infinite set of equations:

$$(n = 1) \quad a_1 - 3a_0 = 1$$

$$(n = 2) \quad a_2 - 3a_1 = 2$$

$$(n = 3) \quad a_3 - 3a_2 = 3$$

$$\vdots \qquad \vdots \qquad \vdots$$

Multiplying the first of these equations by x, the second by x^2, the third by x^3, etc., we obtain

$$(n = 1) \quad a_1 x^1 - 3a_0 x^1 = 1x^1$$

$$(n = 2) \quad a_2 x^2 - 3a_1 x^2 = 2x^2$$

$$(n = 3) \quad a_3x^3 - 3a_2x^3 = 3x^3$$

$$\vdots \qquad \vdots \qquad \vdots$$

Adding this second set of equations we find that

$$\sum_{n=1}^{\infty} a_n x^n - 3 \sum_{n=1}^{\infty} a_{n-1} x^n = \sum_{n=1}^{\infty} n x^n. \tag{1}$$

We want to solve for a_n in terms of n. To accomplish this, let $f(x) = \sum_{n=0}^{\infty} a_n x^n$ be the (ordinary) generating function for the sequence $a_0, a_1, a_2, \ldots$. Then Eq. (1) can be rewritten as

$$(f(x) - a_0) - 3x \sum_{n=1}^{\infty} a_{n-1} x^{n-1} = \sum_{n=1}^{\infty} n x^n \left(= \sum_{n=0}^{\infty} n x^n \right). \tag{2}$$

As $\sum_{n=1}^{\infty} a_{n-1} x^{n-1} = \sum_{n=0}^{\infty} a_n x^n = f(x)$, the left-hand side of Eq. (2) becomes $(f(x) - 1) - 3xf(x)$.

Before we can proceed, we need the generating function for the sequence 0, 1, 2, 3, ... Recall from Chapter 10 that

$$\frac{1}{(1-x)^2} = \frac{d}{dx}\left(\frac{1}{1-x}\right) = 1 + 2x + 3x^2 + \cdots$$

Consequently, $x/(1-x)^2 = x + 2x^2 + 3x^3 + \cdots$, and

$$(f(x) - 1) - 3xf(x) = \frac{x}{(1-x)^2}, \quad \text{so} \quad f(x) = \frac{1}{(1-3x)} + \frac{x}{(1-x)^2(1-3x)}.$$

Using a partial fraction decomposition,

$$\frac{x}{(1-x)^2(1-3x)} = \frac{A}{(1-x)} + \frac{B}{(1-x)^2} + \frac{C}{(1-3x)},$$

or $\qquad x = A(1-x)(1-3x) + B(1-3x) + C(1-x)^2.$

From the following assignments for x we get

$$(x = 1): \quad 1 = B(-2), \qquad B = -\frac{1}{2}.$$

$$\left(x = \frac{1}{3}\right): \quad \frac{1}{3} = C\left(\frac{2}{3}\right)^2, \qquad C = \frac{3}{4}.$$

$$(x = 0): \quad 0 = A + B + C, \qquad A = -(B + C) = -\frac{1}{4}.$$

Therefore, $\qquad f(x) = \dfrac{1}{1-3x} + \dfrac{(-1/4)}{(1-x)} + \dfrac{(-1/2)}{(1-x)^2} + \dfrac{(3/4)}{(1-3x)}$

$$= \frac{(7/4)}{(1-3x)} + \frac{(-1/4)}{(1-x)} + \frac{(-1/2)}{(1-x)^2}.$$

We find a_n by determining the coefficient of x^n in each of the three summands.

a) $(7/4)/(1 - 3x) = (7/4)[1/(1 - 3x)]$
$$= (7/4)[1 + (3x) + (3x)^2 + (3x)^3 + \cdots],$$
and the coefficient of x^n is $(7/4)3^n$.

b) $(-1/4)/(1 - x) = (-1/4)[1 + x + x^2 + \cdots]$, and the coefficient of x^n here is $(-1/4)$.

c) $(-1/2)/(1 - x)^2 = (-1/2)(1 - x)^{-2}$
$$= (-1/2)[\tbinom{-2}{0} + \tbinom{-2}{1}(-x) + \tbinom{-2}{2}(-x)^2 + \tbinom{-2}{3}(-x)^3 + \cdots]$$
with the coefficient of x^n given by $(-1/2)\tbinom{-2}{n}(-1)^n = (-1/2)(-1)^n\tbinom{2+n-1}{n} \cdot (-1)^n = (-1/2)(n + 1)$.

Therefore $a_n = (7/4)3^n - (1/2)n - (3/4)$, $n \geq 0$. (Notice that there is no special concern here with $a_n^{(p)}$.)   □

We consider a second example, which has a familiar result.

EXAMPLE 11.26   Let $n \in \mathbf{N}$. For $r \geq 0$, let $a(n, r)$ = the number of ways we can select, with repetition, r objects from a set of n distinct objects.

For $n \geq 1$, let $\{b_1, b_2, \ldots, b_n\}$ be the set of these objects, and consider object b_1. Exactly two things can happen.

a) The object b_1 is never selected. Hence the r objects are selected from $\{b_2, \ldots, b_n\}$. This we can do in $a(n - 1, r)$ ways.

b) The object b_1 is selected at least once. Then we must select $r - 1$ objects from $\{b_1, b_2, \ldots, b_n\}$, so we can continue to select b_1 in addition to the one selection of it we've already made. There are $a(n, r - 1)$ ways to accomplish this.

Then $a(n, r) = a(n - 1, r) + a(n, r - 1)$, since these two cases cover all possibilities and are mutually disjoint.

Let $f_n = \sum_{r=0}^{\infty} a(n, r)x^r$ be the generating function for the sequence $a(n, 0)$, $a(n, 1)$, $a(n, 2)$, ... From $a(n, r) = a(n - 1, r) + a(n, r - 1)$, $n \geq 1$, $r \geq 1$, it follows that $a(n, r)x^r = a(n - 1, r)x^r + a(n, r - 1)x^r$ and

$$\sum_{r=1}^{\infty} a(n, r)x^r = \sum_{r=1}^{\infty} a(n - 1, r)x^r + \sum_{r=1}^{\infty} a(n, r - 1)x^r.$$

Realizing that $a(n, 0) = 1$, $n \geq 0$, and $a(0, r) = 0$ for $r > 0$, we write

$$f_n - a(n, 0) = f_{n-1} - a(n - 1, 0) + x\sum_{r=1}^{\infty} a(n, r - 1)x^{r-1},$$

so $f_n - 1 = f_{n-1} - 1 + xf_n$. Therefore, $f_n - xf_n = f_{n-1}$, or $f_n = f_{n-1}/(1 - x)$.
If $n = 5$, for example, then

$$f_5 = \frac{f_4}{(1 - x)} = \frac{1}{(1 - x)} \cdot \frac{f_3}{(1 - x)} = \frac{f_3}{(1 - x)^2} = \frac{f_2}{(1 - x)^3} = \frac{f_1}{(1 - x)^4}$$

$$= \frac{f_0}{(1 - x)^5} = \frac{1}{(1 - x)^5},$$

since $f_0 = a(0, 0) + a(0, 1)x + a(0, 2)x^2 + \cdots = 1 + 0 + 0 + \cdots$.

In general, $f_n = 1/(1 - x)^n = (1 - x)^{-n}$, so $a(n, r)$ is the coefficient of x^r in $(1 - x)^{-n}$ which is $\binom{-n}{r}(-1)^r = \binom{n+r-1}{r}$.

(Here we dealt with a recurrence relation for $a(n, r)$, $n, r \geq 0$, a discrete function of two variables.) $\square$

Our last example shows how generating functions may be used to solve a system of recurrence relations.

EXAMPLE 11.27   This example provides an approximate model for the propagation of high- and low-energy neutrons as they strike the nuclei of fissionable material (e.g., uranium) and are absorbed. Here we deal with a fast reactor where there is no moderator (such as water). (In reality, all the neutrons have fairly high energy and there are not just two energy levels. There is a continuous spectrum of energy levels and those neutrons at the upper end of the spectrum are called the high-energy neutrons. The higher energy neutrons tend to produce more new neutrons than the lower energy ones.)

Consider the reactor at time 0 and suppose one high-energy neutron is injected into the system. During each time interval thereafter (about 1 microsecond, or 10^{-6} second) the following occur:

a)   When a high-energy neutron interacts with a nucleus (of fissionable material), upon absorption this results (one microsecond later) in two new high-energy neutrons and one low-energy one.

b)   For a low-energy neutron only one neutron of each level is produced.

Assuming that all free neutrons interact with nuclei one microsecond after their creation, find functions of n so that

a_n = the number of high-energy neutrons,

b_n = the number of low-energy neutrons,

in the reactor after n microseconds, $n \geq 0$.

Here we have $a_0 = 1$, $b_0 = 0$ and the *system* of recurrence relations

$$a_{n+1} = 2a_n + b_n \tag{1}$$

$$b_{n+1} = a_n + b_n . \tag{2}$$

Let $f(x) = \sum_{n=0}^{\infty} a_n x^n$, $g(x) = \sum_{n=0}^{\infty} b_n x^n$ be the generating functions for the sequences $\{a_n \mid n \geq 0\}$, $\{b_n \mid n \geq 0\}$, respectively.

From Eq. (1) and (2), when $n \geq 0$

$$a_{n+1}x^{n+1} = 2a_n x^{n+1} + b_n x^{n+1} \tag{1$'$}$$

$$b_{n+1}x^{n+1} = a_n x^{n+1} + b_n x^{n+1}. \tag{2$'$}$$

Summing Eq. (1)$'$ over all $n \geq 0$, we have

$$\sum_{n=0}^{\infty} a_{n+1}x^{n+1} = 2x \sum_{n=0}^{\infty} a_n x^n + x \sum_{n=0}^{\infty} b_n x^n. \tag{1$''$}$$

In similar fashion, Eq. (2)′ yields

$$\sum_{n=0}^{\infty} b_{n+1}x^{n+1} = x \sum_{n=0}^{\infty} a_n x^n + x \sum_{n=0}^{\infty} b_n x^n. \qquad (2)''$$

Introducing the generating functions at this point we get

$$f(x) - a_0 = 2xf(x) + xg(x) \qquad (1)''$$

$$g(x) - b_0 = xf(x) + xg(x), \qquad (2)''$$

a system of equations relating the generating functions. Solving this system we find

$$f(x) = \frac{1-x}{x^2 - 3x + 1} = \left(\frac{5+\sqrt{5}}{10}\right)\left(\frac{1}{\alpha - x}\right) + \left(\frac{5-\sqrt{5}}{10}\right)\left(\frac{1}{\beta - x}\right)$$

$$g(x) = \frac{x}{x^2 - 3x + 1} = \left(\frac{-5 - 3\sqrt{5}}{10}\right)\left(\frac{1}{\alpha - x}\right) + \left(\frac{-5 + 3\sqrt{5}}{10}\right)\left(\frac{1}{\beta - x}\right),$$

where

$$\alpha = \frac{3+\sqrt{5}}{2}, \qquad \beta = \frac{3-\sqrt{5}}{2}.$$

Consequently

$$a_n = \left(\frac{5+\sqrt{5}}{10}\right)\left(\frac{3-\sqrt{5}}{2}\right)^{n+1} + \left(\frac{5-\sqrt{5}}{10}\right)\left(\frac{3+\sqrt{5}}{2}\right)^{n+1}$$

$$b_n = \left(\frac{-5 - 3\sqrt{5}}{10}\right)\left(\frac{3-\sqrt{5}}{2}\right)^{n+1} + \left(\frac{-5 + 3\sqrt{5}}{10}\right)\left(\frac{3+\sqrt{5}}{2}\right)^{n+1}, \qquad n \geq 0. \quad \square$$

11.5 ■ A SPECIAL KIND OF NONLINEAR RECURRENCE RELATION

Thus far our study of recurrence relations has dealt with linear relations with constant coefficients. The study of nonlinear recurrence relations and relations with variable coefficients is not a topic we shall pursue except for one special nonlinear relation that lends itself to the method of generating functions.

We shall develop the method in a counting problem on data structures. Before doing so, we observe that if $f(x) = \sum_{i=0}^{\infty} a_i x^i$ is the generating function for a_0, a_1, a_2, ..., then $[f(x)]^2$ generates $a_0 a_0$, $a_0 a_1 + a_1 a_0$, $a_0 a_2 + a_1 a_1 + a_2 a_0$, ..., $a_0 a_n + a_1 a_{n-1} + a_2 a_{n-2} + \cdots + a_{n-1} a_1 + a_n a_0$, ...

EXAMPLE 11.28   In Section 3.1 we introduced the idea of a tree diagram. As we mentioned in Chapter 6, trees, in general, are connected graphs that contain no cycles (or loops). Here we examine a binary rooted tree.

In Fig. 11.6 we see two such trees, where the circled vertex denotes the *root*. These trees are called *binary* because from each vertex there are at most two edges (called *branches*) descending from that vertex.

Figure 11.6

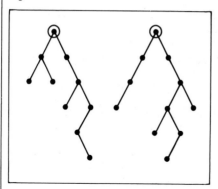

Figure 11.7

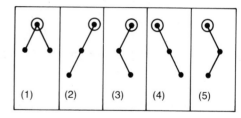

(1)   (2)   (3)   (4)   (5)

In particular these rooted binary trees are *ordered* in the sense that a left branch descending from a vertex is considered different from a right branch descending from that vertex. For the case of three vertices, the five possible ordered rooted binary trees are shown in Fig. 11.7. (If no attention were paid to order, then the last four rooted trees would be the same structure.)

Our objective is to count, for $n \geq 0$, the number b_n of rooted ordered binary trees on n vertices. Assuming that we know the values of b_i, $0 \leq i \leq n$, to obtain b_{n+1}, we select one vertex as the root and note, as in Fig. 11.8, that the substructures descending on the left and right of the root are (rooted ordered binary) subtrees whose total number of vertices is n. Among these possible subtrees is the empty subtree, of which there is only $1(= b_0)$.

Figure 11.8

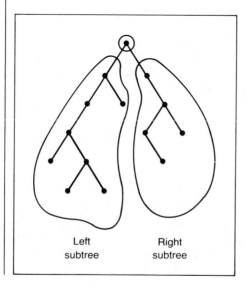

Left   Right
subtree   subtree

Now consider how the n vertices in these two subtrees can be divided up.

(1) 0 vertices on the left; n vertices on the right. This results in $b_0 b_n$ overall substructures to be counted in b_{n+1}.

(2) 1 vertex on the left, $n - 1$ vertices on the right, yielding $b_1 b_{n-1}$ rooted ordered binary trees on $n + 1$ vertices.

. . . . . .

$(i + 1)$ i vertices on the left, $n - i$ on the right for a count of $b_i b_{n-i}$ toward b_{n+1}.

. . . . . .

$(n + 1)$ n vertices on the left and none on the right, contributing $b_n b_0$ of the trees.

Hence, for all $n \geq 0$, we have

$$b_{n+1} = b_0 b_n + b_1 b_{n-1} + b_2 b_{n-2} + \cdots + b_{n-1} b_1 + b_n b_0 ,$$

and $$\sum_{n=0}^{\infty} b_{n+1} x^{n+1} = \sum_{n=0}^{\infty} (b_0 b_n + b_1 b_{n-1} + \cdots + b_{n-1} b_1 + b_n b_0) x^{n+1} . \qquad (1)$$

Now let $f(x) = \sum_{n=0}^{\infty} b_n x^n$ be the generating function for $b_0, b_1, b_2, \ldots .$ We rewrite Eq. (1) as

$$(f(x) - b_0) = x \sum_{n=0}^{\infty} (b_0 b_n + b_{n-1} b_1 + \cdots + b_n b_0) x^n = x[f(x)]^2 .$$

This brings us to the quadratic (in $f(x)$)

$$x[f(x)]^2 - f(x) + 1 = 0, \quad \text{so} \quad f(x) = [1 \pm \sqrt{1 - 4x}]/(2x) .$$

$\sqrt{1 - 4x} = (1 - 4x)^{1/2} = \binom{1/2}{0} + \binom{1/2}{1}(-4x) + \binom{1/2}{2}(-4x)^2 + \cdots$, where the coefficient of x^n, $n \geq 1$, is

$$\binom{1/2}{n}(-4)^n = \frac{(1/2)((1/2) - 1)((1/2) - 2) \cdots ((1/2) - n + 1)}{n!}(-4)^n$$

$$= (-1)^{n-1} \frac{(1/2)(1/2)(3/2) \cdots ((2n - 3)/2)}{n!}(-4)^n$$

$$= \frac{(-1)2^n(1)(3) \cdots (2n - 3)}{n!}$$

$$= \frac{(-1)2^n(n!)(1)(3) \cdots (2n - 3)(2n - 1)}{(n!)(n!)(2n - 1)}$$

$$= \frac{(-1)(2)(4) \cdots (2n)(1)(3) \cdots (2n - 1)}{(2n - 1)(n!)(n!)} = \frac{(-1)}{(2n - 1)}\binom{2n}{n} .$$

In $f(x)$ we select the negative radical; otherwise, we'll have negative values for the b_n's. Then

$$f(x) = \frac{1}{2x}\left[1 - \left[1 - \sum_{n=1}^{\infty} \frac{1}{(2n - 1)}\binom{2n}{n} x^n\right]\right],$$

and b_n, the coefficient of x^n in $f(x)$, is half the coefficient of x^{n+1} in

$$\sum_{n=1}^{\infty} \frac{1}{(2n-1)} \binom{2n}{n} x^n.$$

So

$$b_n = \frac{1}{2}\left[\frac{1}{2(n+1)-1}\right]\binom{2(n+1)}{n+1} = \frac{(2n)!}{(n+1)!(n!)} = \frac{1}{(n+1)}\binom{2n}{n}.$$

(This sequence is called the *Catalan numbers* after Eugene Catalan (1814–1894) who used them in determining the number of ways to parenthesize the expression $x_1 x_2 x_3 \ldots x_n$.)   □

EXERCISES

1. Solve the following recurrence relations by the method of generating functions.
   a) $a_{n+1} - a_n = 3^n$, $n \geq 0$, $a_0 = 1$.
   b) $a_{n+1} - a_n = n^2$, $n \geq 0$, $a_0 = 1$.
   c) $a_{n+2} - 2a_{n+1} + a_n = 2^n$, $n \geq 0$, $a_0 = 1$, $a_1 = 2$.
   d) $a_{n+2} - 2a_{n+1} + a_n = n$, $n \geq 0$, $a_0 = 1$, $a_1 = 2$.

2. For n distinct objects, let $a(n,r)$ denote the number of ways we can select, without repetition, r of the n objects, $0 \leq r \leq n$. $a(n,r) = 0$ when $r > n$. Use the recurrence relation $a(n,r) = a(n-1,r-1) + a(n-1,r)$, $n \geq 1$, $r \geq 1$, to show that $f(x) = (1+x)^n$ generates $a(n,r)$, $r \geq 0$.

3. Solve the following systems of recurrence relations.
   a) $a_{n+1} = -2a_n - 4b_n$
       $b_{n+1} = 4a_n + 6b_n$
       $n \geq 0$, $a_0 = 1$, $b_0 = 0$
   b) $a_{n+1} = 2a_n - b_n + 2$
       $b_{n+1} = -a_n + 2b_n - 1$
       $n \geq 0$, $a_0 = 0$, $b_0 = 1$

4. Verify that

$$\frac{1}{2}\left(\frac{1}{2n+1}\right)\binom{2n+2}{n+1} = \left(\frac{1}{n+1}\right)\binom{2n}{n},$$

   for all $n \geq 0$.

5. For the rooted ordered binary trees of Example 11.28, calculate b_4 and draw all of these four-vertex structures.

6. For a convex polygon of n sides, let t_n denote the number of ways the interior of the polygon can be partitioned into triangular regions by drawing non-intersecting diagonals.
   a) Show that $t_3 = 1$, $t_4 = 2$, $t_5 = 5$.
   b) Define $t_0 = t_1 = 0$, $t_2 = 1$. Verify that $t_{n+1} = t_2 t_n + t_3 t_{n-1} + \cdots + t_{n-1} t_3 + t_n t_2$.
   c) Find t_n as a function of n.

11.6 ■ SUMMARY AND HISTORICAL REVIEW

In this chapter the recurrence relation has emerged as another tool for solving combinatorial problems. In these problems we analyze a given situation and then express the result a_n in terms of the results for certain smaller nonnegative integers. Once the recurrence relation is determined, one can solve for any value of a_n (within reason). With access to a computer, such relations are particularly valuable, especially if they cannot be solved explicitly.

The study of recurrence relations can be traced back to the Fibonacci relation $F_{n+2} = F_{n+1} + F_n$, $n \geq 0$, $F_0 = 0$, $F_1 = 1$, which was investigated by Leonardo of Pisa (c. 1175–1250) in 1202. In his *Liber Abaci*, he deals with a problem concerning the number of rabbits that result in one year if one starts with a single pair that produces another pair at the end of each month. Each new pair starts to breed similarly one month after its birth, and we assume no rabbits die during the given year. Hence, at the end of the first month there are two pairs of rabbits; three pairs after two months; five pairs after three months, and so on. (As mentioned in the summary of Chapter 10, this result was solved explicitly by the method of generating functions by Abraham DeMoivre (1667–1754) in 1718.)

This same sequence appears in the work of the German mathematician Johannes Kepler (1571–1630), who used it in his studies on how the leaves of a plant or flower are arranged about its stem. In 1844 the French mathematician Gabriel Lamé (1795–1870) used the sequence in his analysis of the efficiency of the Euclidean algorithm. Later, François Lucas (1842–1891) derived many properties of this sequence and was the first to call these numbers the Fibonacci sequence. The UMAP article by Roger V. Jean [6] gives many applications dealing with this sequence.

Comparable coverage of the material presented here can be found in Chapter 3 of C. Liu [9]. For more on the theoretical development of linear recurrence relations with constant coefficients examine Chapter 9 of N. Finizio and G. Ladas [4]. This text contains an application in optics where a recurrence relation is used to determine the path of a ray passing through a series of equally spaced thin lenses.

Applications in probability theory dealing with recurrent events, random walks, and ruin problems can be found in Chapters XIII and XIV of the classic text by W. Feller [3]. The UMAP module by D. Sherbert [13] introduces difference equations and includes an application in economics known as the *Cobweb Theorem*. The text by S. Goldberg [5] has more on applications in the social sciences.

Recursive techniques in the generation of permutations, combinations, and partitions of integers are developed in Chapter 3 of R. Brualdi [2] and Chapter 5 of E. Page and L. Wilson [10]. The algorithm presented in Section 11.1 for the permutations of $\{1, 2, 3, \ldots, n\}$ first appeared in the work of H. Steinhaus [12] and is often referred to as the *adjacent mark ordering algorithm*. This result was rediscovered later, independently by H. Trotter [14], and S. Johnson [7]. Efficient sorting methods for permutations and other combinatorial structures are analyzed in detail in the text by D. Knuth [8]. The work of E. Reingold, J. Nievergelt, and N. Deo [11] also deals with such algorithms and their computer implementation.

Finally, for those who enjoyed the rooted ordered binary trees in Section 11.5, Chapter 3 of A. Aho, J. Hopcroft, and J. Ullman [1] should prove interesting.

REFERENCES

1. Aho, Alfred V., Hopcroft, John E., and Ullman, Jeffrey D., *Data Structures and Algorithms,* Addison-Wesley, Reading, Massachusetts, 1983.

2. Brualdi, Richard A., *Introductory Combinatorics,* Elsevier North-Holland, Inc., New York, 1977.

3. Feller, William, *An Introduction to Probability Theory and Its Applications,* Volume I, 3rd ed., John Wiley & Sons, New York, 1968.

4. Finizio, N., and Ladas, G., *An Introduction to Differential Equations (with Difference Equations, Fourier Series and Partial Differential Equations),* Wadsworth Publishing Company, Belmont, California, 1982.

5. Goldberg, Samuel, *Introduction to Difference Equations (with Illustrative Examples from Economics, Psychology, and Sociology),* John Wiley & Sons, New York, 1958.

6. Jean, Roger V., "The Fibonacci Sequence," *The UMAP Journal,* Volume 5, No. 1 (1984), pp. 23–47.

7. Johnson, Selmer M., "Generation of Permutations by Adjacent Transposition," *Mathematics of Computation,* Volume 17 (1963), pp. 282–285.

8. Knuth, Donald E., *The Art of Computer Programming/Volume 3 Sorting and Searching,* Addison-Wesley, Reading, Massachusetts, 1973.

9. Liu, C. L., *Introduction to Combinatorial Mathematics,* McGraw-Hill, New York, 1968.

10. Page, E. S., and Wilson, L. B., *An Introduction to Computational Combinatorics,* Cambridge University Press, Cambridge, England, 1979.

11. Reingold, E. M., Nievergelt, J., and Deo, N., *Combinatorial Algorithms: Theory and Practice,* Prentice-Hall, Englewood Cliffs, New Jersey, 1977.

12. Steinhaus, Hugo D., *One Hundred Problems in Elementary Mathematics,* Basic Books, New York, 1964.

13. Sherbert, Donald R., *Difference Equations with Applications,* UMAP Module 322, Birkhauser Boston, Inc., Cambridge, Massachusetts, 1980.

14. Trotter, H. F., "ACM Algorithm 115 — Permutations," *Communications of the ACM,* Volume 5 (1962), pp. 434–435.

■ MISCELLANEOUS EXERCISES

1. For $n \in \mathbf{Z}^+$ and $n \geq k + 1 \geq 1$, verify algebraically the recursion formula

$$\binom{n}{k + 1} = \left(\frac{n - k}{k + 1}\right)\binom{n}{k}$$

for binomial coefficients.

2. a) For $n \geq 0$, let B_n denote the number of partitions of $\{1, 2, 3, \ldots, n\}$. Set $B_0 = 1$ for the partitions of $\emptyset$. Verify that for all $n \geq 0$,

$$B_{n+1} = \sum_{i=0}^{n} \binom{n}{n-i} B_i = \sum_{i=0}^{n} \binom{n}{i} B_i .$$

   (The numbers B_i, $0 \leq i$, are referred to as the *Bell numbers* after Eric Temple Bell (1883–1960).)

   b) How are the Bell numbers related to the Stirling numbers of the second kind?

3. Let $n, k \in \mathbf{Z}^+$, and define $p(n, k)$ to be the number of partitions of n into exactly k summands. Prove that $p(n, k) = p(n-1, k-1) + p(n-k, k)$.

4. Find and solve a recurrence relation for the number of quaternary $(0, 1, 2, 3)$ sequences of length n that have no consecutive identical symbols.

5. For $n \in \mathbf{Z}^+$, d_n denotes the number of derangements of $\{1, 2, 3, \ldots, n\}$, as discussed in Section 7.3.

   a) If $n > 2$, show that d_n satisfies the recurrence relation

$$d_n = (n-1)(d_{n-1} + d_{n-2}), \qquad d_2 = 1, \qquad d_1 = 0 .$$

   b) How can we define d_0 so that the result in (a) is valid for $n \geq 2$?

   c) Rewrite the result in (a) as $d_n - nd_{n-1} = -[d_{n-1} - (n-1)d_{n-2}]$. How can $d_n - nd_{n-1}$ be expressed in terms of d_{n-2}, d_{n-3}?

   d) Show that $d_n - nd_{n-1} = (-1)^n$.

   e) Let $f(x) = \sum_{n=0}^{\infty} (d_n x^n)/n!$. Multiplying both sides of the equation in (d) by $x^n/n!$ and summing for $n \geq 2$, verify that $f(x) = (e^{-x})/(1-x)$. Hence

$$d_n = n! \left[1 - \frac{1}{1!} + \frac{1}{2!} - \frac{1}{3!} + \cdots + \frac{(-1)^n}{n!} \right] .$$

6. For $n \geq 0$, let $m = \lfloor (n+1)/2 \rfloor$. Prove that $F_{n+2} = \sum_{k=0}^{m} \binom{n-k+1}{k}$. (You may want to look back at Examples 10.14 and 11.8.)

7. For $n \geq 0$, draw n ovals in the plane so that each oval intersects each of the others in exactly two points and no three ovals are coincident. If a_n denotes the number of regions in the plane that results from these n ovals, find and solve a recurrence relation for a_n.

8. a) Divide the plane into three regions by drawing two parallel lines. For $n \geq 0$, draw n additional lines so that each line intersects all other lines including the two parallel lines with which we started. No three of these $n + 2$ lines are coincident. Find and solve a recurrence relation for the number of regions that result in the plane from these $n + 2$ lines.

   b) Answer part (a) if there are k parallel lines, $k \geq 2$.

9. In a certain population model the probability that a couple will have n children satisfies the recurrence relation $p_n = 0.6p_{n-1}$, $n \geq 1$. (a) Find p_n in terms of p_0. (b) Since $\sum_{i=0}^{\infty} p_i = 1$, determine p_0.

10. When considering certain geographic restrictions the recurrence relation in Exercise 9 is replaced by $p_n = (1/n)p_{n-1}$, $n \geq 1$. Find p_n in terms of p_0 and determine p_0.

11. For $n \geq 0$, let us toss a coin $2n$ times.

    a) If a_n is the number of sequences of $2n$ tosses where n heads and n tails occur, find a_n in terms of n.

    b) Find constants r, s, and t so that $(r + sx)^t = f(x) = \sum_{n=0}^{\infty} a_n x^n$.

    c) Let b_n denote the number of sequences of $2n$ tosses where the number of heads and tails are equal for the first time only after all $2n$ tosses have been made. (For example, if $n = 3$, HHHTTT, HHTHTT are counted in b_n, HTHHTT and HHTTHT are not.)
       Define $b_0 = 0$ and show that for all $n \geq 1$, $a_n = a_0 b_n + a_1 b_{n-1} + \cdots + a_{n-1}b_1 + a_n b_0$.

    d) Let $g(x) = \sum_{n=0}^{\infty} b_n x^n$. Show that $g(x) = 1 - 1/f(x)$, and then solve for b_n, $n \geq 1$.

12. (Gambler's Ruin)

    a) When Cathy and Jill play checkers each has probability $\frac{1}{2}$ of winning. There is never a tie and the games are independent in the sense that no matter how many games the girls have played each girl still has probability $\frac{1}{2}$ of winning the next game. After each game the loser gives the winner a quarter. If Cathy has $2.00 to play with and Jill has $2.50 and they play until one of them is broke, what is the probability that Cathy gets wiped out?

    b) Answer part (a) if Jill is tired so that she only has probability $\frac{1}{4}$ of winning each game.

12

Groups,
Coding Theory,
and Polya's Method
of Enumeration

In the study of algebraic structures we examine properties shared by particular mathematical systems. Then we generalize our findings in order to study the underlying structure common to these particular examples.

In Chapter 8 we did this with the ring structure, which depended upon two binary operations. Now we turn to a structure dependent on one binary operation. This structure is called a *group*.

Our study of groups will examine many ideas comparable to those for rings. However, here we shall dwell primarily on those aspects of the structure that are needed for applications in coding theory and a counting method developed by George Polya.

12.1 ■ DEFINITION, EXAMPLES, AND ELEMENTARY PROPERTIES

Definition 12.1 ▶ If G is a nonempty set and $\circ$ is a binary operation on G, then $(G, \circ)$ is called a *group* if the following conditions are satisfied.

1. For $a, b \in G$, $a \circ b \in G$. (Closure of G under $\circ$)
2. For $a, b, c \in G$, $a \circ (b \circ c) = (a \circ b) \circ c$. (The Associative Property)
3. There exists $e \in G$ where $a \circ e = e \circ a = a$, for all $a \in G$. (The Existence of an Identity)
4. For each $a \in G$ there is an element $b \in G$ such that $a \circ b = b \circ a = e$. (Existence of Inverses)

If, in addition, $a \circ b = b \circ a$ for all $a, b \in G$, then G is called a *commutative*, or *abelian*, group. (The adjective "abelian" honors the Norwegian mathematician Niels Henrik Abel (1802–1829).)

EXAMPLE 12.1 Under ordinary addition, each of $\mathbf{Z}, \mathbf{Q}, \mathbf{R}, \mathbf{C}$ is an abelian group. None of these are groups under multiplication, since 0 has no multiplicative inverse. However, $\mathbf{Q}^*, \mathbf{R}^*, \mathbf{C}^*$ are multiplicative abelian groups.

If $(R, +, \cdot)$ is a ring, then $(R, +)$ is an abelian group; the nonzero elements of a *field* form a multiplicative abelian group. □

EXAMPLE 12.2 For $n \in \mathbf{Z}^+$, $(\mathbf{Z}_n, +)$ is an abelian group. When p is a prime, $(\mathbf{Z}_p^*, \cdot)$ is an abelian group. Tables 12.1 and 12.2 demonstrate this for $n = 6$ and $p = 7$. (Recall that in $\mathbf{Z}_n$ we often write a for $[a] = \{a + kn \mid k \in \mathbf{Z}\}$.) □

Table 12.1

+	0	1	2	3	4	5
0	0	1	2	3	4	5
1	1	2	3	4	5	0
2	2	3	4	5	0	1
3	3	4	5	0	1	2
4	4	5	0	1	2	3
5	5	0	1	2	3	4

Table 12.2

·	1	2	3	4	5	6
1	1	2	3	4	5	6
2	2	4	6	1	3	5
3	3	6	2	5	1	4
4	4	1	5	2	6	3
5	5	3	1	6	4	2
6	6	5	4	3	2	1

Definition 12.2

▶ When G is a finite group, the number of elements in G is called the *order* of G and is denoted $|G|$.

EXAMPLE 12.3

For any $n \in \mathbf{Z}^+$, $|(\mathbf{Z}_n, +)| = n$, while $|(\mathbf{Z}_p^*, \cdot)| = p - 1$ for any prime p. □

From here on the group operation will be written multiplicatively, unless it is given otherwise. So $a \circ b$ now becomes ab.

We examine the following properties shared by all groups.

Theorem 12.1

▶ For any group G

a) the identity of G is unique.

b) the inverse of each element of G is unique.

c) if $a, b, c \in G$ and $ab = ac$, then $b = c$.

d) if $a, b, c \in G$ and $ba = ca$, then $b = c$.

e) G is abelian if and only if $(ab)^2 = a^2 b^2$ for all $a, b \in G$.

Proof

a) If e_1, e_2 are both identities in G, then $e_1 = e_1 e_2 = e_2$. (Justify each equality.)

b) Let $a \in G$ and suppose that b, c are both inverses of a. Then $b = be = b(ac) = (ba)c = ec = c$. (Justify each equality.)

The proofs of (c)–(e) are left for the reader. ∎

On the basis of the result in Theorem 12.1(b) the unique inverse of a will be designated by a^{-1}. When the group is written additively, $-a$ is used.

As in the case of multiplication in a ring, we have powers of elements in a group. We define $a^0 = e$, $a^1 = a$, $a^2 = a \cdot a$, and in general $a^{n+1} = a^n \cdot a$, $n \geq 0$. Since each group element has an inverse, here the case of a^{-n} must also be considered. For $n \in \mathbf{Z}^+$, we define $a^{-n} = (a^{-1})^n$. Then a^n is defined for all $n \in \mathbf{Z}$ and it can be shown that for $m, n \in \mathbf{Z}$, $a^m \cdot a^n = a^{m+n}$; and, $(a^m)^n = a^{mn}$.

If the group operation is addition, then multiples replace powers and for any m, $n \in \mathbf{Z}$, $a, b \in G$,

$$ma + na = (m + n)a \qquad m(na) = (mn)a \qquad m(a + b) = ma + mb.$$

EXAMPLE 12.4

Let $G = (\mathbf{Z}_6, +)$. If $H = \{0, 2, 4\}$, then H is a nonempty subset of G. Table 12.3 shows that $(H, +)$ is also a group under the binary operation of G.

Table 12.3

+	0	2	4
0	0	2	4
2	2	4	0
4	4	0	2

□

This situation gives rise to the following.

Definition
12.3

▶ Let G be a group and $\emptyset \neq H \subseteq G$. If H is a group under the binary operation of G, we call H a *subgroup* of G.

EXAMPLE 12.5

a) Every group G has $\{e\}$, G as subgroups. These are the *trivial* subgroups of G. All others are termed *nontrivial*, or *proper*.

b) In addition to $H = \{0, 2, 4\}$, the subset $K = \{0, 3\}$ is also a (proper) subgroup of $G = (\mathbf{Z}_6, +)$.

c) The group $(\mathbf{Z}, +)$ is a subgroup of $(\mathbf{Q}, +)$, which is a subgroup of $(\mathbf{R}, +)$. Yet $(\mathbf{Z}^*, \cdot)$ is not a subgroup of $(\mathbf{Q}^*, \cdot)$. (Why not?)   □

For a group G and $\emptyset \neq H \subseteq G$, the following tells us when H is a subgroup of G.

Theorem
12.2

▶ If H is a nonempty subset of a group G, then H is a subgroup of G iff (a) for all $a, b \in H$, $ab \in H$; and (b) for all $a \in H$, $a^{-1} \in H$.

Proof

If H is a subgroup of G, then by Definition 12.3 H is a group under the same binary operation. Hence it satisfies all the group conditions, including the two mentioned here. Conversely, let $\emptyset \neq H \subseteq G$ with H satisfying (a) and (b). For $a, b, c \in H$, $(ab)c = a(bc)$ in G, so $(ab)c = a(bc)$ in H. (We say that H "inherits" the associative property from G.) Finally, as $H \neq \emptyset$, let $a \in H$. By (b), $a^{-1} \in H$ and by (a), $aa^{-1} = e \in H$, so H contains the identity element, and is a group.   ∎

A finiteness condition improves the situation.

Theorem
12.3

▶ If G is a group and $\emptyset \neq H \subseteq G$, with H *finite*, then H is a subgroup of G iff H is closed under the binary operation of G.

Proof

As in the proof of Theorem 12.2, if H is a subgroup of G, then H is closed under the binary operation of G. Conversely, let H be a finite nonempty subset of G that is so closed. If $a \in H$, then $aH = \{ah \mid h \in H\} \subseteq H$ because of the closure condition. By cancellation in G, $ah_1 = ah_2 \Rightarrow h_1 = h_2$, so $|aH| = |H|$. With $aH \subseteq H$ and $|aH| = |H|$, it follows from H being *finite* that $aH = H$. As $a \in H$, there exists $b \in H$ with $ab = a$. But (in G) $ab = ae$, so $b = e$ and H contains the identity. As $e \in H = aH$, there is an element $c \in H$ such that $ac = e$. Then $(ca)^2 = (c(ac))a = (ce)a = ca = (ca)e$, so $ca = e$, and $c = a^{-1} \in H$. Consequently, by Theorem 12.2, H is a subgroup of G.   ∎

The finiteness condition in Theorem 12.3 is crucial. Both $\mathbf{Z}^+$ and $\mathbf{N}$ are nonempty closed subsets of the group $(\mathbf{Z}, +)$, yet neither has the additive inverses needed for the group structure.

The next example provides a nonabelian group.

EXAMPLE 12.6

Consider the first equilateral triangle in Fig. 12.1(a). If this triangle is rotated counterclockwise through 120° about an axis perpendicular to its plane and passing through its center C, we obtain the second triangle shown in Fig. 12.1(a). As a result, the vertex originally labeled 1 in Fig. 12.1(a) is now in the position that was labeled

3. Likewise, 2 is now in the position originally occupied by 1; 3 has moved to where 2 was. This can be described by the function $\pi_1: \{1, 2, 3\} \rightarrow \{1, 2, 3\}$, where $\pi_1(1) = 3$, $\pi_1(2) = 1$, $\pi_1(3) = 2$. A more compact notation, $\left(\begin{smallmatrix} 1 & 2 & 3 \\ 3 & 1 & 2 \end{smallmatrix}\right)$, where we write $\pi_1(i)$ below i for each $1 \le i \le 3$, emphasizes that π_1 is a permutation of $\{1, 2, 3\}$. If π_2 denotes the counterclockwise rotation through $240°$ then $\pi_2 = \left(\begin{smallmatrix} 1 & 2 & 3 \\ 2 & 3 & 1 \end{smallmatrix}\right)$. For the identity π_0, i.e., the rotation through $n(360°)$, $n \in \mathbf{Z}$, $\pi_0 = \left(\begin{smallmatrix} 1 & 2 & 3 \\ 1 & 2 & 3 \end{smallmatrix}\right)$. These rotations are called *rigid motions* of the triangle. They are two-dimensional motions that keep the center C fixed and preserve the shape of the triangle. Hence the triangle looks the same as when we started, except for a possible rearrangement of some of its vertices.

Figure 12.1

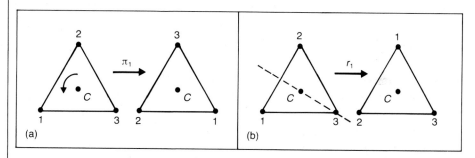

(a) (b)

In addition to these rotations the triangle can be reflected along an axis passing through a vertex and the midpoint of the opposite side. For the diagonal axis that bisects the base angle on the right, the reflection gives the result in Fig. 12.1(b). This we represent by $r_1 = \left(\begin{smallmatrix} 1 & 2 & 3 \\ 2 & 1 & 3 \end{smallmatrix}\right)$. A similar reflection about the axis bisecting the left base angle yields the permutation $r_2 = \left(\begin{smallmatrix} 1 & 2 & 3 \\ 1 & 3 & 2 \end{smallmatrix}\right)$. When the triangle is reflected about its vertical axis, we have $r_3 = \left(\begin{smallmatrix} 1 & 2 & 3 \\ 3 & 2 & 1 \end{smallmatrix}\right)$. Each r_i, $1 \le i \le 3$, is a three-dimensional motion.

Let $G = \{\pi_0, \pi_1, \pi_2, r_1, r_2, r_3\}$, the set of rigid motions (in space) of the equilateral triangle. We make G into a group by defining the rigid motion $\alpha\beta$, for α, $\beta \in G$, as that motion obtained by first applying α and then β. Hence, for example, $\pi_1 r_1 = r_3$. We can see this geometrically, but it will be easier to consider the permutations as follows: $\pi_1 r_1 = \left(\begin{smallmatrix} 1 & 2 & 3 \\ 3 & 1 & 2 \end{smallmatrix}\right)\left(\begin{smallmatrix} 1 & 2 & 3 \\ 2 & 1 & 3 \end{smallmatrix}\right)$, where, for example, $\pi_1(1) = 3$ and $r_1(3) = 3$ and we write $1 \xrightarrow{\pi_1} 3 \xrightarrow{r_1} 3$. So $1 \xrightarrow{\pi_1 r_1} 3$ in the product $\pi_1 r_1$. (Note that the order in which we write the product $\pi_1 r_1$ here is the opposite from the order of their composite function as defined in Section 3.5. The notation of Section 3.5 occurs in analysis, while in algebra there is a tendency toward this opposite order.) Also, $2 \xrightarrow{\pi_1} 1 \xrightarrow{r_1} 2$ and $3 \xrightarrow{\pi_1} 2 \xrightarrow{r_1} 1$, so $\pi_1 r_1 = \left(\begin{smallmatrix} 1 & 2 & 3 \\ 3 & 2 & 1 \end{smallmatrix}\right) = r_3$.

Table 12.4 verifies that under this binary operation G is closed, with identity π_0. Also $\pi_1^{-1} = \pi_2$, $\pi_2^{-1} = \pi_1$, and every other element is its own inverse. Since the elements of G are actually functions, the associative property follows from Theorem 3.5 (although in reverse order).

We computed $\pi_1 r_1$ as r_3, but from Table 12.4 we see that $r_1 \pi_1 = r_2$. With $\pi_1 r_1 = r_3 \ne r_2 = r_1 \pi_1$, G is nonabelian.

Table 12.4

$\cdot$	π_0	π_1	π_2	r_1	r_2	r_3
π_0	π_0	π_1	π_2	r_1	r_2	r_3
π_1	π_1	π_2	π_0	r_3	r_1	r_2
π_2	π_2	π_0	π_1	r_2	r_3	r_1
r_1	r_1	r_2	r_3	π_0	π_1	π_2
r_2	r_2	r_3	r_1	π_2	π_0	π_1
r_3	r_3	r_1	r_2	π_1	π_2	π_0

This group can also be obtained as the group of all permutations of the set $\{1, 2, 3\}$. As such it is denoted by S_3 (the *symmetric* group on three symbols).   □

EXAMPLE 12.7   The symmetric group S_4 consists of the 24 permutations of $\{1, 2, 3, 4\}$. Here $\pi_0 = \left(\begin{smallmatrix} 1 & 2 & 3 & 4 \\ 1 & 2 & 3 & 4 \end{smallmatrix}\right)$ is the identity. If $\alpha = \left(\begin{smallmatrix} 1 & 2 & 3 & 4 \\ 2 & 1 & 4 & 3 \end{smallmatrix}\right)$, $\beta = \left(\begin{smallmatrix} 1 & 2 & 3 & 4 \\ 3 & 1 & 2 & 4 \end{smallmatrix}\right)$, then $\alpha\beta = \left(\begin{smallmatrix} 1 & 2 & 3 & 4 \\ 1 & 3 & 4 & 2 \end{smallmatrix}\right)$, $\beta\alpha = \left(\begin{smallmatrix} 1 & 2 & 3 & 4 \\ 4 & 2 & 1 & 3 \end{smallmatrix}\right)$, so S_4 is nonabelian. Also, $\beta^{-1} = \left(\begin{smallmatrix} 1 & 2 & 3 & 4 \\ 2 & 3 & 1 & 4 \end{smallmatrix}\right)$ and $\alpha^2 = \pi_0 = \beta^3$. Within S_4 there is a subgroup of order 8 that represents the group of rigid motions for a square.   □

We turn now to a construction for making larger groups out of smaller ones.

Theorem 12.4   ▶ Let $(G, \circ)$ and $(H, *)$ be groups. Define the binary operation $\cdot$ on $G \times H$ by $(g_1, h_1) \cdot (g_2, h_2) = (g_1 \circ g_2, h_1 * h_2)$. Then $(G \times H, \cdot)$ is a group, called the *direct product* of G and H.

Proof   The verification of the group postulates for $(G \times H, \cdot)$ is left to the reader.   ■

EXAMPLE 12.8   Consider the groups $(\mathbf{Z}_2, +)$, $(\mathbf{Z}_3, +)$. On $G = \mathbf{Z}_2 \times \mathbf{Z}_3$ define $(a_1, b_1) \cdot (a_2, b_2) = (a_1 + a_2, b_1 + b_2)$. Then G is a group of order 6 where the identity is $(0, 0)$, and the inverse, for example, of the element $(1, 2)$ is $(1, 1)$.   □

EXERCISES

1. For each of the following sets, determine whether or not the set is a group under the stated operation. If so, determine its identity and the inverse of each element. If it is not a group, state the condition(s) of the definition that it violates.

   a) $\{-1, 1\}$ under multiplication.   b) $\{-1, 1\}$ under addition.

   c) $\{-1, 0, 1\}$ under addition.   d) $\{10n \,|\, n \in \mathbf{Z}\}$ under addition.

   e) The set of all functions $f: A \to A$, where $A = \{1, 2, 3, 4\}$, under function composition.

   f) The set of all one-to-one functions $g: A \to A$, where $A = \{1, 2, 3, 4\}$, under function composition.

   g) $\{a/2^n \,|\, a, n \in \mathbf{Z}, n \geq 0\}$ under addition.

2. Prove parts (c)–(e) of Theorem 12.1.

3. If G is a group, prove that for all $a, b \in G$,
   a) $(a^{-1})^{-1} = a$.   b) $(ab)^{-1} = b^{-1}a^{-1}$.

4. Prove that a group G is abelian iff for all $a, b \in G$, $(ab)^{-1} = a^{-1}b^{-1}$.

5. Find all subgroups in each of the groups.
   a) $(\mathbf{Z}_{12}, +)$.   b) $(\mathbf{Z}_{11}^*, \cdot)$.   c) S_3.

6. a) How many rigid motions are there for a square?
   b) Make a group table for these rigid motions like the one in Table 12.4 for the equilateral triangle. What is the identity for this group? Describe the inverse of each element geometrically.

7. a) How many rigid motions are there for a regular pentagon? Describe them geometrically.
   b) Answer (a) for a regular n-gon, $n \geq 3$.

8. In the group S_5, let

$$\alpha = \begin{pmatrix} 1 & 2 & 3 & 4 & 5 \\ 2 & 3 & 1 & 4 & 5 \end{pmatrix}, \qquad \beta = \begin{pmatrix} 1 & 2 & 3 & 4 & 5 \\ 2 & 1 & 5 & 3 & 4 \end{pmatrix}.$$

Determine $\alpha\beta$, $\beta\alpha$, α^3, β^4, α^{-1}, β^{-1}, $(\alpha\beta)^{-1}$, $(\beta\alpha)^{-1}$, $\beta^{-1}\alpha^{-1}$.

9. a) Prove Theorem 12.4.
   b) Extending the idea developed in Theorem 12.4 and Example 12.8 to the group $\mathbf{Z}_6 \times \mathbf{Z}_6 \times \mathbf{Z}_6 = \mathbf{Z}_6^3$, answer the following.
      i) What is the order of this group?
      ii) Find a subgroup of $\mathbf{Z}_6^3$ of order 6, one of order 12, and one of order 36.
      iii) Determine the inverse of each of the elements $(2, 3, 4)$, $(4, 0, 2)$, $(5, 1, 2)$.

10. If H, K are subgroups of a group G, prove that $H \cap K$ is also a subgroup of G.

12.2 ■ HOMOMORPHISMS, ISOMORPHISMS, AND CYCLIC GROUPS

We turn our attention once again to functions that preserve structure.

EXAMPLE 12.9   Let $G = (\mathbf{Z}, +)$ and $H = (\mathbf{Z}_4, +)$. Define $f: G \to H$ by

$$f(x) = [x] = \{x + 4k \,|\, k \in \mathbf{Z}\}.$$

For any $x, y \in G$,

$$f(x + y) = [x + y] = [x] + [y] = f(x) + f(y).$$
$$\uparrow \qquad\qquad\qquad\qquad\qquad\qquad\quad \uparrow$$

The operation in G   The operation in H

Here f preserves the group operations and is an example of a group homomorphism.   □

Definition
12.4

▶ If $(G, \circ)$ and $(H, *)$ are groups and $f: G \rightarrow H$, f is called a *group homomorphism* if for all $a, b \in G$, $f(a \circ b) = f(a) * f(b)$.

When we know that the given structures are groups the function f is simply called a homomorphism.

Some properties of homomorphisms are given in the following.

Theorem
12.5

▶ Let $(G, \circ)$, $(H, *)$ be groups with respective identities e_G, e_H. If $f: G \rightarrow H$ is a homomorphism, then

a) $f(e_G) = e_H$;   b) $f(a^{-1}) = [f(a)]^{-1}$ for any $a \in G$; and

c) $f(S)$ is a subgroup of H for any subgroup S of G.

Proof

a) $e_H * f(e_G) = f(e_G) = f(e_G \circ e_G) = f(e_G) * f(e_G)$, so by Theorem 12.1(d), $f(e_G) = e_H$.

b) The proof of this part is left for the reader.

c) If S is a subgroup of G, then $S \neq \emptyset$, so $f(S) \neq \emptyset$. Let $x, y \in f(S)$. Then $x = f(a)$, $y = f(b)$, for $a, b \in S$. Since S is a subgroup of G, $a \circ b \in S$, and $x * y = f(a) * f(b) = f(a \circ b) \in f(S)$. Finally $x^{-1} = [f(a)]^{-1} = f(a^{-1}) \in f(S)$ since $a^{-1} \in S$ when $a \in S$. Consequently, by Theorem 12.2, $f(S)$ is a subgroup of H.   ∎

Definition
12.5

▶ If $f: (G, \circ) \rightarrow (H, *)$ is a homomorphism, we call f an *isomorphism* if it is one-to-one and onto. In this case G, H are said to be *isomorphic groups*.

EXAMPLE 12.10

Let $f: (\mathbf{R}^+, \cdot) \rightarrow (\mathbf{R}, +)$ where $f(x) = \log_{10}(x)$. This function is both one-to-one and onto. (Verify these properties.) For $a, b \in \mathbf{R}^+$, $f(ab) = \log_{10}(ab) = \log_{10} a + \log_{10} b = f(a) + f(b)$. Therefore, f is an isomorphism and the group of positive real numbers under multiplication is abstractly the same as the group of all real numbers under addition. Here the function f translates a problem in the multiplication of real numbers (a somewhat difficult problem without a calculator) into a problem dealing with the addition of real numbers (an easier arithmetic consideration). This was a major reason behind the use of logarithms before the advent of calculators.   □

EXAMPLE 12.11

Let G be the group of complex numbers $\{1, -1, i, -i\}$ under multiplication. Table 12.5 shows the group. With $H = (\mathbf{Z}_4, +)$, consider $f: G \rightarrow H$ defined by

$$f(1) = [0]   f(-1) = [2]   f(i) = [1]   f(-i) = [3].$$

Then $f((i)(-i)) = f(1) = [0] = [1] + [3] = f(i) + f(-i)$, and $f((-1)(-i)) = f(i) = [1] = [2] + [3] = f(-1) + f(-i)$.

Although we have not checked all possible cases, the function is an isomorphism. Note that the image under f of the subgroup $\{1, -1\}$ of G is $\{[0], [2]\}$, a subgroup of H.

Let us take a closer look at the group G. Here $i^1 = i$, $i^2 = -1$, $i^3 = -i$, $i^4 = 1$, so every element of G is a power of i, and we say that i *generates* G. This is denoted by $G = \langle i \rangle$. (It is also true that $G = \langle -i \rangle$. Verify this.)   □

Table 12.5

·	1	−1	i	−i
1	1	−1	i	−i
−1	−1	1	−i	i
i	i	−i	−1	1
−i	−i	i	1	−1

This example leads us to the following.

Definition 12.6 ▶ A group G is called *cyclic* if there is an element $x \in G$ such that for all $a \in G$, $a = x^n$ for some $n \in \mathbf{Z}$.

EXAMPLE 12.12 The group $H = (\mathbf{Z}_4, +)$ is cyclic. Here the operation is addition so we have multiples instead of powers. We find that both [1] and [3] generate H. For the case of [3] we have $1 \cdot [3] = [3]$, $2 \cdot [3] = [2]$, $3 \cdot [3] = [1]$, $4 \cdot [3] = [0]$. Hence $H = \langle[3]\rangle = \langle[1]\rangle$. □

The concept of a cyclic group leads to a related idea. Given a group G, if $a \in G$ consider the set $S = \{a^k \mid k \in \mathbf{Z}\}$. From Theorem 12.2 it follows that S is a subgroup of G. This subgroup is called the *subgroup generated by a* and is designated by $\langle a \rangle$. In Example 12.11 $\langle i \rangle = \langle -i \rangle = G$; also, $\langle -1 \rangle = \{-1, 1\}$ and $\langle 1 \rangle = \{1\}$.

Definition 12.7 ▶ If G is a group and $a \in G$, the *order of a*, denoted $o(a)$, is $|\langle a \rangle|$. (If $|\langle a \rangle|$ is infinite we say that a has infinite order.)

In Example 12.11, $o(1) = 1$, $o(-1) = 2$, while i and $-i$ each have order 4.

Let us take a second look at the idea of order for $|\langle a \rangle|$ finite. With $a \in G$ and $\langle a \rangle = \{\ldots, a^{-1}, e, a, a^2, a^3, \ldots\}$ finite, by the pigeonhole principle there must be repetitions. So there is a *least* positive integer s where $a^s = a^t$, $1 \le t < s$. Then $a^{s-t} = e$ and $\langle a \rangle = \{a, a^2, \ldots, a^{s-t}\}$. Therefore $o(a)$ can be defined as the *smallest positive integer n for which $a^n = e$.*

Theorem 12.6 ▶ Let $a \in G$ with $o(a) = n$. If $k \in \mathbf{Z}$ and $a^k = e$, then $n \mid k$.

Proof By the division algorithm $k = qn + r$, $0 \le r < n$, and $e = a^k = a^{qn+r} = (a^n)^q a^r = (e^q)(a^r) = a^r$. If $0 < r < n$, we contradict the definition of n as $o(a)$. Hence $r = 0$ and $k = qn$. ∎

We now examine some further results on cyclic groups.

Theorem 12.7 ▶ Let G be a cyclic group.

a) If G is infinite, then G is isomorphic to $(\mathbf{Z}, +)$.

b) If $|G| = n$, then G is isomorphic to $(\mathbf{Z}_n, +)$.

Proof
a) For $G = \langle a \rangle = \{a^k \mid k \in \mathbf{Z}\}$, let $f: G \to \mathbf{Z}$ be defined by $f(a^k) = k$. (Could we have $a^k = a^t$ with $k \neq t$? If so, f would not be a function.) For $a^m, a^n \in G$, $f(a^m \cdot a^n) = f(a^{m+n}) = m + n = f(a^m) + f(a^n)$, so f is a homomorphism. We leave to the reader the verification that f is one-to-one and onto.

b) If $G = \langle a \rangle = \{a, a^2, \ldots, a^{n-1}, a^n = e\}$, the function $f: G \to \mathbf{Z}_n$ defined by $f(a^k) = [k]$ is an isomorphism. (Verify this.)   ∎

EXAMPLE 12.13   If $G = \langle g \rangle$, since $g^m \cdot g^n = g^{m+n} = g^{n+m} = g^n \cdot g^m$ for all $m, n \in \mathbf{Z}$, G is abelian. The converse, however, is false. The group H of Table 12.6 is abelian, and $o(e) = 1$, $o(a) = o(b) = o(c) = 2$. Since no element of H has order 4, H cannot be cyclic. (The group H is the smallest noncyclic group and is known as the *Klein Four* group.)   □

Table 12.6

·	e	a	b	c
e	e	a	b	c
a	a	e	c	b
b	b	c	e	a
c	c	b	a	e

Our last result concerns the structure of subgroups in a cyclic group.

Theorem 12.8   ▶ Any subgroup of a cyclic group is cyclic.

Proof   Let $G = \langle a \rangle$. If H is a subgroup of G, each element of H has the form a^k, $k \in \mathbf{Z}$. For $H \neq \{e\}$, let t be the smallest positive integer such that $a^t \in H$. (How do we know such an integer t exists?) We claim that $H = \langle a^t \rangle$. Since $a^t \in H$, by the closure property for the subgroup H, $\langle a^t \rangle \subseteq H$. For the opposite inclusion, let $b \in H$ with $b = a^s$, $s \in \mathbf{Z}$. If $t \nmid s$, by the division algorithm $s = qt + r$, $q, r \in \mathbf{Z}$, $0 < r < t$. Consequently, $a^s = a^{qt+r}$ and $a^r = a^{-qt}a^s = (a^t)^{-q}b$, with $a^t, b \in H$. But if $a^r \in H$, we contradict the minimality of t. Hence $r = 0$ and $b = a^{qt} = (a^t)^q \in \langle a^t \rangle$, so $H = \langle a^t \rangle$, a cyclic group.   ∎

EXERCISES

1. Prove Theorem 12.5(b).
2. If $f: G \to H$, $g: H \to K$ are homomorphisms, prove that the composite function $g \circ f: G \to K$, where $(g \circ f)(x) = g(f(x))$, is a homomorphism.
3. If $G = (\mathbf{Z}_6, +)$, $H = (\mathbf{Z}_3, +)$, $K = (\mathbf{Z}_2, +)$, find an isomorphism for the groups $H \times K$ and G.
4. Let $f: G \to H$ be a group homomorphism onto H. If G is abelian, prove that H is abelian.
5. Find the order of each element in the group of rigid motions of (a) the equilateral triangle; and (b) the square.

6. In S_5 find an element of order n, $2 \leq n \leq 5$. Also determine the cyclic subgroups of S_5 that these elements generate.

7. Verify that $(\mathbf{Z}_p^*, \cdot)$ is cyclic for the primes 5, 7, and 11.

8. For a group G, prove that the function $f: G \to G$ defined by $f(a) = a^{-1}$, is an isomorphism iff G is abelian.

9. a) Find all generators of the cyclic groups $(\mathbf{Z}_{12}, +)$, $(\mathbf{Z}_{16}, +)$, and $(\mathbf{Z}_{24}, +)$.

   b) Let $G = \langle a \rangle$ with $o(a) = n$. Prove that a^k, $k \in \mathbf{Z}^+$, generates G iff k and n are relatively prime.

   c) If G is a cyclic group of order n, how many distinct generators does it have?

10. Let $f: G \to H$ be a group homomorphism. If $a \in G$ with $o(a) = n$, and $o(f(a)) = k$, prove that $k \mid n$.

12.3 ■ COSETS AND LAGRANGE'S THEOREM

In the last two sections, for any finite group G and subgroup H of G, we had $|H|$ dividing $|G|$. In this section we'll see that this was not mere chance, but is true in general. To prove this we need one new idea.

Definition 12.8

▶ If H is a subgroup of G, for any $a \in G$ the set $aH = \{ah \mid h \in H\}$ is called a *left coset* of H in G. $Ha = \{ha \mid h \in H\}$ is a *right coset*.

If the operation in G is addition, we write $a + H$ in place of aH, where $a + H = \{a + h \mid h \in H\}$.

When the term coset is used in this chapter it will refer to left coset. For abelian groups there is no need to distinguish between left and right cosets. This is not so for nonabelian groups.

EXAMPLE 12.14

If G is the group of Example 12.6 and $H = \{\pi_0, \pi_1, \pi_2\}$, the coset $r_1 H = \{r_1 \pi_0, r_1 \pi_1, r_1 \pi_2\} = \{r_1, r_2, r_3\}$. Likewise we have $r_2 H = r_3 H = \{r_1, r_2, r_3\}$, while $\pi_0 H = \pi_1 H = \pi_2 H = H$.

We see that for any $\alpha \in G$, $|\alpha H| = |H|$, and that $G = H \cup r_1 H$ is a partition of G.

For the subgroup $K = \{\pi_0, r_1\}$, we find $r_2 K = \{r_2, \pi_2\}$, $r_3 K = \{r_3, \pi_1\}$. Again a partition of G arises: $G = K \cup r_2 K \cup r_3 K$. (*Note:* $Kr_2 = \{\pi_0 r_2, r_1 r_2\} = \{r_2, \pi_1\} \neq r_2 K$) □

EXAMPLE 12.15

For $G = (\mathbf{Z}_{12}, +)$ and $H = \{[0], [4], [8]\}$

$$[0] + H = \{[0], [4], [8]\} = [4] + H = [8] + H = H$$
$$[1] + H = \{[1], [5], [9]\} = [5] + H = [9] + H$$
$$[2] + H = \{[2], [6], [10]\} = [6] + H = [10] + H$$
$$[3] + H = \{[3], [7], [11]\} = [7] + H = [11] + H,$$

and $H \cup ([1] + H) \cup ([2] + H) \cup ([3] + H)$ is a partition of G. □

These examples lead us to the following results.

Lemma
12.1

▶ If H is a subgroup of the finite group G, then for any a, $b \in G$, (a) $|aH| = |H|$; and (b) $aH = bH$ or $aH \cap bH = \emptyset$.

Proof

a) Since $aH = \{ah \mid h \in H\}$, $|aH| \leq |H|$. If $|aH| < |H|$, we have $ah_i = ah_j$ with h_i, h_j distinct elements of H. By left cancellation in G we get the contradiction $h_i = h_j$, so $|aH| = |H|$.

b) If $aH \cap bH \neq \emptyset$, let $c = ah_1 = bh_2$, h_1, $h_2 \in H$. If $x \in aH$, then $x = ah$, $h \in H$, and so $x = (bh_2h_1^{-1})h = b(h_2h_1^{-1}h) \in bH$, and $aH \subseteq bH$. Similarly, $y \in bH \Rightarrow y = bh_3$, $h_3 \in H \Rightarrow y = (ah_1h_2^{-1})h_3 = a(h_1h_2^{-1}h_3) \in aH$, so $bH \subseteq aH$. Therefore aH and bH are either disjoint or identical.

   (We observe here that if $g \in G$, then $g \in gH$ since $e \in H$. Also, by (b), G can be partitioned into mutually disjoint cosets.)   ∎

We are now ready to prove the main result of this section.

Theorem
12.9

▶ (*Lagrange's Theorem*) If G is a group of order n with H a subgroup of order m, then m divides n.

Proof

If $H = G$ the result follows. Otherwise $m < n$ and there exists an element $a \in G - H$. Since $a \notin H$, $aH \neq H$, so $aH \cap H = \emptyset$. If $G = aH \cup H$, $|G| = |aH| + |H| = 2|H|$ and the theorem follows. If not, there is an element $b \in G - (H \cup aH)$, with $bH \cap H = \emptyset = bH \cap aH$ and $|bH| = |H|$. If $G = bH \cup aH \cup H$, we have $|G| = 3|H|$. Otherwise we're back to an element $c \in G$ with $c \notin bH \cup aH \cup H$. Since G is finite this process terminates and we find that $G = a_1 H \cup a_2 H \cup \ldots \cup a_k H$. Therefore, $|G| = k|H|$ and m divides n.   ∎

We close with the statements of two corollaries. Their proofs are requested in the section exercises.

Corollary
12.1

▶ If G is finite and $a \in G$, then $\sigma(a)$ divides $|G|$.

Corollary
12.2

▶ Any group of prime order is cyclic.

EXERCISES

1. Let $G = S_4$. (a) For $\alpha = \left(\begin{smallmatrix} 1 & 2 & 3 & 4 \\ 2 & 3 & 4 & 1 \end{smallmatrix}\right)$, find the subgroup $H = \langle\alpha\rangle$. (b) Determine the left cosets of H in G.

2. Answer Exercise 1 for $\beta = \left(\begin{smallmatrix} 1 & 2 & 3 & 4 \\ 2 & 3 & 1 & 4 \end{smallmatrix}\right)$.

3. If $\gamma = \left(\begin{smallmatrix} 1 & 2 & 3 & 4 \\ 2 & 1 & 4 & 3 \end{smallmatrix}\right) \in S_4$, how many cosets does $\langle\gamma\rangle$ determine?

4. For $G = (\mathbf{Z}_{24}, +)$, find the cosets determined by the subgroup $H = \langle[3]\rangle$. Do likewise for the subgroup $K = \langle[4]\rangle$.

5. Let $G = \mathbf{R} \times \mathbf{R}$, with group operation $+$ defined by $(a, b) + (c, d) = (a + c, b + d)$, $a, b, c, d \in \mathbf{R}$.

   a) If $H = \{(a, 0) \mid a \in \mathbf{R}\}$, prove that H is a subgroup of G.

   b) Give a geometric interpretation of the cosets of H in G.

6. a) Let R be a ring with unity u. Prove that the units of R form a group under the multiplication of the ring.

   b) If $R = (\mathbf{Z}_n, +, \cdot)$, how many elements are in its (multiplicative) group of units?

7. If G is a group of order n and $a \in G$, prove that $a^n = e$.

8. a) *(Fermat's Theorem)* If p is a prime, prove that $a^p \equiv a \pmod{p}$ for any $a \in \mathbf{Z}$. (How is this related to Exercise 12(a) of Section 8.3?)

   b) *(Euler's Theorem)* For any $n \in \mathbf{Z}^+$, $a \in \mathbf{Z}$, prove that if $(a, n) = 1$, then $a^{\phi(n)} \equiv 1 \pmod{n}$.

   c) How are these two theorems related?

   d) Is there any connection between these two theorems and the results in Exercises 6 and 7?

9. Let p be a prime. (a) If G has order $2p$, prove that every proper subgroup of G is cyclic. (b) If G has order p^2, prove that G has a subgroup of order p.

10. Prove Corollaries 12.1 and 12.2.

12.4 ∎ ELEMENTS OF CODING THEORY

In this and the next four sections we introduce an area of applied mathematics called *algebraic coding theory*. This theory was inspired by the fundamental paper of Claude Shannon (1948) along with results by M. Golay (1949) and Richard Hamming (1950). Since that time it has become an area of great interest where algebraic structures, probability, and combinatorics all play a role.

Our coverage will be held to an introductory level as we seek to model the transmission of information represented by strings of the signals 0 and 1.

In digital communications, when information is transmitted in the form of strings of 0's and 1's, certain problems arise. As a result of "noise" in the channel, when a certain signal is transmitted a different signal may be received, thus causing a wrong decision to be made by the receiver. Hence we want to develop techniques to help us detect, and perhaps even correct, transmission errors. However, we can only improve the chances of correct transmission; there are no guarantees.

Our model uses a *binary symmetric channel*, as shown in Fig. 12.2. The adjective "binary" appears since an individual signal is represented by one of the bits 0 or 1. When a transmitter sends the signal 0 or 1 in such a channel, associated with either signal is a (constant) probability p for incorrect transmission. When that probability p is the same for both signals the channel is called *symmetric*. Here, for example, we have probability p of sending 0 and having 1 received. The probability of sending signal 0 and having it received correctly is then $1 - p$. All possibilities are considered in Fig. 12.2.

Figure 12.2

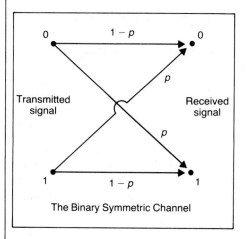

The Binary Symmetric Channel

EXAMPLE 12.16 Consider the string $c = 10110$. We regard c as an element of the group $\mathbf{Z}_2^5$, formed from the direct product of five copies of $(\mathbf{Z}_2, +)$. To shorten notation we write 10110 instead of $(1, 0, 1, 1, 0)$. When sending each bit (individual signal) of c through the binary symmetric channel, we assume the probability of incorrect transmission is $p = 0.05$.

The probability of transmitting c with no errors is $(.95)^5 \doteq 0.77$. Here, and throughout our discussion of coding theory, we assume that the transmission of any signal does not depend in any way on the transmissions of the prior signals. Consequently, the probability of the occurrence of all of these *independent* events (in their prescribed order) is given by the product of their individual probabilities.

What is the probability that the party receiving the five-bit message receives the string $r = 00110$, i.e., the original message with an error in the first position? The probability of incorrect transmission for the first bit is 0.05, so with the assumption of independent events, $(0.05)(0.95)^4 \doteq 0.041$ is the probability of sending $c = 10110$ and receiving $r = 00110$. With $e = 10000$, we can write $c + e = r$, and interpret r as the result of the sum of the original message c and the particular *error pattern* $e = 10000$. Since $c, r, e \in \mathbf{Z}_2^5$ and $-1 = 1$ in $\mathbf{Z}_2$, we also have $c + r = e$ and $r + e = c$.

In transmitting $c = 10110$, the probability of receiving $r = 00100$ is

$$(0.05)(0.95)^2(0.05)(0.95) \doteq 0.002,$$

so this multiple error is not very likely to occur.

Finally if we transmit $c = 10110$, what is the probability that r differs from c in exactly two places? To answer this we sum the probabilities for each error pattern consisting of two 1's and three 0's. Each such pattern has probability 0.002. Since there are $\binom{5}{2}$ such patterns, the probability of two errors in transmission is

$$\binom{5}{2}(0.05)^2(0.95)^3 \doteq 0.021.$$ □

These results lead us to the following theorem.

Theorem 12.10

▶ Let $c \in \mathbf{Z}_2^n$. For the transmission of c through a binary symmetric channel with probability p of incorrect transmission

a) the probability of receiving $r = c + e$, where e is a *particular* error pattern consisting of k 1's and $(n - k)$ 0's, is $p^k(1 - p)^{n-k}$.

b) the probability that k errors are made in the transmission is $\binom{n}{k}p^k(1 - p)^{n-k}$.

In Example 12.16, the probability of making at most one error in the transmission of $c = 10110$ is $(0.95)^5 + \binom{5}{1}(0.05)(0.95)^4 \doteq 0.977$. Thus the chance for multiple errors in transmission will be considered negligible throughout the discussion in this chapter. Such an assumption is valid when p is small. In actuality, a binary symmetric channel is considered "good" when $p < 10^{-5}$. However, no matter what else we stipulate, we always want $p < 1/2$.

To improve the accuracy of transmission in a binary symmetric channel, certain types of coding schemes can be used where extra signals are provided.

For $m, n \in \mathbf{Z}^+$, let $n > m$. Consider $\emptyset \neq W \subseteq \mathbf{Z}_2^m$. The set W consists of the *messages* to be transmitted. We add to each $w \in W$ extra signals to form the *code word* c, where $c \in \mathbf{Z}_2^n$. This process is called *encoding* and is represented by the function $E\colon W \to C$. Then $E(w) = c$ and $E(W) = C \subseteq \mathbf{Z}_2^n$. To keep the code words for the messages distinct, E must be one-to-one. Upon transmission c is transformed into $T(c)$, where $T(c) \in \mathbf{Z}_2^n$. Unfortunately, T is not a function, since $T(c)$ may be different at different transmission times. (See Fig. 12.3.)

Figure 12.3

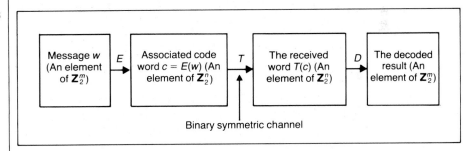

Binary symmetric channel

Upon receiving $T(c)$ we want to apply a decoding function $D\colon \mathbf{Z}_2^n \to \mathbf{Z}_2^m$ to remove the extra signals and, we hope, obtain the original message w. Ideally $D \circ T \circ E$ should be the identity function on W, with $D\colon C \to W$. Since this cannot be achieved, we seek functions E and D such that there is a high probability of decoding the received word $T(c)$ and recapturing the original message w. In addition we want the ratio m/n to be as large as possible so that an excessive number of signals are not added to w in getting $E(w)$. This ratio m/n measures the *efficiency* of our scheme and is called the *rate* of the code. Finally, the functions E and D should be more than theoretical results; they must be practical in the sense that they can be implemented electronically.

In such a scheme, the functions E and D are called the *encoding* and *decoding* functions, respectively, of an (n, m) *block code*.

We illustrate these ideas in the following two examples.

EXAMPLE 12.17   Consider the $(m + 1, m)$ block code for $m = 8$. Let $W = \mathbf{Z}_2^8$. For any $w = w_1 w_2 \ldots w_8 \in W$, define $E : \mathbf{Z}_2^8 \rightarrow \mathbf{Z}_2^9$ by $E(w) = w_1 w_2 \ldots w_8 w_9$, where $w_9 = \sum_{i=1}^{8} w_i$, with the addition performed modulo 2. For example, $E(11001101) = 110011011$, and $E(00110011) = 001100110$.

For all $w \in \mathbf{Z}_2^8$, $E(w)$ contains an even number of 1's. So for $w = 11010110$ and $E(w) = 110101101$, if we receive $T(c) = T(E(w))$ as 100101101, from the odd number of 1's in $T(c)$ we know that a mistake has occurred in transmission. Hence we are able to *detect* single errors in transmission. But we have no way to correct such errors.

The probability of sending the code word 110101101 and making at most one error in transmission is

$$\underbrace{(1 - p)^9}_{} + \underbrace{\binom{9}{1} p (1 - p)^8}_{}.$$

All nine bits are   One bit is changed in
correctly transmitted.   transmission and an error is detected.

For $p = 0.001$ this gives $(0.999)^9 + \binom{9}{1}(0.001)(0.999)^8 \doteq 0.999964$.

If we detect an error and we are able to relay a signal back to the transmitter to repeat the transmission of the code word, and continue this until the received word has an even number of 1's, then the probability of sending the code word 110101101 and receiving the correct transmission is approximately 0.999964.[†]

Should an even number of errors occur in transmission, $T(c)$ is unfortunately accepted as the correct code word and we interpret its first eight components as the original message. This scheme is called the $(m + 1, m)$ *parity-check code* and is only appropriate when multiple errors are not likely to occur.

If we send the message 11010110 through the channel we have probability $(0.999)^8 = 0.992028$ of correct transmission. By using this parity-check code the chances of getting the correct message increases to (approximately) 0.999964. However, an extra signal is sent (and perhaps additional transmissions are needed) and the rate of the code has decreased from 1 to 8/9.

But suppose that instead of sending eight bits we sent 160 bits, in successive strings of length 8. The chances of receiving the correct message without any coding scheme would be $(0.999)^{160} \doteq 0.852076$. With the parity-check method we send 180 bits but the chance for correct transmission increases to $(0.999964)^{20} \doteq 0.999280$. □

EXAMPLE 12.18   The $(3m, m)$ *triple repetition code* is one where we can *detect* and *correct* single errors in transmission. With $m = 8$ and $W = \mathbf{Z}_2^8$, we define $E : \mathbf{Z}_2^8 \rightarrow \mathbf{Z}_2^{24}$ by $E(w_1 w_2 \ldots w_7 w_8) = w_1 w_2 \ldots w_8 w_1 w_2 \ldots w_8 w_1 w_2 \ldots w_8$.

Hence if $w = 10110111$, $c = E(w) = 101101111011011110110111$.

[†]The probability of sending the code word 110101101 and receiving the correct transmission under the conditions described is, for $p = 0.001$, $(0.999)^9 + [\binom{9}{1}(0.001)(0.999)^8](0.999)^9 + [\binom{9}{1}(0.001) \cdot (0.999)^8]^2(0.999)^9 + \cdots = (0.999)^9(1 + x + x^2 + \cdots) = (0.999)^9(1/(1 - x))$, where $x = \binom{9}{1}(0.001)(0.999)^8$. This probability is 0.999964 (to six decimal places).

The decoding function $D: \mathbf{Z}_2^{24} \to \mathbf{Z}_2^8$ is carried out by the majority rule. For example, if $T(c) = 101001110011011110110110$ we have three errors. We decode $T(c)$ by examining the first, ninth, and seventeenth positions to see which signal appears more times. Here it is 1, so we decode the first entry in the decoded message as 1. Continuing with the entries in the second, tenth, and eighteenth positions, the second entry of the decoded message is 0. As we proceed, we recapture the correct message, 10110111.

Although we have more than one transmission error here, all is well unless two errors occur with the second error eight or sixteen spaces after the first, i.e., if two or more incorrect transmissions occur for the same bit of the original message.

Now how does this scheme compare with the other methods we have? With $p = 0.001$, the probability of correctly decoding a single bit is $(0.999)^3 + \binom{3}{1} \cdot (0.001)(0.999)^2 \doteq 0.999997$. So the probability of receiving and correctly decoding the eight-bit message is $(0.999997)^8 = 0.999976$, just slightly better than the result from the parity-check method. But we transmit 24 signals for this message, so our rate is now $\frac{1}{3}$. For this increased accuracy and the ability to detect and correct single errors, we pay with an increase in transmission time.   □

EXERCISES

1. Let C be a set of code words, where $C \subseteq \mathbf{Z}_2^7$. In each of the following, two of e (error pattern), r (received word) and c (code word) are given, with $r = c + e$. Determine the third term.

   a) $c = 1010110$, $r = 1011111$ b) $c = 1010110$, $e = 0101101$

   c) $e = 0101111$, $r = 0000111$

2. A binary symmetric channel has probability $p = 0.05$ of incorrect transmission. If the code word $c = 011011101$ is transmitted, what is the probability that (a) we receive $r = 011111101$? (b) we receive $r = 111011100$? (c) a single error occurs? (d) a double error occurs? (e) a triple error occurs? (f) three errors occur with no two of them consecutive?

3. Let $E: \mathbf{Z}_2^3 \to \mathbf{Z}_2^9$ be the encoding function for the $(9, 3)$ triple repetition code.

   a) If $D: \mathbf{Z}_2^9 \to \mathbf{Z}_2^3$ is the corresponding decoding function, apply D to decode the received words

       i) 111101100;   ii) 000100011;   iii) 010011111.

   b) Find three different received words r for which $D(r) = 000$.

   c) For any $w \in \mathbf{Z}_2^3$, what is $|D^{-1}(w)|$?

4. The $(5m, m)$ five-times repetition code has encoding function $E: \mathbf{Z}_2^m \to \mathbf{Z}_2^{5m}$, where $E(w) = wwwww$. Decoding with $D: \mathbf{Z}_2^{5m} \to \mathbf{Z}_2^m$ is accomplished by the majority rule. (Here we are able to correct single and double errors made in transmission.)

   a) With $p = 0.05$, what is the probability for the transmission and correct decoding of the signal 0?

   b)   Answer (a) for the message 110.

   c)   For $m = 2$, decode the received word $r = 0111001001$.

   d)   If $m = 2$, find three received words r where $D(r) = 00$.

   e)   For $m = 2$ and $D: \mathbf{Z}_2^{10} \rightarrow \mathbf{Z}_2^2$, what is $|D^{-1}(w)|$ for $w \in \mathbf{Z}_2^2$?

12.5 ■ THE HAMMING METRIC

In this section we develop the general principles for discussing the error-detecting and error-correcting capabilities of a coding scheme. These ideas are due to Richard Hamming.

We start by considering a code $C \subseteq \mathbf{Z}_2^4$, with $c_1 = 0111$, $c_2 = 1111 \in C$. Now both the transmitter and the receiver know the elements of C. So if the transmitter sends c_1 but the person receiving the code word receives $T(c_1)$ as 1111 then he or she feels that c_2 was transmitted and makes whatever decision (a wrong one) c_2 implies. Consequently, although only one transmission error was made the results could be unpleasant. Why is this? Unfortunately we have two code words that are almost the same. They are rather *close* to each other, for they differ in only one component.

We describe this notion of closeness more precisely as follows.

Definition 12.9   ▶   For any element $x = x_1 x_2 \ldots x_n \in \mathbf{Z}_2^n$, $n \in \mathbf{Z}^+$, the *weight of x*, denoted $wt(x)$, is the number of components x_i, of x, $1 \leq i \leq n$, where $x_i = 1$. If $y \in \mathbf{Z}_2^n$, the *distance between x and y*, denoted $d(x, y)$, is the number of components where $x_i \neq y_i$, $1 \leq i \leq n$.

EXAMPLE 12.19   For $n = 5$, let $x = 01001$ and $y = 11101$. Then $wt(x) = 2$, $wt(y) = 4$, and $d(x, y) = 2$. In addition, $x + y = 10100$, so $wt(x + y) = 2$. Is it just by chance that $d(x, y) = wt(x + y)$? For $1 \leq i \leq 5$, $x_i + y_i$ contributes a count of 1 to $wt(x + y) \Leftrightarrow x_i \neq y_i \Leftrightarrow x_i, y_i$ contribute a count of 1 to $d(x, y)$. (This is true for any $n \in \mathbf{Z}^+$ so $wt(x + y) = d(x, y)$ for all $x, y \in \mathbf{Z}_2^n$.)   □

When $x, y \in \mathbf{Z}_2^n$, we write $d(x, y) = \sum_{i=1}^n d(x_i, y_i)$ where,

$$\text{for } 1 \leq i \leq n, \qquad d(x_i, y_i) = \begin{cases} 0 & \text{if } x_i = y_i \\ 1 & \text{if } x_i \neq y_i. \end{cases}$$

Lemma 12.2   ▶   For any $x, y \in \mathbf{Z}_2^n$, $wt(x + y) \leq wt(x) + wt(y)$.

Proof   This lemma is proved by examining the individual components x_i, y_i, $x_i + y_i$, $1 \leq i \leq n$, of x, y, $x + y$, respectively. Only one situation would cause this inequality to be false: if $x_i + y_i = 1$ while $x_i = 0$ and $y_i = 0$, for some $1 \leq i \leq n$. But this never occurs since $x_i + y_i = 1$ implies that exactly one of x_i and y_i is 1.   ■

Theorem 12.11   ▶   The distance function d defined on $\mathbf{Z}_2^n \times \mathbf{Z}_2^n$ satisfies the following for all $x, y, z \in \mathbf{Z}_2^n$.

a) $d(x, y) \geq 0$;

b) $d(x, y) = 0 \Leftrightarrow x = y$;

c) $d(x, y) = d(y, x)$; and

d) $d(x, z) \leq d(x, y) + d(y, z)$.

Proof We leave the first three parts for the reader and prove (d).

In $\mathbf{Z}_2^n$, $y + y = 0$, so $d(x, z) = wt(x + z) = wt(x + (y + y) + z) = wt((x + y) + (y + z)) \leq wt(x + y) + wt(y + z)$, by Lemma 12.2. With $wt(x + y) = d(x, y)$, $wt(y + z) = d(y, z)$, the result follows. (This property is generally called the *Triangle Inequality*.) ∎

When a function satisfies the four properties listed here it is called a *distance function* and we call $(\mathbf{Z}_2^n, d)$ a *metric space*. Hence d is often referred to as the *Hamming metric*.

Definition 12.10 ▶ For $n, k \in \mathbf{Z}^+$, $x \in \mathbf{Z}_2^n$, the *sphere* of radius k centered at x is defined as $S(x, k) = \{y \in \mathbf{Z}_2^n \mid d(x, y) \leq k\}$.

EXAMPLE 12.20 For $n = 3$ and $x = 110 \in \mathbf{Z}_2^3$, $S(x, 1) = \{110, 010, 100, 111\}$, $S(x, 2) = \{110, 010, 100, 111, 000, 101, 011\}$. □

With these preliminaries we turn now to the two major results of this section.

Theorem 12.12 ▶ Let $E: W \to C$ be an encoding function with the set of messages $W \subseteq \mathbf{Z}_2^m$ and the set of code words $E(W) = C \subseteq \mathbf{Z}_2^n$, $m < n$. For $k \in \mathbf{Z}^+$, we can detect transmission errors of weight $\leq k$ iff the minimum distance between code words is at least $k + 1$.

Proof The set C is known to both the transmitter and the receiver, so if $w \in W$ is the message and $c = E(w)$ is transmitted, let $c \neq T(c) = r$. If the minimum distance between code words is at least $k + 1$ then the transmission of c can result in as many as k errors and r will not be listed in C. Hence we can detect all errors e where $wt(e) \leq k$. Conversely, let c_1, c_2 be code words with $d(c_1, c_2) < k + 1$. Then $c_2 = c_1 + e$ where $wt(e) \leq k$. If we send c_1 and $T(c_1) = c_2$ we would feel that c_2 had been sent, thus failing to detect an error of weight $\leq k$. ∎

What can we say about error-correcting capability?

Theorem 12.13 ▶ With E, W, C as in Theorem 12.12, and $k \in \mathbf{Z}^+$, we can construct a decoding function $D: \mathbf{Z}_2^n \to W$ that corrects all transmission errors of weight $\leq k$ iff the minimum distance between code words is at least $2k + 1$.

Proof For $c \in C$, consider $S(c, k) = \{x \in \mathbf{Z}_2^n \mid d(c, x) \leq k\}$. Define $D: \mathbf{Z}_2^n \to W$ as follows. If $r \in \mathbf{Z}_2^n$ and $r \in S(c, k)$ for some code word c, $D(r) = w$ where $E(w) = c$. If $r \notin S(c, k)$ for all $c \in C$, define $D(r) = w_0$, where w_0 is some arbitrary message that remains fixed once it is chosen. The only problem we could face here is that D might not be a function. This will happen if there is an element r in $\mathbf{Z}_2^n$ with r in both $S(c_1, k)$ and $S(c_2, k)$ for distinct code words c_1, c_2. But $r \in S(c_1, k) \Rightarrow d(r, c_1) \leq k$, and $r \in S(c_2, k) \Rightarrow d(r, c_2) \leq k$, so $d(c_1, c_2) \leq d(c_1, r) + d(r, c_2) \leq k + k < 2k + 1$. Consequently, if the minimum distance between code words is at least $2k + 1$, D is a function, and it will decode all possible received words, correcting any transmission error of weight $\leq k$. Conversely, if c_1, $c_2 \in C$ and $d(c_1, c_2) \leq 2k$, then c_2 can be obtained from c_1 by making at most $2k$ changes. Starting at code word c_1 we make approximately half (exactly, $\lfloor d(c_1, c_2)/2 \rfloor$) of these changes. This brings us to $r = c_1 + e_1$ with $wt(e_1) \leq k$. Continuing from r we make the remaining changes to get to c_2 and find $r + e_2 = c_2$ with $wt(e_2) \leq k$. But then $r = c_2 + e_2$. Now with $c_1 + e_1 = r = c_2 + e_2$, and $wt(e_1), wt(e_2) \leq k$, how can one decide on the code word from which r arises? As a result we have an error of weight $\leq k$ that cannot be corrected. ∎

EXAMPLE 12.21   With $W = \mathbf{Z}_2^2$ let $E: W \to \mathbf{Z}_2^6$ be given by

$$E(00) = 000000 \quad E(10) = 101010 \quad E(01) = 010101 \quad E(11) = 111111.$$

Then the minimum distance between code words is 3, so we can detect double errors and correct single ones.

With $S(000000, 1) = \{x \in \mathbf{Z}_2^6 \mid d(000000, x) \leq 1\} = \{000000, 100000, 010000, 001000, 000100, 000010, 000001\}$, the decoding function $D: \mathbf{Z}_2^6 \to W$ gives $D(x) = 00$ for all $x \in S(000000, 1)$.

Similarly, $S(010101, 1) = \{x \in \mathbf{Z}_2^6 \mid d(010101, x) \leq 1\} = \{010101, 110101, 000101, 011101, 010001, 010111, 010100\}$, and here $D(x) = 01$ for each $x \in S(010101, 1)$. At this point our definition of D accounts for 14 of the elements in $\mathbf{Z}_2^6$. Continuing to define D for the 14 elements in $S(101010, 1)$ and $S(111111, 1)$, there remain 36 other elements to account for. We define $D(x) = 00$ (or any other message) for these 36 other elements and have a decoding function that will correct single errors.

With regard to detection, if $c = 010101$ and $T(c) = r = 111101$, we can detect this double error. But if $T(c) = r_1 = 111111$, a triple error has occurred, so we think that $c = 111111$ and incorrectly decode r_1 as 11 instead of the correct message, 01. □

12.6 ■ THE PARITY-CHECK AND GENERATOR MATRICES

In this section we introduce an example where encoding and decoding functions are given by matrices over $\mathbf{Z}_2$.

EXAMPLE 12.22   Let

$$G = \begin{bmatrix} 1 & 0 & 0 & 1 & 1 & 0 \\ 0 & 1 & 0 & 0 & 1 & 1 \\ 0 & 0 & 1 & 1 & 0 & 1 \end{bmatrix}$$

be a 3×6 matrix over $\mathbf{Z}_2$. The first three columns of G form the 3×3 identity matrix I_3. Letting A denote the matrix formed from the last three columns of G we write $G = [I_3 | A]$ to denote its structure. The matrix G is called a *generator matrix*. We use G to define an encoding function $E: \mathbf{Z}_2^3 \to \mathbf{Z}_2^6$ as follows. For $w \in \mathbf{Z}_2^3$, $E(w) = wG$ is the element in $\mathbf{Z}_2^6$ obtained by multiplying w, considered as a three-dimensional row vector, by the matrix G on its right. Unlike the results on matrix multiplication in Chapter 5, in the calculations here we have $1 + 1 = 0$, not $1 + 1 = 1$.

(If the set W of messages is not all of $\mathbf{Z}_2^3$, we'll assume that all of $\mathbf{Z}_2^3$ is encoded. The transmitter and receiver will both know the real messages of importance and their corresponding code words.)

We find here, for example, that

$$E(110) = (110)G = [110]\begin{bmatrix} 1 & 0 & 0 & 1 & 1 & 0 \\ 0 & 1 & 0 & 0 & 1 & 1 \\ 0 & 0 & 1 & 1 & 0 & 1 \end{bmatrix} = [110101],$$

and

$$E(010) = (010)G = [010]\begin{bmatrix} 1 & 0 & 0 & 1 & 1 & 0 \\ 0 & 1 & 0 & 0 & 1 & 1 \\ 0 & 0 & 1 & 1 & 0 & 1 \end{bmatrix} = [010011].$$

Note that $E(110)$ can be obtained by adding the first two rows of G while $E(010)$ is simply the second row of G.

The set of code words is $C = \{000000, 100110, 010011, 001101, 110101, 101011, 011110, 111000\} \subseteq \mathbf{Z}_2^6$, so decoding is accomplished by simply dropping the last three components of the code word. In addition, the minimum distance between code words is 3 so we can detect errors of weight ≤ 2 and correct single errors.

For any $w = w_1 w_2 w_3 \in \mathbf{Z}_2^3$, $E(w) = w_1 w_2 w_3 w_4 w_5 w_6 \in \mathbf{Z}_2^6$. Since

$$E(w) = [w_1 w_2 w_3]\begin{bmatrix} 1 & 0 & 0 & 1 & 1 & 0 \\ 0 & 1 & 0 & 0 & 1 & 1 \\ 0 & 0 & 1 & 1 & 0 & 1 \end{bmatrix}$$

$$= [w_1 w_2 w_3 (w_1 + w_3)(w_1 + w_2)(w_2 + w_3)],$$

we have $w_4 = w_1 + w_3$, $w_5 = w_1 + w_2$, $w_6 = w_2 + w_3$, and these equations are called the *parity-check equations*. With each $w_i \in \mathbf{Z}_2$, $1 \leq i \leq 6$, $w_i = -w_i$, so the equations can be rewritten as

$$\begin{aligned} w_1 \quad\quad\ + w_3 + w_4 \quad\quad\quad\quad &= 0 \\ w_1 + w_2 \quad\quad\quad\ + w_5 \quad\ &= 0 \\ w_2 + w_3 \quad\quad\quad\ + w_6 &= 0. \end{aligned}$$

Thus we find that

$$\begin{bmatrix} 1 & 0 & 1 & 1 & 0 & 0 \\ 1 & 1 & 0 & 0 & 1 & 0 \\ 0 & 1 & 1 & 0 & 0 & 1 \end{bmatrix}\begin{bmatrix} w_1 \\ w_2 \\ w_3 \\ w_4 \\ w_5 \\ w_6 \end{bmatrix} = H \cdot (E(w))^{\mathrm{tr}} = \begin{bmatrix} 0 \\ 0 \\ 0 \end{bmatrix},$$

where $(E(w))^{tr}$ denotes the transpose of $E(w)$. Consequently, if $r = r_1r_2\ldots r_6 \in \mathbf{Z}_2^6$, we can identify r as a code word if

$$H \cdot r^{tr} = \begin{bmatrix} 0 \\ 0 \\ 0 \end{bmatrix}.$$

Writing $H = [B\,|\,I_3]$, we notice that if the rows and columns of B are interchanged we get A. Hence $B = A^{tr}$.

In general, if $m < n$ and $E: \mathbf{Z}_2^m \to \mathbf{Z}_2^n$ is given by a *generator matrix* $G = [I_m\,|\,A]$, then A is an m by $(n-m)$ matrix and the associated *parity-check matrix* H has the form $[A^{tr}\,|\,I_{n-m}]$. Matrix H can also be used to define the encoding function E, for if $w = w_1w_2\ldots w_m \in \mathbf{Z}_2^m$, $E(w) = w_1w_2\ldots w_mw_{m+1}\ldots w_n$, where $w_{m+1}, \ldots, w_n$ can be determined from the set of $n-m$ (parity-check) equations that arise from $H \cdot (E(w))^{tr} = \mathbf{0}$, the column vector of $n-m$ 0's.

From the theory developed earlier on error correction, since the minimum distance between the code words of this example is 3 we should be able to develop a decoding function that corrects single errors.

Suppose we receive $r = 110110$. If we have a long list of code words to check r against, we would be better off to examine $H \cdot r^{tr}$, which is called the *syndrome* of r. Here

$$H \cdot r^{tr} = \begin{bmatrix} 1 & 0 & 1 & 1 & 0 & 0 \\ 1 & 1 & 0 & 0 & 1 & 0 \\ 0 & 1 & 1 & 0 & 0 & 1 \end{bmatrix} \begin{bmatrix} 1 \\ 1 \\ 0 \\ 1 \\ 1 \\ 0 \end{bmatrix} = \begin{bmatrix} 0 \\ 1 \\ 1 \end{bmatrix},$$

so r is not a code word. Hence we at least detect an error. Looking back at the list of code words we see that $d(100110, r) = 1$. For all other $c \in C$, $d(r, c) \geq 2$. Writing $r = c + e = 100110 + 010000$, we find that the transmission error (of weight 1) occurs in the second component of r. Is it a coincidence that the syndrome $H \cdot r^{tr}$ produced the second column of H? If not, we can use this to realize that if a single transmission error occurred it took place at the second component. Changing the second component of r we get c; the message w comprises the first three components of c.

Let $r = c + e$, where c is a code word and e is an error pattern of weight 1. Suppose that 1 is in the ith component of e, $1 \leq i \leq 6$. Then

$$H \cdot r^{tr} = H \cdot (c + e)^{tr} = H \cdot (c^{tr} + e^{tr}) = H \cdot c^{tr} + H \cdot e^{tr}.$$

With c a code word, $H \cdot c^{tr} = \mathbf{0}$, so $H \cdot r^{tr} = H \cdot e^{tr} = i$th column of matrix H.

Since we are primarily concerned with transmissions where multiple errors are rare, this technique is of definite value. If we ask for more however, we find ourselves expecting too much.

For suppose that we receive $r = 000111$. Computing the syndrome

$$H \cdot r^{tr} = \begin{bmatrix} 1 & 0 & 1 & 1 & 0 & 0 \\ 1 & 1 & 0 & 0 & 1 & 0 \\ 0 & 1 & 1 & 0 & 0 & 1 \end{bmatrix} \begin{bmatrix} 0 \\ 0 \\ 0 \\ 1 \\ 1 \\ 1 \end{bmatrix} = \begin{bmatrix} 1 \\ 1 \\ 1 \end{bmatrix},$$

we obtain a result that is not one of the columns of H. Yet $H \cdot r^{tr}$ can be obtained as the sum of two columns from H. If $H \cdot r^{tr}$ came from the first and sixth columns of H, correcting these components in r results in the code word 100110. If we sum the third and fifth columns of H to get this syndrome, upon changing the third and fifth components of r we get a second code word, 001101. So we cannot expect H to correct multiple errors. This is no surprise, since the minimum distance between code words is 3.   □

We summarize the results of Example 12.22 for the general situation. For $m < n$, let $E: \mathbf{Z}_2^m \to \mathbf{Z}_2^n$ be given by $E(w) = wG$, where $G = [I_m | A]$ is an $m \times n$ matrix over $\mathbf{Z}_2$. The matrix G is called the *generator matrix* for the code $C = E(\mathbf{Z}_2^m)$. Then $H = [A^{tr} | I_{n-m}]$ is the unique *parity-check matrix* associated with G. Matrix H provides a decoding scheme that corrects single errors in transmissions if:

a) H does not contain a column of 0's. (If the ith column of H had all 0's and $H \cdot r^{tr} = \mathbf{0}$ for a received word r, we couldn't decide whether r were a code word or a received word whose ith component was incorrectly transmitted. We do not want to compare r with all code words when C is large.)

b) No two columns of H are the same. (If the ith and jth columns of H are the same and $H \cdot r^{tr}$ equals this repeated column, how would we decide on which component of r to change?)

When H satisfies these two conditions we get the following decoding algorithm. For any $r \in \mathbf{Z}_2^n$, if $T(c) = r$, then

1. With $H \cdot r^{tr} = \mathbf{0}$, we feel that the transmission was correct and that r is the code word that was transmitted. The decoded message is then the first m components of r.

2. With $H \cdot r^{tr}$ equal to the ith column of H, we feel that there has been a single error in transmission and change the ith component of r in order to get the code word c. Here the first m components of c yield the original message.

3. If neither case 1 nor case 2 occurs, we feel that there has been more than one transmission error and we cannot provide a reliable way to decode in this situation.

We close by recalling an earlier comment on the use of H to define E. If we start with a parity-check matrix $H = [B | I_{n-m}]$ and use it in this manner to define E, we obtain the same set of code words that is generated by the unique associated generator matrix $G = [I_m | B^{tr}]$.

1. For Example 12.21, list the elements in $S(101010, 1)$ and $S(111111, 1)$.

2. Decode each of the following received words for Example 12.21.

   a) 110101   b) 101011   c) 001111   d) 110000.

3. a) If $x \in \mathbf{Z}_2^{10}$, determine $|S(x, 1)|$, $|S(x, 2)|$, $|S(x, 3)|$.

   b) For $n, k \in \mathbf{Z}^+$, $1 \le k \le n$, if $x \in \mathbf{Z}_2^n$, what is $|S(x, k)|$?

4. Let $E: \mathbf{Z}_2^5 \to \mathbf{Z}_2^{25}$ be an encoding function where the minimum distance between code words is 9. What is the largest value of k so that we can detect errors of weight $\le k$? If we wish to correct errors of weight $\le n$, what is the maximum value for n?

5. For each of the following encoding functions, find the minimum distance between the code words. Discuss the error-detecting and error-correcting capabilities of each code.

   a) $E: \mathbf{Z}_2^2 \to \mathbf{Z}_2^5$

   $00 \to 00001$ $01 \to 01010$

   $10 \to 10100$ $11 \to 11111$

   b) $E: \mathbf{Z}_2^2 \to \mathbf{Z}_2^{10}$

   $00 \to 0000000000$ $01 \to 0000011111$

   $10 \to 1111100000$ $11 \to 1111111111$

   c) $E: \mathbf{Z}_2^3 \to \mathbf{Z}_2^6$

   $000 \to 000111$ $001 \to 001001$

   $010 \to 010010$ $011 \to 011100$

   $100 \to 100100$ $101 \to 101010$

   $110 \to 110001$ $111 \to 111000$

   d) $E: \mathbf{Z}_2^3 \to \mathbf{Z}_2^8$

   $000 \to 00011111$ $001 \to 00111010$

   $010 \to 01010101$ $011 \to 01110000$

   $100 \to 10001101$ $101 \to 10101000$

   $110 \to 11000100$ $111 \to 11100011$

6. a) Use the parity-check matrix H of Example 12.22 to decode the following received words.

   i) 111101   ii) 110101   iii) 001111   iv) 100100

   v) 110001   vi) 111111   vii) 111100   viii) 010100

   b) Are all the results in part (a) uniquely determined?

7. The encoding function $E: \mathbf{Z}_2^2 \to \mathbf{Z}_2^5$ is given by the generator matrix

$$G = \begin{bmatrix} 1 & 0 & 1 & 1 & 0 \\ 0 & 1 & 0 & 1 & 1 \end{bmatrix}.$$

   a) Determine all code words. What can we say about the error-detection capability of this code? What about its error-correction capability?

   b) Find the associated parity-check matrix H.

    c)   Use H to decode each of the following received words.

        i)   11011   ii)   10101   iii)   11010

        iv)   00111   v)   11101   vi)   00110

8. Define the encoding function $E: \mathbf{Z}_2^3 \to \mathbf{Z}_2^6$ by means of the parity-check matrix

$$H = \begin{bmatrix} 1 & 0 & 1 & 1 & 0 & 0 \\ 1 & 1 & 0 & 0 & 1 & 0 \\ 1 & 0 & 1 & 0 & 0 & 1 \end{bmatrix}.$$

    a)   Determine all code words.

    b)   Does this code correct all single errors in transmission?

9. Find the generator and parity-check matrices for the $(9, 8)$ single parity-check coding scheme of Example 12.17.

10. a)   Show that the 1×9 matrix $G = [1 \quad 1 \quad 1 \quad \ldots \quad 1]$ is the generator matrix for the $(9, 1)$ nine-times repetition code.

     b)   What is the associated parity-check matrix H in this case?

11. For an (n, m) code C with generator matrix $G = [I_m | A]$ and parity-check matrix $H = [A^{tr} | I_{n-m}]$, the $(n, n - m)$ code C^d with generator matrix $[I_{n-m} | A^{tr}]$ and parity-check matrix $[A | I_m]$ is called the *dual code* of C. Show that the codes in each of Exercises 9 and 10 constitute a pair of dual codes.

12. Given $n \in \mathbf{Z}^+$, let $M(n, k) \subseteq \mathbf{Z}_2^n$ contain the maximum number of code words of length n, where the minimum distance between code words is $2k + 1$. Prove that

$$\frac{2^n}{\sum_{i=0}^{2k} \binom{n}{i}} \leq |M(n, k)| \leq \frac{2^n}{\sum_{i=0}^{k} \binom{n}{i}}.$$

(The upper bound on $|M(n, k)|$ is called the *Hamming bound*, while the lower bound is referred to as the *Gilbert bound*.)

12.7 ■ GROUP CODES: DECODING WITH COSET LEADERS

Now that we've seen some introductory material on coding theory it is time to see how the group structure enters the picture.

Definition 12.11 ▶ Let $E: \mathbf{Z}_2^m \to \mathbf{Z}_2^n$ be an encoding function. The code $C = E(\mathbf{Z}_2^m)$ is called a *group code* if C is a subgroup of $\mathbf{Z}_2^n$.

When the code words form a group, it will be easier to compute the minimum distance between code words.

Theorem 12.14 ▶ In a group code the minimum distance between code words is the minimum of the weights of the nonzero elements of the code.

Proof Let $a, b, c \in C$ where $d(a, b)$ is minimum and c is nonzero with minimum weight. By closure in the group C, $a + b$ is a code word. Since $d(a, b) = \text{wt}(a + b)$, by the choice of c, $d(a, b) \geq \text{wt}(c)$. (Why is $d(a, b) > 0$?) Conversely, $\text{wt}(c) = d(c, \mathbf{0})$, where $\mathbf{0}$ is a code word since C is a group. Then $d(c, \mathbf{0}) \geq d(a, b)$ by the choice of a, b, so $\text{wt}(c) \geq d(a, b)$. Consequently, $d(a, b) = \text{wt}(c)$. ∎

If C is a set of code words and $|C| = 1024$, we have to compute $\binom{1024}{2} = 523{,}776$ distances to find the minimum distance between code words. But if we can recognize that C possesses a group structure, we need only compute the weights of the 1023 nonzero elements of C.

Is there some way to guarantee a group structure on the code words? By Theorem 12.5(c), if $E: \mathbf{Z}_2^m \to \mathbf{Z}_2^n$ is a group homomorphism, then $C = E(\mathbf{Z}_2^m)$ will be a subgroup of $\mathbf{Z}_2^n$.

Theorem 12.15 ▶ Let $E: \mathbf{Z}_2^m \to \mathbf{Z}_2^n$ be an encoding function given by a generator matrix G or the associated parity-check matrix H. Then $C = E(\mathbf{Z}_2^m)$ is a group code.

Proof We establish these results by proving that the function E arising from G or H is a group homomorphism.

If $x, y \in \mathbf{Z}_2^m$, then $E(x + y) = (x + y)G = xG + yG = E(x) + E(y)$. Hence E is a homomorphism and $C = E(\mathbf{Z}_2^m)$ is a group code.

For the case of H, if x is a message, then $E(x) = x_1 x_2 \ldots x_m x_{m+1} \ldots x_n$, where $x = x_1 x_2 \ldots x_m \in \mathbf{Z}_2^m$ and $H \cdot (E(x))^{\text{tr}} = \mathbf{0}$. In particular, $E(x)$ is uniquely determined by these two properties. If y is also a message, then $x + y$ is likewise, and $E(x + y)$ has $(x_1 + y_1), (x_2 + y_2), \ldots, (x_m + y_m)$ as its first m components, as does $E(x) + E(y)$. Further, $H \cdot (E(x) + E(y))^{\text{tr}} = H \cdot (E(x)^{\text{tr}} + E(y)^{\text{tr}}) = H \cdot E(x)^{\text{tr}} + H \cdot E(y)^{\text{tr}} = \mathbf{0} + \mathbf{0} = \mathbf{0}$. Since $E(x + y)$ is the unique element of $\mathbf{Z}_2^n$ with $(x_1 + y_1)$, $(x_2 + y_2), \ldots, (x_m + y_m)$ as its first m components and with $H \cdot (E(x + y))^{\text{tr}} = \mathbf{0}$, it follows that $E(x + y) = E(x) + E(y)$. So E is a group homomorphism and $C = \{c \in \mathbf{Z}_2^n \mid H \cdot c^{\text{tr}} = \mathbf{0}\}$ is a group code. ∎

Now we use the group structure of C together with its cosets in $\mathbf{Z}_2^n$ to develop a scheme for decoding. Our example uses the code developed in Example 12.22, but the procedure applies for any group code.

EXAMPLE 12.23 We develop a table for decoding as follows.

1. First list in a row the elements of the group code C, starting with the identity.

000000 100110 010011 001101 110101 101011 011110 111000 .

2. Next select an element x of $\mathbf{Z}_2^6$ ($\mathbf{Z}_2^n$, in general) where x does not appear anywhere in the table developed so far and x has minimum weight. Then list the elements of the coset $x + C$, with $x + c$ directly below c for each $c \in C$. For $x = 100000$ we have

000000 100110 010011 001101 110101 101011 011110 111000

100000 000110 110011 101101 010101 001011 111110 011000 .

3. Repeat step 2 until the cosets provide a partition of $\mathbf{Z}_2^6$. This results in the *decoding table* shown in Table 12.7.

4. Once the decoding table is constructed, for any received word r we find the column containing r and use the first three components of the code word c at the top of the column to decode r.

Table 12.7 Decoding table for the code of Example 12.22

000000	100110	010011	001101	110101	101011	011110	111000
100000	000110	110011	101101	010101	001011	111110	011000
010000	110110	000011	011101	100101	111011	001110	101000
001000	101110	011011	000101	111101	100011	010110	110000
000100	100010	010111	001001	110001	101111	011010	111100
000010	100100	010001	001111	110111	101001	011100	111010
000001	100111	010010	001100	110100	101010	011111	111001
010100	110010	000111	011001	100001	111111	001010	101100

From the table we find the code words for the received words

$$r_1 = 101001 \quad r_2 = 111010 \quad r_3 = 001001 \quad r_4 = 111011$$

are, respectively

$$c_1 = 101011 \quad c_2 = 111000 \quad c_3 = 001101 \quad c_4 = 101011 .$$

From these results the respective messages are

$$w_1 = 101 \quad w_2 = 111 \quad w_3 = 001 \quad w_4 = 101 .$$

The entries in the first column of Table 12.7 are called the *coset leaders*. For the first seven rows the coset leaders will be the same in all tables, with some permutations of rows possible. However, for the last row, 100001 or 001010 could have been used in place of 010100 since they also have minimum weight 2. So the table need not be unique.

How do the coset leaders help us? It seems that the code words in the first row are what we used to decode r_1, r_2, r_3, and r_4 above.

Consider the received words $r_1 = 101001$ and $r_2 = 111010$ in the sixth row, where the coset leader is $x = 000010$. Computing syndromes, we find that

$$H \cdot (r_1)^{\text{tr}} = \begin{bmatrix} 0 \\ 1 \\ 0 \end{bmatrix} = H \cdot (r_2)^{\text{tr}} = H \cdot x^{\text{tr}}.$$

This is not just a coincidence.   □

Theorem 12.16 ▶ Let $C \subseteq \mathbf{Z}_2^n$ be a group code for a parity-check matrix H, and let $r_1, r_2 \in \mathbf{Z}_2^n$. For the table of cosets of C in $\mathbf{Z}_2^n$, r_1 and r_2 are in the same coset of C iff $H \cdot (r_1)^{\text{tr}} = H \cdot (r_2)^{\text{tr}}$.

Proof If r_1 and r_2 are in the same coset, then $r_1 = x + c_1$, $r_2 = x + c_2$, where x is the coset leader, and c_1 and c_2 are the code words at the tops of the respective columns for r_1, r_2. Then $H \cdot (r_1)^{\text{tr}} = H \cdot (x + c_1)^{\text{tr}} = H \cdot x^{\text{tr}} + H \cdot (c_1)^{\text{tr}} = H \cdot x^{\text{tr}} + \mathbf{0} = H \cdot x^{\text{tr}}$, since c_1 is a code word. Likewise, $H \cdot (r_2)^{\text{tr}} = H \cdot x^{\text{tr}}$, so r_1, r_2 have the same syndrome. Conversely, $H \cdot (r_1)^{\text{tr}} = H \cdot (r_2)^{\text{tr}} \Rightarrow H \cdot (r_1 + r_2)^{\text{tr}} = \mathbf{0} \Rightarrow r_1 + r_2$ is a code word c. Hence $r_1 + r_2 = c$, so $r_1 = r_2 + c$ and $r_1 \in r_2 + C$. Since $r_2 \in r_2 + C$, we have r_1, r_2 in the same coset. ∎

In decoding received words, when Table 12.7 is used we must search through 64 elements to find a given received word. For $C \subseteq \mathbf{Z}_2^{12}$ there are 4096 strings, each with 12 bits. Such a searching process is tedious. So perhaps we should be thinking about having a computer do the searching. Presently it appears that this means storing the entire table: $6 \times 64 = 384$ bits of storage for Table 12.7; $12 \times 4096 = 49,152$ bits for $C \subseteq \mathbf{Z}_2^{12}$. We should like to improve this situation. Before things get better, however, they'll look worse as we enlarge Table 12.7, as shown in Table 12.8. This new table includes to the left of the coset leaders (the transposes of) the syndromes for each row.

Table 12.8 Decoding table of Table 12.7 with syndromes

000	000000	100110	010011	001101	110101	101011	011110	111000
110	100000	000110	110011	101101	010101	001011	111110	011000
011	010000	110110	000011	011101	100101	111011	001110	101000
101	001000	101110	011011	000101	111101	100011	010110	110000
100	000100	100010	010111	001001	110001	101111	011010	111100
010	000010	100100	010001	001111	110111	101001	011100	111010
001	000001	100111	010010	001100	110100	101010	011111	111001
111	010100	110010	000111	011001	100001	111111	001010	101100

Now we can decode a received word r by the following procedure.

1. Compute the syndrome $H \cdot r^{\text{tr}}$.
2. Find the coset leader x to the right of $H \cdot r^{\text{tr}}$.
3. Add x to r to get c. (The code word c that we are seeking at the top of the column containing r satisfies $c + x = r$, or $c = x + r$.)

Consequently, all that is needed from Table 12.8 are the first two columns, which will require $(3)(8) + (6)(8) = 72$ storage bits. With 18 more storage bits for H we can store what we need for this decoding process, called *decoding by coset leaders,* in 90 storage bits, as opposed to the original estimate of 384 bits.

Applying this procedure to $r = 110110$, we find the syndrome

$$H \cdot r^{\text{tr}} = \begin{bmatrix} 0 \\ 1 \\ 1 \end{bmatrix}.$$

Since 011 is to the left of the coset leader $x = 010000$, the code word $c = 010000 + 110110 = 100110$, and we recapture the original message, 100.

Since the code here is a group code where the minimum weight of the nonzero code words is 3, we expected to be able to find a decoding scheme that corrected single errors. Here this is accomplished because the error patterns of weight 1 are all coset leaders. We cannot correct all double errors since only one error pattern of weight 2 is a coset leader. All error patterns of weight 1 or 2 would have to be coset leaders before our decoding scheme could correct both single and double errors in transmission.

Unlike the situation in Example 12.22 where syndromes were also used for decoding, things here are a bit different. Decoding by coset leaders gives an answer for *all* received words, not just for those that are codewords or have syndromes that appear among the columns of the parity-check matrix H. In addition, as our last result will affirm, this method provides a decoding scheme that is as good as any other.

Theorem 12.17

▶ When decoding by coset leaders, if $r \in \mathbf{Z}_2^n$ is a received word and r is decoded as the code word c^* (which we then decode to get the message), then $d(c^*, r) \le d(c, r)$ for all code words c.

Proof

Let x be the coset leader for the coset containing r. Then $r = c^* + x$, or $r + c^* = x$, so $d(c^*, r) = \text{wt}(r + c^*) = \text{wt}(x)$. If c is any code word, $d(c, r) = \text{wt}(c + r)$, and $c + r = c + (c^* + x) = (c + c^*) + x$. Since C is a group code, $c + c^* \in C$ and $c + r$ is in the coset $x + C$. Among the elements in the coset $x + C$, the coset leader x is chosen to have minimum weight, so $\text{wt}(c + r) \ge \text{wt}(x)$. Consequently, $d(c^*, r) = \text{wt}(x) \le \text{wt}(c + r) = d(c, r)$. ■

12.8 ■ HAMMING MATRICES

We found the parity-check matrix H helpful in correcting single errors in transmission when (a) H had no column of 0's; and (b) no two columns of H were alike. For the matrix

$$H = \begin{bmatrix} 1 & 1 & 0 & 1 & 1 & 0 & 0 \\ 1 & 0 & 1 & 1 & 0 & 1 & 0 \\ 0 & 1 & 1 & 1 & 0 & 0 & 1 \end{bmatrix}$$

we find that H satisfies these two conditions and that for the number of rows, $r = 3$, in H we have the maximum number of columns possible. If any additional column is added, H will no longer be useful for correcting single errors.

The generator matrix G associated with H is

$$G = \begin{bmatrix} 1 & 0 & 0 & 0 & 1 & 1 & 0 \\ 0 & 1 & 0 & 0 & 1 & 0 & 1 \\ 0 & 0 & 1 & 0 & 0 & 1 & 1 \\ 0 & 0 & 0 & 1 & 1 & 1 & 1 \end{bmatrix}.$$

Consequently we have a $(7,4)$ group code. The encoding function $E: \mathbf{Z}_2^4 \to \mathbf{Z}_2^7$ encodes four-bit messages into seven-bit code words. We realize that since H is determined by three parity-check equations, we have now maximized the number of bits we can have in the messages. (In addition, the columns of H, read from top to bottom, are the binary equivalents of the integers from 1 to 7.)

In general, if we start with r parity-check equations, then the parity-check matrix H can have as many as $2^r - 1$ columns and still be used to correct single errors. Under these circumstances $H = [A \,|\, I_r]$, where A is an $r \times 2^r - 1 - r$ matrix, and $G = [I_m \,|\, A^{tr}]$ with $m = 2^r - 1 - r$. The parity-check matrix H associated with a $(2^r - 1,\ 2^r - 1 - r)$ group code in this way is called a *Hamming matrix,* and the code is referred to as a *Hamming code.*

EXAMPLE 12.24   If $r = 4$, then $2^r - 1 = 15$, $2^r - 1 - r = 11$ and one possible Hamming matrix H for $r = 4$ is

$$\begin{bmatrix} 1 & 1 & 1 & 1 & 1 & 1 & 1 & 0 & 0 & 0 & 0 & 1 & 0 & 0 & 0 \\ 1 & 1 & 1 & 1 & 0 & 0 & 0 & 1 & 1 & 1 & 0 & 0 & 1 & 0 & 0 \\ 1 & 1 & 0 & 0 & 1 & 1 & 0 & 1 & 1 & 0 & 1 & 0 & 0 & 1 & 0 \\ 1 & 0 & 1 & 0 & 1 & 0 & 1 & 1 & 0 & 1 & 1 & 0 & 0 & 0 & 1 \end{bmatrix}.$$

(Once again the columns of H contain the binary equivalents of the integers from 1 to $15\ (= 2^4 - 1)$.)

This matrix H is the parity-check matrix of a Hamming $(15, 11)$ code whose rate is $11/15$. □

With regard to the rate of these Hamming codes, for any $r \geq 2$, the rate m/n of such a code is given by

$$m/n = (2^r - 1 - r)/(2^r - 1) = 1 - [r/(2^r - 1)].$$

As r increases, $r/(2^r - 1)$ goes to 0 and the rate approaches 1.

We close our discussion on coding theory with one final observation. In Section 12.6 we presented G (and H) in what is called the *systematic form.* Other arrangements of the rows and columns of these matrices are also possible and these yield *equivalent codes.* (More on this can be found in the text by L. Dornhoff and F. Hohn [2].) We mention this here since it is often common practice to list the columns in a Hamming matrix of r rows so that the binary representations of 1 to $2^r - 1$ appear as the columns of H are read from left to right. For the Hamming $(7, 4)$ code, the matrix H mentioned at the start of this section would take the (equivalent) form

$$H_1 = \begin{bmatrix} 0 & 0 & 0 & 1 & 1 & 1 & 1 \\ 0 & 1 & 1 & 0 & 0 & 1 & 1 \\ 1 & 0 & 1 & 0 & 1 & 0 & 1 \end{bmatrix}.$$

Here the identity appears in the first, second, and fourth columns instead of the last three. Consequently, we would use these components for the parity checks and find that if we send the message $w = w_1 w_2 w_3 w_4$, the corresponding code word $E(w)$ is $c_1 c_2 w_1 c_3 w_2 w_3 w_4$, where

$$
\begin{aligned}
c_1 &= w_1 + w_2 && + w_4 \\
c_2 &= w_1 && + w_3 + w_4 \\
c_3 &= && w_2 + w_3 + w_4,
\end{aligned}
$$

so that $H_1 \cdot (E(w))^{\text{tr}} = \mathbf{0}$.

EXERCISES

1. Let $E: \mathbf{Z}_2^8 \to \mathbf{Z}_2^{12}$ be the encoding function for a code C. How many calculations are needed to find the minimum distance between code words? How many calculations are needed if E is a group homomorphism?

2. a) Use Table 12.8 to decode the following received words.

   000011   100011   111110   100001   001100   011110   001111   111100

   b) Do any of the results in (a) change if a different set of coset leaders is used?

3. a) Construct a decoding table (with syndromes) for the group code given by the generator matrix

$$
G = \begin{bmatrix} 1 & 0 & 1 & 1 & 0 \\ 0 & 1 & 0 & 1 & 1 \end{bmatrix}.
$$

   b) Use the table from part (a) to decode the received words:

   11110   11101   11011   10100   10011   10101   11111   01100

   c) Does this code correct single errors in transmission?

4. Let

$$
H = \begin{bmatrix} 1 & 1 & 0 & 1 & 1 & 0 & 0 \\ 1 & 0 & 1 & 1 & 0 & 1 & 0 \\ 0 & 1 & 1 & 1 & 0 & 0 & 1 \end{bmatrix}
$$

   be the parity-check matrix for a Hamming $(7, 4)$ code.

   a) Encode the messages: 1000   1100   1011   1110   1001   1111.

   b) Decode the received words: 1100001   1110111   0010001   0011100.

   c) Construct a decoding table consisting of the syndromes and coset leaders for this code.

   d) Use the result in part (c) to decode the received words in part (b).

5. a) What are the dimensions of the generator matrix for the Hamming $(63, 57)$ code? What are the dimensions for the associated parity-check matrix H?

   b) What is the rate of this code?

6. Compare the rates of the Hamming $(7, 4)$ code and the $(3, 1)$ triple-repetition code.

7. a) Let $p = 0.01$ be the probability of incorrect transmission for a binary symmetric channel. If the message 1011 is sent, using the Hamming $(7, 4)$ code, what is the probability for correct decoding?

   b) Answer (a) for a 20-bit message sent in five blocks of length 4.

8. A standard practice in dealing with Hamming codes is to write the columns in normal binary order (going from left to right), as we did with the matrix H_1 at the end of Section 12.8. If we do this for the Hamming $(15, 11)$ code, in which columns do we find the identity matrix? Where is the identity for the Hamming $(31, 26)$ code? Where is it for the Hamming $(2^r - 1, 2^r - 1 - r)$ code?

12.9 ■ COUNTING AND EQUIVALENCE: BURNSIDE'S THEOREM

In this section and the next two we shall develop a counting technique known as Polya's method of enumeration. Our development will not be very rigorous. Often we shall only state the general results of the theory as seen in the solution of a specific problem.

EXAMPLE 12.25   We have a set of sticks, all of the same length and color, and a second set of round plastic disks. Each disk contains two holes, as shown in Fig. 12.4, and the sticks can be inserted into these holes in order to form different shapes, such as a square. If each disk is either red or white, how many distinct squares can we form?

Figure 12.4

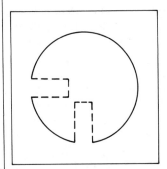

If the square is considered stationary, then the four disks are located at four distinct locations; a red or white disk is used at each location. Thus there are $2^4 = 16$ different configurations, as shown in Fig. 12.5, where a dark circle indicates a red disk. The configurations have been split into six classes, $c\ell(1)$, $c\ell(2)$, ..., $c\ell(6)$, according to the number and relative location of the red disks.

Now suppose that the square is not fixed, but that it can be moved about in space. Unless the vertices (disks) are marked somehow, certain configurations in Fig. 12.5 are indistinguishable when we move them about.

Figure 12.5

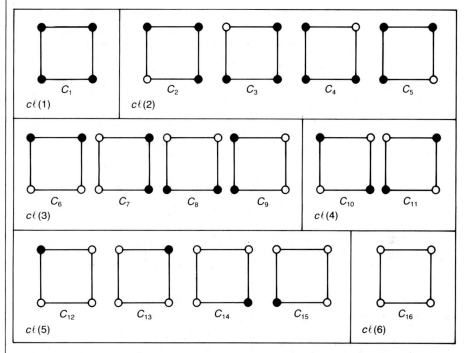

To place these notions in a more mathematical setting, we use the nonabelian group of three-dimensional rigid motions of a square to define an equivalence relation on the configurations in Fig. 12.5. Since this group will be used throughout this section and the next two sections, we now give a detailed description of its elements.

In Fig. 12.6 we have the group $G = \{\pi_0, \pi_1, \pi_2, \pi_3, r_1, r_2, r_3, r_4\}$ for the rigid motions of the square in (a). Parts (b)–(i) of the figure show how each element of G is applied. We have expressed each group element as a permutation of $\{1, 2, 3, 4\}$ and in a new form called a *product of disjoint cycles*. For example, in (b) we find $\pi_1 = (1234)$. The cycle (1234) indicates that if we start with the square in (a), after applying π_1, 1 has moved to the position originally occupied by 2, 2 to that of 3, 3 to that of 4, and 4 to that of 1. In general, if xy appears in a cycle, then x moves to the position originally occupied by y. Also, for a cycle where x and y appear as $(x \dots y)$, y moves into the position originally occupied by x when the motion described by this cycle is applied. Note that $(1234) = (2341) = (3412) = (4123)$. We say that each of these cycles has *length* 4, the number of elements in the cycle. In the case of r_1 in part (f) of the figure, starting with 1 we find that r_1 sends 1 to 4 so we have $(14 \dots)$ as the start of our first cycle in this *decomposition* of r_1. However, here r_1 sends 4 to 1 so we have completed a portion, namely (14), of the complete decomposition. We then select a vertex that has not yet appeared, for example, vertex 2. Since r_1 sends 2 to 3 and 3, in turn, to 2, we get a second cycle (23). This exhausts all vertices and so $(14)(23) = r_1$, where these cycles have no vertex in common. Here $(14)(23) =$

$(23)(14) = (23)(41) = (32)(41)$ all provide a representation of r_1 as a product of disjoint cycles, each of length two. Lastly, for the group element $r_3 = (13)(2)(4)$, the cycle (2) indicates that 2 is fixed by, or *invariant* under, the permutation r_3. When the number of vertices involved is known, we sometimes write $r_3 = (13)$ where the missing elements are understood to be fixed. However, we shall write all cycles in our decompositions, for this will be useful later in our discussion.

Figure 12.6

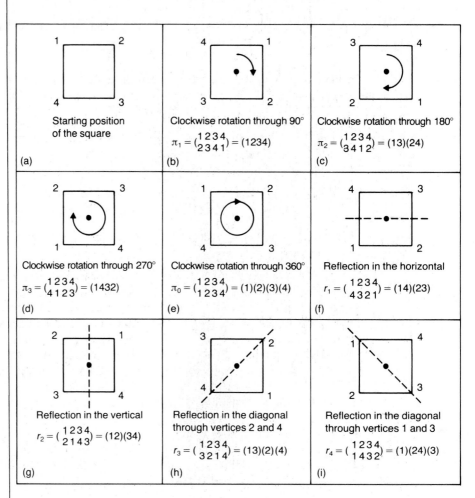

Before continuing with the main discussion, let us examine one more result on disjoint cycles.

In the group S_6 of all permutations of $\{1, 2, 3, 4, 5, 6\}$, let $\pi = \begin{pmatrix} 1 & 2 & 3 & 4 & 5 & 6 \\ 2 & 3 & 1 & 4 & 6 & 5 \end{pmatrix}$. As a product of disjoint cycles

$$\pi = (123)(4)(56) = (56)(4)(123) = (4)(231)(65).$$

If $\sigma \in S_6$, with $\sigma = \left(\begin{smallmatrix} 1 & 2 & 3 & 4 & 5 & 6 \\ 2 & 4 & 5 & 1 & 6 & 3 \end{smallmatrix}\right)$, then

$$\sigma = (124)(356) = \begin{pmatrix} 1 & 2 & 3 & 4 & 5 & 6 \\ 2 & 4 & 3 & 1 & 5 & 6 \end{pmatrix} \begin{pmatrix} 1 & 2 & 3 & 4 & 5 & 6 \\ 1 & 2 & 5 & 4 & 6 & 3 \end{pmatrix},$$

so each cycle can be thought of as an element of S_6.

Finally, if $\alpha = (124)(3)(56)$ and $\beta = (13)(245)(6)$ are elements of S_6, then

$$\alpha\beta = (124)(3)(56)(13)(245)(6) = (143)(256),$$

while

$$\beta\alpha = (13)(245)(6)(124)(3)(56) = (132)(465).$$

Returning to the 16 configurations, or colorings, in Fig. 12.5, we now examine how each element of the group G, in Fig. 12.6, acts upon these configurations. For example, $\pi_1 = \left(\begin{smallmatrix} 1 & 2 & 3 & 4 \\ 2 & 3 & 4 & 1 \end{smallmatrix}\right)$ permutes the numbers $\{1, 2, 3, 4\}$, according to a 90° clockwise rotation for the square in Fig. 12.6(a), yielding the result in Fig. 12.6(b). How does such a rotation act on $S = \{C_1, C_2, \ldots, C_{16}\}$, our set of colorings? We use π_1^* to distinguish between the 90° clockwise rotation for $\{1, 2, 3, 4\}$ and the same rotation when applied to $S = \{C_1, C_2, \ldots, C_{16}\}$. We find that

$\pi_1^* =$

$$\begin{pmatrix} C_1 & C_2 & C_3 & C_4 & C_5 & C_6 & C_7 & C_8 & C_9 & C_{10} & C_{11} & C_{12} & C_{13} & C_{14} & C_{15} & C_{16} \\ C_1 & C_3 & C_4 & C_5 & C_2 & C_7 & C_8 & C_9 & C_6 & C_{11} & C_{10} & C_{13} & C_{14} & C_{15} & C_{12} & C_{16} \end{pmatrix}.$$

As a product of disjoint cycles,

$$\pi_1^* = (C_1)(C_2 C_3 C_4 C_5)(C_6 C_7 C_8 C_9)(C_{10} C_{11})(C_{12} C_{13} C_{14} C_{15})(C_{16}).$$

We note that under the action of π_1^* no configuration is changed into one that is in another class.

As a second example, consider the reflection r_3 in Fig. 12.6(h). The action of this rigid motion on S is given by

$r_3^* =$

$$\begin{pmatrix} C_1 & C_2 & C_3 & C_4 & C_5 & C_6 & C_7 & C_8 & C_9 & C_{10} & C_{11} & C_{12} & C_{13} & C_{14} & C_{15} & C_{16} \\ C_1 & C_2 & C_5 & C_4 & C_3 & C_7 & C_6 & C_9 & C_8 & C_{10} & C_{11} & C_{14} & C_{13} & C_{12} & C_{15} & C_{16} \end{pmatrix}$$

$$= (C_1)(C_2)(C_3 C_5)(C_4)(C_6 C_7)(C_8 C_9)(C_{10})(C_{11})(C_{12} C_{14})(C_{13})(C_{15})(C_{16}).$$

Once again no configuration is taken by r_3^* into one that is outside the class that it was in originally.

Using the idea of *the group G acting on the set S* we define a relation $\mathcal{R}$ on S as follows. For colorings $C_i, C_j \in S$, $1 \le i, j \le 16$, $C_i \mathcal{R} C_j$ if there is a permutation $\sigma \in G$ such that $\sigma^*(C_i) = C_j$, i.e., as σ^* acts on the 16 configurations in S, C_i is transformed into C_j. This relation $\mathcal{R}$ is an equivalence relation as we now verify.

a) (Reflexive Property) For all $C_i \in S$, $1 \le i \le 16$, $C_i \mathcal{R} C_i$ since G contains the identity permutation. ($\pi_0^*(C_i) = C_i$ for all $1 \le i \le 16$.)

b) (Symmetric Property) If $C_i \, \mathcal{R} \, C_j$ for $C_i, C_j \in S$, then $\sigma^*(C_i) = C_j$, for $\sigma \in G$. Since G is a group, $\sigma^{-1} \in G$, and we find that $(\sigma^*)^{-1} = (\sigma^{-1})^*$. (Verify this for two choices of $\sigma \in G$.) Hence $C_i = (\sigma^{-1})^*(C_j)$, and $C_j \, \mathcal{R} \, C_i$.

c) (Transitive Property) Let $C_i, C_j, C_k \in S$ with $C_i \, \mathcal{R} \, C_j$, $C_j \, \mathcal{R} \, C_k$. Then $C_j = \sigma^*(C_i)$, $C_k = \tau^*(C_j)$, for $\sigma, \tau \in G$. By closure in G, $\sigma\tau \in G$, and we find $(\sigma\tau)^* = \sigma^*\tau^*$, where σ is applied first in $\sigma\tau$, σ^* first in $\sigma^*\tau^*$. (Verify this for two specific permutations $\sigma, \tau \in G$.) Then $C_k = (\sigma\tau)^*(C_i)$ and $\mathcal{R}$ is transitive. (The reader may have noticed that $C_k = \tau^*(C_j) = \tau^*(\sigma^*(C_i))$ and felt that we should have written $(\sigma\tau)^* = \tau^*\sigma^*$. Once again there has been a change in the notation for the composite function as we first defined it in Chapter 3. Here we write $\sigma^*\tau^*$ for $(\sigma\tau)^*$ and σ^* is applied first.)

Since $\mathcal{R}$ is an equivalence relation on S, $\mathcal{R}$ partitions S into equivalence classes, which are precisely the classes $c\ell(1), c\ell(2), \ldots, c\ell(6)$ of Fig. 12.5. Consequently, there are six nonequivalent configurations under the group action. So among the original 16 colorings only 6 are really distinct.

What has happened in this example generalizes as follows. With S a set of configurations, let G be a group of permutations that act on S. If the relation $\mathcal{R}$ is defined on S by $x \, \mathcal{R} \, y$ if $\pi^*(x) = y$, for some $\pi \in G$, then $\mathcal{R}$ is an equivalence relation.

With only red and white disks to connect the sticks, the answer to this example could have been determined from the results in Fig. 12.5. However, we developed quite a bit of mathematical overkill to answer the question. Referring to S as the set of 2-colorings of the vertices of a square, we start to wonder about the role of 2 and seek the number of nonequivalent configurations if the disks come in three or more colors.

In addition we might notice that the function $f(r, w) = r^4 + r^3w + 2r^2w^2 + rw^3 + w^4$ is the generating function (of two variables) for the number of nonequivalent configurations from S. Here the coefficient of r^iw^{4-i}, $0 \le i \le 4$, yields the number of distinct 2-colorings that have i red disks and $(4 - i)$ white ones. The coefficient of r^2w^2 is 2, because of the two equivalence classes $c\ell(3)$ and $c\ell(4)$. Finally, $f(1, 1) = 6$, the number of equivalence classes. This generating function $f(r, w)$ is called the *pattern inventory* for the configurations. We shall examine it in detail in the next two sections.   □

For now we extend our present results in the following theorem. (A proof of this result is given on pages 136–137 of C. L. Liu [7].)

Theorem 12.18 ▶ (*Burnside's Theorem*) Let S be a set of configurations on which a group G of permutations acts. The number of equivalence classes into which S is partitioned by the action of G is then given by

$$\frac{1}{|G|} \sum_{\pi \in G} \psi(\pi^*),$$

where $\psi(\pi^*)$ is the number of configurations in S fixed by π^*.

To convince ourselves of the truth of this theorem we first examine two examples where we already know the answers.

EXAMPLE 12.26

In Example 12.25 we find that $\psi(\pi_1^*) = 2$, since only C_1 and C_{16} are fixed, or *invariant*, under π_1^*. For $r_3 \in G$, however, $\psi(r_3^*) = 8$, since $C_1, C_2, C_4, C_{10}, C_{11}, C_{13}, C_{15}, C_{16}$ remain fixed under this group action. In like manner $\psi(\pi_2^*) = 4$, $\psi(\pi_3^*) = 2$, $\psi(\pi_0^*) = 16$, $\psi(r_1^*) = \psi(r_2^*) = 4$, $\psi(r_4^*) = 8$. With $|G| = 8$, Burnside's theorem implies that the number of equivalence classes, or nonequivalent configurations, is $(1/8)(16 + 2 + 4 + 2 + 4 + 4 + 8 + 8) = (1/8)(48) = 6$, the original answer. □

EXAMPLE 12.27

In how many ways can six people be arranged around a circular table if two arrangements are considered equivalent when one can be obtained from the other by means of a clockwise rotation through $i \cdot 60°$, $0 \le i \le 5$?

Figure 12.7

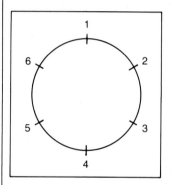

Here the six distinct people are to be placed in six chairs located at a table, as in Fig. 12.7. Our permutation group G consists of the clockwise rotations π_i through $i \cdot 60°$, $0 \le i \le 5$. Here reflections are not meaningful. The situation is two-dimensional, for we can only rotate the circle (representing the table) in the plane; the circle never lifts off the plane. The total number of possible configurations is 6! We find that $\psi(\pi_0^*) = 6!$, while $\psi(\pi_i^*) = 0$, for $1 \le i \le 5$. (It's impossible to move different people and simultaneously have them stay in a fixed location.) Consequently, the total number of nonequivalent seating arrangements is

$$\left(\frac{1}{|G|}\right) \sum_{\sigma \in G} \psi(\sigma^*) = \left(\frac{1}{6}\right)(6! + 0 + 0 + 0 + 0 + 0) = 5!,$$

as we found in Example 1.14 of Chapter 1. □

We now examine a situation where the power of this theorem is made apparent.

EXAMPLE 12.28

In how many ways can the vertices of a square be 3-colored, if the square can be moved about in three dimensions?

Now we have the sticks of Example 12.25 along with red, white, and blue disks. Considering the group in Fig. 12.6, we find the following:

$\psi(\pi_0^*) = 3^4$, since the identity fixes all 81 configurations in the set S of possible configurations.

$\psi(\pi_1^*) = \psi(\pi_3^*) = 3$, for each of π_1^*, π_3^* leaves invariant only those configurations with all vertices the same color.

$\psi(\pi_2^*) = 9$, for π_2^* can only fix those configurations where the opposite (diagonally) vertices have the same color. Consider a square like the one in Fig. 12.8. There are three choices for placing a colored disk at vertex 1, and then one choice for matching it at vertex 3. Likewise there are three choices for colors at vertex 2, and then one for vertex 4. Consequently there are nine configurations invariant under π_2^*.

$\psi(r_1^*) = \psi(r_2^*) = 9$. In the case of r_1^*, for the square in Fig. 12.8, we have three choices for coloring each of the vertices 1 and 2, and then we must match the color of vertex 4 with the color of vertex 1, and the color of vertex 3 with that of vertex 2.

Finally, $\psi(r_3^*) = \psi(r_4^*) = 27$. For r_3^*, we have nine choices for coloring the two vertices at 2 and 4, and three choices for vertex 1. Then there is only one choice for vertex 3, since we must match the color of vertex 1.

By Burnside's theorem the number of nonequivalent configurations is $(1/8)(3^4 + 3 + 3^2 + 3 + 3^2 + 3^2 + 3^3 + 3^3) = 21$.

Figure 12.8

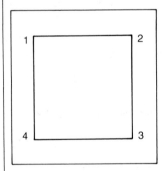

EXERCISES

1. For the configurations in Fig. 12.5
   a) determine π_2^*, π_3^*, r_2^*, r_4^*.
   b) verify that $(\pi_1^{-1})^* = (\pi_1^*)^{-1}$ and $(r_3^{-1})^* = (r_3^*)^{-1}$.
   c) verify that $(\pi_1 r_1)^* = \pi_1^* r_1^*$ and $(\pi_3 r_4)^* = \pi_3^* r_4^*$.

2. Express each of the following elements of S_7 as a product of disjoint cycles.

$$\alpha = \begin{pmatrix} 1 & 2 & 3 & 4 & 5 & 6 & 7 \\ 2 & 4 & 6 & 7 & 1 & 5 & 3 \end{pmatrix} \quad \beta = \begin{pmatrix} 1 & 2 & 3 & 4 & 5 & 6 & 7 \\ 3 & 6 & 5 & 2 & 1 & 7 & 4 \end{pmatrix}$$

$$\gamma = \begin{pmatrix} 1 & 2 & 3 & 4 & 5 & 6 & 7 \\ 2 & 3 & 1 & 7 & 5 & 4 & 6 \end{pmatrix} \quad \delta = \begin{pmatrix} 1 & 2 & 3 & 4 & 5 & 6 & 7 \\ 4 & 2 & 7 & 1 & 3 & 6 & 5 \end{pmatrix}$$

3. a) Determine the order of each of the elements in Exercise 2.

   b) State a general result about the order of an element in S_n in terms of the lengths of the cycles in its decomposition as a product of disjoint cycles.

4. a) Determine the number of distinct ways one can color the vertices of an equilateral triangle using the colors red and white, if the triangle is free to move in three dimensions.

   b) Answer part (a) if the color blue is also available.

5. Answer the questions in Exercise 4 for a regular pentagon.

6. a) How many distinct ways are there to paint the *edges* of a square with three different colors?

   b) Answer part (a) for the edges of a regular pentagon.

7. We make a child's bracelet by symmetrically placing four beads about a circular wire. The colors of the beads are red, white, blue, and green, and there are at least four beads of each color.

   a) How many distinct bracelets can we make in this way, if the bracelets can be rotated but not reflected?

   b) Answer part (a) if the bracelets can be rotated and reflected.

8. A baton is painted with three cylindrical bands of color (not necessarily distinct), with each band of the same length.

   a) How many distinct paintings can be made if there are three colors of paint available? How many for four colors?

   b) Answer part (a) for batons with four cylindrical bands.

   c) Answer part (a) for batons with n cylindrical bands.

   d) Answer parts (a) and (b) if adjacent cylindrical bands are to have different colors.

Figure 12.9

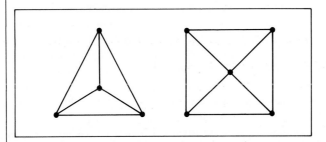

9. In how many ways can we 2-color the vertices of the configurations in Fig. 12.9 if they are free to move in

   a) two dimensions?

   b) three dimensions?

10. A pyramid has a square base and four faces that are equilateral triangles. If we can move the pyramid about (in three dimensions), how many nonequivalent ways are there to paint its five faces, if we have paint of four different colors? How many if the color of the base must be different from the color(s) of the triangular faces?

11. a) In how many ways can we paint the cells of a 3×3 chessboard using red and blue paint? (The back of the chessboard is black.)

    b) In how many ways can we construct a 3×3 chessboard by joining (with paste) the edges of nine 1×1 plastic squares that are transparent and tinted red or blue? (There are nine squares of each color available.)

12. Answer Exercise 11 for a 4×4 chessboard. (Replace each "nine" in part (b) with "sixteen.")

13. In how many ways can we paint the seven horses on a carousel using black, brown, and white paint?

14. a) Let S be a set of configurations and G a group of permutations that act on S. If $x \in S$, prove that $\{\pi \in G \mid \pi^*(x) = x\}$ is a subgroup of G.

    b) Determine the respective subgroups in part (a) for the configurations C_7 and C_{15} in Fig. 12.5.

12.10 ■ THE CYCLE INDEX

In applying Burnside's theorem we have been faced with computing $\psi(\pi^*)$ for each $\pi \in G$, where G is a permutation group acting on a set S of configurations. As the number of available colors increases and the configurations get more complex, such computations can get a bit involved. In addition, it seems that if we can determine the number of 2-colorings for a set S of configurations, we should be able to use some of the work in this case to determine the number of 3-colorings, 4-colorings, etc. We shall now find some assistance as we return to the solution of Example 12.25. This time more attention will be paid to the representation of each permutation $\pi \in G$ as a product of disjoint cycles. Our results are summarized in Table 12.9.

For π_0, the identity of G, we write $\pi_0 = (1)(2)(3)(4)$, a product of four disjoint cycles. We shall represent this cycle structure algebraically by x_1^4 where x_1 indicates a cycle of length 1. The term x_1^4 is called the *cycle structure representation* of π_0. Here we interpret disjoint as independent, in the sense that whatever color is used to paint the vertices in one cycle has no bearing on the choice of color for the vertices in another cycle. As long as all the vertices in a given cycle have the same color we shall find configurations that are invariant under π_0^*. (Admittedly, this seems like mathematical overkill again, since π_0^* fixes all 2-colorings of the square.) In addition, since we can paint the vertices in each cycle either red or white, we have 2^4 configurations, and we find that $(r + w)^4 = r^4 + 4r^3w + 6r^2w^2 + 4rw^3 + w^4$ generates these 16 configurations. For example, from the term $6r^2w^2$ we find that there are six configurations with two red and two white vertices, as found in classes (3) and (4) of Fig. 12.5.

Turning to π_1, we find $\pi_1 = (1234)$, a cycle of length 4. This cycle structure is represented by x_4, and here there are only two invariant configurations. As the cycle structure for π_1 has only one cycle this tells us that for a configuration to be invariant under π_1^* every vertex in this cycle must be painted the same color. With two colors to choose from there are only two possible configurations, C_1 and C_{16}. In this case the term $(r^4 + w^4)$ generates these configurations.

Continuing with r_1, we have $r_1 = (14)(23)$, a product of two disjoint cycles of length 2; the term x_2^2 represents this cycle structure. For a configuration to be fixed by r_1^*, the vertices at 2 and 3 must be the same color, i.e., we have two choices for coloring the vertices in (23). We also have two choices for coloring the vertices in (14). Consequently, we get 2^2 invariant configurations: $C_1(r^4)$, $C_7(r^2w^2)$, $C_9(r^2w^2)$, $C_{16}(w^4)$. $[(r^2 + w^2)^2 = r^4 + 2r^2w^2 + w^4.]$

Finally, in the case of $r_3 = (13)(2)(4)$, $x_2x_1^2$ indicates its decomposition into one cycle of length 2 and two of length 1. The vertices at 1 and 3 must be painted the same color if the configuration is to be fixed by r_3^*. With three cycles and two choices of color for each cycle we find 2^3 invariant configurations. These are $C_1(r^4)$, $C_2(r^3w)$, $C_4(r^3w)$, $C_{10}(r^2w^2)$, $C_{11}(r^2w^2)$, $C_{13}(rw^3)$, $C_{15}(rw^3)$, $C_{16}(w^4)$. The generating function for these configurations is $(r^2 + w^2)(r + w)^2$, for when we consider the cycle (13) we have two choices: both vertices red (r^2) or both vertices white (w^2). This gives us $(r^2 + w^2)$. For each single vertex in the two cycles of length 1, $r + w$ provides the

Table 12.9

Rigid Motions π (Elements of G)	Configurations in S that are Invariant under π^*	Cycle Structure Representation of π	Inventory of Configurations that are Invariant under π^*			
$\pi_0 = (1)(2)(3)(4)$	2^4: All configurations in S	x_1^4	$(r + w)^4$	$= r^4 + 4r^3w + 6r^2w^2 + 4rw^3 + w^4$		
$\pi_1 = (1234)$	2: C_1, C_{16}	x_4	$r^4 + w^4$	$= r^4$		$+ w^4$
$\pi_2 = (13)(24)$	2^2: C_1, C_{10}, C_{11}, C_{16}	x_2^2	$(r^2 + w^2)^2$	$= r^4$	$+ 2r^2w^2$	$+ w^4$
$\pi_3 = (1432)$	2: C_1, C_{16}	x_4	$r^4 + w^4$	$= r^4$		$+ w^4$
$r_1 = (14)(23)$	2^2: C_1, C_7, C_9, C_{16}	x_2^2	$(r^2 + w^2)^2$	$= r^4$	$+ 2r^2w^2$	$+ w^4$
$r_2 = (12)(34)$	2^2: C_1, C_6, C_8, C_{16}	x_2^2	$(r^2 + w^2)^2$	$= r^4$	$+ 2r^2w^2$	$+ w^4$
$r_3 = (13)(2)(4)$	2^3: C_1, C_2, C_4, C_{10}, C_{11}, C_{13}, C_{15}, C_{16}	$x_2x_1^2$	$(r^2 + w^2)(r + w)^2 = r^4 + 2r^3w + 2r^2w^2 + 2rw^3 + w^4$			
$r_4 = (1)(24)(3)$	2^3: C_1, C_3, C_5, C_{10}, C_{11}, C_{12}, C_{14}, C_{16}	$x_2x_1^2$	$(r^2 + w^2)(r + w)^2 = r^4 + 2r^3w + 2r^2w^2 + 2rw^3 + w^4$			
	$P_G(x_1, x_2, x_3, x_4) = \frac{1}{8}(x_1^4 + 2x_4 + 3x_2^2 + 2x_2x_1^2)$		Complete Inventory $\left.\right\}$ $= 8r^4 + 8r^3w + 16r^2w^2 + 8rw^3 + 8w^4$			

choices for each cycle, $(r + w)^2$ the choices for the two. By the independence of choice of colors as we go from one cycle to another, $(r^2 + w^2)(r + w)^2$ generates the 2^3 configurations invariant under r_3^*.

Similar arguments provide the information in Table 12.9 for the permutations π_2, π_3, r_2, and r_4.

At this point we see that what determines the number of configurations invariant under π^*, for $\pi \in G$, depends upon the cycle structure of π. Within each cycle the same color must be used, but that color can be selected from the two or more choices made available. For r_1, we had two cycles (of length 2), and 2^2 configurations. If three colors were available, the number of invariant configurations would have been 3^2. For m colors, the number is m^2. Adding these terms for all the cycle structures that arise gives $\sum_{\pi \in G} \psi(\pi^*)$.

Since we now wish to place more emphasis on cycle structures we define the *cycle index*, P_G, for *group G* as

$$P_G(x_1, x_2, x_3, x_4) = \frac{1}{|G|} \sum_{\pi \in G} \text{(cycle structure representation of } \pi).$$

In this example,

$$P_G(x_1, x_2, x_3, x_4) = (1/8)(x_1^4 + 2x_4 + 3x_2^2 + 2x_2x_1^2).$$

When each occurrence of x_1, x_2, x_3, x_4 is replaced by 2, we have the number of nonequivalent 2-colorings equal to

$$P_G(2,2,2,2) = (1/8)(2^4 + 2(2) + 3(2^2) + 2(2)(2^2)) = 6.$$

We summarize our present findings in the following.

Theorem 12.19 ▶ Let S be a set of configurations that are acted upon by a permutation group G. (G is a subgroup of S_n, the group of all permutations of $\{1, 2, 3, \ldots, n\}$, and the cycle index $P_G(x_1, x_2, x_3, \ldots, x_n)$ of G is $(1/|G|)\sum_{\pi \in G}$ (cycle structure representation of π).) The number of nonequivalent m-colorings of S is then $P_G(m, m, m, \ldots, m)$.

We close this section with an example that uses this theorem.

EXAMPLE 12.29 In how many distinct ways can we 4-color the vertices of a regular hexagon that is free to move in space?

For a regular hexagon there are twelve rigid motions: (a) the six clockwise rotations through 0°, 60°, 120°, 180°, 240°, 300°; (b) the three reflections in diagonals through opposite vertices; and (c) the three reflections through lines passing through the midpoints of opposite edges.

In Fig. 12.10 we have listed each group element as a product of disjoint cycles, together with its cycle structure representation. Here

$$P_G(x_1, x_2, x_3, x_4, x_5, x_6) = (1/12)(x_1^6 + 2x_6 + 2x_3^2 + 4x_2^3 + 3x_1^2x_2^2),$$

and there are

$$P_G(4,4,4,4,4,4) = (1/12)(4^6 + 2(4) + 2(4^2) + 4(4^3) + 3(4^2)(4^2)) = 430$$

Figure 12.10

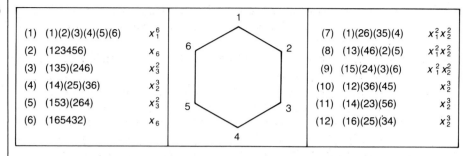

(1)	(1)(2)(3)(4)(5)(6)	x_1^6		(7)	(1)(26)(35)(4)	$x_1^2 x_2^2$
(2)	(123456)	x_6		(8)	(13)(46)(2)(5)	$x_1^2 x_2^2$
(3)	(135)(246)	x_3^2		(9)	(15)(24)(3)(6)	$x_1^2 x_2^2$
(4)	(14)(25)(36)	x_2^3		(10)	(12)(36)(45)	x_2^3
(5)	(153)(264)	x_3^2		(11)	(14)(23)(56)	x_2^3
(6)	(165432)	x_6		(12)	(16)(25)(34)	x_2^3

nonequivalent 4-colorings of a regular hexagon. (*Note:* Even though neither x_4 nor x_5 occurs in a cycle structure representation, we may list these variables among the arguments of P_G.) □

EXERCISES

1. In how many ways can we 5-color the vertices of a square that is free to move in (a) two dimensions? (b) three dimensions?

2. Answer Exercise 1 for a regular pentagon.

3. Find the number of nonequivalent 4-colorings of the vertices in the configurations of Fig. 12.11, when they are free to move in (a) two dimensions; and (b) three dimensions.

Figure 12.11

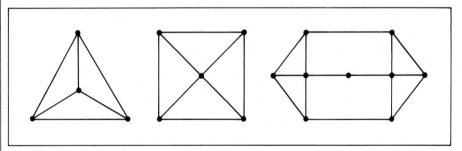

4. a) In how many ways can we 3-color the vertices of a regular hexagon that is free to move in space?

   b) Give a combinatorial argument to show that for all $m \in \mathbf{Z}^+$, $(m^6 + 2m + 2m^2 + 4m^3 + 3m^4)$ is divisible by 12.

5. a) In how many ways can we 5-color the vertices of a regular hexagon that is free to move in two dimensions?

   b) Answer part (a) if the hexagon is free to move in three dimensions.

   c) Find two 5-colorings that are equivalent in case (b) but distinct for case (a).

6. In how many distinct ways can we 3-color the *edges* in the configurations of Fig. 12.11, if they are free to move in (a) two dimensions; and (b) three dimensions?

7. a)   In how many distinct ways can we 3-color the *edges* of a square that is free to move in three dimensions?

   b)   In how many distinct ways can we 3-color both the vertices and the edges of such a square?

   c)   For a square that can move in three dimensions, let k, m, n denote the number of distinct ways we can 3-color its vertices (alone), its edges (alone), and both its vertices and edges, respectively. Does $n = km$? (Give a geometric explanation.)

12.11 ■ THE PATTERN INVENTORY: POLYA'S METHOD OF ENUMERATION

In this final section we return to Example 12.25, concentrating now on the pattern inventory and how it can be derived from the cycle index.

For $\pi_0 \in G$, every configuration in S is invariant. The cycle structure (representation) for π_0 is given by x_1^4, where for each cycle of length 1 we have a choice of coloring the vertex in that cycle red (r) or white (w). Using $+$ to represent *or*, we write $(r + w)$ to denote the two choices for that vertex (cycle of length 1). With four such cycles, $(r + w)^4$ generates the patterns of the 16 configurations.

In the case of $\pi_1 = (1234)$, x_4 denotes the cycle structure and here all four vertices must be the same color for the configuration to remain fixed under π_1^*. Consequently, we have all four vertices red or all four vertices white, and we express this algebraically by $r^4 + w^4$.

At this point we notice that for each of the permutations we have considered, the number of factors in the pattern inventory for that permutation equals the number of factors in its cycle structure (representation). Is this just a coincidence?

Continuing with $r_1 = (14)(23)$, we find the cycle structure x_2^2. For the cycle (14) we must color both of the vertices 1 and 4 either red or white. These choices are represented by $r^2 + w^2$. Since there are two such cycles of length 2, the generating function for the pattern inventory of configurations of S fixed by r_1^* is $(r^2 + w^2)^2$. Once again the number of factors in the cycle structure equals the number of factors in the corresponding term of the generating function.

Lastly, for $r_3 = (13)(2)(4)$, the cycle structure is $x_1^2 x_2$. For each of the cycles (2) and (4), $r + w$ represents the choices for each of these vertices, while $(r + w)^2$ accounts for all four colorings of the pair. The cycle (13) indicates that vertices 1 and 3 must have the same color; $r^2 + w^2$ accounts for the two possibilities. Hence the summand in the generating function for these configurations is $(r + w)^2(r^2 + w^2)$, and we find three factors in both the cycle structure and the corresponding summand of the generating function. But even more comes to light here.

Looking at the terms in the cycle structures we see that, for $1 \le i \le n$, the factor x_i in the cycle structure corresponds with the term $r^i + w^i$ in the associated summand of the generating function.

Continuing with the cycle structures for π_2, π_3, r_2, and r_4, we find that the pattern inventory can be obtained by replacing each x_i in $P_G(x_1, x_2, x_3, x_4)$ with

$r^i + w^i$, $1 \le i \le 4$. Consequently,

$$P_G(r + w, r^2 + w^2, r^3 + w^3, r^4 + w^4) = r^4 + r^3 w + 2r^2 w^2 + r w^3 + w^4.$$

(This result is $(1/8)$-th of the complete inventory listed in Table 12.9.)

If we had three colors, red, white, and blue, the replacement for x_i would be $r^i + w^i + b^i$, $1 \le i \le 4$.

We generalize these observations in the following.

Theorem 12.20 ▶ (*Polya's Method of Enumeration*) Let S be a set of configurations that are acted upon by a permutation group G, where G is a subgroup of S_n and has cycle index $P_G(x_1, x_2, \ldots, x_n)$. Then the generating function for the pattern inventory of non-equivalent m-colorings of S is given by

$$P_G\left(\sum_{i=1}^{m} c_i, \sum_{i=1}^{m} c_i^2, \ldots, \sum_{i=1}^{m} c_i^n \right),$$

where $c_1, c_2, \ldots, c_m$ denote the m colors that are available.

We apply this theorem in the following examples.

EXAMPLE 12.30 A child's bracelet is formed by placing three beads — red, white, and blue — on a circular piece of wire. Bracelets are considered equivalent if one can be obtained from the other by a (planar) rotation. Find the pattern inventory for these bracelets.

Here G is the group of rotations of an equilateral triangle, so $G = \{(1)(2)(3), (123), (132)\}$, where 1, 2, 3 denote the vertices of the triangle. Then $P_G(x_1, x_2, x_3) = (1/3)(x_1^3 + 2x_3)$ and the pattern inventory is given by $(1/3)[(r + w + b)^3 + 2(r^3 + w^3 + b^3)] = (1/3)[3r^3 + 3r^2 w + 3r^2 b + 3rw^2 + 6rwb + 3rb^2 + 3w^3 + 3w^2 b + 3wb^2 + 3b^3] = r^3 + r^2 w + r^2 b + rw^2 + 2rwb + rb^2 + w^3 + w^2 b + wb^2 + b^3$.

If the bracelets can also be reflected, then G becomes $\{(1)(2)(3), (123), (132), (1)(23), (2)(13), (3)(12)\}$, and the pattern inventory here is the same as the one above, with one exception. Here we have rwb, instead of $2rwb$, because the nonequivalent (for rotations) patterns in Fig. 12.12 become equivalent when reflections are allowed.

Figure 12.12

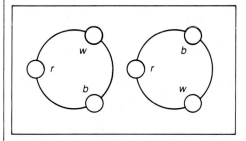

□

EXAMPLE 12.31 Consider the 3-colorings of the configurations in Example 12.25. If the three colors are red, white, and blue, how many nonequivalent configurations have exactly two red vertices?

Given that $P_G(x_1, x_2, x_3, x_4) = (1/8)(x_1^4 + 2x_4 + 3x_2^2 + 2x_2x_1^2)$, the answer is the sum of the coefficients of r^2w^2, r^2b^2, and r^2wb in $(1/8)[(r + w + b)^4 + 2(r^4 + w^4 + b^4) + 3(r^2 + w^2 + b^2)^2 + 2(r^2 + w^2 + b^2)(r + w + b)^2]$.

In $(r + w + b)^4$, we find the term $6r^2w^2 + 6r^2b^2 + 12r^2wb$. For $3(r^2 + w^2 + b^2)^2$, we are interested in the term $6r^2w^2 + 6r^2b^2$, while $4r^2w^2 + 4r^2b^2 + 4r^2bw$ arises in $2(r^2 + w^2 + b^2)(r + w + b)^2$.

Then $(1/8)[6r^2w^2 + 6r^2b^2 + 12r^2wb + 6r^2w^2 + 6r^2b^2 + 4r^2w^2 + 4r^2b^2 + 4r^2bw] = 2r^2w^2 + 2r^2b^2 + 2r^2bw$, the inventory of the six nonequivalent configurations that contain exactly two red vertices.   □

Our next example deals with the pattern inventory for the 2-colorings of the vertices of a cube. (The colors are red and white.)

Figure 12.13

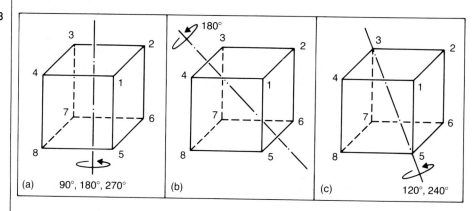

(a)   90°, 180°, 270°   (b)   (c)   120°, 240°

EXAMPLE 12.32   For the cube in Fig. 12.13, we find that its group G of rigid motions consists of the following.

1. The identity transformation with cycle structure x_1^8.

2. Rotations through 90°, 180°, 270° about an axis through the centers of two opposite faces: From Fig. 12.13(a) we have

    90° rotation: (1234)(5678)   Cycle structure: x_4^2
    180° rotation: (13)(24)(57)(68)   Cycle structure: x_2^4
    270° rotation: (1432)(5876)   Cycle structure: x_4^2.

    Since there are two other pairs of opposite faces, these nine rotations account for the term $3x_2^4 + 6x_4^2$ in the cycle index.

3. Rotations through 180° about an axis through the midpoints of two opposite edges: As in Fig. 12.13(b), we have the permutation (17)(28)(34)(56), whose cycle structure is given by x_2^4. With six pairs of opposite edges these rotations contribute the term $6x_2^4$ to the cycle index.

4. Rotations through 120° and 240° about an axis through two diagonally opposite vertices: From part (c) of the figure we have

$\qquad$ 120° rotation: $(168)(274)(3)(5)$ $\qquad$ Cycle structure: $x_1^2 x_3^2$

$\qquad$ 240° rotation: $(186)(247)(3)(5)$ $\qquad$ Cycle structure: $x_1^2 x_3^2$.

Here there are four such pairs of vertices and these result in the term $8x_1^2 x_3^2$ in the cycle index.

Therefore, $P_G(x_1, x_2, \ldots, x_8) = (1/24)(x_1^8 + 9x_2^4 + 6x_4^2 + 8x_1^2 x_3^2)$, and the pattern inventory for these configurations is

$$(1/24)[(r + w)^8 + 9(r^2 + w^2)^4 + 6(r^4 + w^4)^2 + 8(r + w)^2(r^3 + w^3)^2]$$
$$= r^8 + r^7 w + 3r^6 w^2 + 3r^5 w^3 + 7r^4 w^4 + 3r^3 w^5 + 3r^2 w^6 + rw^7 + w^8.$$

Replacing r and w by 1, we find 23 nonequivalent configurations here. $\qquad\square$

Since Polya's Method of Enumeration was first developed to count isomers of organic compounds, we close this section with an application that deals with a certain class of organic compounds. This is based on an example by C. L. Liu. (See pp. 152–154 of reference [7].)

Figure 12.14

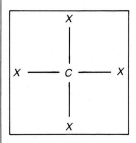

EXAMPLE 12.33

Here we are concerned with organic molecules of the form shown in Fig. 12.14, where C is a carbon atom and X denotes any of the following components: Br (bromine), H (hydrogen), CH_3 (methyl), or C_2H_5 (ethyl). (For example, if each X is replaced by H the compound CH_4 (methane) results.) Figure 12.14 should not be allowed to mislead us. The structure of these organic compounds is three-dimensional. Consequently we turn to the regular tetrahedron in order to model this structure. We place the carbon atom at the center of the tetrahedron and then place our selections for X at vertices 1, 2, 3, 4.

The group G acting on these configurations is given as follows:

1. The identity transformation $(1)(2)(3)(4)$ with cycle structure x_1^4.

2. Rotations through 120°, 240° about an axis through a vertex and the center of the opposite face: As seen in Fig. 12.15(a) we have

$\qquad$ 120° rotation: $(1)(243)$ with cycle structure $x_1 x_3$

$\qquad$ 240° rotation: $(1)(234)$ with cycle structure $x_1 x_3$.

Figure 12.15

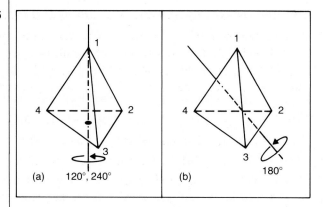

(a)   120°, 240°   (b)   180°

By symmetry there are three other pairs of vertices and opposite faces, so these rigid motions account for the term $8x_1x_3$ in $P_G(x_1, x_2, x_3, x_4)$.

3. Rotations of 180° about an axis through the midpoints of two opposite edges: The case shown in part (b) of the figure is the permutation (14)(23) whose cycle structure is x_2^2. With three pairs of opposite edges we get the term $3x_2^2$ in $P_G(x_1, x_2, x_3, x_4)$.

Hence $P_G(x_1, x_2, x_3, x_4) = (1/12)[x_1^4 + 8x_1x_3 + 3x_2^2]$, and $P_G(4, 4, 4, 4) = (1/12)[4^4 + 8(4^2) + 3(4^2)] = 36$, so there are 36 distinct organic compounds that can be formed in this way.

Lastly, if we wish to know how many of these compounds have exactly two bromine atoms, we let w, x, y, z represent the "colors" Br, H, CH_3, C_2H_5, respectively, and find the sum of the coefficients of w^2x^2, w^2y^2, w^2z^2, w^2xy, w^2xz, w^2yz in the pattern inventory

$$(1/12)[(w + x + y + z)^4 + 8(w + x + y + z)(w^3 + x^3 + y^3 + z^3)$$
$$+ 3(w^2 + x^2 + y^2 + z^2)^2].$$

For $(w + x + y + z)^4$ the term is $6w^2x^2 + 6w^2y^2 + 6w^2z^2 + 12w^2xy + 12w^2xz + 12w^2yz$. The middle summand of the pattern inventory does not give rise to any of these situations, while in $3(w^2 + x^2 + y^2 + z^2)^2$ we find $6w^2x^2 + 6w^2y^2 + 6w^2z^2$.

Consequently the pattern inventory for the compounds containing exactly two bromine atoms is

$$(1/12)[12w^2x^2 + 12w^2y^2 + 12w^2z^2 + 12w^2xy + 12w^2xz + 12w^2yz]$$

and there are six such organic compounds.   □

EXERCISES

1. a) Find the pattern inventory for the 2-colorings of the edges of a square that is free to move in two dimensions; three dimensions. (Let the colors be red and white.)

    b) Answer part (a) for 3-colorings, with the colors red, white, and blue.

2. If a regular pentagon is free to move in space and we can color its vertices with red, white, and blue paint, how many nonequivalent configurations have exactly three red vertices? How many have two red, one white, and two blue vertices?

3. Suppose that in Example 12.32 we 2-color the faces of the cube, which is free to move in space.

    a) How many distinct 2-colorings are there for this situation?

    b) If the available colors are red and white, determine the pattern inventory.

    c) How many nonequivalent colorings have three red and three white faces?

4. For the organic compounds in Example 12.33, how many have at least one bromine atom? How many have exactly three hydrogen atoms?

5. Find the pattern inventories for the 2-colorings of the vertices in the configurations in Fig. 12.11, when they are free to move in space. (Let the colors be green and gold.)

6. a) In how many ways can the seven (identical) horses on a carousel be painted with black, brown, and white paint, so that there are three black, two brown, and two white horses?

    b) In how many ways are there equal numbers of black and brown horses?

    c) Give a combinatorial argument to verify that for all $n \in \mathbf{Z}^+$, $n^7 + 6n$ is divisible by 7.

7. a) In how many ways can we paint the eight squares of a 2 × 4 chessboard using the colors red and white? (The back of the chessboard is black cardboard.)

    b) Find the pattern inventory for the colorings in part (a).

    c) How many of the colorings in part (a) have four red and four white squares? How many have six red and two white squares?

8. a) In how many ways can we 2-color the eight regions of the pinwheel in Fig. 12.16, using the colors black and gold, if the back of each region remains grey?

Figure 12.16

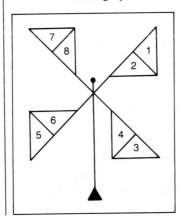

b)   Answer part (a) for the possible 3-colorings, using black, gold, and blue paints to color the regions.

c)   For the colorings in part (b), how many have four black, two gold, and two blue regions?

9. Let $m, n \in \mathbf{Z}^+$ with $n \geq 3$. How many distinct summands appear in the pattern inventory for the m-colorings of the vertices of a regular polygon of n sides?

12.12 ■ SUMMARY AND HISTORICAL REVIEW

Although the notion of a group of transformations evolved slowly in the study of geometry, the major thrust in the development of the group concept came from the study of polynomial equations.

Methods for solving quadratic equations were known to the ancient Greeks. Then in the sixteenth century advances were made toward solving cubic and quartic equations over $\mathbf{Q}$. Continuing with polynomials of fifth and higher degree, attempts were made by both Leonhard Euler (1707–1783) and Joseph-Louis Lagrange (1736–1813) to solve the general quintic. Lagrange realized there had to be a connection between the degree n of a polynomial equation and the permutation group S_n. However, it was Neils Henrik Abel (1802–1829) who finally proved that it was not possible to find a formula for solving the general quintic using only addition, subtraction, multiplication, division, and root extraction. During this same period, the existence of a necessary and sufficient condition for when a polynomial of degree $n \geq 5$ with rational coefficients can be solved by radicals was investigated and solved by the illustrious French mathematician Evariste Galois (1811–1832). Since his work utilizes the structures of both groups and fields, we shall say more about him in the summary of Chapter 13.

Following the accomplishments of Galois, group theory established itself further in many areas of mathematics. During the late nineteenth century, for example, the German mathematician Felix Klein (1849–1929), in what has come to be known as the *Erlanger Programm,* attempted to codify all existing geometries according to the group of transformations under which the properties of the geometry were invariant.

Many other mathematicians, such as Augustin-Louis Cauchy (1789–1857), Arthur Cayley (1821–1895), Ludwig Sylow (1832–1918), Richard Dedekind (1831–1916), and Leopold Kronecker (1823–1891) contributed to the further development of certain types of groups. However, it was not until 1900 that sets of defining conditions were given for the general abstract group.

During the twentieth century a great deal of research has been carried on in the attempt to analyze the structure of finite groups. For finite abelian groups it is known that any such group is isomorphic to a direct product of cyclic groups of prime power order. In the case of the finite nonabelian group a great deal more was needed. Starting with the work of Galois one finds particular attention paid to a special type of subgroup called a normal subgroup. For any group G, a subgroup H is called *normal,* if for all $g \in G$ and $h \in H$, $ghg^{-1} \in H$. In an abelian group every subgroup is normal, but not so for nonabelian groups. In every group G, both $\{e\}$ and G are normal subgroups, but if G has no other normal subgroups it is called *simple*. During the past five decades mathematicians have sought to find all the finite simple groups and explain their role in the structure of all finite groups. Among the prime movers in this development are Professors Walter Feit, John Thompson, Daniel Gorenstein, Michael Aschbacher, and Robert Griess, Jr.

There are many texts one can turn to for further study in the theory of groups. At the introductory level the text by V. Larney [6] provides further coverage beyond the introduction given in this chapter. The text by I. Herstein [5] is an excellent source and includes material on Galois theory.

The beginnings of algebraic coding theory can be traced to 1941 when Claude Elwood Shannon began his investigations of problems in communications. These problems were prompted by the needs of the war effort. His research resulted in many new ideas and principles which were later published in 1948 in the journal article [12]. As a result of this work Shannon is acknowledged as the founder of information theory. After this publication, results by M. Golay [3] and R. Hamming [4] soon followed. These provided further impetus for research in this area. If one consults the bibliography at the end of Volume II of the texts by F. MacWilliams and N. Sloane [8], the 1478 references listed there should provide some idea of the activity in this area that followed after 1950.

Our coverage of coding theory followed the development in Chapter 5 of the text by L. Dornhoff and F. Hohn [2]. The text by V. Pless [9] provides a nice coverage of topics at a fairly intermediate level. More advanced work in coding can be found in the books by F. MacWilliams and N. Sloane [8], and A. Street and W. Wallis [13]. An interesting application on the use of the pigeonhole principle in coding theory is given in Chapter XI of [13].

In the last three sections of the chapter we came upon an enumeration technique whose development is attributed to the contemporary mathematician George Polya. His article [10] provided the fundamental techniques for counting equivalence classes of chemical isomers, graphs, and trees. (To some extent the ideas in this work were anticipated by J. Redfield [11].) Since then these techniques have been found invaluable for counting problems in such areas as the electronic realizations of Boolean functions. Polya's fundamental theorem was first generalized in the article by N. DeBruijn [1] and other extensions of these ideas can be found in the literature.

Our coverage of this topic follows the presentation given in the article by A. Tucker [14]. A more rigorous presentation of this method can be found in Chapter 5 of the text by C. Liu [7].

REFERENCES

1. De Bruijn, Nicolaas Govert, "Polya's Theory of Counting," Chapter 5 in *Applied Combinatorial Mathematics,* edited by Edwin F. Beckenbach, John Wiley, New York, 1964.

2. Dornhoff, Larry L., and Hohn, Franz E., *Applied Modern Algebra,* Macmillan, New York, 1978.

3. Golay, Marcel J. E., "Notes on Digital Coding," *Proceedings of the IRE 37,* 1949, p. 657.

4. Hamming, Richard W., "Error Detecting and Error Correcting Codes," *Bell System Technical Journal 29,* 1950, pp. 147–160.

5. Herstein, I. N., *Topics in Algebra,* 2nd ed., Xerox College Publishing, Lexington, Massachusetts, 1975.

6. Larney, Violet H., *Abstract Algebra: A First Course,* Prindle, Weber & Schmidt, Boston, Massachusetts, 1975.

7. Liu, C. L., *Introduction to Combinatorial Mathematics,* McGraw-Hill, New York, 1968.

8. MacWilliams, F. Jessie, and Sloane, Neil J. A., *The Theory of Error-Correcting Codes,* Volumes I and II, North-Holland, Amsterdam, 1977.

9. Pless, Vera, *Introduction to the Theory of Error-Correcting Codes,* John Wiley, New York, 1982.

10. Polya, George, "Kombinatorische Anzahlbestimmungen für Gruppen, Graphen und Chemishe Verbindungen," *Acta Mathematica 68,* 1937, pp. 145–254.

11. Redfield, J. Howard, "The Theory of Group Reduced Distributions," *American Journal of Mathematics 49,* 1927, pp. 433–455.

12. Shannon, Claude E., "The Mathematical Theory of Communication," *Bell System Technical Journal 27,* 1948, pp. 379–423, 623–656. Reprinted in C. E. Shannon and W. Weaver, *The Mathematical Theory of Communication,* University of Illinois Press, Urbana, Illinois, 1949.

13. Street, Anne Penfold, and Wallis, W. D., *Combinatorial Theory: An Introduction,* The Charles Babbage Research Center, Winnipeg, Canada, 1977.

14. Tucker, Alan, "Polya's Enumeration Formula by Example," *Mathematics Magazine 47,* 1974, pp. 248–256.

■ MISCELLANEOUS EXERCISES

1. Let $f: G \to H$ be a group homomorphism with e_H the identity in H. Prove that
   a) $K = \{x \in G \mid f(x) = e_H\}$ is a subgroup of G.
   b) for all $g \in G$, $x \in K$, $gxg^{-1} \in K$.

2. If $G, H,$ and K are groups and $G = H \times K$, prove that G contains subgroups that are isomorphic to H and K.

3. Let G be a group where $a^2 = e$ for all $a \in G$. Prove that G is abelian.

4. If G is a group of even order, prove that there is an element $a \in G$ with $a \neq e$ and $a = a^{-1}$.

5. Let $f: G \to H$ be a group homomorphism onto H. If T is a subgroup of H, prove that $f^{-1}(T) = \{a \in G \mid f(a) \in T\}$ is a subgroup of G.

6. Considering the possible orders of elements, prove that the groups $\mathbf{Z}_9$ and $\mathbf{Z}_3 \times \mathbf{Z}_3$ are not isomorphic.

7. a) Regard $\mathbf{Z}_2^n$ as a group code. Given that a, b, $c \in \mathbf{Z}_2^n$, prove that $d(a, b) = d(a + c, b + c)$.

 b) For the group code $C \subseteq \mathbf{Z}_2^n$, let a, b be distinct code words. With $x \in C$, define $d(x, C) = \min\{d(x, c) \mid c \in C, c \neq x\}$. Prove that $d(a, C) = d(b, C)$.

8. Let $C \subseteq \mathbf{Z}_2^n$ be a k–error-correcting group code, where $4k + 2 > n$.

 a) Show that if we list the code words (in a column), any column in this list contains all 0's or half 0's and half 1's.

 b) Verify that the total number of 1's in the list is at least $(2k + 1)[|C| - 1]$.

 c) Prove that $\dfrac{2(2k + 1)}{4k + 2 - n} \geq |C|$.

9. a) Define the relation $\mathcal{R}$ on $\mathbf{Z}_2^7$ by $a_1 a_2 \ldots a_7 \ \mathcal{R} \ b_1 b_2 \ldots b_7$ if $b_1 b_2 \ldots b_7$ is a cyclic shift of $a_1 a_2 \ldots a_7$. (For example, $a_1 a_2 \ldots a_7 = 0110011$ yields $1100110 = a_2 a_3 \ldots a_7 a_1$ after one cyclic shift, while 1001101 results from $a_1 a_2 \ldots a_7$ after two cyclic shifts. Each of these examples is referred to as a cyclic shift.) Prove that $\mathcal{R}$ is an equivalence relation on $\mathbf{Z}_2^7$.

 b) How many equivalence classes result in the partition of $\mathbf{Z}_2^7$ induced by $\mathcal{R}$?

10. Let $C \subseteq \mathbf{Z}_2^7$ where C contains the elements 0000000, 0110100, 1001011, 1111111 and all their resulting cyclic shifts.

 a) Prove that C is a group code.

 b) What are the error-correcting and error-detecting capabilities of the code?

11. A projectile undergoes testing by being fired at a target. Assume that the probability of a hit is 0.75 for any single test and that the results of successive firings are independent.

 a) If four projectiles are fired, what is the probability of (i) exactly two hits? (ii) at least two hits?

 b) How many projectiles must we fire in order for the probability of at least one hit to be at least 0.95?

12. a) In how many ways can Nicole paint the eight regions of the square shown in Fig. 12.17, if five colors are available?

 b) In how many ways if she uses only four of the colors?

Figure 12.17

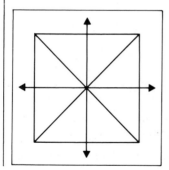

13

Finite Fields and Combinatorial Designs

It is time now to recall the ring structure of Chapter 8 as we examine rings of polynomials and their role in the construction of finite fields. We know that for any prime p, $(\mathbf{Z}_p, +, \cdot)$ is a finite field, but here we shall find other finite fields. Just as the order of a finite Boolean algebra is restricted to powers of 2, for finite fields we'll find that the possible orders are p^n, where p is a prime and $n \in \mathbf{Z}^+$. Applications of these finite fields will include a discussion of such combinatorial designs as Latin squares. Finally we shall investigate the structure of a finite geometry and discover how these geometries and combinatorial designs are interrelated.

13.1 ■ POLYNOMIAL RINGS

We recall that a ring $(R, +, \cdot)$ consists of a nonempty set R, where $(R, +)$ is an abelian group, $(R, \cdot)$ is closed under the associative operation $\cdot$, and the two operations are related by the distributive laws: $a(b + c) = ab + ac$, $(b + c)a = ba + ca$, for $a, b, c \in R$. (We write ab for $a \cdot b$.)

Let x denote an indeterminate, i.e., a formal symbol that is not an element of the ring R. We use x to define the following.

Definition 13.1

▶ Given a ring $(R, +, \cdot)$, an expression of the form $f(x) = a_n x^n + a_{n-1} x^{n-1} + \cdots + a_1 x^1 + a_0 x^0$, where $a_i \in R$, $0 \le i \le n$, is called a *polynomial in the indeterminate x over R*.

If a_n is not the zero element of R, then a_n is called *the leading coefficient of $f(x)$* and we say that $f(x)$ has *degree n*. Hence the degree of a polynomial is the highest power of x that occurs in a summand of the polynomial. The term $a_0 x^0$ is called the *constant*, or *constant term*, of $f(x)$.

If $g(x) = b_m x^m + b_{m-1} x^{m-1} + \cdots + b_1 x^1 + b_0 x^0$ is also a polynomial in x over R, then $f(x) = g(x)$ if $m = n$ and $a_i = b_i$ for all $1 \le i \le n$.

EXAMPLE 13.1

a) Over the ring $R = (\mathbf{Z}_6, +, \cdot)$, $5x^2 + 3x^1 - 2x^0$ is a polynomial of degree 2, with leading coefficient 5 and constant term $-2x^0$. Here we are using a to denote $[a]$. This polynomial can also be expressed as $5x^2 + 3x^1 + 4x^0$ since $[4] = [-2]$ in $\mathbf{Z}_6$.

b) If 0 is the zero element of ring R, then $0x^0 = 0$ is said to have *no degree* and no leading coefficient. A polynomial over R that is the zero element or of degree 0 is called a *constant polynomial*. The polynomial $5x^0$ over $\mathbf{Z}_7$ has degree 0 and is a constant polynomial.   □

We introduce operations of addition and multiplication for these polynomials in order to obtain a new ring. For a ring $(R, +, \cdot)$, let

$$f(x) = a_n x^n + a_{n-1} x^{n-1} + \cdots + a_1 x^1 + a_0 x^0$$

$$g(x) = b_m x^m + b_{m-1} x^{m-1} + \cdots + b_1 x^1 + b_0 x^0,$$

where $a_i, b_j \in R$ for $0 \le i \le n$, $0 \le j \le m$.

Assume that $n \geq m$. We define

$$f(x) + g(x) = \sum_{i=0}^{n} (a_i + b_i)x^i, \tag{1}$$

where $b_i = 0$ for $i > m$, and

$$f(x)g(x) = (a_n b_m)x^{n+m} + (a_n b_{m-1} + a_{n-1}b_m)x^{n+m-1} + \cdots$$
$$+ (a_1 b_0 + a_0 b_1)x^1 + (a_0 b_0)x^0. \tag{2}$$

In the definition of $f(x) + g(x)$, each coefficient $(a_i + b_i)$, $0 \leq i \leq n$, is obtained from the addition in R. For $f(x)g(x)$, the coefficient of x^t is $\sum_{k=0}^{t} a_{t-k}b_k$, where all additions and multiplications occur in R, and $0 \leq t \leq n + m$.

These operations are needed in the following result.

Theorem 13.1 ▶ Given a ring R, let $R[x]$ denote the set of all polynomials in the indeterminate x with coefficients in R. Under the operations of addition and multiplication given above in Eqs. (1) and (2), $(R[x], +, \cdot)$ is a ring, called the *polynomial ring*, or *ring of polynomials, over R*.

Proof The ring properties for $R[x]$ follow from those of R. Consequently, we shall prove the associative law of multiplication here, and leave the proofs of the other properties to the reader. Let $h(x) = \sum_{k=0}^{p} c_k x^k$, with $f(x)$, $g(x)$ as defined earlier. A typical summand in $(f(x)g(x))h(x)$ has the form Ax^t where $0 \leq t \leq (m + n) + p$ and A is the sum of all products of the form $(a_i b_j)c_k$, with $0 \leq i \leq n$, $0 \leq j \leq m$, $0 \leq k \leq p$, and $i + j + k = t$. In $f(x)(g(x)h(x))$ the coefficient of x^t is the sum of all products $a_i(b_j c_k)$, again with $0 \leq i \leq n$, $0 \leq j \leq m$, $0 \leq k \leq p$, and $i + j + k = t$. Since R is associative under multiplication, $(a_i b_j)c_k = a_i(b_j c_k)$ for each of these terms and the coefficient of x^t in $(f(x)g(x))h(x)$ is the same as it is in $f(x)(g(x)h(x))$. Hence $(f(x)g(x))h(x) = f(x)(g(x)h(x))$. ■

Corollary 13.1 ▶ Let $R[x]$ be a polynomial ring.

a) If R is commutative, then $R[x]$ is commutative.

b) If R is a ring with unity, then $R[x]$ is a ring with unity.

c) $R[x]$ is an integral domain iff R is an integral domain.

Proof The proof of this corollary is left for the reader. ■

From this point on we shall write x instead of x^1. If R has unity u, we define $x^0 = u$ and for any $r \in R$ we write rx^0 as r.

EXAMPLE 13.2 Let $f(x)$, $g(x) \in \mathbf{Z}_8[x]$ with $f(x) = 4x^2 + 1$, $g(x) = 2x + 3$. Then $f(x)$ has degree 2 and $g(x)$ has degree 1. From our past experiences with polynomials we expect the degree of $f(x)g(x)$ to be 3, the sum of the degrees of $f(x)$ and $g(x)$. Here, however, $f(x)g(x) = (4x^2 + 1)(2x + 3) = 8x^3 + 12x^2 + 2x + 3 = 4x^2 + 2x + 3$, since $[8] = [0]$ in $\mathbf{Z}_8$. So the degree of $f(x)g(x) = 2 < 3 = $ degree $f(x) + $ degree $g(x)$. □

What causes the result in Example 13.2 is the existence of zero divisors in the ring $\mathbf{Z}_8$. This observation leads us to the following.

Theorem 13.2

▶ Let $(R, +, \cdot)$ be a commutative ring with unity. (We let 0 denote the zero of R and 1 the unity.) Then R is an integral domain if and only if for all $f(x), g(x) \in R[x]$,

$$\text{degree } f(x)g(x) = \text{degree } f(x) + \text{degree } g(x).$$

Proof

Let $f(x) = \sum_{i=0}^{n} a_i x^i$, $g(x) = \sum_{j=0}^{m} b_j x^j$, with $a_n \neq 0$, $b_m \neq 0$. If R is an integral domain, then $a_n b_m \neq 0$, so the degree of $f(x)g(x) = n + m = \text{degree } f(x) + \text{degree } g(x)$. Conversely, if R is not an integral domain, let $a, b \in R$ with $a \neq 0$, $b \neq 0$, $ab = 0$. The polynomials $f(x) = ax + 1$, $g(x) = bx + 1$ each have degree 1, but $f(x)g(x) = (a + b)x + 1$ and degree $f(x)g(x) \leq 1 < 2 = \text{degree } f(x) + \text{degree } g(x)$. ∎

In this chapter we shall be concerned primarily with polynomial rings $F[x]$, where F is a field (and $F[x]$ is an integral domain). Consequently, we shall not dwell any further on situations where degree $f(x)g(x) < \text{degree } f(x) + \text{degree } g(x)$. In addition, unless it is stated otherwise, we shall denote the zero element of a field by 0 and use 1 to denote its unity.

Let R be a ring with $f(x) = a_n x^n + \cdots + a_1 x + a_0 \in R[x]$. If $r \in R$, then $f(r) = a_n r^n + \cdots + a_1 r + a_0 \in R$. We are especially interested in those values of r for which $f(r) = 0$.

Definition 13.2

▶ Let $f(x) \in R[x]$ and $r \in R$. If $f(r) = 0$, then r is called a *root* of the polynomial $f(x)$.

EXAMPLE 13.3

a) If $f(x) = x^2 - 2 \in \mathbf{R}[x]$, then $f(x)$ has $\sqrt{2}$ and $-\sqrt{2}$ as roots since $(\sqrt{2})^2 - 2 = 0 = (-\sqrt{2})^2 - 2$. In addition we can write $f(x) = (x - \sqrt{2})(x + \sqrt{2})$, with $x - \sqrt{2}, x + \sqrt{2} \in \mathbf{R}[x]$. However, if we regard $f(x)$ as an element of $\mathbf{Q}[x]$, then $f(x)$ has no roots, since $\sqrt{2}$ and $-\sqrt{2}$ are irrational numbers. Consequently, the existence of roots for a polynomial is dependent on the underlying ring of coefficients.

b) For $f(x) = x^2 + 3x + 2 \in \mathbf{Z}_6[x]$, we find that

$$f(0) = (0)^2 + 3(0) + 2 = 2 \qquad f(3) = (3)^2 + 3(3) + 2 = 20 = 2$$

$$f(1) = (1)^2 + 3(1) + 2 = 6 = 0 \qquad f(4) = (4)^2 + 3(4) + 2 = 30 = 0$$

$$f(2) = (2)^2 + 3(2) + 2 = 12 = 0 \qquad f(5) = (5)^2 + 3(5) + 2 = 42 = 0$$

Consequently, $f(x)$ has four roots: 1, 2, 4, 5. This is more than we expected. In our prior experiences a polynomial of degree two had at most two roots. □

As a result of Example 13.3(b), we shall now develop the concepts needed to find out when a polynomial of degree n has at most n roots.

Definition 13.3

▶ Let F be a field. For $f(x)$, $g(x) \in F[x]$, $f(x)$ is called a *divisor* (or *factor*) of $g(x)$ if there exists $h(x) \in F[x]$ with $f(x)h(x) = g(x)$. In this situation we also say that $f(x)$ *divides* $g(x)$ and that $g(x)$ is a *multiple* of $f(x)$.

This leads us to the *Division Algorithm* for polynomials.

Theorem 13.3

▶ (*Division Algorithm*) Let $f(x)$, $g(x) \in F[x]$ with $f(x) \neq 0$, the zero polynomial. There exist unique polynomials $q(x)$, $r(x) \in F[x]$ such that $g(x) = q(x)f(x) + r(x)$, where $r(x) = 0$ or degree $r(x) <$ degree $f(x)$.

Proof

Let $S = \{g(x) - t(x)f(x) \mid t(x) \in F[x]\}$.

If $0 \in S$, then $0 = g(x) - t(x)f(x)$ for some $t(x) \in F[x]$. Then with $q(x) = t(x)$ and $r(x) = 0$ we have $g(x) = q(x)f(x) + r(x)$.

If $0 \notin S$, consider the degrees of the elements of S, and let $r(x) = g(x) - q(x)f(x)$ be an element in S of minimum degree. Since $r(x) \neq 0$, the result follows if degree $r(x) <$ degree $f(x)$. If not, let

$$r(x) = a_n x^n + a_{n-1} x^{n-1} + \cdots + a_2 x^2 + a_1 x + a_0, \qquad a_n \neq 0,$$

$$f(x) = b_m x^m + b_{m-1} x^{m-1} + \cdots + b_2 x^2 + b_1 x + b_0, \qquad b_m \neq 0,$$

with $n \geq m$. Define

$$h(x) = r(x) - [a_n b_m^{-1} x^{n-m}]f(x) = (a_n - a_n b_m^{-1} b_m)x^n + (a_{n-1} - a_n b_m^{-1} b_{m-1})x^{n-1}$$
$$+ \cdots + (a_{n-m} - a_n b_m^{-1} b_0)x^{n-m} + a_{n-m-1} x^{n-m-1} + \cdots + a_1 x + a_0.$$

Then $h(x)$ has degree less than n, the degree of $r(x)$. More important, $h(x) = [g(x) - q(x)f(x)] - [a_n b_m^{-1} x^{n-m}]f(x) = g(x) - [q(x) + a_n b_m^{-1} x^{n-m}]f(x)$, so $h(x) \in S$ and this contradicts the choice of $r(x)$ as having minimum degree. Consequently, degree $r(x) <$ degree $f(x)$ and we have the existence part of the theorem.

For uniqueness, let $g(x) = q_1(x)f(x) + r_1(x) = q_2(x)f(x) + r_2(x)$ where $r_1(x) = 0$ or degree $r_1(x) <$ degree $f(x)$, and $r_2(x) = 0$ or degree $r_2(x) <$ degree $f(x)$. Then $[q_2(x) - q_1(x)]f(x) = r_1(x) - r_2(x)$, and if $q_2(x) - q_1(x) \neq 0$, then degree $([q_2(x) - q_1(x)]f(x)) \geq$ degree $f(x)$ while $r_1(x) - r_2(x) = 0$ or degree $[r_1(x) - r_2(x)] <$ degree $f(x)$. Consequently, $q_1(x) = q_2(x)$, and $r_1(x) = r_2(x)$. ■

The Division Algorithm provides the following results on roots and factors.

Theorem 13.4

▶ (*The Remainder Theorem*) For $f(x) \in F[x]$ and $a \in F$, the remainder in the division of $f(x)$ by $x - a$ is $f(a)$.

Proof

From the Division Algorithm, $f(x) = q(x)(x - a) + r(x)$, with $r(x) = 0$ or degree $r(x) <$ degree $(x - a) = 1$. Hence $r(x) = r$ is an element of F. Substituting a for x we find $f(a) = q(a)(a - a) + r(a) = 0 + r = r$. ■

Theorem 13.5

▶ (*The Factor Theorem*) If $f(x) \in F[x]$ and $a \in F$, then $x - a$ is a factor of $f(x)$ iff a is a root of $f(x)$.

Proof | If $x - a$ is a factor of $f(x)$, then $f(x) = q(x)(x - a)$. With $f(a) = q(a)(a - a) = 0$, a is a root of $f(x)$. Conversely, suppose that a is a root of $f(x)$. By the Division Algorithm, $f(x) = q(x)(x - a) + r$, where $r \in F$. Since $f(a) = 0$ it follows that $r = 0$, so $f(x) = q(x)(x - a)$, and $x - a$ is a factor of $f(x)$. ∎

We use these results to establish the last major idea in this section.

Theorem 13.6

▶ If $f(x) \in F[x]$ has degree $n \geq 1$, then $f(x)$ has at most n distinct roots in F.

Proof | The proof is by mathematical induction on the degree of $f(x)$. If $f(x)$ has degree 1, then $f(x) = ax + b$, $a, b \in F$, $a \neq 0$. With $f(-a^{-1}b) = 0$, $f(x)$ has at least one root in F. If c_1 and c_2 are both roots, then $f(c_1) = ac_1 + b = 0 = ac_2 + b = f(c_2)$. By cancellation in a ring, $ac_1 + b = ac_2 + b \Rightarrow ac_1 = ac_2$. Since F is a field and $a \neq 0$, $ac_1 = ac_2 \Rightarrow c_1 = c_2$, so $f(x)$ has only one root in F.

Now assume the result of the theorem is true for all polynomials of degree k in $F[x]$. Consider a polynomial $f(x)$ of degree $k + 1$. If $f(x)$ has no roots in F, the theorem follows. Otherwise, let $r \in F$ with $f(r) = 0$. By the Factor Theorem, $f(x) = (x - r)g(x)$ where $g(x)$ has degree k. Consequently, by the induction hypothesis, $g(x)$ has at most k distinct roots in F and $f(x)$, in turn, has at most $k + 1$ distinct roots. ∎

EXAMPLE 13.4

a) Let $f(x) = x^2 - 6x + 9 \in \mathbf{R}[x]$. Then $f(x)$ has the roots 3, 3 and $f(x) = (x - 3)(x - 3)$, a factorization into first degree, or *linear*, factors. Here we say that 3 is a root of *multiplicity* 2.

b) For $g(x) = x^2 + 4 \in \mathbf{R}[x]$, $g(x)$ has no real roots, but Theorem 13.6 is not contradicted. (Why?) In $\mathbf{C}[x]$, $g(x)$ has the roots $2i$, $-2i$ and can be factored as $g(x) = (x - 2i)(x + 2i)$.

c) If $h(x) = x^2 + 2x + 6 \in \mathbf{Z}_7[x]$, then $h(2) = 0$, $h(3) = 0$ and these are the only roots. Also, $h(x) = (x - 2)(x - 3) = x^2 - 5x + 6 = x^2 + 2x + 6$, since $[-5] = [2]$ in $\mathbf{Z}_7$.

d) As we saw in Example 13.3(b), the polynomial $x^2 + 3x + 2$ has four roots. This is not a contradiction to Theorem 13.6, since $\mathbf{Z}_6$ is not a field. Also, $x^2 + 3x + 2 = (x + 1)(x + 2) = (x + 4)(x + 5)$, two distinct factorizations. □

We close with one final remark, without proof, on the idea of factorization in $F[x]$. If $f(x) \in F[x]$ has degree n, and $r_1, r_2, \ldots, r_n$ are the roots of $f(x)$ in F (where it is possible for a root to be repeated) then $f(x) = a_n(x - r_1)(x - r_2) \cdots (x - r_n)$, where a_n is the leading coefficient of $f(x)$. This representation of $f(x)$ is unique up to the order of the first-degree factors.

EXERCISES

1. How many polynomials are there of degree 2 in $\mathbf{Z}_{11}[x]$? How many have degree 3? degree 4? degree n, $n \in \mathbf{N}$?

2. a) Find two nonzero polynomials $f(x)$, $g(x)$ in $\mathbf{Z}_{12}[x]$ where $f(x)g(x) = 0$.

    b) Find polynomials $h(x)$, $k(x) \in \mathbf{Z}_{12}[x]$ such that degree $h(x) = 5$, degree $k(x) = 2$ and degree $h(x)k(x) = 3$.

3. Complete the proofs of Theorem 13.1 and Corollary 13.1.

4. If $f(x) = ax^3 + bx^2 + cx + d$, $g(x) = 5x^2 + 3x - 7 \in \mathbf{Z}[x]$ and $f(x) = (3x + 1)g(x)$, find a, b, c, and d.

5. a) If $f(x) = x^4 - 16$, find its roots and factorization in $\mathbf{Q}[x]$.

    b) Answer part (a) for $f(x) \in \mathbf{R}[x]$.

    c) Answer part (a) for $f(x) \in \mathbf{C}[x]$.

    d) Answer parts (a), (b), (c) for $f(x) = x^4 - 25$.

6. For each of the following pairs $f(x)$, $g(x)$, find $q(x)$, $r(x)$ so that $g(x) = q(x)f(x) + r(x)$, where $r(x) = 0$ or degree $r(x) <$ degree $f(x)$.

    a) $f(x)$, $g(x) \in \mathbf{Q}[x]$, $f(x) = x^4 - 5x^3 + 7x$, $g(x) = x^5 - 2x^2 + 5x - 3$

    b) $f(x)$, $g(x) \in \mathbf{Z}_2[x]$, $f(x) = x^2 + 1$, $g(x) = x^4 + x^3 + x^2 + x + 1$

    c) $f(x)$, $g(x) \in \mathbf{Z}_5[x]$, $f(x) = x^2 + 3x + 1$, $g(x) = x^4 + 2x^3 + x + 4$

7. Find the remainder when $f(x) = x^8 + 7x^5 - 4x^4 + 3x^3 + 5x^2 - 4$ is divided by $x - 3$. ($f(x)$, $x - 3 \in \mathbf{Q}[x]$.)

8. a) Find all roots of $f(x) = x^2 + 4x$ if $f(x) \in \mathbf{Z}_{12}[x]$.

    b) Find four distinct linear polynomials $g(x)$, $h(x)$, $s(x)$, $t(x) \in \mathbf{Z}_{12}[x]$ so that $f(x) = g(x)h(x) = s(x)t(x)$.

9. How many units are there in the ring $\mathbf{Z}_5[x]$? How many in $\mathbf{Z}_7[x]$? How many in $\mathbf{Z}_p[x]$, p a prime?

10. For $n \in \mathbf{Z}^+$, let $f(x) \in \mathbf{Z}_n[x]$. Prove that if a, $b \in \mathbf{Z}$ and $a \equiv b \pmod{n}$, then $f(a) \equiv f(b) \pmod{n}$.

13.2 ■ IRREDUCIBLE POLYNOMIALS: FINITE FIELDS

We now wish to construct finite fields other than $(\mathbf{Z}_p, +, \cdot)$. The construction will use the following special polynomials.

Definition 13.4

▶ Let $f(x) \in F[x]$, with F a field and degree $f(x) \geq 2$. We call $f(x)$ *reducible* if there exist $g(x)$, $h(x) \in F[x]$, where $f(x) = g(x)h(x)$ and degree $g(x)$, $h(x) \geq 1$. If $f(x)$ is not reducible it is called *irreducible*, or *prime*.

Theorem 13.7 contains some useful observations about irreducible polynomials.

Theorem 13.7

▶ For polynomials in $F[x]$,

a) any nonzero polynomial of degree ≤ 1 is irreducible.

b) if $f(x) \in F[x]$ with degree $f(x) = 2$ or 3, then $f(x)$ is reducible if and only if $f(x)$ has a root in F.

Proof | The proof is left for the reader. ■

EXAMPLE 13.5

a) The polynomial $x^2 + 1$ is irreducible in $\mathbf{Q}[x]$ or $\mathbf{R}[x]$, but in $\mathbf{C}[x]$ we find $x^2 + 1 = (x + i)(x - i)$.

b) Let $f(x) = x^4 + 2x^2 + 1 \in \mathbf{R}[x]$. Although $f(x)$ has no real roots, it is reducible since $(x^2 + 1)^2 = x^4 + 2x^2 + 1$. Hence part (b) of Theorem 13.7 is not applicable for polynomials of degree > 3.

c) In $\mathbf{Z}_2[x]$, $f(x) = x^3 + x^2 + x + 1$ is reducible since $f(1) = 0$. But $g(x) = x^3 + x + 1$ is irreducible since $g(0) = g(1) = 1$.

d) Let $h(x) = x^4 + x^3 + x^2 + x + 1 \in \mathbf{Z}_2[x]$. Is $h(x)$ reducible in $\mathbf{Z}_2[x]$? Since $h(0) = h(1) = 1$, $h(x)$ has no first-order factors, but perhaps we can find $a, b, c, d \in \mathbf{Z}_2$ such that $(x^2 + ax + b)(x^2 + cx + d) = x^4 + x^3 + x^2 + x + 1$.

By comparing coefficients of like powers of x, we find $a + c = 1$, $ac + b + d = 1$, $ad + bc = 1$, $bd = 1$. With $bd = 1$, it follows that $b = 1$ and $d = 1$, so $ac + b + d = 1 \Rightarrow ac = 1 \Rightarrow a = c = 1 \Rightarrow a + c = 0$. This contradicts $a + c = 1$. Consequently $h(x)$ is irreducible in $\mathbf{Z}_2[x]$. □

The next results awaken memories of Chapters 6 and 8.

Definition 13.5 ▶ A polynomial $f(x) \in F[x]$ is *monic* if its leading coefficient is 1, the unity of F.

Definition 13.6 ▶ If $f(x), g(x) \in F[x]$, then $h(x) \in F[x]$ is a *greatest common divisor* of $f(x), g(x)$

a) if $h(x)$ divides each of $f(x), g(x)$; and

b) if $k(x) \in F[x]$ and $k(x)$ divides both $f(x), g(x)$, then $k(x)$ divides $h(x)$.

We now state the following results on the existence and uniqueness of the greatest common divisor, which we write once again as g.c.d. Furthermore, there is a method for finding this g.c.d. that is called the Euclidean Algorithm for Polynomials. A proof for the first result is outlined in the section exercises.

Theorem 13.8 ▶ Let $f(x), g(x) \in F[x]$, with at least one of $f(x), g(x)$ not the zero polynomial. Then any polynomial of minimum degree that can be written as a linear combination of $f(x), g(x)$, i.e., $s(x)f(x) + t(x)g(x)$, $s(x), t(x) \in F[x]$, will be a greatest common divisor of $f(x), g(x)$. If we require the g.c.d. to be monic it will be unique.

Theorem 13.9 ▶ (*Euclidean Algorithm for Polynomials*) Let $f(x), g(x) \in F[x]$ with degree $f(x) \le$ degree $g(x)$ and $f(x) \neq 0$. Applying the Division Algorithm we write

$$g(x) = q(x)f(x) + r(x), \qquad \text{degree } r(x) < \text{degree } f(x)$$
$$f(x) = q_1(x)r(x) + r_1(x), \qquad \text{degree } r_1(x) < \text{degree } r(x)$$
$$r(x) = q_2(x)r_1(x) + r_2(x), \qquad \text{degree } r_2(x) < \text{degree } r_1(x)$$
$$\vdots \qquad\qquad \vdots$$
$$r_{k-2}(x) = q_k(x)r_{k-1}(x) + r_k(x), \qquad \text{degree } r_k(x) < \text{degree } r_{k-1}(x)$$
$$r_{k-1}(x) = q_{k+1}(x)r_k(x) + r_{k+1}(x), \qquad r_{k+1}(x) = 0.$$

Then $r_k(x)$, the last nonzero remainder, is a constant multiple of the greatest common divisor of $f(x)$, $g(x)$. (Multiplying $r_k(x)$ by the inverse of its leading coefficient allows us to obtain the unique monic polynomial we call the greatest common divisor.)

Definition 13.7

▶ If $f(x)$, $g(x) \in F[x]$ and their g.c.d. is 1, then $f(x)$ and $g(x)$ are called *relatively prime*.

The last results we need to construct our new finite fields provide the analog of a construction we developed in Section 8.3.

Theorem 13.10

▶ Let $s(x) \in F[x]$, $s(x) \neq 0$. Define relation $\mathcal{R}$ on $F[x]$ by $f(x) \mathcal{R} g(x)$ if $f(x) - g(x) = t(x)s(x)$, $t(x) \in F[x]$, i.e. $s(x)$ divides $f(x) - g(x)$. Then $\mathcal{R}$ is an equivalence relation on $F[x]$.

Proof

The verification of the reflexive, symmetric, and transitive properties of $\mathcal{R}$ is left for the reader. ∎

When the situation in Theorem 13.10 occurs, we say that $f(x)$ is *congruent* to $g(x)$ *modulo* $s(x)$, and write $f(x) \equiv g(x)$ (mod $s(x)$). The relation $\mathcal{R}$ is referred to as *congruence modulo* $s(x)$.

Let us examine the equivalence classes for such a relation.

EXAMPLE 13.6

Let $s(x) = x^2 + x + 1 \in \mathbf{Z}_2[x]$. Then

a) $[0] = [x^2 + x + 1]$
$$= \{0, x^2 + x + 1, x^3 + x^2 + x, (x + 1)(x^2 + x + 1), \ldots\}$$
$$= \{t(x)(x^2 + x + 1) \mid t(x) \in \mathbf{Z}_2[x]\}.$$

b) $[1] = \{1, x^2 + x, x(x^2 + x + 1) + 1, (x + 1)(x^2 + x + 1) + 1, \ldots\}$
$$= \{t(x)(x^2 + x + 1) + 1 \mid t(x) \in \mathbf{Z}_2[x]\}.$$

c) $[x] = \{x, x^2 + 1, x(x^2 + x + 1) + x, (x + 1)(x^2 + x + 1) + x, \ldots\}$
$$= \{t(x)(x^2 + x + 1) + x \mid t(x) \in \mathbf{Z}_2[x]\}.$$

d) $[x + 1] = \{x + 1, x^2, x(x^2 + x + 1) + (x + 1), (x + 1)(x^2 + x + 1)$
$$+ (x + 1), \ldots\} = \{t(x)(x^2 + x + 1) + (x + 1) \mid t(x) \in \mathbf{Z}_2[x]\}.$$

Are these all of the equivalence classes? If $f(x) \in \mathbf{Z}_2[x]$, by the Division Algorithm $f(x) = q(x)s(x) + r(x)$ where $r(x) = 0$ or degree $r(x) <$ degree $s(x)$. Since $f(x) - r(x) = q(x)s(x)$, $f(x) \equiv r(x)$ (mod $s(x)$), so $f(x) \in [r(x)]$. Consequently, to determine all the equivalence classes we consider the possibilities for $r(x)$. Here $r(x) = 0$ or degree $r(x) < 2$, so $r(x) = ax + b$, where a, $b \in \mathbf{Z}_2$. With only two choices for each of a, b, there are four possible choices for $r(x)$: 0, 1, x, $x + 1$. □

We now place a ring structure on these equivalence classes. Considering how this was accomplished in Chapter 8 for $\mathbf{Z}_n$, we define addition by $[f(x)] + [g(x)] = [f(x) + g(x)]$. Since degree $(f(x) + g(x)) \leq \max\{\text{degree } f(x), \text{degree } g(x)\}$, we can find the equivalence class for $[f(x) + g(x)]$ without too much trouble. Here, for example, $[x] + [x + 1] = [x + (x + 1)] = [2x + 1] = [1]$ since $2 = 0$ in $\mathbf{Z}_2$.

In defining the multiplication of these equivalence classes we run into a little more difficulty. For instance, what is $[x][x]$ in Example 13.6? If, in general, we define $[f(x)][g(x)] = [f(x)g(x)]$, it is possible for degree $f(x)g(x) \geq$ degree $s(x)$, so we may not readily find $[f(x)g(x)]$ in the list of equivalence classes. However, if degree $f(x)g(x) \geq$ degree $s(x)$, then using the Division Algorithm we can write $f(x)g(x) = q(x)s(x) + r(x)$, where $r(x) = 0$ or degree $r(x) <$ degree $s(x)$. With $f(x)g(x) = q(x)s(x) + r(x)$, $f(x)g(x) \equiv r(x) \pmod{s(x)}$, and we define $[f(x)g(x)] = [r(x)]$, where $[r(x)]$ does occur in the list of equivalence classes.

From these observations we construct Tables 13.1 and 13.2 for the addition and multiplication, respectively, of $\{[0], [1], [x], [x + 1]\}$. (In these tables we write a for $[a]$).

Table 13.1

+	0	1	x	x + 1
0	0	1	x	x + 1
1	1	0	x + 1	x
x	x	x + 1	0	1
x + 1	x + 1	x	1	0

Table 13.2

·	0	1	x	x + 1
0	0	0	0	0
1	0	1	x	x + 1
x	0	x	x + 1	1
x + 1	0	x + 1	1	x

From the multiplication table (Table 13.2) we find that not only do these equivalence classes form a ring but a field, where $[1]^{-1} = [1]$, $[x]^{-1} = [x + 1]$ and $[x + 1]^{-1} = [x]$. This field of order 4 is denoted by $\mathbf{Z}_2[x]/(x^2 + x + 1)$. In addition, for the nonzero elements of this field we find that $[x]^1 = [x]$, $[x]^2 = [x + 1]$, $[x]^3 = [1]$, so we have a cyclic group of order 3. But since the nonzero elements of any field form a group under multiplication, and any group of order 3 is cyclic, why bother with this observation? In general, the nonzero elements of *any* finite field form a cyclic group under multiplication. (A proof for this can be found in Chapter 12 of reference [9].)

The construction above is summarized in the following theorem. An outline of the proof is given in the section exercises.

Theorem 13.11

▶ Let $s(x)$ be a nonzero polynomial in $F[x]$.

a) The equivalence classes of $F[x]$ for the relation of congruence modulo $s(x)$ form a commutative ring with unity under the operations

$$[f(x)] + [g(x)] = [f(x) + g(x)], \quad [f(x)][g(x)] = [f(x)g(x)] = [r(x)],$$

where $r(x)$ is the remainder obtained upon dividing $f(x)g(x)$ by $s(x)$. This ring is denoted by $F[x]/(s(x))$.

b) If $s(x)$ is irreducible in $F[x]$, then $F[x]/(s(x))$ is a field.

We now examine another example of a finite field that arises by virtue of Theorem 13.11.

EXAMPLE 13.7   In $\mathbf{Z}_3[x]$ the polynomial $s(x) = x^2 + x + 2$ is irreducible since $s(0) = 2$, $s(1) = 1$, $s(2) = 2$. Consequently, $\mathbf{Z}_3[x]/(s(x))$ is a field containing all equivalence classes of the form $[ax + b]$, where $a, b \in \mathbf{Z}_3$. These are the possible remainders when a polynomial $f(x) \in \mathbf{Z}_3[x]$ is divided by $s(x)$. The nine equivalence classes are $[0]$, $[1]$, $[2]$, $[x]$, $[x + 1]$, $[x + 2]$, $[2x]$, $[2x + 1]$, $[2x + 2]$.

Instead of constructing a complete multiplication table, we examine four sample multiplications.

a) $[2x][x] = [2x^2] = [2x^2 + 0] = [2x^2 + (x^2 + x + 2)] = [3x^2 + x + 2] = [x + 2]$ since $3 = 0$ in $\mathbf{Z}_3$.

b) $[x + 1][x + 2] = [x^2 + 3x + 2] = [x^2 + 2] = [x^2 + 2 + 2(x^2 + x + 2)] = [2x]$.

c) $[2x + 2]^2 = [4x^2 + 8x + 4] = [x^2 + 2x + 1] = [(-x - 2) + (2x + 1)]$ since $x^2 \equiv (-x - 2) \pmod{s(x)}$. Consequently, $[2x + 2]^2 = [x - 1] = [x + 2]$.

d)   Often we write the equivalence classes without brackets and concentrate on the coefficients of the powers of x. For example, 11 is written for $[x + 1]$ and 21 represents $[2x + 1]$. Consequently, $(21) \cdot (12) = [2x + 1][x + 2] = [2x^2 + 5x + 2] = [2x^2 + 2x + 2] = [2(-x - 2) + 2x + 2] = [-4 + 2] = [-2] = [1]$, so $(21)^{-1} = (12)$.

We make one more observation about this field.

$$[x]^1 = [x] \qquad\qquad [x]^3 = [2x + 2] \qquad [x]^5 = [2x] \qquad\qquad [x]^7 = [x + 1]$$
$$[x]^2 = [2x + 1] \qquad [x]^4 = [2] \qquad\qquad\quad [x]^6 = [x + 2] \qquad [x]^8 = [1]$$

Therefore the nonzero elements of $\mathbf{Z}_3[x]/(s(x))$ form a cyclic group under multiplication.   □

We complete this section by characterizing the order of a finite field. To accomplish this we need the following idea.

Definition 13.8   ▶   Let $(R, +, \cdot)$ be a ring. If there is a least positive integer n such that $nr = z$, the zero of R, for all $r \in R$, we say that R has *characteristic n,* and write $\text{char}(R) = n$. If no such integer exists, R is said to have *characteristic* 0.

EXAMPLE 13.8   a)   The ring $(\mathbf{Z}_3, +, \cdot)$ has characteristic 3; $(\mathbf{Z}_4, +, \cdot)$ has characteristic 4; in general, $(\mathbf{Z}_n, +, \cdot)$ has characteristic n.

b)   The rings $(\mathbf{Z}, +, \cdot)$, $(\mathbf{Q}, +, \cdot)$ have characteristic 0.

c)   A ring can be infinite and still have positive characteristic. For example, $\mathbf{Z}_3[x]$ is an infinite ring but it has characteristic 3.

d)   The ring in Example 13.6 has characteristic 2. In Example 13.7 the characteristic of the ring is 3. Unlike the examples in (a), the order of a finite ring can be different from its characteristic.

Examples 13.6 and 13.7, however, are more than just rings. They are fields with prime characteristic. Could this property be true for all finite fields?   □

Theorem 13.12

▶ Let $(F, +, \cdot)$ be a field. If char$(F) > 0$, then char(F) must be prime.

Proof

In this proof we write the unity of F as u so that it is distinct from the positive integer 1. Let char$(F) = n > 0$. If n is not prime, we write $n = mk$, where $m, k \in \mathbf{Z}^+$ and $1 < m, k < n$. By the definition of characteristic, $nu = z$, the zero of F. Hence $(mk)u = z$. But

$$(mk)u = \underbrace{(u + u + \cdots + u)}_{mk \text{ summands}} = \underbrace{(u + u + \cdots + u)}_{m \text{ summands}}\underbrace{(u + u + \cdots + u)}_{k \text{ summands}} = (mu)(ku).$$

With F a field, $(mu)(ku) = z \Rightarrow (mu) = z$ or $(ku) = z$. Assume without loss of generality that $ku = z$. Then for any $r \in F$, $kr = k(ur) = (ku)r = zr = z$, contradicting the choice of n as the characteristic of F. Consequently, char(F) is prime. ■

(The proof of Theorem 13.12 actually requires only that F be an integral domain.)
If F is a finite field and $m = |F|$, then $ma = z$ for all $a \in F$, since $(F, +)$ is an additive group of order m. (See Exercise 7 of Section 12.3.) Consequently, F has positive characteristic and by Theorem 13.12 this characteristic is prime. This leads us to the following.

Theorem 13.13

▶ Any finite field F has order p^t, where p is a prime and $t \in \mathbf{Z}^+$.

Proof

Since F is a finite field, let char$(F) = p$, a prime, and let u denote the unity and z the zero element. Then $S_0 = \{u, 2u, 3u, \ldots, pu = z\}$ is a set of p distinct elements in F. If not, $mu = nu$ for $1 \leq m < n \leq p$ and $(n - m)u = z$, with $0 < n - m < p$. This contradicts char$(F) = p$. If $F = S_0$, then $|F| = p^1$ and the result follows. If not, let $a \in F - S_0$. Then $S_1 = \{ma + nu \mid 0 < m, n \leq p\}$ is a subset of F with $|S_1| \leq p^2$. If $|S_1| < p^2$, then $m_1a + n_1u = m_2a + n_2u$, with $0 < m_1, m_2, n_1, n_2 \leq p$ and at least one of $m_1 - m_2, n_2 - n_1 \neq 0$. Should $m_1 - m_2 = 0$, then $(m_1 - m_2)a = z = (n_2 - n_1)u$, with $0 < |n_2 - n_1| < p$. Consequently, for all $x \in F$, $|n_2 - n_1|x = |n_2 - n_1|ux = zx$ with $0 < |n_2 - n_1| < p = $ char(F). Therefore, $(m_1 - m_2)a = (n_2 - n_1)u \neq z$. Choose $k \in \mathbf{Z}^+$ such that $0 < k < p$ and $k(m_1 - m_2) \equiv 1 \pmod{p}$. Then $a = k(m_1 - m_2)a = k(n_2 - n_1)u$, and $a \in S_0$, another contradiction. Hence $|S_1| = p^2$, and if $F = S_1$ the theorem is proved. If not, continue this process with an element $b \in F - S_1$. Then $S_2 = \{\ell b + ma + nu \mid 0 < \ell, m, n \leq p\}$ will have order p^3. (Prove this.) Since F is finite we get to a point where $F = S_{t-1}$ for $t \in \mathbf{Z}^+$, and $|F| = |S_{t-1}| = p^t$. ■

As a result of this theorem there can be no finite fields with such orders as 6, 10, 12, ... In addition, for each prime p and each $t \in \mathbf{Z}^+$, there is really only one field of order p'. Any two finite fields of the same order are isomorphic. These fields were discovered by the French mathematician Evariste Galois (1811–1832) in his work on the nonexistence of formulas for solving general polynomial equations of degree ≥ 5 over **Q**. As a result, a finite field of order p' is denoted by $GF(p')$ where the letters *GF* stand for *Galois field*.

EXERCISES

1. Determine whether or not each of the following polynomials is irreducible over the given fields. If it is reducible, provide a factorization into irreducible factors.

   a) $x^2 + 3x - 1$ over **Q, R, C**   b) $x^4 - 2$ over **Q, R, C**

   c) $x^2 + x + 1$ over $\mathbf{Z}_3, \mathbf{Z}_5, \mathbf{Z}_7$ d) $x^4 + x^3 + 1$ over $\mathbf{Z}_2$

2. Give an example of a polynomial $f(x) \in \mathbf{R}[x]$ where $f(x)$ has degree 6, is reducible, but has no real roots.

3. How many monic polynomials in $\mathbf{Z}_7[x]$ have degree 5?

4. Prove Theorem 13.7.

5. Below is an outline for a proof of Theorem 13.8.

   a) Let $S = \{s(x)f(x) + t(x)g(x) \mid s(x), t(x) \in F[x]\}$. Select an element $m(x)$ of minimum degree in S. (Recall that the zero polynomial has no degree so it is not selected.) Can we guarantee that $m(x)$ is monic?

   b) Show that if $h(x) \in F[x]$ and $h(x)$ divides both $f(x)$, $g(x)$ then $h(x)$ divides $m(x)$.

   c) Show that $m(x)$ divides $f(x)$. If not, use the Division Algorithm and write $f(x) = q(x)m(x) + r(x)$, where $r(x) \neq 0$ and degree $r(x) <$ degree $m(x)$. Then show that $r(x) \in S$ and obtain a contradiction.

   d) Repeat the argument in (c) to show that $m(x)$ divides $g(x)$.

6. Prove Theorems 13.9 and 13.10.

7. Use the Euclidean Algorithm for Polynomials to find the g.c.d. of each pair of polynomials, over the designated field F. Then write the g.c.d. as $s(x)f(x) + t(x)g(x)$, where $s(x), t(x) \in F[x]$.

   a) $f(x) = x^2 + x - 2, \quad g(x) = x^5 - x^4 + x^3 + x^2 - x - 1 \quad$ in $\mathbf{Q}[x]$.

   b) $f(x) = x^4 + x^3 + 1, \quad g(x) = x^2 + x + 1 \quad$ in $\mathbf{Z}_2[x]$.

   c) $f(x) = x^4 + 2x^2 + 2x + 2, \quad g(x) = 2x^3 + 2x^2 + x + 1 \quad$ in $\mathbf{Z}_3[x]$.

8. For Example 13.6, determine which equivalence class contains each of the following:

   a) $x^4 + x^3 + x + 1$ b) $x^3 + x^2 + 1$ c) $x^4 + x^3 + x^2 + 1$

9. Below is an outline for the proof of Theorem 13.11.

   a) Prove that the operations defined in Theorem 13.11(a) are well-defined by showing that if $f(x) \equiv f_1(x) \pmod{s(x)}$, and $g(x) \equiv g_1(x) \pmod{s(x)}$, then $f(x) + g(x) \equiv f_1(x) + g_1(x) \pmod{s(x)}$ and $f(x)g(x) \equiv f_1(x)g_1(x) \pmod{s(x)}$.

   b) Verify the ring properties for the equivalence classes in $F[x]/(s(x))$.

c) Let $f(x) \in F[x]$, with $f(x) \neq 0$ and degree $f(x) <$ degree $s(x)$. If $s(x)$ is irreducible in $F[x]$, why does it follow that 1 is the g.c.d. of $f(x)$ and $s(x)$?

d) Use (c) to prove that if $s(x)$ is irreducible in $F[x]$, then $F[x]/(s(x))$ is a field.

10. a) Show that $s(x) = x^2 + 1$ is reducible in $\mathbf{Z}_2[x]$.

b) Find the equivalence classes for the ring $\mathbf{Z}_2[x]/(s(x))$.

c) Is $\mathbf{Z}_2[x]/(s(x))$ an integral domain?

11. For the field in Example 13.7, find each of the following:

a) $[x + 2][2x + 2] + [x + 1]$ b) $[2x + 1] - [x + 1][2x + 1]$

c) $[2x + 1]^2[x + 2]$ d) $(22)^{-1} = [2x + 2]^{-1}$.

12. Let $s(x) = x^4 + x^3 + 1$, an irreducible polynomial in $\mathbf{Z}_2[x]$.

a) What is the order of the field $\mathbf{Z}_2[x]/(s(x))$?

b) Find $[x^2 + x + 1]^{-1}$ in this field. (*Hint:* Find a, b, c, $d \in \mathbf{Z}_2$ so that $[x^2 + x + 1][ax^3 + bx^2 + cx + d] = [1]$.)

c) Determine $[x^3 + x + 1][x^2 + 1]$ in this field.

13. For p a prime, let $s(x)$ be irreducible of degree n in $\mathbf{Z}_p[x]$.

a) How many elements are there in the field $\mathbf{Z}_p[x]/(s(x))$?

b) How many elements in $\mathbf{Z}_p[x]/(s(x))$ generate the multiplicative group of nonzero elements of this field?

14. For Theorem 13.13, prove that $|S_2| = p^3$.

15. Find the orders n for all fields $GF(n)$, where $100 \leq n \leq 150$.

13.3 ■ LATIN SQUARES

Our first application deals with the structure called a Latin square. Such configurations arise in the study of combinatorial designs and play a role in statistics in the design of experiments. We introduce the structure in the following example.

EXAMPLE 13.9 A petroleum corporation is interested in testing four types of gasoline additives to see their effects on mileage. To do so, a research team designs an experiment where four different automobiles, denoted A, B, C, and D, are run on a fixed track in a laboratory. Each run uses the same prescribed amount of fuel with one of the additives present. In order to see how each additive affects each type of auto, the schedule in Table 13.3, where the additives are numbered 1, 2, 3, and 4, is followed. This schedule provides a way to test each additive thoroughly in each type of auto. If one additive produces the best results in all four auto types, the experiment will recognize its superior capability.

The same corporation is also interested in testing four other additives developed for cleaning engines. A similar schedule for these tests is shown in Table 13.4, where these engine-cleaning additives are also denoted as 1, 2, 3, and 4.

The research team is interested in the combined effect of both types of additives. It requires 16 days to test the 16 possible pairs of additives (one for improved mileage,

Table 13.3

Auto	Mon	Tues	Wed	Thurs
A	1	2	3	4
B	2	1	4	3
C	3	4	1	2
D	4	3	2	1

(Day spanning Mon–Thurs)

Table 13.4

Auto	Mon	Tues	Wed	Thurs
A	1	2	3	4
B	3	4	1	2
C	4	3	2	1
D	2	1	4	3

(Day spanning Mon–Thurs)

the other for cleaning engines) in every automobile. If they need the results in four days the research team must design the schedules so that every pair is tested once by some auto. Since there are 16 ordered pairs in $\{1, 2, 3, 4\} \times \{1, 2, 3, 4\}$ this can be done in the allotted time if the schedules in Tables 13.3 and 13.4 are superimposed to obtain the schedule in Table 13.5. Here, for example, the entry $(4, 3)$ indicates that on Tuesday, auto C is used to test the combined effect of the fourth additive for improved mileage and the third additive for maintaining a clean engine.

Table 13.5

Auto	Mon	Tues	Wed	Thurs
A	(1, 1)	(2, 2)	(3, 3)	(4, 4)
B	(2, 3)	(1, 4)	(4, 1)	(3, 2)
C	(3, 4)	(4, 3)	(1, 2)	(2, 1)
D	(4, 2)	(3, 1)	(2, 4)	(1, 3)

(Day spanning Mon–Thurs)

□

What has happened here leads us to the following concepts.

Definition 13.9 ▶ An $n \times n$ *Latin square* is a square array of symbols, usually 1, 2, 3, ..., n, where each symbol appears exactly once in each row and each column of the array.

EXAMPLE 13.10
a) Tables 13.3 and 13.4 are examples of 4×4 Latin squares.
b) For any $n \geq 2$, using n instead of 0, we can obtain an $n \times n$ Latin square from the table of the group $(\mathbf{Z}_n, +)$. □

From the two Latin squares in Example 13.9 we were able to produce all of the ordered pairs in $S \times S$, for $S = \{1, 2, 3, 4\}$. We now question whether or not we can do this for $n \times n$ Latin squares in general.

Definition 13.10 ▶ Let $L_1 = (a_{ij})$, $L_2 = (b_{ij})$ be two $n \times n$ Latin squares, where $1 \leq i, j \leq n$ and each a_{ij}, $b_{ij} \in \{1, 2, 3, ..., n\}$. If the n^2 ordered pairs (a_{ij}, b_{ij}), $1 \leq i, j \leq n$, are distinct, then L_1, L_2 are a *pair of orthogonal Latin squares*.

EXAMPLE 13.11
a) There is no pair of 2×2 orthogonal Latin squares, since the only possibilities are

$$L_1: \begin{matrix} 1 & 2 \\ 2 & 1 \end{matrix} \quad \text{and} \quad L_2: \begin{matrix} 2 & 1 \\ 1 & 2 \end{matrix}.$$

b)   In the 3 × 3 case we find the orthogonal pair:

$$
L_1:\ \begin{matrix} 1 & 2 & 3 \\ 2 & 3 & 1 \\ 3 & 1 & 2 \end{matrix}
\quad \text{and} \quad
L_2:\ \begin{matrix} 1 & 2 & 3 \\ 3 & 1 & 2 \\ 2 & 3 & 1 \end{matrix}
$$

c)   The two 4 × 4 Latin squares in Example 13.9 form an orthogonal pair. The 4 × 4 Latin square shown here in Table 13.6 is orthogonal to each of the Latin squares in that example.

Table 13.6

$$
\begin{matrix}
1 & 2 & 3 & 4 \\
4 & 3 & 2 & 1 \\
2 & 1 & 4 & 3 \\
3 & 4 & 1 & 2
\end{matrix}
$$

□

We could continue listing some larger Latin squares, but we've seen enough of them at this point to ask the following questions:

1. What is the first value of $n > 2$ for which there is no pair of orthogonal $n \times n$ Latin squares?
2. For $n > 1$, what can we say about the number of $n \times n$ Latin squares that can be constructed so that each pair of them is orthogonal?
3. Is there a method to assist us in constructing a pair of orthogonal $n \times n$ Latin squares for certain values of $n > 2$?

Before we can examine these questions we need to standardize some of our results.

Definition 13.11

▶   If L is an $n \times n$ Latin square, L is said to be in *standard form* if its first row is 1   2   3   … n.

Except for L_2 in Example 13.11(a), all the Latin squares we've seen in this section are in standard form. If a Latin square is not in standard form it can be put in that form by interchanging some of the symbols.

EXAMPLE 13.12

The 5 × 5 Latin square (a) is not in standard form. If, however, we replace each occurrence of 4 by 1, each occurrence of 5 by 4, and each occurrence of 1 by 5, the result is the (standard) 5 × 5 Latin square (b).

$$
\begin{matrix}
4 & 2 & 3 & 5 & 1 \\
1 & 3 & 5 & 4 & 2 \\
3 & 4 & 2 & 1 & 5 \\
2 & 5 & 1 & 3 & 4 \\
5 & 1 & 4 & 2 & 3
\end{matrix}
\qquad
\begin{matrix}
1 & 2 & 3 & 4 & 5 \\
5 & 3 & 4 & 1 & 2 \\
3 & 1 & 2 & 5 & 4 \\
2 & 4 & 5 & 3 & 1 \\
4 & 5 & 1 & 2 & 3
\end{matrix}
$$
$$
\text{(a)} \qquad\qquad\qquad \text{(b)}
$$

□

It is often convenient to deal with Latin squares in standard form. But will this affect our results on orthogonal pairs in any way?

Theorem 13.14

▶ Let L_1, L_2 be an orthogonal pair of $n \times n$ Latin squares. If L_1, L_2 are standardized as L_1^*, L_2^*, then L_1^*, L_2^* are orthogonal.

Proof

The proof of this result is left for the reader.   ■

These ideas are needed for the main results of this section.

Theorem 13.15

▶ If $n \in \mathbf{Z}^+$, $n > 2$, then the largest number of $n \times n$ Latin squares that are orthogonal in pairs is $n - 1$.

Proof

Let L_1, L_2, ..., L_k be k distinct $n \times n$ Latin squares that are in standard form and orthogonal in pairs. We write $a_{ij}^{(m)}$ to denote the entry in the ith row and jth column of L_m, where $1 \le i, j \le n$, $1 \le m \le k$. Since these Latin squares are in standard form, $a_{11}^{(m)} = 1$, $a_{12}^{(m)} = 2$, ..., $a_{1n}^{(m)} = n$ for all $1 \le m \le k$. Now consider $a_{21}^{(m)}$, $1 \le m \le k$. These entries in the second row and first column are below $a_{11}^{(m)} = 1$. Thus, $a_{21}^{(m)} \ne 1$, $1 \le m \le k$, or the configuration is not a Latin square. Further, if there exists $1 \le \ell < m \le k$ with $a_{21}^{(\ell)} = a_{21}^{(m)}$ then the pair L_ℓ, L_m could not be an orthogonal pair. (Why not?) Consequently, there are at best $n - 1$ choices for the a_{21} entries in any of our $n \times n$ Latin squares, and the result follows from this observation.   ■

This theorem places an upper bound on the number of $n \times n$ Latin squares that are orthogonal in pairs. We now find that for certain values of n this upper bound can be attained. In addition, our next result provides a method for constructing these Latin squares, though initially not in standard form. The construction uses the structure of a finite field.

Theorem 13.16

▶ Let $n \in \mathbf{Z}^+$, $n > 2$. If p is a prime and $n = p^t$, $t \in \mathbf{Z}^+$, then there are $n - 1$ Latin squares that are $n \times n$ and orthogonal in pairs.

Proof

Let $F = GF(p^t)$, the Galois field of order $p^t = n$. Consider $F = \{f_1, f_2, \ldots, f_n\}$, where f_1 is the unity and f_n is the zero element.

We construct $n - 1$ Latin squares as follows.

For any $1 \le k \le n - 1$, let L_k be the $n \times n$ array $(a_{ij}^{(k)})$, $1 \le i, j \le n$, where $a_{ij}^{(k)} = f_k f_i + f_j$.

First we show that each L_k is a Latin square. If not, there are two identical elements of F in the same row or column of L_k. Suppose that a repetition occurs in a column, i.e., $a_{rj}^{(k)} = a_{sj}^{(k)}$ for $1 \le r, s \le n$. Then $a_{rj}^{(k)} = f_k f_r + f_j = f_k f_s + f_j = a_{sj}^{(k)}$. This implies that $f_k f_r = f_k f_s$, by the cancellation for addition in F. Since $k \ne n$, $f_k \ne f_n$, the zero of F. Consequently, f_k is invertible, so $f_r = f_s$ and $r = s$. A similar proof shows that there are no repetitions in any row of L_k. (This proof is left as an exercise.)

At this point we have $n - 1$ Latin squares, $L_1, L_2, \ldots, L_{n-1}$. Now we shall prove that they are orthogonal in pairs. If not, let $1 \le k, m \le n - 1$ with

$$a_{ij}^{(k)} = a_{rs}^{(k)}, \quad a_{ij}^{(m)} = a_{rs}^{(m)}, \quad 1 \le i, j, r, s \le n, \quad (i, j) \ne (r, s).$$

(Then the same ordered pair occurs twice when we superimpose L_k, L_m.)

$$a_{ij}^{(k)} = a_{rs}^{(k)} \Leftrightarrow f_k f_i + f_j = f_k f_r + f_s$$

$$a_{ij}^{(m)} = a_{rs}^{(m)} \Leftrightarrow f_m f_i + f_j = f_m f_r + f_s.$$

Subtracting these equations we find that $(f_k - f_m)f_i = (f_k - f_m)f_r$. With $k \neq m$, $(f_k - f_m)$ is not the zero of F, so it is invertible and we have $f_i = f_r$. Putting this back into either of the prior equations, it follows that $f_j = f_s$. Consequently, $i = r$, $j = s$. Therefore for $k \neq m$, L_k, L_m form an orthogonal pair.   ∎

The first value of n that is not a power of a prime is 6. The existence of a pair of 6×6 orthogonal Latin squares was first investigated by Leonhard Euler (1707–1783) when he sought a solution to the "problem of the 36 officers." This problem deals with six different regiments where six officers, each with a different rank, are selected from each regiment. (There are only six possible ranks.) The objective is to arrange the 36 officers in a 6×6 array so that in any row or column of the array every rank and every regiment is represented exactly once. Hence each officer in the square array corresponds to an ordered pair (i, j) where $1 \leq i, j \leq 6$, with i for his regiment and j for his rank. In 1782 Euler conjectured that the problem could not be solved, i.e., that there is no pair of 6×6 orthogonal Latin squares. He went further and conjectured that for any $n \in \mathbf{Z}^+$, if $n \equiv 2 \pmod 4$ then there is no pair of $n \times n$ orthogonal Latin squares. In 1900 G. Tarry verified Euler's conjecture for $n = 6$ by a systematic enumeration of all possible 6×6 Latin squares. However, it was not until 1960, through the combined efforts of R. C. Bose, S. S. Shrikhande, and E. T. Parker that the remainder of Euler's conjecture was proved false. They showed that if $n \in \mathbf{Z}^+$ with $n \equiv 2 \pmod 4$ and $n > 6$, then there exists a pair of $n \times n$ orthogonal Latin squares.

For more on this result and Latin squares in general, the reader should consult the chapter references.

EXERCISES

1. a) Rewrite this 4×4 Latin square in standard form.

$$
\begin{array}{cccc}
1 & 3 & 4 & 2 \\
3 & 1 & 2 & 4 \\
2 & 4 & 3 & 1 \\
4 & 2 & 1 & 3
\end{array}
$$

   b) Find a 4×4 Latin square in standard form that is orthogonal to the result in part (a).

   c) Apply the reverse of the process in part (a) to the result in part (b). Show that your answer is orthogonal to the given 4×4 Latin square.

2. Prove Theorem 13.14.

3. Complete the proof of the first part of Theorem 13.16.

4. The three 4×4 Latin squares in Tables 13.3, 13.4, and 13.6 are orthogonal in pairs. Can you find another 4×4 Latin square that is orthogonal to each of these three?

5. Use the method of Theorem 13.16 to obtain four 5×5 Latin squares that are orthogonal in pairs. Rewrite these results in standard form.

6. Find three 7×7 Latin squares that are orthogonal in pairs. Rewrite these results in standard form.

7. Extend the experiment in Example 13.9 so that the research team needs three 4×4 Latin squares that are orthogonal in pairs.

8. A Latin square L is called *self-orthogonal* if L and its transpose L^{tr} form an orthogonal pair.

   a) Show that there is no 3×3 self-orthogonal Latin square.

   b) Give an example of a 4×4 Latin square that is self-orthogonal.

   c) If $L = (a_{ij})$ is an $n \times n$ self-orthogonal Latin square, prove that the elements a_{ii}, $1 \le i \le n$, must all be distinct.

13.4 ■ FINITE GEOMETRIES AND AFFINE PLANES

In the Euclidean geometry of the real plane we find that (a) two distinct points determine a unique line; and (b) if ℓ is a line in the plane, and P a point not on ℓ, then there is a unique line ℓ' that contains P and is parallel to ℓ. During the eighteenth and nineteenth centuries non-Euclidean geometries were developed when alternatives to condition (b) were investigated. Yet all of these geometries contained infinitely many points and lines. The notion of a finite geometry did not appear until the end of the nineteenth century in the work of Gino Fano (*Giornale di Matematiche*, 1892).

How can we construct such a geometry? To do so, we return to the more familiar Euclidean geometry. In order to describe points and lines in this plane algebraically we introduced a set of coordinate axes and identified each point P by an ordered pair (c, d) of real numbers. This description set up a one-to-one correspondence between the points in the plane and the set $\mathbf{R} \times \mathbf{R}$. By using the idea of slope we could represent uniquely each line in this plane by either (1) $x = a$, where the slope is infinite; or (2) $y = mx + b$, where m is the slope; a, m, and b are any real numbers. We also found that two distinct lines are parallel if and only if they have the same slope. When their slopes are distinct the lines intersect in a unique point.

Instead of using real numbers for the points (c, d) and the lines $x = a$, $y = mx + b$, we turn to a comparable *finite* structure, the finite field. Our objective is to construct what is called a (finite) affine plane.

Definition 13.12 ▶ Let $\mathscr{P}$ be a finite set of points, and $\mathscr{L}$ a set of subsets of $\mathscr{P}$, called lines. A *(finite) affine plane* on the sets $\mathscr{P}$ and $\mathscr{L}$ is a finite structure satisfying the following conditions.

A1) Two distinct points of $\mathscr{P}$ are in only one element of $\mathscr{L}$, i.e., on only one line.

A2) For any $\ell \in \mathcal{L}$, and $P \in \mathcal{P}$ with $P \notin \ell$, there exists a unique element $\ell' \in \mathcal{L}$ where $P \in \ell'$ and ℓ, ℓ' have no point in common.

A3) There are four points in $\mathcal{P}$, no three of which are collinear.

The reason for condition (A3) is to avoid uninteresting situations like the one in Fig. 13.1. If only conditions (A1) and (A2) were considered, then this system would be an affine plane.

Figure 13.1

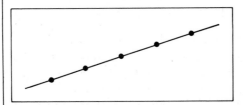

We return now to our construction. Let $F = GF(n)$, where $n = p^t$ for some prime p, $t \in \mathbf{Z}^+$. In constructing our affine plane, denoted by $AP(F)$, we let $\mathcal{P} = \{(c, d) \mid c, d \in F\}$. Thus we have n^2 points.

How many lines do we have for the set $\mathcal{L}$?

The lines fall into two categories. For a line of infinite slope the equation is $x = a$, where $a \in F$. Thus we have n such "vertical lines." The other lines are given algebraically by $y = mx + b$, where $m, b \in F$. With n choices for each of m and b there are n^2 lines that are not "vertical." Hence $|\mathcal{L}| = n^2 + n$.

Before we verify that $AP(F)$, with $\mathcal{P}$ and $\mathcal{L}$ as constructed, is an affine plane, we make two other observations.

First, for any line $\ell \in \mathcal{L}$, if ℓ is given by $x = a$, then there are n choices for y on $\ell = \{(a, y) \mid y \in F\}$. Thus ℓ contains exactly n points. If ℓ is given by $y = mx + b$, for $m, b \in F$, then for each choice of x we have y uniquely determined, and ℓ consists of n points.

Now consider any point $(c, d) \in \mathcal{P}$. This point is on the line $x = c$. Furthermore, on any line $y = mx + b$ of finite slope m, $d - mc$ uniquely determines b. With n choices for m, (c, d) is on the n lines of the form $y = mx + (d - mc)$. Overall, (c, d) is on $n + 1$ lines.

Thus far in our construction of $AP(F)$ we have a set $\mathcal{P}$ of points and a set $\mathcal{L}$ of lines where (a) $|\mathcal{P}| = n^2$; (b) $|\mathcal{L}| = n^2 + n$; (c) each $\ell \in \mathcal{L}$ contains n points; and (d) each point in $\mathcal{P}$ is on exactly $n + 1$ lines. We now prove that $AP(F)$ satisfies the three conditions to be an affine plane.

A1) Let (c, d), $(e, f) \in \mathcal{P}$. Using the two-point formula for the equation of a line we have

$$(e - c)(y - d) = (f - d)(x - c) \tag{1}$$

as a line on which we find both (c, d) and (e, f). Each of these points is on $n + 1$ lines. Could there be a second line containing both of them?

The point (c, d) is on the line $x = c$. If (e, f) is also on that line then $e = c$, but $f \neq d$ since the points are distinct. With $e = c$, Eq. (1) reduces to $0 = (f - d)(x - c)$, or $x = c$ since $f - d \neq 0$, and we do not have a second line. With $c \neq e$, if (c, d), (e, f) are on a second line $y = mx + b$, then $d = mc + b$, $f = me + b$ and $(f - d) = m(e - c)$. Since our coefficients are from a field and $e \neq c$, $m = (f - d)(e - c)^{-1}$ and $b = d - (f - d)(e - c)^{-1}c$. Consequently this second line containing (c, d) and (e, f) is

$$y = (f - d)(e - c)^{-1}x + [d - (f - d)(e - c)^{-1}c],$$

or $(e - c)(y - d) = (f - d)(x - c)$, which is Eq. (1). Thus two points from $\mathcal{P}$ are on only one line and condition (A1) is satisfied.

A2) To verify this condition, consider point P and line ℓ as shown in Fig. 13.2. Since there are n points on any line, let $P_1, P_2, \ldots, P_n$ be the points of ℓ. Since P is not on ℓ, P and P_i determine a unique line ℓ_i, $1 \leq i \leq n$. We showed earlier that each point is on $n + 1$ lines so there is one additional line ℓ' with P on ℓ' and ℓ' not intersecting ℓ.

Figure 13.2

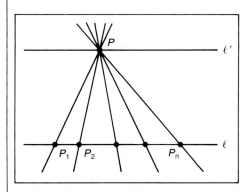

A3) The last condition uses the field F. Since $|F| \geq 2$ there is a unity 1 and zero element 0 in F. Considering the points $(0, 0)$, $(1, 0)$, $(0, 1)$, $(1, 1)$, if line ℓ contains any three of these points, then two of the points have the form (c, c), (c, d). Consequently the equation for ℓ is given by $x = c$, which is not satisfied by either (d, c) or (d, d). Hence no three of these points are collinear.

We have now shown the following.

Theorem 13.17 ▶ If F is a finite field, the system based on the set $\mathcal{P}$ of points and the set $\mathcal{L}$ of lines, as described above, is an affine plane denoted by $AP(F)$.

Some particular examples will indicate a connection between these finite geometries, or affine planes, and the Latin squares of the previous section.

EXAMPLE 13.13 For $F = (\mathbf{Z}_2, +, \cdot)$, $n = |F| = 2$. The affine plane in Fig. 13.3 has $n^2 = 4$ points and $n^2 + n = 6$ lines. For example, the line $\ell_4 = \{(1, 0), (1, 1)\}$ and ℓ_4 contains no other

Figure 13.3

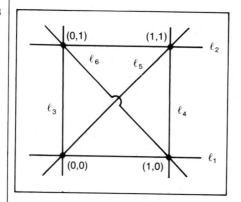

points that the figure might suggest. Furthermore, ℓ_5 and ℓ_6 are parallel lines in this finite geometry since they do not intersect.　　□

EXAMPLE 13.14

Let $F = GF(2^2)$ where we have $F = \{00, 01, 10, 11\}$ with addition and multiplication given by Table 13.7. We use this field to construct a finite geometry with $n^2 = 16$ points and $n^2 + n = 20$ lines. The 20 lines can be partitioned into five *parallel classes* of four lines each.

Table 13.7

+	00	01	10	11
00	00	01	10	11
01	01	00	11	10
10	10	11	00	01
11	11	10	01	00

·	00	01	10	11
00	00	00	00	00
01	00	01	10	11
10	00	10	11	01
11	00	11	01	10

Class 1: Here we have the lines of infinite slope. These four "vertical" lines are given by the equations $x = 00$, $x = 01$, $x = 10$, and $x = 11$.

Class 2: For the "horizontal" class, or class of slope 0, we have the four lines $y = 00$, $y = 01$, $y = 10$, and $y = 11$.

Class 3: The lines with slope 01 are those whose equations are $y = 01x + 00$, $y = 01x + 01$, $y = 01x + 10$, and $y = 01x + 11$.

Class 4: This class consists of the lines with equations $y = 10x + 00$, $y = 10x + 01$, $y = 10x + 10$, and $y = 10x + 11$.

Class 5: The last class contains the four lines given by $y = 11x + 00$, $y = 11x + 01$, $y = 11x + 10$, and $y = 11x + 11$.

Since each line in $AP(F)$ contains four points and each parallel class contains four lines, we shall see now how three of these parallel classes partition the 16 points of $AP(F)$.

For the class with $m = 01$, there are four lines: (1) $y = 01x + 00$; (2) $y = 01x + 01$; (3) $y = 01x + 10$; and (4) $y = 01x + 11$. Above each point in $AP(F)$

Figure 13.4

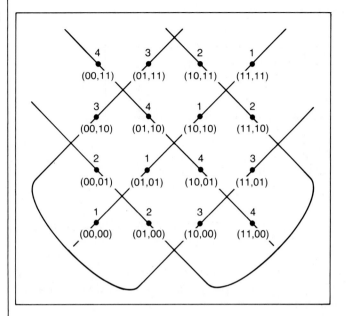

we write the number corresponding to the line it is on (see Fig. 13.4). This configuration can be given by the Latin square:

$$4 \quad 3 \quad 2 \quad 1$$
$$3 \quad 4 \quad 1 \quad 2$$
$$2 \quad 1 \quad 4 \quad 3$$
$$1 \quad 2 \quad 3 \quad 4$$

If this process is repeated for classes 4 and 5, we get the partitions shown in Figs. 13.5 and 13.6, respectively. In each class the lines are listed, for the given

Figure 13.5

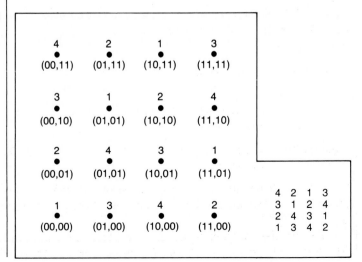

Figure 13.6

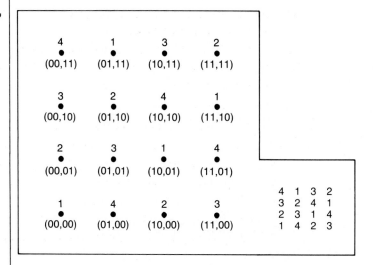

slope, in the same order as for Fig. 13.4. Next to each figure is its corresponding Latin square.

These figures give us three 4 × 4 Latin squares that are orthogonal in pairs.   □

Based on this example we close this section with the following theorem.

Theorem
13.18

▶   Let $F = GF(n)$, where $n \geq 3$, $n = p^t$, p a prime. The Latin squares that arise from $AP(F)$ for the $n - 1$ parallel classes where the slope is neither 0 nor infinite are orthogonal in pairs.

Proof   A proof of this result is outlined in the section exercises.   ■

EXERCISES

1. Complete the following table dealing with affine planes.

Field	Number of points	Number of lines	Number of points on a line	Number of lines on a point
	25			
GF(3^2)				
		56		
				17
			31	

2. How many parallel classes do each of the affine planes in Exercise 1 determine? How many lines are in each class?

3. Construct the affine plane $AP(\mathbf{Z}_3)$. Determine its parallel classes and the corresponding Latin squares for the classes of finite nonzero slope.

4. Answer Exercise 3 replacing $\mathbf{Z}_3$ by $\mathbf{Z}_5$.

5. Determine each of the following lines.

   a) The line in $AP(\mathbf{Z}_7)$ that is parallel to $y = 4x + 2$ and contains $(3, 6)$.

   b) The line in $AP(\mathbf{Z}_{11})$ that is parallel to $2x + 3y + 4 = 0$ and contains $(10, 7)$.

   c) The line in $AP(F)$, where $F = GF(2^2)$, that is parallel to $10y = 11x + 01$ and contains $(11, 01)$. (See Table 13.7.)

6. Suppose we try to construct an affine plane $AP(\mathbf{Z}_6)$ as we did in this section.

   a) Determine which of the conditions (A1), (A2), (A3) fail in this situation.

   b) Find how many lines contain a given point P and how many points are on a given line ℓ, for this "geometry."

7. The following provides an outline for a proof of Theorem 13.18.

   a) Consider a parallel class of lines given by $y = mx + b$, where $m \in F$, $m \neq 0$. Show that each line in this class intersects each "vertical" line and each "horizontal" line in exactly one point of $AP(F)$. Thus the configuration obtained by labeling the points of $AP(F)$, as in Figs. 13.4, 13.5, and 13.6, is a Latin square.

   b) To show that the Latin squares corresponding to two different classes, other than the classes of slope 0 or infinite slope, are orthogonal, assume that an ordered pair (i, j) appears more than once when one square is superimposed upon the other. How does this lead to a contradiction?

13.5 ■ BLOCK DESIGNS AND PROJECTIVE PLANES

In this final section we examine a type of combinatorial design and see how it is related to the structure of a finite geometry.

EXAMPLE 13.15
Dick (d) and his wife Mary (m) go to New York City with their five children Richard (r), Peter (p), Christopher (c), Brian (b), and Julie (j). While staying in the city they receive three passes each day, for a week, to visit the Empire State Building. Can we make up a schedule for this family so that everyone gets to visit this attraction the same number of times?

The following schedule is one possibility.

(1) b, c, d (2) b, j, r (3) b, m, p (4) c, j, m

(5) c, p, r (6) d, j, p (7) d, m, r

Here the result was obtained by trial and error. For a problem of this size such a technique is feasible. However, it is not very good for all problems of this type. Furthermore, in asking for a certain schedule we may be asking for something that doesn't exist. In the problem above, for example, each pair of family members is together on only one visit. If the family had received four passes each day we would not be able to construct a table that maintained this property.   □

The situation in this example generalizes as follows.

Definition 13.13

▶ Let V be a set with v elements. A collection $\{B_1, B_2, \ldots, B_b\}$ of subsets of V is called a *balanced incomplete block design*, or (v, b, r, k, λ)-*design*, if the following conditions are satisfied:

a) For $1 \le i \le b$, B_i contains k elements, where k is a fixed constant and $k < v$.

b) Each element $x \in V$ is in r ($\le b$) of the subsets B_i, $1 \le i \le b$.

c) Every pair x, y of elements of V appear together in λ ($\le b$) of the subsets B_i, $1 \le i \le b$.

The elements of V are often called *varieties* because of early applications in the design of experiments that dealt with tests on fertilizers and plants. The b subsets B_1, $B_2, \ldots, B_b$ of V, are called *blocks,* where each block contains k varieties. The number r is referred to as the *replication number* of the design. Finally, λ is termed the *covalency* for the design. This parameter makes the design balanced in the following sense. For general block designs we have a number λ_{xy} for each pair x, $y \in V$; if λ_{xy} is the same for all pairs of elements from V, then λ represents this common measure and the design is called balanced. In this text we deal only with balanced designs.

EXAMPLE 13.16

a) The schedule in Example 13.15 is an example of a $(7, 7, 3, 3, 1)$-design.

b) For $V = \{1, 2, 3, 4, 5, 6\}$, the ten blocks

1 2 4	1 3 4	1 5 6	2 3 6	3 4 6
1 2 6	1 3 5	2 3 5	2 4 5	4 5 6

constitute a $(6, 10, 5, 3, 2)$-design.

c) If F is a finite field with $|F| = n$, then the affine plane $AP(F)$ yields an $(n^2, n^2 + n, n + 1, n, 1)$-design. Here the varieties are the n^2 points in $AP(F)$; the $n^2 + n$ lines are the blocks of the design.   □

At this point there are five parameters determining our design. We now examine how these parameters are related.

Theorem 13.19

▶ For a (v, b, r, k, λ)-design (1) $vr = bk$; and (2) $\lambda(v - 1) = r(k - 1)$.

Proof

(1) With b blocks in the design and k elements per block, listing all the elements of the blocks we get bk symbols. This collection of symbols consists of the elements of V with each element appearing r times, for a total of vr symbols. Hence $vr = bk$. (2) For this property we introduce the *pairwise incidence matrix* A for the design. With $|V| = v$, let $t = \binom{v}{2}$, the number of pairs of elements in V. We construct the $t \times b$ matrix $A = (a_{ij})$ by defining $a_{ij} = 1$ if the ith pair of elements from V is in the jth block of the design; if not, $a_{ij} = 0$.

$$
\begin{array}{c@{\qquad}ccccc}
 & B_1 & B_2 & \cdots & B_b \\[4pt]
x_1x_2 & \begin{bmatrix} a_{11} \end{bmatrix} & a_{12} & \cdots & a_{1b} \\
x_1x_3 & a_{21} & a_{22} & \cdots & a_{2b} \\
\vdots & \vdots & \vdots & \vdots & \vdots \\
x_1x_n & a_{n-1,1} & a_{n-1,2} & \cdots & a_{n-1,b} \\
x_2x_3 & a_{n,1} & a_{n,2} & \cdots & a_{n,b} \\
\vdots & \vdots & \vdots & \vdots & \vdots \\
x_{n-1}x_n & a_{t,1} & a_{t,2} & \cdots & a_{t,b}
\end{array}
$$

We now count the number of 1's in matrix A in two ways.

a) Consider the rows. Since each pair x_i, x_j, $1 \le i < j \le n$, appears in λ blocks, it follows that each row contains λ 1's. With t rows in the matrix the number of 1's is then $\lambda t = \lambda v(v - 1)/2$.

b) Now consider the columns. As each block contains k elements, this determines $\binom{k}{2} = k(k - 1)/2$ pairs, and this is the number of 1's in each column of matrix A. With b columns the total number of 1's is $bk(k - 1)/2$.

Then, $\lambda v(v - 1)/2 = bk(k - 1)/2 = vr(k - 1)/2$, so $\lambda(v - 1) = r(k - 1)$. ∎

When n is a power of a prime, the $(n^2, n^2 + n, n + 1, n, 1)$-design can be obtained from the affine plane $AP(F)$, where $F = GF(n)$. Here the points are the varieties while the lines are the blocks. We shall now introduce a construction that enlarges $AP(F)$ to what is called a finite projective plane. From this projective plane we can construct the $(n^2 + n + 1, n^2 + n + 1, n + 1, n + 1, 1)$-design. First let us see how these two kinds of planes compare.

Definition 13.14 ▶ If $\mathscr{P}'$ is a finite set of points and $\mathscr{L}'$ a set of lines formed from subsets of $\mathscr{P}'$, the (finite) plane based on $\mathscr{P}'$ and $\mathscr{L}'$ is called a *projective plane* if the following conditions are satisfied.

P1) Two distinct points of $\mathscr{P}'$ are on only one line.

P2) Any two lines from $\mathscr{L}'$ intersect in a unique point.

P3) There are four points in $\mathscr{P}'$, no three of which are collinear.

The difference between the affine and projective planes lies in the condition dealing with the existence of parallel lines. Here the parallel lines of the affine plane, based on $\mathscr{P}$, $\mathscr{L}$, will intersect when the given system is enlarged to the projective plane based on $\mathscr{P}'$, $\mathscr{L}'$.

The construction proceeds as follows.

EXAMPLE 13.17 Start with an affine plane $AP(F)$ where $F = GF(n)$. For each point $(x, y) \in \mathscr{P}$, rewrite the point as $(x, y, 1)$. We then think of the points as ordered triples (x, y, z) where $z = 1$. Rewrite the equations of the lines $x = c$, $y = mx + b$ in $AP(F)$ as

$x = cz$, $y = mx + bz$, where $z = 1$. We still have our original $AP(F)$, but with a change of notation.

Add the set of points $\{(1,0,0)\} \cup \{(x,1,0) \mid x \in F\}$ to $\mathcal{P}$ to get the set $\mathcal{P}'$. Then $|\mathcal{P}'| = n^2 + n + 1$. Let ℓ_∞ be the subset of $\mathcal{P}'$ consisting of these new points. This new line can be given by the equation $z = 0$, with the stipulation that we never have $x = y = z = 0$. Hence $(0,0,0) \notin \mathcal{P}'$.

Let us examine these ideas for the affine plane $AP(\mathbf{Z}_2)$. Here $\mathcal{P} = \{(0,0), (1,0), (0,1), (1,1)\}$, so

$$\mathcal{P}' = \{(0,0,1), (1,0,1), (0,1,1), (1,1,1)\} \cup \{(1,0,0), (0,1,0), (1,1,0)\}.$$

The six lines in $\mathcal{L}$ were originally

$\quad x = 0$: $\{(0,0), (0,1)\}$ $\quad$ $y = 0$: $\{(0,0), (1,0)\}$ $\quad$ $y = x$: $\{(0,0), (1,1)\}$

$\quad x = 1$: $\{(1,0), (1,1)\}$ $\quad$ $y = 1$: $\{(0,1), (1,1)\}$ $\quad$ $y = x + 1$: $\{(0,1), (1,0)\}$

We now rewrite these as

$$x = 0 \qquad y = 0 \qquad y = x$$
$$x = z \qquad y = z \qquad y = x + z,$$

and add a new line ℓ_∞ defined by $z = 0$. These constitute the set of lines $\mathcal{L}'$ for our projective plane. At this point we consider z as a *variable*. Consequently, the line $x = z$ consists of the points $(0,1,0)$, $(1,0,1)$, and $(1,1,1)$. In fact, each line of $\mathcal{L}$ that consisted of two points will now contain three points when considered in $\mathcal{L}'$. The set $\mathcal{L}'$ consists of the following seven lines.

$\quad x = 0$: $\{(0,0,1), (0,1,0), (0,1,1)\}$ $\quad$ $y = z$: $\{(1,0,0), (0,1,1), (1,1,1)\}$

$\quad y = 0$: $\{(0,0,1), (1,0,0), (1,0,1)\}$ $\quad$ $y = x$: $\{(0,0,1), (1,1,0), (1,1,1)\}$

$\quad x = z$: $\{(0,1,0), (1,0,1), (1,1,1)\}$ $\quad$ $y = x + z$: $\{(0,1,1), (1,1,0), (1,0,1)\}$

$\quad z = 0$ (ℓ_∞): $\{(0,1,0), (1,0,0), (1,1,0)\}$

In the original affine plane the lines $x = 0$ and $x = 1$ were parallel, since no point in this plane satisfied both equations simultaneously. Here in this new system $x = 0$ and $x = z$ intersect in the point $(0,1,0)$, so they are no longer parallel in the sense of $AP(\mathbf{Z}_2)$. Likewise, $y = x$ and $y = x + 1$ were parallel in $AP(\mathbf{Z}_2)$, whereas here the lines $y = x$ and $y = x + z$ intersect at $(1,1,0)$. We depict this projective plane based on $\mathcal{P}'$ and $\mathcal{L}'$ as in Fig. 13.7. Here the "circle" through $(1,0,1)$, $(1,1,0)$, $(0,1,1)$ is the line $y = x + z$. Note that every line intersects ℓ_∞, which is often called the *line at infinity*. This line consists of the three *points at infinity*. We define two lines to be parallel in the projective plane when they intersect in a point at infinity, or on ℓ_∞.

This projective plane provides us with a $(7,7,3,3,1)$-design like the one we developed by trial and error in Example 13.15.   □

We generalize the results of Example 13.17 in the following.

Let n be a power of a prime. The affine plane $AP(F)$, for $F = GF(n)$, provides an example of an $(n^2, n^2 + n, n + 1, n, 1)$-design. In $AP(F)$ the $n^2 + n$ lines fall into $n + 1$ parallel classes. For each parallel class we add a point at infinity to $AP(F)$. The line at infinity, ℓ_∞, is then defined as the set of points at infinity. In this way we obtain

Figure 13.7

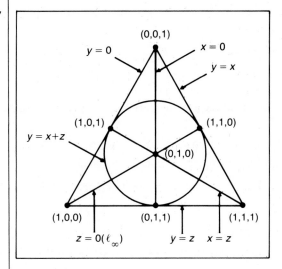

the projective plane over $GF(n)$, which has $n^2 + n + 1$ points and $n^2 + n + 1$ lines. Here each point is on $n + 1$ lines, and each line contains $n + 1$ points. Furthermore, any two points in this plane are on only one line. Consequently, we have an example of an $(n^2 + n + 1, n^2 + n + 1, n + 1, n + 1, 1)$-design.

EXERCISES

1. Let $V = \{1, 2, \ldots, 9\}$. Determine the values of v, b, r, k, and λ for the design given by the blocks

1 2 6	1 4 7	2 3 4	2 7 9	3 7 8	4 6 8
1 3 5	1 8 9	2 5 8	3 6 9	4 5 9	5 6 7

2. Find an example of a $(4, 4, 3, 3, \lambda)$-design.

3. Find an example of a $(7, 7, 4, 4, \lambda)$-design.

4. Complete the following table so that the parameters v, b, r, k, λ in any row are possible in a balanced incomplete block design.

v	b	r	k	λ
4			3	2
9	12		3	
10		9		2
13		4	4	
	30	10		3

5. Is it possible to have a (v, b, r, k, λ)-design where (a) $b = 28$, $r = 4$, $k = 3$? (b) $v = 17$, $r = 8$, $k = 5$?

6. Given a (v, b, r, k, λ)-design with $b = v$, prove that if v is even then λ is even.

7. A (v, b, r, k, λ)-design is called a *triple system* if $k = 3$. When $k = 3$ and $\lambda = 1$ we call the design a *Steiner triple system*.

    a) Prove that in every triple system $\lambda(v - 1)$ is even and $\lambda v(v - 1)$ is divisible by 6.

    b) Prove that in every Steiner triple system v is congruent to 1 or 3 modulo 6.

8. Verify that the following blocks constitute a Steiner triple system on nine varieties.

$$
\begin{array}{cccccc}
1\ 2\ 8 & 1\ 4\ 7 & 2\ 3\ 4 & 2\ 7\ 9 & 3\ 8\ 9 & 4\ 6\ 8 \\
1\ 3\ 5 & 1\ 6\ 9 & 2\ 5\ 6 & 3\ 6\ 7 & 4\ 5\ 9 & 5\ 7\ 8
\end{array}
$$

9. In a programming class Professor Madge has n students and she wants to assign m students to each of p computer projects. If each student is assigned to the same number of projects, (a) in how many projects will each individual student be involved? (b) in how many projects will each pair of students be involved?

10. In constructing the projective plane from $AP(\mathbf{Z}_2)$ in Example 13.17, why didn't we want to include the point $(0, 0, 0)$ in the set $\mathcal{P}'$?

11. Determine the values of v, b, r, k, and λ for the balanced incomplete block design associated with the projective plane that arises from $AP(F)$ for the following choices of F: (a) $\mathbf{Z}_5$; (b) $\mathbf{Z}_7$; and (c) $GF(8)$.

12. a) List the points and lines in $AP(\mathbf{Z}_3)$. How many parallel classes are there for this finite geometry? What are the parameters for the associated balanced incomplete block design?

    b) List the points and lines for the projective plane that arises from $AP(\mathbf{Z}_3)$. Determine the points on ℓ_∞ and use them to determine the "parallel" classes for this geometry. What are the parameters for the associated balanced incomplete block design?

13.6 ■ SUMMARY AND HISTORICAL REVIEW

The structure of a field was first developed in Chapter 8. Here we examined polynomial rings and their role in the structure of finite fields. In this study attention was directed to applications in finite geometries and combinatorial designs.

In Chapter 9 we saw that the order of a finite Boolean algebra could be only a power of 2. Now we find that for a finite field the order can be only a power of a prime, and that for any prime p, $n \in \mathbf{Z}^+$, there is only one field, up to isomorphism, of order p^n. This field is denoted by $GF(p^n)$, in honor of the French mathematician Evariste Galois (1811–1832).

The finite fields $(\mathbf{Z}_p, +, \cdot)$, for p a prime, were obtained in Chapter 8 by means of the equivalence relation, congruence modulo p, defined on $\mathbf{Z}$. Using these finite fields, we developed here the integral domains $\mathbf{Z}_p[x]$. Then, with $s(x)$ an irreducible polynomial of degree n in $\mathbf{Z}_p[x]$, a similar equivalence relation, namely congruence modulo $s(x)$, gave us a set of p^n equivalence classes, denoted $\mathbf{Z}_p[x]/(s(x))$. These p^n equivalence classes became the elements of the field $GF(p^n)$. (Although we did not prove every possible result in general, it can be shown that over any finite field $\mathbf{Z}_p$ there is an irreducible polynomial of degree n for any $n \in \mathbf{Z}^+$.)

The theory of finite fields was developed by Galois in his work on the solutions of polynomial equations. As we mentioned in the summary of Chapter 12, the study of polynomial equations was an area of mathematical research that challenged many mathematicians from the sixteenth to the nineteenth centuries. In the nineteenth century, Neils Henrik Abel (1802–1829) first showed that the solution of the general quintic could not be given by radicals. Galois showed that for any polynomial of degree n over a field F, there is a corresponding group G that is isomorphic to a subgroup of S_n, the group of permutations of $\{1, 2, 3, \ldots, n\}$. The essence of Galois' work is that such a polynomial equation can be solved by (addition, subtraction, multiplication, division, and) radicals if its corresponding group is *solvable*. Now what makes a group solvable? We say that a group G is solvable if it has a chain of subgroups $G = K_1 \supset K_2 \supset K_3 \supset \ldots \supset K_t = \{e\}$, where for all $2 \le i \le t$, K_i is a normal subgroup of K_{i-1} (i.e., $xyx^{-1} \in K_i$ for all $y \in K_i$, $x \in K_{i-1}$), and $|K_{i-1}| \div |K_i|$ is a prime. One finds that all subgroups of S_i, $1 \le i \le 4$, are solvable, but for $n \ge 5$ there are subgroups of S_n that are not solvable.

Though it seems that Galois theory is concerned predominantly with groups, there is a great deal more on the theory of fields that we have not mentioned. As a consequence of Galois' work, the areas of field theory and finite group theory became topics of great mathematical interest.

For more on *Galois theory,* the reader will find Chapter 6 of the text by V. Larney [7] and Chapter 12 in the book by N. H. McCoy and T. Berger [9] good places to start from. Chapter 5 of I. N. Herstein [5] has more on the topic. A detailed presentation can be found in the text by O. Zariski and P. Samuel [13]. Appendix E in the text by V. Larney [7] includes an interesting short account of the life of Galois; more on his life can be found in the somewhat fictional account by L. Infeld [6]. The article by T. Rothman [10] provides a contemporary discussion of the inacurracies and myths surrounding the life, and especially the death, of Galois.

The Latin squares, combinatorial designs, and finite geometries of the later sections of the chapter showed us how the finite field structure entered into problems of design. Dating back to the time of Leonhard Euler (1707–1783) and the problem of the "thirty-six officers," the study of orthogonal Latin squares has been developed considerably since 1900, and especially since 1960 with the work of R. C. Bose, S. S. Shrikhande, and E. T. Parker. Chapter 7 of the monograph by H. Ryser [11] provides the details of their accomplishments. The text by C. L. Liu [8] includes ideas from coding theory in its discussion of Latin squares.

The study of finite geometries can be traced back to the work of Gino Fano who, in 1892, considered a finite three-dimensional geometry consisting of 15 points, 35 lines, and 15 planes. However, it was not until 1906 that these geometries gained any notice, when O. Veblen and W. Bussey began their study of finite projective geometries. For more on this topic the reader should find the texts by A. Albert and R. Sandler [1] and H. Dorwart [3] very interesting. The text by P. Dombowski [2] provides an extensive coverage for those seeking more advanced study.

Finally, the notion of designs was first studied by statisticians in the area called the design of experiments. Through the research of R. A. Fisher and his followers, this area has come to play an important role in the modern theory of statistical analysis. In our development we examined conditions under which a (v, b, r, k, λ)-design could exist and how such designs were related to affine planes and finite projective planes. The text by M. Hall, Jr. [4] provides more on this topic as does the work by A. Street and W. Wallis [12]. Chapter XIII of [12] includes material relating designs and coding theory.

REFERENCES

1. Albert, A. Adrian, and Sandler, R., *An Introduction to Finite Projective Planes,* Holt, Rinehart, and Winston, New York, 1968.

2. Dombowski, Peter, *Finite Geometries,* Springer-Verlag, New York, 1968.

3. Dorwart, Harold L., *The Geometry of Incidence,* Prentice-Hall, Englewood Cliffs, New Jersey, 1966.

4. Hall, Marshall, Jr., *Combinatorial Theory,* Blaisdell, Waltham, Massachusetts, 1967.

5. Herstein, I. N., *Topics in Algebra,* 2nd ed., Xerox College Publishing, Lexington, Massachusetts, 1975.

6. Infeld, Leopold, *Whom the Gods Love,* McGraw-Hill, New York, 1948.

7. Larney, Violet H., *Abstract Algebra: A First Course,* Prindle, Weber & Schmidt, Boston, Massachusetts, 1975.

8. Liu, C. L., *Topics in Combinatorial Mathematics,* Mathematical Association of America, 1972.

9. McCoy, Neal H., and Berger, Thomas R., *Algebra: Groups, Rings, and Other Topics,* Allyn and Bacon, Boston, Massachusetts, 1977.

10. Rothman, Tony, "Genius and Biographers: The Fictionalization of Evariste Galois," *The American Mathematical Monthly,* Vol. 89, No. 2, 1982, pp. 84–106.

11. Ryser, Herbert J., *Combinatorial Mathematics,* Carus Mathematical Monographs, Number 14, Mathematical Association of America, 1963.

12. Street, Anne Penfold, and Wallis, W. D., *Combinatorial Theory: An Introduction,* The Charles Babbage Research Center, Winnipeg, Canada, 1977.

13. Zariski, Oscar, and Samuel, Pierre, *Commutative Algebra,* Vol. I, Van Nostrand, New York, 1958.

■ MISCELLANEOUS EXERCISES

1. Determine n if over $GF(n)$ there are 6561 monic polynomials of degree 5 with no constant term.

2. a) Let $f(x) = a_n x^n + \cdots + a_1 x + a_0 \in \mathbf{Z}[x]$. If $r/s \in \mathbf{Q}$, with $(r, s) = 1$ and $f(r/s) = 0$, prove that $s \mid a_n$ and $r \mid a_0$.

   b) Find the rational roots, if any exist, of the following polynomials over $\mathbf{Q}$. Factor $f(x)$ in $\mathbf{Q}[x]$.

   i) $f(x) = 2x^3 + 3x^2 - 2x - 3$

   ii) $f(x) = 3x^3 + 5x^2 + 5x + 2$

   iii) $f(x) = x^5 - (3/2)x^4 + x - (3/2)$

   c) Show that the polynomial $f(x) = x^{100} - x^{50} + x^{20} + x^3 + 1$ has no rational root.

3. a) Verify that $s(x) = x^2 + 1$ is irreducible in $\mathbf{R}[x]$.

   b) Show that the field $\mathbf{R}[x]/(s(x))$ is isomorphic to the field $\mathbf{C}$ of complex numbers.

4. Verify that the polynomial $f(x) = x^4 + x^3 + x + 1$ is reducible over any field F (finite or infinite).

5. Four of the seven blocks in a $(7, 7, 3, 3, 1)$-design are $\{1, 3, 7\}$, $\{1, 5, 6\}$, $\{2, 6, 7\}$, and $\{3, 4, 6\}$. Determine the other three blocks.

6. A projective plane is coordinatized with the elements of a field F. If this plane contains 91 lines, what are $|F|$ and $\text{char}(F)$?

7. Let $V = \{x_1, x_2, \ldots, x_v\}$ be the set of varieties and $\{B_1, B_2, \ldots, B_b\}$ the collection of blocks for a (v, b, r, k, λ)-design. We define the incidence matrix A for the design by

$$A = (a_{ij})_{v \times b}, \qquad \text{where } a_{ij} = \begin{cases} 1, & \text{if } x_i \in B_j \\ 0, & \text{otherwise}. \end{cases}$$

   a) How many 1's are there in each row and column of A?

   b) Let $J_{m \times n}$ be the $m \times n$ matrix where every entry is 1. For $J_{n \times n}$ we write J_n. Prove that for the incidence matrix A, $A \cdot J_b = r \cdot J_{v \times b}$ and $J_v \cdot A = k \cdot J_{v \times b}$.

   c) Show that

$$A \cdot A^{\text{tr}} = \begin{bmatrix} r & \lambda & \lambda & \ldots & \lambda \\ \lambda & r & \lambda & \ldots & \lambda \\ \lambda & \lambda & r & \ldots & \lambda \\ \ldots & \ldots & \ldots & \ldots & \ldots \\ \lambda & \lambda & \lambda & \ldots & r \end{bmatrix} = (r - \lambda)I_v + \lambda J_v,$$

   where I_v is the $v \times v$ (multiplicative) identity.

   d) Prove that $\det(A \cdot A^{\text{tr}}) = (r - \lambda)^{v-1}[r + (v - 1)\lambda] = (r - \lambda)^{v-1} rk$.

14

An Introduction
to Graph Theory

With this chapter we start to develop the last major topic of this text. Unlike other areas in mathematics, the theory of graphs has a definite starting place, in a paper published in 1736 by the Swiss mathematician Leonhard Euler (1707–1783). The main idea behind this work grew out of a now popular situation known as the seven bridges of Königsberg. We shall examine the solution of this problem from which Euler developed the fundamental concepts for the theory of graphs.

Unlike the continuous graphs of early algebra courses, the graphs we examine here are finite in structure and can be used to analyze relationships and applications in many different settings. We have seen some examples of applications of graph theory in earlier chapters (3–7, 9 and 11). However, the development here will be independent of these prior discussions.

14.1 ■ DEFINITIONS AND EXAMPLES

When we use a road map we are often concerned with seeing how to get from one town to another by means of the roads indicated on the map. Consequently we are dealing with two distinct sets of objects: towns and roads. As we have seen many times before, such sets of objects can be used to define a relation. If V denotes the set of towns and E the set of roads, we can define a relation $\mathcal{R}$ on V by $a \mathcal{R} b$ if there is a system of roads that takes us from a to b. If these are all two-way roads, then if $a \mathcal{R} b$ we also have $b \mathcal{R} a$. Should all the roads under consideration be two-way we have a symmetric relation.

One way to represent any relation is by listing the ordered pairs that are its elements. Here, however, it is more convenient to use a picture as in Fig. 14.1. This figure demonstrates the possible ways of traveling among six towns using the eight roads indicated. It shows that there is at least one system of roads connecting any two towns (identical or distinct). This pictorial representation is a lot easier to work with than the 36 ordered pairs of the relation $\mathcal{R}$.

At the same time, Fig. 14.1 would be appropriate for representing six communication centers with the eight "roads" interpreted as communication links. If each link provides two-way communication we should be quite concerned about the vulnerability of center a due to such factors as equipment breakdowns or enemy

Figure 14.1

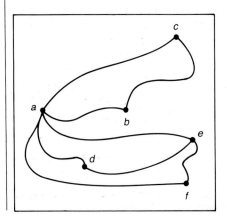

attacks. Without center a, neither b nor c can communicate with any of d, e, or f. From these observations we consider the following concepts.

Definition 14.1

▶ Let V be a nonempty set and $E \subseteq V \times V$. The pair (V, E) is then called a *directed graph* (on V), or *digraph*[†] (on V), where V is the set of *vertices*, or *nodes*, while E is its set of *edges*. We write $G = (V, E)$ to denote such a graph.

Figure 14.2

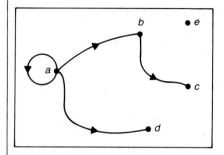

Figure 14.2 provides an example of a directed graph on $V = \{a, b, c, d, e\}$ with $E = \{(a, a), (a, b), (a, d), (b, c)\}$. The direction of an edge is indicated by placing a directed arrow on the edge, as shown here. For any edge, such as (b, c), we say that the edge is *incident* with the vertices b, c; b is said to be *adjacent to* c, while c is *adjacent from* b. In addition b is called the *origin*, or *source*, of the edge while c is the *terminus*, or *terminating vertex*. The edge (a, a) is an example of a *loop* and the vertex e that is not incident with any edge is called an *isolated* vertex.

When there is no concern about the direction of an edge the graph is called *undirected*, as in Fig. 14.3(a). This graph is a more compact way of describing the directed graph in Fig. 14.3(b). In an undirected graph there are undirected edges such

Figure 14.3

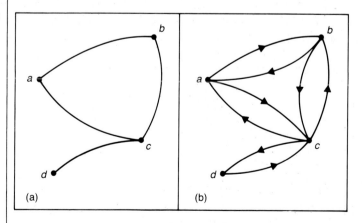

(a)   (b)

as $\{a,b\}$, $\{b,c\}$, $\{a,c\}$, $\{c,d\}$ in Fig. 14.3(a). An edge such as $\{a,b\}$ stands for $\{(a,b),(b,a)\}$. Although $(a,b) = (b,a)$ only when $a = b$, we do have $\{a,b\} = \{b,a\}$ for any a, b. Lastly, we write $\{a,a\}$ to denote a loop in an undirected graph; $\{a,a\}$ is considered the same as (a,a).

In general, if a graph G is not specified as directed or undirected, it is assumed to be undirected. If it contains no loops it is called *loop-free*.

Definition 14.2 ► Let $G = (V,E)$ be an undirected graph. For $x, y \in V$, we say there is a *path* in G from x to y if there exists a finite nonempty sequence of distinct edges from E such as $\{x,x_1\}, \{x_1,x_2\}, \dots, \{x_{i-1},x_i\}, \dots, \{x_{n-1},x_n\}, \{x_n,y\}$. When $x = y$ we call the path a *cycle*. The number of edges in a path (cycle) is called the *length* of the path (cycle).

Figure 14.4

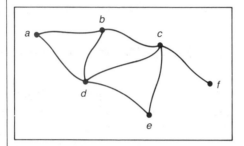

In Fig. 14.4, $\{a,b\}$, $\{b,c\}$, and $\{c,f\}$ provide a path from a to f. The sequence $\{a,b\}, \{b,c\}, \{c,e\}, \{e,d\}, \{d,c\}, \{c,f\}$ also provides a path. Here, however, we have passed through vertex c twice. When a path between two vertices does not pass through any vertex more than once, the path is called *simple*. The same applies to a cycle, except that the first and last vertices are the same. In Fig. 14.4, $\{a,b\}$, $\{b,d\}$, $\{d,a\}$, is a simple cycle, while $\{a,b\}, \{b,d\}, \{d,c\}, \{c,e\}, \{e,d\}, \{d,a\}$ is a cycle that is not simple.

For a directed graph we use the terms *directed paths* and *directed cycles*.

The notion of a path is needed in the following graph property.

Definition 14.3 ► Let $G = (V,E)$ be an undirected graph. We call G *connected* if there is a path between any two distinct vertices of G.

Let $G = (V,E)$ be a directed graph. Its associated undirected graph is the graph obtained from G by ignoring the directions on the edges. If this associated graph is connected we consider G connected.

A graph that is not connected is called *disconnected*.

The graphs in Fig. 14.1, 14.3, and 14.4 are connected. In Fig. 14.2 the graph is not connected, since, for example, there is no path from a to e.

EXAMPLE 14.1   In Fig. 14.5 we have an undirected graph on $V = \{a,b,c,d,e,f,g\}$. This graph is not connected since, for example, there is no path from a to e. However, the graph is composed of pieces that are connected, and these pieces are called the *components* of the graph. Hence a graph is connected iff it has only one component.   □

Figure 14.5

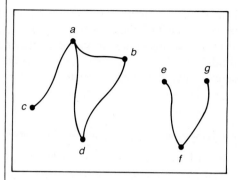

Considering how many concepts we have introduced, it is time to prove a first result in this new theory.

Theorem 14.1

▶ If $G = (V, E)$ is a connected graph, then there is a simple path between any two vertices in G.

Proof

Since G is connected, for any $a, b \in V$ there is a path from a to b. Select one of shortest length, say $\{a, x_1\}, \{x_1, x_2\}, \ldots, \{x_n, b\}$. If this path is not simple we have the situation $\{a, x_1\}, \{x_1, x_2\}, \ldots, \{x_{k-1}, x_k\}, \{x_k, x_{k+1}\}, \{x_{k+1}, x_{k+2}\}, \ldots, \{x_{m-1}, x_m\}, \{x_m, x_{m+1}\}, \ldots, \{x_n, b\}$, where $x_k = x_m$. But then $\{a, x_1\}, \{x_1, x_2\}, \ldots, \{x_{k-1}, x_k\}, \{x_m, x_{m+1}\}, \ldots, \{x_n, b\}$ is a shorter path. ■

Before closing this first section we extend our concept of a graph. When dealing with a road map it is definitely possible for two towns to be connected by more than one road. Thus far we have allowed at most one edge for two vertices; we now consider an extension.

Definition 14.4

▶ A graph $G = (V, E)$ is called a *multigraph* if for some $a, b \in V$, $a \neq b$, there are two or more edges of the form (a) (a, b) (for a directed graph); or (b) $\{a, b\}$ (for an undirected graph).

Figure 14.6

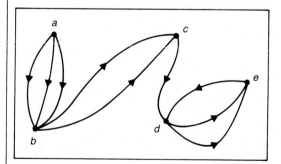

Figure 14.6 shows an example of a directed multigraph. Since there are three edges from a to b, we say that the edge (a, b) has *multiplicity* 3. The edges (b, c) and (d, e) each have multiplicity 2.

For $n \in \mathbf{Z}^+$, a multigraph is called an *n-graph* if no edge in the graph has multiplicity greater than n. The graph in Fig. 14.6 is a directed 3-graph.

We shall need the idea of a multigraph later in the chapter when we solve the problem of the seven bridges of Königsberg. Whenever we are dealing with a multigraph G, we shall state explicitly that G is a multigraph.

EXERCISES

1. List three situations, different from those in this section, where a graph could prove useful.

2. For the graph in Fig. 14.7, determine (a) a path from b to d that is simple; (b) a path from b to d that is not simple; (c) a cycle from b to b that is simple; and (d) a cycle from b to b that is not simple.

3. For the graph in Fig. 14.7, how many simple paths are there from b to f?

Figure 14.7

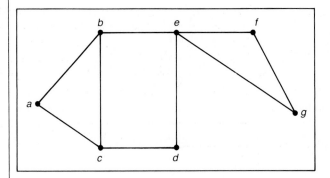

4. If a, b are vertices in a connected undirected graph G, the *distance* from a to b is defined to be the length of a shortest path from a to b. For the graph in Fig. 14.8, find the distance from d to every other vertex in G.

Figure 14.8

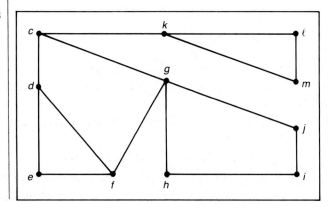

5. Seven towns a, b, c, d, e, f, and g are connected by a system of highways as follows: (1) I-22 goes from a to c, passing through b; (2) I-33 goes from c to d and then passes through b as it continues to f; (3) I-44 goes from d through e to a; (4) I-55 goes from f to b, passing through g; and (5) I-66 goes from g to d.

   a) Using vertices for towns and directed edges for segments of highways between towns, draw a directed graph that models this situation.

   b) List the simple paths from g to a.

   c) What is the smallest number of highway segments that would have to be closed down so that travel from b to d is disrupted?

   d) Is it possible to leave town c and return there, visiting each of the other towns only once?

   e) What is the answer to (d) if we are not required to return to c?

   f) Is it possible to start at some town and drive over all of these highways exactly once? (You are allowed to visit a town more than once and you need not return to the town from which you started.)

6. Figure 14.9 shows an undirected graph representing a section of a department store. The vertices indicate where cashiers are located, while the edges denote unblocked aisles between cashiers. The department store wants to set up a security system where (plain clothes) guards are placed at certain cashier locations so that each cashier either has a guard at his or her location or there is a guard with a cashier one aisle away. What is the smallest number of guards needed?

Figure 14.9

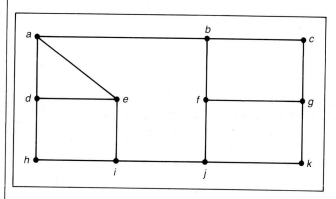

7. Let $G = (V, E)$ be a connected undirected graph, and let $\{a, b\}$ be an edge of G. Prove that $\{a, b\}$ is part of a cycle iff its removal (the vertices a and b are left) does not disconnect G.

8. Give an example of a connected graph G where removing any edge of G results in a disconnected graph.

9. Let G be a graph that satisfies the condition in Exercise 8.

   a) Must G be loop-free?   b) Could G be a multigraph?

   c) If G has n vertices, can we determine how many edges it has?

10. a)   If $G = (V, E)$ is an undirected graph with $|V| = v$, $|E| = e$, and no loops, prove that $2e \leq v^2 - v$.

    b)   State the corresponding inequality for the case when G is directed.

11. Let $G = (V, E)$ be an undirected graph. Define a relation $\mathcal{R}$ on V by $a \,\mathcal{R}\, b$ if $a = b$ or there is a path in G from a to b. Prove that $\mathcal{R}$ is an equivalence relation. Describe the partition of V induced by $\mathcal{R}$.

14.2 ■ SUBGRAPHS, COMPLEMENTS, AND GRAPH ISOMORPHISM

As we've done in the study of other discrete structures, we focus now on the following two ideas.

a)   What type of substructure is there to study for a graph?

b)   Is it possible to draw two graphs that appear distinct but have the same underlying structure?

To answer (a) we introduce the following.

Definition 14.5   ▶ If $G = (V, E)$ is a graph (directed or undirected), $G' = (V', E')$ is called a *subgraph* of G if $\emptyset \neq V' \subseteq V$ and $E' \subseteq E$, where each edge in E' is incident with vertices in V'.

Figure 14.10

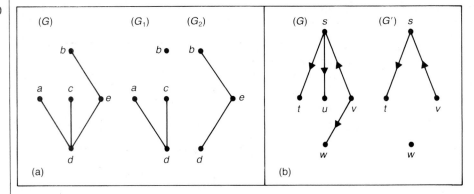

Figure 14.10(a) provides us with an undirected graph G and two of its subgraphs, G_1 and G_2. The vertex b is isolated in subgraph G_1. Part (b) of the figure provides a directed example.

The idea of a subgraph gives us a way to develop the complement of an undirected graph. Before doing so however, we define a type of graph that is maximal in size for a given number of vertices.

Definition
14.6

▶ Let V be a set of n vertices. The *complete graph* on V, denoted K_n, is a loop-free undirected graph where for any a, $b \in V$, $a \neq b$, there is an edge $\{a, b\}$.

Figure 14.11

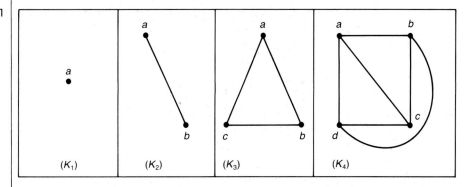

(K_1) (K_2) (K_3) (K_4)

Figure 14.11 provides the complete graphs K_n, $1 \le n \le 4$. We shall see, when we examine the idea of graph isomorphism, that these are the only possible complete graphs for the number of vertices given.

In determining the complement of a set in Chapter 2, we needed to know the universal set under consideration. The complete graph plays a role similar to a universal set.

Definition
14.7

▶ Let G be a loop-free undirected graph on n vertices. The *complement of G*, denoted $\overline{G}$, is the subgraph of K_n consisting of the n vertices in G and all edges that are not in G. (If $G = K_n$, $\overline{G}$ is a graph consisting of n vertices and no edges. Such a graph is called a *null* graph.)

Figure 14.12

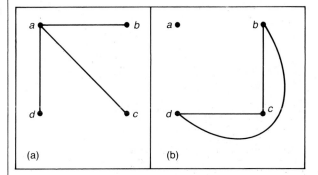

(a)   (b)

In Fig. 14.12(a) there is an undirected graph on four vertices. Its complement is shown in part (b) of the figure. In the complement, vertex a is isolated.

Once again we have reached a point where many new ideas have been defined. To demonstrate why these ideas are important we apply them now to the solution of an interesting puzzle.

EXAMPLE 14.2

(*Instant Insanity*) The game of Instant Insanity is played with four cubes. Each of the six faces on a cube is painted with one of the colors red (R), white (W), blue (B), or yellow (Y). The object of the game is to place the cubes in a column of four such that on each of the four sides of the column all four colors appear.

Figure 14.13

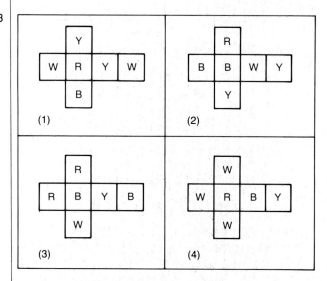

Consider the cubes in Fig. 14.13 and number them as shown. (These cubes are only one example of this game. Many others exist.) First we shall estimate the number of arrangements that are possible here. If we place cube 1 at the bottom of the column there are at most three different ways in which we can do this. In Fig. 14.13 cube 1 is unfolded and we see that it makes no difference whether we place the red face on the table or the opposite white face on the table. We are only concerned with the other four faces at the base of our column. With three pairs of opposite faces there will be at most three ways to place the first cube for the base of the column. Now consider cube 2. Although some colors are repeated, no pair of opposite faces has the same color. Hence we have six ways to place the second cube on top of the first. We can then rotate the second cube without changing either the face on the top of the first cube or the face on the bottom of the second cube. With four possible rotations we may place the second cube on top of the first in as many as 24 different ways. Continuing the argument, we find that there can be as many as $(3)(24)(24)(24) = 41{,}472$ possibilities to consider. And there may not even be a solution!

In solving the puzzle we realize that it is difficult to keep track of colors on opposite faces of cubes, and columns of colors. A graph (actually a labeled multi-graph) helps us to visualize the situation. In Fig. 14.14 we have a graph on four vertices R, W, B, and Y. As we consider each cube we examine its three pairs of opposite faces. For example, cube 1 has a pair of opposite faces painted yellow and blue, so we draw an edge connecting Y and B and label it 1 (for cube 1). The other two edges in the figure that are labeled with 1 account for the pairs of opposite faces

Figure 14.14

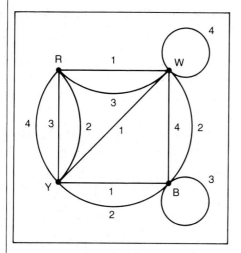

that are white and yellow, and red and white. Doing likewise for the other cubes we arrive at the graph in the figure. A loop, such as the one at B, with label 3, indicates a pair of opposite faces with the same color.

From the graph we see that there is a total of 12 edges falling into four sets of 3, according to the labels for the cubes. At each vertex the number of edges incident to (or from) the vertex counts the number of faces on the four cubes that have that color. (We count a loop twice.) Hence Fig. 14.14 tells us that for our four cubes we have five red faces, seven white ones, six blue ones, and six that are yellow.

With the four cubes stacked in a column we examine two opposite sides of the column. This gives us four edges in the graph of Fig. 14.14, where each label appears once. Since each color is to appear only once on a side of the column, each color must appear twice as an endpoint of these four edges. If we can accomplish the same result for the other two sides of the column, we have solved the puzzle. In Fig. 14.15(a) we see that two opposite sides of our column each have the four colors if the cubes are arranged according to the subgraph shown there. However, to also accomplish this for

Figure 14.15

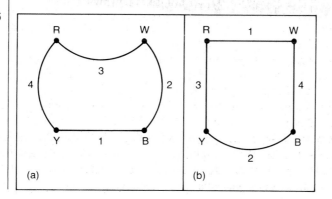

the other two sides of the column we need a second such subgraph that doesn't use any edge in (a). In this case a second such subgraph does exist, as shown in (b) of the figure.

Figure 14.16 shows how to arrange the cubes as indicated by the subgraphs in Fig. 14.15.

Figure 14.16

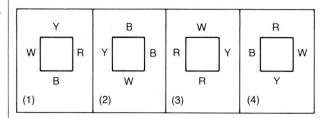

In general, for any four cubes we construct a labeled multigraph and try to find two subgraphs where (1) each subgraph contains all four vertices, and four edges, one for each label; (2) in each subgraph each vertex is incident with exactly two edges (a loop is counted twice); and (3) no edge of the graph appears in both subgraphs. □

Now we turn to the second question posed at the start of the section.

Figure 14.17

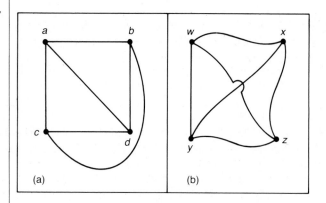

Figure 14.17 shows two undirected graphs on four vertices. Since straight edges and curved edges are considered the same here, each graph represents six adjacent pairs of vertices. In fact we probably feel that these graphs are both examples of the graph K_4. We make this feeling mathematically rigorous in the following idea.

Definition 14.8

▶ Let $G_1 = (V_1, E_1)$ and $G_2 = (V_2, E_2)$ be two undirected graphs. A function $f: V_1 \rightarrow V_2$ is called a *graph isomorphism* if (a) f is one-to-one and onto; and (b) for all $a, b \in V_1$, $\{a, b\} \in E_1$ iff $\{f(a), f(b)\} \in E_2$. When such a function exists, G_1 and G_2 are called *isomorphic graphs*.

The vertex correspondence of a graph isomorphism preserves adjacencies. In this way the structure of the graphs is preserved.

For the graphs in Fig. 14.17 the function f defined by $f(a) = w$, $f(b) = x$, $f(c) = y$, $f(d) = z$ provides an isomorphism. Consequently, as far as structure is concerned these graphs are considered the same.

Let us examine this idea in a more difficult situation.

Figure 14.18

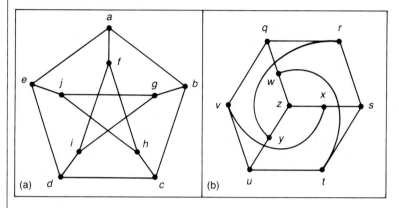

(a)

(b)

EXAMPLE 14.3 In Fig. 14.18 we have two graphs, each on ten vertices. Unlike the graphs in Fig. 14.17, it is not immediate if these graphs are isomorphic.

One finds that the correspondence given by

$$a \to q \qquad c \to u \qquad e \to r \qquad g \to x \qquad i \to z$$
$$b \to v \qquad d \to y \qquad f \to w \qquad h \to t \qquad j \to s$$

preserves all adjacencies. For example, $\{f, h\}$ is an edge in graph (a) with $\{w, t\}$ the corresponding edge in graph (b). But how did we come up with the correspondence?

We note that since an isomorphism preserves adjacencies it will then preserve graph substructures such as paths and cycles. In graph (a) the edges $\{a, f\}$, $\{f, i\}$, $\{i, d\}$, $\{d, e\}$, $\{e, a\}$ constitute a (simple) cycle of length 5. Hence we must preserve this as we try to find an isomorphism. The corresponding edges in graph (b) are $\{q, w\}$, $\{w, z\}$, $\{z, y\}$, $\{y, r\}$, $\{r, q\}$, which also provide a (simple) cycle of length 5. In addition, starting at vertex a in graph (a) we find a path that will "visit" each vertex only once. We express this path by $a \to f \to h \to c \to b \to g \to j \to e \to d \to i$. For the graphs to be isomorphic there must be a corresponding path in graph (b). Here the path described by $q \to w \to t \to u \to v \to x \to s \to r \to y \to z$ is the counterpart.   ☐

The observations above are some of the ways we can determine whether two graphs are isomorphic. Other considerations will be discussed throughout the chapter.

We close this section with one more example on graph isomorphism.

Figure 14.19

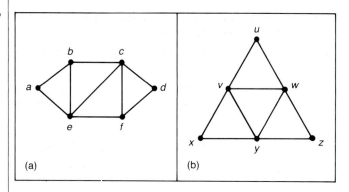

(a)

(b)

EXAMPLE 14.4 The two graphs in Fig. 14.19 each have six vertices and nine edges. Therefore it is reasonable to ask if they are isomorphic.

In graph (a) vertex a is adjacent to two other vertices of the graph. Consequently, if we try to construct an isomorphism between these graphs, we should correspond vertex a with a comparable one in graph (b), say vertex u. A similar situation exists for vertex d and either vertex x or z. But no matter which of the vertices x or z we use, there remains one vertex in graph (b) that is adjacent to two other vertices. And there is no other such vertex in graph (a) to continue our one-to-one structure preserving correspondence. Consequently, these graphs are not isomorphic.

Furthermore, in graph (b) it is possible to start at any vertex and find a cycle that includes every edge of the graph. For example, if we start at vertex u, the cycle $u \rightarrow w \rightarrow v \rightarrow y \rightarrow w \rightarrow z \rightarrow y \rightarrow x \rightarrow v \rightarrow u$ possesses this property. This does not happen in graph (a) where the only paths that include each edge start at either b or f and then terminate at f or b, respectively. □

EXERCISES

1. Each of the labeled multigraphs in Fig. 14.20 arises in the analysis of a set of four blocks for the game of Instant Insanity. In each case determine a solution to the puzzle, if possible.

Figure 14.20

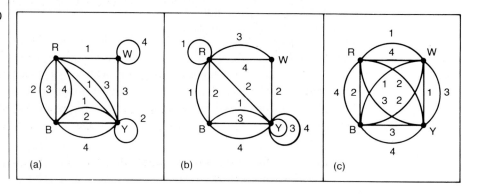

(a)

(b)

(c)

2. Find all (loop-free) nonisomorphic undirected graphs with four vertices. How many of these graphs are connected?

3. For each pair of graphs in Fig. 14.21, determine whether or not the graphs are isomorphic.

Figure 14.21

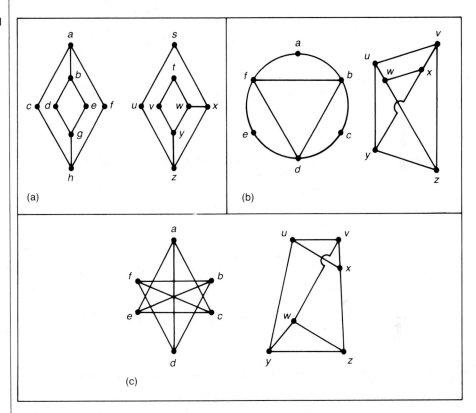

(a)

(b)

(c)

4. a) Coach Rizzo must make up a schedule for the five softball teams in her league. If each team plays two other teams, model a possible schedule by using a graph.

   b) Although it is possible to have different schedules in part (a), show that these schedules are the same except for a permutation of the names of the teams.

5. Let G be an undirected (loop-free) graph with v vertices and e edges. How many edges are there in $\overline{G}$?

6. Let $f: G_1 \to G_2$ be a graph isomorphism. If there is a path of length 3 between the vertices a and b in G_1, prove that in G_2 there is a path of length 3 between the vertices $f(a)$ and $f(b)$.

7. a) If G_1, G_2 are (loop-free) undirected graphs, prove that G_1, G_2 are isomorphic iff $\overline{G}_1$, $\overline{G}_2$ are isomorphic.

b) Determine if the graphs in Fig. 14.22 are isomorphic.

Figure 14.22

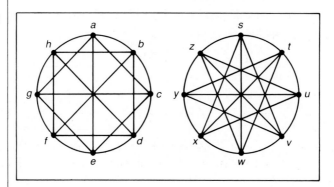

8. a) Let G be an undirected graph with n vertices. If G is isomorphic to its own complement $\overline{G}$, how many edges must G have? (Such a graph is called *self-complementary*.)

   b) Find an example of a self-complementary graph on four vertices and one on five vertices.

   c) If G is a self-complementary graph on n vertices, prove that n is congruent to 0 or 1 modulo 4.

9. Let G be a simple cycle on n vertices. Prove that G is self-complementary iff $n = 5$.

10. a) Find a graph G where both G and $\overline{G}$ are connected.

    b) If G is a graph on n vertices, $n \geq 2$, and G is not connected, prove that $\overline{G}$ is connected.

11. a) Extend Definition 14.8 to directed graphs.

    b) Determine, up to isomorphism, all (loop-free) directed graphs on three vertices.

    c) Determine if the directed graphs in Fig. 14.23 are isomorphic.

Figure 14.23

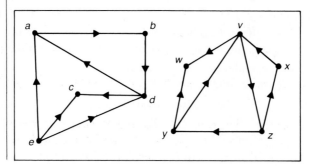

12. a) How many subgraphs $H = (V, E)$ of K_6 satisfy $|V| = 3$? (If two subgraphs are isomorphic but have different vertex sets, consider them distinct.)

   b) How many subgraphs $H = (V, E)$ of K_6 satisfy $|V| = 4$?

   c) How many subgraphs does K_6 have?

   d) For $n \geq 3$, how many subgraphs does K_n have?

14.3 ■ VERTEX DEGREE: EULER PATHS AND CYCLES

In Example 14.4 we found that the number of edges incident with a vertex could be used to show that two undirected graphs were not isomorphic. We now find this idea even more helpful.

Definition 14.9 ▶ Let G be an undirected graph or multigraph. For any vertex v of G, the *degree of v*, written $\deg(v)$, is the number of edges in G that are incident with v.

Figure 14.24

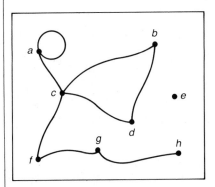

EXAMPLE 14.5 For the graph in Fig. 14.24, $\deg(b) = \deg(d) = \deg(f) = \deg(g) = 2$, $\deg(c) = 4$, $\deg(e) = 0$, $\deg(h) = 1$. For vertex a we have $\deg(a) = 3$, since we count a loop twice. Since h has degree 1, it is called a *pendant* vertex. □

Using the idea of vertex degree we have the following result.

Theorem 14.2 ▶ If $G = (V, E)$ is an undirected graph or multigraph, then $\sum_{v \in V} \deg(v) = 2|E|$.

Proof As we consider each edge $\{a, b\}$ in graph G, the edge contributes a count of 1 to each of $\deg(a)$, $\deg(b)$, and consequently a count of 2 to $\sum_{v \in V} \deg(v)$. Thus $2|E|$ accounts for $\deg(v)$, for all $v \in V$, and $\sum_{v \in V} \deg(v) = 2|E|$. ■

This theorem provides some insight into the number of odd vertices that can exist in a graph.

Corollary 14.1 ▶ For any undirected graph or multigraph the number of vertices of odd degree must be even.

Proof We leave the proof for the reader. ∎

We apply Theorem 14.2 in the following.

EXAMPLE 14.6 An undirected graph (or multigraph) where each vertex has the same degree is called a *regular* graph. Is it possible to have a regular graph where each vertex has degree 4 and there are 10 edges?

From Theorem 14.2, $2|E| = 20 = 4|V|$, so we have five vertices of degree 4. Figure 14.25 provides two nonisomorphic examples that satisfy the requirements.

Figure 14.25

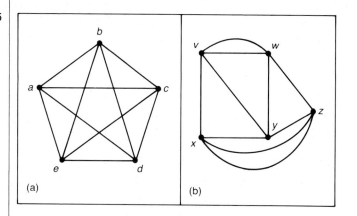

(a) (b)

If we want each vertex to have degree 4, with 15 edges in the graph, we find that $2|E| = 30 = 4|V|$, from which it follows that no such graph is possible. □

We turn now to the reason Euler developed the idea of the degree of a vertex, namely, the solution of the problem dealing with the seven bridges of Königsberg.

Figure 14.26

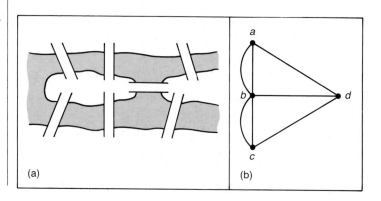

(a) (b)

EXAMPLE 14.7   (*The Seven Bridges of Königsberg*) During the eighteenth century the city of Königsberg (in East Prussia) was divided into four sections (including the island of Kneiphof) by the Pregel river. Seven bridges connected these regions, as shown in Fig. 14.26(a). It was said that residents spent their Sunday walks trying to find a starting point so that they could walk about the city, cross each bridge exactly once, and return to the starting point.

In order to determine whether or not such a cycle existed, Euler represented the four sections of the city and the seven bridges by the multigraph in Fig. 14.26(b). Here he found four vertices with $\deg(a) = \deg(c) = \deg(d) = 3$ and $\deg(b) = 5$. He also found that the existence of such a cycle depended on the number of vertices of odd degree in the graph.   □

Before proving the general result we give the following definition.

Definition 14.10   ▶ Let $G = (V, E)$ be an undirected graph or multigraph. G is said to have an *Euler cycle* if there is a cycle in G that passes through every $v \in V$ and traverses every edge of the graph exactly once. If there is a path from a to b in G and this path passes through every vertex of G and traverses each edge in G exactly once the path is called an *Euler path*.

The problem of the seven bridges is now settled as we characterize the graphs that have an Euler cycle.

Theorem 14.3   ▶ Let $G = (V, E)$ be an undirected graph or multigraph. Then G has an Euler cycle iff G is connected and every vertex in G has even degree.

Proof   If G has an Euler cycle, then for any $a, b \in V$ there is a path from a to b, namely that part of the cycle that starts at a and terminates at b. Hence G is connected.

Let s be the starting vertex of the Euler cycle. For any other vertex v of G, each time the cycle comes to v it then departs from the vertex. Thus the cycle has traversed either two (new) edges that are incident with v or a (new) loop at v. In either case a count of 2 is contributed to $\deg(v)$. Since v is not the starting point, a count of 2 is obtained each time the cycle passes through v, so $\deg(v)$ is even. As for the starting vertex s, the first edge of the cycle must be distinct from the last edge, and as any other visit to s results in a count of 2 for $\deg(s)$, we have $\deg(s)$ even.

Conversely, let G be connected with every vertex of even degree. If the number of edges in G is 1 or 2, then G must be as shown in Fig. 14.27. Euler cycles are immediate in these cases. We proceed now by induction and assume the result true for

Figure 14.27

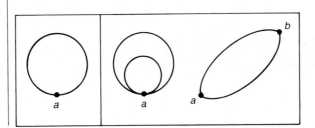

all situations where there are fewer than n edges. If G has n edges, select a vertex s in G as a starting point to build an Euler cycle. Since G is connected and each vertex has even degree, we can at least construct a cycle C on s. (Verify this.) Should the cycle contain every edge of G, we are finished. If not, remove the edges of the cycle from G, making sure to remove any vertex that would become isolated. The remaining subgraph K has all vertices of even degree, but it may not be connected. However, each component of K is connected and will have an Euler cycle. (Why?) In addition, each of these Euler cycles has a vertex that is on C. Consequently, starting at s we travel on C until we arrive at a vertex s_1 that is on the Euler cycle of a component C_1 of K. Then we traverse this Euler cycle and, returning to s_1, continue on C until we reach a vertex s_2 that is on the Euler cycle of component C_2 of K. Since the graph G is finite, as we continue this process we construct an Euler cycle for G.   ∎

Should G be connected and not have too many vertices of odd degree, we can at least find an Euler path in G.

Corollary 14.2

▶ If G is an undirected graph or multigraph, we can construct an Euler path in G iff G is connected and has only two vertices of odd degree.

Proof

If G is connected and a and b are the vertices of G that have odd degree, add an additional edge $\{a, b\}$ to G. We now have a graph G_1 that is connected with every vertex of even degree. Hence G_1 has an Euler cycle C and when the edge $\{a, b\}$ is removed from C we obtain an Euler path for G. (Thus the Euler path starts at one of the vertices of odd degree and terminates at the other odd vertex.) We leave the details of the converse for the reader.   ∎

Returning now to the seven bridges of Königsberg, Figure 14.26(b) is a connected multigraph, but it has four vertices of odd degree. Consequently there is no Euler path or cycle for this situation.

EXERCISES

1. Determine $|V|$ for the following graphs or multigraphs, G.
   a) G has nine edges and all vertices have degree 3.
   b) G is regular with 15 edges.
   c) G has 10 edges with two vertices of degree 4 and all others of degree 3.

2. If $G = (V, E)$ is a connected graph with $|E| = 17$ and $\deg(v) \geq 3$ for all $v \in V$, what is the maximum value for $|V|$?

3. If G is an undirected graph with n vertices and e edges, let $\delta = \min_{v \in V}\{\deg(v)\}$ and $\Delta = \max_{v \in V}\{\deg(v)\}$. Prove that $\delta \leq 2(e/n) \leq \Delta$.

4. Let $G = (V, E)$, $H = (V', E')$ be undirected graphs with $f: V \to V'$ establishing an isomorphism between the graphs.
   a) Prove that $f^{-1}: V' \to V$ is also an isomorphism for G and H.
   b) If $a \in V$, prove that $\deg(a)$ (in G) $= \deg(f(a))$ (in H).

   c) Prove that vertex a is isolated (pendant) in G iff $f(a)$ is isolated (pendant) in H.

5. Complete the proofs of Corollaries 14.1 and 14.2.

6. a) Find an Euler cycle for the graph in Fig. 14.28.

   b) If the edge $\{d, e\}$ is removed from this graph, find an Euler path for the resulting subgraph.

Figure 14.28

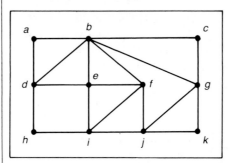

7. Determine the value(s) of n for which the complete graph K_n has an Euler cycle. For which n does K_n have an Euler path but not an Euler cycle?

8. For the graph in Fig. 14.26(b), what is the smallest number of bridges that must be removed so that the resulting subgraph has an Euler path but not an Euler cycle? Which bridge(s) should we remove?

9. When visiting a chamber of horrors, Paul and David try to figure out whether they can travel through the seven rooms and surrounding corridor of the attraction without passing through any door more than once. If they must start from the starred position in the corridor shown in Fig. 14.29, can they accomplish their goal?

Figure 14.29

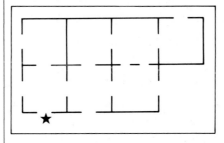

10. Let G be a loop-free undirected graph on n (≥ 3) vertices. If G has only one vertex of even degree, how many vertices in $\overline{G}$ have even degree?

11. If $G = (V, E)$ is a directed graph or multigraph, we define the *incoming*, or *in*, *degree* of a vertex $v \in V$ as the number of edges in G that are incident into v. The number of edges that are incident from v determines the *outgoing*, or *out*, *degree*

of v. We denote these numbers by $\deg^+(v)$ and $\deg^-(v)$, respectively.

   a) Prove that in any directed graph or multigraph $G = (V, E)$, $\sum_{v \in V} \deg^+(v) = \sum_{v \in V} \deg^-(v)$.

   b) Let G be a directed graph on n vertices. If the associated undirected graph for G is K_n, prove that $\sum_{v \in V} [\deg^+(v)]^2 = \sum_{v \in V} [\deg^-(v)]^2$.

12. a) Prove that a directed graph or multigraph $G = (V, E)$ has a directed Euler cycle iff G is connected and $\deg^+(v) = \deg^-(v)$ for all $v \in V$.

   b) If $G = (V, E)$ is a directed graph or multigraph, prove that G has a directed Euler path iff (i) G is connected; (ii) $\deg^+(v) = \deg^-(v)$ for all but two vertices x, y in V; and (iii) $\deg^+(x) = \deg^-(x) + 1$, $\deg^-(y) = \deg^+(y) + 1$.

   c) Find a directed Euler cycle for the graph in Fig. 14.30.

Figure 14.30

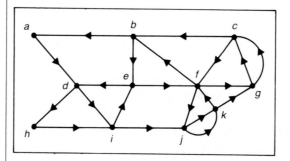

   d) A directed graph is called *strongly connected* if there is a directed path from a to b for all vertices a, b. Prove that if a directed graph has a directed Euler cycle, then it is strongly connected. Is the converse true?

13. Let $V = \{000, 001, 010, \ldots, 110, 111\}$. For any four-bit sequence $b_1 b_2 b_3 b_4$ draw an edge from the element $b_1 b_2 b_3$ to the element $b_2 b_3 b_4$ in V.

   a) Draw the graph $G = (V, E)$ as described.

   b) Find a directed Euler cycle for G.

   c) Equally space eight 0's and eight 1's around the edge of a rotating (clockwise) disk so that these 16 bits form a circular sequence where the (consecutive) subsequences of length 4 provide the binary representations of 0, 1, 2, $\ldots$, 14, 15, in some order.

14. If $G = (V, E)$ is an undirected graph with $|V| = n$, $|E| = k$, the following matrices are used to represent G.

Let $V = \{v_1, v_2, \ldots, v_n\}$. Define the *adjacency matrix* $A = (a_{ij})_{n \times n}$ where $a_{ij} = 1$ if $\{v_i, v_j\} \in E$, otherwise $a_{ij} = 0$.

If $E = \{e_1, e_2, \ldots, e_k\}$, the *incidence matrix* I is the $n \times k$ matrix $(b_{ij})_{n \times k}$ where $b_{ij} = 1$ if v_i is a vertex on the edge e_j, otherwise $b_{ij} = 0$.

   a) Find the adjacency and incidence matrices associated with the graph in Fig. 14.31.

Figure 14.31

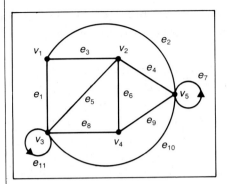

b) Calculating A^2 and using the Boolean operations where $0 + 0 = 0, 0 + 1 = 1 + 0 = 1 + 1 = 1, 0 \cdot 0 = 0 \cdot 1 = 1 \cdot 0 = 0, 1 \cdot 1 = 1$, prove that the entry in row i and column j of A^2 is 1 iff there is a path of length two between the ith and jth vertices of V.

c) If we calculate A^2 using ordinary addition and multiplication, what do the entries in the matrix reveal about G?

d) What is the column sum for each column of A? Why?

e) What is the column sum for each column of I? Why?

14.4 ■ PLANAR GRAPHS

On a road map it is usually the case that the lines indicating the roads and highways intersect only at junctions or towns. Sometimes roads seem to intersect when one road is located above another, as in the case of an overpass. In this case the two roads are at different levels, or planes.

Definition
14.11

▶ A graph G is called *planar* if G can be drawn in the plane with its edges intersecting only at vertices of G.

EXAMPLE 14.8   The graphs in Fig. 14.32 are planar. The first is a 3-regular graph, since each vertex has degree 3; it is planar since no edges intersect except at the vertices. In graph (b)

Figure 14.32

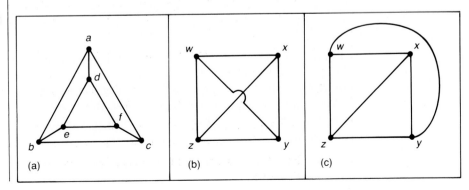

(a)   (b)   (c)

it appears that we have a *nonplanar* graph since the edges $\{x, z\}$ and $\{w, y\}$ overlap at a point other than a vertex. However, we can redraw this graph as in (c) of the figure. Consequently K_4 is planar.   □

EXAMPLE 14.9   Just as K_4 is planar, so are the graphs K_1, K_2, K_3.

Figure 14.33

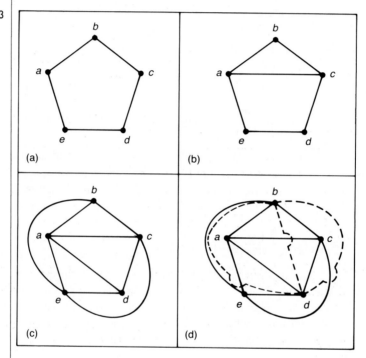

(a)

(b)

(c)

(d)

The construction of K_5 is shown in Fig. 14.33. Start with the pentagon in part (a) of the figure. Since a complete graph contains an edge for every pair of distinct vertices, we add edge $\{a, c\}$ as shown in (b). This edge is contained entirely within the interior of the pentagon in (a). (We could have drawn the edge in the exterior region determined by the pentagon. The reader will be asked in the exercises to show that the same conclusion arises in this case.) Continuing to part (c), we add in the edges $\{a, d\}$, $\{c, e\}$, and $\{b, e\}$. Now we consider the vertices b and d. The edge $\{b, d\}$ is needed in order to have K_5. Vertex d is inside the region formed by the cycle edges $\{a, c\}$, $\{c, e\}$, and $\{e, a\}$, while b is outside of the region; thus in drawing the edge $\{b, d\}$ we must intersect one of the existing edges at least once (as shown by the dotted edges in (d)). Consequently, K_5 is nonplanar. (Since this proof appeals to a diagram, the reader may feel that it lacks rigor. We will prove K_5 nonplanar by another method, later in this section.)   □

Before we can characterize all nonplanar graphs we need to examine another class of graphs.

Definition 14.12

▶ A graph $G = (V, E)$ is called *bipartite* if $V = V_1 \cup V_2$ with $V_1 \cap V_2 = \emptyset$, and every edge of G is of the form $\{a, b\}$ with one of the vertices a, b in V_1 and the other vertex in V_2. If each vertex in V_1 is joined with every vertex in V_2 we have a *complete bipartite* graph; when $|V_1| = m$, $|V_2| = n$, the graph is denoted by $K_{m,n}$.

Figure 14.34

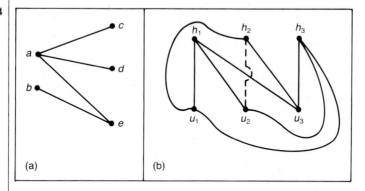

(a) (b)

Figure 14.34 shows two bipartite graphs. The graph in (a) satisfies the definition for $V_1 = \{a, b\}$, $V_2 = \{c, d, e\}$. If we add the edges $\{b, d\}$ and $\{b, c\}$ the result is the complete bipartite graph $K_{2,3}$, which is planar. Graph (b) of the figure is $K_{3,3}$. Let $V_1 = \{h_1, h_2, h_3\}$ and $V_2 = \{u_1, u_2, u_3\}$, and interpret V_1 as a set of houses and V_2 as a set of utilities. Then $K_{3,3}$ is called the *utility graph*. Can we hook up the houses with each of the utilities and avoid having overlapping utility lines? In Fig. 14.34(b) it appears that this is not possible and that $K_{3,3}$ is nonplanar. (Once again we deduce the nonplanarity of a graph from a diagram. However, we shall verify $K_{3,3}$ nonplanar by another method, later in the section.)

We shall see that when dealing with nonplanar graphs either K_5 or $K_{3,3}$ will be the source of the problem. Before stating the general result however, we need one final new idea.

Definition 14.13

▶ The graphs $G_1 = (V_1, E_1)$, $G_2 = (V_2, E_2)$ are called *homeomorphic* if G_2 can be obtained from G_1 by the insertion or deletion of vertices of degree 2.

The three graphs in Fig. 14.35 are all homeomorphic. One can think of homeomorphic graphs as being isomorphic up to vertices of degree 2. Any graph is

Figure 14.35

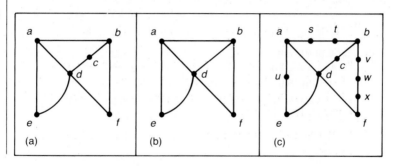

(a) (b) (c)

homeomorphic to itself. (Insert a vertex of degree 2 and then delete it.) Finally, if two graphs are homeomorphic they are simultaneously planar (or nonplanar).

These preliminaries lead us to the following result.

Theorem 14.4

▶ (*Kuratowski's Theorem*) A graph is nonplanar iff it contains a subgraph that is homeomorphic to either K_5 or $K_{3,3}$.

Proof

(This theorem was first proved by the Polish mathematician Kasimir Kuratowski in 1930.) If a graph G has a subgraph homeomorphic to either K_5 or $K_{3,3}$, it is clear that G is nonplanar. The converse of this theorem, however, is much more difficult to prove. (A proof can be found in Chapter 8 of C. L. Liu [13].) ∎

We demonstrate the use of Kuratowski's Theorem in the following.

Figure 14.36

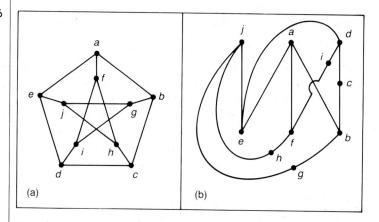

(a)   (b)

EXAMPLE 14.10

Figure 14.36(a) is a familiar graph called the *Petersen graph*. Part (b) of the figure provides a subgraph of the Petersen graph that is homeomorphic to $K_{3,3}$. (Deletion of vertices c, g, h, i leaves $K_{3,3}$.) Hence the Petersen graph is nonplanar. □

When a graph is planar and connected we find the following relation, due to Euler.

Theorem 14.5

▶ Let $G = (V, E)$ be a connected planar graph with $|V| = v$, $|E| = e$. Let r be the number of regions in the plane determined by G; one of these regions has infinite area and is called *the infinite region*. Then $v - e + r = 2$.

Proof

The proof is by induction on e. If $e = 1$, then G is isomorphic to one of the graphs in Fig. 14.37. For graph (a), $v = 1$, $e = 1$, and $r = 2$. Graph (b) has $v = 2$, $e = 1$, and $r = 1$. In both cases $v - e + r = 2$.

Assume the result is true for all connected planar graphs with k edges. Let G be connected and planar with v vertices, r regions, and $e = k + 1$ edges. We consider two cases.

Figure 14.37

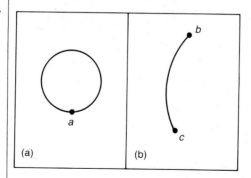

(a) (b)

Case 1: G has a pendant vertex a, i.e., $\deg(a) = 1$. Let H be the subgraph of G obtained by deleting vertex a and the edge incident with it. Then H has $v - 1$ vertices, k edges, and r regions. By the induction hypothesis applied to H, $(v - 1) - k + r = 2$. Hence $[(v - 1) + 1] - (k + 1) + r = 2 = v - e + r$, so Euler's theorem holds for G.

Case 2: If G has no pendant vertex, let $\{a, b\}$ be an edge on the boundary of a finite region, i.e., a region that encloses a finite area. (Such a region must exist. If not, G is a path with v vertices and e edges, with $e = v - 1$, and G has a pendant vertex.) Removing edge $\{a, b\}$ from G yields a subgraph H with v vertices, $e - 1$ edges and $r - 1$ regions. Considering graph H, the induction hypothesis implies that $v - (e - 1) + (r - 1) = 2$, from which we have $v - e + r = 2$. Thus Euler's theorem follows for G. ∎

Corollary 14.3

▶ Let $G = (V, E)$ be a loop-free connected planar graph with $|V| = v$, $|E| = e > 2$, and r regions. Then $3r \le 2e$ and $e \le 3v - 6$.

Proof Since G is loop-free, and not a multigraph, the boundary of each region (including the infinite region) contains at least three edges. Each edge is used twice as part of the boundary of a region, so $2e \ge 3r$. From Euler's theorem $2 = v - e + r \le v - e + (2/3)e = v - (1/3)e$, so $6 \le 3v - e$, or $e \le 3v - 6$. ∎

We now consider what this corollary does and does not imply. If $G = (V, E)$ is a loop-free connected graph with $|E| > 2$, then if $e > 3v - 6$, G is not planar. However, if $e \le 3v - 6$ we cannot conclude that G is planar.

EXAMPLE 14.11 The graph K_5 is loop-free and connected with 10 edges and five vertices. Consequently, $3v - 6 = 15 - 6 = 9 < 10 = e$, and K_5 is nonplanar. □

EXAMPLE 14.12 The graph $K_{3,3}$ is loop-free and connected with nine edges and six vertices. Here $3v - 6 = 18 - 6 = 12 \ge 9 = e$, so we cannot conclude that $K_{3,3}$ is planar unless we make the mistake of arguing by the converse. But we did claim earlier that $K_{3,3}$ is nonplanar.

If $K_{3,3}$ were planar, since each region in the graph is bounded by at least four edges, we have $4r \le 2e$ as in the proof of Corollary 14.3. From Euler's theorem, $v - e + r = 2$, or $r = e - v + 2 = 9 - 6 + 2 = 5$, so $20 = 4r \le 2e = 18$. From this contradiction we have $K_{3,3}$ nonplanar.	□

Figure 14.38

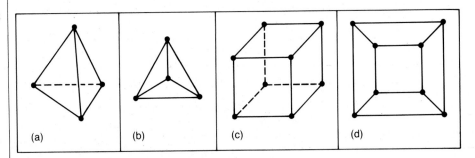

(a)	(b)	(c)	(d)

EXAMPLE 14.13

We use Euler's theorem to characterize the *Platonic solids*. (For these solids all faces are congruent and all (interior) solid angles are equal.) In Fig. 14.38 we have two of these solids. Part (a) of the figure shows the regular tetrahedron, which has four faces, each an equilateral triangle. Concentrating on the edges of the tetrahedron, we focus on its underlying framework. As we view this framework from a point directly above the center of one of the faces we picture the planar representation in (b). This planar graph determines four regions (corresponding to the four faces); three regions meet at each of the four vertices. Part (c) of the figure provides another Platonic solid, the cube. Its associated planar graph is given in (d). In this graph there are six regions with three regions meeting at each vertex.

Based on our observations for the regular tetrahedron and cube, we shall determine the other Platonic solids by means of their associated planar graphs. In these graphs $G = (V, E)$ we have $v = |V|$; $e = |E|$; $r =$ the number of planar regions determined by G; $m =$ the number of edges in the boundary of each region; and $n =$ the number of regions that meet at each vertex. Thus the constants $m, n \ge 3$. Since each edge is used in the boundary of two regions and there are r regions, each with m edges, it follows that $2e = mr$. Counting the endpoints of the edges we get $2e$. But all these endpoints can also be counted by considering what happens at each vertex. Since n regions meet at each vertex, n edges meet there, so there are n endpoints of edges to count at each of the v vertices. This totals nv endpoints of edges, so $2e = nv$. From Euler's theorem we have

$$0 < 2 = v - e + r = \frac{2e}{n} - e + \frac{2e}{m} = e\left(\frac{2m - mn + 2n}{mn}\right).$$

With $e, m, n > 0$, $2m - mn + 2n > 0 \Rightarrow mn - 2m - 2n < 0 \Rightarrow mn - 2m - 2n + 4 < 4 \Rightarrow (m - 2)(n - 2) < 4$.

Since $m, n \ge 3$, we have $(m - 2), (n - 2) \in \mathbf{Z}^+$, and there are five cases to consider:

1. $(m - 2) = (n - 2) = 1; m = n = 3$ (The regular tetrahedron)
2. $(m - 2) = 2, n - 2 = 1; m = 4, n = 3$ (The cube)
3. $(m - 2) = 1, n - 2 = 2; m = 3, n = 4$ (The octahedron)
4. $(m - 2) = 3, n - 2 = 1; m = 5, n = 3$ (The dodecahedron)
5. $(m - 2) = 1, n - 2 = 3; m = 3, n = 5$ (The icosahedron).

The planar graphs for cases 3–5 are shown in Fig. 14.39.   ☐

Figure 14.39

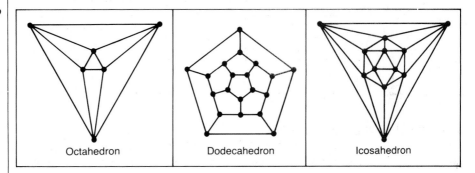

Octahedron   Dodecahedron   Icosahedron

The last idea we shall discuss for planar graphs is the notion of a *dual* graph. This concept is also valid for planar graphs with loops and planar multigraphs. To construct a dual for a planar graph or multigraph G with $V = \{a, b, c, d, e, f\}$, place a point (vertex) inside each region, including the infinite region, determined by the graph. For each edge shared by two regions, draw an edge connecting the vertices inside these regions. In Fig. 14.40 G^d is a dual for the graph $G = (V, E)$. From this example we make the following observations:

1. A vertex of degree 2 in G yields a pair of edges in G^d that connect two vertices. Hence G^d may be a multigraph. (Here vertex e provides the edges $\{a, e\}, \{e, f\}$ in G that brought about the two edges connecting v and z in G^d.)

Figure 14.40

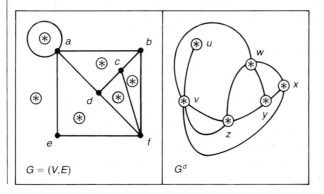

$G = (V, E)$ G^d

2. A loop in G yields a pendant vertex in G^d. (It is also true that a pendant vertex in G yields a loop in G^d.)

3. The degree of a vertex in G^d is the number of edges in the boundary of the region in G that contains that vertex.

(Why is G^d called *a* dual of G instead of *the* dual of G? The section exercises will show that it is possible to have isomorphic graphs G_1 and G_2 with respective duals G_1^d, G_2^d that are not isomorphic.)

In order to examine further the relationship between a graph G and a dual G^d of G we introduce the following.

Definition 14.14

▶ Let $G = (V, E)$ be an undirected graph or multigraph. A subset E' of E is called a *cut-set* of G if by removing the edges (but not the vertices) in E' from G we increase the number of components of G, and we cannot accomplish this for any proper subset of E'.

Figure 14.41

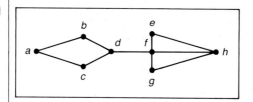

EXAMPLE 14.14

In the graph in Fig. 14.41, note that each of the sets $\{\{a, b\}, \{a, c\}\}$, $\{\{a, b\}, \{c, d\}\}$, $\{\{e, h\}, \{f, h\}, \{g, h\}\}$, and $\{\{d, f\}\}$ is a cut-set.   □

We return now to the graphs in Fig. 14.40, redrawing them as in Fig. 14.42 in order to emphasize relationships between their edges.

Figure 14.42

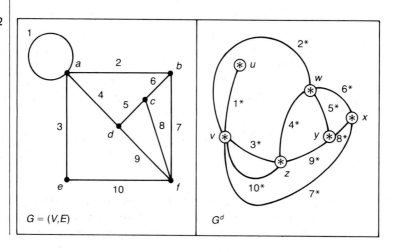

Here the edges in G are labeled 1, 2, ..., 10. The numbering scheme for G^d is obtained as follows. The edge labeled 4*, for example, connects the vertices w and z in G^d. We drew this edge because edge 4 in G was a common edge of the regions containing these vertices. Likewise, edge 7 is common to the region containing x and the infinite region containing v. Hence we label the edge in G^d that connects x and v with 7*.

In graph G the set of edges labeled 6, 7, 8 constitute a simple cycle. What about the edges labeled 6*, 7*, 8* in G^d? If they are removed from G^d, then vertex x becomes isolated and G^d is disconnected. Thus these edges form a cut-set in G^d. In similar fashion edges 2, 4, 10 form a cut-set G in while in G^d the edges 2*, 4*, 10* yield a simple cycle.

In general, there is a one-to-one correspondence between the simple cycles (cut-sets) of a planar graph G and the cut-sets (simple cycles) of a dual G^d of G.

All these theoretical observations are interesting, but let us stop here and see how we might apply the idea of a dual.

EXAMPLE 14.15   If we consider the five finite regions in Fig. 14.43(a) as countries on a map and we construct the subgraph (since we do not use the infinite region) of a dual as shown in part (b), then we find the following relationship.

Figure 14.43

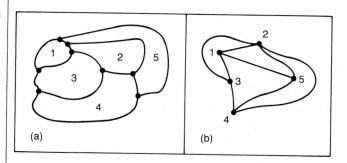

(a)   (b)

Suppose we want to color the five regions of the map in (a) so two countries that share a common border are colored with different colors. This idea can be translated into the dual notion of coloring the vertices in (b) so that adjacent vertices are colored with different colors. (Such coloring problems will be examined further in Section 14.6.)   □

The final result for this section provides us with an application for an electrical network. This material is based on Example 8.6 on pp. 227–230 of the text by C. L. Liu [13].

EXAMPLE 14.16   In Fig. 14.44 we see an electrical network with nine contacts (switches) that control the excitation of a light. We want to construct a dual network where a second light will go on (off) whenever the light in our given network is off (on).

Figure 14.44

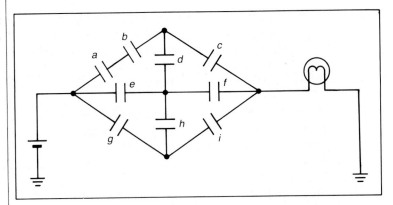

The contacts (switches) are of two types: normally open (as shown in Fig. 14.44) or normally closed. We use a and a' as in Fig. 14.45 to represent the normally open and closed contacts, respectively.

Figure 14.45

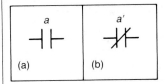

In Fig. 14.46(a) a *one-terminal-pair-graph* represents the network in Fig. 14.44. Here the special pair of vertices are labeled 1 and 2, and are called the *terminals* of the graph. Also each edge is labeled according to its corresponding contact in Fig. 14.44.

A one-terminal-pair-graph G is called a *planar-one-terminal-pair-graph* if G is planar, and the resulting graph is also planar when an edge connecting the terminals is added to G. Figure 14.46(b) shows this situation. Constructing a dual of (b) we obtain the graph in (c) of the figure. Removal of the dotted edge results in the terminals 1*, 2* for this dual, which is a one-terminal-pair-graph. This graph provides the dual network in Fig. 14.46(d).

We make two observations in closing.

1. When the contacts at a, b, c are closed in the original network (Fig. 14.44), the light is on. In Fig. 14.46(b) the edges a, b, c, j are a simple cycle that includes the terminals. In part (c) of the figure a^*, b^*, c^*, j^* form a cut-set disconnecting the terminals 1*, 2*. Lastly, with a', b', c' open in (d) of the figure, no current gets past the first level of contacts (switches) and the light is off.

2. In like manner, the edges c, f, i, j form a cut-set that separates the terminals in Fig. 14.46(b). (When the contacts at c, f, i are open in Fig. 14.44 the light is off.) Figure 14.46(c) shows how c^*, f^*, i^*, j^* form a simple cycle that connects 1*, 2*. If c', f', i' are closed in (d), current flows through the dual network and the light is on.   □

Figure 14.46

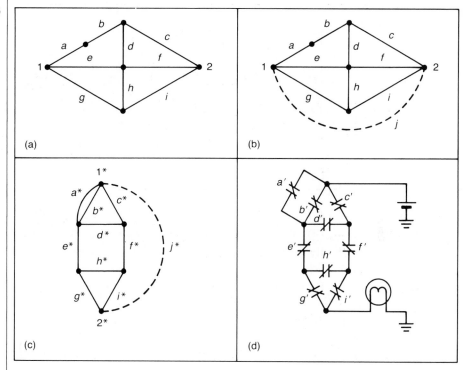

(a)   (b)   (c)   (d)

EXERCISES

1. Verify that the conclusion in Example 14.9 is unchanged if Fig. 14.33(b) has edge $\{a, c\}$ drawn in the exterior of the pentagon.

2. Show that if any edge is removed from K_5 the resulting subgraph is planar. Is this true for the graph $K_{3,3}$?

3. a) How many vertices and how many edges are there in the complete bipartite graphs $K_{4,7}$, $K_{7,11}$, $K_{m,n}$?

   b) If the graph $K_{m,12}$ has 72 edges, what is m?

4. Can a bipartite graph contain a cycle of odd length? Explain.

5. Let $G = (V, E)$ be a loop-free connected graph with $|V| = v$. If $|E| > (v/2)^2$ prove that G cannot be bipartite.

6. Determine whether or not the graphs in Fig. 14.47 are planar. If planar, redraw the graph with no edges overlapping. If nonplanar, find a subgraph homeomorphic to either K_5 or $K_{3,3}$.

7. Let $G = (V, E)$ be a loop-free connected planar graph with $\deg(v) = 4$, for all $v \in V$. If $|E| = 16$, how many regions are there in a planar depiction of G?

8. Suppose that $G = (V, E)$ is a loop-free planar graph with $|V| = v$, $|E| = e$, and $\omega(G)$ = the number of connected components of G. (a) State and prove an

Figure 14.47

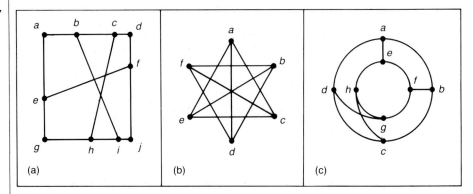

extension of Euler's theorem for such a graph. (b) Prove that Corollary 14.3 remains valid for G loop-free and planar, but not connected.

9. If G is a loop-free connected planar graph, prove that there is a vertex v of G with $\deg(v) < 6$.

10. Let $G = (V, E)$ be a loop-free connected graph with $|V| \geq 11$. Prove that either G or its complement $\overline{G}$ must be nonplanar.

11. a)   Let $k \in \mathbf{Z}^{+}$, $k \geq 3$. If $G = (V, E)$ is a connected planar graph with $|V| = v$,

$|E| = e$, and each cycle of length at least k, prove that $e \leq \left(\dfrac{k}{k-2} \right)(v-2)$.

   b)   What is the minimal cycle length in $K_{3,3}$?

   c)   Use parts (a) and (b) to conclude that $K_{3,3}$ is nonplanar.

   d)   Use part (a) to prove that the Petersen graph is nonplanar.

12. Find a dual graph for each of the planar graphs in Fig. 14.48.

Figure 14.48

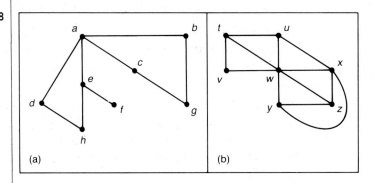

13. Find duals for the planar graphs that correspond with the five Platonic solids.

14. a)   Show that the graphs in Fig. 14.49 are isomorphic.

   b)   Draw a dual for each graph.

   c)   Show that the duals obtained in part (b) are not isomorphic.

Figure 14.49

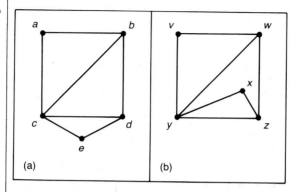

(a)   (b)

d) Two graphs G and H are called *2-isomorphic* if one can be obtained from the other by applying either or both of the following procedures a finite number of times.

1. In Fig. 14.50 we split a vertex, namely r, of G and obtain the graph H, which is disconnected. (When the splitting, or separation, of a vertex in a graph results in a new graph with more components, that vertex is called a *cut-vertex* or *articulation point*.)

Figure 14.50

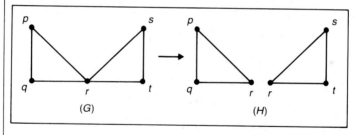

(G)   (H)

2. In Fig. 14.51 (next page) we obtain graph (d) from (a) by
   i) first splitting the graph at two distinct vertices j and q;
   ii) then reflecting one subgraph about the horizontal axis; and
   iii) then joining vertex $j(q)$ in one subgraph with vertex $q(j)$ in the other subgraph.

   Prove that the dual graphs obtained in part (c) are 2-isomorphic.

e) For the cut-set $\{\{a, b\}, \{c, b\}, \{d, b\}\}$ in part (a) of Fig. 14.49, find the corresponding simple cycle in its dual. In the dual of the graph in Fig. 14.49(b), find the cut-set that corresponds with the cycle $\{w, z\}, \{z, x\}, \{x, y\}, \{y, w\}$ in the given graph.

15. For the graph in Fig. 14.51(a) find two cut-sets that contain the edge $\{k, m\}$.

16. a) If $G = (V, E)$ is a loop-free 3-regular connected graph with 8 vertices, what is $|E|$?

    b) Find two graphs G_1 and G_2 that satisfy the conditions in part (a) with G_1 planar and G_2 nonplanar.

Figure 14.51

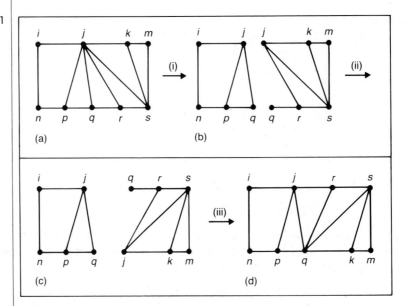

17. Find the dual network for the electrical network in Fig. 14.52.

Figure 14.52

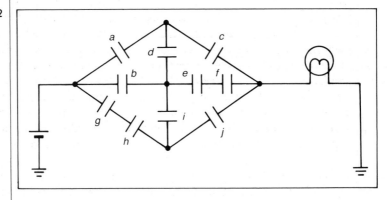

14.5 ■ HAMILTON PATHS AND CYCLES

In 1859 the Irish mathematician Sir William Hamilton (1805–1865) developed a game that he sold to a Dublin toy manufacturer. The game consisted of a wooden regular dodecahedron with the 20 corner points (vertices) labeled with the names of prominent cities. The objective of the game was to find a simple cycle along the edges of the solid so that each city was on the cycle exactly once. Figure 14.53 is the planar graph for this Platonic solid; such a cycle is designated by the darkened edges. This illustration leads us to the following.

Figure 14.53

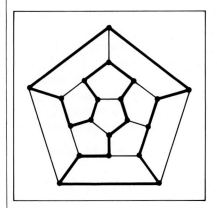

Definition 14.15

▶ If $G = (V, E)$ is a graph or multigraph, we say that G has a *Hamilton cycle* if there is a simple cycle in G that contains every vertex in V. A *Hamilton path* is a simple path in G that contains each vertex.

Given a graph with a Hamilton cycle, the deletion of any edge in the cycle results in a Hamilton path. It is possible, however, for a graph to have a Hamilton path without having a Hamilton cycle.

It may seem that the existence of a Hamilton cycle (path) and the existence of an Euler cycle (path) for a graph are similar problems. The Hamilton cycle (path) is designed to visit each vertex in a graph only once; the Euler cycle (path) traverses the graph so that each edge is traveled exactly once. Unfortunately, there is no helpful connection between the two ideas, and unlike the situation for Euler cycles (paths), there do not exist necessary and sufficient conditions on a graph G that guarantee the existence of a Hamilton cycle (path). If a graph has a Hamilton cycle, then it will at least be connected. Many theorems exist that establish either necessary or sufficient conditions for a connected graph to have a Hamilton cycle or path. We shall investigate two of these results later. When confronted with particular graphs we shall often resort to trial and error, with a few helpful observations.

EXAMPLE 14.17

If G is the graph in Fig. 14.54, the edges $\{a, b\}$, $\{b, c\}$, $\{c, f\}$, $\{f, e\}$, $\{e, d\}$, $\{d, g\}$, $\{g, h\}$, $\{h, i\}$ yield a Hamilton path for G. But does G have a Hamilton cycle?

Figure 14.54

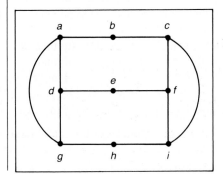

Since G has nine vertices, if there is a Hamilton cycle in G it must contain nine edges. Let us start at vertex b and try to build a Hamilton cycle. With regard to the symmetry in the graph it doesn't matter whether we go from b either to c or to a. We'll go to c. At c we can go either to f or to i. Using symmetry again, we go to f. Then we delete edge $\{c, i\}$ from further consideration since we cannot return to vertex c. In order to include vertex i in our cycle we must now go from f to i (to h to g). With edges $\{c, f\}$, $\{f, i\}$ in the cycle we delete edge $\{e, f\}$. But now once we get to e we are stuck. Hence there is no Hamilton cycle for the graph.   □

This example indicates a few helpful hints for trying to find a Hamilton cycle in a graph $G = (V, E)$.

1. If G has a Hamilton cycle, then for all $v \in V$, $\deg(v) \geq 2$.

2. If $a \in V$ and $\deg(a) = 2$ then the two edges incident with vertex a must appear in any Hamilton cycle for G.

3. If $a \in V$ and $\deg(a) > 2$, then as we try to build a Hamilton cycle, once we pass through vertex a, any unused edges incident with a are deleted from consideration.

4. In building a Hamilton cycle for G we cannot obtain a simple cycle for a subgraph of G unless it contains all the vertices of G.

Our next example provides an interesting technique for trying to decide whether or not a particular graph has a Hamilton path.

Figure 14.55

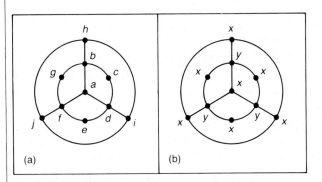

(a)   (b)

EXAMPLE 14.18

In Fig. 14.55(a) we have a connected graph G and we wish to know if G contains a Hamilton path. Part (b) of the figure provides the same graph with a set of labels x, y. This labeling was accomplished as follows. First we label vertex a with the letter x. Those vertices adjacent to a, namely b, d, and f, are then labeled with the letter y. Then we label the unlabeled vertices adjacent to b, d, and f with x. At this point all vertices in G are labeled. Now if G is to have a Hamilton path there must be an alternating sequence of five x's and five y's. (Why?) Since only three vertices are labeled with y, this is impossible. Hence G has no Hamilton path (or cycle).   □

Our next example provides an application that calls for Hamilton cycles in a complete graph.

EXAMPLE 14.19

At Professor Alfred's science camp, 17 students have lunch together each day at a circular table. Since they are trying to get to know each other better, they make an effort each afternoon to sit next to two different colleagues. For how many afternoons can they do this? How can they arrange themselves on these occasions?

In order to answer this problem we consider the graph K_n, where $n \geq 3$ and is odd. This graph has n vertices and $\binom{n}{2} = n(n-1)/2$ edges. A Hamilton cycle in K_n has n edges so we can have at most $(n-1)/2$ Hamilton cycles with no two having an edge in common.

Figure 14.56

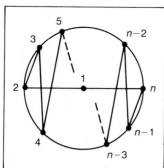

Figure 14.57

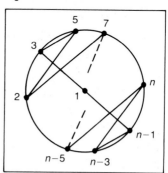

Consider the circle in Fig. 14.56 and the subgraph of K_n consisting of the n vertices and the n edges $\{1, 2\}, \{2, 3\}, \ldots, \{n-1, n\}, \{n, 1\}$. Keep the vertices on the circumference fixed and rotate this Hamilton cycle clockwise through the angle $[1/(n-1)](2\pi)$. This gives us the Hamilton cycle (Fig. 14.57) made up of edges $\{1, 3\}, \{3, 5\}, \{5, 2\}, \{2, 7\}, \ldots, \{n, n-3\}, \{n-3, n-1\}, \{n-1, 1\}$. This Hamilton cycle has no edge in common with the first cycle. If we continue to rotate the cycle in Fig. 14.56 in this way through angles $[k/(n-1)](2\pi)$, $2 \leq k \leq (n-3)/2$, we obtain a set of Hamilton cycles where no two cycles have an edge in common.

Therefore the 17 students at the science camp can dine for $[(17-1)/2] = 8$ days before they must sit next to the same student twice. Using Fig. 14.56 with $n = 17$, eight such possible arrangements can be obtained. □

We turn now to some further results on Hamilton paths.

Theorem 14.6

▶ Let K_n^* be a complete directed graph, i.e., K_n^* has n vertices and for any distinct pair x, y of vertices, at least one of the edges (x, y) or (y, x) is in K_n^*. Such a graph always contains a Hamilton path.

Proof

Let $m \geq 2$ with p_m a simple path containing the $m-1$ edges (v_1, v_2), (v_2, v_3), $\ldots$, (v_{m-1}, v_m). If $m = n$, we're finished. If not, let v be a vertex that doesn't appear in p_m.

If (v, v_1) is an edge in K_n^* we can extend p_m by adjoining this edge. If not, then (v_1, v) must be an edge. Now suppose that (v, v_2) is in the graph. Then we have the larger path: (v_1, v), (v, v_2), (v_2, v_3), $\ldots$, (v_{m-1}, v_m). If (v, v_2) is not an edge in K_n^*, then (v_2, v) must be. As we continue this process there are only two possibilities:

(a) For some $1 \le k \le m - 1$ the edges (v_k, v), (v, v_{k+1}) are in K_n^* and we replace (v_k, v_{k+1}) with this pair of edges; or (b) (v_m, v) is in K_n^* and we add this edge to p_m. Either case results in a simple path p_{m+1} that includes $m + 1$ vertices and has m edges. This process can be repeated until we have such a path on n vertices. ∎

EXAMPLE 14.20 In a round-robin tournament each player plays every other player exactly once. We want to somehow rank the players according to the results of the tournament. Since we could have players a, b, and c where a beats b, b beats c, but c beats a, it is not always possible to have a ranking where a player in a certain position has beaten all of the opponents in later positions. Representing the players by vertices, construct a directed graph G on these vertices by drawing edge (x, y) if x beats y. Then by Theorem 14.6 it is possible to list the players so that each has beaten the next player on the list. □

Theorem 14.7 ▶ Let $G = (V, E)$ be a loop-free graph with $|V| = n$. If $\deg(x) + \deg(y) \ge n - 1$ for all $x, y \in V$, $x \ne y$, then G has a Hamilton path.

Proof First we prove that G is connected. If not, let C_1, C_2 be two components of G, with x, $y \in V$, $x \in C_1$, $y \in C_2$. Let C_i have n_i vertices, $i = 1, 2$. Then $\deg(x) \le n_1 - 1$, $\deg(y) \le n_2 - 1$ and $\deg(x) + \deg(y) \le (n_1 + n_2) - 2 \le n - 2$, contradicting the condition given in the theorem. Consequently, G is connected.

Now we build a Hamilton path for G. For $m \ge 2$, let p_m be the path $\{v_1, v_2\}$, $\{v_2, v_3\}$, ..., $\{v_{m-1}, v_m\}$ of length $m - 1$. (We relabel vertices if necessary.) Such a path exists, since for $m = 2$ all that is needed is one edge. If v_1 is adjacent to any vertex v other than $v_2, v_3, \ldots, v_m$, we add the edge $\{v, v_1\}$ to p_m to get p_{m+1}. The same type of procedure is carried out if v_m is adjacent to a vertex other than $v_1, v_2, \ldots,$ v_{m-1}. If we are able to enlarge p_m to p_n in this way we get a Hamilton path. Otherwise the path p_m: $\{v_1, v_2\}$, ..., $\{v_{m-1}, v_m\}$ has v_1, v_m adjacent only to vertices in p_m, and $m < n$. When this happens we claim that G contains a simple cycle on these vertices. If v_1 and v_m are adjacent, the cycle is $\{v_1, v_2\}$, $\{v_2, v_3\}$, ..., $\{v_{m-1}, v_m\}$, $\{v_m, v_1\}$. If v_1 and v_m are not adjacent, then v_1 is adjacent to a subset S of the vertices in $\{v_2, v_3, \ldots, v_{m-1}\}$. If there is a vertex $v_t \in S$ such that v_m is adjacent to v_{t-1} then we can get the cycle by adding $\{v_1, v_t\}$, $\{v_{t-1}, v_m\}$ to p_m, and deleting $\{v_{t-1}, v_t\}$ as shown in Fig. 14.58. If not, let $|S| = k < m - 1$. Then $\deg(v_1) = k$ and $\deg(v_m) \le (m - 1) - k$, and we have the contradiction $\deg(v_1) + \deg(v_m) \le m - 1 < n - 1$. Hence there is a simple cycle connecting $v_1, v_2, \ldots, v_m$.

Now consider a vertex $v \in V$ that is not found on this cycle. Since G is connected, there is a path from v to a first vertex v_r in the cycle, as in Fig. 14.59(a).

Figure 14.58

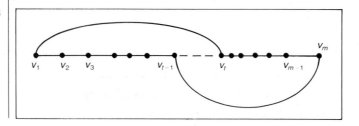

Figure 14.59

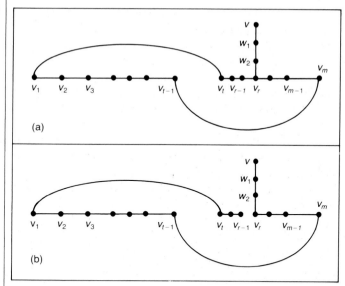

(a)

(b)

Removing the edge $\{v_{r-1}, v_r\}$ (or $\{v_1, v_t\}$ if $r = t$) we get the path (longer than the original p_m) shown in Fig. 14.59(b). Applying the process applied to p_m to the path in Fig. 14.59(b), we continue to increase the length of the path until it includes every vertex of G. ∎

Corollary 14.4

▶ Let $G = (V, E)$ be a loop-free graph with n vertices. If $\deg(v) \geq (n - 1)/2$ for all $v \in V$, then G has a Hamilton path.

Proof

The proof is left as an exercise for the reader. ∎

A problem that is related to the search for Hamilton cycles in a graph is the *traveling salesman problem*. Here a traveling salesman leaves his home and visits certain locations before returning. The objective is to find a way to make his trip most efficient (perhaps, in terms of total distance traveled or total cost). The problem can be modeled with a labeled (edges have distances or costs associated with them) graph where the most efficient Hamilton cycle is sought. Unfortunately, no general method presently exists for solving such a problem. The references by Bellman, Cooke, and Lockett [3], Bellmore and Nemhauser [4], Elsayed [9], and Elsayed and Stern [9] should prove interesting to the reader who wants to learn more about this important optimization problem.

EXERCISES

1. Give an example of a connected graph that has:
   a) Neither an Euler cycle nor a Hamilton cycle.
   b) An Euler cycle but no Hamilton cycle.

c)   A Hamilton cycle but no Euler cycle.

d)   Both a Hamilton cycle and an Euler cycle.

2. Characterize the type of graph in which an Euler path (cycle) is also a Hamilton path (cycle).

3. Find a Hamilton cycle, if one exists, for each of the graphs or multigraphs in Fig. 14.60. If the graph has no Hamilton cycle, determine if it has a Hamilton path.

Figure 14.60

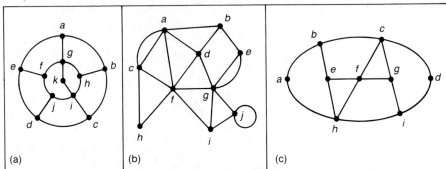

(a)   (b)   (c)

4. a)   Show that the Petersen graph (Fig. 14.36(a)) contains no Hamilton cycle but that it has a Hamilton path.

  b)   Show that if any vertex (and the edges incident to it) is removed from the Petersen graph, then the resulting subgraph has a Hamilton cycle.

5. a)   For $n \geq 3$, how many different Hamilton cycles are there in the complete graph K_n?

  b)   How many edge-disjoint Hamilton cycles are there in K_{21}?

  c)   Nineteen students in a nursery school play a game each day where they hold hands to form a circle. For how many days can they do this with no student holding hands with the same playmate twice?

6. a)   Let $G = (V, E)$ be a bipartite undirected graph with V partitioned as $V_1 \cup V_2$. Prove that if $|V_1| \neq |V_2|$, then G cannot have a Hamilton cycle.

  b)   If the graph G in part (a) has a Hamilton path, prove that $|V_1| - |V_2| = \pm 1$.

7. Find a counterexample to the converse of Theorem 14.7.

8. a)   Let $G = (V, E)$ be a loop-free graph with $|V| = n \geq 3$. If $\deg(x) + \deg(y) \geq n$ for all nonadjacent $x, y \in V$, prove that G contains a Hamilton cycle.

  b)   Helen and Dominic invite 10 friends to dinner. In this group of 12 people everyone knows at least 6 others. Prove that the 12 can be seated around a circular table so that each person is acquainted with the persons sitting on either side.

9. Prove Corollary 14.4.

10. Let $n = 2^k$, $k \in \mathbf{Z}^+$. We use the n k-bit sequences (of 0's and 1's) to represent $1, 2, 3, \ldots, n$, so that for two consecutive integers $i, i + 1$, the corresponding

k-bit sequences differ in exactly one component. This representation is called a *Gray Code*.

a)   For $k = 3$, use a graph model with $V = \{000, 001, 010, \ldots, 111\}$ to find such a code for 1, 2, 3, $\ldots$, 8. How is this related to the concept of a Hamilton path?

b)   Answer (a) for $k = 4$.

11.   If $G = (V, E)$ is an undirected graph, a subset I of V is called *independent* if no two vertices in I are adjacent. An independent set I is called *maximal* if no vertex v can be added to I with $I \cup \{v\}$ independent. The *independence number* of G, denoted $\beta(G)$, is the size of any largest maximal independent set in G.

Figure 14.61

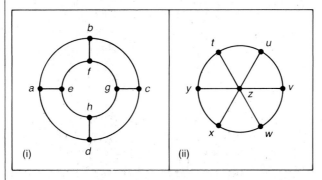

(i)   (ii)

a)   For each graph in Fig. 14.61 find two maximal independent sets with different sizes.

b)   Find $\beta(G)$ for each graph in part (a).

c)   Let I be an independent set in $G = (V, E)$. What type of subgraph does I determine in $\overline{G}$?

12.   Let $G = (V, E)$ be an undirected graph with I an independent set. For each $a \in I$ and any Hamilton cycle C for G there will be $\deg(a) - 2$ edges in E that are incident with a and not in C. Therefore there are $\sum_{a \in I} [\deg(a) - 2] = \sum_{a \in I} \deg(a) - 2|I|$ edges in E that do not appear in C.

a)   Why are these $\sum_{a \in I} \deg(a) - 2|I|$ edges distinct?

b)   Let $v = |V|$, $e = |E|$. Prove that if $e - \sum_{a \in I} \deg(a) + 2|I| < v$, then G has no Hamilton cycle.

c)   Select a suitable independent set I and use part (b) to show that the graph in Fig. 14.62 (known as the Herschel graph) has no Hamilton cycle.

Figure 14.62

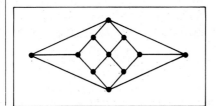

14.6 ■ GRAPH COLORING AND CHROMATIC POLYNOMIALS

At the J. & J. Chemical Company, Jeannette is in charge of the storage of chemical compounds in the company warehouse. Since certain types of compounds (such as acids and bases) should not be kept in the same vicinity, she decides to have her partner Jack partition the warehouse into separate storage areas where incompatible chemical reagents can be stored in separate compartments. How can she determine the number of storage compartments that Jack will have to build?

If this company sells 25 chemical compounds, let $\{c_1, c_2, \ldots, c_{25}\} = V$, a set of vertices. For any $1 \le i < j \le 25$, we draw the edge $\{c_i, c_j\}$ if c_i and c_j must be stored in separate compartments. This gives us an undirected graph $G = (V, E)$.

We now introduce the following concept.

Definition 14.16 ▶ If $G = (V, E)$ is an undirected graph, a *proper coloring* of G occurs when we color the vertices of G so that if $\{a, b\}$ is an edge in G, then a and b are colored with different colors. (Hence adjacent vertices have different colors.) The minimum number of colors needed to properly color G is called the *chromatic number* of G and is written $\gamma(G)$.

Returning to assist Jeannette at the warehouse, we find that the number of storage compartments Jack must build is equal to $\gamma(G)$ for the graph we constructed on $V = \{c_1, c_2, \ldots, c_{25}\}$. But how do we compute $\gamma(G)$? Before we present any work on how to determine the chromatic number of a graph, we turn to the following related idea.

In Example 14.15 we mentioned the connection between coloring the regions in a planar map (with neighboring regions having different colors) and properly coloring the vertices in an associated graph. Determining the smallest number of colors needed to color planar maps in this way has been a problem of interest for over a century.

In about 1850, Francis Guthrie (1831–1899) became interested in the general problem after showing how to color the counties on a map of England with only four colors. Shortly after, he showed the "Four-color Problem" to his younger brother Frederick. At that time Frederick Guthrie (1833–1866) was a student of Augustus DeMorgan (1806–1871), who then communicated the problem (in 1852) to William Hamilton (1805–1865). The problem did not interest Hamilton and lay dormant for about 25 years. Then in 1878, the scientific community was made aware of the problem through an announcement by Arthur Cayley (1821–1895) at a meeting of the London Mathematical Society. In 1879, Cayley stated the problem in the first volume of the *Proceedings of the Royal Geographical Society*. Shortly thereafter, the British mathematician, Sir Alfred Kempe (1849–1922), devised a proof that remained unquestioned for over a decade. Then in 1890, Percy John Heawood (1861–1955), another British mathematician, found a mistake in Kempe's work.

The problem remained unsolved until 1976, when it was finally established by Kenneth Appel and Wolfgang Haken. Their proof employs a very intricate computer analysis of 1936 (reducible) configurations.

Now although only four colors are needed to properly color the regions in a planar map, we need more than four colors to properly color the vertices of some nonplanar graphs.

We start with some small examples. Then we shall find a way to determine $\gamma(G)$ from smaller subgraphs of G. We shall also obtain what is called the chromatic polynomial for G and see how it is used in computing $\gamma(G)$.

Figure 14.63

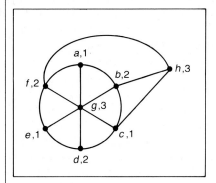

EXAMPLE 14.21 For the graph G in Fig. 14.63, we start at vertex a and next to each vertex write the number of colors needed to properly color the vertices of G considered up to that point. Going to vertex b, the 2 indicates the need for a second color. Proceeding alphabetically to f we find that two colors are needed to properly color $\{a, b, c, d, e, f\}$. For vertex g a third color is needed; this third color can also be used for vertex h, since $\{g, h\}$ is not an edge in G. Thus $\gamma(G) = 3$.   □

EXAMPLE 14.22
a) For any $n \geq 1$, $\gamma(K_n) = n$.

b) The chromatic number of the Herschel graph (Fig. 14.62) is 2.

c) If G is the Petersen graph, $\gamma(G) = 3$.   □

   We turn now to a method of determining $\gamma(G)$. Our coverage follows the development in the survey article [16] by R. C. Read.

   Let G be an undirected graph and let λ be a positive integer representing the number of colors that are available for properly coloring the vertices of G. Our objective is to find a polynomial $P(G, \lambda)$, in the variable λ, which will tell us in how many different ways we can properly color the vertices of G using at most λ colors.

EXAMPLE 14.23
a) If $G = (V, E)$ and $|V| = n$, $E = \emptyset$, then G consists of n isolated points and by the rule of product $P(G, \lambda) = \lambda^n$.

b) If $G = K_n$, then in order to properly color G at least n colors must be available. Here, by the rule of product, $P(G, \lambda) = \lambda(\lambda - 1)(\lambda - 2) \cdots (\lambda - n + 1)$, which we denote by $\lambda^{(n)}$. For $\lambda < n$, $P(G, \lambda) = 0$ and there are no ways to properly color K_n. $P(G, \lambda) > 0$ for the first time when $\lambda = n = \gamma(G)$.

c) For each path in Fig. 14.64, we consider the number of choices at each successive vertex. Proceeding alphabetically, we find that $P(G_1, \lambda) = \lambda(\lambda - 1)^3$ and $P(G_2, \lambda) = \lambda(\lambda - 1)^4$. Since $P(G_1, 1) = 0 = P(G_2, 1)$, but $P(G_1, 2) = 2 = P(G_2, 2)$, $\gamma(G_1) = \gamma(G_2) = 2$. If five colors are available we can properly color G_1 in $5(4)^3 = 320$ ways; G_2 can be so colored in $5(4)^4 = 1280$ ways.

Figure 14.64

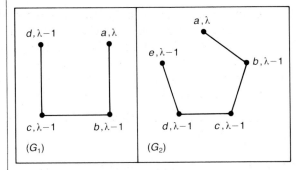

(G_1) (G_2)

In general, if G is a simple path on n vertices, then $P(G, \lambda) = \lambda(\lambda - 1)^{n-1}$.

d) If G is made up of components $C_1, C_2, \ldots, C_k$, again by the rule of product it follows that $P(G, \lambda) = P(C_1, \lambda) \cdot P(C_2, \lambda) \cdots P(C_k, \lambda)$.   □

As a result of Example 14.23(d), we shall concentrate on connected graphs. In many instances in discrete mathematics, methods have been employed to solve problems in large cases by breaking these down into two or more smaller cases. Once again we use this method of attack. To do so the following ideas and notation are needed.

Let $G = (V, E)$ be an undirected graph. For $e = \{a, b\} \in E$, let G_e denote the subgraph of G obtained by deleting e from G (without removing vertices a and b). From G_e a second subgraph of G is obtained by coalescing the vertices a and b. This second subgraph is denoted by G'_e.

Figure 14.65

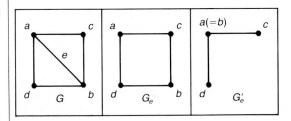

EXAMPLE 14.24

Figure 14.65 shows G_e and G'_e for graph G with the edge e as specified. Note how the coalescing of a and b in G'_e results in the coalescing of the two pairs of edges $\{d, b\}$, $\{d, a\}$ and $\{a, c\}$, $\{b, c\}$.   □

Using these special subgraphs we turn now to the main result.

Theorem
14.8

▶   (*Decomposition Theorem for Chromatic Polynomials*) If $G = (V, E)$ is a connected graph and $e \in E$, then

$$P(G_e, \lambda) = P(G, \lambda) + P(G'_e, \lambda).$$

Proof | Let $e = \{a, b\}$. The number of ways to properly color the vertices in G_e with (at most) λ colors is $P(G_e, \lambda)$. Those colorings where a and b have different colors are proper colorings of G. The colorings of G_e that are not proper colorings of G occur when a and b have the same color. But each of these colorings correspond with a proper coloring for G'_e. This partition of the $P(G_e, \lambda)$ colorings into the two disjoint subsets described results in the equation $P(G_e, \lambda) = P(G, \lambda) + P(G'_e, \lambda)$. ■

When calculating chromatic polynomials we shall place brackets about a graph to indicate its chromatic polynomial.

EXAMPLE 14.25 | The following calculations yield $P(G, \lambda)$ for G a simple cycle of length 4.

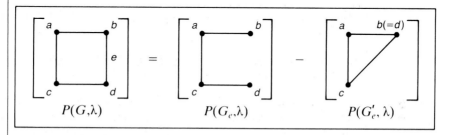

$$P(G,\lambda) \qquad P(G_e,\lambda) \qquad P(G'_e, \lambda)$$

From Example 14.23(c), $P(G_e, \lambda) = \lambda(\lambda - 1)^3$. With $G'_e = K_3$, $P(G'_e, \lambda) = \lambda^{(3)}$. Therefore,

$$P(G, \lambda) = \lambda(\lambda - 1)^3 - \lambda(\lambda - 1)(\lambda - 2) = \lambda(\lambda - 1)[(\lambda - 1)^2 - (\lambda - 2)]$$

$$= \lambda(\lambda - 1)[\lambda^2 - 3\lambda + 3] = \lambda^4 - 4\lambda^3 + 6\lambda^2 - 3\lambda.$$

Since $P(G, 1) = 0$ while $P(G, 2) = 2 > 0$, $\gamma(G) = 2$. □

EXAMPLE 14.26 | Here we find a second application of Theorem 14.8.

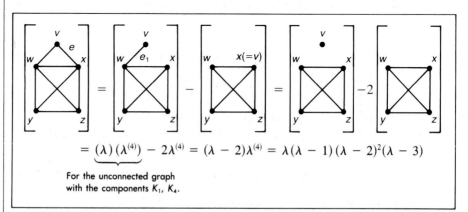

$$= (\lambda)(\lambda^{(4)}) - 2\lambda^{(4)} = (\lambda - 2)\lambda^{(4)} = \lambda(\lambda - 1)(\lambda - 2)^2(\lambda - 3)$$

For the unconnected graph with the components K_1, K_4.

For $1 \le \lambda \le 3$, $P(G, \lambda) = 0$, while $P(G, \lambda) > 0$ for $\lambda \ge 4$. Consequently the given graph has chromatic number 4. □

The chromatic polynomials in Examples 14.25 and 14.26 suggest the following results.

Theorem 14.9

▶ For any graph G, the constant term in $P(G, \lambda)$ is 0.

Proof

For any graph G, since $V \neq \emptyset$, $\gamma(G) > 0$. If $P(G, \lambda)$ has constant term a, then $P(G, 0) = a > 0$. This implies that there are a ways to color G properly with 0 colors. ■

Theorem 14.10

▶ Let $G = (V, E)$ with $|E| > 0$. Then the sum of the coefficients in $P(G, \lambda)$ is 0.

Proof

Since $|E| \geq 1$, $\gamma(G) \geq 2$. Consequently $P(G, 1) = 0 =$ the sum of the coefficients in $P(G, \lambda)$. ■

Since the chromatic polynomial of a complete graph is easy to determine, an alternative method for finding $P(G, \lambda)$ can be obtained. Theorem 14.8 reduced the problem to smaller graphs. Here edges are added to a given graph until we reach complete graphs.

Theorem 14.11

▶ Let $G = (V, E)$, with $a, b \in V$ but $\{a, b\} = e \notin E$. We write G_e^+ for the graph obtained from G by adding the edge $e = \{a, b\}$. Coalescing the vertices a and b in G gives us the subgraph G_e^{++} of G. Under these circumstances $P(G, \lambda) = P(G_e^+, \lambda) + P(G_e^{++}, \lambda)$.

Proof

This result follows from Theorem 14.8, since $P(G_e^+, \lambda) = P(G, \lambda) - P(G_e^{++}, \lambda)$. ■

EXAMPLE 14.27

We apply Theorem 14.11 in the following.

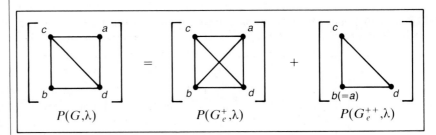

$$P(G, \lambda) \qquad\qquad P(G_e^+, \lambda) \qquad\qquad P(G_e^{++}, \lambda)$$

Here $P(G, \lambda) = \lambda^{(4)} + \lambda^{(3)} = \lambda(\lambda - 1)(\lambda - 2)^2$, so $\gamma(G) = 3$. In addition, if six colors are available the vertices in G can be properly colored in $6(5)(4)^2 = 480$ ways. □

Our last result again uses complete graphs.

Theorem 14.12

▶ Let G be an undirected graph with subgraphs G_1, G_2. If $G = G_1 \cup G_2$ and $G_1 \cap G_2 = K_n$, then

$$P(G, \lambda) = [P(G_1, \lambda) \cdot P(G_2, \lambda)]/\lambda^{(n)}.$$

Proof

Since $G_1 \cap G_2 = K_n$, K_n is a subgraph of G_1, G_2, and $\gamma(G_1)$, $\gamma(G_2) \geq n$. There are $\lambda^{(n)}$ proper colorings of K_n. For each of these $\lambda^{(n)}$ colorings there are $P(G_1, \lambda)/\lambda^{(n)}$ ways to properly color the remaining vertices in G_1. Likewise, there are $P(G_2, \lambda)/\lambda^{(n)}$ ways for the remaining vertices in G_2. By the rule of product,

$$P(G, \lambda) = P(K_n, \lambda) \cdot \frac{P(G_1, \lambda)}{\lambda^{(n)}} \cdot \frac{P(G_2, \lambda)}{\lambda^{(n)}} = \frac{P(G_1, \lambda) \cdot P(G_2, \lambda)}{\lambda^{(n)}}.$$ ∎

EXAMPLE 14.28

Consider the graph in Example 14.26. Let G_1 be the subgraph consisting of the vertices w, x, y, z and all edges in G that connect them. Let G_2 be the simple cycle determined by v, w, and x. Then $G_1 \cap G_2$ is the edge $\{w, x\}$, so $G_1 \cap G_2 = K_2$. Therefore

$$P(G, \lambda) = \frac{P(G_1, \lambda) \cdot P(G_2, \lambda)}{\lambda^{(2)}} = \frac{\lambda^{(4)} \cdot \lambda^{(3)}}{\lambda^{(2)}}$$

$$= [\lambda^2(\lambda - 1)^2(\lambda - 2)^2(\lambda - 3)]/[\lambda(\lambda - 1)]$$

$$= \lambda(\lambda - 1)(\lambda - 2)^2(\lambda - 3),$$

as shown in Example 14.26. □

A great deal more can be said about chromatic polynomials. In addition there are many unanswered questions. For example, no one has found a set of conditions that indicate if a given polynomial in λ is the chromatic polynomial for some graph. More about this topic is introduced in the article by R. C. Read [16].

EXERCISES

1. A pet-shop owner receives a shipment of tropical fish. Among the different species in the shipment are certain pairs where one species feeds upon the other. These pairs must consequently be kept in different aquaria. Model this problem as a graph-coloring problem and tell how to determine the smallest number of aquaria needed to preserve all the fish in the shipment.

2. As chairwoman of church committees, Mrs. Blasi is faced with scheduling the meeting times for 15 committees. Each committee meets for one hour each week. If someone is on more than one committee, then those committees must be scheduled at different times. Model this problem as a graph-coloring problem and tell how to determine the least number of meeting times Mrs. Blasi has to consider for scheduling the 15 committee meetings.

3. Find the chromatic number of the following graphs.
   a) The complete bipartite graph $K_{m,n}$.
   b) A simple cycle on n vertices, $n \geq 3$.
   c) The graphs in Figs. 14.38(d), 14.41, and 14.61.

4. If G is a graph with components $C_1, C_2, \ldots, C_k$, how is $\gamma(G)$ related to $\gamma(C_i)$, $1 \leq i \leq k$?

Figure 14.66

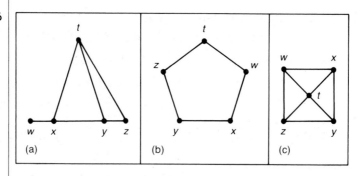

5. a) Determine the chromatic polynomials for the graphs in Fig. 14.66.

b) Find $\gamma(G)$ for each graph.

c) If five colors are available, in how many ways can the vertices of each graph be properly colored?

Figure 14.67

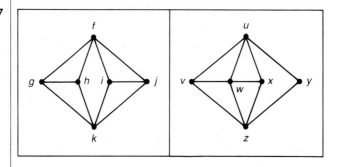

6. a) Determine whether the graphs in Fig. 14.67 are isomorphic.

b) Find $P(G, \lambda)$ for each graph.

c) Comment on the results found in parts (a) and (b).

7. For $n \geq 3$, let C_n denote the simple cycle of length n.

a) What is $P(C_3, \lambda)$?

b) If $n \geq 4$, show that $P(C_n, \lambda) = P(L_n, \lambda) - P(C_{n-1}, \lambda)$, where L_n denotes the simple path of length $n - 1$.

c) Verify that $P(L_n, \lambda) = \lambda(\lambda - 1)^{n-1}$, for any $n \geq 2$.

d) Establish the relations

$$P(C_n, \lambda) - (\lambda - 1)^n = (\lambda - 1)^{n-1} - P(C_{n-1}, \lambda), \qquad n \geq 4,$$

$$P(C_n, \lambda) - (\lambda - 1)^n = P(C_{n-2}, \lambda) - (\lambda - 1)^{n-2}, \qquad n \geq 5.$$

e) Prove that for any $n \geq 3$, $P(C_n, \lambda) = (\lambda - 1)^n + (-1)^n(\lambda - 1)$.

8. For $n \geq 3$, the *wheel graph*, W_n, is obtained from a simple cycle of length n by placing a new vertex within the cycle and adding edges (*spokes*) from this new

vertex to each vertex of the cycle.

a) What relationship is there between $\gamma(C_n)$ and $\gamma(W_n)$?

b) Use part (e) of Exercise 7 to show that

$$P(W_n, \lambda) = \lambda(\lambda - 2)^n + (-1)^n \lambda(\lambda - 2).$$

9. Let $G = (V, E)$ be a loop-free undirected graph with chromatic polynomial $P(G, \lambda)$ and $|V| = n$. Use Theorem 14.11 to prove that $P(G, \lambda)$ has degree n and leading coefficient 1.

10. Let $G = (V, E)$ be a loop-free undirected graph.

a) For each such graph where $|V| \leq 3$, find $P(G, \lambda)$ and show that it is a sum of consecutive powers of λ. Also show that the coefficients of these consecutive powers alternate in sign.

b) Now consider $G = (V, E)$ where $|V| = n \geq 4$, $|E| = k$. Prove by induction that $P(G, \lambda)$ is a sum of consecutive powers of λ and that the coefficients of these consecutive powers alternate in sign. (For the induction hypothesis, assume that the result is true for $G = (V, E)$ where either (i) $|V| = n - 1$ or (ii) $|V| = n$, but $|E| = k - 1$.)

c) Prove that if $|V| = n$, then the coefficient of λ^{n-1} in $P(G, \lambda)$ is the negative of $|E|$.

Figure 14.68

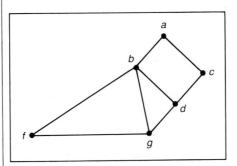

11. Use Theorem 14.12 to find $P(G, \lambda)$ for the graph G in Fig. 14.68.

12. If G is a planar graph, prove that if $P(G, \lambda) = \lambda(\lambda^2 - 3\lambda - 2)f(\lambda)$ with the degree of $f(\lambda) \geq 2$, then $f(\lambda)$ has a root in **Z**. What is the chromatic number of G?

14.7 ■ SUMMARY AND HISTORICAL REVIEW

Unlike other areas in mathematics, the beginnings of graph theory can be traced to a definite time and place: the problem of the seven bridges of Königsberg, which was solved in 1736 by Leonhard Euler (1707–1783). In 1752 we find Euler's theorem for planar graphs. (This result was originally presented in terms of polyhedra.) However, after these developments, little was accomplished in this area for almost a century.

Then in 1847 Gustav Kirchhoff (1824–1887) examined a special type of graph called a tree. (A *tree* is an undirected graph that is connected but contains no cycles.) Kirchhoff used this concept in applications dealing with electrical networks in his extension of Ohm's laws for electrical flow. Ten years later Arthur Cayley (1821–1895) developed this same type of graph in order to count the distinct isomers of the saturated hydrocarbons C_nH_{2n+2}, $n \in \mathbf{Z}^+$.

This period also saw two other major ideas come to light. The *four-color conjecture* was first investigated by Francis Guthrie (1831–1899) in about 1850. In Section 14.6 we related some of the history of this problem, which was solved by an intricate computer analysis in 1976 by Kenneth Appel and Wolfgang Haken.

The second major idea was the Hamilton cycle. This cycle is named for Sir William Hamilton (1805–1865), who used the idea in 1859 for an intriguing puzzle that used the edges on a regular dodecahedron. A solution to this puzzle is not very difficult to find but mathematicians still search for necessary and sufficient conditions to characterize those undirected graphs that possess a Hamilton path or cycle.

Following these developments, we find little activity until after 1920. One problem that was solved at this time was the characterization of planar graphs, by Kasimir Kuratowski in 1930. In 1936 we find the publication of the first book on graph theory, written by Denses König, a prominent researcher in the field. Since then there has been a great deal of activity in the area, the computer providing assistance in the last three decades. Among the many contemporary researchers (not mentioned in the chapter references) in this and related fields one finds Claude Berge, V. Chvátal, Paul Erdös, Laszlo Lovász, W. T. Tutte, and Hassler Whitney.

Comparable coverage of the material presented in this chapter is contained in Chapters 6, 8, and 9 of C. L. Liu [13]. More advanced work is found in the works by J. Bondy and U. Murty [6] and N. Deo [7]. A proof of Kuratowski's Theorem appears in Chapter 8 of C. L. Liu [13]. Recent developments in the field are given in the two-volume *MAA Studies in Mathematics,* edited by D. Fulkerson [10].

Applications of graph theory in electrical networks can be found in S. Seshu and M. Reed [18]. In the text by N. Deo [7], applications in coding theory, electrical networks, operations research, computer programming, and chemistry occupy Chapters 12–15. The text by F. Roberts [17] applies the methods of graph theory to the social sciences.

More on chromatic polynomials can be found in the survey article by R. C. Read [16] and in Chapter VI of C. L. Liu [14]. The role of Polya's theory in graphical enumeration is examined in Chapter 10 of N. Deo [7]. A thorough coverage of this topic is found in the text by F. Harary and E. Palmer [12].

Additional coverage on the historical development of graph theory is given in N. Biggs, E. Lloyd, and R. Wilson [5].

Many applications in graph theory involve large graphs that require the decision-making capability of a computer in conjunction with the ingenuity of mathematical methods. Chapter 11 of N. Deo [7] presents computer algorithms dealing with several of the graph-theoretic properties we have studied here. Along the same line, the text by A. Aho, J. Hopcroft, and J. Ullman [1] provides even more for the reader interested in computer science.

As mentioned at the end of Section 14.5, the problem of the traveling salesman is closely related to the search for a Hamilton cycle in a graph. This is a graph-theoretic problem of interest in both operations research and computer science. The article by M. Bellmore and G. L. Nemhauser [4] provides a good introductory survey of results on this problem. The text by R. Bellman, K. L. Cooke, and J. A. Lockett [3] includes an algorithmic treatment of this problem along with other graph problems. Recent developments on applications, where a robot visits different locations in an automated warehouse in order to fill a given order, are examined in the articles by E. A. Elsayed [8] and by E. A. Elsayed and R. G. Stern [9].

Considering its impact on the scientific community, the solution of the four-color problem can be examined further by starting with the paper by K. Appel and W. Haken [2]. An appreciation for the development of this work can be gained from a study of the earlier article by W. Haken [11]. The proof uses a computer analysis to handle a large number of cases; the article by T. Tymoczko [19] examines the role of such techniques in pure mathematics.

Finally, the article by A. Ralston [15] demonstrates the connections between coding theory, combinatorics, graph theory, and computer science.

REFERENCES

1. Aho, Alfred V., Hopcroft, John E., and Ullman, Jeffrey D., *Data Structures and Algorithms,* Addison-Wesley, Reading, Massachusetts, 1983.

2. Appel, Kenneth, and Haken, Wolfgang, "Every Planar Map Is Four Colorable," *Bulletin of the American Mathematical Society* 82, (1976), pp. 711–712.

3. Bellman, R., Cooke, K. L., and Lockett, J. A., *Algorithms, Graphs, and Computers,* Academic Press, New York, 1970.

4. Bellmore, M., and Nemhauser, G. L., "The Traveling Salesman Problem: A Survey," *Operations Research,* Vol. 16, 1968, pp. 538–558.

5. Biggs, N., Lloyd, E. K., and Wilson, R. J., *Graph Theory (1736–1936),* Clarendon Press, Oxford, England, 1976.

6. Bondy, J. A., and Murty, U. S. R., *Graph Theory with Applications,* Elsevier North-Holland, New York, 1976.

7. Deo, Narsingh, *Graph Theory with Applications to Engineering and Computer Science,* Prentice-Hall, Englewood Cliffs, New Jersey, 1974.

8. Elsayed, E. A., "Algorithms for Optimal Material Handling in Automatic Warehousing Systems," *Int. J. Prod. Res.,* Vol. 19, 1981, pp. 525–535.

9. Elsayed, E. A., and Stern, R. G., "Computerized Algorithms for Order Processing in Automated Warehousing Systems," *Int. J. Prod. Res.,* Vol. 21, 1983, pp. 579–586.

10. Fulkerson, D. R., ed., *Studies in Graph Theory,* Parts I, II, *MAA Studies in Mathematics,* Vol. 11 and 12, The Mathematical Association of America, 1975.

11. Haken, Wolfgang, "An Attempt to Understand the Four-Color Problem," *Journal of Graph Theory,* Vol. 1, No. 3, (1977).

12. Harary, Frank, and Palmer, Edgar M., *Graphical Enumeration,* Academic Press, New York, 1973.

13. Liu, C. L., *Introduction to Combinatorial Mathematics,* McGraw-Hill, New York, 1968.

14. Liu, C. L., *Topics in Combinatorial Mathematics,* Mathematical Association of America, 1972.

15. Ralston, Anthony, "De Bruijn Sequences — A Model Example of the Interaction of Discrete Mathematics and Computer Science," *Mathematics Magazine,* Vol. 55, No. 3, May 1982, pp. 131–143.

16. Read, R. C., "An Introduction to Chromatic Polynomials," *Journal of Combinatorial Theory,* Vol. 4, 1968, pp. 52–71.

17. Roberts, Fred S., *Discrete Mathematical Models,* Prentice-Hall, Englewood Cliffs, New Jersey, 1976.

18. Seshu, S., and Reed, M. B., *Linear Graphs and Electrical Networks,* Addison-Wesley, Reading, Massachusetts, 1961.

19. Tymoczko, Thomas, "Computers, Proofs and Mathematicians: A Philosophical Investigation of the Four-Color Proof," *Mathematics Magazine,* Vol. 53, No. 3, May 1980, pp. 131–138.

■ MISCELLANEOUS EXERCISES

1. Let G be a loop-free undirected graph on n vertices. If G has 56 edges and $\overline{G}$ has 80 edges, what is n?

2. Let G_1 and G_2 be two loop-free undirected connected graphs, each on n vertices. If every vertex in both G_1 and G_2 has degree 2, prove that G_1 and G_2 are isomorphic.

3. An n-cube is an undirected graph $G = (V, E)$ where the vertices of G are the (ordered) n-tuples of 0's and 1's. Edges in G are of the form $\{v, w\}$ where v and w differ in exactly one component.

   a) Prove that G is connected.

   b) Find $|V|$, $|E|$.

4. Let G be an undirected graph with C a simple cycle in G. If e_1 and e_2 are any two edges in C, prove that G contains a cut-set S such that $C \cap S = \{e_1, e_2\}$.

5. a) If the edges of K_6 are painted either red or blue, prove that there is a red triangle or blue triangle that is a subgraph.

   b) Prove that in any group of six people there must be three who are total strangers to one another or three who are mutual friends.

6. If $G = (V, E)$ is an undirected loop-free graph, the *line graph* of G, denoted $L(G)$, is a graph with the set E as vertices, where we join two vertices e_1, e_2 in $L(G)$ iff e_1, e_2 are adjacent edges in G.

Figure 14.69

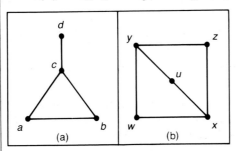

(a)   (b)

a) Find $L(G)$ for each of the graphs in Fig. 14.69.

b) Assuming that $|V| = n$, $|E| = e$, show that $L(G)$ has e vertices and $(1/2) \sum_{v \in V} \deg(v)[\deg(v) - 1] = [(1/2) \sum_{v \in V} [\deg(v)]^2] - e = \sum_{v \in V} \binom{\deg(v)}{2}$ edges.

c) Prove that if G has an Euler cycle, then $L(G)$ has both an Euler cycle and a Hamilton cycle.

d) If $G = K_4$, examine $L(G)$ to show that the converse of (c) is false.

e) Prove that if G has a Hamilton cycle, then so does $L(G)$.

f) Examine $L(G)$ for the graph in Fig. 14.69(b) to show that the converse of (e) is false.

g) Verify that $L(G)$ is nonplanar for $G = K_5$ and $G = K_{3,3}$.

h) Give an example of a graph G where G is planar but $L(G)$ is not.

7. If $G = (V, E)$ is an undirected graph, a subset K of V is called a *covering* of G if for every edge $\{a, b\}$ of G either a or b is in K. The number of vertices in a minimum covering is called the *covering number* of G.

a) Prove that if $I \subseteq V$, then I is an independent set in G iff $V - I$ is a covering of G.

b) Verify that $|V|$ is the sum of the independence number of G and its covering number.

8. If $G = (V, E)$ is an undirected graph, a subset D of V is called a *dominating set* if for all $v \in V$, either $v \in D$ or v is adjacent to a vertex in D. (See Section 9.3 for more on dominating sets.) If D is a dominating set and no proper subset of D has this property, D is called *minimal*. The size of any smallest minimal dominating set in G is denoted $\alpha(G)$ and is called the *domination number* of G.

a) If G has no isolated points, prove that if D is a minimal dominating set, then $V - D$ is a dominating set.

b) If $I \subseteq V$ and I is independent, prove that I is a dominating set iff I is maximal independent.

c) Show that $\alpha(G) \le \beta(G)$, and $|V| \le \beta(G)\gamma(G)$.

Figure 14.70

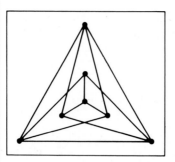

9. For the graph G in Fig. 14.70, answer the following questions.

   a) What are $\alpha(G)$, $\beta(G)$, $\gamma(G)$?

   b) Does G have an Euler or Hamilton cycle?

   c) Is G bipartite? Is it planar?

10. Explain why each of the following polynomials in λ cannot be a chromatic polynomial.

   a) $\lambda^4 - 5\lambda^3 + 7\lambda^2 - 6\lambda + 3$ b) $3\lambda^3 - 4\lambda^2 + \lambda$

   c) $\lambda^4 - 3\lambda^3 + 5\lambda^2 - 4\lambda$ d) $\lambda^4 - 5\lambda^2 + 4\lambda$

11. If $G = (V, E)$ is an undirected graph, any subgraph of G that is a complete graph is called a *clique* in G. The number of vertices in a largest clique in G is called the *clique number* for G and is denoted by $\text{cl}(G)$.

   a) How are $\gamma(G)$ and $\text{cl}(G)$ related?

   b) If G is made up of components C_1, C_2, ..., C_k, how is $\text{cl}(G)$ related to $\text{cl}(C_1)$, $\text{cl}(C_2)$, ..., $\text{cl}(C_k)$?

   c) Can a graph G have 25 edges and clique number 8?

   d) Is there any relationship between $\text{cl}(G)$ and $\beta(\overline{G})$?

12. Just as we defined $\lambda^{(n)}$ when dealing with the chromatic polynomial for K_n, for $n \geq 1$, we define $x^{(n)} = x(x - 1)(x - 2)\cdots(x - n + 1)$. Multiplying this out results in a polynomial in x of degree n. We write this polynomial as $\sum_{i=0}^{n} s(n, i)x^i$, and call each coefficient $s(n, i)$, $0 \leq i \leq n$, a *Stirling number of the first kind*.

   For all n, $s(n, n) = 1$ and $s(n, 0) = 0$. Also, for $i < 0$ or $i > n$, $s(n, i) = 0$. In addition $\sum_{i=0}^{n+1} s(n + 1, i)x^i = x^{(n+1)} = x^{(n)}(x - n) = [\sum_{i=0}^{n} s(n, i)x^i](x - n)$. Consequently, for any $1 \leq i \leq n$, the comparison of coefficients of like powers of x results in the recurrence relation

$$s(n + 1, i) = s(n, i - 1) - ns(n, i).$$

   a) Use the recurrence relation to find a table for $s(n, i)$, $1 \leq n \leq 5$, $0 \leq i \leq n$.

   b) Write a computer program to extend the table in part (a) to all $n \leq 15$, $0 \leq i \leq n$.

   c) For $m, n \in \mathbf{Z}^+$, we define the delta function $\delta(m, n)$ to be 1 when $m = n$, and 0 when $m \neq n$. If $k = \max\{m, n\}$ then the Stirling numbers of the first and second kinds are related by

$$\sum_{i=0}^{k} s(m, i) S(i, n) = \delta(m, n) = \sum_{i=0}^{k} S(m, i) s(i, n).$$

Verify these results for the cases where $m = n = 4$, and where $m = 5$, $n = 4$.

13. a) The complete bipartite graph $K_{m,n}$ contains 16 edges, and satisfies $m \leq n$. Determine m, n so that $K_{m,n}$ possesses (i) an Euler cycle but not a Hamilton cycle; (ii) both a Hamilton cycle and an Euler cycle.

  b) Generalize the results of part (a).

15

Trees

Continuing our study of graph theory we focus on a special type of graph called a tree. First used in 1847 by Gustav Kirchhoff (1824–1887) in his work on electrical networks, trees were later redeveloped and named by Arthur Cayley (1821–1895). In 1857 Cayley used these graphs to enumerate the different isomers of the saturated hydrocarbons C_nH_{2n+2}, $n \in \mathbf{Z}^+$.

With the advent of digital computers, many new applications were found for trees. Special types of trees are prominent in the study of data structures, sorting, coding theory, and in certain optimization problems.

15.1 ■ DEFINITIONS, PROPERTIES AND EXAMPLES

Definition 15.1

▶ Let $G = (V, E)$ be an undirected graph. G is called a *tree*[†] if G is connected and contains no cycles.

Since a loop is a cycle of length one, a tree is loop-free.

Figure 15.1

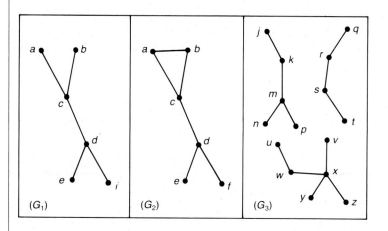

(G_1) (G_2) (G_3)

In Fig. 15.1 the graph G_1 is a tree but the graph G_2 is not because of the cycle $\{a, b\}, \{b, c\}, \{c, a\}$. Graph G_3 is not connected, so it cannot be a tree. However, each component of G_3 is a tree, and we call G_3 a *forest*.

When a graph is a tree we write T instead of G to emphasize this structure.

In Fig. 15.1 we see that G_1 is a subgraph of G_2 where G_1 contains all the vertices of G_2 and G_1 is a tree. In this situation G_1 is called a *spanning tree* for G_2. We think of a spanning tree as providing minimal connectivity for the graph and as a minimal skeletal framework holding the vertices together.

We now examine some properties of trees.

Theorem 15.1

▶ If a, b are vertices in a tree $T = (V, E)$, then there is a unique path that connects these vertices.

[†]As in the case of graphs, the terminology in the study of trees is not standard and the reader may find some differences from one textbook to another.

Proof We first note that in a tree all paths are simple. (Why?) Since T is connected there is at least one path in T that connects a and b. If there were more, from two such paths some of the edges could be combined to form a cycle. But T has no cycles. ∎

Theorem 15.2 ▶ If $G = (V, E)$ is an undirected graph, then G is connected iff G has a spanning tree.

Proof If G has a spanning tree T, then for every pair a, b of vertices in V a subset of the edges in T provides a (unique) path between a and b, so G is connected. Conversely, if G is connected and G is not a tree, remove all loops from G. If the resulting subgraph G_1 is not a tree, then G_1 must contain a (longer) cycle. For each such cycle remove a distinct edge of G_1. The resulting subgraph G_2 contains all the vertices in G, is connected, and contains no cycles. Hence G_2 is a spanning tree for G. ∎

Figure 15.2

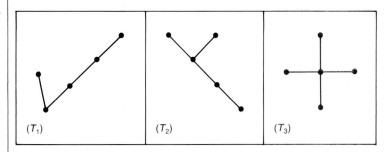

(T_1) (T_2) (T_3)

Figure 15.2 shows three nonisomorphic trees that exist for five vertices. Although they are not isomorphic they all have the same number of edges, namely four. This leads us to the following general result.

Theorem 15.3 ▶ In any tree $T = (V, E)$, $|V| = |E| + 1$.

Proof This result was given earlier as Theorem 6.3. ∎

As we examine the trees in Fig. 15.2 we also see that each tree has at least two pendant vertices, i.e., vertices of degree 1. This is also true in general.

Theorem 15.4 ▶ For any tree $T = (V, E)$, if $|V| \geq 2$ then T has at least two pendant vertices.

Proof Let $|V| = n \geq 2$. From Theorem 15.3, $|E| = n - 1$, and by Theorem 14.2, $2(n - 1) = 2n - 2 = 2|E| = \sum_{v \in V} \deg(v)$. If $\deg(v) \geq 2$ for all $v \in V$, then $\sum_{v \in V} \deg(v) \geq 2|V| = 2n$. From the contradiction $2n - 2 \geq 2n$, it follows that there are at least two vertices v for which $\deg(v) < 2$. Since there are no isolated vertices, these vertices must be pendant. ∎

Figure 15.3

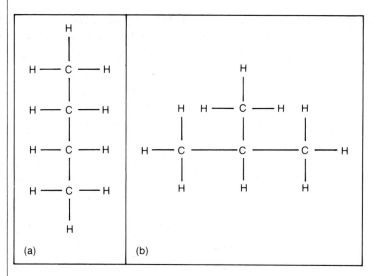

(a)   (b)

EXAMPLE 15.1

In Fig. 15.3 we have two trees, each with 14 vertices (labeled with C's and H's) and 13 edges. Each vertex has degree 4 (C, carbon atom) or degree 1 (H, hydrogen atom). Part (b) of the figure has a carbon atom (C) at the center of the tree. This carbon atom is adjacent to four vertices, three of which have degree 4. Since there is no vertex (C atom) in part (a) that possesses this property, the two trees are not isomorphic. They serve as models for the two chemical isomers that correspond with the saturated hydrocarbon C_4H_{10}. Part (a) represents butane; part (b) represents isobutane.   □

We close this section with a second result from chemistry.

EXAMPLE 15.2

If a saturated hydrocarbon has n carbon atoms, show that it has $2n + 2$ hydrogen atoms.

Considering the saturated hydrocarbon as a tree $T = (V, E)$, let k equal the number of pendant vertices, or hydrogen atoms, in the tree. Then with a total of $n + k$ vertices, where the n carbon atoms each have degree 4, we find that $4n + k = \sum_{v \in V} \deg(v) = 2|E| = 2(|V| - 1) = 2(n + k - 1)$, and $4n + k = 2(n + k - 1) \Rightarrow k = 2n + 2$.   □

EXERCISES

1. a)   Draw the graphs of all nonisomorphic trees on six vertices.

   b)   How many isomers does hexane (C_6H_{14}) have?

2. a)   Verify that all trees are planar.

   b)   Derive Theorem 15.3 from part (a) and Euler's theorem.

3. a)   If $G = (V, E)$ is a forest with $|V| = v$, $|E| = e$, and ω components (trees), what relationship exists among v, e, and ω?

   b)   What is the smallest number of edges we must add to G in order to get a tree?

4. a) If a tree has four vertices of degree 2, one vertex of degree 3, two of degree 4, and one of degree 5, how many pendant vertices does it have?

   b) If a tree $T = (V, E)$ has v_2 vertices of degree 2, v_3 vertices of degree 3, ..., and v_m vertices of degree m, what are $|V|$ and $|E|$?

5. If $G = (V, E)$ is an undirected graph, prove that G is a tree if there is a unique path between any two vertices of G.

6. The connected undirected graph $G = (V, E)$ has 30 edges. What is the maximum value that $|V|$ can have?

7. Let $G = (V, E)$ be a loop-free connected undirected graph with no isolated vertices. Let H be a subgraph of G. The *complement of H in G* is the subgraph of G made up of those edges in G that are not in H (along with the vertices incident to these edges).

   a) If T is a spanning tree of G, prove that the complement of T in G does not contain a cut-set.

   b) If C is a cut-set in G, prove that the complement of C in G does not contain a spanning tree.

8. Prove that the following statements are equivalent for an undirected graph $G = (V, E)$.

   a) G is a tree;

   b) G is connected but the removal of any edge from G disconnects G into two subgraphs that are trees;

   c) G contains no cycles and $|V| = |E| + 1$;

   d) G is connected and $|V| = |E| + 1$; and,

   e) G contains no cycles and if a, $b \in V$ with $\{a, b\} \notin E$, then the graph obtained by adding edge $\{a, b\}$ to G has precisely one cycle.

9. A labeled tree is one where the vertices are labeled. If the tree has n vertices, $\{1, 2, 3, \ldots, n\}$ is used as the set of labels. We find that two trees that are isomorphic without labels may become nonisomorphic when labeled. In Fig. 15.4 the first two trees are isomorphic as labeled trees. The third tree is isomorphic to

Figure 15.4

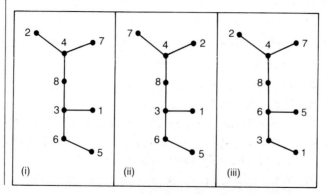

(i) (ii) (iii)

the other two if we ignore the labels; as a labeled tree, however, it is not isomorphic to either of the other two.

The number of nonisomorphic trees with n labeled vertices can be counted by setting up a one-to-one correspondence between these trees and the n^{n-2} sequences (with repetitions allowed) $x_1, x_2, \ldots, x_{n-2}$ from $\{1, 2, 3, \ldots, n\}$. If T is one such labeled tree, we use the following algorithm to establish the one-to-one correspondence.

Step 1: Set the counter i to 1.

Step 2: Set $T(i) = T$.

Step 3: Since a tree has at least two pendant vertices, select the pendant vertex in $T(i)$ with the smallest label y_i. Now remove the edge $\{x_i, y_i\}$ from $T(i)$ and use x_i for the ith component of the sequence.

Step 4: If $i = n - 2$, we have the sequence corresponding to the given labeled tree $T(1)$. If $i \neq n - 2$, increase i by 1, set $T(i)$ equal to the resulting subtree obtained in step (3) and return to step (3).

a) Find the six-digit sequence for trees (i) and (iii) in Fig. 15.4

b) If v is a vertex in T, show that the number of times the label on v appears in the sequence $x_1, x_2, \ldots, x_{n-2}$ is $\deg(v) - 1$.

c) Reconstruct the labeled tree on eight vertices that is associated with the sequence 2, 6, 5, 5, 5, 5.

d) Develop an algorithm for reconstructing a tree from a given sequence $x_1, x_2, \ldots, x_{n-2}$.

15.2 ■ ROOTED TREES

We turn now to directed trees. We find a variety of applications for a special type of directed tree called a *rooted tree*.

Definition 15.2 ▶ If G is a directed graph, G is called a *directed tree* if the undirected graph associated with G is a tree. When G is a directed tree, G is called a *rooted tree* if there is a unique vertex r, called the *root*, in G with the in degree of $r = \deg^+(r) = 0$, and for all other vertices v, $\deg^+(v)$, the in degree of v, is 1.

The tree in Fig. 15.5(a) is directed but not rooted; the tree in (b) is rooted with root r.

We draw rooted trees as in Fig. 15.5(b) but with the directions understood as going from the upper level to the lower level, so that the arrows aren't needed.

In a rooted tree, a vertex v with the out degree of $v = \deg^-(v) = 0$ is called a *leaf* (or *terminal vertex*). Vertices u, v, x, y, z are leaves in Fig. 15.5(b). All other vertices are called *branch nodes* (or *internal vertices*).

Consider the vertex s in this rooted tree (Fig. 15.5(b)). The path from the root, r, to s is of length 2, so we say that s is at *level* 2 in the tree, or that s has *level number*

Figure 15.5

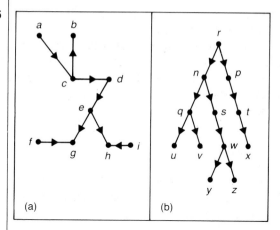

(a)　　　(b)

2. Similarly, x is at level 3, while y has level number 4. We call s a *son* of n and n the *father* of s. Vertices w, y, z are considered *descendants* of s, n, and r; s, n, and r are called *ancestors* of w, y, and z. In general, if v_1 and v_2 are vertices in a rooted tree and v_1 has the smaller level number, then v_1 is an ancestor of v_2, (or v_2 is a descendant of v_1) if there is a path from v_1 to v_2. Two vertices with a common father are referred to as *brothers*. Such is the case for vertices q and s, whose father is vertex n. Lastly, if v_1 is any vertex of the tree, the *subtree at v_1* is the subgraph made up of the root v_1 and all of its descendants (there may be none).

EXAMPLE 15.3　In Fig. 15.6(a) a rooted tree is used to represent the table of contents of a three chapter ($C1$, $C2$, $C3$) book. Vertices with level number 2 are for sections within a chapter, while those at level 3 represent subsections within a section. □

　　　The tree in Fig. 15.6(a) suggests an order for the vertices if we examine the subtrees at $C1$, $C2$, and $C3$ from left to right. We now consider a second example that provides such an order.

Figure 15.6

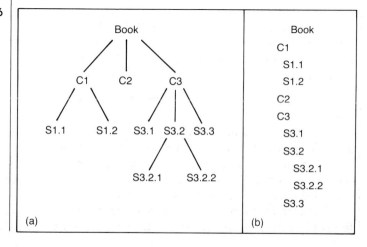

(a)　　　(b)

Figure 15.7

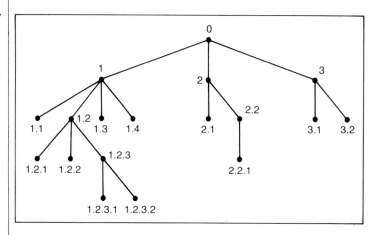

EXAMPLE 15.4   In the tree T of Fig. 15.7, the edges (branches) leaving each internal vertex are *ordered* from left to right. Hence T is called an *ordered rooted tree*.

We label the vertices for this tree by the following algorithm.

Step 1:   First assign the root the label (or *address*) 0.

Step 2:   Next assign the positive integers 1, 2, 3, ... to the vertices at level 1, going from left to right.

Step 3:   Now let v be an internal vertex at level $n \geq 1$, and let $v_1, v_2, \ldots, v_k$ denote the sons of v (going from left to right). If a is the label assigned to vertex v, assign the labels $a.1, a.2, \ldots, a.k$ to the sons $v_1, v_2, \ldots, v_k$, respectively.

Consequently ,each vertex in T, other than the root, has a label of the form $a_1.a_2.a_3. \ldots .a_n$ iff that vertex has level number n. This is known as the *universal address system*.

This system provides a way to *order* all vertices in T. If u and v are two vertices in T with addresses b and c, respectively, we define $b < c$ if (a) $b = a_1.a_2. \ldots .a_m$, $c = a_1.a_2. \ldots .a_m.a_{m+1}. \ldots .a_n$, with $m < n$; or, (b) $b = a_1.a_2. \ldots .a_m.x_1. \ldots .y$, $c = a_1.a_2. \ldots .a_m.x_2. \ldots .z$, where $x_1, x_2 \in \mathbf{Z}^+$ and $x_1 < x_2$.

For the tree under consideration this ordering yields

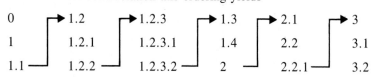

Since this is like the alphabetical ordering in a dictionary, the order is called the *lexicographic*, or *dictionary*, *order*.   □

We consider one further application of a rooted tree in the study of computer algorithms.

Figure 15.8

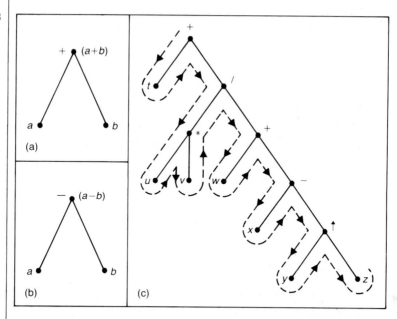

(a) $+ \bullet \ (a+b)$

a b

(a)

(b) $- \bullet \ (a-b)$

a b

(b)

(c)

EXAMPLE 15.5   A rooted binary tree can represent a binary operation as in (a) and (b) of Fig. 15.8. To avoid confusion when dealing with a noncommutative operation $\circ$ we label the root as $\circ$ and require the result to be $a \circ b$ where a is the left son and b the right son of the root.

In evaluating $t + (uv)/(w + x - y^z)$ in a procedural language like FORTRAN, we write the expression in the form $t + (u * v)/(w + x - y \uparrow z)$, where $*$ denotes multiplication and $\uparrow$ exponentiation. When the computer evaluates this expression it performs the binary operations (within each parenthesized part) according to a hierarchy of operations where exponentiation precedes multiplication and division, which in turn precede addition and subtraction. In Fig. 15.9 we number the operations in the order they are performed by the computer. In order for the computer to evaluate this expression it must somehow scan the expression to perform the operations in the order specified.

Figure 15.9

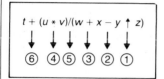

$t + (u * v)/(w + x - y \uparrow z)$

⑥ ④⑤ ③ ② ①

Instead of scanning back and forth continuously, however, the expression is converted by the machine into a notation that is independent of parentheses. This is known as Polish notation, in honor of the Polish (actually Ukrainian) logician Jan Lukasiewicz (1878–1956). Here the *infix* notation, $a \circ b$, for a binary operation $\circ$, becomes ∘ *a b*, the *prefix* (or Polish) notation. The advantage now is that the expression in Fig. 15.9 can be rewritten without parentheses as $+ \; t / * \; uv + w - x \uparrow yz$, where the evaluation proceeds from right to left. When a binary operation is encountered it is performed on the two operands to its right. The result is then treated as one of the operands for the next binary operation encountered as we continue to the left.

The use of Polish notation is important for compilation and can be attained by representing a given expression by a rooted tree, as in Fig. 15.8(c). Here each variable (or constant) is used to label a leaf of the tree. Each internal vertex is labeled by a binary operation whose left and right operands are the left and right subtrees it determines. Starting at the root, as we traverse the tree from top to bottom and left to right, as shown in Fig. 15.8(c), we find the Polish notation by writing down the labels of the vertices in the order in which they are visited.   □

In these last two examples we have seen the importance of order. Several methods exist for systematically ordering the vertices in a tree. The three most prevalent in the study of data structures are the preorder, inorder, and postorder, which are defined recursively in the following.

**Definition
15.3**
▶   Let $T = (V, E)$ be a rooted tree with root r. If T has no other vertices, then the root by itself constitutes the *preorder, inorder,* and *postorder traversal* of T. If $|V| > 1$, let $T_1, T_2, T_3, \ldots, T_k$ denote the subtrees of T as we go from left to right (as in Fig. 15.10).

Figure 15.10

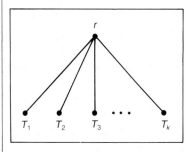

a)   The *preorder traversal* of T starts at r and then visits the vertices of T_1 in preorder, then the vertices in T_2 in preorder, and so on until the vertices in T_k are visited in preorder.

b)   The *inorder traversal* of T first visits the vertices in T_1 in inorder, then visits the root r, and then visits in inorder the vertices of the subtrees $T_2, T_3, \ldots, T_k$.

c)   The *postorder traversal* of T visits in postorder the vertices of the subtrees $T_1, T_2, \ldots, T_k$, and then visits the root.

We demonstrate these ideas in the following example.

EXAMPLE 15.6   Consider the rooted tree in Fig. 15.11.

Figure 15.11

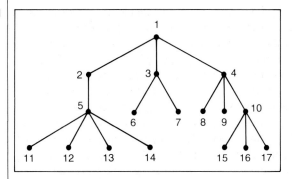

a) *Preorder:* Starting at vertex 1 we visit the subtree T_1 rooted at vertex 2. After visiting vertex 2 we proceed to the subtree rooted at vertex 5 and after visiting vertex 5 we then go to the subtree rooted at vertex 11. Since this subtree has no other vertices, we visit vertex 11 and then return to vertex 5 from which we visit, in succession, vertices 12, 13, and 14. Following this we *backtrack* (14 to 5 to 2 to 1) to the root and then visit the vertices in the subtree T_2 in the preorder 3, 6, 7. Finally, after returning to the root for the last time, we traverse the subtree T_3 in the preorder 4, 8, 9, 10, 15, 16, 17. Hence the preorder listing of the vertices in this tree is 1, 2, 5, 11, 12, 13, 14, 3, 6, 7, 4, 8, 9, 10, 15, 16, 17.

In this ordering we start at the root and build a path as far as we can. At each level we go to the leftmost vertex not previously visited at the next level, until we reach a leaf ℓ. Then we backtrack to the father f of this leaf ℓ and visit its brother b (and the subtree b determines) directly on its right. If no such brother b exists, we backtrack to the grandfather g of the leaf ℓ and visit, if it exists, a vertex u that is the brother of f directly to its right in the tree. Eventually we generate a set of paths from the root to each of the leaves (going from left to right) in the tree.

The vertices in Figs. 15.7 and 15.8(c) are visited in preorder.

b) *Inorder:* Starting from vertex 1 we visit in inorder the vertices in the subtree T_1 rooted at vertex 2. This takes us to vertex 5 and then to vertex 11, a leaf. Hence, in visiting the vertices of the subtree rooted at vertex 5 in inorder, we start listing the vertices visited as 11, 5, 12, 13, 14. We next visit vertex 2, the root of subtree T_1, and then the root at vertex 1. To complete this traversal the vertices in subtrees T_2 and T_3 must now be visited in inorder. For T_2 this takes us from vertex 6 to vertex 3 (the root of T_2), and then to vertex 7. Finally, for subtree T_3 with root at vertex 4 we first visit vertex 8, then the root at vertex 4, and then vertex 9. After that we traverse the subtree rooted at vertex 10; this gives the inorder listing 15, 10, 16, 17. Consequently, the inorder traversal of the tree in Fig. 15.11 determines the (inorder) sequence 11, 5, 12, 13, 14, 2, 1, 6, 3, 7, 8, 4, 9, 15, 10, 16, 17 for the vertices.

c) *Postorder:* For the postorder traversal of a tree we start at the root r and build the longest path, going to the leftmost son at each internal vertex arrived at. When we arrive at a leaf ℓ we visit this vertex and then backtrack to its father f. However, we do not visit f until after all of its descendants are visited. The next vertex we visit is found by applying the same procedure at f that was originally applied at r in obtaining ℓ. And at no time is any vertex visited more than once or before any of its descendants.

For the given tree, the postorder traversal starts with a postorder traversal of the subtree T_1 rooted at vertex 2. This yields the listing 11, 12, 13, 14, 5, 2. Proceeding to the subtree T_2 the postorder listing continues with 6, 7, 3. Then for T_3 we find 8, 9, 15, 16, 17, 10, 4 as the postorder listing. Consequently, for this tree, the postorder traversal visits the vertices in the order $11, 12, 13, 14, 5, 2, 6,$ $7, 3, 8, 9, 15, 16, 17, 10, 4, 1.$ □

The following spanning tree for a connected graph is related to the notion of preorder.

Let $G = (V, E)$ be a connected undirected graph with $r \in V$. Starting from r we construct a simple path in G. If this path includes every vertex in V we are finished. If not, let x and y be the last two vertices visited along this path, with y the last vertex. We then return, or *backtrack*, to the vertex x and construct a second simple path in G that starts at x and doesn't include any vertex already visited. If no such path exists, backtrack to the father f of x and see how far it is possible to branch off from f, building a simple path (with no previously visited vertices) to a new leaf y_1. Should all edges from the vertex f lead to vertices already encountered, backtrack one level higher, and continue the process. Since the graph is finite and connected, this technique, which is called *backtracking*, or *depth-first search*, will eventually determine a spanning tree T for G, where r is regarded as the root of T. Using T, the vertices of G can then be ordered in a preorder listing.

The depth-first search serves as a framework around which many algorithms can be designed to test for certain graph properties. In Section 15.4 one such algorithm will be examined in detail.

EXAMPLE 15.7   The spanning trees T_1, T_2 (both rooted at a) in Fig. 15.12 each provide a depth-first search for the vertices of $G = (V, E)$. □

Figure 15.12

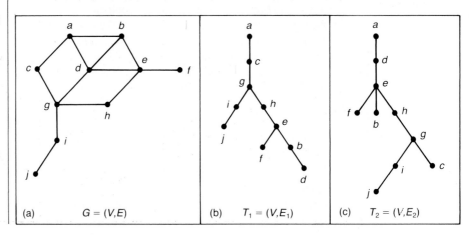

(a) $G = (V,E)$ (b) $T_1 = (V, E_1)$ (c) $T_2 = (V, E_2)$

A second method for searching the vertices of a connected undirected graph is the *breadth-first search*. Here we designate one vertex as the root and fan out to all sons of the root. From each son we then fan out to those vertices (not previously visited) that are adjacent to one of these sons. Continuing this process a vertex is never listed twice, so no cycle is constructed, and with G finite the process eventually terminates. The spanning trees T_3 and T_4 of Fig. 15.13 each provide a breadth-first search of graph G in Fig. 15.12(a). The tree T_3 is rooted at a; T_4 has d as its root.

Figure 15.13

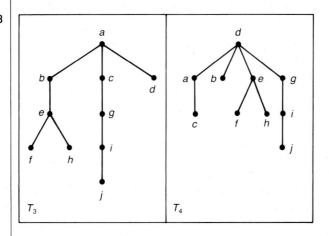

We apply the ideas of graph searching in the following example.

EXAMPLE 15.8 Let $G = (V, E)$ be an undirected graph on the vertices v_1, v_2, ..., v_7. If Fig. 15.14(a) is the adjacency matrix $A(G)$ for G, how can this representation of G be used to determine if G is connected, without drawing the graph?

Figure 15.14

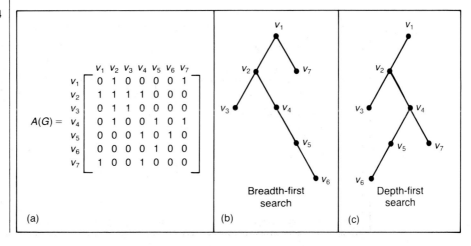

Using v_1 as the root, in part (b) of the figure the graph is searched by means of its adjacency matrix, using a breadth-first search. First we visit the sons of v_1, listing them in ascending order according to the subscripts on the v's in $A(G)$. Continuing the search, as all vertices in G are reached G is shown to be connected.

The same conclusion follows from the depth-first search in part (c). The tree here also has v_1 as its root. As the tree branches out to search the graph, it does so by listing the first vertex found adjacent to v_1 according to the row in $A(G)$ for v_1. Likewise, from v_2 the first new vertex in this search is found from $A(G)$ to be v_3. The vertex v_3 is a leaf in this tree since no new vertex can be visited from v_3. Backtracking to v_2, row 2 of $A(G)$ indicates that v_4 can now be visited from v_2. Continuing this process, the connectedness of G follows from part (c) of the figure.   □

It is time now to return to the main discussion on rooted trees.

Definition 15.4

▶ If $T = (V, E)$ is a rooted tree and $m \in \mathbf{Z}^+$, then T is called an *m-ary tree* if $\deg^-(v)$, the out degree of v, is m for all internal vertices v. When $m = 2$ we say the tree is a *binary tree.*[†]

The tree in Fig. 15.8(c) is an example of a rooted binary tree.

Some properties of these trees are considered in the following.

Theorem 15.5

▶ Let $T = (V, E)$ be an *m*-ary tree with $|V| = n$. If T has ℓ leaves and i internal vertices then (a) $n = mi + 1$; (b) $\ell = (m - 1)i + 1$; (c) $i = (\ell - 1)/(m - 1) = (n - 1)/m$.

Proof   This proof is left for the section exercises.   ∎

EXAMPLE 15.9   The Wimbledon tennis championship is a single-elimination tournament where a player (or doubles team) is eliminated after a single loss. If 27 women compete in the singles championship, how many matches must be played to determine the number one female player?

Figure 15.15

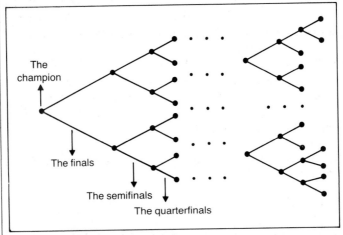

The champion

The finals

The semifinals

The quarterfinals

[†]The type of binary tree used in this chapter differs from the one defined in Section 11.5. (Unfortunatel; both types appear in the literature as binary trees.)

Consider the tree in Fig. 15.15. With 27 women competing there are 27 leaves in this binary tree, so from Theorem 15.5(c) the number of internal vertices (which is the number of matches) is $i = (\ell - 1)/(m - 1) = (27 - 1)/(2 - 1) = 26$. □

EXAMPLE 15.10 A classroom contains 25 minicomputers that must be connected to a wall socket that has four outlets. Connections are made by using extension cords that each have four outlets. What is the least number of cords needed to get these computers set up for class use?

The wall socket is considered as the root of an m-ary tree for $m = 4$. The minicomputers are the leaves of this tree, so $\ell = 25$. Each internal vertex corresponds with an extension cord, so by part (c) of Theorem 15.5, we need $(\ell - 1)/(m - 1) = (25 - 1)/(4 - 1) = 8$ extension cords. □

Definition 15.5 ▶ If $T = (V, E)$ is a rooted tree and h is the largest level number achieved by a leaf of T, then T is said to have *height* h. A rooted tree T of height h is said to be *balanced* if the level number of every leaf in T is $h - 1$ or h.

The rooted tree in Fig. 15.11 is a balanced tree of height 3. Tree T_1 in Fig. 15.12 has height 6 but is not balanced. (Why?)

The tree for the tournament in Example 15.9 must be balanced in order for the tournament to be as fair as possible. If it is not balanced, some competitor will receive more than one bye (an opportunity to advance without playing a match).

Before stating our next theorem let us recall that for any $x \in \mathbf{R}$, $\lfloor x \rfloor$ denotes the greatest integer in x, while $\lceil x \rceil$, the *ceiling of* x, is defined by $\lceil x \rceil = \lfloor x \rfloor = x$, for $x \in \mathbf{Z}$; $\lceil x \rceil = \lfloor x \rfloor + 1$, $x \notin \mathbf{Z}$.

Theorem 15.6 ▶ Let $T = (V, E)$ be an m-ary tree of height h. If T has ℓ leaves, then $\ell \le m^h$, and $h \ge \lceil \log_m \ell \rceil$.

Proof The proof is by induction on h. When $h = 1$, T is a tree with a root and m sons. In this case $\ell = m = m^h$, and the result is true. Assume the result for a tree of height $h - 1$, and consider a tree T with height h and ℓ leaves. (The level numbers that are possible for these leaves are 1, 2, . . . , h, with at least m of the leaves at level h.) The ℓ leaves of T are also the ℓ leaves (total) for the m subtrees T_i, $1 \le i \le m$, of T rooted at each of the sons of the root. For $1 \le i \le m$, let ℓ_i be the number of leaves in subtree T_i. (In the case where leaf and root coincide, $\ell_i = 1$. But since $m \ge 1$ and $h - 1 \ge 1$, we have $m^{h-1} \ge 1 = \ell_i$.) By the induction hypothesis, $\ell_i \le m^{h-1}$, so $\ell = \ell_1 + \ell_2 + \cdots + \ell_m \le m(m^{h-1}) = m^h$. With $\ell \le m^h$, $\log_m \ell \le \log_m (m^h) = h$, and since $h \in \mathbf{Z}^+$, $h \ge \lceil \log_m \ell \rceil$. ∎

Corollary 15.1 ▶ Let T be an m-ary tree with ℓ leaves and height h. If T is balanced, then $h = \lceil \log_m \ell \rceil$.

Proof This proof is left as an exercise. ∎

EXAMPLE 15.11 (*Decision Trees*) Given eight coins (identical in appearance) and a pan balance, determine the heavier counterfeit coin.

Let the coins be labeled 1, 2, 3, . . . , 8. In using the pan balance there are three outcomes to consider: (a) the two sides balance to indicate that the coins in the two pans are not counterfeit; (b) the left pan of the balance goes down, indicating that the counterfeit coin is in the left pan; or (c) the right pan goes down, indicating that it holds the counterfeit coin.

Figure 15.16

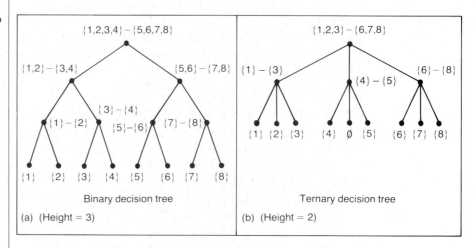

Binary decision tree

(a) (Height = 3)

Ternary decision tree

(b) (Height = 2)

In Fig. 15.16(a), we search for the counterfeit coin by first balancing coins $1, 2, 3, 4$ against $5, 6, 7, 8$. If the balance tips to the right we follow the right branch from the root to then analyze coins $5, 6$ against $7, 8$. Otherwise the balance tips to the left and coins $1, 2$ are tested against $3, 4$. At each successive level we have half as many coins to test, so that at level 3 the heavier counterfeit coin has been identified.

The tree in part (b) of the figure finds the heavier coin in two weighings. The first weighing balances coins $1, 2, 3$ against $6, 7, 8$. Three possible outcomes can occur: (i) the balance tips to the right, indicating that the heavier coin is 6, 7, or 8, and we follow the right branch from the root; (ii) the balance tips to the left and we follow the left branch to find which of $1, 2, 3$ is the heavier, or, (iii) the pans balance and we follow the center branch to find which of $4, 5$ is heavier. At each internal vertex the label indicates which coins are being compared. Unlike part (a), a conclusion may be deduced in part (b) when a coin is not included in a weighing. Finally, when comparing coins 4 and 5, since equality cannot take place, the center leaf is labeled with $\emptyset$.

In this particular problem the height of the ternary tree used must be at least 2. With eight coins involved, the tree will have at least eight leaves. Consequently, with $\ell \geq 8$, it follows from Theorem 15.6 that $h \geq \lceil \log_3 \ell \rceil \geq \lceil \log_3 8 \rceil = 2$, so at least two weighings are needed. If n coins are involved, the ternary tree will have ℓ leaves where $\ell \geq n$, and its height h satisfies $h \geq \lceil \log_3 n \rceil$.   □

We conclude this section with an example on a sorting procedure called the *merge sort*. Before doing this, however, a few preliminaries are needed.

In Example 11.4 (Chapter 11) the bubble sort was introduced in the study of linear recurrence relations. There we found that the number of comparisons needed to sort a list of n items is $n(n - 1)/2$. Consequently, this algorithm determines a function $f: \mathbf{Z}^+ \to \mathbf{R}$ defined by $f(n) = n(n - 1)/2$. This function f is dominated by the function $g: \mathbf{Z}^+ \to \mathbf{R}$ defined by $g(n) = n^2$, the *time complexity function* of the algorithm. The bubble sort is said to require $\mathcal{O}(n^2)$ comparisons, i.e., for large n, the number of comparisons equals cn^2, where c is a proportionality constant that is generally not specified, since it depends on such factors as the compiler and the computer that are used. (See Miscellaneous Exercise 28 in Chapter 3 and Miscellaneous Exercise 11 in Chapter 5.) This function g is used as a measure of the time needed to execute this algorithm on a problem of size n. To compare algorithms for the solution of a given problem of size n, one then compares their corresponding complexity functions.

We shall find that the merge sort requires $\mathcal{O}(n \log_2 n)$ comparisons.

Given a list of n items to sort in ascending order, the *merge sort* recursively splits the given list and all subsequent sublists in half (or as close as possible to half) until each sublist contains a single element. Then the procedure merges these sublists in ascending order until the original n items have been so sorted. The splitting and merging processes can best be described by a pair of balanced binary rooted trees, as in the next example.

Figure 15.17

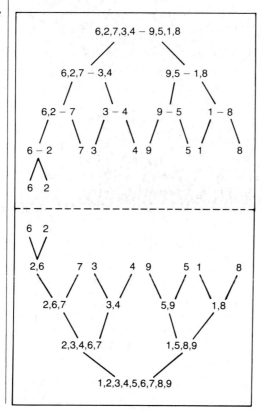

EXAMPLE 15.12

(*Merge Sort*) Using the merge sort, Fig. 15.17 sorts the list $6, 2, 7, 3, 4, 9, 5, 1, 8$. The tree at the top of the figure shows how the process first splits the given list into sublists of size 1. The merging process is then outlined by the tree at the bottom of the figure. □

To compare the merge sort to the bubble sort, we want to determine its complexity function. The following lemma will be needed for this task.

Lemma 15.1

▶ Let L_1 and L_2 be two sorted lists of ascending numbers, where L_i contains n_i elements, $i = 1, 2$. Then L_1 and L_2 can be merged into one ascending list L using at most $n_1 + n_2 - 1$ comparisons.

Proof

To merge L_1, L_2 into list L we perform the following algorithm.

Step 1: Set L equal to the empty list $\emptyset$.

Step 2: Compare the first elements in L_1, L_2. Remove the smaller of the two from the list it is in and place it at the end of L.

Step 3: For the present lists L_1, L_2 (one change has been made) there are two considerations.

a) If either of L_1, L_2 is empty, then the other list is concatenated to the end of L. This completes the merging process.

b) If not, return to step 2.

Each comparison of a number from L_1 with one from L_2 results in an element being added to list L, so there cannot be more than $n_1 + n_2$ comparisons. When one of the lists L_1 or L_2 becomes empty, no further comparisons are needed, so the maximum number of comparisons needed is $n_1 + n_2 - 1$. ■

To determine the complexity function of the merge sort, consider a list of n elements. We do not treat the general problem here, assuming instead that $n = 2^h$. (The result is true for the general situation, however.) In the splitting process, the list of 2^h elements is first split into two sublists of size 2^{h-1}. (These are the level 1 vertices in the tree representing the splitting process.) Continuing the process, each successive list of size 2^{h-k}, $h > k$, is at level k and splits into two sublists of size $(1/2)(2^{h-k}) = 2^{h-k-1}$. At level h the sublists each contain $2^{h-h} = 1$ element.

Reversing the process, we first merge the $n = 2^h$ leaves into 2^{h-1} ordered sublists of size 2. These sublists are at level $h - 1$ and require $(1/2)(2^h) = 2^{h-1}$ comparisons (one per pair). Continuing this merging process, at each of the 2^k vertices at level k, $1 \le k < h$, there is a sublist of size 2^{h-k}, obtained from merging the two sublists of size 2^{h-k-1} at its sons (on level $k + 1$). From Lemma 15.1, this merging requires at most $2^{h-k-1} + 2^{h-k-1} - 1 = 2^{h-k} - 1$ comparisons. When the sons of the root are reached there are two sublists of size 2^{h-1} (at level 1). To merge these sublists into the final list requires at most $2^{h-1} + 2^{h-1} - 1 = 2^h - 1$ comparisons.

Consequently, for $1 \leq k \leq h$, at level k there are 2^{k-1} pairs of vertices. At each of these vertices is a sublist of size 2^{h-k}, so it takes at most $2^{h-k+1} - 1$ comparisons to merge each pair of sublists. With 2^{k-1} pairs of vertices at level k, the total number of comparisons at level k is at most $2^{k-1}(2^{h-k+1} - 1)$. Summing over all levels k, $1 \leq k \leq h$, the total number of comparisons is at most

$$\sum_{k=1}^{h} 2^{k-1}(2^{h-k+1} - 1) = \sum_{k=0}^{h-1} 2^k(2^{h-k} - 1) = \sum_{k=0}^{h-1} 2^h - \sum_{k=0}^{h-1} 2^k = h \cdot 2^h - (2^h - 1).$$

With $n = 2^h$, we have $h = \log_2 n$ and

$$h \cdot 2^h - (2^h - 1) = n \log_2 n - (n - 1) = n \log_2 n - n + 1,$$

where $n \log_2 n$ is the dominating term for large n. Thus the complexity function for this sorting procedure is $g(n) = n \log_2 n$. Hence the number of comparisons needed to merge sort a list of n items is $\mathbb{O}(n \log_2 n)$, (i.e. equal to $dn \log_2 n$ for some proportionality constant d, and all $n \geq n_0$, for some particular $n_0 \in \mathbf{Z}^+$).

Although $n \log_2 n \leq n^2$ for all $n \in \mathbf{Z}^+$, it does *not* follow that since the bubble sort is $\mathbb{O}(n^2)$ and the merge sort is $\mathbb{O}(n \log_2 n)$ that the merge sort is more efficient than the bubble sort for all $n \in \mathbf{Z}^+$. For $n \leq 15$ the bubble sort requires less programming effort and generally takes less time than the merge sort. However, as n increases, the ratio of the running times, as measured by $(cn^2)/(dn \log_2 n) = (c/d)(n/\log_2 n)$, gets arbitrarily large. Consequently, as the input list increases in size, the $\mathbb{O}(n^2)$ algorithm (bubble sort) will take significantly more time than the $\mathbb{O}(n \log_2 n)$ algorithm (merge sort).

For more on function complexity and sorting algorithms, the reader should examine [1], [4], and [5] in the chapter references.

EXERCISES

1. Answer the following questions for the tree in Fig. 15.18

Figure 15.18

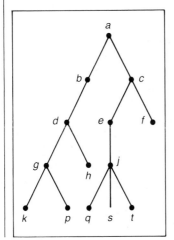

a)   Which vertices are the leaves?

b)   Which vertex is the root?

c)   Which vertex is the father of g?

d)   Which vertices are the descendants of c?

e)   Which vertices are the brothers of s?

f)   What is the level number of vertex f?

g)   Which vertices have level number 4?

h)   What is the height of the tree?

i)   Is the tree balanced?

2. Let $T = (V, E)$ be a rooted tree ordered by a universal address system.

   a)   If vertex v in T has address 2.1.3.6, what is the smallest number of brothers that v must have?

   b)   For the vertex v in part (a), find the address of its father.

   c)   How many ancestors does the vertex v in part (a) have?

   d)   With the presence of v in T, what other addresses must there be in the system?

3. a)   Write the expression $(w + x - y)/(\pi * z^3)$ in Polish notation, using a rooted tree.

   b)   What is the value of the expression (in Polish notation) $/ \uparrow a - bc + d * ef$, if $a = c = d = e = 2$, $b = f = 4$?

   c)   Extend the structure of the rooted tree in Fig. 15.8(c) to include monary (or unary) operations such as the trigonometric functions and multiplicative inverse. Find the rooted trees for the following.

   i) $(\sin \pi x) + (y - z)^{-1}$ ii) $\sin(\pi x + y) - z^{-1}$

4. List the vertices in the tree of Fig. 15.19 when they are visited in a preorder traversal, an inorder traversal, and a postorder traversal.

Figure 15.19

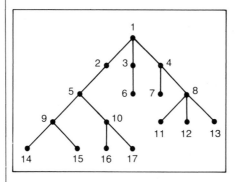

5. The rooted binary tree $T = (V, E)$ has $V = \{a, b, c, \ldots, i, j, k\}$. The preorder traversal of T visits the vertices in the order $a, b, d, e, c, f, h, i, g, j, k$;

the postorder listing yields $d, e, b, h, i, f, j, k, g, c, a$. From this information draw T.

6. a) Let $T = (V, E)$ be a rooted tree, and let preorder (v) denote the position of vertex v in a preorder listing of V. Also, let desc(v) denote the number of descendants of v. For $v_1, v_2 \in V$, prove that v_1 is a descendant of v_2 iff preorder$(v_2) <$ preorder$(v_1) \leq$ preorder$(v_2) +$ desc(v_2).

   b) State and prove a comparable result using postorder in place of preorder.

7. Find three different depth-first search spanning trees (two with the same root) for (a) K_6; (b) the graph in Fig. 14.48(a); and, (c) the graph in Fig. 14.55(a).

8. Find a breadth-first spanning tree for each graph in Exercise 7.

9. Let $G = (V, E)$ be an undirected graph with adjacency matrix $A(G)$ as shown here.

$$
\begin{array}{c c c c c c c c c}
 & v_1 & v_2 & v_3 & v_4 & v_5 & v_6 & v_7 & v_8 \\
v_1 & 0 & 1 & 0 & 0 & 0 & 0 & 1 & 0 \\
v_2 & 1 & 1 & 0 & 1 & 1 & 0 & 1 & 0 \\
v_3 & 0 & 0 & 0 & 1 & 0 & 1 & 0 & 1 \\
v_4 & 0 & 1 & 1 & 0 & 0 & 0 & 0 & 0 \\
v_5 & 0 & 1 & 0 & 0 & 0 & 0 & 1 & 0 \\
v_6 & 0 & 0 & 1 & 0 & 0 & 1 & 0 & 0 \\
v_7 & 1 & 1 & 0 & 0 & 1 & 0 & 0 & 0 \\
v_8 & 0 & 0 & 1 & 0 & 0 & 0 & 0 & 0 \\
\end{array}
$$

Use a breadth-first search based on $A(G)$ to determine if G is connected.

10. a) For the graph in Fig. 15.20 find a breadth-first spanning tree rooted at vertex a.

    b) How many different breadth-first spanning trees are rooted at vertex a?

Figure 15.20

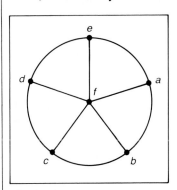

11. Prove Theorem 15.5 and Corollary 15.1.

12. With m, n, i, ℓ as in Theorem 15.5, prove that

    (a) $n = (m\ell - 1)/(m - 1)$; and   (b) $\ell = [(m - 1)n + 1]/m$.

13. For $m \geq 3$, an m-ary tree can be transformed into a binary tree as shown in Fig. 15.21.

Figure 15.21

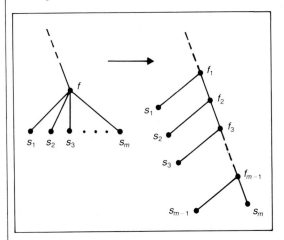

a) Use this technique to transform the ternary decision tree in Fig. 15.16(b).

b) If T is a quaternary-rooted tree of height 3, what is the height of T after it is transformed into a binary tree?

c) Answer part (b) if T is an m-ary tree of height h.

14. a) At a men's singles tennis tournament 25 players each bring a can of tennis balls. When a match is played, one can of balls is opened and used, then kept by the loser. The winner takes the unopened can on to his next match. How many cans of tennis balls will be opened during this tournament? How many matches are played in the tournament?

b) In how many matches did the tournament champion play?

c) If a match is won by the first opponent to win three sets, what is the maximum number of sets that could have been played (by all entrants) during the tournament?

15. What is the maximum number of internal vertices that a quaternary tree of height 8 can have? What is the number for an m-ary tree of height h?

16. On the first Sunday of 1984 Rizzo and Frenchie start a chain letter with each of them sending five letters (to ten different friends). Each person receiving the letter is to send out five copies to five new people on the Sunday following the letter's arrival. After the first seven Sundays have passed, what is the total number of chain letters that have been mailed? How many were mailed on the last three Sundays?

17. Use a ternary decision tree to repeat Example 15.11 for a set of 12 coins, where one of the coins is a heavier counterfeit.

18. Apply the merge sort to each of the following lists. Draw the splitting and merging trees for each application of the procedure.

a) $-1, 0, 2, -2, 3, 6, -3, 5, 1, 4$ b) $-1, 7, 4, 11, 5, -8, 15, -3, -2, 6, 10, 3$.

19. Related to the merge sort is a somewhat more efficient procedure called the *quick sort*. Here we start with a list L: $a_1, a_2, \ldots, a_n$, and use a_1 as a pivot to develop two sublists L_1 and L_2 as follows. For $i > 1$, if $a_i < a_1$, place a_i at the end of the first list being developed (this is L_1 at the end of the process); otherwise, place a_i at the end of the second list.

After all a_i, $i > 1$, have been processed, place a_1 at the end of the first list. Now apply quick sort recursively to each of the lists L_1, L_2 to obtain sublists $L_{11}, L_{12}, L_{21}, L_{22}$. Continue the process until the resulting sublists each contain one element. The sublists are then ordered and their concatenation gives the ordering sought for the original list L.

Apply quick sort to each list in Exercise 18.

15.3 ■ WEIGHTED TREES AND PREFIX CODES

In Chapter 12 we saw an application on the transmission of information represented by sequences of 0's and 1's. This application used the structure of a group to develop encoding and decoding schemes.

At this point we want to develop a different type of coding scheme for the transmission of a set of symbols, such as the letters in the alphabet. Since there are 26 letters we should be able to encode these symbols in terms of sequences of five binary bits, since $2^4 < 26 < 2^5$. However, in reading a page of a newspaper one realizes that not all letters occur with the same frequency. Consequently, it would be more efficient to use binary sequences of different lengths, with the most frequently occurring letters (such as e, i, t) represented by the shortest possible sequences. For example, consider $S = \{a, e, n, r, t\}$, a subset of the alphabet. Represent the elements of S by the binary sequences:

$$a: 01 \qquad e: 0 \qquad n: 101 \qquad r: 10 \qquad t: 1.$$

If the message "*ata*" is to be transmitted, the binary sequence 01101 is sent. Unfortunately, this sequence is also transmitted for the messages "*etn*", "*atet*", and "*an*".

Consider a second encoding scheme given by:

$$a: 111 \qquad e: 0 \qquad n: 1100 \qquad r: 1101 \qquad t: 10.$$

Here the message "*ata*" is represented by the sequence 11110111 and there are no other possibilities to confuse the situation. What's more, the labeled binary tree in Fig. 15.22 can be used to decode the sequence 11110111. Starting at the root, traverse the edge labeled 1 to the right son (of the root). Continuing along the next two edges labeled with 1, we arrive at the leaf labeled a. Hence the unique path from the root to the vertex at a is unambiguously determined by the first three 1's in the sequence 11110111. Returning to the root, the next two symbols in the sequence, namely 10, determine the unique path along the edge from the root to its right son, followed by the edge from that son to its left son. This terminates at the vertex labeled t. Again returning to the root, the final three bits of the sequence determine the letter a for a second time. Hence the tree "decodes" 11110111 as *ata*.

Figure 15.22

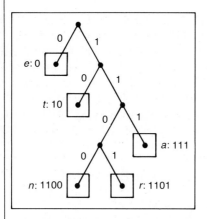

Now why did the second encoding scheme work out so readily while the first led to ambiguities? In the first scheme, r is represented as 10 and n as 101. If we encounter the symbols 10, how can we determine if the symbols represent r or the first two symbols of 101, which represents n? The problem is that the sequence for r is a prefix of the sequence for n. This dilemma does not occur in the second encoding scheme, which suggests the following.

Definition 15.6 ▶ A set P of binary sequences (representing a set of symbols) is called a *prefix code* if no sequence in P is the prefix of any other sequence in P.

Consequently, the binary sequences 111, 0, 1100, 1101, 10 constitute a prefix code for the letters a, e, n, r, t, respectively. But how did the binary tree of Fig. 15.22 come about? To deal with this problem, the following concept is needed.

Definition 15.7 ▶ If T is a rooted binary tree of height h, then T is called a *full* binary tree if all the leaves in T are at level h.

Figure 15.23

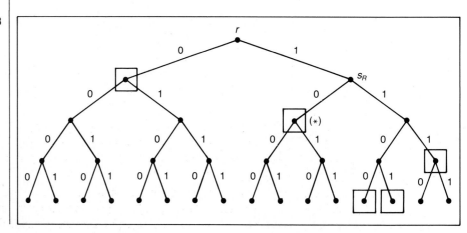

EXAMPLE 15.13.

For the prefix code $P = \{111, 0, 1100, 1101, 10\}$, the longest binary sequence has length 4. Draw the labeled full binary tree of height 4, as in Fig. 15.23. The elements of P are assigned to the vertices of this tree as follows. For example, the sequence 10 traces the path from the root r to its right son s_R. Then it continues to the left son of s_R where the box (marked with the asterisk) indicates completion of the sequence. Returning to the root, the other four sequences are traced out in similar fashion, resulting in the five boxed vertices. For each boxed vertex remove the subtree (except for the root) that it determines. The resulting pruned tree is the tree of Fig. 15.22. □

We turn now to a method of determining a labeled tree, where the frequency of occurrence of each symbol is taken into account. Once again a prefix code is needed where short sequences are used for the more frequently occurring symbols. If there are many symbols, such as all 26 letters of the alphabet, a trial and error method for constructing such a tree is not efficient. Instead there is an elegant construction developed by David A. Huffman that provides a recursive technique for constructing such trees.

The general problem now arises as follows.

Let $w_1, w_2, \ldots, w_n$ be a set of positive numbers called *weights*, where $w_1 \le w_2 \le \ldots \le w_n$. If $T = (V, E)$ is a binary tree with n leaves, assign these weights (in any one-to-one manner) to the n leaves. The result is called a *binary tree for the weights* $w_1, w_2, \ldots, w_n$. The *weight of the tree*, denoted $W(T)$, is defined as $\sum_{i=1}^{n} w_i \ell(w_i)$ where, for $1 \le i \le n$, $\ell(w_i)$ is the level number of the leaf assigned the weight w_i. The objective is to assign the weights so that $W(T)$ is as small as possible. A binary tree T_1 for these weights is said to be an *optimal* tree if $W(T_1) \le W(T)$ for any other binary tree T for the weights.

Figure 15.24

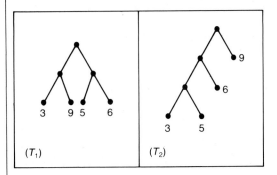

(T_1) (T_2)

Figure 15.24 shows two binary trees for the weights 3, 5, 6, and 9. For tree T_1, $W(T_1) = \sum_{i=1}^{4} w_i \ell(w_i) = (3 + 9 + 5 + 6) \cdot 2 = 46$, since each leaf has level number 2. In the case of T_2, $W(T_2) = 3 \cdot 3 + 5 \cdot 3 + 6 \cdot 2 + 9 \cdot 1 = 45$, which we shall find is optimal.

The major idea behind Huffman's construction is that in order to obtain an optimal tree T for the n weights $w_1, w_2, w_3, \ldots, w_n$, one considers an optimal tree T_1 for the $n - 1$ weights $w_1 + w_2, w_3, \ldots, w_n$. (It cannot be assumed that $w_1 + w_2 \le w_3$).

In particular, the tree T_1 is transformed into T by replacing the leaf with weight $w_1 + w_2$ by a tree of height 1 with left son of weight w_1 and right son of weight w_2. Assuming the tree T_2 in Fig. 15.24 is optimal for the four weights $1 + 2, 5, 6, 9$, then the tree in Fig. 15.25 will be optimal for the five weights $1, 2, 5, 6, 9$.

Figure 15.25

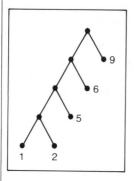

To establish these claims, the following lemma is needed.

Lemma 15.2

▶ If T is an optimal tree for the n weights $w_1 \leq w_2 \leq \ldots \leq w_n$, then there exists an optimal tree T' in which the leaves of weights w_1 and w_2 are brothers.

Proof   Let v be an internal vertex of T where the level number of v is maximal for all internal vertices. Let w_x and w_y be the weights assigned to the sons x, y of vertex v, with $w_x \leq w_y$. By the choice of vertex v, $\ell(w_x) = \ell(w_y) \geq \ell(w_1), \ell(w_2)$. Consider the case of $w_1 < w_x$. (If $w_1 = w_x$, then w_1 and w_x can be interchanged and we would consider the case of $w_2 < w_y$. Applying the following proof to this case, we would find that w_y and w_2 can be interchanged.)

If $\ell(w_x) > \ell(w_1)$, let $\ell(w_x) = \ell(w_1) + j$, $j \in \mathbf{Z}^+$. Then $w_1\ell(w_1) + w_x\ell(w_x) = w_1\ell(w_1) + w_x[\ell(w_1) + j] = w_1\ell(w_1) + w_xj + w_x\ell(w_1) > w_1\ell(w_1) + w_1j + w_x\ell(w_1) = w_1\ell(w_x) + w_x\ell(w_1)$. So $W(T) = w_1\ell(w_1) + w_x\ell(w_x) + \sum_{i \neq 1, x} w_i\ell(w_i) > w_1\ell(w_x) + w_x\ell(w_1) + \sum_{i \neq 1, x} w_i\ell(w_i)$. Consequently, by interchanging the locations of the weights w_1 and w_x, a tree of smaller weight is obtained. But this contradicts the choice of T as an optimal tree. Therefore $\ell(w_x) = \ell(w_1) = \ell(w_y)$. In a similar manner it can be shown that $\ell(w_y) = \ell(w_2)$, so $\ell(w_x) = \ell(w_y) = \ell(w_1) = \ell(w_2)$. Interchanging the locations of the pairs w_1, w_x, and w_2, w_y, we obtain an optimal tree T_1, where w_1, w_2 are brothers. ■

(From this lemma we see that small weights will be at the bottom level in an optimal tree.)

Theorem 15.7

▶ Let T be an optimal tree for the weights $w_1 + w_2, w_3, \ldots, w_n$, where $w_1 \leq w_2 \leq w_3 \leq \ldots \leq w_n$. At the leaf with weight $w_1 + w_2$ place a binary tree of height 1 and assign the weights w_1, w_2 to the sons (leaves) of this former leaf. The new binary tree T_1 so constructed is then optimal for the weights $w_1, w_2, w_3, \ldots, w_n$.

Proof | Let T_2 be an optimal tree for the weights $w_1, w_2, \ldots, w_n$, where the leaves for weights w_1, w_2 are brothers. Remove the leaves of weights w_1, w_2 and assign the weight $w_1 + w_2$ to their father (now a leaf). This binary tree is denoted T_3 and $W(T_2) = W(T_3) + w_1 + w_2$. Also, $W(T_1) = W(T) + w_1 + w_2$. Since T is optimal, $W(T) \leq W(T_3)$. If $W(T) < W(T_3)$, then $W(T_1) < W(T_2)$, contradicting the choice of T_2 as optimal. Hence $W(T_1) = W(T_2)$ and T_1 is optimal for the weights $w_1, w_2, \ldots, w_n$. ∎

Remark. The proof above started with an optimal tree T_2 whose existence rests on the fact that there is only a finite number of ways we can assign n weights to a rooted binary tree with n leaves. Consequently, with a finite number of assignments there is at least one where $W(T)$ is minimal. But finite numbers can be large. This proof establishes the existence of an optimal tree for a set of weights, and develops a recursive construction for such a tree. The following example will use such a construction.

EXAMPLE 15.14 | Construct an optimal prefix code for the symbols a, o, q, u, y, z which occur (in a given sample) with frequencies $20, 28, 4, 17, 12, 7$, respectively.

Figure 15.26

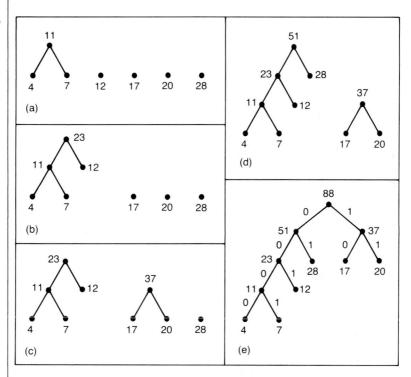

Figure 15.26 shows the construction that follows Huffman's recursive procedure. In part (a), weights 4 and 7 are combined so that we then consider the construction for the weights $11, 12, 17, 20, 28$. At each step we create a tree with subtrees rooted at the

two smallest weights (these two smallest weights belong to vertices each of which is originally either isolated (tree with just a root) or the root of a tree obtained earlier in the construction). From the last result, the prefix code is determined as

$$a: 11 \quad o: 01 \quad q: 0000 \quad u: 10 \quad y: 001 \quad z: 0001 \qquad \square$$

EXERCISES

1. For the prefix code given in Fig. 15.22, decode the sequences (a) 1001111101; (b) 10111100110001101; and (c) 1101111110010.

2. A code for $\{a, b, c, d, e\}$ is given by $a: 00$ $b: 01$ $c: 101$ $d: x10$ $e: yz1$, where $x, y, z \in \{0, 1\}$. Determine x, y, and z so that the given code is a prefix code.

3. Construct an optimal prefix code for the symbols a, b, c, $\ldots$, i, j that occur (in a given sample) with respective frequencies $78, 16, 30, 35, 125, 31, 20, 50, 80, 3$.

4. Let L_i, $1 \le i \le 4$, be four lists of numbers, each sorted in ascending order. The numbers of entries in these lists are 75, 40, 110, and 50, respectively.

    a) How many comparisons are needed to merge these four lists by merging L_1 and L_2, merging L_3 and L_4, and then merging the two resulting lists?

    b) How many comparisons are needed if we first merge L_1 and L_2, then merge the result with L_3, and finally merge this result with L_4?

    c) In order to minimize the total number of comparisons in this merging of the four lists, what order should the merging follow?

    d) Extend the result in part (c) to n sorted lists $L_1, L_2, \ldots, L_n$.

15.4 ■ BICONNECTED COMPONENTS AND ARTICULATION POINTS

Let $G = (V, E)$ be the connected undirected graph in Fig. 15.27(a), where each vertex represents a communication center. Here an edge $\{x, y\}$ indicates the existence of a communication link between the centers at x and y.

Figure 15.27

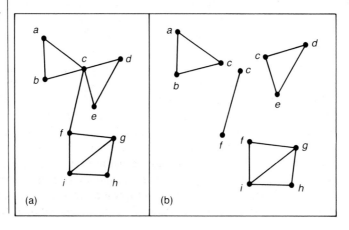

(a)   (b)

By splitting the vertices at c and f we obtain the collection of subgraphs in part (b) of the figure. These vertices are examples of the following.

Definition 15.8

▶ A vertex v in an undirected graph $G = (V, E)$ is called an *articulation point* if the splitting of v results in a graph with more components than in G. A graph with no articulation points is called *biconnected*. If a graph has one or more articulation points, the biconnected subgraphs that result upon the complete splitting of these vertices are called the *biconnected components* of the graph.

In terms of communication centers and links, the articulation points of the graph indicate where the system is most vulnerable. Without articulation points such a system is more likely to survive disruptions at a communication center, whether by the breakdown of a technical device or by external forces.

The problem of finding the articulation points in a connected graph provides an application for the depth-first spanning tree. The objective here is the development of an algorithm that determines the articulation points of a connected undirected graph. If no such points exist, then the graph is biconnected. Should such vertices exist, the resulting biconnected components provide information about such properties as the planarity and chromatic number of the given graph.

The following preliminaries are needed for developing this algorithm.

Lemma 15.3

▶ Let $G = (V, E)$ be a connected undirected graph with $T = (V, E')$ a depth-first spanning tree of G. If $\{a, b\} \in E$ but $\{a, b\} \notin E'$ then a is either an ancestor or descendant of b.

Proof

Suppose that a is neither an ancestor nor a descendant of b. From the depth-first spanning tree T we obtain a preorder listing for the vertices in V. For any $v \in V$ let $\text{dfi}(v)$ denote the depth-first index of vertex v, i.e., the position of v in the preorder listing. Assume that $\text{dfi}(a) < \text{dfi}(b)$. Consequently, a is encountered before b in the preorder traversal of T. But then with $\{a, b\} \in E$ it was possible for the depth-first search to go from a to b and use the edge $\{a, b\}$ in T.   ∎

If $G = (V, E)$ is a connected undirected graph, let $T = (V, E')$ be a depth-first spanning tree for G, as shown in Fig. 15.28. By Lemma 15.3, the dotted edge $\{a, b\}$, which is not part of T, indicates an edge that may exist in G. Such an edge is called a *back edge* of T, and here a is an ancestor of b. (The $\text{dfi}(a) = 3$ while $\text{dfi}(b) = 6$.) A dotted edge, such as $\{b, c\}$ in the figure, cannot exist in G, also because of Lemma 15.3. Thus all edges of G are either edges in T or back edges of T.

Let $x \in V$. For G as above, let $T_{x,s}$ be the subtree consisting of edge $\{x, s\}$ (s a son of x) together with the tree T_s rooted at s. If there is no back edge from a descendant of x in $T_{x,s}$ to an ancestor of x, then the splitting of vertex x results in the separation of $T_{x,s}$ from G, and x is an articulation point. If no other articulation points of G occur in $T_{x,s}$, then the addition to $T_{x,s}$ of all other edges in G determined by the vertices in $T_{x,s}$ results in a biconnected component of G. (Since a root has no ancestors, the root is an articulation point if it has more than one son.)

The depth-first spanning tree preorders the vertices of G. Let $\text{dfi}(x)$ denote the depth-first index of x in that preorder. If y is a descendant of x, then $\text{dfi}(x) < \text{dfi}(y)$.

Figure 15.28

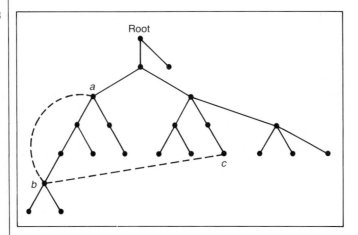

For y an ancestor of x, $\mathrm{dfi}(x) > \mathrm{dfi}(y)$. Define $\mathrm{low}(x) = \min\{\mathrm{dfi}(y)\,|\,y$ is connected in G to either x or a descendant of $x\}$. If z is the father of x (in T), then there are two possibilities to consider:

1. $\mathrm{low}(x) = \mathrm{dfi}(z)$: In this case T_x, the subtree rooted at x, contains no vertex that is adjacent to an ancestor of z by means of a back edge of T. Hence z is an articulation point of G. If T_x contains no articulation points, then T_x together with edge $\{z, x\}$ is a biconnected component of G. (Now remove T_x and edge $\{z, x\}$ from T and apply this idea to the remaining subtree of T.)

2. $\mathrm{low}(x) < \mathrm{dfi}(z)$: Here there is a descendant of z that can be joined (by a back edge) to an ancestor of z.

To deal in an efficient manner with these ideas we develop the following algorithm. Let $G = (V, E)$ be a connected undirected graph defined in terms of its adjacency matrix $A(G)$.

Step 1: Find a depth-first spanning tree T for G. Let $y_1, y_2, \ldots, y_n$ be the vertices of G preordered by T. Then $\mathrm{dfi}(y_j) = j$, $1 \le j \le n$.

Step 2: Start with y_n and continue back to $y_{n-1}, y_{n-2}, \ldots, y_4, y_3$, determining $\mathrm{low}(y_j)$, $3 \le j \le n$, recursively as follows:

a) $\mathrm{low}'(y_j) = \min\{\mathrm{dfi}(z)\,|\,z$ is adjacent in G to $y_j\}$.

b) If $s_1, s_2, \ldots, s_m$ are the sons of y_j, then $\mathrm{low}(y_j) = \min\{\mathrm{low}'(y_j), \mathrm{low}(s_1), \mathrm{low}(s_2), \ldots, \mathrm{low}(s_m)\}$. (No problem arises here, for the vertices are examined in the reverse order to the given preorder. Consequently, if s is the son of f, $\mathrm{low}(s)$ is determined before $\mathrm{low}(f)$.)

Step 3: Let w_j be the father of y_j. If $\mathrm{low}(y_j) = \mathrm{dfi}(w_j)$ then w_j is an articulation point of G (and T). Moreover, the subtree rooted at y_j together with the edge $\{w_j, y_j\}$ is part of a biconnected component of G.

EXAMPLE 15.15

We apply this algorithm to the graph $G = (V, E)$ given by the adjacency matrix $A(G)$ shown here.

	x_1	x_2	x_3	x_4	x_5	x_6	x_7	x_8	x_9	x_{10}	
x_1	0	1	1	1	1	1	0	0	1	0	y_1
x_2	1	0	0	0	0	0	0	0	0	0	y_2
x_3	1	0	0	0	1	0	0	0	0	0	y_3
x_4	1	0	0	0	0	1	1	1	0	0	y_5
x_5	1	0	1	0	0	0	0	0	0	0	y_4
x_6	1	0	0	1	0	0	0	0	0	0	y_6
x_7	0	0	0	1	0	0	0	1	0	0	y_7
x_8	0	0	0	1	0	0	1	0	1	1	y_8
x_9	1	0	0	0	0	0	0	1	0	0	y_9
x_{10}	0	0	0	0	0	0	0	1	0	0	y_{10}
	y_1	y_2	y_3	y_5	y_4	y_6	y_7	y_8	y_9	y_{10}	

Figure 15.29

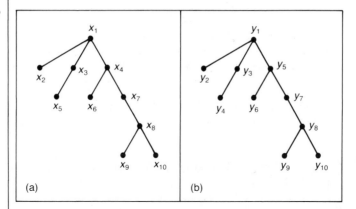

(a)   (b)

The 10 vertices of G may be traversed in the preorder listing given in Figure 15.29(a). Part (b) of that figure provides a re-indexing of these vertices so that for the depth-first spanning tree shown, dfi(y_j) = j, $1 \le j \le 10$. These are the y_j's, $1 \le j \le 10$, that appear at the ends of the rows and columns in the adjacency matrix $A(G)$. Starting with vertex y_{10}, from $A(G)$ we find low'(y_{10}) = 8, since y_8 is adjacent to y_{10}, but for $1 \le i \le 7$, y_i is not adjacent to y_{10}. With y_{10} a leaf, if follows that low(y_{10}) = 8. Since low(y_{10}) = dfi(y_8), and y_8 is the father of y_{10}, y_8 is an articulation point of G; the edge $\{y_8, y_{10}\}$ is the biconnected component. In Fig. 15.30(a), for each y_j, $3 \le j \le 10$, the ordered pair next to y_j is (low'(y_j), low(y_j)). These are computed by the algorithm as we proceed backwards from y_{10} to y_9, to y_8, and so forth, to y_3. In computing y_8, for example, from $A(G)$ we find that low'(y_8) = 5, since y_5 is adjacent to y_8 and for any $1 \le i \le 4$, y_i is not adjacent to y_8. Since y_9 is a son of y_8 and low(y_9) = 1, it follows that low(y_8) = 1. For both y_3 and y_5, low(y_3) = low(y_5) = 1 = dfi(y_1), with y_1 the father of both y_3 and y_5. Hence y_1 is an articulation point and the following subtrees

Figure 15.30

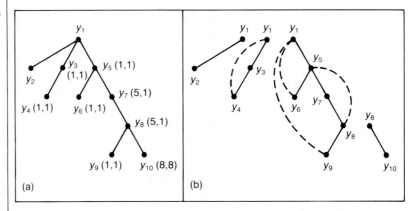

(a)   (b)

give rise to two more biconnected components: (a) the subtree rooted at y_3 together with the edge $\{y_1, y_3\}$; and (b) the subtree rooted at y_5, with the edge $\{y_8, y_{10}\}$ deleted and the edge $\{y_1, y_5\}$ adjoined. Finally, since y_2 is a leaf, the edge $\{y_1, y_2\}$ constitutes one more biconnected component.

Figure 15.30(b) shows the articulation points y_1 and y_8 for G. These determine the four biconnected components of G shown in the figure.   □

EXERCISES

1. Find the articulation points and biconnected components for the graph in Fig. 15.31.

Figure 15.31

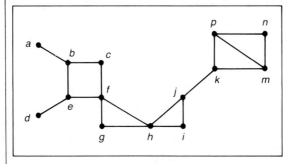

2. Let $G = (V, E)$ be a connected undirected graph. If $z \in V$ prove that z is an articulation point of G if and only if there exist $x, y \in V$ such that every path from x to y contains the vertex z.

3. Let $T = (V, E)$ be a tree with $|V| = n \geq 2$.

   a) What is the smallest and largest number of articulation points that T can have? Describe the trees for each of these cases.

   b) How many biconnected components does T have?

4. a) Let $T = (V, E)$ be a tree. If $v \in V$, prove that v is an articulation point of T if and only if $\deg(v) > 1$.

   b) Let $G = (V, E)$ be a loop-free connected undirected graph with $|E| \geq 1$. Prove that G has at least two vertices that are not articulation points.

5. If $B_1, B_2, \ldots, B_k$ are the biconnected components of a connected undirected graph G, how is $\gamma(G)$ related to $\gamma(B_i)$, $1 \leq i \leq k$?

6. For a connected undirected graph G, explain why the planarity of G depends on the planarity of its biconnected components.

7. A connected undirected graph G has biconnected components B_1, B_2, and B_3. A second connected undirected graph G' has biconnected components B_1', B_2', and B_3'. If B_i is isomorphic to B_i', $1 \leq i \leq 3$, does it follow that G and G' are isomorphic?

8. The following adjacency matrices each represent a connected undirected graph G. In each case:

   a) Determine a depth-first spanning tree for G rooted at x_1.

   b) Using the result in part (a), apply the algorithm developed in this section to find the articulation points and biconnected components of G.

	x_1	x_2	x_3	x_4	x_5	x_6	x_7	x_8
x_1	0	1	1	1	1	0	0	0
x_2	1	0	0	1	1	0	0	0
x_3	1	0	0	0	0	1	1	0
x_4	1	1	0	0	1	0	0	0
x_5	1	1	0	1	0	0	0	0
x_6	0	0	1	0	0	0	1	0
x_7	0	0	1	0	0	1	0	1
x_8	0	0	0	0	0	0	1	0

	x_1	x_2	x_3	x_4	x_5	x_6	x_7	x_8
x_1	0	1	1	1	1	0	0	0
x_2	1	0	1	1	0	0	0	0
x_3	1	1	0	1	0	1	0	0
x_4	1	1	1	0	0	0	0	0
x_5	1	0	0	0	0	0	1	1
x_6	0	0	1	0	0	0	0	0
x_7	0	0	0	0	1	0	0	1
x_8	0	0	0	0	1	0	1	0

9. In step 2 of the algorithm for articulation points, why wasn't it necessary to compute $\text{low}(y_1)$ and $\text{low}(y_2)$?

15.5 ■ SUMMARY AND HISTORICAL REVIEW

The structure now called a tree first appeared in 1847 in the work of Gustav Kirchhoff (1824–1877) on electrical networks. The concept also appeared at this time in the book *Geometrie die Lage* by Karl von Staudt (1798–1867). In 1857 trees were redeveloped by Arthur Cayley (1821–1895), who was unaware of these earlier developments. Cayley was the first to call the structure a "tree," and he used it in applications dealing with chemical isomers. He also investigated the enumeration of certain classes of trees. In his first work on trees Cayley enumerated unlabeled rooted trees. This was then followed by the enumeration of unlabeled ordered trees. Two of Cayley's contemporaries who also studied trees were Carl Borchardt (1817–1880) and Marie Ennemond Jordan (1838–1922).

The formula n^{n-2} for the number of labeled trees on n vertices (Exercise 9 at the end of Section 15.1) was discovered in 1860 by Carl Borchardt. Cayley later gave an independent development of the formula in 1889. Since then, there have been other derivations. These are surveyed in the book by J. W. Moon [7].

The paper by G. Polya [8] is a pioneering work on the enumeration of trees and other combinatorial structures. Polya's theory of enumeration, which we saw in Chapter 12, was developed in this work. For more on the enumeration of trees the reader should see Chapter 15 of F. Harary [2].

The high-speed digital computer has proved to be a constant impetus for applications of trees. The first application of these structures was in the manipulation of algebraic formulae. This dates back to 1951 in the work of Grace Hopper. Since then computer applications of trees have been widely investigated. In the beginning, particular results appeared only in the documentation of specific algorithms. The first general survey of the applications of trees was made in 1961 by Kenneth Iverson as part of a broader survey on data structures. Such ideas as preorder and postorder can be traced to the early 1960's, as evidenced in the work of Zdzislaw Pawlak, Lyle Johnson, and Kenneth Iverson. Additional material on these orders and the procedures for their implementation on a computer can be found in Chapter 6 of the text by A. Aho, J. Hopcroft, and J. Ullman [1]. More on the properties and computer applications of trees is given in Section 3 of Chapter 2 in the work by D. Knuth [4]. Sorting techniques and their use of trees can be further studied in Chapter 11 of A. Aho, J. Hopcroft, and J. Ullman [1]. An extensive investigation will warrant the coverage found in volume 3 of D. Knuth [5].

The technique in Section 15.3 for designing prefix codes is based on methods developed by D. Huffman [3]. The algorithm for articulation points is discussed on Pages 144–146 of A. Aho, J. Hopcroft, and J. Ullman [1].

Finally, Chapter 7 of C. L. Liu [6] deals with trees, simple cycles, and cut-sets, and the vector spaces associated with these ideas. The reader with a background in linear or abstract algebra should find this material of interest.

REFERENCES

1. Aho, Alfred V., Hopcroft, John E., and Ullman, Jeffrey D., *Data Structures and Algorithms,* Addison-Wesley, Reading, Massachusetts, 1983.

2. Harary, Frank, *Graph Theory*. Addison-Wesley, Reading, Massachusetts, 1969.

3. Huffman, David A., "A Method for the Construction of Minimum Redundancy Codes," *Proceedings of the IRE,* Vol. 40, 1952, pp. 1098–1101.

4. Knuth, Donald E., *The Art of Computer Programming,* Vol. 1, 2nd ed., Addison-Wesley, Reading, Massachusetts, 1973.

5. Knuth, Donald E., *The Art of Computer Programming,* Vol. 3, Addison-Wesley, Reading, Massachusetts, 1973.

6. Liu, C.L., *Introduction to Combinatorial Mathematics,* McGraw-Hill, New York, 1968.

7. Moon, John Wesley, *Counting Labelled Trees,* Canadian Mathematical Congress, Montreal, Canada, 1970.

8. Polya, George, "Kombinatorische Anzahlbestimmungen für Gruppen, Graphen und Chemische Verbindungen," *Acta Mathematica,* Vol. 68, 1937, pp. 145–234.

MISCELLANEOUS EXERCISES

1. a) Let $G = (V, E)$ be a loop-free undirected graph with $|V| = n$. Prove that G is a tree iff $P(G, \lambda) = \lambda(\lambda - 1)^{n-1}$.

   b) Prove that $\gamma(G) = 2$ for any tree with two or more vertices.

   c) If $G = (V, E)$ is a connected undirected graph with $|V| = n$, prove that for any integer $\lambda \geq 0$, $P(G, \lambda) \leq \lambda(\lambda - 1)^{n-1}$.

2. The forest $G = (V, E)$ contains 26 vertices and 21 edges. How many components does G have?

3. Let $T = (V, E)$ be an ordered rooted tree, as in Example 15.4. Assume vertex v has address $a_1.a_2.a_3. \ldots .a_{10}$.

   a) At what level is vertex v in the tree?

   b) What are the addresses of the vertices on the path from the root of T to vertex v?

4. Let $G = (V, E)$ be an undirected graph where $V = \{v_2, v_3, v_4, \ldots, v_{30}\}$; $E = \{\{v_i, v_j\} \mid 2 \leq i < j \leq 30$ and i, j have a common divisor greater than 1$\}$.

   a) Determine $\omega(G)$.

   b) Determine a spanning forest for G.

5. Let $G = (V, E)$ be a connected undirected graph. If $H = (V, E')$ is a subgraph of G, prove that H is a Hamilton path in G iff H is a spanning tree of G where (in H) each vertex has degree at most 2.

6. A telephone communication system is set up at a company where 125 executives are employed. The system is initialized by the president who calls her four vice-presidents. Each vice-president then calls four other executives, who in turn call four others, and so on.

   a) How many calls are made in reaching all 125 executives?

b)   How many executives, aside from the president, are required to make calls?

7. If T is an m-ary tree, the *total path length* of T is the sum of the lengths of all paths in the tree from the root to each of its vertices.

   a)   For any nonnegative integer h, let x_h denote the minimal total path length for an m-ary tree of height h. Show that x_h satisfies the recurrence relation $x_{h+1} = x_h + m(h + 1)$, with initial condition $x_0 = 0$.

   b)   Solve the recurrence relation in part (a) for x_h.

8. In an m-ary tree T, the *external path length* of T is the sum of the lengths of all paths in T from the root to each of the leaves.

   a)   If $h \geq 0$, let y_h denote the minimal external path length for an m-ary tree of height h. Show that y_h satisfies the recurrence relation $y_{h+1} = y_h - h + m(h + 1)$, with the initial condition $y_0 = 0$.

   b)   Solve the recurrence relation in part (a) for y_h.

9. Let T be a binary tree with the vertices of T ordered by a preorder traversal. This traversal assigns the label 1 to all internal vertices of T and the label 0 to each leaf. The sequence of 0's and 1's that results from the preorder traversal of T is called the tree's *characteristic sequence*.

   a)   Find the characteristic sequence for the binary tree in Fig. 15.8(c).

   b)   Determine the binary trees for the characteristic sequences: (i) 1011001010100; and (ii) 10111100001010111000.

   c)   What are the last two symbols in the characteristic sequence for all binary trees? Why?

10. For $k \in \mathbf{Z}^+$, let $n = 2^k$, and consider the list L: $a_1, a_2, a_3, \ldots, a_n$. To sort L in ascending order, first compare the entries a_i and $a_{i+(n/2)}$, $1 \leq i \leq n/2$. For the resulting 2^{k-1} ordered pairs, merge sort the ith and $(i + (n/4))$-th ordered pairs, $1 \leq i \leq n/4$. Now do a merge sort on the ith and $(i + (n/8))$-th ordered quadruples, $1 \leq i \leq n/4$. Continue the process until the elements of L are in ascending order.

   a)   Apply this sorting procedure to the list

   $$L:\ 11, 3, 4, 6, -5, 7, 35, -2, 1, 23, 9, 15, 18, 2, -10, 5.$$

   b)   If $n = 2^k$, how many comparisons at most does this procedure require?

11. a)   If T is a full binary tree of height 5, how many leaves does T have? How many internal vertices? How many edges (branches)?

    b)   Answer part (a) for a full binary tree of height h, where $h \in \mathbf{Z}^+$.

12. Let T be a full binary tree of height h. If $v_1, v_2, \ldots, v_m$ are the leaves of T, evaluate $\sum_{i=1}^{m} 2^{-d_i}$, where d_i is the level number of leaf v_i, $1 \leq i \leq m$.

16

Optimization
and Matching

Using the structures of trees and graphs, this final chapter introduces techniques that arise in the area of mathematics called *operations research*. These techniques optimize certain results by considering graphs and multigraphs that have a nonnegative real number, called a weight, associated with each edge of the graph. These numbers relate information such as the distance between the vertices that are the endpoints of the edge, or perhaps the amount of material that can be shipped from one vertex to another along an edge that represents a highway or air route. With the graphs providing the framework, the optimization methods are developed in an algorithmic manner to facilitate their implementation on a computer. Among the problems we analyze are the determinations of

1. a spanning tree for a given graph or multigraph where the sum of the weights of the edges in the tree is minimal; and,

2. the maximum amount of material that can be transported from a starting point (the source) to a terminating point (the sink), where the weight of an edge indicates an allowable capacity for the material being transported.

16.1 ■ MINIMAL SPANNING TREES: THE ALGORITHMS OF KRUSKAL AND PRIM

Figure 16.1

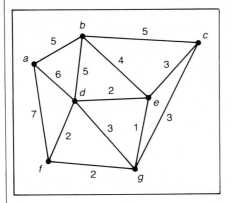

A loosely coupled computer network is to be set up for a system of seven computers. The graph G in Fig. 16.1 models the situation. The computers are represented by the vertices in the graph; the edges represent transmission lines that are being considered for linking certain pairs of computers. Associated with each edge e in G is a nonnegative real number wt(e), called the *weight of e*. The weight of an edge indicates the projected cost for constructing that particular transmission line. The objective is to link all the computers while minimizing the total cost of construction. To do so requires a spanning tree T, where the sum of the weights of the edges in T is minimal. The construction of such an *optimal* spanning tree can be accomplished by algorithms developed by Joseph Kruskal and Robert Prim.

These algorithms are called *greedy* algorithms. When such an algorithm is used, at each step of the process an optimal (maximal or minimal) choice is made from the remaining available data. If what appears to be the best choice *locally* (e.g., for a vertex c and the vertices near c) turns out to be the best choice *globally* (e.g., for all vertices of the graph), then the greedy algorithm will lead to an optimal solution.

We first consider Kruskal's algorithm. This algorithm is given recursively as follows.

Let $G = (V, E)$ be an undirected connected graph where $|V| = n$ and each edge e is assigned a nonnegative real number wt(e). To find an optimal (minimal) spanning tree for G apply the following.

Kruskal's Algorithm

Step 1: Set the counter $i = 1$ and select an edge $e_1 \in E$ where wt(e_1) is as small as possible.

Step 2: For $1 \le i \le n - 2$, if edges $e_1, e_2, \ldots, e_i$ have been selected, then select edge e_{i+1} from the remaining edges in G so that (a) wt(e_{i+1}) is as small as possible; and (b) the subgraph of G determined by the edges $e_1, e_2, \ldots, e_i, e_{i+1}$ (and the vertices they are adjacent to) contains no cycles.

Step 3: Replace i by $i + 1$.
If $i = n - 1$, the subgraph of G determined by edges $e_1, e_2, \ldots, e_{n-1}$ is connected with n vertices and $n - 1$ edges, so it is an optimal spanning tree for G.
If $i < n - 1$, return to step 2.

Before establishing the validity of the algorithm, we consider the following example.

EXAMPLE 16.1 Apply Kruskal's algorithm to the graph in Fig. 16.1.

Initialization: ($i = 1$) Since there is a unique edge $\{e, g\}$ of weight 1, start with $T = \{\{e, g\}\}$. (T starts as a tree with one edge, and after each iteration it grows into a larger tree or forest. After the last iteration the subgraph T is an optimal spanning tree for the given graph G.)

First Iteration: Among the remaining edges in G, three have weight 2. Select $\{d, f\}$, which satisfies the conditions in step 2. Now T is the forest $\{\{e, g\}, \{d, f\}\}$, and i is increased to 2. With $i = 2 < 6$, return to step 2.

Second Iteration: Two remaining edges have weight 2. Selecting $\{d, e\}$, T is now the tree $\{\{e, g\}, \{d, f\}, \{d, e\}\}$, and i increases to 3. But since $3 < 6$, the algorithm directs us back to step 2.

Third Iteration: Among the edges of G that are not in T, edge $\{f, g\}$ has minimal weight 2. However, if this edge is added to T, the result contains a cycle, which destroys the tree structure being sought. Consequently, the edges $\{c, e\}$, $\{c, g\}$, and $\{d, g\}$ are considered. Edge $\{d, g\}$ brings about a cycle, but either $\{c, e\}$ or $\{c, g\}$ satisfies the conditions in step 2. Selecting $\{c, e\}$, T grows to $\{\{e, g\}, \{d, f\}, \{d, e\}, \{c, e\}\}$ and i is increased to 4. Returning to step 2, the fourth and fifth iterations provide the following.

Fourth Iteration: $T = \{\{e, g\}, \{d, f\}, \{d, e\}, \{c, e\}, \{b, e\}\}$; i increases to 5.

Fifth Iteration: $T = \{\{e, g\}, \{d, f\}, \{d, e\}, \{c, e\}, \{b, e\}, \{a, b\}\}$. The counter i now becomes 6 = (number of vertices in G) − 1. So T is an optimal tree for graph G and has weight $1 + 2 + 2 + 3 + 4 + 5 = 17$.

Figure 16.2

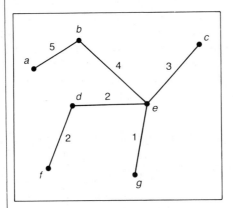

Figure 16.2 shows this spanning tree of minimal weight.   □

Example 16.1 demonstrates that Kruskal's algorithm does generate a spanning tree. Also, the algorithm is greedy, for it selects from the remaining edges an edge of minimal weight that doesn't create a cycle. The following result guarantees that the spanning tree obtained is optimal.

Theorem 16.1

▶ Let $G = (V, E)$ be a weighted connected undirected graph. Any spanning tree for G that is obtained by Kruskal's algorithm is optimal.

Proof

Let $|V| = n$, and let T be a spanning tree for G obtained by Kruskal's algorithm. The edges in T are labeled as $e_1, e_2, \ldots, e_{n-1}$, according to the order in which they are generated by the algorithm. For any optimal tree T' of G, define $d(T') = k$ if k is the smallest positive integer such that T and T' both contain $e_1, e_2, \ldots, e_{k-1}$, but $e_k \notin T'$.

Let T_1 be an optimal tree for which $d(T_1) = r$ is maximal. If $r = n$, then $T = T_1$ and the result follows. Otherwise, $r \le n - 1$ and adding edge e_r (of T) to T_1 produces the cycle C, where there exists an edge e'_r of C that is in T_1 but not in T.

Start with tree T_1. Adding e_r to T_1 and deleting e_r', we obtain a connected graph with n vertices and $n - 1$ edges. This graph is a tree, T_2. The weights of T_1 and T_2 satisfy $\text{wt}(T_2) = \text{wt}(T_1) + \text{wt}(e_r) - \text{wt}(e_r')$.

Following the selection of $e_1, e_2, \ldots, e_{r-1}$ in Kruskal's algorithm, the edge e_r is chosen so that $\text{wt}(e_r)$ is minimal and no cycle results when e_r is added to the subgraph H of G determined by $e_1, e_2, \ldots, e_{r-1}$. Since e_r' produces no cycle when added to the subgraph H, by the minimality of $\text{wt}(e_r)$, $\text{wt}(e_r') \geq \text{wt}(e_r)$. Hence $\text{wt}(e_r) - \text{wt}(e_r') \leq 0$, so $\text{wt}(T_2) \leq \text{wt}(T_1)$. But with T_1 optimal, $\text{wt}(T_2) = \text{wt}(T_1)$, so T_2 is optimal.

The tree T_2 is optimal and has the edges $e_1, e_2, \ldots, e_{r-1}, e_r$ in common with T, so $d(T_2) \geq r + 1 > r = d(T_1)$, contradicting the choice of T_1. Consequently, $T_1 = T$ and Kruskal's algorithm produces the optimal tree T. ∎

A second technique for constructing an optimal tree was developed by Robert Prim. In this greedy algorithm, the vertices in the graph are partitioned into two sets: processed and not processed. At first only one vertex is in the set P of processed vertices, and all other vertices are in the set N of vertices to be processed. Each iteration of the algorithm increases the set P by one vertex while the size of set N decreases by one. The algorithm is summarized as follows.

Let $G = (V, E)$ be a weighted connected undirected graph. To obtain an optimal tree T for G, apply the given recursive procedure.

Prim's Algorithm

Step 1: Set the counter $i = 1$ and place a vertex $v_1 \in V$ into set P. Define $N = V - \{v_1\}$ and $T = \emptyset$.

Step 2: For $1 \leq i \leq n - 1$, let $P = \{v_1, v_2, \ldots, v_i\}$, $T = \{e_1, e_2, \ldots, e_{i-1}\}$, $N = V - P$. Add to T the shortest edge (i.e., the edge of minimal weight) in G that connects a vertex x in P with a vertex $y \ (= v_{i+1})$ in N. Place y in P and delete it from N.

Step 3: Increase the counter by 1.
  If $i = n$, the subgraph of G determined by the edges $e_1, e_2, \ldots, e_{n-1}$ is connected with n vertices and $n - 1$ edges, and is an optimal tree for G.
  If $i < n$, go back to step 2.

We use this algorithm to find an optimal tree for the graph in Fig. 16.1.

EXAMPLE 16.2 Prim's Algorithm generates an optimal tree as follows.

Initialization: $i = 1$; $P = \{a\}$; $N = \{b, c, d, e, f, g\}$; $T = \emptyset$.

First Iteration: $T = \{\{a, b\}\}$; $P = \{a, b\}$; $N = \{c, d, e, f, g\}$; $i = 2$.

Second Iteration: $T = \{\{a, b\}, \{b, e\}\}$; $P = \{a, b, e\}$; $N = \{c, d, f, g\}$; $i = 3$.

Third Iteration:	$T = \{\{a, b\}, \{b, e\}, \{e, g\}\}$; $P = \{a, b, e, g\}$; $N = \{c, d, f\}$; $i = 4$.
Fourth Iteration:	$T = \{\{a, b\}, \{b, e\}, \{e, g\}, \{d, e\}\}$; $P = \{a, b, e, g, d\}$; $N = \{c, f\}$; $i = 5$.
Fifth Iteration:	$T = \{\{a, b\}, \{b, e\}, \{e, g\}, \{d, e\}, \{f, g\}\}$; $P = \{a, b, e, g, d, f\}$; $N = \{c\}$; $i = 6$.
Sixth Iteration:	$T = \{\{a, b\}, \{b, e\}, \{e, g\}, \{d, e\}, \{f, g\}, \{c, g\}\}$; $P = \{a, b, e, g, d, f, c\} = V$; $N = \emptyset$; $i = 7 = |V|$. Hence T is an optimal spanning tree of weight 17 for G, as seen in Fig. 16.3.

Figure 16.3

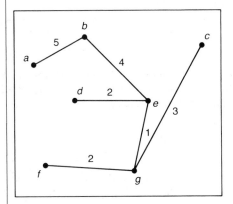

We conclude this section with a statement on the validity of Prim's algorithm. The proof is similar to that given for Theorem 16.1 and is left for the reader.

Theorem 16.2 ▶ Let $G = (V, E)$ be a weighted connected undirected graph. Any spanning tree for G that is obtained by Prim's algorithm is optimal.

EXERCISES

1. Apply Kruskal's and Prim's algorithms to determine minimal spanning trees for the graph in Fig. 16.4.

Figure 16.4

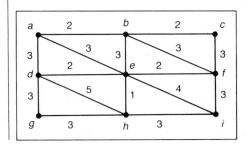

2. The following table provides information on the distance (in miles) between pairs of cities in the state of Indiana.

Table 16.1

	Bloomington	Evansville	Fort Wayne	Gary	Indianapolis	South Bend
Evansville	119	—	—	—	—	—
Fort Wayne	174	290	—	—	—	—
Gary	198	277	132	—	—	—
Indianapolis	51	168	121	153	—	—
South Bend	198	303	79	58	140	—
Terre Haute	58	113	201	164	71	196

   A system of highways that connects these seven cities is to be constructed. Determine which highways should be constructed so that the cost of construction is minimal. (Assume that the cost of construction of a mile of highway is the same between any pair of cities.)

3. a) Answer Exercise 2 under the additional requirement that the system includes a highway directly linking Evansville and Indianapolis.

   b) If there must be a direct link between Fort Wayne and Gary in addition to the one connecting Evansville and Indianapolis, find the smallest number of miles of highway that must be constructed.

4. Modify Kruskal's algorithm to find a minimal spanning tree that includes one or more prescribed edges.

5. a) Modify Kruskal's algorithm to determine an optimal tree of *maximal* weight.

   b) Interpret the information of Exercise 2 in terms of the number of calls that can be placed between pairs of cities by the adoption of certain new telephone transmission lines. (Cities that are not directly linked must communicate through one or more intermediate cities.) How can the seven cities be minimally connected and allow a maximum number of calls to be placed?

6. Prove Theorem 16.2.

7. Let $G = (V, E)$ be a weighted connected undirected graph where for distinct edges e_1, $e_2 \in E$, $\text{wt}(e_1) \neq \text{wt}(e_2)$. Prove that G has only one minimal spanning tree.

16.2 ■ TRANSPORT NETWORKS: THE MAX-FLOW MIN-CUT THEOREM

This section provides an application for weighted directed graphs to the flow of a commodity from a source to a prescribed destination. Such commodities may be gallons of oil that flow through pipelines or numbers of telephone calls transmitted in a communication system. In modeling such situations, the weight of an edge in the directed graph is interpreted as a capacity that places an upper limit, for example, on the amount of oil that can flow through a certain part of a system of pipelines. These ideas are defined formally in the following.

Definition 16.1

▶ Let $N = (V, E)$ be a loop-free connected directed graph. Then N is called a *network*, or *transport network*, if the following conditions are satisfied:

a) There exists a unique vertex $a \in V$ with $\deg^+(a)$, the in degree of a, equal to 0. This vertex a is called the *source*.

b) There is a unique vertex $z \in V$, called the *sink*, where $\deg^-(z)$, the out degree of z, equals 0.

c) The graph N is weighted, so there is a function from E to the set of nonnegative real numbers that assigns to each edge $e = (v, w) \in E$ a *capacity*, denoted by $c(e) = c(v, w)$.

Figure 16.5

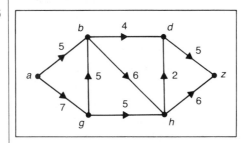

EXAMPLE 16.3

The graph in Fig. 16.5 is a transport network. Here vertex a is the source; the sink is at vertex z. Since $c(a, b) + c(a, g) = 5 + 7 = 12$, the amount of the commodity being transported from a to z cannot exceed 12. With $c(d, z) + c(h, z) = 5 + 6 = 11$, the amount is further restricted to be no greater than 11. To determine the maximum amount that can be transported from a to z, we must consider the capacities of all edges in the network.   □

The following is introduced to assist in this problem.

Definition 16.2

▶ If $N = (V, E)$ is a transport network, a function f from E to the nonnegative real numbers is called a *flow* for N if

a) $f(e) \leq c(e)$ for each edge $e \in E$; and

b) for any $v \in V$, other than the source a or the sink z, $\sum_{w \in V} f(w, v) = \sum_{w \in V} f(v, w)$.

(If there is no edge (v, w), $f(v, w) = 0$.)

The first condition specifies that the amount of material transported along a given edge cannot exceed the capacity of that edge. Condition (b) requires that the amount of material flowing into a vertex v must equal the amount that flows out from this vertex. This is so for all vertices except the source and sink.

Figure 16.6

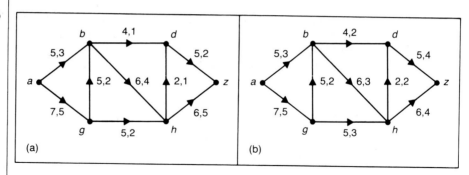

(a)   (b)

EXAMPLE 16.4   For each network in Fig. 16.6 the label x, y on edge e is determined so that $x = c(e)$ and y is the value assigned for a possible flow f. The label on each edge e satisfies $f(e) \leq c(e)$. In part (a) of the figure, the "flow" into vertex g is 5 while the "flow" out from that vertex is $2 + 2 = 4$. Hence the function f is not a flow in this case. The function f for part (b) does satisfy both conditions, so it is a flow for the given network.   □

Definition 16.3

▶   Let f be a flow for a transport network $N = (V, E)$.

a)   An edge e of the network is called *saturated* if $f(e) = c(e)$. When $f(e) < c(e)$ the edge is called *unsaturated*.

b)   If a is the source of N, then val$(f) = \sum_{v \in V} f(a, v)$ is called the *value of the flow*.

EXAMPLE 16.5   For the network in Fig. 16.6(b), only the edge (h, d) is saturated. All other edges are unsaturated. The value of the flow in this network is

$$\text{val}(f) = \sum_{v \in V} f(a, v) = f(a, b) + f(a, g) = 3 + 5 = 8 \,.$$

But is there another flow f_1 such that val$(f_1) > 8$? The determination of a *maximal flow*, i.e. a flow that achieves the greatest possible value, is the objective of the remainder of this section. To accomplish this we observe that in the network of Fig. 16.6(b),

$$\sum_{v \in V} f(a, v) = 8 = 4 + 4 = f(d, z) + f(d, h) = \sum_{v \in V} f(v, z) \,.$$

Consequently, the total flow leaving the source a equals the total flow into the sink z.   □

The last remark in Example 16.5 seems like a reasonable circumstance, but will it occur in general? To prove the result for any network, we need the following special type of cut-set.

Definition 16.4

▶   If $N = (V, E)$ is a transport network and C is a cut-set for the undirected graph associated with N, then C is called a *cut*, or *a-z cut*, if the removal of the edges in C from the network results in the separation of a and z.

EXAMPLE 16.6

Each of the dotted curves in Fig. 16.7 indicates a cut for the given network. The cut C_1 consists of the undirected edges $\{a, g\}$, $\{b, d\}$, $\{b, g\}$, and $\{b, h\}$. This cut partitions the vertices of the network into two sets $P = \{a, b\}$ and its complement $\overline{P} = \{d, g, h, z\}$, so C_1 is denoted as $(P, \overline{P})$. The *capacity of a cut,* denoted $c(P, \overline{P})$, is defined by

$$c(P, \overline{P}) = \sum_{\substack{v \in P \\ w \in \overline{P}}} c(v, w),$$

the sum of the capacities of all edges (v, w) where $v \in P$ and $w \in \overline{P}$. In this example, $c(P, \overline{P}) = c(a, g) + c(b, d) + c(b, h) = 7 + 4 + 6 = 17$. (Considering the *directed* edges of the cut $C_1 = (P, \overline{P})$, namely (a, g), (b, d), (b, h), the removal of these edges does not result in a subgraph with two components. However, these three edges are minimal in the sense that their removal eliminates any possible directed path from a to z.)

Figure 16.7

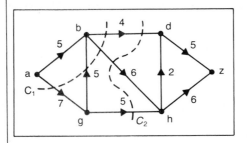

The cut C_2 induces the vertex partition $Q = \{a, b, g\}$, $\overline{Q} = \{d, h, z\}$ and has capacity $c(Q, \overline{Q}) = c(b, d) + c(b, h) + c(g, h) = 4 + 6 + 5 = 15$.

A third cut of interest is the one that induces the vertex partition $S = \{a, b, d, g, h\}$, $\overline{S} = \{z\}$. (What are the edges in this cut?) Its capacity is 11.   □

Using the idea of the capacity of a cut, this next result provides an upper bound for the value of a flow in a network.

Theorem 16.3

▶ Let f be a flow in a network $N = (V, E)$. If $C = (P, \overline{P})$ is any cut in N, then $\mathrm{val}(f)$ cannot exceed $c(P, \overline{P})$.

Proof

Let vertex a be the source in N and vertex z the sink. Since $\deg^+(a) = 0$, for any $w \in V$, $f(w, a) = 0$; consequently,

$$\mathrm{val}(f) = \sum_{v \in V} f(a, v) = \sum_{v \in V} f(a, v) - \sum_{w \in V} f(w, a).$$

By condition (b) in the definition of a flow, for any $x \in P$, $x \neq a$, $\sum_{v \in V} f(x, v) - \sum_{w \in V} f(w, x) = 0$.

Adding the results in the above equations we have

$$\text{val}(f) = \left[\sum_{v \in V} f(a, v) - \sum_{w \in V} f(w, a) \right] + \sum_{\substack{x \in P \\ x \neq a}} \left[\sum_{v \in V} f(x, v) - \sum_{w \in V} f(w, x) \right]$$

$$= \sum_{\substack{x \in P \\ v \in V}} f(x, v) - \sum_{\substack{x \in P \\ w \in V}} f(w, x)$$

$$= \left[\sum_{\substack{x \in P \\ v \in P}} f(x, v) + \sum_{\substack{x \in P \\ v \in \overline{P}}} f(x, v) \right] - \left[\sum_{\substack{x \in P \\ w \in P}} f(w, x) + \sum_{\substack{x \in P \\ w \in \overline{P}}} f(w, x) \right].$$

Since

$$\sum_{\substack{x \in P \\ v \in P}} f(x, v) \quad \text{and} \quad \sum_{\substack{x \in P \\ w \in P}} f(w, x)$$

are summed over the same set of all ordered pairs in $P \times P$, these summations are equal. Consequently,

$$\text{val}(f) = \sum_{\substack{x \in P \\ v \in \overline{P}}} f(x, v) - \sum_{\substack{x \in P \\ w \in \overline{P}}} f(w, x).$$

For all $x, w \in V$, $f(w, x) \geq 0$, so

$$\sum_{\substack{x \in P \\ w \in \overline{P}}} f(w, x) \geq 0 \quad \text{and} \quad \text{val}(f) \leq \sum_{\substack{x \in P \\ v \in \overline{P}}} f(x, v) \leq \sum_{\substack{x \in P \\ v \in \overline{P}}} c(x, v) = c(P, \overline{P}). \qquad \blacksquare$$

From Theorem 16.3 we find that in a network N the value for *any* flow is less than or equal to the capacity of *any* cut in that network. Hence the value of the maximum flow cannot exceed the minimum capacity over all cuts in a network. For the network in Fig. 16.7 it can be shown that the cut consisting of edges (d, z) and (h, z) has minimum capacity 11. Consequently, the maximum flow f for the network satisfies $\text{val}(f) \leq 11$. It will turn out that the value of the maximum flow is 11. How to construct such a flow and why its value equals the minimum capacity among all cuts will be dealt with in the main theorem of this section.

However, before we deal with this construction let us note that in the proof of Theorem 16.3, the value of a flow is given by

$$\text{val}(f) = \sum_{\substack{x \in P \\ v \in \overline{P}}} f(x, v) - \sum_{\substack{x \in P \\ w \in \overline{P}}} f(w, x),$$

where $(P, \overline{P})$ is *any* cut in N. Therefore, once a flow is constructed in a network, then for any cut $(P, \overline{P})$ in the network, the value of the flow equals the sum of the flows in the edges directed from the vertices in P to those in $\overline{P}$ minus the sum of the flows in the edges directed from the vertices in $\overline{P}$ to those in P.

This observation leads to the following result.

Corollary 16.1

▶ If f is a flow in a transport network $N = (V, E)$, then the value of the flow from the source a is equal to the value of the flow into the sink z.

Proof Let $P = \{a\}$, $\overline{P} = V - \{a\}$, and $Q = V - \{z\}$, $\overline{Q} = \{z\}$. From the above observation

$$\sum_{\substack{x \in P \\ v \in \overline{P}}} f(x, v) - \sum_{\substack{x \in P \\ w \in \overline{P}}} f(w, x) = \text{val}(f) = \sum_{\substack{y \in Q \\ v \in \overline{Q}}} f(y, v) - \sum_{\substack{y \in Q \\ w \in \overline{Q}}} f(w, y).$$

With $P = \{a\}$ and $\deg^+(a) = 0$, $\sum_{x \in P, w \in \overline{P}} f(w, x) = \sum_{w \in \overline{P}} f(w, a) = 0$. Similarly, for $\overline{Q} = \{z\}$ and $\deg^-(z) = 0$, $\sum_{y \in Q, w \in \overline{Q}} f(w, y) = \sum_{y \in Q} f(z, y) = 0$. Consequently,

$$\sum_{\substack{x \in P \\ v \in \overline{P}}} f(x, v) = \sum_{v \in \overline{P}} f(a, v) = \text{val}(f) = \sum_{\substack{y \in Q \\ v \in \overline{Q}}} f(y, v) = \sum_{y \in Q} f(y, z),$$

and this establishes the corollary. ■

We turn now to the main result of the section.

Theorem 16.4

▶ *(The Max-Flow Min-Cut Theorem)* For a transport network $N = (V, E)$ the maximum flow that can be attained in N is equal to the minimum capacity over all cuts in the network.

Proof By Theorem 16.3, if $(P, \overline{P})$ is a cut of minimum capacity in N, then the value of any flow f in N satisfies $\text{val}(f) \leq c(P, \overline{P})$. To verify the existence of a flow f for which $\text{val}(f) = c(P, \overline{P})$, the following algorithm, called the *labeling procedure*, provides the necessary steps.

The Labeling Procedure:

Step 1: Given a network N, define an initial flow f in N by $f(e) = 0$ for every e in E. (This function f satisfies the conditions in the definition of a flow.)

Step 2: Label the source a with $(-, \infty)$. (This label indicates that from the source a as much material is available as is needed to achieve a maximum flow.)

Step 3: For any vertex x that is adjacent from a, label x as follows:

a) If $c(a, x) - f(a, x) > 0$, define $\Delta(x) = c(a, x) - f(a, x)$ and label x with $(a^+, \Delta(x))$.

b) If $c(a, x) - f(a, x) = 0$, leave vertex x unlabeled.

(The label $(a^+, \Delta(x))$ indicates that the present flow from a to x can be increased by the amount $\Delta(x)$, with the $\Delta(x)$ additional units supplied from the source a.)

Step 4: Let $x \in V$, $x \neq a$, and suppose that x is labeled.
If $(x, y) \in E$ and y is not labeled, label vertex y as follows:

a) If $c(x, y) - f(x, y) > 0$, define $\Delta(y) = \min\{\Delta(x), c(x, y) - f(x, y)\}$, and label y as $(x^+, \Delta(y))$.

b) If $c(x, y) - f(x, y) = 0$, vertex y is left unlabeled.

(The label $(x^+, \Delta(y))$ indicates that the present flow into vertex y can be increased by an amount $\Delta(y)$ taken from vertex x.)

Step 5: Let $x \in V$, with x labeled.
   If $(y, x) \in E$, and y is not labeled, label vertex y as follows:

a) If $f(y, x) > 0$, label vertex y as $(x^-, \Delta(y))$, where $\Delta(y) = \min\{\Delta(x), f(y, x)\}$.

b) If $f(y, x) = 0$, vertex y is left unlabeled.

(The label $(x^-, \Delta(y))$ tells us that by decreasing the flow from y to x, the total flow out of y to the labeled vertices can be decreased by $\Delta(y)$. These $\Delta(y)$ units may then be used to increase the total flow from y to the unlabeled vertices.)

Since a vertex y may be adjacent to or from more than one labeled vertex, the results of this procedure need not be unique. In addition, if x is labeled, the network could include both of the edges (x, y) and (y, x), perhaps resulting in two possible labels for y. But the procedure is designed to create a maximum flow and there may be more than one. Nonetheless, when a vertex can be labeled in more than one way, an arbitrary choice must be made.

As we apply the labeling procedure to the vertices of the given network, there are two possible cases to consider.

Case 1. If the sink z is labeled as $(x^+, \Delta(z))$, the flow in the edge (x, z) can be increased from $f(x, z)$ to $f(x, z) + \Delta(z)$, as indicated by the label.

Considering vertex x, it can be labeled as either $(v^+, \Delta(x))$ or $(v^-, \Delta(x))$, where $\Delta(x) \geq \Delta(z)$. For the label $(v^+, \Delta(x))$, we may regard vertex v as the source for increasing the flow in edge (x, z) by the amount $\Delta(z)$. In this case we likewise increase the present flow in edge (v, x) from $f(v, x)$ to $f(v, x) + \Delta(z)$ (not $f(v, x) + \Delta(x)$). Should x have the label $(v^-, \Delta(x))$, the flow in the edge (x, v) is changed from $f(x, v)$ to $f(x, v) - \Delta(z)$ in order to provide the additional flow of Δz units from x to z.

As this process continues back to source a, each directed edge along a path from a to z has its flow increased by Δz units. When this is accomplished, all vertex labels, except for $(-, \infty)$ on the source, are removed; the process is repeated to see if it is possible to increase the flow even more.

Case 2: If the labeling procedure is carried out as far as possible and the sink z is unlabeled, then the maximum flow has been attained. Let P be the vertices in V that are labeled, and $\overline{P} = V - P$. Since z is not labeled, the flows in the edges (x, y), $x \in P$, $y \in \overline{P}$, satisfy $f(x, y) = c(x, y)$. Also, for any edge (w, v), $w \in \overline{P}$, $v \in P$, $f(w, v) = 0$. Consequently, there is a flow for the given network, where the value of the flow is equal to the capacity of the cut $(P, \overline{P})$. From Theorem 16.3 this flow is a maximum. ∎

Figure 16.8

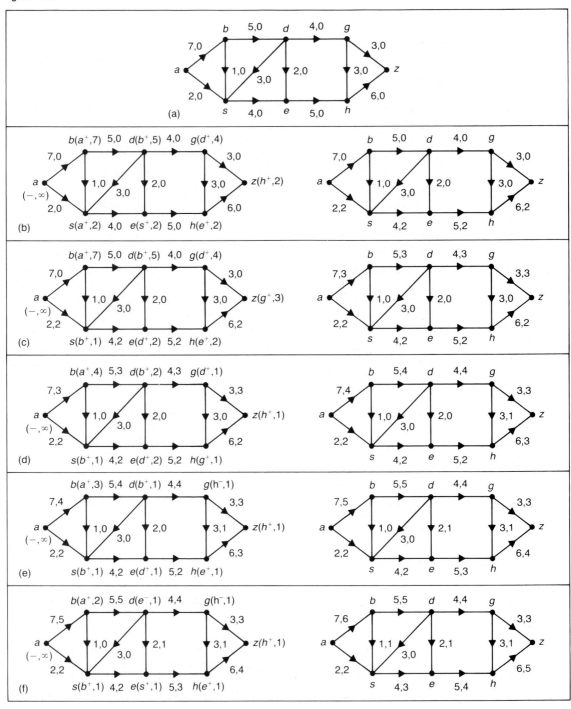

Before demonstrating the labeling procedure we state one last corollary, whose proof is left as an exercise.

Corollary 16.2

▶ Let $N = (V, E)$ be a transport network where for each $e \in E$, $c(e)$ is a positive integer. Then there is a maximum flow f for N where $f(e)$ is a nonnegative integer for each edge e.

EXAMPLE 16.7

Use the labeling procedure to find a maximum flow for the transport network in Fig. 16.8(a).

In this network, each edge is labeled with an ordered pair x, y, where x is the capacity of the edge and $y = 0$ indicates an initial flow for the network. Figure 16.8(b) demonstrates the first application of the labeling procedure. Here a choice is made in labeling the sink, z. We chose $(h^+, 2)$ as the label instead of $(g^+, 3)$. (What other label could have been used for vertex e?) Backtracking from z to h to e to s to a, and increasing the flow in each edge by $\Delta(z) = 2$, we obtain the new flow in Fig. 16.8(b). Parts (c), (d), (e), and (f) of the figure show a second, third, fourth, and fifth application of the labeling procedure. Note how vertex g has a negative label in parts (e) and (f). Also, part (f) provides a second instance of a negative label, this time for vertex d. Applying the labeling procedure one last time, the transport network is labeled as in Fig. 16.9. Here the sink z is unlabeled and the second case in the procedure is now followed. Letting $P = \{a, b\}$, $\overline{P} = \{s, d, e, g, h, z\}$, $c(P, \overline{P}) = c(b, d) + c(b, s) + c(a, s) = 5 + 1 + 2 = 8 = 5 + 3 = f(h, z) + f(g, z)$, the flow into z. The edges in N that are crossed by the dotted curve are the edges of the (undirected) cut-set associated with the cut $(P, \overline{P})$. It consists of all edges of the form $\{x, y\}$ where $x \in P$, $y \in \overline{P}$.

Figure 16.9

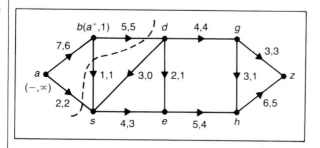

□

We close this section with three examples that are modeled with the concepts of the transport network. After setting up the models, the final solution of each example is left to the section exercises.

EXAMPLE 16.8

Computer chips are manufactured (in units of a thousand) at three companies c_1, c_2, and c_3. These chips are then distributed to two computer manufacturers, m_1 and m_2, through the "transport network" in Fig. 16.10(a), where there are the three sources c_1, c_2, c_3, and the two sinks m_1 and m_2. Company c_1 can produce up to 15

Figure 16.10

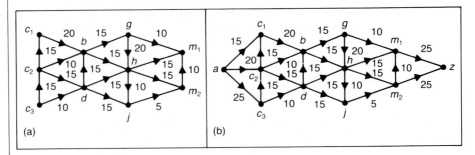

(a)   (b)

units, company c_2 up to 20 units, and company c_3 up to 25 units. If each manufacturer needs 25 units, how many units should each of the companies produce so that they can meet the demand of the manufacturers, or at least supply them with as many units as the network will allow?

In order to model this example with a transport network, a source a and sink z are introduced as in Fig. 16.10(b). The manufacturing capabilities of the three companies are then used to label edges (a, c_1), (a, c_2), and (a, c_3). For the edges (m_1, z) and (m_2, z), the demands are used as labels. To answer the question posed here, one applies the labeling procedure to this network to find the value of a maximum flow. □

Figure 16.11

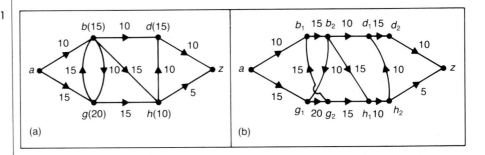

(a)   (b)

EXAMPLE 16.9

The transport network of Fig. 16.11(a) has an added restriction, for there are capacities assigned to vertices other than the source and sink. Part (b) of the figure shows how to redraw the network to obtain one where the labeling procedure can be applied. For any vertex v other than a or z, split v into vertices v_1 and v_2. Draw an edge from v_1 to v_2 and label it with the capacity originally assigned to v. An edge of the form (v, w), where $v \neq a$, $w \neq z$, then becomes the edge (v_2, w_1), maintaining the capacity of (v, w). Edges of the form (a, v) become (a, v_1) with capacity $c(a, v)$. An edge such as (w, z) is replaced by the edge (w_2, z), with capacity $c(w, z)$.

The maximum flow for the given network is now determined by applying the labeling procedure to the network in Fig. 16.11(b). □

EXAMPLE 16.10

During the practice of war games, messengers must deliver information from headquarters (vertex a) to a field command station (vertex z). Since certain roads may be blocked or destroyed, how many messengers should be sent out so that each travels along a path that has no edge in common with any other path taken?

Figure 16.12

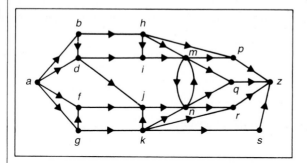

Since the distances between vertices is not relevant here, the graph of Fig. 16.12 has no capacities assigned to its edges. The problem here is to determine the maximum number of edge-disjoint paths from a to z. Assigning each edge a capacity of 1, the problem is converted into a maximum flow problem, where the number of edge-disjoint paths (from a to z) equals the value of a maximum flow for the network. ☐

EXERCISES

1. a) For the network in Fig. 16.13, let the capacity of each edge be 10. If each edge e in the figure is labeled by a function f, as shown, determine the values of s, t, w, x, and y so that f is a flow in the network.

Figure 16.13

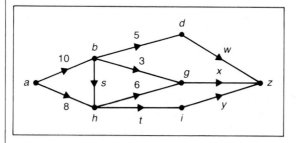

   b) What is the value of this flow?

   c) Find three cuts $(P, \overline{P})$ in this network that have capacity 30.

2. If $N = (V, E)$ is a transport network, let f be a flow in N and $(P, \overline{P})$ a cut. Prove that the value of the flow f equals $c(P, \overline{P})$ iff

   a) $f(e) = c(e)$ for each edge $e = (x, y)$, $x \in P$, $y \in \overline{P}$; and

   b) $f(e) = 0$ for each edge $e = (v, w)$, $v \in \overline{P}$, $w \in P$.

3. Find a maximum flow and the corresponding minimum cut for each transport network in Fig. 16.14.

Figure 16.14

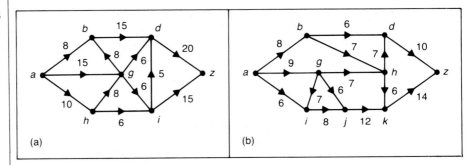

(a)   (b)

4. Apply the labeling procedure to find a maximum flow in Examples 16.8, 16.9, and 16.10.

5. Prove Corollary 16.2.

6. In each of the following "transport networks" two companies c_1 and c_2 produce a certain product that is used by two manufacturers m_1 and m_2. For the network in part (a) of Fig. 16.15, c_1 produces 8 units while c_2 produces 7 units; manufacturer m_1 requires 7 units and manufacturer m_2 needs 6 units. In the network of Fig. 16.15(b), each company produces 7 units and each manufacturer needs 6 units. In which situation(s) can the supply meet the manufacturers' demands?

Figure 16.15

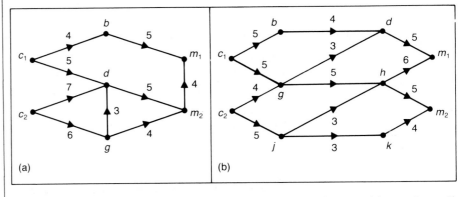

(a)   (b)

7. Find a maximum flow for the network in Fig. 16.16. The capacities on the undirected edges indicate the capacity is the same in either direction.

Figure 16.16

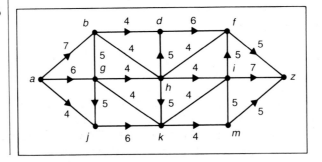

16.3 ■ MATCHING THEORY

The Villa school district must hire four teachers to teach classes in the following subjects: mathematics (s_1), computer science (s_2), chemistry (s_3), physics (s_4), and biology (s_5). Four candidates who are interested in teaching in this district are Miss Carelli (c_1), Mr. Ritter (c_2), Ms. Camille (c_3), and Mrs. Lewis (c_4). Miss Carelli is certified in mathematics and computer science; Mr. Ritter in mathematics and physics; Ms. Camille in biology; Mrs. Lewis in chemistry, physics and computer science. If the district hires all four candidates, can each teacher be assigned to teach a (different) subject in which he or she is certified?

This problem is an example of a general situation called the *assignment problem*. Using the principle of inclusion and exclusion in conjunction with the rook polynomial (see Sections 7.4 and 7.5), one can determine in how many ways, if any, the four teachers may be assigned to teach a different subject for which he or she is qualified. However, these techniques do not provide a means of setting up any of these assignments. In Fig. 16.17 the problem is modeled by means of a bipartite graph $G = (V, E)$, where V is partitioned as $X \cup Y$ with $X = \{c_1, c_2, c_3, c_4\}$ and $Y = \{s_1, s_2, s_3, s_4, s_5\}$. The edges $\{c_1, s_2\}, \{c_2, s_4\}, \{c_3, s_5\}, \{c_4, s_3\}$ demonstrate such an assignment of X into Y.

Figure 16.17

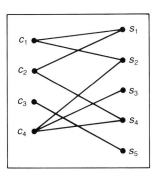

To examine this idea further, the following concepts are introduced.

Definition 16.5 ▶ Let $G = (V, E)$ be a bipartite graph with V partitioned as $X \cup Y$. (Each edge of E has the form $\{x, y\}$ with $x \in X$ and $y \in Y$.)

a) A *matching* in G is a subset of E such that no two edges share a common vertex in X or Y.

b) A *complete matching* of X into Y is a matching in G such that every $x \in X$ is the endpoint of an edge.

In terms of functions, a matching is a function that establishes a one-to-one correspondence between a subset of X and a subset of Y. When the matching is complete, a one-to-one function from X into Y is defined. The example in Fig. 16.17 contains such a function and a complete matching.

For a bipartite graph $G = (V, E)$ with V partitioned as $X \cup Y$, a complete matching of X into Y requires $|X| \leq |Y|$. If $|X|$ is large, then the construction of such a matching cannot be accomplished just by observation or trial and error. The following theorem, due to the English mathematician Philip Hall (1935), provides necessary and sufficient conditions for the existence of such a matching. The proof of the theorem, however, is not that given by Hall. A constructive proof that uses the material developed on transport networks is given.

Theorem 16.5   ▶ Let $G = (V, E)$ be bipartite with V partitioned as $X \cup Y$. A complete matching of X into Y exists iff for every subset A of X, $|A| \leq |R(A)|$, where $R(A)$ is the subset of Y consisting of those vertices that are adjacent to the vertices in A.

Before proving the theorem we illustrate its use in the following example.

EXAMPLE 16.11   a)   The bipartite graph in Fig. 16.18(a) has no complete matching. Any attempt to construct such a matching must include $\{x_1, y_1\}$ and either $\{x_2, y_3\}$ or $\{x_3, y_3\}$. If $\{x_2, y_3\}$ is included, there is no match for x_3. Likewise, if $\{x_3, y_3\}$ is included, we are not able to match x_2. If $A = \{x_1, x_2, x_3\} \subseteq X$, then $R(A) = \{y_1, y_2\}$. With $|A| = 3 > 2 = |R(A)|$, it follows from Theorem 16.5 that no complete matching can exist.

Figure 16.18

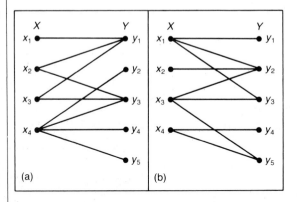

(a)   (b)

b)   For the graph in part (b) of the figure, consider the exhaustive listing in Table 16.2. Assuming the validity of Theorem 16.5, this listing indicates that the graph contains a complete matching.   □

We turn now to a proof of the theorem.

Proof   With V partitioned as $X \cup Y$, let $X = \{x_1, x_2, \ldots, x_m\}$ and $Y = \{y_1, y_2, \ldots, y_n\}$. Construct a transport network N that extends graph G by introducing two new vertices a (the source) and z (the sink). For each vertex x_i, $1 \leq i \leq m$, draw edge (a, x_i); for each vertex y_j, $1 \leq j \leq n$, draw edge (y_j, z). Each new edge is given a capacity of 1. Let M be any positive number that exceeds $|X|$. Assign each edge in G the capacity M.

Table 16.2

A	R(A)	|A|	|R(A)|
$\emptyset$	$\emptyset$	0	0
$\{x_1\}$	$\{y_1, y_2, y_3\}$	1	3
$\{x_2\}$	$\{y_2\}$	1	1
$\{x_3\}$	$\{y_2, y_3, y_5\}$	1	3
$\{x_4\}$	$\{y_4, y_5\}$	1	2
$\{x_1, x_2\}$	$\{y_1, y_2, y_3\}$	2	3
$\{x_1, x_3\}$	$\{y_1, y_2, y_3, y_5\}$	2	4
$\{x_1, x_4\}$	Y	2	5
$\{x_2, x_3\}$	$\{y_2, y_3, y_5\}$	2	3
$\{x_2, x_4\}$	$\{y_2, y_4, y_5\}$	2	3
$\{x_3, x_4\}$	$\{y_2, y_3, y_4, y_5\}$	2	4
$\{x_1, x_2, x_3\}$	$\{y_1, y_2, y_3, y_5\}$	3	4
$\{x_1, x_2, x_4\}$	Y	3	5
$\{x_1, x_3, x_4\}$	Y	3	5
$\{x_2, x_3, x_4\}$	$\{y_2, y_3, y_4, y_5\}$	3	4
X	Y	4	5

The original graph G and its associated network N appear as in Fig. 16.19. Using this network, it follows that a complete matching exists in G iff there is a maximum flow in N that uses all edges (a, x_i), $1 \le i \le m$. Then the value of such a maximum flow is $m = |X|$.

Figure 16.19

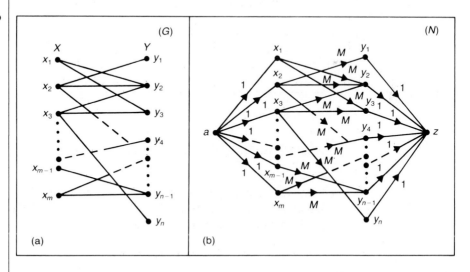

(a)   (b)

If $|A| \le |R(A)|$ for all $A \subseteq X$, we shall prove that there is a complete matching in G by showing that for any cut $(P, \overline{P})$ in N, $c(P, \overline{P}) \ge |X|$. Let $A \subseteq X$, where $A = \{x_1, x_2, \ldots, x_i\}$, for some $0 \le i \le m$. (Elements of X are relabeled, if necessary, so that the subscripts on the elements of A are consecutive. When $i = 0$, $A = \emptyset$.) Let

Figure 16.20

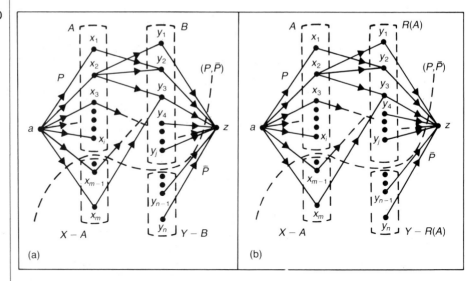

(a)

(b)

P be the source a together with the vertices in A and the set $B \subseteq Y$, as shown in Fig. 16.20(a). (Elements of Y are also relabeled if necessary.) Then $\overline{P} = (X - A) \cup (Y - B) \cup \{z\}$. If there is an edge $\{x, y\}$ with $x \in A$ and $y \in (Y - B)$, then the capacity of that edge is a summand in $c(P, \overline{P})$ and $c(P, \overline{P}) \geq M > |X|$. Should no such edge exist, then $c(P, \overline{P})$ is determined by the capacities of (1) the edges from the source a to the vertices in $X - A$, and (2) the edges from the vertices in B to the sink z. Since each of these edges has capacity 1, $c(P, \overline{P}) = |X - A| + |B| = |X| - |A| + |B|$. With $B \supseteq R(A)$, we have $|B| \geq |R(A)|$, and since $|R(A)| \geq |A|$, it follows that $|B| \geq |A|$. Consequently $c(P, \overline{P}) = |X| + (|B| - |A|) \geq |X|$. Therefore, since every cut in network N has capacity at least $|X|$, by Theorem 16.4, any maximum flow for N has value $|X|$. Such a flow will result in exactly $|X|$ edges from X to Y having flow 1 and this flow provides a complete matching of X into Y.

Conversely, let $A \subseteq X$ with $|A| > |R(A)|$. Let $(P, \overline{P})$ be the cut shown for the network in Fig. 16.20(b), with $P = \{a\} \cup A \cup R(A)$, $\overline{P} = (X - A) \cup (Y - R(A)) \cup \{z\}$. Then $c(P, \overline{P})$ is determined by (1) the edges from the source a to the vertices in $X - A$, and (2) the edges from the vertices in $R(A)$ to the sink z. Hence $c(P, \overline{P}) = |X - A| + |R(A)| = |X| - (|A| - |R(A)|) < |X|$, since $|A| > |R(A)|$. Since the network has a cut of capacity less than $|X|$, once again by Theorem 16.4 it follows that any maximum flow in the network has value smaller than $|X|$. Therefore there is no complete matching from X into Y for the given bipartite graph G.   ■

EXAMPLE 16.12   Five students s_1, s_2, s_3, s_4, and s_5 are members of three committees c_1, c_2, and c_3. The bipartite graph of Fig. 16.21(a) indicates the committee memberships. Each committee is to select a student representative to meet with the school president. Can a selection be made so that each committee has a distinct representative?

Although this problem is small enough to solve by inspection, we use the ideas developed in the proof of Theorem 16.5. Figure 16.21(b) provides the network for

Figure 16.21

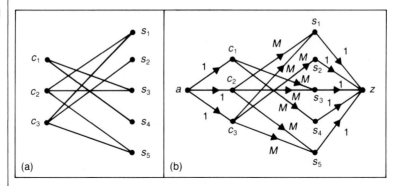

(a) (b)

the given bipartite graph. In Fig. 16.22(a) the labeling procedure (Theorem 16.4) is applied for the first time and indicates the edge (c_3, s_5) as a possible start for a complete matching. (Many edge labels are omitted in the figures in order to simplify the diagrams. Any unlabeled edge that starts at a or terminates at z has the label $1, 0$; all other unlabeled edges are labeled $M, 0$.) Two further applications of the procedure provide the network in Fig. 16.22(b). Here the maximum flow indicates that (c_1, s_4), (c_2, s_3), (c_3, s_5) is a possible solution to the choice of representatives.

Figure 16.22

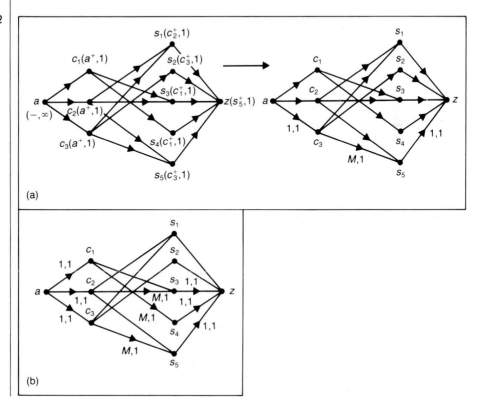

(a)

(b)

This example is a particular instance of a problem studied by Philip Hall. He considered a collection of sets $A_1, A_2, \ldots, A_n$ where the elements $a_1, a_2, \ldots, a_n$ were called a *system of distinct representatives* for the collection if (a) $a_i \in A_i$, $1 \leq i \leq n$; and (b) $a_i \neq a_j$, $1 \leq i < j \leq n$. Rewording Theorem 16.5 in this context we find that the collection $A_1, A_2, \ldots, A_n$ has a system of distinct representatives iff, for all $1 \leq i \leq n$, the union of any i of the sets $A_1, A_2, \ldots, A_n$ contains at least i elements.　□

Although the condition in Theorem 16.5 may be very tedious to check, the following corollary provides sufficient conditions for the existence of a complete matching.

Corollary 16.3　▶ Let $G = (V, E)$ be a bipartite graph with V partitioned as $X \cup Y$. There is a complete matching of X into Y if, for some $k \in \mathbf{Z}^+$, $\deg(x) \geq k \geq \deg(y)$ for all vertices $x \in X$ and $y \in Y$.

Proof　This proof is left for the section exercises.　■

EXAMPLE 16.13
a) Corollary 16.3 is applicable to the graph in Fig. 16.21(a). Here the appropriate value of k is 2.

b) There are 50 students (25 females and 25 males) in the senior class at Bell High School. If each female in the class is appreciated by exactly five of the males, and each male enjoys the company of exactly five of the females in the class, then it is possible for each male to go to the class party with a female he likes, and each female will attend with a male who likes her. (As a result of problems of this type, the condition in Theorem 16.5 is often referred to in the literature as *Hall's Marriage Condition*.)　□

For problems such as Example 16.11(a), where a complete matching does not exist, the following type of matching is often of interest.

Definition 16.6　▶ If $G = (V, E)$ is a bipartite graph with V partitioned as $X \cup Y$, a *maximal matching* in G is one that matches as many vertices in X as possible with the vertices in Y.

To investigate the existence and construction of a maximal matching, the following new idea is presented.

Definition 16.7　▶ Let $G = (V, E)$ be a bipartite graph where V is partitioned as $X \cup Y$. If $A \subseteq X$, then $\delta(A) = |A| - |R(A)|$ is called the *deficiency of A*. The *deficiency of graph G*, denoted $\delta(G)$, is given by $\delta(G) = \max\{\delta(A) | A \subseteq X\}$.

For $\emptyset \subseteq X$, we have $R(\emptyset) = \emptyset$, so $\delta(\emptyset) = 0$ and $\delta(G) \geq 0$. If $\delta(G) > 0$ there is a subset A of X with $|A| - |R(A)| > 0$, so $|A| > |R(A)|$ and by Theorem 16.5 there is no complete matching of X into Y.

EXAMPLE 16.14　The graph in Fig. 16.23(a) has no complete matching. (See Example 16.11(a).) For $A = \{x_1, x_2, x_3\}$, $R(A) = \{y_1, y_3\}$ and $\delta(A) = 3 - 2 = 1$. As a result of this subset A,

Figure 16.23

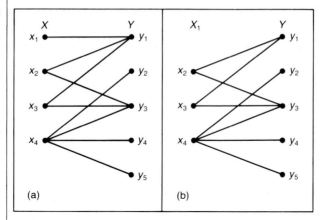

(a)   (b)

$\delta(G) = 1$. Removing one of the vertices from A (and the edges incident with it) we obtain the subgraph in part (b) of the figure. This (bipartite) subgraph provides a complete matching from $X_1 = \{x_2, x_3, x_4\}$ into Y. The edges $\{x_2, y_1\}$, $\{x_3, y_3\}$, and $\{x_4, y_4\}$ indicate one such matching which is also a maximal matching of X into Y. □

The ideas developed in Example 16.14 lead to the following.

Theorem 16.6

▶ Let $G = (V, E)$ be bipartite with V partitioned as $X \cup Y$. The maximum number of vertices in X that can be matched with those in Y is $|X| - \delta(G)$. Moreover, a matching of size $|X| - \delta(G)$ exists.

Proof

We provide a constructive proof, using transport networks as in the proof of Theorem 16.5. As in Figure 16.19, let N be the network associated with the bipartite graph G. The result will follow by showing that (a) the capacity of any cut $(P, \overline{P})$ in N is greater than or equal to $|X| - \delta(G)$; and (b) there exists a cut with capacity $|X| - \delta(G)$.

Let $(P, \overline{P})$ be any cut in N, where P is made up of the source a, the vertices in $A \subseteq X$, and the vertices in $B \subseteq Y$. (See Fig. 16.20(a)). (As in the proof of Theorem 16.5, the subsets A, B may be $\emptyset$.)

1. If edge (x, y) is in N with $x \in A$ and $y \in Y - B$, then $c(x, y)$ is a summand in $c(P, \overline{P})$. Since $c(x, y) = M > |X|$, $c(P, \overline{P}) > |X| \geq |X| - \delta(G)$.

2. If no such edge as in (1) exists, then $c(P, \overline{P})$ is determined by the $|X - A|$ edges from a to $X - A$ and the $|B|$ edges from B to z. Since each of these edges has capacity 1, $c(P, \overline{P}) = |X - A| + |B| = |X| - |A| + |B|$. As no edge connects a vertex in A with a vertex in $Y - B$, $R(A) \subseteq B$ and $|R(A)| \leq |B|$. Consequently, $c(P, \overline{P}) = (|X| - |A|) + |B| \geq (|X| - |A|) + |R(A)| = |X| - (|A| - |R(A)|) = |X| - \delta(A) \geq |X| - \delta(G)$.

Therefore in either case, $c(P, \overline{P}) \geq |X| - \delta(G)$, for any cut $(P, \overline{P})$ in N.

To complete the proof we must establish the existence of a cut with capacity $|X| - \delta(G)$. Since $\delta(G) = \max\{\delta(A) | A \subseteq X\}$, select a subset A of X with $\delta(G) =$

$\delta(A)$. Examining Fig. 16.20(b), let $P = \{a\} \cup A \cup R(A)$. Then $\overline{P} = (X - A) \cup (Y - R(A)) \cup \{z\}$. Since there is no edge between the vertices in A and those in $Y - R(A)$, $c(P, \overline{P}) = |X - A| + |R(A)| = |X| - (|A| - |R(A)|) = |X| - \delta(A) = |X| - \delta(G)$. ∎

We close this section with two examples that deal with these concepts.

EXAMPLE 16.15

Let $G = (V, E)$ be bipartite with V partitioned as $X \cup Y$. For each $x \in X$, $\deg(x) \geq 4$ and for each $y \in Y$, $\deg(y) \leq 5$. If $|X| \leq 15$, find an upper bound for $\delta(G)$.

Let $\emptyset \neq A \subseteq X$ and let $E_1 \subseteq E$ where $E_1 = \{\{a, b\} \mid a \in A, b \in R(A)\}$. Since $\deg(a) \geq 4$ for all $a \in A$, $|E_1| \geq 4|A|$. With $\deg(b) \leq 5$ for all $b \in R(A)$, $|E_1| \leq 5|R(A)|$. Hence $4|A| \leq 5|R(A)|$ and $\delta(A) = |A| - |R(A)| \leq |A| - (4/5)|A| = (1/5)|A|$. Since $A \subseteq X$, $|A| \leq 15$, so $\delta(A) \leq (1/5)(15) = 3$. Consequently, $\delta(G) = \max\{\delta(A) \mid A \subseteq X\} \leq 3$. □

EXAMPLE 16.16

At a certain college each of 110 freshmen is to be advised by a faculty advisor when she or he arrives for registration. These freshmen fall into four groups: 20 chemistry majors, 20 physics majors, 35 mathematics majors, and 35 computer science majors. Each prospective chemist may select any one of 12 faculty advisors (not necessarily the same 12 for each prospective chemist); each physics major any one of 8 such advisors. The mathematics majors each have a choice of 5 advisors but the future computer scientists are limited to a choice of only 3. There are 60 faculty advisors and each can handle up to 7 students at one time. On registration day, all of the students are waiting in the auditorium when the registrar arrives. If students are assigned according to their preferences as described above, what is the largest crowd of students that should have to wait for a later meeting with an advisor of their choice?

Let X be the set of 110 students and Y the set of 60 faculty members. The objective then is to find a maximum matching of X into Y. For any $x \in X$, $y \in Y$, draw an edge $\{x, y\}$ if student x selects faculty advisor y. This then yields the bipartite graph $G = (V, E)$, where V is partitioned as $X \cup Y$. To answer the question we need to compute $\delta(G)$.

Suppose that $A \subseteq X$ where the numbers of chemistry, physics, mathematics, and computer science majors in A are n_1, n_2, n_3, and n_4, respectively. Then $\delta(A) = |A| - |R(A)| = (n_1 + n_2 + n_3 + n_4) - |R(A)|$. With the restrictions on the number of possible advisor selections for each major and the number of students an advisor can assist at one time, $7|R(A)| \geq 12n_1 + 8n_2 + 5n_3 + 3n_4$. Consequently, $\delta(A) \leq (n_1 + n_2 + n_3 + n_4) - (12n_1 + 8n_2 + 5n_3 + 3n_4)/7 = (-5n_1/7) + (-n_2/7) + (2n_3/7) + (4n_4/7)$. Since A was chosen arbitrarily, this upper bound on $\delta(A)$ is maximal when $n_1 = n_2 = 0$, $n_3 = 35$, and $n_4 = 35$. Hence $\delta(G) \leq (70 + 140)/7 = 30$, so no matter how the students are assigned (following the procedures), there should be at least $|X| - \delta(G) = 80$ who are assigned, and no more than 30 who are kept waiting. □

EXERCISES

1. For the graph in Fig. 16.17, if four edges are selected at random, what is the probability that they provide a complete matching of X into Y?

2. Cathy is liked by Albert, Joseph, and Robert; Janice by Joseph and Dennis; Theresa by Albert and Joseph; Nettie by Dennis, Joseph, and Frank; and Karen by Albert, Joseph, and Robert. (a) Set up a bipartite graph to model the matching problem where each man is paired with a woman he likes. (b) Draw the associated network for the graph in part (a) and determine a maximum flow for this network. What complete matching does this determine? (c) Is there a complete matching that pairs Janice with Dennis and Nettie with Frank? (d) Is it possible to determine two complete matchings where each man is paired with two different women?

3. At Rydell High School the senior class is represented on six school committees by Annemarie (A), Gary (G), Jill (J), Kenneth (K), Michael (M), Norma (N), Paul (P), and Rosemary (R). The senior members of these committees are $\{A, G, J, P\}$, $\{G, J, K, R\}$, $\{A, M, N, P\}$, $\{A, G, M, N, P\}$, $\{A, G, K, N, R\}$, and $\{G, K, N, R\}$.

   a) The student government calls a meeting that requires the presence of a senior member from each committee. Find a selection that maximizes the number of seniors involved.

   b) Before the meeting the finances of each committee are to be reviewed by a senior who is not on that committee. Can this be accomplished? If so, how?

4. Prove Corollary 16.3.

5. Brenda is in charge of assigning students to part-time jobs at the college where she works. She has 25 student applications and there are 25 different part-time jobs available on the campus. Each applicant is qualified for at least four of the jobs, but each job can be performed by at most four of the applicants. Can Brenda assign all the students to jobs for which they are qualified? Explain.

6. For each of the following collections of sets, determine if possible a system of distinct representatives. If no such system exists, explain why.

   a) $A_1 = \{2, 3, 4\}$, $A_2 = \{3, 4\}$, $A_3 = \{1\}$, $A_4 = \{2, 3\}$.

   b) $A_1 = A_2 = A_3 = \{2, 4, 5\}$, $A_4 = A_5 = \{1, 2, 3, 4, 5\}$.

   c) $A_1 = \{1, 2\}$, $A_2 = \{2, 3, 4\}$, $A_3 = \{2, 3\}$, $A_4 = \{1, 3\}$, $A_5 = \{2, 4\}$.

7. a) Determine all systems of distinct representatives for the collection of sets: $A_1 = \{1, 2\}$, $A_2 = \{2, 3\}$, $A_3 = \{3, 4\}$, $A_4 = \{4, 1\}$.

   b) Given the collection of sets: $A_1 = \{1, 2\}$, $A_2 = \{2, 3\}$, $A_3 = \{3, 4\}$, ..., $A_n = \{n, 1\}$, how many different systems of distinct representatives exist for the collection?

8. Let $A_1, A_2, \ldots, A_n$ be a collection of sets where $A_1 = A_2 = \ldots = A_n$ and $|A_i| = k > 0$, $1 \le i \le n$. (a) Prove that the given collection has a system of distinct representatives iff $n \le k$. (b) When $n \le k$, how many different systems exist for the collection?

9. At each of seven universities there are six graduates with training in both computer science and Japanese. Six international corporations wish to hire 10, 9, 8, 7, 7, 7 graduates with these qualifications. Company policy (at all six corporations) dictates that no two newly hired professionals with this training may

be from the same university. Explain in terms of Theorem 16.5 whether or not all of these graduates will be hired by these six corporations.

10. Let $G = (V, E)$ be bipartite with V partitioned as $X \cup Y$. For all $x \in X$, $\deg(x) \geq 3$, and for all $y \in Y$, $\deg(y) \leq 7$. If $|X| \leq 50$, find an upper bound on $\delta(G)$.

11. a) Let $G = (V, E)$ be the bipartite graph in Fig. 16.24, with V partitioned as $X \cup Y$. Determine $\delta(G)$ and a maximal matching of X into Y.

    b) For any bipartite graph $G = (V, E)$, with V partitioned as $X \cup Y$, if $\beta(G)$ denotes the independence number of G, show that $|Y| = \beta(G) - \delta(G)$.

    c) Determine a largest maximal independent set of vertices for the graphs in Fig. 16.23(a) and 16.24.

Figure 16.24

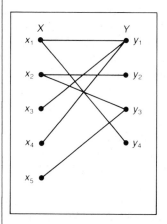

12. An aircraft company hires 100 new employees: 10 mathematicians, 10 computer scientists, 20 civil engineers, 25 mechanical engineers, and 35 electrical engineers. At present the company is undertaking 25 projects. Each mathematician has a choice of 4 projects on which to work, each civil engineer or computer scientist a choice of 5, each mechanical engineer a choice of 12, and each electrical engineer a choice of 15. If there is not enough work for more than 10 people on any one of these projects, what is the largest number of new employees that should have to wait for the start of new projects before they can begin working on a project?

16.4 ■ SUMMARY AND HISTORICAL REVIEW

This last chapter has provided us with a sample of how graph theory enters into an area of mathematics called operations research. Each topic was presented in an algorithmic manner that can be used in the computer implementation needed for solving each type of problem. Comparable coverage of this material can be found in Chapters 10 and 11 of the text by C. L. Liu [13] and in Chapter 9 of A. Tucker [19].

Chapters 4 and 5 of E. Lawler [12] have an extensive coverage that includes many of the more recent developments on networks and matching. This text provides a wide variety of applications and includes references for additional reading.

Section 16.1 provided two techniques for finding a minimal spanning tree in a weighted undirected graph. These techniques were developed in the late 1950's by Joseph Kruskal [11] and Robert Prim [16]. Computer implementation of these techniques is given in Chapters 6 and 7 of A. Aho, J. Hopcroft, and J. Ullman [1]. This reference also discusses the efficiency and speed of these algorithms. An interesting application of the concept of the minimal spanning tree in a physical science setting is provided in the article by D. Shier [18].

As found in Section 16.2, problems dealing with the allocation of resources or the shipment of goods can be modeled by means of transport networks. The fundamental work by G. Dantzig, L. Ford, and D. Fulkerson that brought about the labeling procedure for these networks and the Max-Flow Min-Cut Theorem can be found in their pioneering articles [5, 6, 7, 8]. The classic text by L. Ford and D. Fulkerson [9] provides excellent coverage of this topic. In addition, the reader may wish to examine Chapter 8 of the text by C. Berge [2] or Chapter 7 of the book by R. Busacker and T. Saaty [4]. Chapter 10 in C. L. Liu [13] includes coverage on an extension to networks where the flow in each edge is restricted by a lower as well as an upper capacity. For more applications the reader should examine the article by D. Fulkerson on pages 139–171 of reference [10].

The last topic discussed here dealt with matching in a bipartite graph. The theory behind this was first developed by Philip Hall in 1935, but here the ideas on transport networks were used to provide an algorithm for a solution. Chapter 7 of the text by O. Ore [15] provides a very readable introduction to this topic along with some applications. For more on systems of representatives the reader should examine Chapter 5 of the monograph by H. Ryser [17]. A second method for finding a maximal matching in a bipartite graph is called the *Hungarian method*. This is given in Chapter 5 of the text by J. Bondy and U. Murty [3] and in Chapter 10 of the book by C. Berge [2]. In addition to its application in solving the assignment problem, matching theory has many interesting combinatorial implications. One may learn more about these in the survey article by L. Mirsky and H. Perfect [14].

REFERENCES

1. Aho, Alfred V., Hopcroft, John E., and Ullman, Jeffrey D., *Data Structures and Algorithms,* Addison-Wesley, Reading, Massachusetts, 1983.

2. Berge, Claude, *The Theory of Graphs and Its Applications,* John Wiley & Sons, New York, 1962.

3. Bondy, J. A., and Murty, U. S. R., *Graph Theory with Applications,* Elsevier North Holland, New York, 1976.

4. Busacker, Robert G., and Saaty, Thomas L., *Finite Graphs and Networks,* McGraw-Hill, New York, 1965.

5. Dantzig, G. B., and Fulkerson, D. R., *Computation of Maximal Flows in Networks,* The RAND Corporation, P-677, 1955.

6. Dantzig, G. B., and Fulkerson, D. R., *On the Max Flow Min Cut Theorem,* The RAND Corporation, RM-1418-1, 1955.

7. Ford, L. R., Jr., *Network Flow Theory,* The RAND Corporation, P-923, 1956.

8. Ford, L. R., Jr., and Fulkerson, D. R., "Maximal flow through a network," *Canadian Journal of Mathematics,* Vol. 8, 1956, pp. 399–404.

9. Ford, L. R., Jr., and Fulkerson, D. R., *Flows in Networks,* Princeton University Press, Princeton, New Jersey, 1962.

10. Fulkerson, D. R., ed., *Studies in Graph Theory,* Part I, *MAA Studies in Mathematics,* Vol. 11, The Mathematical Association of America, 1975.

11. Kruskal, Joseph B., "On the shortest spanning subtree of a graph and the traveling salesman problem," *Proceedings of the AMS,* Vol. 1, Number 1, 1956, pp. 48–50.

12. Lawler, Eugene, *Combinatorial Optimization: Networks and Matroids,* Holt, Rinehart and Winston, New York, 1976.

13. Liu, C. L., *Introduction to Combinatorial Mathematics,* McGraw-Hill, New York, 1968.

14. Mirsky, L., and Perfect, H., "Systems of representatives," *Journal of Mathematical Analysis and Applications,* Vol. 3, 1966, pp. 520–568.

15. Ore, Oystein, *Theory of Graphs,* American Mathematical Society, Providence, Rhode Island, 1962.

16. Prim, Robert C., "Shortest connection networks and some generalizations," *Bell System Technical Journal,* Vol. 36, 1957, pp. 1389–1401.

17. Ryser, Herbert J., *Combinatorial Mathematics,* Carus Mathematical Monographs, Number 14, Mathematical Association of America, 1963.

18. Shier, Douglas R., "Testing for homogeneity using minimum spanning trees," *The UMAP Journal,* Vol. 3, No. 3, 1982, pp. 273–283.

19. Tucker, Alan, *Applied Combinatorics,* John Wiley & Sons, New York, 1980.

■ MISCELLANEOUS EXERCISES

1. a) Let $G = (V, E)$ be a weighted connected undirected graph. If $e_1 \in E$ with $\text{wt}(e_1) < \text{wt}(e)$ for all other edges $e \in E$, prove that edge e_1 is part of every minimal spanning tree for G.

   b) With G as in part (a), suppose that there are edges e_1, $e_2 \in E$ with $\text{wt}(e_1) < \text{wt}(e_2) < \text{wt}(e)$ for all other edges $e \in E$. Prove or disprove: Edge e_2 is part of every minimal spanning tree for G.

2. a) Let $G = (V, E)$ be a weighted connected undirected graph where every edge e of G is part of a simple cycle. Prove that if $e_1 \in E$ with

$\text{wt}(e_1) > \text{wt}(e)$ for all other edges $e \in E$, then any spanning tree for G that contains e_1 cannot be minimal.

b) With G as in part (a), suppose that $e_1, e_2 \in E$ with $\text{wt}(e_1) > \text{wt}(e_2) > \text{wt}(e)$ for all other edges $e \in E$. Prove or disprove: Edge e_2 is not part of any minimal spanning tree for G.

3. Use the concept of flow in a transport network to construct a directed multigraph $G = (V, E)$, with $V = \{u, v, w, x, y\}$ and $\deg^+(u) = 1$, $\deg^-(u) = 3$; $\deg^+(v) = 3$, $\deg^-(v) = 3$; $\deg^+(w) = 3$, $\deg^-(w) = 4$; $\deg^+(x) = 5$, $\deg^-(x) = 4$; $\deg^+(y) = 4$, $\deg^-(y) = 2$.

4. A set of words $\{qs, tq, ut, pqr, srt\}$ is to be transmitted using a binary code for each letter.

a) Show that it is possible to select one letter from each word as a system of distinct representatives for these words.

b) If a letter is selected at random from each of the five words, what is the probability that the selection is a system of distinct representatives for the words?

5. For $n \in \mathbf{Z}^+$ and for each $1 \le i \le n$, let $A_i = \{1, 2, 3, \ldots, n\} - \{i\}$. How many different systems of distinct representatives exist for the collection A_1, A_2, A_3, $\ldots$, A_n?

6. This exercise outlines a proof of the Birkhoff–von Neumann Theorem.

a) For $n \in \mathbf{Z}^+$, an $n \times n$ matrix is called a *permutation* matrix if there is exactly one 1 in each row and column, and all other entries are 0. How many 5×5 permutation matrices are there? How many $n \times n$?

b) An $n \times n$ matrix B is called *doubly stochastic* if $b_{ij} \ge 0$ for all $1 \le i, j \le n$, and the sum of the entries in any row or column is 1.

If $B = \begin{bmatrix} 0.2 & 0.1 & 0.7 \\ 0.4 & 0.5 & 0.1 \\ 0.4 & 0.4 & 0.2 \end{bmatrix}$, verify that B is doubly stochastic.

c) Find four positive real numbers c_1, c_2, c_3, and c_4 and four permutation matrices P_1, P_2, P_3, and P_4 such that $c_1 + c_2 + c_3 + c_4 = 1$ and $B = c_1 P_1 + c_2 P_2 + c_3 P_3 + c_4 P_4$.

d) Part (c) is a special case of the Birkhoff–von Neumann theorem: If B is any $n \times n$ doubly stochastic matrix, then there exist positive real numbers c_1, c_2, $\ldots$, c_k, and permutation matrices P_1, P_2, $\ldots$, P_k such that $\sum_{i=1}^k c_i = 1$ and $\sum_{i=1}^k c_i P_i = B$. To prove this result, proceed as follows: Construct a bipartite graph $G = (V, E)$ with V partitioned as $X \cup Y$, where $X = \{x_1, x_2, \ldots, x_n\}$, $Y = \{y_1, y_2, \ldots, y_n\}$. The vertex x_i, $1 \le i \le n$, corresponds with the ith row of B; vertex y_j, $1 \le j \le n$, corresponds with the jth column of B. The edges of G are of the form $\{x_i, y_j\}$ if and only if $b_{ij} > 0$. We claim that there is a complete matching of X into Y.

If not, there is a subset A of X with $|A| > |R(A)|$, i.e., there is a set of r rows of B having positive entries in s columns and $r > s$. What is the sum of these r rows of B? Yet the sum of these same entries, when added column by column, is less than or equal to s. (Why?) Consequently, we have a contradiction.

As a result of the complete matching of X into Y there are n positive entries in B that occur so that no two are in the same row or column. (Why?) If c_1 is the smallest of these entries, then we may write $B = c_1 P_1 + B_1$ where P_1 is an $n \times n$ permutation matrix where the 1's are located according to the positive entries in B that came about from the complete matching. What are the sums of the entries in the rows and columns of B_1?

e) How is the proof completed?

Answers

Chapter 1

Sections 1.1 and 1.2—p. 8

1. a) 288 b) 24 c) 8
3. a) 136,080 b) 9×10^5
  i) a) 68,880 b) 450,000
  ii) a) 28,560 b) 180,000
  iii) a) 33,600 b) 225,000
5. 2^9 7. $26 \sum_{i=0}^{7} (36)^i$
11. $P(n + 1, r) = [(n + 1)!]/[(n + 1 - r)!] = [(n + 1)/(n + 1 - r)] [(n!)/((n - r)!)]$
  $= [(n + 1)/(n + 1 - r)] \cdot P(n, r)$.
13. a) 6720 b) 6!
15. a) 26^3 (for five or six letters) b) 15,600
17. $(14!)/[(7!)(7!)]$; $(14!)/[(7!)(7!)]$
  General statement: For real numbers a, b, and nonnegative integers m, n, the number of paths from (a, b) to $(a + m, b + n)$ is $(m + n)!/(m! n!)$.
19. a) 6! b) 2(5!)
21. The number of ways one can arrange the n objects $x_1, x_1, x_1, x_2, x_2, x_2, \ldots, x_k, x_k, x_k$ (for $n = 3k$) is $(n!)/(3!)^k$, an integer.

Section 1.3—p. 16

1. a) $\binom{20}{12}$ b) $\binom{10}{6}\binom{10}{6}$ c) $\sum_{i=1}^{5} \binom{10}{2i}\binom{10}{12-2i}$
  d) $\sum_{i=7}^{10} \binom{10}{i}\binom{10}{12-i}$ e) $\sum_{i=8}^{10} \binom{10}{i}\binom{10}{12-i}$

3. $\binom{2n}{n} + \binom{2n}{n-1} = \dfrac{(2n)!}{n! \, n!} + \dfrac{(2n)!}{(n-1)!(n+1)!}$

  $= \dfrac{(2n)!(n+1)^2}{(n+1)!(n+1)!} + \dfrac{(2n)!(n)(n+1)}{(n+1)!(n+1)!}$

  $= \left(\dfrac{1}{2}\right) \left[\dfrac{(2n)!(2n+2)(n+1) + (2n)!(2n+2)n}{(n+1)!(n+1)!} \right]$

  $= \left(\dfrac{1}{2}\right) \binom{2n+2}{n+1}$.

5. a) 120 b) 56 c) 110
7. a) $\binom{12}{3}\binom{9}{3}\binom{6}{3}\binom{3}{3}$ or $(12!)/(3!)^4$ b) $(12!)/(4!)^2(2!)^2$
9. 17 11. $\binom{n}{3}$; $\binom{n}{3} - n(n - 4) - n$, $n \geq 4$.
15. a) 2^3 b) 2^{10} c) 3^{10} d) 4^5 e) 4^{10}
17. Consider the expansions of (a) $[(1 + x) - x]^n$; (b) $[(2 + x) - (x + 1)]^n$; and (c) $[(2 + x) - x]^n$.

Section 1.4—p. 22

1. a) $\binom{14}{10}$ b) $\binom{9}{5}$ c) $\binom{12}{8}$
3. $\binom{23}{20}$ 5. a) 2^5 b) 2^n
7. a) $\binom{35}{32}$ b) $\binom{31}{28}$ c) $\binom{11}{8}$
  d) 1 e) $\binom{43}{40}$ f) $\binom{31}{28} - \binom{5}{2}$

9. a) $\binom{14}{5}$　　　　　　　　b) $\binom{11}{5} + 3\binom{10}{4} + 3\binom{9}{3} + \binom{8}{2}$

11. a) $\binom{7}{4}$　　　　　　　　b) $\sum_{i=0}^{3}\binom{9-2i}{7-2i}$

13. a) $\binom{16}{12}$　　　　　　　b) 5^{12}

15. a) $(5^3)\binom{29}{25}$

　　b) $[4\binom{23}{20} + 4^2\binom{18}{15} + 4^2\binom{13}{10} + 4^3\binom{8}{5}] + [4^2\binom{28}{25} + 4\binom{18}{15} + 4^2\binom{13}{10} + 4^2\binom{8}{5} + 4^3]$

17. a) $\binom{8}{6}\binom{34}{31}$　　　　　　b) $\binom{5}{3}\binom{34}{31}$

19. For the first object there are r choices. For the second object there are $r + 1$ choices, for if this object is placed in the same container as the first object it can be placed above or below it. In similar fashion there are $r + 2$ choices for the third object, $\ldots$, $r + (n - 1)$ choices for the nth object. By the rule of product the number of distributions is

$$(r)(r + 1)(r + 2)\cdots(r + n - 1) = [(r + n - 1)!]/(r - 1)! = P(r + n - 1, r - 1).$$

Miscellaneous Exercises — p. 25

1. $\binom{4}{1}\binom{7}{2} + \binom{4}{2}\binom{7}{4} + \binom{4}{3}\binom{7}{6}$

3. a) $10!/[2!\,1!\,7!]$　　　　　b) $10!/[2!\,1!\,7!]$

　c) Let a, b, c be any real numbers and let m, n, p be nonnegative integers. The number of paths from (a, b, c) to $(a + m, b + n, c + p)$ is $(m + n + p)!/[m!\,n!\,p!]$.

　d) $([9!/7!\,2!] + (9!/7!)) + ([7!/5!\,2!] + (7!/5!)) + ([5!/3!\,2!] + (5!/3!)) + ((3!/2!) + (3!/1!))$.

5. a) 10^{25}　　　　　　　b) $(10)(11)(12)\cdots(34) = 34!/9!$

　c) $(25!)\binom{24}{9}$

7. a) $C(12, 8)$　　　　　　b) $P(12, 8)$

9. $[(7!/2!)\binom{6}{1}](6)$　　　　　　　　11. $(1/10)[10!/(4!\,3!\,3!)]$

13. a) i) $\binom{5}{4} + \binom{5}{2}\binom{4}{2} + \binom{4}{4}$　　　　　　　　ii) $\binom{8}{4} + \binom{6}{2}\binom{5}{2} + \binom{7}{4}$

　　iii) $\binom{8}{4} + \binom{6}{2}\binom{5}{2} + \binom{7}{4} - 9$

　b) i) $\binom{5}{1}\binom{4}{3} + \binom{5}{3}\binom{4}{1}$　　　　　　ii) and iii) $\binom{5}{1}\binom{5}{3} + \binom{7}{3}\binom{4}{1}$

15. a) $(5)(9!)$　　　　　　b) $(3)(8!)$

17. $17 + (10)(2) + \binom{6}{2} + 5\binom{3}{2} + \binom{5}{4}$　　　　　　19. $\binom{16}{12} - 5\binom{8}{4}$

21. $2\binom{n+r-k-1}{n-k} + (n + r - k - 1)\binom{n+r-k-2}{n-k}$

23. $0 = (1 + (-1))^n = \binom{n}{0} - \binom{n}{1} + \binom{n}{2} - \binom{n}{3} + \cdots + (-1)^n\binom{n}{n}$,
　　so $\binom{n}{0} + \binom{n}{2} + \binom{n}{4} + \cdots = \binom{n}{1} + \binom{n}{3} + \binom{n}{5} + \cdots$

Chapter 2　Section 2.1 — p. 36

1. a) $r \to q$　　b) $q \to p$　　c) $r \wedge q$　　d) $(s \wedge r) \to q$

3. (a), (d), (e), (f), (g)

p	q		a) $p \to (p \vee q)$		d) $(p \to q) \to (q \to p)$
0	0		1		1
0	1		1		0
1	0		1		1
1	1		1		1

e) $[p \wedge (p \to q)] \to q$ f) $(p \wedge q) \to p$ g) $q \leftrightarrow (\bar{p} \vee \bar{q})$

e)	f)	g)
1	1	0
1	1	1
1	1	0
1	1	0

(b), (c), (h)

p	q	r
0	0	0
0	0	1
0	1	0
0	1	1
1	0	0
1	0	1
1	1	0
1	1	1

b) $p \to (q \to r)$ c) $(p \to q) \to r$

b)	c)
1	0
1	1
1	0
1	1
1	1
1	1
0	0
1	1

h) $[(p \to q) \wedge (q \to r)] \to (p \to r)$

1
1
1
1
1
1
1
1

5. a)

p	q	$p \to q$	$\bar{q} \to \bar{p}$	$q \to p$	$\bar{p} \to \bar{q}$
0	0	1	1	1	1
0	1	1	1	0	0
1	0	0	0	1	1
1	1	1	1	1	1

b)   Converse: If quadrilateral $ABCD$ is a rectangle, then quadrilateral $ABCD$ is a square. (FALSE)

Inverse: If quadrilateral $ABCD$ is not a square, then quadrilateral $ABCD$ is not a rectangle. (FALSE)

Contrapositive: If quadrilateral $ABCD$ is not a rectangle, then quadrilateral $ABCD$ is not a square. (TRUE)

7. a) 2^5 b) 2^n

9. a)

p	q	$p \wedge q$	$(\bar{p} \vee q) \wedge (p \wedge (p \wedge q))$
0	0	0	0
0	1	0	0
1	0	0	0
1	1	1	1

b) $(\bar{p} \wedge q) \vee (p \vee (p \vee q)) \Leftrightarrow (p \vee q)$.

11. a) $(p \uparrow p)$ b) $(p \uparrow p) \uparrow (q \uparrow q)$ c) $(p \uparrow q) \uparrow (p \uparrow q)$

d) $p \uparrow (q \uparrow q)$

e) $(r \uparrow s) \uparrow (r \uparrow s)$, where r stands for $p \uparrow (q \uparrow q)$ and s for $q \uparrow (p \uparrow p)$.

13.

p	q	$\overline{(p \downarrow q)}$	$(\bar{p} \uparrow \bar{q})$	$(\bar{p} \uparrow q)$	$(\bar{p} \downarrow \bar{q})$
0	0	0	0	0	0
0	1	1	1	0	0
1	0	1	1	0	0
1	1	1	1	1	1

15. 27

Section 2.2 — p. 43

1. They are all the same set.

3. (b) and (d) are false; the remaining parts are true.

5. $|A| = 6$

7. (a) $|A| = 7$; (b) If A has 2^n subsets of odd cardinality, then $|A| = n + 1$.

9. a)   31   b)   30   c)   28

11. Let $W = \{1\}$, $X = \{\{1\}, 2\}$, $Y = \{X, 3\}$

13. a)   Let $x \in A$. Since $A \subseteq B$, $x \in B$. Then with $B \subseteq C$, $x \in C$. Hence $x \in A \Rightarrow x \in C$
    and $A \subseteq C$.

    b)   Let $x \in A$. $A \subset B \Rightarrow x \in B$. $B \subseteq C \Rightarrow x \in C$. Hence $A \subseteq C$. Since $A \subset B$, there
    is a $y \in B$ where $y \notin A$. With $B \subseteq C$, $y \in C$. Consequently, $A \subseteq C$ and $y \in C$
    with $y \notin A$, so $A \subset C$.

    c)   If $x \in A$, then $A \subseteq B \Rightarrow x \in B$, and $B \subset C \Rightarrow x \in C$. Hence $A \subseteq C$. Since $B \subset C$
    there exists $y \in C$ with $y \notin B$. Also, $A \subseteq B$ and $y \notin B \Rightarrow y \notin A$. Consequently,
    $A \subseteq C$ and $y \in C$ with $y \notin A \Rightarrow A \subset C$.

    d)   Since $A \subset B$, it follows that $A \subseteq B$. The result then follows from (c).

15. $n = 20$

17. Let $A = \{x, y, a_1, a_2, \ldots, a_n\}$. The number of subsets of A of size r is $\binom{n+2}{r}$. These fall
into four classes: (1) the $\binom{n}{r}$ subsets containing neither x nor y; (2) the $\binom{n}{r-1}$ subsets
containing x but not y; (3) the $\binom{n}{r-1}$ subsets containing y but not x; (4) the $\binom{n}{r-2}$ subsets
containing both x and y.

Section 2.3 — p. 52

1. a) $\{1, 2, 3, 5\}$ b) A c) and d) $\mathcal{U} - \{2\}$

    e) $\{4, 8\}$ f) $\{1, 2, 3, 4, 5, 8\}$ g) $\emptyset$

    h) $\{2, 4, 8\}$ i) $\{1, 3, 4, 5, 8\}$

3. a)   (i)   b)   (iii)   c)   (ii)   d)   (iv)

5. a)   Let $\mathcal{U} = \{1, 2, 3\}$, $A = \{1\}$, $B = \{2\}$, $C = \{3\}$. Then $A \cap C = B \cap C = \emptyset$ but
    $A \neq B$.

    b)   For $\mathcal{U} = \{1, 2\}$, $A = \{1\}$, $B = \{2\}$, $C = \mathcal{U}$, $A \cup C = B \cup C$ but $A \neq B$.

c) $x \in A \Rightarrow x \in A \cup C \Rightarrow x \in B \cup C$. If $x \in B$, then $A \subseteq B$. If $x \in C$, then $x \in A \cap C = B \cap C$ and $x \in B$. In either case $A \subseteq B$. Likewise, $y \in B \Rightarrow y \in B \cup C = A \cup C$, so $y \in A$ or $y \in C$. If $y \in C$, then $y \in B \cap C = A \cap C$. In either case $y \in A$ and $B \subseteq A$. Hence $A = B$.

d) Let $x \in A$. Consider two cases: (1) $x \in C \Rightarrow x \notin A \bigtriangleup C \Rightarrow x \notin B \bigtriangleup C \Rightarrow x \in B$. (2) $x \notin C \Rightarrow x \in A \bigtriangleup C \Rightarrow x \in B \bigtriangleup C \Rightarrow x \in B$ (since $x \notin C$). In either case $x \in B$, so $A \subseteq B$. In a similar way it follows that $B \subseteq A$ and $A = B$.

7. 7; 1

9. a) $\emptyset = (A \cup B) \cap (A \cup \overline{B}) \cap (\overline{A} \cup B) \cap (\overline{A} \cup \overline{B})$

   b) $A = A \cup (A \cap B)$

   c) $A \cap B = (A \cup B) \cap (A \cup \overline{B}) \cap (\overline{A} \cup B)$

   d) $A = (A \cap B) \cup (A \cap \mathcal{U})$.

11. a) Let $\mathcal{U} = \{1, 2, 3\}$, $A = \{1\}$, $B = \{2\}$. Then $\{1, 2\} \in \mathcal{P}(A \cup B)$ but $\{1, 2\} \notin \mathcal{P}(A) \cup \mathcal{P}(B)$.

   b) $X \in \mathcal{P}(A \cap B) \Leftrightarrow X \subseteq A \cap B \Leftrightarrow X \subseteq A$ and $X \subseteq B \Leftrightarrow X \in \mathcal{P}(A)$ and $X \in \mathcal{P}(B) \Leftrightarrow X \in \mathcal{P}(A) \cap \mathcal{P}(B)$, so $\mathcal{P}(A \cap B) = \mathcal{P}(A) \cap \mathcal{P}(B)$.

13. a) 2^6 b) 2^n

   c) In the membership table, $A \subseteq B$ if the columns for A, B are such that whenever a 1 occurs in the column for A there is a corresponding 1 in the column for B.

   d)

A	B	C	$A \cup \overline{B}$	$(A \cap B) \cup (\overline{B} \cap C)$
0	0	0	1	1
0	0	1	1	1
0	1	0	0	1
0	1	1	0	0
1	0	0	1	1
1	0	1	1	1
1	1	0	1	1
1	1	1	1	1

15. a) $[-6, 9]$ b) $[-8, 12]$ c) $\emptyset$

   d) $[-8, -6) \cup (9, 12]$ e) $[-14, 21]$ f) $[-2, 3]$

   g) $\mathbf{R}$ h) $[-2, 3]$.

17. $x \in A \cup (\bigcap_{i \in I} B_i) \Leftrightarrow x \in A$ or $(x \in B_i$ for all $i \in I) \Leftrightarrow$ For all $i \in I$, $x \in A$ or $x \in B_i \Leftrightarrow$ For all $i \in I$, $x \in A \cup B_i \Leftrightarrow x \in \bigcap_{i \in I} (A \cup B_i)$. $[A \cap (\bigcup_{i \in I} B_i) = \bigcup_{i \in I} (A \cap B_i)$ now follows by duality]

Sections 2.4 and 2.5 — p. 58

1. a) $24! + 24! - 22!$ b) $26! - [24! + 24! - 23!]$

3. $9! + 9! - 8!$ 5. a) $2/9$ b) $4/9$

7. a) $[13!/(2!)^3] - 3[12!/(2!)^2] + 3(11!/2!) - 10!$

   b) Divide the result in (a) by $[13!/(2!)^3]$

9. $Pr(A) = 1/3$, $Pr(B) = 7/15$, $Pr(A \cap B) = 2/15$, $Pr(A \cup B) = 2/3$; $Pr(A \cup B) = 2/3 = 1/3 + 7/15 - 2/15 = Pr(A) + Pr(B) - Pr(A \cap B)$

11. $3/28$

Miscellaneous Exercises — p. 62

1. a)   Valid   b)   Invalid   c)   Invalid

3. a)   128   b)   8

5. a)   126 (if teams wear different uniforms); 63 (if teams are not distinguishable).

      112 (if teams wear different uniforms); 56 (if teams are not distinguishable).

   b) $2^n - 2$; $(1/2)(2^n - 2)$. $2^n - 2 - 2n$; $(1/2)(2^n - 2 - 2n)$.

7. a) $\binom{r+1}{m}$ $(m \le r + 1)$ b) $\binom{n-k+1}{k}$ $(2k \le n + 1)$.

9. a) $[0, 14/3]$ b) $(0, 9/5]$ c) $\{0\}$

   d) $\{0\} \cup (6, 12]$ e) $[0, +\infty)$ f) $(0, +\infty)$

   g) $\{0\}$ h) $\varnothing$

11. a)

A	B	$A \cap B$
0	0	0
0	1	0
1	0	0
1	1	1

Since $A \subseteq B$, consider only rows 1, 2, 4. For these rows $A \cap B = A$.

   b)

A	B	C	$A \cap B$	$B \cup C$	$A \cup B \cup C$
0	0	0	0	0	0
0	0	1	0	1	1
0	1	0	0	1	1
0	1	1	0	1	1
1	0	0	0	0	1
1	0	1	0	1	1
1	1	0	1	1	1
1	1	1	1	1	1

Since $A \cap B = A$ and $B \cup C = C$, consider only rows 1, 2, 4, 8. For these four rows $A \cup B \cup C = C$.

   c)

A	B	C	$(A \cap \bar{B}) \cup (B \cap \bar{C})$	$A \cap \bar{C}$
0	0	0	0	0
0	0	1	0	0
0	1	0	1	0
0	1	1	0	0
1	0	0	1	1
1	0	1	0	0
1	1	0	1	1
1	1	1	0	0

For $C \subseteq B \subseteq A$ consider only rows 1, 5, 7, 8. Here $(A \cap \bar{B}) \cup (B \cap \bar{C}) = A \cap \bar{C}$.

d)

A	B	C	$A \triangle B$	$A \triangle C$	$B \triangle C$
0	0	0	0	0	0
0	0	1	0	1	1
0	1	0	1	0	1
0	1	1	1	1	0
1	0	0	1	1	0
1	0	1	1	0	1
1	1	0	0	1	1
1	1	1	0	0	0

When $A \triangle B = C$ we consider rows 1, 4, 6, 7. In these cases $A \triangle C = B$ and $B \triangle C = A$.

13. $7^{15} - 3(3^{15}) + 3$

15. $2/(7!)$ 17. a) $\binom{15}{12}/\binom{19}{16}$ b) $\binom{11}{8}/\binom{19}{16}$ c) $1/\binom{19}{16}$.

Chapter 3

Section 3.1 — p. 70

1. $A \times B = \{(1, 2), (2, 2), (3, 2), (4, 2), (1, 5), (2, 5), (3, 5), (4, 5)\}$
    $B \times A = \{(2, 1), (2, 2), (2, 3), (2, 4), (5, 1), (5, 2), (5, 3), (5, 4)\}$
    $A \cup (B \times C) = \{1, 2, 3, 4, (2, 3), (2, 4), (2, 7), (5, 3), (5, 4), (5, 7)\}$
    $(A \cup B) \times C = \{(1, 3), (2, 3), (3, 3), (4, 3), (5, 3), (1, 4), (2, 4), (3, 4), (4, 4), (5, 4), (1, 7),$
    $(2, 7), (3, 7), (4, 7), (5, 7)\} = (A \times C) \cup (B \times C)$.

3. a) 9   b) 2^9 c) 2^9 d) 2^7 e) $\binom{9}{5}$ f) $\binom{9}{7} + \binom{9}{8} + \binom{9}{9}$.

5. $\mathscr{R} = \{(2, 10), (2, 12), (2, 14), (3, 12), (4, 12), (5, 10), (6, 12), (7, 14)\}$.

7.

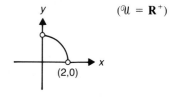

9. b) Follows from (a) by duality.

    c) $(x, y) \in (A \cap B) \times C \Leftrightarrow x \in A \cap B, y \in C \Leftrightarrow (x \in A$ and $x \in B)$ and $y \in C \Leftrightarrow$
        $(x \in A$ and $y \in C)$ and $(x \in B$ and $y \in C) \Leftrightarrow (x, y) \in A \times C$ and $(x, y) \in$
        $B \times C \Leftrightarrow (x, y) \in (A \times C) \cap (B \times C)$

    d) Follows from (c) by duality.

11. $(x, y) \in A \times (B - C) \Leftrightarrow x \in A, y \in B - C \Leftrightarrow x \in A$ and $(y \in B$ and $y \notin C) \Leftrightarrow$
    $(x \in A$ and $y \in B)$ and $(x \in A$ and $y \notin C) \Leftrightarrow (x, y) \in A \times B$ and $(x, y) \notin A \times C \Leftrightarrow$
    $(x, y) \in (A \times B) - (A \times C)$.

Section 3.2 — p. 74

1. a) Function. Range $= \{7, 8, 11, 16, 23, \ldots\}$

    b) Relation, not a function   c) Function. Range $= \mathbf{R}$.

    d) and e) Relation, not a function

3. a) 1) $\{(1, x), (2, x), (3, x), (4, x)\}$

        2) $\{(1, y), (2, y), (3, y), (4, y)\}$

　　　　3)　$\{(1, z), (2, z), (3, z), (4, z)\}$

　　　　4)　$\{(1, x), (2, y), (3, x), (4, y)\}$

　　　　5)　$\{(1, x), (2, y), (3, z), (4, x)\}$

　　b)　3^4　　c)　0　　d)　4^3　　e)　24　　f)　3^3　　g)　3^2　　h)　3^2

5. a)　One-to-one; the range is the set of all odd integers.

　　b)　One-to-one; the range is $\mathbf{Q}$.

　　c)　Not one-to-one; the range is $\{0, \pm 6, \pm 24, \pm 60, \ldots\} = \{n^3 - n \mid n \in \mathbf{Z}\}$.

　　d)　One-to-one; the range is $\mathbf{R}$.

　　e)　One-to-one; the range is $[-1, 1]$.

　　f)　Not one-to-one; the range is $[0, 1]$.

7. 4^2

9. a)　$f(A_1 \cup A_2) = \{y \in B \mid y = f(x),\ x \in A_1 \cup A_2\} = \{y \in B \mid y = f(x),\ x \in A_1$ or $x \in A_2\} = \{y \in B \mid y = f(x),\ x \in A_1\} \cup \{y \in B \mid y = f(x),\ x \in A_2\} = f(A_1) \cup f(A_2)$.

　　c)　From (b), $f(A_1 \cap A_2) \subseteq f(A_1) \cap f(A_2)$. Conversely, $y \in f(A_1) \cap f(A_2) \Rightarrow y = f(x_1) = f(x_2),\ x_1 \in A_1,\ x_2 \in A_2 \Rightarrow y = f(x_1)$ and $x_1 = x_2$ (since f is injective) $\Rightarrow y \in f(A_1 \cap A_2)$. So f injective $\Rightarrow f(A_1 \cap A_2) = f(A_1) \cap f(A_2)$.

Section 3.3 — p. 79

1. a)　$A = \{1, 2, 3, 4\}$, $B = \{v, w, x, y, z\}$, $f = \{(1, v), (2, v), (3, w), (4, x)\}$

　　b)　A, B as in (a), $f = \{(1, v), (2, x), (3, z), (4, y)\}$

　　c)　$A = \{1, 2, 3, 4, 5\}$, $B = \{w, x, y, z\}$, $f = \{(1, w), (2, w), (3, x), (4, y), (5, z)\}$

　　d)　$A = \{1, 2, 3, 4\}$, $B = \{w, x, y, z\}$, $f = \{(1, w), (2, x), (3, y), (4, z)\}$

3. (a), (b), (c), (f) are one-to-one and onto.

　　d)　Neither one-to-one nor onto; range $= [0, +\infty)$

　　e)　Neither one-to-one nor onto; range $= [-\frac{1}{4}, +\infty)$

5. (For the case $n = 5$, $m = 3$):

$$\sum_{k=0}^{5} (-1)^k \binom{5}{5-k}(5 - k)^3 = (-1)^0 \binom{5}{5}5^3 + (-1)^1 \binom{5}{4}4^3 + (-1)^2 \binom{5}{3}3^3 + (-1)^3 \binom{5}{2}2^3$$
$$+ (-1)^4 \binom{5}{1}1^3 + (-1)^5 \binom{5}{0}0^3$$
$$= 125 - 5(64) + 10(27) - 10(8) + 5 = 0$$

7. For any $r \in \mathbf{R}$ there is at least one $a \in \mathbf{R}$ such that $a^5 - 2a^2 + a - r = 0$ because the polynomial $x^5 - 2x^2 + x - r$ has odd degree and real coefficients. Consequently, f is onto. However, $f(0) = 0 = f(1)$, so f is not one-to-one.

9.

$m \backslash n$	1	2	3	4	5	6	7	8	9	10
9	1	255	3025	7770	6951	2646	462	36	1	
10	1	511	9330	34105	42525	22827	5880	750	45	1

Section 3.4 — p. 82

1. The pigeons are the socks; the pigeonholes are the colors.

3. 7

5.

In triangle ABC, divide each side into three equal parts and form the nine congruent triangles shown in the figure. Let R_1 be the interior of triangle ADE together with the points on segment DE, excluding D, E. Region R_2 is the interior of triangle DFG together with the points on segments DG, FG, excluding D, F. Regions $R_3, \ldots, R_9$ are defined similarly so that the interior of $\triangle ABC$ is the union of these nine regions and $R_i \cap R_j = \emptyset$, for $i \neq j$. Then if 10 points are chosen in the interior of $\triangle ABC$, at least two of these points are in R_i for some $1 \leq i \leq 9$, and these two points are at a distance less than $\frac{1}{3}$ from each other.

7. Consider the subsets A of S where $1 \leq |A| \leq 3$. Since $|S| = 5$, there are $\binom{5}{1} + \binom{5}{2} + \binom{5}{3} = 25$ such subsets A. Let s_A denote the sum of the elements in A. Then $1 \leq s_A \leq 7 + 8 + 9 = 24$. So by the pigeonhole principle, there are two subsets of S whose elements yield the same sum.

9. For $1 \leq i \leq 42$, let x_i be the total number of resumés Brace has sent out from the start of his senior year to the end of the ith day. Then $1 \leq x_1 < x_2 < \ldots < x_{42} \leq 60$, and $x_1 + 23 < x_2 + 23 < \ldots < x_{42} + 23 \leq 83$. We have 42 distinct numbers $x_1, x_2, \ldots, x_{42}$, and 42 other distinct numbers $x_1 + 23, x_2 + 23, \ldots, x_{42} + 23$, all between 1 and 83 inclusive. By the pigeonhole principle, $x_i = x_j + 23$ for some $1 \leq j < i \leq 42$ and $x_i - x_j = 23$.

Section 3.5 — p. 92

1. a) 25   b) 5^{25} c) 5^{25} d) 5^{15}

3. (a), (b), and (d) are commutative and associative; (c) is neither commutative nor associative

5. $f(z_1, z_2) = z_2$ because z_1 is an identity for f.

7. a) 5

b)

A_3	A_4	A_5
25	25	6
25	2	4
60	40	20
25	40	10

c) A_1, A_2

9. h is onto $\Leftrightarrow$ for all $b \in B$, $d \in D$, there exists $a \in A$, $c \in C$ with $h(a, c) = (b, d) \Leftrightarrow$ for all $b \in B$, $d \in D$, there exists $a \in A$, $c \in C$ with $f(a) = b$, $g(c) = d \Leftrightarrow f$, g are onto. h is one-to-one $\Leftrightarrow$ [for all a, $a_1 \in A$, c, $c_1 \in C$, $h(a, c) = h(a_1, c_1) \Rightarrow a = a_1$, $c = c_1$] $\Leftrightarrow$ [for all a, $a_1 \in A$, c, $c_1 \in C$, $f(a) = f(a_1) \Rightarrow a = a_1$, and $g(c) = g(c_1) \Rightarrow c = c_1$] $\Leftrightarrow f$, g are one-to-one.

11. $g^2(A) = g(T \cap (S \cup A)) = T \cap (S \cup [T \cap (S \cup A)])$
$= T \cap [(S \cup T) \cap (S \cup (S \cup A))] = T \cap [(S \cup T) \cap (S \cup A)]$
$= [T \cap (S \cup T)] \cap (S \cup A) = T \cap (S \cup A) = g(A)$.

13. $a = 3, b = -1; a = -3, b = 2$

15. a) $(b, a) \in (\mathcal{R}_1 \cup \mathcal{R}_2)^c \Leftrightarrow (a, b) \in \mathcal{R}_1 \cup \mathcal{R}_2 \Leftrightarrow (a, b) \in \mathcal{R}_1$ or $(a, b) \in \mathcal{R}_2 \Leftrightarrow (b, a) \in \mathcal{R}_1^c$ or $(b, a) \in \mathcal{R}_2^c \Leftrightarrow (b, a) \in \mathcal{R}_1^c \cup \mathcal{R}_2^c$

b) $(b, a) \in (\mathcal{R}_1 \cap \mathcal{R}_2)^c \Leftrightarrow (a, b) \in \mathcal{R}_1 \cap \mathcal{R}_2 \Leftrightarrow (a, b) \in \mathcal{R}_1$ and $(a, b) \in \mathcal{R}_2 \Leftrightarrow (b, a) \in \mathcal{R}_1^c$ and $(b, a) \in \mathcal{R}_2^c \Leftrightarrow (b, a) \in \mathcal{R}_1^c \cap \mathcal{R}_2^c$

c) $(a, b) \in (\mathcal{R}_1^c)^c \Leftrightarrow (b, a) \in (\mathcal{R}_1^c) \Leftrightarrow (a, b) \in \mathcal{R}_1$.

17. a) $f^{-1}(x) = (1/2)(\ln x - 5)$

b)   For $x \in \mathbf{R}^{+}$,

$$(f \circ f^{-1})(x) = f((1/2)(\ln x - 5)) = e^{2((1/2)(\ln x - 5)) + 5} = e^{\ln x - 5 + 5} = e^{\ln x} = x;$$

for $x \in \mathbf{R}$,

$$(f^{-1} \circ f)(x) = f^{-1}(e^{2x+5}) = (1/2)[\ln(e^{2x+5}) - 5] = (1/2)[2x + 5 - 5] = x.$$

c)

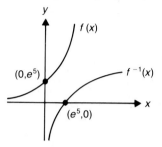

19. f, g invertible $\Rightarrow$ each of f, g is both one-to-one and onto $\Rightarrow g \circ f$ is one-to-one and onto $\Rightarrow g \circ f$ invertible. Since $(g \circ f) \circ (f^{-1} \circ g^{-1}) = 1_C$ and $(f^{-1} \circ g^{-1}) \circ (g \circ f) = 1_A$, $f^{-1} \circ g^{-1}$ is an inverse of $g \circ f$. By uniqueness of inverses $f^{-1} \circ g^{-1} = (g \circ f)^{-1}$.

21.   a) $[0, 2)$ b) $[-1, 2)$ c) $[0, 1)$

d) $[0, 2)$ e) $[-1, 2)$ f) $[0, 2)$

g) $[-1, 3)$ h) $[-1, 0) \cup [2, 4)$

23.   a) $a \in f^{-1}(B_1 \cap B_2) \Leftrightarrow f(a) \in B_1 \cap B_2 \Leftrightarrow f(a) \in B_1$ and $f(a) \in B_2 \Leftrightarrow a \in f^{-1}(B_1)$ and $a \in f^{-1}(B_2) \Leftrightarrow a \in f^{-1}(B_1) \cap f^{-1}(B_2)$

c) $a \in f^{-1}(\overline{B_1}) \Leftrightarrow f(a) \in \overline{B_1} \Leftrightarrow f(a) \notin B_1 \Leftrightarrow a \notin f^{-1}(B_1) \Leftrightarrow a \in \overline{f^{-1}(B_1)}$

25.   a)   i) $f(x) = 2x;$ ii) $f(x) = \lfloor x/2 \rfloor$

b)   No. The set $\mathbf{Z}$ is not finite.

Miscellaneous Exercises — p. 96

1.   a) $(a, b) \in A \times B \Rightarrow a \in A$, $b \in B \Rightarrow a \in C$, $b \in D$, since $A \subseteq C$, $B \subseteq D \Rightarrow$ $(a, b) \in C \times D$.

b)   If either A or B is $\emptyset$, the result need not follow. For example, let $A = \{1, 2\}$, $B = \emptyset$, $C = \{1\}$, $D = \{1\}$, then $A \times B = \emptyset \subseteq C \times D$ but $A \not\subseteq C$. If neither A nor B is $\emptyset$, let $a \in A$, $b \in B$. Then $(a, b) \in A \times B \subseteq C \times D \Rightarrow a \in C$, $b \in D$. Hence $A \subseteq C$ and $B \subseteq D$.

3.   a) (B)   b) (B)   c) $2^{12} - 4^3$

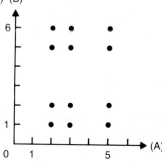

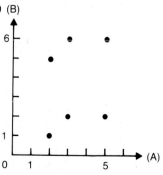

5. a) $(x, y) \in (A \cap B) \times (C \cap D) \Leftrightarrow x \in A \cap B, \, y \in C \cap D \Leftrightarrow (x \in A, y \in C)$ and $(x \in B, y \in D) \Leftrightarrow (x, y) \in A \times C$ and $(x, y) \in B \times D \Leftrightarrow (x, y) \in (A \times C) \cap (B \times D)$

b) $(x, y) \in (A \cup B) \times (C \cup D) \Leftrightarrow x \in A \cup B, \, y \in C \cup D \Leftrightarrow (x \in A \text{ or } x \in B)$ and $(y \in C \text{ or } y \in D) \Leftrightarrow (x \in A \text{ and } y \in C)$ or $(x \in B \text{ and } y \in D)$ or $(x \in A \text{ and } y \in D)$ or $(x \in B \text{ and } y \in C) \Leftrightarrow ((x, y) \in A \times C)$ or $((x, y) \in B \times D)$ or $((x, y) \in A \times D)$ or $((x, y) \in B \times C) \Leftrightarrow (x, y) \in (A \times C) \cup (B \times D) \cup (A \times D) \cup (B \times C)$

7. $A(2, 3) = 9$ 9. a) $(7!)/[2(7^5)]$

11. For $1 \le i \le 10$, let x_i be the number of letters typed on day i. Then $x_1 + x_2 + x_3 + \cdots + x_8 + x_9 + x_{10} = 84$, or $x_3 + \cdots + x_8 = 54$. Suppose that $x_1 + x_2 + x_3 < 25$, $x_2 + x_3 + x_4 < 25$, ..., $x_8 + x_9 + x_{10} < 25$. Then $x_1 + 2x_2 + 3(x_3 + \cdots + x_8) + 2x_9 + x_{10} < 8(25) = 200$, or $3(x_3 + \cdots + x_8) < 160$. Consequently, $54 = x_3 + \cdots + x_8 < \frac{160}{3} = 53\frac{1}{3}$.

13. Let the n distinct objects be $x_1, x_2, \ldots, x_n$. Place x_n in a container. Now there are two *distinct* containers. For each of $x_1, x_2, \ldots, x_{n-1}$ there are two choices and this gives 2^{n-1} distributions. Among these there is one where $x_1, x_2, \ldots, x_{n-1}$ are in the container with x_n, so we remove this distribution and find $S(n, 2) = 2^{n-1} - 1$.

15. 42   17. Any $n \in \mathbf{Z}^+$ where n is odd.

19. $x < y \Rightarrow f(x) < f(y)$ (since f is increasing) $\Rightarrow g(f(x)) < g(f(y))$ (since g is increasing.) Hence $g \circ f$ is increasing.

21. $f \circ g = \{(x, z), (y, y), (z, x)\}$; $g \circ f = \{(x, x), (y, z), (z, y)\}$; $f^{-1} = \{(x, z), (y, x), (z, y)\}$; $g^{-1} = \{(x, y), (y, x), (z, z)\}$; $(g \circ f)^{-1} = \{(x, x), (y, z), (z, y)\} = f^{-1} \circ g^{-1}$; $g^{-1} \circ f^{-1} = \{(x, z), (y, y), (z, x)\}$.

23. a) 4   b) 10

25. a) $a \in A \Rightarrow f(a) \in f(A) \Rightarrow a \in f^{-1}(f(A))$. If f is one-to-one, $A = f^{-1}(f(A))$.

b) $b \in f(f^{-1}(B)) \Rightarrow b = f(a)$ for some $a \in f^{-1}(B) \Rightarrow b = f(a)$ where $f(a) \in B \Rightarrow b \in B$. If f is onto, $B = f(f^{-1}(B))$.

27. a) $(\pi \circ \sigma)(x) = (\sigma \circ \pi)(x) = x$

b) $\pi^n(x) = x - n$; $\sigma^n(x) = x + n$ $(n \ge 2)$.

c) $\pi^{-n}(x) = x + n$; $\sigma^{-n}(x) = x - n$ $(n \ge 2)$.

29. $(9901)(121)(200 \times 10^{-9})$ seconds $\doteq 0.24$ seconds.

31. a) and b) $m! \, S(n, m)$

33. For $\prod_{k=1}^n (k - i_k)$ to be odd, $(k - i_k)$ must be odd for all $1 \le k \le n$, that is, one of k, i_k must be even and the other odd. Since n is odd, $n = 2m + 1$ and in the list $1, 2, \ldots, n$ there are m even integers and $m + 1$ odd integers. Let $1, 3, 5, \ldots, n$ be the pigeons and $i_1, i_3, i_5, \ldots, i_n$ the pigeonholes. At most m of the pigeonholes can be even integers, so $(k - i_k)$ must be even for at least one $k = 1, 3, 5, \ldots, n$. Consequently, $\prod_{k=1}^n (k - i_k)$ is even.

Chapter 4

Section 4.1 — p. 106

1. a) 25; 125   b) 3906   3. 12   5. 780

7. a) $x \in AC \Rightarrow x = ac, a \in A, c \in C \Rightarrow x \in BD$, since $A \subseteq B, C \subseteq D$.

b) If $A\emptyset \ne \emptyset$, let $x \in A\emptyset$. $x \in A\emptyset \Rightarrow x = yz, y \in A, z \in \emptyset$. But $z \in \emptyset$ is impossible. Hence $A\emptyset = \emptyset$. [In like manner $\emptyset A = \emptyset$].

9. (Theorem 4.1): a) $x \in A \Rightarrow x = x\lambda \in A\{\lambda\} \Rightarrow A \subseteq A\{\lambda\}$. $y \in A\{\lambda\} \Rightarrow y = ab$, $a \in A$, $b \in \{\lambda\}$, so $b = \lambda$. Hence $y = a\lambda = a \in A$ and $A\{\lambda\} \subseteq A$. Consequently $A\{\lambda\} = A$.

   c) $x \in A(B \cup C) \Rightarrow x = ay$, $a \in A$, $y \in B$ or $y \in C$. So $x \in AB$ or $x \in AC$, and $A(B \cup C) \subseteq AB \cup AC$. Conversely, let $z \in AB \cup AC$. Then $z \in AB$ or $z \in AC$. Assume, without loss of generality, that $z \in AB$. $z \in AB \Rightarrow z = ab$, $a \in A$, $b \in B \Rightarrow z = ab$, $a \in A$, $b \in B \cup C \Rightarrow z \in A(B \cup C) \Rightarrow AB \subseteq A(B \cup C)$. In like manner, $AC \subseteq A(B \cup C)$, so $AB \cup AC \subseteq A(B \cup C)$.

   e) $x \in A(B \cap C) \Rightarrow x = ay$, $a \in A$, $y \in B \cap C \Rightarrow x = ay$, $a \in A$, $y \in B$ and $y \in C \Rightarrow x \in AB$ and $x \in AC \Rightarrow x \in AB \cap AC \Rightarrow A(B \cap C) \subseteq AB \cap AC$. Let $A = \{y, yy\}$, $B = \{x, xx, y\}$, $C = \{y, yx\}$. Then $yyx \in AB \cap AC$ but $yyx \notin A(B \cap C)$, so $A(B \cap C) \subset AB \cap AC$ can occur.

   (Theorem 4.2): a) For any $a \in A$, $a = a\lambda$ where $\lambda \in B^*$, so $A \subseteq AB^*$.

   d) Let $a \in A^*$. If $a = \lambda$, then $a \in B^*$. $a \neq \lambda \Rightarrow a \in A^+ \Rightarrow a \in B^+$ (by part (c)) $\Rightarrow a \in B^*$. Hence $A \subseteq B \Rightarrow A^* \subseteq B^*$.

   e) $x \in AA^* \Rightarrow x = a_1 a_2$, $a_1 \in A$, $a_2 \in A^k \subseteq A^*$, $k \geq 0 \Rightarrow x \in A^{k+1} \subseteq A^*$, so $AA^* \subseteq A^+$. Conversely, $y \in A^+ \Rightarrow y \in A^k$, $k \geq 1 \Rightarrow y = y_1 y_2$, $y_1 \in A$, $y_2 \in A^* \Rightarrow y \in AA^*$, so $A^+ \subseteq AA^*$. [The proof for $A^*A = A^+$ is similar.]

11. a) $x \in A(\bigcup_{i \in I} B_i) \Leftrightarrow x = ab$, $a \in A$, $b \in \bigcup_{i \in I} B_i \Leftrightarrow x = ab$, $a \in A$, $b \in B_i$, for some $i \in I \Leftrightarrow x \in AB_i$, for some $i \in I \Leftrightarrow x \in \bigcup_{i \in I} AB_i$.

   b) $y \in (\bigcup_{i \in I} B_i)A \Leftrightarrow y = dc$, $d \in \bigcup_{i \in I} B_i$, $c \in A \Leftrightarrow y = dc$, $d \in B_i$, for some $i \in I$, $c \in A \Leftrightarrow y \in B_i A$ for some $i \in I \Leftrightarrow y \in \bigcup_{i \in I} B_i A$.

Section 4.2 — p. 113

1. a) 0010101; s_1 b) 0000000; s_1 c) 001000000; s_0

3. a) 010110   b)

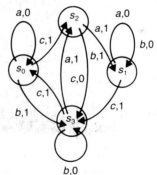

5. a) 010000; s_2

   b) (s_1) 100000; s_2

       (s_2) 000000; s_2

       (s_3) 110010; s_2

   d) s_1

   e) $x = 101$ (unique)

c)

	ν		ω	
	0	1	0	1
s_0	s_0	s_1	0	0
s_1	s_1	s_2	1	1
s_2	s_2	s_2	0	0
s_3	s_0	s_3	0	1
s_4	s_2	s_3	0	1

7. a)   i) 15   ii) 3^{15} iii) 2^{15} b) 6^{15}

Section 4.3 — p. 120

1. a) 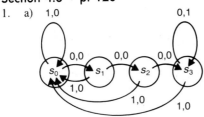   b)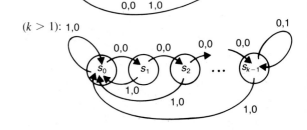

3. $(k = 1)$: 1,0 $(k > 1)$: 1,0

7. Both 110 and 111 are transfer sequences from s_2 to s_5.

Miscellaneous Exercises — p. 122

1. Let $x \in \Sigma$ and $A = \{x\}$. Then $A^2 = \{xx\}$ and $(A^2)^* = \{\lambda, x^2, x^4, \ldots\}$. $A^* = \{\lambda, x, x^2, x^3, \ldots\}$ and $(A^*)^2 = A^*$, so $(A^2)^* \neq (A^*)^2$.

3. $A = \{0, 1, 01, 011, 0111, \ldots, 11, 111, 1111, \ldots\} = B$

5. $\mathbb{O}_{02} = \{1\}^*\{0\} (\{0\}\{1\}^*\{0\})^*$; $\mathbb{O}_{22} = \{0\}(\{00\}^*\{1\}^*)^*\{0\}$; $\mathbb{O}_{11} = \emptyset$; $\mathbb{O}_{00} = (\{1\}^*\{00\}^*)^*$; $\mathbb{O}_{10} = \{1\}(\{1\}^*\{00\}^*)^*$

9. Both 1110 and 1111 are transfer sequences from s_0 to s_5.

Chapter 5

Section 5.1 — p. 129

1. a) $\{(1, 1), (2, 2), (3, 3), (4, 4), (1, 2), (2, 1), (2, 3), (3, 2)\}$

    b) $\{(1, 1), (2, 2), (3, 3), (4, 4), (1, 2)\}$

    c) $\{(1, 1), (2, 2), (1, 2), (2, 1)\}$

3. a)   reflexive, antisymmetric, transitive

    b)   reflexive, transitive

    c)   reflexive, symmetric, transitive

    d)   symmetric

    e)   (even)-reflexive, symmetric, transitive; (odd)-symmetric

    f)   (even)-reflexive, symmetric, transitive; (odd)-symmetric

    g)   reflexive, symmetric

h) reflexive, transitive

5. a) For all $x \in A$, $(x,x) \in \mathcal{R}_1, \mathcal{R}_2$, so $(x,x) \in \mathcal{R}_1 \cap \mathcal{R}_2$ and $\mathcal{R}_1 \cap \mathcal{R}_2$ is reflexive.

   b) i) $(x,y) \in \mathcal{R}_1 \cap \mathcal{R}_2 \Rightarrow (x,y) \in \mathcal{R}_1, \mathcal{R}_2 \Rightarrow (y,x) \in \mathcal{R}_1, \mathcal{R}_2 \Rightarrow (y,x) \in \mathcal{R}_1 \cap \mathcal{R}_2$ and $\mathcal{R}_1 \cap \mathcal{R}_2$ is symmetric.

       ii) $(x,y), (y,x) \in \mathcal{R}_1 \cap \mathcal{R}_2 \Rightarrow (x,y), (y,x) \in \mathcal{R}_1, \mathcal{R}_2$. By the antisymmetry of $\mathcal{R}_1$ (or $\mathcal{R}_2$), $x = y$ and $\mathcal{R}_1 \cap \mathcal{R}_2$ is antisymmetric.

       iii) $(x,y), (y,z) \in \mathcal{R}_1 \cap \mathcal{R}_2 \Rightarrow (x,y), (y,z) \in \mathcal{R}_1, \mathcal{R}_2 \Rightarrow (x,z) \in \mathcal{R}_1, \mathcal{R}_2$ (transitive property) $\Rightarrow (x,z) \in \mathcal{R}_1 \cap \mathcal{R}_2$, so $\mathcal{R}_1 \cap \mathcal{R}_2$ is transitive.

7. a) True

   b) False: let $A = \{1,2\}$ and $\mathcal{R} = \{(1,2),(2,1)\}$.

   c) i) Reflexive: true

       ii) Symmetric: false. Let $A = \{1,2\}$, $\mathcal{R}_1 = \{(1,1)\}$, $\mathcal{R}_2 = \{(1,1),(1,2)\}$.

       iii) Antisymmetric and transitive: false. Let $A = \{1,2\}$, $\mathcal{R}_1 = \{(1,2)\}$, $\mathcal{R}_2 = \{(1,2),(2,1)\}$.

   d) i) Reflexive: false. Let $A = \{1,2\}$, $\mathcal{R}_1 = \{(1,1)\}$, $\mathcal{R}_2 = \{(1,1),(2,2)\}$.

       ii) Symmetric: false. Let $A = \{1,2\}$, $\mathcal{R}_1 = \{(1,2)\}$, $\mathcal{R}_2 = \{(1,2),(2,1)\}$.

       iii) Antisymmetric: true

       iv) Transitive: false. Let $A = \{1,2\}$, $\mathcal{R}_1 = \{(1,2),(2,1)\}$, $\mathcal{R}_2 = \{(1,1),(1,2),(2,1),(2,2)\}$.

   e) True.

9. There may exist an element $a \in A$ such that for all $b \in B$ neither (a,b) nor $(b,a) \in \mathcal{R}$.

11. $r - n$ counts the elements in $\mathcal{R}$ of the form (a,b), $a \neq b$. Since $\mathcal{R}$ is symmetric, $r - n$ is even.

Section 5.2 — p. 139

1. $\mathcal{R} \circ \mathcal{S} = \{(1,3),(1,4)\}$; $\mathcal{S} \circ \mathcal{R} = \{(1,2),(1,3),(1,4),(2,4)\}$; $\mathcal{R}^2 = \mathcal{R}^3 = \{(1,4),(2,4),(4,4)\}$; $\mathcal{S}^2 = \mathcal{S}^3 = \{(1,1),(1,2),(1,3),(1,4)\}$.

3. $(a,d) \in (\mathcal{R}_1 \circ \mathcal{R}_2) \circ \mathcal{R}_3 \Rightarrow (a,c) \in \mathcal{R}_1 \circ \mathcal{R}_2, (c,d) \in \mathcal{R}_3$
   for some $c \in C \Rightarrow (a,b) \in \mathcal{R}_1, (b,c) \in \mathcal{R}_2, (c,d) \in \mathcal{R}_3$
   for some $b \in B$, $c \in C \Rightarrow (a,b) \in \mathcal{R}_1, (b,d) \in \mathcal{R}_2 \circ \mathcal{R}_3 \Rightarrow (a,d) \in \mathcal{R}_1 \circ (\mathcal{R}_2 \circ \mathcal{R}_3)$,
   and $(\mathcal{R}_1 \circ \mathcal{R}_2) \circ \mathcal{R}_3 \subseteq \mathcal{R}_1 \circ (\mathcal{R}_2 \circ \mathcal{R}_3)$.

5. This follows by the pigeonhole principle. Here the pigeons are the $2^{n^2} + 1$ integers between 0 and 2^{n^2}, inclusive, and the pigeonholes are the 2^{n^2} relations on A.

7. 2^{21}

9. d) Let s_{xy} be the entry in row (x) and column (y) of M. Then s_{yx} appears in row (x) and column (y) of M^{tr}. $\mathcal{R}$ is antisymmetric $\Leftrightarrow (s_{xy} = s_{yx} = 1 \Rightarrow x = y) \Leftrightarrow M \cap M^{tr} \leq I_n$.

11. $\mathcal{R}$

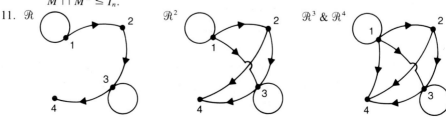

13. a) 2^{25} b) 2^{15}

15. a) $\mathcal{R}_1$: $\begin{bmatrix} 1 & 1 & 0 & 0 & 0 \\ 1 & 1 & 0 & 0 & 0 \\ 0 & 0 & 1 & 1 & 0 \\ 0 & 0 & 1 & 1 & 0 \\ 0 & 0 & 0 & 0 & 1 \end{bmatrix}$ $\mathcal{R}_2$: $\begin{bmatrix} 1 & 1 & 1 & 0 & 0 \\ 1 & 1 & 1 & 0 & 0 \\ 1 & 1 & 1 & 0 & 0 \\ 0 & 0 & 0 & 1 & 1 \\ 0 & 0 & 0 & 1 & 1 \end{bmatrix}$

   b) Given an equivalence relation $\mathcal{R}$ on a finite set A, list the elements of A so that elements in the same cell of the partition (see Section 5.4) are adjacent. The resulting relation matrix will then have square blocks of 1's along the diagonal (from upper left to lower right).

Section 5.3 — p. 147

1.

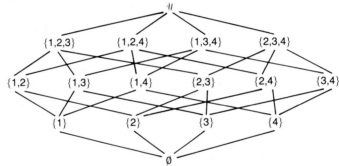

3. For all $a \in A$, $b \in B$, $a\,\mathcal{R}_1\,a$ and $b\,\mathcal{R}_2\,b$ so $(a,b)\,\mathcal{R}\,(a,b)$. $(a,b)\,\mathcal{R}\,(c,d)$, $(c,d)\,\mathcal{R}$ $(a,b) \Rightarrow a\,\mathcal{R}_1\,c$, $c\,\mathcal{R}_1\,a$ and $b\,\mathcal{R}_2\,d$, $d\,\mathcal{R}_2\,b \Rightarrow a = c$, $b = d \Rightarrow (a,b) = (c,d)$. $(a,b)\,\mathcal{R}\,(c,d)$, $(c,d)\,\mathcal{R}\,(e,f) \Rightarrow a\,\mathcal{R}_1\,c$, $c\,\mathcal{R}_1\,e$ and $b\,\mathcal{R}_2\,d$, $d\,\mathcal{R}_2\,f \Rightarrow a\,\mathcal{R}_1\,e$, $b\,\mathcal{R}_2\,f \Rightarrow (a,b)\,\mathcal{R}\,(e,f)$.

5. $\emptyset < \{1\} < \{2\} < \{3\} < \{1,2\} < \{1,3\} < \{2,3\} < \{1,2,3\}$.
   (There are other possibilities.)

7. a)

   b) $3 < 2 < 1 < 4$ or $3 < 1 < 2 < 4$.

   c) 2

11. Let x, y both be least upper bounds. Then $x\,\mathcal{R}\,y$ since y is an upper bound and x is a least upper bound. Likewise, $y\,\mathcal{R}\,x$. $\mathcal{R}$ antisymmetric $\Rightarrow x = y$. (The proof for the glb is similar.)

13. Let $\mathcal{U} = \{1,2\}$, $A = \mathcal{P}(\mathcal{U})$, and $\mathcal{R}$ the inclusion relation. Then $(A, \mathcal{R})$ is a poset but not a total order. Let $B = \{\emptyset, \{1\}\}$. Then $(B \times B) \cap \mathcal{R}$ is a total order.

15. $n + \binom{n}{2}$.

17. a) False. Let $\mathcal{U} = \{1,2\}$, $A = \mathcal{P}(\mathcal{U})$ and $\mathcal{R}$ be the inclusion relation. Then $(A, \mathcal{R})$ is a lattice where for any S, $T \in A$, $\mathrm{lub}\{S, T\} = S \cup T$ and $\mathrm{glb}\{S, T\} = S \cap T$. However, $\{1\}$ and $\{2\}$ are not related, so $(A, \mathcal{R})$ is not a total order.

b)   If $(A, \mathcal{R})$ is a total order, then for any $x, y \in A$, $x \mathcal{R} y$ or $y \mathcal{R} x$. For $x \mathcal{R} y$, lub $\{x, y\} = y$ and glb $\{x, y\} = x$. Consequently, $(A, \mathcal{R})$ is a lattice.

Sections 5.4 and 5.5 — p. 156

1. $\mathcal{R} = \{(1, 1), (1, 2), (2, 1), (2, 2), (3, 3), (3, 4), (4, 3), (4, 4), (5, 5)\}$.

3. $\mathcal{R}$ is not transitive since $1 \mathcal{R} 2$ and $2 \mathcal{R} 3$ but $1 \not{\mathcal{R}} 3$.

5. a)   For all $(x, y) \in A$, $x + y = x + y \Rightarrow (x, y) \mathcal{R} (x, y)$. $(x_1, y_1) \mathcal{R} (x_2, y_2) \Rightarrow x_1 + y_1 = x_2 + y_2 \Rightarrow x_2 + y_2 = x_1 + y_1 \Rightarrow (x_2, y_2) \mathcal{R} (x_1, y_1)$. $(x_1, y_1) \mathcal{R} (x_2, y_2)$, $(x_2, y_2) \mathcal{R} (x_3, y_3) \Rightarrow x_1 + y_1 = x_2 + y_2$, $x_2 + y_2 = x_3 + y_3$, so $x_1 + y_1 = x_3 + y_3$ and $(x_1, y_1) \mathcal{R} (x_3, y_3)$. Since $\mathcal{R}$ is reflexive, symmetric and transitive, it is an equivalence relation.

   b) $[(1, 3)] = \{(1, 3), (2, 2), (3, 1)\}$; $[(2, 4)] = \{(1, 5), (2, 4), (3, 3), (4, 2), (5, 1)\}$; $[(1, 1)] = \{(1, 1)\}$.

   c) $A = \{(1, 1)\} \cup \{(1, 2), (2, 1)\} \cup \{(1, 3), (2, 2), (3, 1)\} \cup$
      $\{(1, 4), (2, 3), (3, 2), (4, 1)\} \cup \{(1, 5), (2, 4), (3, 3), (4, 2), (5, 1)\} \cup$
      $\{(2, 5), (3, 4), (4, 3), (5, 2)\} \cup \{(3, 5), (4, 4), (5, 3)\} \cup \{(4, 5), (5, 4)\} \cup \{(5, 5)\}$.

7. 300

9. a)   Not possible. With $\mathcal{R}$ reflexive, $|\mathcal{R}| \geq 7$.

   b) $\mathcal{R} = \{(x, x) \,|\, x \in \mathbf{Z}, 1 \leq x \leq 7\}$.

   c)   Not possible. With $\mathcal{R}$ symmetric, $|\mathcal{R}| - 7$ must be even.

   d) $\mathcal{R} = \{(x, x) \,|\, x \in \mathbf{Z}, 1 \leq x \leq 7\} \cup \{(1, 2), (2, 1)\}$.

   e) $\mathcal{R} = \{(x, x) \,|\, x \in \mathbf{Z}, 1 \leq x \leq 7\} \cup \{(1, 2), (2, 1)\} \cup \{(3, 4), (4, 3)\}$.

   f) and h)   Not possible with $r - 7$ odd.

   g) and i)   Not possible. See remark at the end of Section 5.4.

11. a) s_2 and s_5 are equivalent.

    b) s_2 and s_5 are equivalent.

    c) s_2 and s_7 are equivalent; s_3 and s_4 are equivalent.

13. a) s_1 and s_7 are equivalent;

       s_4 and s_5 are equivalent.

    b)   i)   0 0 0 0

        ii)   0

        iii)   0 0

	ν		ω	
M:	0	1	0	1
s_1	s_4	s_1	1	0
s_2	s_1	s_2	1	0
s_3	s_6	s_1	1	0
s_4	s_3	s_4	0	0
s_6	s_2	s_1	1	0

Miscellaneous Exercises — p. 159

1. a)   False. Let $A = \{1, 2\}$, $I = \{1, 2\}$, $\mathcal{R}_1 = \{(1, 1)\}$, $\mathcal{R}_2 = \{(2, 2)\}$. Then $\bigcup_{i \in I} \mathcal{R}_i$ is reflexive but neither $\mathcal{R}_1$ nor $\mathcal{R}_2$ is reflexive. Conversely, however, if $\mathcal{R}_i$ is reflexive for all (actually, at least one) $i \in I$, then $\bigcup_{i \in I} \mathcal{R}_i$ is reflexive.

    b)   True. $\bigcap_{i \in I} \mathcal{R}_i$ is reflexive $\Leftrightarrow (a, a) \in \bigcap_{i \in I} \mathcal{R}_i$ for all $a \in A \Leftrightarrow (a, a) \in \mathcal{R}_i$ for all $a \in A$ and $i \in I \Leftrightarrow \mathcal{R}_i$ is reflexive for all $i \in I$.

3. $(a, c) \in \mathcal{R}_2 \circ \mathcal{R}_1 \Rightarrow$ for some $b \in A$, $(a, b) \in \mathcal{R}_2$, $(b, c) \in \mathcal{R}_1$. With $\mathcal{R}_1$, $\mathcal{R}_2$ symmetric, $(b, a) \in \mathcal{R}_2$, $(c, b) \in \mathcal{R}_1$, so $(c, a) \in \mathcal{R}_1 \circ \mathcal{R}_2 \subseteq \mathcal{R}_2 \circ \mathcal{R}_1$. $(c, a) \in \mathcal{R}_2 \circ \mathcal{R}_1 \Rightarrow$ $(c, d) \in \mathcal{R}_2$, $(d, a) \in \mathcal{R}_1$, for some $d \in A$. Then $(d, c) \in \mathcal{R}_2$, $(a, d) \in \mathcal{R}_1$ by symmetry, and $(a, c) \in \mathcal{R}_1 \circ \mathcal{R}_2$, so $\mathcal{R}_2 \circ \mathcal{R}_1 \subseteq \mathcal{R}_1 \circ \mathcal{R}_2$ and the result follows.

5. $(c, a) \in (\mathcal{R}_1 \circ \mathcal{R}_2)^c \Leftrightarrow (a, c) \in \mathcal{R}_1 \circ \mathcal{R}_2 \Leftrightarrow (a, b) \in \mathcal{R}_1$, $(b, c) \in \mathcal{R}_2$, for some $b \in B \Leftrightarrow (b, a) \in \mathcal{R}_1^c$, $(c, b) \in \mathcal{R}_2^c$, for some $b \in B \Leftrightarrow (c, a) \in \mathcal{R}_2^c \circ \mathcal{R}_1^c$.

7. Let $\mathcal{U} = \{1, 2, 3, 4, 5\}$, $A = \mathcal{P}(\mathcal{U}) - \{\mathcal{U}, \emptyset\}$. Under the inclusion relation A is a poset with the five minimal elements $\{x\}$, $1 \leq x \leq 5$, but no least element. Also, A has five maximal elements — the five subsets of $\mathcal{U}$ of size 4 — but no greatest element.

9. $n = 10$

11. a)   reflexive, transitive

   c) $f_1 \in [f] \Rightarrow |f_1(n)| \leq m_1 |f(n)|$ for all $n \geq k_1$; $f_2 \in [f] \Rightarrow |f_2(n)| \leq m_2 |f(n)|$ for all $n \geq k_2$. Then $|(f_1 + f_2)(n)| = |f_1(n) + f_2(n)| \leq |f_1(n)| + |f_2(n)| \leq (m_1 + m_2)|f(n)|$ for all $n \geq \max\{k_1, k_2\}$. Hence $f_1 + f_2 \in [f]$.

13. No. Consider $S = \{q \in \mathbf{Q} \mid 0 < q < 1\}$.

15. a) a b) a c) c d) e e) z f) e g) v
   $(A, \mathcal{R})$ is a lattice with z the greatest (and only maximal) element and a the least (and only minimal) element.

17. a)

Adjacency List		Index List	
1	2	1	1
2	3	2	2
3	1	3	3
4	4	4	5
5	5	5	6
6	3	6	8
7	5		

   b)

Adjacency List		Index List	
1	2	1	1
2	3	2	2
3	1	3	3
4	5	4	4
5	4	5	5
		6	6

   c)

Adjacency List		Index List	
1	2	1	1
2	3	2	2
3	1	3	3
4	4	4	6
5	5	5	7
6	1	6	8
7	4		

19. b)   The cells of the partition are the connected components of G.

21.   One possible order is 10, 3, 8, 6, 7, 9, 1, 4, 5, 2, where program 10 is run first and program 2 last.

Chapter 6

Section 6.1 — p. 172

1. a)   Since $1 = 1^2$ the result is true for $n = 1$. Assume the result for $n = k$: $1 + 3 + 5 + \cdots + (2k - 1) = k^2$. Now consider the case for $n = k + 1$: $[1 + 3 + 5 + \cdots + (2k - 1)] + (2k + 1) = k^2 + (2k + 1) = (k + 1)^2$. Consequently, the result is true for all $n \in \mathbf{Z}^+$ by the principle of finite induction.

   c)   Since $1 \cdot 3 = (1)(2)(9)/6$, the result is true for $n = 1$. Assume the result for $n = k$: $1 \cdot 3 + 2 \cdot 4 + 3 \cdot 5 + \cdots + k(k + 2) = k(k + 1)(2k + 7)/6$. Then consider the case for $n = k + 1$: $[1 \cdot 3 + 2 \cdot 4 + \cdots + k(k + 2)] + (k + 1)(k + 3) = [k(k + 1)(2k + 7)/6] + (k + 1)(k + 3) = [(k + 1)/6][k(2k + 7) + 6(k + 3)] = (k + 1)(2k^2 + 13k + 18)/6 = (k + 1)(k + 2)(2k + 9)/6$. Hence the result follows for all $n \in \mathbf{Z}^+$ by the principle of finite induction.

3. a)   7626   b)   627,874

5. For $n = 11$, $11 - 2 = 9 < 9\frac{1}{6} = (11^2 - 11)/12$. Assume the result true for $n = k(\geq 11)$: $k - 2 < (k^2 - k)/12$. When $n = k + 1$, $(k - 2 < (k^2 - k)/12) \Rightarrow (k - 2) + 1 < ((k^2 - k)/12) + 1 \Rightarrow (k + 1) - 2 < (k^2 - k + 12)/12$. For $k > 6$, $2k > 12$ and $k > 12 - k$, so $(k + 1) - 2 < (k^2 + k)/12 = [(k + 1)^2 - (k + 1)]/12$. Consequently the result follows for all $n \geq 11$ by the principle of induction.

7. For $n = 5$, $2^5 = 32 > 25 = 5^2$. Assume the result for $n = k(\geq 5)$: $2^k > k^2$. For $k > 3$, $k(k - 2) > 1$, or $k^2 > 2k + 1$. $2^k > k^2 \Rightarrow 2^k + 2^k > k^2 + k^2 \Rightarrow 2^{k+1} > k^2 + k^2 > k^2 + (2k + 1) = (k + 1)^2$. Hence the result is true for $n \geq 5$ by the principle of induction.

9. Assume $S(k)$. For $S(k + 1)$, $\sum_{i=1}^{k+1} i = ([k + (1/2)]^2/2) + (k + 1) = (k^2 + k + (1/4) + 2k + 2)/2 = [(k + 1)^2 + (k + 1) + (1/4)]/2 = [(k + 1) + (1/2)]^2/2$. So $S(k) \Rightarrow S(k + 1)$. However, we have no first value of k where $S(k)$ is true: for any $k \geq 1$, $\sum_{i=1}^{k} i = (k)(k + 1)/2$ and $(k)(k + 1)/2 = [k + (1/2)]^2/2 \Rightarrow 0 = 1/4$.

11. Let $T = \{n \in \mathbf{Z}^+ \mid S(n)$ is false$\}$. $S(1)$ true $\Rightarrow 1 \notin T$. If $T \neq \emptyset$, then T has a least element r, since $T \subseteq \mathbf{Z}^+$. However, $S(1), S(2), \ldots, S(r - 1)$ true $\Rightarrow S(r)$ true. Hence $T = \emptyset$ and the result follows.

13. a) For $n = 2$, $A \cap (B_1 \cup B_2) = (A \cap B_1) \cup (A \cap B_2)$ is the distributive law of intersection over union, which is true. Assuming that $A \cap (B_1 \cup B_2 \cup \ldots \cup B_k) = (A \cap B_1) \cup (A \cap B_2) \cup \ldots \cup (A \cap B_k)$ we find that $A \cap (B_1 \cup B_2 \cup \ldots \cup B_k \cup B_{k+1}) = A \cap [(B_1 \cup B_2 \cup \ldots \cup B_k) \cup B_{k+1}] = (A \cap (B_1 \cup B_2 \cup \ldots \cup B_k)) \cup (A \cap B_{k+1}) = (A \cap B_1) \cup (A \cap B_2) \cup \ldots \cup (A \cap B_k) \cup (A \cap B_{k+1})$, and the result follows for all $n \geq 2$ by the principle of induction.

   b) For $n = 2$, $\overline{A_1 \cap A_2} = \overline{A_1} \cup \overline{A_2}$ is one of the DeMorgan laws. Assuming that

$$\overline{A_1 \cap A_2 \cap \ldots \cap A_k} = \overline{A_1} \cup \overline{A_2} \cup \ldots \cup \overline{A_k},$$

we have

$$\overline{A_1 \cap A_2 \cap \ldots \cap A_k \cap A_{k+1}} = \overline{(A_1 \cap A_2 \cap \ldots \cap A_k) \cap A_{k+1}}$$
$$= \overline{(A_1 \cap A_2 \cap \ldots \cap A_k)} \cup \overline{A_{k+1}}$$
$$= \overline{A_1} \cup \overline{A_2} \cup \ldots \cup \overline{A_k} \cup \overline{A_{k+1}},$$

and the result is true for all $n \geq 2$ by the principle of induction.

Section 6.2 — p. 176

1. e) If $a \mid x$, $a \mid y$ then $x = ac$, $y = ad$ for some $c, d \in \mathbf{Z}$. So $z = x - y = a(c - d)$, and $a \mid z$. The proofs for the other cases are similar.

   g) Follows from (f) by mathematical induction.

3. Since q is prime its only positive divisors are 1 and q. With p a prime, $p > 1$. Hence $p \mid q \Rightarrow p = q$.

5. $b \mid a$, $b \mid (a + 2) \Rightarrow b \mid [ax + (a + 2)y]$ for all $x, y \in \mathbf{Z}$. Let $x = -1$, $y = 1$. Then $b > 0$ and $b \mid 2$, so $b = 1$ or 2.

7. Let $a = 2m + 1$, $b = 2n + 1$, $m, n \geq 0$. Then $a^2 + b^2 = 4(m^2 + m + n^2 + n) + 2$, so $2 \mid (a^2 + b^2)$ but $4 \nmid (a^2 + b^2)$.

9. a) $137 = (10001001)_2 = (2021)_4 = (211)_8$

   b) $6243 = (1100001100011)_2 = (1201203)_4 = (14143)_8$

   c) $12,345 = (11000000111001)_2 = (3000321)_4 = (30071)_8$

15. a) Since $2 \mid 10^t$ for all $t \in \mathbf{Z}^+$, $2 \mid n$ iff $2 \mid r_0$.

   b) Follows from the fact that $4 \mid 10^t$ for $t \geq 2$.

   c) Follows from the fact that $8 \mid 10^t$ for $t \geq 3$. In general,

$$2^{t+1} \mid n \text{ iff } 2^{t+1} \mid (r_t \cdot 10^t + \cdots + r_1 \cdot 10 + r_0).$$

Section 6.3 — p. 181

1. a) $(1820, 231) = 7 = 1820(8) + 231(-63)$

   b) $(2597, 1369) = 1 = 2597(534) + 1369(-1013)$

   c) $(4001, 2689) = 1 = 4001(-1117) + 2689(1662)$

   d) $(7983, 7982) = 1 = 7983(1) + 7982(-1)$

3. $(n, n + 1) = 1$; $[n, n + 1] = n(n + 1)$

5. $(a, b) = d \Rightarrow d = ax + by$, for $x, y \in \mathbf{Z}$. $(a, b) = d \Rightarrow a/d, b/d \in \mathbf{Z}$. $1 = (a/d)x + (b/d)y \Rightarrow (a/d, b/d) = 1$.

7. Let $(a, b) = h$, $(b, d) = g$. $(a, b) = h \Rightarrow h \mid a$, $h \mid b \Rightarrow h \mid (a \cdot 1 + bc) \Rightarrow h \mid d$. $h \mid b$, $h \mid d \Rightarrow h \mid g$. $(b, d) = g \Rightarrow g \mid b$, $g \mid d \Rightarrow g \mid (d \cdot 1 + b(-c)) \Rightarrow g \mid a$. $g \mid b$, $g \mid a$, $h = (a, b) \Rightarrow g \mid h$. $h \mid g$, $g \mid h$, $g, h \in \mathbf{Z}^+ \Rightarrow g = h$.

9. Symmetric

11. There is no solution for $c \neq 12, 18$. For $c = 12$, the solutions are $x = 118 - 165k$, $y = -10 + 14k$, $k \in \mathbf{Z}$. For $c = 18$, the solutions are $x = 177 - 165k$, $y = -15 + 14k$, $k \in \mathbf{Z}$.

13. $(a, b) = 1 \Rightarrow ax + by = 1$, $x, y \in \mathbf{Z}$. Then $acx + bcy = c$. $a \mid acx$, $a \mid bcy$ (since $a \mid bc) \Rightarrow a \mid c$.

15. Let $(a, b) = g$, $[a, b] = h$. $(a, b) = g \Rightarrow as + bt = g$, for $s, t \in \mathbf{Z}$. $[a, b] = h \Rightarrow h = ma = nb$, $m, n \in \mathbf{Z}^+$. $hg = has + hbt = nbas + mabt = ab(ns + mt) \Rightarrow ab \mid hg$. $(a, b) = g \Rightarrow g \mid a$, $g \mid b$, so $(a/g)b = (b/g)a$ is a common multiple of a and b. Consequently $h \mid (a/g)b$, and $hx = (a/g)b$, $x \in \mathbf{Z}$, or $ghx = ab$. Hence $gh \mid ab$.

Section 6.4 — p. 184

1. a) $2^2 \cdot 3^3 \cdot 5^3 \cdot 11$ b) $2^4 \cdot 3 \cdot 5^2 \cdot 7^2 \cdot 11^2$ c) $3^2 \cdot 5^3 \cdot 7^2 \cdot 11 \cdot 13$

3. The result is true for $n = 2$ by Lemma 6.2. For $k \geq 2$, assume that $p \mid a_1 a_2 \cdots a_k \Rightarrow p \mid a_i$, for some $1 \leq i \leq k$. Then $p \mid a_1 a_2 \cdots a_k a_{k+1} \Rightarrow p \mid (a_1 a_2 \cdots a_k) a_{k+1} \Rightarrow p \mid a_1 a_2 \cdots a_k$, or $p \mid a_{k+1} \Rightarrow p \mid a_i$, for some $1 \leq i \leq k + 1$. Hence the result follows for all $n \in \mathbf{Z}^+$ by the principle of finite induction.

5. b) Let $\log_{10} p = a/b$, where $a, b \in \mathbf{Z}^+$, $(a, b) = 1$. Then $p = 10^{(a/b)}$ or $p^b = 10^a = 2^a \cdot 5^a$, contradicting the Fundamental Theorem of Arithmetic.

7. a) $\tau(2) = \tau(3) = \tau(5) = 2$; $\tau(2^2) = \tau(3^2) = \tau(5^2) = 3$; $\tau(6) = \tau(10) = \tau(14) = 4$; $\tau(2^4) = \tau(3^4) = \tau(5^4) = 5$; $\tau(12) = \tau(18) = \tau(20) = 6$.

   b) For any $k > 1$ and any prime p, $\tau(p^{k-1}) = k$.

Miscellaneous Exercises — p. 186

1. $a + (a + d) + (a + 2d) + \cdots + (a + (n - 1)d) = na + [(n - 1)nd]/2$. For $n = 1$, $a = a + 0$, and the result is true in this case. Assuming that

$$\sum_{i=1}^{k} [a + (i - 1)d] = ka + [(k - 1)kd]/2,$$

we have
$$\sum_{i=1}^{k+1}[a + (i - 1)d] = (ka + [(k - 1)kd]/2) + (a + kd)$$
$$= (k + 1)a + [k(k + 1)d]/2,$$

so the result follows for all $n \in \mathbf{Z}^+$ by induction.

3. a)

n	$n^2 + n + 41$	n	$n^2 + n + 41$	n	$n^2 + n + 41$
1	43	4	61	7	97
2	47	5	71	8	113
3	53	6	83	9	131

   b)   For $n = 39$, $n^2 + n + 41 = 1601$, a prime. But for $n = 40$, $n^2 + n + 41 = (41)^2$, so $S(39) \not\Rightarrow S(40)$.

5. Fix $m = 1$. For $n = 1$ the result is true. Assume $f \circ f^k = f^k \circ f$ and consider $f \circ f^{k+1}$. $f \circ f^{k+1} = f \circ (f \circ f^k) = f \circ (f^k \circ f) = (f \circ f^k) \circ f = f^{k+1} \circ f$. Hence $f \circ f^n = f^n \circ f$ for all $n \in \mathbf{Z}^+$. Now assume that for $t \geq 1$, $f^t \circ f^n = f^n \circ f^t$. Then $f^{t+1} \circ f^n = (f \circ f^t) \circ f^n = f \circ (f^t \circ f^n) = f \circ (f^n \circ f^t) = (f \circ f^n) \circ f^t = (f^n \circ f) \circ f^t = f^n \circ (f \circ f^t) = f^n \circ f^{t+1}$, so $f^m \circ f^n = f^n \circ f^m$ for all $m, n \in \mathbf{Z}^+$.

7. For $n = 1$, $7^3 + 8^3 = 855 = (57)(15)$. Assuming that $57 | (7^{k+2} + 8^{2k+1})$, $7^{(k+1)+2} + 8^{2(k+1)+1} = 7^{k+3} + 8^{2k+3} = 7(7^{k+2}) + 64(8^{2k+1}) = 64(7^{k+2}) + 64(8^{2k+1}) - 57(7^{k+2})$, so $57 | (7^{k+3} + 8^{2k+3})$ and the result follows by induction.

9. Any selection of size 101 from S must contain two consecutive integers n, $n + 1$ and $(n, n + 1) = 1$.

11. Let $x_1, x_2, \ldots, x_{25}$ denote the numbers (in their order on the wheel), and assume that $x_1 + x_2 + x_3 < 39$, $x_2 + x_3 + x_4 < 39$, $\ldots$, $x_{24} + x_{25} + x_1 < 39$, and $x_{25} + x_1 + x_2 < 39$. Then $\sum_{i=1}^{25} 3x_i < 25(39)$. But $\sum_{i=1}^{25} 3x_i = 3\sum_{i=1}^{25} i = (3)(25)(26)/2 = (39)(25)$.

13. a)
$$r = r_0 + r_1 \cdot 10 + r_2 \cdot 10^2 + \cdots + r_n \cdot 10^n$$
$$= r_0 + r_1(9) + r_1 + r_2(99) + r_2 + \cdots + r_n(\underbrace{99\ldots9}_{n\ 9's}) + r_n$$
$$= [9r_1 + 99r_2 + \cdots + (99\ldots9)r_n] + (r_0 + r_1 + r_2 + \cdots + r_n).$$

   Hence $9 | r$ iff $9 | (r_0 + r_1 + r_2 + \cdots + r_n)$.

   c) $3 | t$ for $x = 1$ or 4 or 7; $9 | t$ for $x = 7$.

15. a)   1, 4, 9.   b)   1, 4, 9, 16, $\ldots$, k where k is the largest square less than or equal to n.

Chapter 7

Section 7.1 — p. 198

1. a)   534   b)   524   c)   10

3. a) $\binom{22}{19}$ b) $\binom{22}{19} - 4\binom{14}{11} + 6\binom{6}{3}$

   c) $\binom{16}{13} - [2\binom{10}{7} + \binom{9}{6} + \binom{11}{8}] + [10 + 2\binom{5}{2}]$.

5. $9!/[(3!)^3] - 3[7!/[(3!)^2]] + 3(5!/3!) - 3!$

7. $26! - [3(23!) + 24!] + (20! + 21!)$

9. a) 2^{n-1} b) $2^{n-1}(p - 1)$

11. a)   1600   b)   4399

13. $[6^8 - \binom{6}{1}5^8 + \binom{6}{2}4^8 - \binom{6}{3}3^8 + \binom{6}{2}2^8 - \binom{6}{1}]/6^8$.   15. No

17. a)   If a number n is divisible by a prime p, the execution of the While Loop divides out every occurrence of p. The first loop is for $p = 2$; the next loop for $p = 3$; and the third loop is for primes $p > 3$.

   b)   In working with primes, once we are past $p = 5$, there is no sense in examining even integers as possible candidates for primes.

Sections 7.2 and 7.3—p. 204

1. $E_0 = 768$; $E_1 = 205$; $E_2 = 40$; $E_3 = 10$; $E_4 = 0$; $E_5 = 1$. $\sum_{i=0}^{5} E_i = 1024 = N$.

3. a) $7! - d_7$ $(d_7 \doteq (7!)e^{-1})$;   b) $d_{26} \doteq (26!)e^{-1}$

5. a)   i) $332{,}640$ ii) $398{,}160$

   b)   i) $60{,}480$ ii) $65{,}520$

7. a)   i) $d_n/n!$ ii) $n(d_{n-1})/n!$ iii) $1 - (d_n/n!)$ iv) $[\binom{n}{r}d_{n-r}]/n!$

   b)   i) e^{-1} ii) e^{-1} iii) $1 - e^{-1}$ iv) $(1/r!)e^{-1}$

9. $\binom{n}{0}(n-1)! - \binom{n}{1}(n-2)! + \binom{n}{2}(n-3)! - \cdots + (-1)^{n-1}\binom{n}{n-1}(0!) + (-1)^n\binom{n}{n}$

11. a) $(11{,}088)/(10!) \doteq 0.003$ b) $(13{,}264)/(10!) \doteq 0.004$

Sections 7.4 and 7.5—p. 211

3. a) $\binom{8}{0} + \binom{8}{1}8x + \binom{8}{2}(8 \cdot 7)x^2 + \binom{8}{3}(8 \cdot 7 \cdot 6)x^3 + \binom{8}{4}(8 \cdot 7 \cdot 6 \cdot 5)x^4 + \cdots$

$$+ \binom{8}{8}(8!)x^8 = \sum_{i=0}^{8} \binom{8}{i}P(8,i)x^i.$$ b) $\sum_{i=0}^{n} \binom{n}{i}P(n,i)x^i$

5. a)   i) $(1 + 2x)^3$ ii) $1 + 8x + 14x^2 + 4x^3$

   iii) $1 + 9x + 25x^2 + 21x^3$ iv) $1 + 8x + 16x^2 + 7x^3$

   b)   If the board C consists of n steps, and each step has k blocks, then $r(C,x) = (1 + kx)^n$.

7. $5! - 8(4!) + 21(3!) - 20(2!) + 6(1!) = 20$

9. a) 20 b) $3/10$ c) $7/10$

Miscellaneous Exercises—p. 214

1. 134 3. $\phi(17) = \phi(32) = \phi(48) = 16$

5. $\sum_{i=0}^{8} (-1)^i \binom{8}{i}(8-i)!$

7. a) $[\sum_{i=0}^{3}(-1)^i\binom{4}{i}\binom{52-13i}{13}]/\binom{52}{13}$

   b) $[\sum_{i=1}^{3}(-1)^{i+1}(i)\binom{4}{i}\binom{52-13i}{13}]/\binom{52}{13}$

   c) $[\binom{4}{2}\binom{26}{13} - 3\binom{4}{3}\binom{13}{13}]/\binom{52}{13}$

9. a) $(d_{10})^2 \doteq (10!)^2 e^{-2}$ b) $\sum_{i=0}^{10}(-1)^i\binom{10}{i}[(10-i)!]^2$

11. $[16!/(4!)^4] - \binom{4}{1}[13!/(4!)^3] + \binom{4}{2}[10!/(4!)^2] - \binom{4}{3}(7!/4!) + (4!)$

13. a) $\binom{n-m}{r-m}$

Chapter 8

Section 8.1—p. 221

1. (Example 8.4): $-a = a$, $-b = e$, $-c = d$, $-d = c$, $-e = b$

   (Example 8.5): $-s = s$, $-t = y$, $-v = x$, $-w = w$, $-x = v$, $-y = t$

3. a) x b) $-s = t, \ -t = s, \ -x = x, \ -y = y$

   c) y d)   Yes   e)   No

   f) s, y

5. b) $1, -1, i, -i$

7. $\begin{bmatrix} a & b \\ c & d \end{bmatrix}^{-1} = (1/(ad - bc)) \begin{bmatrix} d & -b \\ -c & a \end{bmatrix}, \qquad ad - bc \neq 0.$

9. a)   For example, $\oplus$ is associative since $(a \oplus b) \oplus c = (a + b - 1) \oplus c = (a + b - 1 + c) - 1 = (a + b + c) - 2$, and $a \oplus (b \oplus c) = a \oplus (b + c - 1) = (a + b + c - 1) - 1 = (a + b + c) - 2$. Also, 1 is the additive identity and the additive inverse of any $a \in \mathbf{Z}$ is $2 - a$.

   b)   Yes   c)   The unity is 0. The units are 0 and 2.

   d)   The ring is an integral domain, but not a field.

Section 8.2 — p. 227

1. Theorem 8.10 (a). If $(S, +, \cdot)$ is a subring of R, then $a - b, \ ab \in S$ for any $a, b \in S$. Conversely, let $a \in S$. Then $a - a = z \in S$ and $z - a = -a \in S$. Also, if $b \in S$, then $-b \in S$, so $a - (-b) = a + b \in S$, and S is a subring by theorem 8.9.

3. a) $(ab)(b^{-1}a^{-1}) = aua^{-1} = aa^{-1} = u$ and $(b^{-1}a^{-1})(ab) = b^{-1}ub = b^{-1}b = u$, so ab is a unit. Since the multiplicative inverse of a unit is unique, $(ab)^{-1} = b^{-1}a^{-1}$.

   b) $A^{-1} = \begin{bmatrix} 2 & -7 \\ -1 & 4 \end{bmatrix}, \qquad B^{-1} = \begin{bmatrix} 1 & -2 \\ -2 & 5 \end{bmatrix}, \qquad (AB)^{-1} = \begin{bmatrix} 4 & -15 \\ -9 & 34 \end{bmatrix},$

       $(BA)^{-1} = \begin{bmatrix} 16 & -39 \\ -9 & 22 \end{bmatrix}, \qquad B^{-1}A^{-1} = \begin{bmatrix} 4 & -15 \\ -9 & 34 \end{bmatrix}.$

5. $(-a)^{-1} = -(a^{-1})$

7. $z \in S, T \Rightarrow z \in S \cap T \Rightarrow S \cap T \neq \emptyset. \ a, b \in S \cap T \Rightarrow a, b \in S$ and $a, b \in T \Rightarrow a + b, \ ab \in S$ and $a + b, \ ab \in T \Rightarrow a + b, \ ab \in S \cap T. \ a \in S \cap T \Rightarrow a \in S$ and $a \in T \Rightarrow -a \in S$ and $-a \in T \Rightarrow -a \in S \cap T$. So $S \cap T$ is a subring of R.

9. b) $\begin{bmatrix} 1 & 0 \\ 0 & 1 \end{bmatrix}$ c) $\begin{bmatrix} 1 & 0 \\ 0 & 0 \end{bmatrix}$

   d) S is an integral domain while R is a noncommutative ring with unity.

11.   2

13. a) $a = au \in aR$, so $aR \neq \emptyset$. If $ar_1, \ ar_2 \in aR$, then $ar_1 - ar_2 = a(r_1 - r_2) \in aR$. Also, for $ar_1 \in aR, \ r \in R, \ r(ar_1) = (ar_1)r = a(r_1 r) \in aR$. Hence aR is an ideal of R.

   b)   Let $a \in R, \ a \neq z$. Then $a = au \in aR$ so $aR = R$. Since $u \in R = aR, \ u = ar$ for some $r \in R$, and $r = a^{-1}$. Hence R is a field.

15. a) $\binom{4}{2}(49)$ b) 7^4

   c)   Yes, the element (u, u, u, u).   d) 4^4

17. b)   If R has a unity u, define $a^0 = u$, for $a \in R, \ a \neq z$. If a is a unit of R, define a^{-n} as $(a^{-1})^n$, for $n \in \mathbf{Z}^+$.

Section 8.3 — p. 232

1. a) $-6, 1, 8, 15$; b) $-9, 2, 13, 24$; c) $-7, 10, 27, 44$

3. b) No, $2 \, \mathcal{R} \, 3$ and $3 \, \mathcal{R} \, 5$ but $5 \, \mathcal{\not R} \, 8$. Also, $2 \, \mathcal{R} \, 3$ and $2 \, \mathcal{R} \, 5$ but $4 \, \mathcal{\not R} \, 15$.

5. a) $[17]^{-1} = [831]$; b) $[100]^{-1} = [111]$; c) $[777]^{-1} = [735]$.

7. (a) 16 units; 0 proper zero divisors; (b) 72 units; 44 proper zero divisors; (c) 1116 units; 0 proper zero divisors.

9. $[\binom{334}{3} + 2\binom{333}{3} + \binom{334}{1}\binom{333}{1}^2]/\binom{1000}{3}$.

11. Let $g = (a, n)$, $h = (b, n)$. $a \equiv b \pmod{n} \Rightarrow a = b + kn$, $k \in \mathbf{Z} \Rightarrow g \mid b$, $h \mid a$. $g \mid b$, $g \mid n \Rightarrow g \mid h$; $h \mid a$, $h \mid n \Rightarrow h \mid g$. Since $g, h > 0$, $g = h$.

13. a) 112 b) $031 - 43 - 3464$

Section 8.4 — p. 237

1. $s \to 0, t \to 1, v \to 2, w \to 3, x \to 4, y \to 5$

3. a) 4 b) 1 c) No

5. No, $\mathbf{Z}_4$ has two units, while the ring in Example 8.3 has only one unit.

7. Let $(R, +, \cdot)$, $(S, \oplus, \odot)$, $(T, +', \cdot')$ be the rings. For any $a, b \in R$, $(g \circ f)(a + b) = g(f(a + b)) = g(f(a) \oplus f(b)) = g(f(a)) +' g(f(b)) = (g \circ f)(a) +' (g \circ f)(b)$. Also, $(g \circ f)(a \cdot b) = g(f(a \cdot b)) = g(f(a) \odot f(b)) = g(f(a)) \cdot' g(f(b)) = (g \circ f)(a) \cdot' (g \circ f)(b)$. Hence, $g \circ f$ is a ring homomorphism.

9. a) It is better to use the form in the program since matrix multiplication is more accurate than matrix inversion. If the inverse is computed first, then round-off error may be compounded as the inverse is raised to the fourth power.

Miscellaneous Exercises — p. 239

1. a) False. Let $R = \mathbf{Z}$ and $S = \mathbf{Z}^+$.

   b) False. Let $R = \mathbf{Z}$ and $S = \{2x \mid x \in \mathbf{Z}\}$

   c) False. Let $R = M_2(\mathbf{Z})$ and $S = \left\{ \begin{bmatrix} a & 0 \\ 0 & 0 \end{bmatrix} \middle| a \in \mathbf{Z} \right\}$.

   d) and e) True.

   f) False. $(\mathbf{Z}, +, \cdot)$ is a subring (but not a field) in $(\mathbf{Q}, +, \cdot)$

   g) False. For any prime p, $\{a/(p^n) \mid a, n \in \mathbf{Z}, n \geq 0\}$ is a subring in $(\mathbf{Q}, +, \cdot)$.

   h) False. $f(6) = 12 = f(2 \cdot 3)$, but $f(2) \cdot f(3) = (4)(6) = 24$.

   i) False. Consider the field in Table 8.6. j) True.

   k) False. Let $R = (\mathbf{Z}, +, \cdot)$, $S = \{2x \mid x \in \mathbf{Z}\}$, $T = \{3x \mid x \in \mathbf{Z}\}$.

   l) False. In $\mathbf{Z}_5$, $1^2 = 4^2$ but $1 \neq 4$.

3. a) $a + a = (a + a)^2 = a^2 + a^2 + a^2 + a^2 = (a + a) + (a + a) \Rightarrow a + a = 2a = z$. [Hence $-a = a$.]

   b) For any $a \in R$, $a + a = z \Rightarrow a = -a$. For $a, b \in R$, $(a + b) = (a + b)^2 = a^2 + ab + ba + b^2 = a + ab + ba + b \Rightarrow ab + ba = z \Rightarrow ab = -ba = ba$, so R is commutative.

5. b) i) $\mathbf{Z} = \{0\} \cup \{1, -1\} \cup \{2, -2\} \cup \{3, -3\} \cup \dots$

   ii) $\{a + bi \mid a, b \in \mathbf{Z}, i^2 = -1\} = \bigcup_{x, y \in \mathbf{Z}} \{x + yi, -x - yi, -y + xi, y - xi\}$

iii) $\mathbf{Z}_7 = \{0\} \cup \{1, 2, 3, 4, 5, 6\}$

7. Let $x = a_1 + b_1$, $y = a_2 + b_2$, for $a_1, a_2 \in A$, $b_1, b_2 \in B$. Then $x - y = (a_1 - a_2) + (b_1 - b_2) \in A + B$. If $r \in R$, and $a + b \in A + B$, with $a \in A$, $b \in B$, then $ra \in A$, $rb \in B$ and $r(a + b) \in A + B$. Similarly, $(a + b)r \in A + B$, and $A + B$ is an ideal of R.

9. Consider the numbers x_1, $x_1 + x_2$, $x_1 + x_2 + x_3$, $\ldots$, $x_1 + x_2 + x_3 + \cdots + x_n$. If one of these numbers is congruent to 0 modulo n, the result follows. If not, there exist $1 \le i < j \le n$ with $(x_1 + x_2 + \cdots + x_i) \equiv (x_1 + \cdots + x_i + x_{i+1} + \cdots + x_j)$ (mod n). Hence n divides $(x_{i+1} + \cdots + x_j)$.

Chapter 9

Section 9.1 — p. 250

1. a) 2^n b) $2^{(2^n)}$

3. a) 2^{64} b) 2^6 c) 2^6
   d) 2^{49}; 2^7 e) 2^{32}; 2^{32}

5. $m + k = 2^n$

7. a) $y + x\bar{z}$ b) $x + y$ c) $wx + z$
   d) $x_1 + x_2 + x_3 + x_4 + \cdots$

9. a) $f \oplus f = 0; f \oplus \bar{f} = 1; f \oplus 1 = \bar{f}; f \oplus 0 = f$

   b)   i) $f \oplus g = 0 \Leftrightarrow f\bar{g} + \bar{f}g = 0 \Rightarrow f\bar{g} = \bar{f}g = 0.$ $f = 1$, $f\bar{g} = 0 \Rightarrow g = 1.$
          $f = 0, \bar{f}g = 0 \Rightarrow g = 0.$ Hence $f = g$.

         iii) $\bar{f} \oplus \bar{g} = \bar{\bar{f}}\bar{g} + \bar{\bar{\bar{g}}}\bar{f} = \bar{f}g + f\bar{g} = f\bar{g} + \bar{f}g = f \oplus g$

         iv) This is the only result that is not true. When f has value 1, g has value 0 and h value 1 (or g has value 1 and h value 0) then $f \oplus gh$ has value 1 but $(f \oplus g)(f \oplus h)$ has value 0.

         v) $fg \oplus fh = \overline{fg}fh + fg\overline{fh} = (\bar{f} + \bar{g})fh + fg(\bar{f} + \bar{h}) = \bar{f}fh + f\bar{g}h + f\bar{f}g + fg\bar{h} = f\bar{g}h + fg\bar{h} = f(\bar{g}h + g\bar{h}) = f(g \oplus h).$

         vi) $\bar{f} \oplus g = \bar{\bar{f}}\bar{g} + \bar{f}g = f\bar{g} + \bar{f}g = f \oplus \bar{g}.$
              $\overline{f \oplus g} = \overline{f\bar{g} + \bar{f}g} = (\bar{f} + g)(f + \bar{g}) = \bar{f}\bar{g} + fg = \bar{f} \oplus g.$

         vii) $f \oplus g = f \oplus h \Rightarrow f \oplus (f \oplus g) = f \oplus (f \oplus h) \Rightarrow (f \oplus f) \oplus g = (f \oplus f) \oplus h \Rightarrow 0 \oplus g = 0 \oplus h \Rightarrow g = h.$

Section 9.2 — p. 259

1. a) $x \oplus y = (x + y)(\overline{xy})$

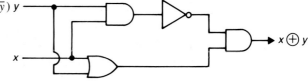

b) $\overline{xy}$

c) $\overline{x + y}$

3. a)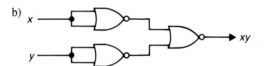

b)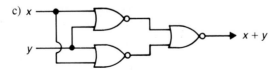

c) x $x + y$

5. $f(w, x, y, z) = \overline{w}\,\overline{x}\,y\,\overline{z} + (w + x + \overline{y})z$

7. $f(w, x, y, z) = (w + y)(\overline{x} + y)(\overline{w} + \overline{y} + z)(x + \overline{y} + z)$

9. a)   64   b)   32

c)   16   d)   8

Section 9.3 — p. 264

1. $uv + wvy + uxz + uyz + wz$

3. Parts (a) and (c) are true.

b)   In Fig. 9.11, let $D_1 = \{c, d\}$, $D_2 = \{a, g\}$.

d)   Let $D_1 = \{a, d\}$, $D_2 = \{a\}$ in Fig. 9.11.

e)   Let $D_1 = \{a\}$, $D_2 = \{d\}$ in Fig. 9.11.

5. a) $f(w, x, y, z) = z$

b) $f(w, x, y, z) = \overline{x}\,\overline{y}\,\overline{z} + x\overline{y}z + xy\overline{z}$

c) $f(v, w, x, y, z) = v\overline{y}\,\overline{z} + \overline{w}\,\overline{x}\,yz + \overline{v}\,\overline{w}\,\overline{z} + \overline{v}x\overline{y}$

Section 9.4 — p. 272

3. a)   30   b)   30   c)   1

d)   21   e)   30   f)   70

5. $y \leq x$

7. From Theorem 9.5(a), with x_1, x_2 distinct atoms, if $x_1 x_2 \neq 0$, then $x_1 = x_1 x_2 = x_2 x_1 = x_2$, a contradiction.

9. No. Let $\mathcal{U} = \{a, b, c\}$, and $\mathcal{B} = (\mathcal{P}(\mathcal{U}), \cup, \cap, \overline{}, \emptyset, \mathcal{U})$. If $\mathcal{B}_1 = \{\emptyset, \mathcal{U}, \{a\}, \{b\}, \{a, b\}\}$, then $\mathcal{B}_1$ satisfies the stated conditions but it is not a subalgebra of $\mathcal{B}$. For example, $\overline{\{a\}} \notin \mathcal{B}_1$.

11. a) $f(0) = f(x\overline{x})$ for any $x \in \mathcal{B}_1$. $f(x\overline{x}) = f(x)f(\overline{x}) = f(x)\overline{f(x)} = 0$.

b)   Follows from (a) by duality.

c) $x \leq y \Leftrightarrow xy = x \Rightarrow f(xy) = f(x) \Rightarrow f(x)f(y) = f(x) \Leftrightarrow f(x) \leq f(y)$.

e) $S_1 \neq \emptyset \Rightarrow f(S_1) \neq \emptyset$. Let x_2, $y_2 \in f(S_1)$ with x_1, $y_1 \in S_1$ and $f(x_1) = x_2$, $f(y_1) = y_2$. Then $f(x_1 + y_1) = f(x_1) + f(y_1) = x_2 + y_2$, and $x_1 + y_1 \in S_1$. Hence $x_2 + y_2 \in f(S_1)$. Also, $\overline{x_2} = \overline{f(x_1)} = f(\overline{x_1})$, and since $\overline{x_1} \in S_1$, it follows that $\overline{x_2} \in f(S_1)$. By exercise 8(d), $f(S_1)$ is a subalgebra of $\mathcal{B}_2$.

13. a) $f(xy) = f(\overline{\overline{x} + \overline{y}}) = \overline{f(\overline{x} + \overline{y})} = \overline{f(\overline{x}) + f(\overline{y})} = \overline{f(\overline{x})} \cdot \overline{f(\overline{y})} = f(\overline{\overline{x}}) \cdot f(\overline{\overline{y}}) = f(x) \cdot f(y)$.

    b) Let $\mathcal{B}_1$, $\mathcal{B}_2$ be Boolean algebras with $f: \mathcal{B}_1 \to \mathcal{B}_2$ one-to-one and onto. Then f is an isomorphism if $f(\overline{x}) = \overline{f(x)}$ and $f(xy) = f(x)f(y)$ for all $x, y \in \mathcal{B}_1$. (Follows from (a) by duality.)

Miscellaneous Exercises — p. 275

1. a) i) For $n = 2$, $\overline{x_1 + x_2} = \overline{x}_1\overline{x}_2$ is true; this is one of the DeMorgan Laws. Assume the result for $n = k(\geq 2)$ and consider the case of $n = k + 1$.

$$\overline{(x_1 + x_2 + \cdots + x_k + x_{k+1})} = \overline{(x_1 + x_2 + \cdots + x_k) + x_{k+1}}$$
$$= \overline{(x_1 + x_2 + \cdots + x_k)}\,\overline{x_{k+1}}$$
$$= \overline{x_1}\overline{x_2}\cdots\overline{x_k}\overline{x_{k+1}}\,.$$

Consequently, the result follows for all $n \geq 2$ by the principle of finite induction.

    ii) Follows from (i) by duality.

    iii) $\left(\sum_{i=1}^{n} x_i\right)\overline{\left(\prod_{i=1}^{n} x_i\right)} = (x_1 + x_2 + \cdots + x_n)(\overline{x_1} + \overline{x_2} + \cdots + \overline{x_n})$. This has value 0 if $x_i = 0$ for all $1 \leq i \leq n$, or if $\overline{x}_i = 0$ for all $1 \leq i \leq n$. In either case $x_1\overline{x_2} + x_2\overline{x_3} + \cdots + x_n\overline{x_1}$ has value 0. In every other case the expression has value 1 and there exist x_i, x_{i+1}, $1 \leq i \leq n - 1$, where one has the value 0 and the other 1, or x_n has value 0(1) while x_1 has value 1(0). In these cases $x_1\overline{x_2} + x_2\overline{x_3} + \cdots + x_n\overline{x_1}$ also has value 1.

    b) $\left(\prod_{i=1}^{n} x_i\right)\overline{\left(\sum_{i=1}^{n} x_i\right)} = (x_1 + \overline{x_2})(x_2 + \overline{x_3})\cdots(x_{n-1} + \overline{x_n})(x_n + \overline{x_1})$

3. She can invite only Nettie and Cathy.

5. a) $f(w,x,y,z) = \overline{w}\,\overline{x} + xy$

    b) $g(v,w,x,y,z) = \overline{v}\,\overline{w}yz + xz + w\overline{y}\,\overline{z} + \overline{x}\,\overline{y}\,\overline{z}$

7. a) $2^{(2^{n-1})}$ b) 2^4; 2^{n+1}

9. a) If $n = 60$, there are 12 divisors and no Boolean algebra contains 12 elements, since 12 is not a power of 2.

    b) If $n = 120$, there are 16 divisors. However, if $x = 4$, then $\overline{x} = 30$ and $x \cdot \overline{x} = (x, \overline{x}) = (4, 30) = 2$, which is not the zero element. So the Inverse Laws are not satisfied.

Chapter 10

Section 10.1 — p. 282

1. a) The coefficient of x^{20} in $(1 + x + x^2 + \cdots + x^7)^4$ or $(1 + x + x^2 + \cdots)^4$.

    b) The coefficient of x^{20} in $(1 + x + x^2 + \cdots + x^{20})^2(1 + x^2 + x^4 + \cdots + x^{20})^2$ or $(1 + x + x^2 + \cdots)^2(1 + x^2 + x^4 + \cdots)^2$

    c) The coefficient of x^{30} in $(x^2 + x^3 + x^4)(x^3 + x^4 + \cdots + x^8)^4$

    d) The coefficient of x^{30} in $(1 + x + x^2 + \cdots + x^{30})^3(1 + x^2 + x^4 + \cdots + x^{30}) \cdot (x + x^3 + x^5 + \cdots + x^{29})$ or $(1 + x + x^2 + \cdots)^3(1 + x^2 + x^4 + \cdots) \cdot (x + x^3 + x^5 + \cdots)$

3. a) The coefficient of x^{10} in $(1 + x + x^2 + x^3 + \cdots)^6$

    b) The coefficient of x^r in $(1 + x + x^2 + x^3 + \cdots)^n$

5. The answer is the coefficient of x^{31} in the generating function

$$(1 + x + x^2 + x^3 + \cdots)^3(1 + x + x^2 + \cdots + x^{10})$$

Section 10.2 — p. 288

1. a) $(1 + x)^8$ b) $8(1 + x)^7$ c) $(1 + x)^{-1}$
    d) $x^3/(1 - x)$ e) $6x^3/(1 + x)$ f) $(1 - x^2)^{-1}$
    g) $(1 - 2x)^{-1}$ h) $x^2/(1 - ax)$

3. a) $\binom{21}{7}$ b) $\binom{n+6}{7}$

5. $\binom{14}{10} - 5\binom{9}{5} + \binom{5}{2}$

7. a) 0 b) $\binom{14}{12} - 5\binom{16}{14}$
    c) $\binom{18}{15} + 4\binom{17}{14} + 6\binom{16}{13} + 4\binom{15}{12} + \binom{14}{11}$

9. $\binom{99}{96} - 4\binom{64}{61} + 6\binom{29}{26}$

11. $[\binom{29}{18} - \binom{12}{1}\binom{23}{12} + \binom{12}{2}\binom{17}{6} - \binom{12}{3}]/(6^{12})$

13. $(1/8)[1 + (-1)^n] + (1/4)\binom{n+1}{1} + (1/2)\binom{n+2}{n}$

15. $(1 - x - x^2 - x^3 - x^4 - x^5 - x^6)^{-1} = [1 - (x + x^2 + \cdots + x^6)]^{-1}$

$$= \underbrace{(x + x^2 + \cdots + x^6)}_{\text{one roll}} + \underbrace{(x + x^2 + \cdots + x^6)^2}_{\text{two rolls}} + \underbrace{(x + x^2 + \cdots + x^6)^3}_{\text{three rolls}} + \cdots$$

17. $a = 1, b = -2, k = -1/5$ 19. $\binom{44}{37}$

Section 10.3 — p. 292

1. 5; 4 + 1; 3 + 2; 3 + 1 + 1; 2 + 2 + 1; 2 + 1 + 1 + 1; 1 + 1 + 1 + 1 + 1.
    7; 6 + 1; 5 + 2; 5 + 1 + 1; 4 + 3; 4 + 2 + 1; 4 + 1 + 1 + 1; 3 + 3 + 1;
    3 + 2 + 2; 3 + 2 + 1 + 1; 3 + 1 + 1 + 1 + 1; 2 + 2 + 2 + 1;
    2 + 2 + 1 + 1 + 1; 2 + 1 + 1 + 1 + 1 + 1; 1 + 1 + 1 + 1 + 1 + 1 + 1.

3. The number of partitions of 6 into 1's, 2's and 3's is 7.

5. (a) and (b)

$$(1 + x^2 + x^4 + x^6 + \cdots)(1 + x^4 + x^8 + \cdots)$$

$$\times (1 + x^6 + x^{12} + \cdots)\cdots = \prod_{i=1}^{\infty} \frac{1}{1 - x^{2i}}$$

7. $(1 + x + x^2 + \cdots + x^{99})(1 + x^5 + x^{10} + \cdots + x^{95})(1 + x^{10} + \cdots + x^{90}) \cdot$
    $(1 + x^{25} + x^{50} + x^{75})(1 + x^{50})$

9. This result follows from the one-to-one correspondence between the Ferrer's graphs with
    summands (rows) not exceeding m and the transpose graphs (also Ferrer's graphs) that
    have m summands (rows).

Section 10.4 — p. 295

1. a) e^{-x} b) e^{2x} c) e^{-ax}
    d) e^{a^2x} e) ae^{a^2x} f) xe^{2x}

3. a) $(1 + x)^2\left(1 + x + \dfrac{x^2}{2}\right)^2$

b) $(1 + x)\left(1 + x + \dfrac{x^2}{2}\right)\left(1 + x + \dfrac{x^2}{2} + \dfrac{x^3}{3!} + \dfrac{x^4}{4!}\right)^2$

c) $(1 + x)^3\left(1 + x + \dfrac{x^2}{2}\right)^4$

5. The answer is the coefficient of $\dfrac{x^{25}}{25!}$ in $\left(\dfrac{x^3}{3!} + \dfrac{x^4}{4!} + \cdots + \dfrac{x^{10}}{10!}\right)^4$

7. a) $(1/2)[3^{20} + 1]/(3^{20})$ b) $(1/4)[3^{20} + 3]/(3^{20})$
    c) $(1/2)[3^{20} - 1]/(3^{20})$ d) $(1/2)[3^{20} - 1]/(3^{20})$
    e) $(1/2)[3^{20} + 1]/(3^{20})$

9. a) $\left(\dfrac{x^5}{5!} + \dfrac{x^6}{6!} + \cdots\right)\left(\dfrac{x^2}{2!} + \dfrac{x^3}{3!} + \cdots\right)^4$

b) $\left(x + \dfrac{x^2}{2!} + \dfrac{x^3}{3!} + \cdots\right)^3\left[(x)\left(\dfrac{x^2}{2!} + \dfrac{x^3}{3!} + \dfrac{x^4}{4!} + \dfrac{x^5}{5!}\right) + \left(\dfrac{x^2}{2!}\right)\left(\dfrac{x^3}{3!} + \dfrac{x^4}{4!} + \dfrac{x^5}{5!}\right)\right.$

$\left. + \left(\dfrac{x^3}{3!}\right)\left(\dfrac{x^4}{4!} + \dfrac{x^5}{5!}\right) + \left(\dfrac{x^4}{4!}\right)\left(\dfrac{x^5}{5!}\right)\right]$

Section 10.5 — p. 297

3. $a_0, a_1 - a_0, a_2 - a_1, a_3 - a_2, \ldots$

Miscellaneous Exercises — p. 299

1. a) $6/(1 - x) + 1/(1 - x)^2$ b) $1/(1 - ax)$
    c) $1/[1 - (1 + a)x]$ d) $1/(1 - x) + 1/(1 - ax)$
5. $[\binom{15}{12} - \binom{4}{1}\binom{9}{6} + \binom{4}{2}]^2$
7. $4^{10} - \binom{4}{1}(10)(3^9) + \binom{4}{2}(10)(9)(2^8) - \binom{4}{3}(10)(9)(8)$
9. a) $1, 5, (5)(7), (5)(7)(9), (5)(7)(9)(11), \ldots$ b) $a = 4, b = -\frac{7}{4}$
11. a) $\binom{19}{8}$ b) $\binom{9}{4}^2/\binom{19}{8}$.

Chapter 11

Section 11.1 — p. 307

1. a) $a_n = 5a_{n-1}, n \geq 1, a_0 = 2$ b) $a_n = -3a_{n-1}, n \geq 1, a_0 = 6$
    c) $a_n = (1/3)a_{n-1}, n \geq 1, a_0 = 1$ d) $a_n = (2/5)a_{n-1}, n \geq 1, a_0 = 7$
3. $c = \pm(3/7)$ 5. 141 months
7. $3(7)^{10/3}$
9. a) 145   b) 45
11. a) 21345   b) 52143, 52134   c) 21534, 21354, 21345
    d) 21543 is 113th, 35421 is 67th, 31524 is 43rd.

Section 11.2 — p. 315

1. a) $a_n = (3/7)(-1)^n + (4/7)(6)^n, n \geq 0$
    b) $a_n = 4(1/2)^n - 2(5)^n, n \geq 0$

c) $a_n = 4 + 3(-1/3)^n, n \geq 0$

d) $a_n = 3 \sin(n\pi/2), n \geq 0$

e) $a_n = 2^n[\cos(n\pi/2) + (1/2) \sin(n\pi/2)], n \geq 0$

f) $a_n = (5 - n)3^n, n \geq 0$

g) $a_n = (\sqrt{2})^n[\cos(3\pi n/4) + 4 \sin(3\pi n/4)], n \geq 0$

h) $a_n = 3(5)^n, n \geq 0$

i) $a_n = (-2/3)^n(1 - 7n), n \geq 0$

j) $a_n = [(-5 + 6n)/9](-1)^n + (14/9)(2^n), n \geq 0$

3. $a_n = (1/10)[7^n - (-3)^n], n \geq 0$

5. a)
$$F_1 = F_2 - F_0$$
$$F_3 = F_4 - F_2$$
$$F_5 = F_6 - F_4$$
$$\vdots$$
$$\underline{\qquad F_{2n-1} = F_{2n} - F_{2n-2} \qquad}$$
$$F_1 + F_3 + F_5 + \cdots + F_{2n-1} = F_{2n} - F_0 = F_{2n}$$

b)
$$F_2 = F_3 - F_1$$
$$F_4 = F_5 - F_3$$
$$F_6 = F_7 - F_5$$
$$\vdots$$
$$\underline{\qquad F_{2n} = F_{2n+1} - F_{2n-1} \qquad}$$
$$F_2 + F_4 + \cdots + F_{2n} = F_{2n+1} - F_1$$
$$F_0 + F_2 + F_4 + \cdots + F_{2n} = F_{2n+1} - 1 \ (F_0 = 0; F_1 = 1).$$

c)
$$F_1 + F_3 + \cdots + F_{2n-1} = F_{2n}$$
$$\underline{F_2 + F_4 + \cdots + F_{2n} \qquad = F_{2n+1} - 1}$$
$$F_1 - F_2 + F_3 - F_4 + \cdots + F_{2n-1} - F_{2n} = (F_{2n} - F_{2n+1}) + 1 = -F_{2n-1} + 1$$

7. $a_n = (1/\sqrt{5})[((1 + \sqrt{5})/2)^{n+1} - ((1 - \sqrt{5})/2)^{n+1}], n \geq 0.$

9. $a_n = [(5 + \sqrt{21})/(2\sqrt{21})][(3 + \sqrt{21})/2]^n - [(5 - \sqrt{21})/(2\sqrt{21})][(3 - \sqrt{21})/2]^n$

11. $a_n = \sqrt{51(4^n) - 35}, n \geq 0.$

Section 11.3 — p. 321

1. a) $a_n = (n + 1)^2, n \geq 0$ b) $a_n = 3 + n(n - 1)^2, n \geq 0$

    c) $a_n = 6(2^n) - 5, n \geq 0$ d) $a_n = 2^n + n(2^{n-1}), n \geq 0.$

3. a) $a_n = a_{n-1} + n, n \geq 1, a_0 = 1$

        $a_n = 1 + [n(n + 1)]/2, n \geq 0.$

    b) $b_n = b_{n-1} + 2, n \geq 2, b_1 = 2$

        $b_n = 2n, n \geq 1, b_0 = 1.$

5. a) $a_n = (3/4)(-1)^n - (4/5)(-2)^n + (1/20)(3)^n, n \geq 0$

    b) $a_n = (2/9)(-2)^n - (5/6)(n)(-2)^n + (7/9), n \geq 0$

    c) $a_n = (-55/54)n(-2)^n + (n^2/9) - (4n/27), n \geq 0$

    d) $a_n = (5/4) - (1/4)(-1)^n - (1/2) \sin(n\pi/2)$

7. $a_n = A + Bn + Cn^2 - (3/4)n^3 + (5/24)n^4$

9. $P = \$117.68$

11. a) $a_n = [(3/4)(3)^n - 5(2)^n + (7n/2) + (21/4)]^{1/2}$, $n \geq 0$

   b) $a_n = (1/2)[(-1)^n + 1]n!$, $n \geq 0$ c) $a_n = 2$, $n \geq 0$

Sections 11.4 and 11.5 — p. 329

1. a) $a_n = (1/2)[1 + 3^n]$, $n \geq 0$ b) $a_n = 1 + [n(n-1)(2n-1)]/6$, $n \geq 0$

   c) $a_n = 2^n$, $n \geq 0$ d) $a_n = [n^3 - 3n^2 + 8n + 6]/6$

3. a) $a_n = 2^n(1 - 2n)$, $b_n = n(2^{n+1})$, $n \geq 0$

   b) $a_n = (-3/4) + (1/2)(n + 1) + (1/4)(3^n)$,

   $b_n = (3/4) + (1/2)(n + 1) - (1/4)(3^n)$, $n \geq 0$

5. $b_4 = (8!)/[(5!)(4!)] = 14$

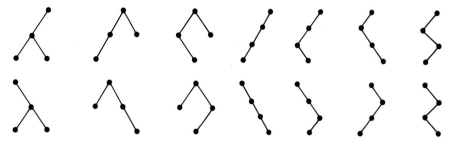

Miscellaneous Exercises — p. 331

1. $\dbinom{n}{k+1} = \dfrac{n!}{(k+1)!(n-k-1)!} = \dfrac{(n-k)}{(k+1)} \cdot \dfrac{n!}{k!(n-k)!} = \dbinom{n-k}{k+1}\dbinom{n}{k}$

3. There are two cases to consider. Case 1 (1 is a summand): Here there are $p(n - 1, k - 1)$ ways to partition $n - 1$ into exactly $k - 1$ summands. Case 2 (1 is not a summand): Here each summand $s_1, s_2, \ldots, s_k > 1$. For $1 \leq i \leq k$, let $t_i = s_i - 1 \geq 1$. Then $t_1, t_2, \ldots, t_k$ provide a partition of $n - k$ into exactly k summands. These cases are exhaustive and disjoint, so by the rule of sum $p(n, k) = p(n - 1, k - 1) + p(n - k, k)$.

5. a) For any derangement, 1 is placed in position i, $2 \leq i \leq n$. Two things then occur. Case 1 (i is in position 1): Here the other $n - 2$ integers are deranged in d_{n-2} ways. With $n - 1$ choices for i this results in $(n - 1)d_{n-2}$ such derangements. Case 2 (i is not in position 1 (or position i)): Here we consider 1 as the new natural position for i, so there are $n - 1$ elements to derange. With $n - 1$ choices for i we have $(n - 1)d_{n-1}$ derangements. Since the two cases are exhaustive and disjoint, the result follows from the rule of sum.

   b) $d_0 = 1$ c) $d_n - nd_{n-1} = d_{n-2} - (n - 2)d_{n-3}$.

7. $a_n = n^2 - n + 2$

9. a) $p_n = (0.6)^n p_0$, $n \geq 0$ b) $p_0 = 2/5$

11. a) $a_n = \binom{2n}{n}$, $n \geq 0$ b) $r = 1$, $s = -4$, $t = -1/2$

   d) $b_n = (1/(2n - 1))\binom{2n}{n}$, $n \geq 1$; $b_0 = 0$

Chapter 12

Section 12.1 — p. 340

1. a)   Yes. The identity is 1 and each element is its own inverse.

   b)   No. The set is not closed under addition and there is no identity.

   c)   No. The set is not closed under addition.

   d)   Yes. The identity is 0; the inverse of $10n$ is $10(-n)$ or $-10n$.

   e)   No. The function $f: A \rightarrow A$ where $f(x) = 1$ for all $x \in A$ is not invertible.

   f)   Yes. The identity is 1_A and the inverse of $g: A \rightarrow A$ is $g^{-1}: A \rightarrow A$.

   g)   Yes. The identity is 0; the inverse of $a/(2^n)$ is $(-a)/(2^n)$.

3. a)   The result follows from Theorem 12.1(b) since both $(a^{-1})^{-1}$ and a are inverses of a^{-1}.

   b) $(b^{-1}a^{-1})(ab) = b^{-1}(a^{-1}a)b = b^{-1}(e)b = b^{-1}b = e$ and $(ab)(b^{-1}a^{-1})$
   $$= a(bb^{-1})a^{-1} = a(e)a^{-1} = aa^{-1} = e.$$
   So $b^{-1}a^{-1}$ is an inverse of ab, and by Theorem 12.1(b), $(ab)^{-1} = b^{-1}a^{-1}$.

5. a) $\{0\}; \{0,6\}; \{0,4,8\}; \{0,3,6,9\}; \{0,2,4,6,8,10\}; \mathbf{Z}_{12}$.

   b) $\{1\}; \{1,10\}; \{1,3,4,5,9\}; \mathbf{Z}_{11}^*$.

   c) $\{\pi_0\}; \{\pi_0, \pi_1, \pi_2\}; \{\pi_0, r_1\}; \{\pi_0, r_2\}; \{\pi_0, r_3\}; S_3$.

7. a)   There are 10: five rotations through $i(72°)$, $0 \leq i \leq 4$, and five reflections about lines containing a vertex and the midpoint of the opposite side.

   b)   For a regular n-gon ($n \geq 3$) there are $2n$ rigid motions. There are the n rotations through $i(360°/n)$, $0 \leq i \leq n - 1$. There are n reflections. For n odd, each reflection is about a line through a vertex and the midpoint of the opposite side. For n even, there are $n/2$ reflections about lines through opposite vertices and $n/2$ reflections about lines through the midpoints of opposite sides.

9. b)   i)   216

   ii) $H_1 = \{(x,0,0) | x \in \mathbf{Z}_6\}$ is a subgroup of order 6;
        $H_2 = \{(x,y,0) | x,y \in \mathbf{Z}_6, y = 0, 3\}$ is a subgroup of order 12;
        $H_3 = \{(x,y,0) | x,y \in \mathbf{Z}_6\}$ has order 36.

   iii) $-(2,3,4) = (4,3,2)$; $-(4,0,2) = (2,0,4)$; $-(5,1,2) = (1,5,4)$.

Section 12.2 — p. 344

1. $f(a^{-1}) \cdot f(a) = f(a^{-1} \cdot a) = f(e_G) = e_H$ and $f(a) \cdot f(a^{-1}) = f(a \cdot a^{-1}) = f(e_G) = e_H$, so $f(a^{-1})$ is an inverse of $f(a)$. By the uniqueness of inverses (Theorem 12.1(b)) it follows that $f(a^{-1}) = [f(a)]^{-1}$.

3. $f(0) = (0,0)$ $f(1) = (1,1)$ $f(2) = (2,0)$
   $f(3) = (0,1)$ $f(4) = (1,0)$ $f(5) = (2,1)$

5. a) $o(\pi_0) = 1$, $o(\pi_1) = o(\pi_2) = 3$, $o(r_1) = o(r_2) = o(r_3) = 2$

   b)   (See Fig. 12.6) $o(\pi_0) = 1$, $o(\pi_1) = o(\pi_3) = 4$, $o(\pi_2) = o(r_1) = o(r_2) = o(r_3) = o(r_4) = 2$.

7. $\mathbf{Z}_5^* = \langle 2 \rangle = \langle 3 \rangle$; $\mathbf{Z}_7^* = \langle 3 \rangle = \langle 5 \rangle$; $\mathbf{Z}_{11}^* = \langle 2 \rangle = \langle 6 \rangle = \langle 7 \rangle = \langle 8 \rangle$.

9. a) $(\mathbf{Z}_{12}, +) = \langle 1 \rangle = \langle 5 \rangle = \langle 7 \rangle = \langle 11 \rangle$
        $(\mathbf{Z}_{16}, +) = \langle 1 \rangle = \langle 3 \rangle = \langle 5 \rangle = \langle 7 \rangle = \langle 9 \rangle = \langle 11 \rangle = \langle 13 \rangle = \langle 15 \rangle$
        $(\mathbf{Z}_{24}, +) = \langle 1 \rangle = \langle 5 \rangle = \langle 7 \rangle = \langle 11 \rangle = \langle 13 \rangle = \langle 17 \rangle = \langle 19 \rangle = \langle 23 \rangle$

   b) Let $G = \langle a^k \rangle$. Since $G = \langle a \rangle$, $a = (a^k)^s$ for $s \in \mathbf{Z}$. Then $a^{1-ks} = e$, so $1 - ks = tn$ since $\circ(a) = n$. $1 - ks = tn \Rightarrow 1 = ks + tn \Rightarrow (k, n) = 1$. Conversely, let $G = \langle a \rangle$ where $a^k \in G$ and $(k, n) = 1$. Then $\langle a^k \rangle \subseteq G$. $(k, n) = 1 \Rightarrow 1 = ks + tn$, $s, t \in \mathbf{Z} \Rightarrow a = a^1 = a^{ks+nt} = (a^k)^s(a^n)^t = (a^k)^s(e)^t = (a^k)^s \in \langle a^k \rangle$. Hence $G \subseteq \langle a^k \rangle$. So $G = \langle a^k \rangle$, or a^k generates G.

   c) $\phi(n)$.

Section 12.3 — p. 346

1. a) $\left\{\left(\begin{smallmatrix}1&2&3&4\\2&3&4&1\end{smallmatrix}\right), \left(\begin{smallmatrix}1&2&3&4\\3&4&1&2\end{smallmatrix}\right), \left(\begin{smallmatrix}1&2&3&4\\4&1&2&3\end{smallmatrix}\right), \left(\begin{smallmatrix}1&2&3&4\\1&2&3&4\end{smallmatrix}\right)\right\}$

   b) $\left(\begin{smallmatrix}1&2&3&4\\2&1&4&3\end{smallmatrix}\right)H = \left\{\left(\begin{smallmatrix}1&2&3&4\\3&2&1&4\end{smallmatrix}\right), \left(\begin{smallmatrix}1&2&3&4\\4&3&2&1\end{smallmatrix}\right), \left(\begin{smallmatrix}1&2&3&4\\1&4&3&2\end{smallmatrix}\right), \left(\begin{smallmatrix}1&2&3&4\\2&1&4&3\end{smallmatrix}\right)\right\}$

   $\left(\begin{smallmatrix}1&2&3&4\\4&2&3&1\end{smallmatrix}\right)H = \left\{\left(\begin{smallmatrix}1&2&3&4\\1&2&3&4\end{smallmatrix}\right), \left(\begin{smallmatrix}1&2&3&4\\2&4&1&3\end{smallmatrix}\right), \left(\begin{smallmatrix}1&2&3&4\\3&1&2&4\end{smallmatrix}\right), \left(\begin{smallmatrix}1&2&3&4\\4&2&3&1\end{smallmatrix}\right)\right\}$

   $\left(\begin{smallmatrix}1&2&3&4\\1&2&4&3\end{smallmatrix}\right)H = \left\{\left(\begin{smallmatrix}1&2&3&4\\2&3&1&4\end{smallmatrix}\right), \left(\begin{smallmatrix}1&2&3&4\\3&4&2&1\end{smallmatrix}\right), \left(\begin{smallmatrix}1&2&3&4\\4&1&3&2\end{smallmatrix}\right), \left(\begin{smallmatrix}1&2&3&4\\1&2&4&3\end{smallmatrix}\right)\right\}$

   $\left(\begin{smallmatrix}1&2&3&4\\1&3&2&4\end{smallmatrix}\right)H = \left\{\left(\begin{smallmatrix}1&2&3&4\\2&4&3&1\end{smallmatrix}\right), \left(\begin{smallmatrix}1&2&3&4\\3&1&4&2\end{smallmatrix}\right), \left(\begin{smallmatrix}1&2&3&4\\4&2&1&3\end{smallmatrix}\right), \left(\begin{smallmatrix}1&2&3&4\\1&3&2&4\end{smallmatrix}\right)\right\}$

   $\left(\begin{smallmatrix}1&2&3&4\\1&4&2&3\end{smallmatrix}\right)H = \left\{\left(\begin{smallmatrix}1&2&3&4\\2&1&3&4\end{smallmatrix}\right), \left(\begin{smallmatrix}1&2&3&4\\3&2&4&1\end{smallmatrix}\right), \left(\begin{smallmatrix}1&2&3&4\\4&3&1&2\end{smallmatrix}\right), \left(\begin{smallmatrix}1&2&3&4\\1&4&2&3\end{smallmatrix}\right)\right\}$

   $\left(\begin{smallmatrix}1&2&3&4\\1&2&3&4\end{smallmatrix}\right)H = H$

3. 12

5. a) Since $(0, 0) \in H$, $H \neq \emptyset$. Let $(a, 0), (b, 0) \in H$.
    Then $(a, 0) + (b, 0) = (a + b, 0) \in H$. Also $(-a, 0) \in H$ for any $(a, 0) \in H$.
    Hence H is a subgroup.

   b) Each coset consists of the points on a horizontal line.

7. Let $\circ(a) = k$. Then $|\langle a \rangle| = k$, so by Lagrange's theorem k divides n. Hence $a^n = a^{km} = (a^k)^m = e^m = e$.

9. a) If H is a proper subgroup of G, then by Lagrange's theorem $|H|$ is 2 or p. If $|H| = 2$, then $H = \{e, x\}$ where $x^2 = e$, so $H = \langle x \rangle$. If $|H| = p$, let $y \in H$, $y \neq e$. Then $\circ(y) = p$, so $H = \langle y \rangle$.

   b) Let $x \in G$, $x \neq e$. Then $\circ(x) = p$ or $\circ(x) = p^2$. If $\circ(x) = p$, then $|\langle x \rangle| = p$. If $\circ(x) = p^2$, then $G = \langle x \rangle$ and $\langle x^p \rangle$ is a subgroup of G of order p.

Section 12.4 — p. 351

1. a) $e = 0001001$ b) $r = 1111011$ c) $c = 0101000$

3. a) i) $D(111101100) = 101$ ii) $D(000100011) = 000$
    iii) $D(010011111) = 011$
   b) $000000000, 000000001, 100000000$ c) 64

Sections 12.5 and 12.6 — p. 358

1. $S(101010, 1) = \{101010, 001010, 111010, 100010, 101110, 101000, 101011\}$
    $S(111111, 1) = (111111, 011111, 101111, 110111, 111011, 111101, 111110\}$

3. a) $|S(x, 1)| = 11$; $|S(x, 2)| = 56$; $|S(x, 3)| = 176$
   b) $|S(x, k)| = 1 + \binom{n}{1} + \binom{n}{2} + \cdots + \binom{n}{k} = \sum_{i=0}^{k} \binom{n}{i}$

5. a) The minimum distance between code words is 3. The code can detect all errors of weight ≤ 2 and can correct all single errors.

   b) The minimum distance between code words is 5. The code can detect all errors of weight ≤ 4 and can correct all errors of weight ≤ 2.

c)   The minimum distance is 2. The code detects all single errors but has no correction capability.

d)   The minimum distance is 3. The code detects all errors of weight ≤ 2 and corrects all single errors.

7. a) $C = \{00000, 10110, 01011, 11101\}$. The minimum distance between code words is 3, so the code can detect all errors of weight ≤ 2 and can correct all single errors.

b) $H = \begin{bmatrix} 1 & 0 & 1 & 0 & 0 \\ 1 & 1 & 0 & 1 & 0 \\ 0 & 1 & 0 & 0 & 1 \end{bmatrix}$

c)   i) 01   ii) 11   v) 11   vi) 10

For (iii) and (iv) the syndrome is $(111)^{tr}$ which is not a column of H. Assuming a double error, if $(111)^{tr} = (110)^{tr} + (001)^{tr}$, then the decoded received word is 01 (for (iii)) and 10 (for (iv)). If $(111)^{tr} = (011)^{tr} + (100)^{tr}$, we get 10 (for (iii)) and 01 (for (iv)).

9. $G = [I_8 | A]$ where I_8 is the 8×8 multiplicative identity matrix and A is a column of eight 1's. $H = [A^{tr} | 1] = [11111111 | 1]$.

11. Compare the generator (parity-check) matrix in Exercise 9 with the parity-check (generator) matrix in Exercise 10.

Sections 12.7 and 12.8 — p. 365

1. $\binom{256}{2}$; 255

3. a)
Syndrome	Coset Leader			
000	00000	10110	01011	11101
110	10000	00110	11011	01101
011	01000	11110	00011	10101
100	00100	10010	01111	11001
010	00010	10100	01001	11111
001	00001	10111	01010	11100
101	11000	01110	10011	00101
111	01100	11010	00111	10001

(The last two rows are not unique.)

b)
Received Word	Code Word	Decoded Message
11110	10110	10
11101	11101	11
11011	01011	01
10100	10110	10
10011	01011	01
10101	11100	11
11111	11101	11
01100	00000	00

5. a) G is 57×63; H is 6×63

b)   The rate is $\frac{57}{63}$

7. a) $(0.99)^4 + \binom{4}{1}(0.99)^3(0.01)$

b) $[(0.99)^4 + \binom{4}{1}(0.99)^3(0.01)]^5$

Section 12.9 — p. 372

1. a) $\pi_2^* =$

$$\begin{pmatrix} C_1 & C_2 & C_3 & C_4 & C_5 & C_6 & C_7 & C_8 & C_9 & C_{10} & C_{11} & C_{12} & C_{13} & C_{14} & C_{15} & C_{16} \\ C_1 & C_4 & C_5 & C_2 & C_3 & C_8 & C_9 & C_6 & C_7 & C_{10} & C_{11} & C_{14} & C_{15} & C_{12} & C_{13} & C_{16} \end{pmatrix}$$

$r_4^* =$

$$\begin{pmatrix} C_1 & C_2 & C_3 & C_4 & C_5 & C_6 & C_7 & C_8 & C_9 & C_{10} & C_{11} & C_{12} & C_{13} & C_{14} & C_{15} & C_{16} \\ C_1 & C_4 & C_3 & C_2 & C_5 & C_9 & C_8 & C_7 & C_6 & C_{10} & C_{11} & C_{12} & C_{15} & C_{14} & C_{13} & C_{16} \end{pmatrix}$$

  b) $(\pi_1^{-1})^* =$

$$\begin{pmatrix} C_1 & C_2 & C_3 & C_4 & C_5 & C_6 & C_7 & C_8 & C_9 & C_{10} & C_{11} & C_{12} & C_{13} & C_{14} & C_{15} & C_{16} \\ C_1 & C_5 & C_2 & C_3 & C_4 & C_9 & C_6 & C_7 & C_8 & C_{11} & C_{10} & C_{15} & C_{12} & C_{13} & C_{14} & C_{16} \end{pmatrix}$$

$$= (\pi_1^*)^{-1}$$

  c) $\pi_3^* r_4^* =$

$$\begin{pmatrix} C_1 & C_2 & C_3 & C_4 & C_5 & C_6 & C_7 & C_8 & C_9 & C_{10} & C_{11} & C_{12} & C_{13} & C_{14} & C_{15} & C_{16} \\ C_1 & C_5 & C_4 & C_3 & C_2 & C_6 & C_9 & C_8 & C_7 & C_{11} & C_{10} & C_{13} & C_{12} & C_{15} & C_{14} & C_{16} \end{pmatrix}$$

$$= (\pi_3 r_4)^*$$

3. a) $o(\alpha) = 7$; $o(\beta) = 12$; $o(\gamma) = 3$; $o(\delta) = 6$.

  b)   Let $\alpha \in S_n$, with $\alpha = c_1 c_2 \ldots c_k$, a product of disjoint cycles. Then $o(\alpha)$ is the l.c.m. of $\ell(c_1), \ell(c_2), \ldots, \ell(c_k)$, where $\ell(c_i) = $ length of c_i, $1 \leq i \leq k$.

5. a)   8   b)   39

7. a)   70   b)   90

9. Triangular Figure: a)   8   b)   8
    Square Figure: a)   12   b)   12

11. a)   140   b)   102

13. 315

Section 12.10 — p. 377

1. a)   165   b)   120

3. (Triangular Figure):   a)   96   b)   80
    (Square Figure):   a)   280   b)   220
    (Hexagonal Figure):   a)   131,584   c)   70,144

5. a)   2635   b)   1505

  c)

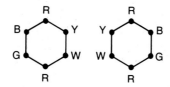

7. a)   21   b)   954

c) No, $k = 21$, $m = 21$, so $km = 441 \neq 954 = n$. Here the location of a certain edge must be considered relative to the location of the vertices. For example,

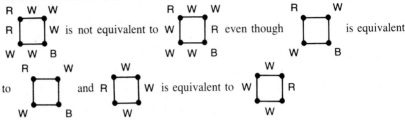

Section 12.11 — p. 382

1. a) i) and ii) $r^4 + w^4 + r^3w + 2r^2w^2 + rw^3$

   b) i) $(1/4)\,[(r + b + w)^4 + 2(r^4 + b^4 + w^4) + (r^2 + b^2 + w^2)^2]$

   ii) $(1/8)\,[(r + b + w)^4 + 2(r^4 + b^4 + w^4) + 3(r^2 + b^2 + w^2)^2$
   $+ 2(r + b + w)^2(r^2 + b^2 + w^2)]$

3. a) 10

   b) $(1/24)\,[(r + w)^6 + 6(r + w)^2(r^4 + w^4) + 3(r + w)^2(r^2 + w^2)^2 + 6(r^2 + w^2)^3$
   $+ 8(r^3 + w^3)^2]$

   c) 2

5. Let g = green and y = gold.

   (Triangular Figure): $(1/6)\,[(g + y)^4 + 2(g + y)(g^3 + y^3) + 3(g + y)^2(g^2 + y^2)]$

   (Square Figure): $(1/8)\,[(g + y)^5 + 2(g + y)(g^4 + y^4) + 3(g + y)(g^2 + y^2)^2$
   $+ 2(g + y)^3(g^2 + y^2)]$

   (Hexagonal Figure): $(1/4)\,[(g + y)^9 + 2(g + y)(g^2 + y^2)^4 + (g + y)^5(g^2 + y^2)^2]$

7. a) 136

   b) $(1/2)\,[(r + w)^8 + (r^2 + w^2)^4]$

   c) 38; 16

9. $\binom{m+n-1}{n}$

Miscellaneous Exercises — p. 386

1. a) Since $f(e_G) = e_H$, $e_G \in K$ and $K \neq \emptyset$. If $x, y \in K$, then $f(x) = f(y) = e_H$ and $f(xy) = f(x)f(y) = e_He_H = e_H$, so $xy \in K$. Also, for $x \in K$, $f(x^{-1}) = [f(x)]^{-1} = e_H^{-1} = e_H$, so $x^{-1} \in K$. Hence K is a subgroup of G.

   b) If $x \in K$, then $f(x) = e_H$. For all $g \in G$,

   $$f(gxg^{-1}) = f(g)f(x)f(g^{-1}) = f(g)e_Hf(g^{-1}) = f(g)f(g^{-1}) = f(gg^{-1}) = f(e_G) = e_H.$$

   Hence, for all $x \in K$, $g \in G$, $gxg^{-1} \in K$.

3. Let $a, b \in G$. Then $a^2b^2 = ee = e = (ab)^2 = abab$. $a^2b^2 = abab \Rightarrow aabb = abab \Rightarrow ab = ba$, so G is abelian.

5. If T is a subgroup of H, then $e_H \in T$ and $f(e_G) = e_H$, so $e_G \in f^{-1}(T)$. Hence $f^{-1}(T) \neq \emptyset$. If $x, y \in f^{-1}(T)$, then $f(x), f(y) \in T$. T a subgroup of $H \Rightarrow f(x)f(y) = f(xy) \in T \Rightarrow xy \in f^{-1}(T)$. Also, $x \in f^{-1}(T) \Rightarrow f(x) \in T \Rightarrow [f(x)]^{-1} \in T \Rightarrow f(x^{-1}) \in T \Rightarrow x^{-1} \in f^{-1}(T)$. Consequently, $f^{-1}(T)$ is a subgroup of G.

7. a) $d(a + c, b + c) = wt(a + c + b + c) = wt(a + b + (c + c))$
$$= wt(a + b + \mathbf{0}) = wt(a + b) = d(a, b).$$

   b) Let c_1, $c_2 \in C$ with $d(a, c_1)$, $d(b, c_2)$ minimal.
   $d(a, c_1) = wt(a + c_1) = wt(b + a + b + c_1) = d(b, a + b + c_1) \geq d(b, c_2)$.
   $d(b, c_2) = wt(b + c_2) = wt(a + b + a + c_2) = d(a, b + a + c_2) \geq d(a, c_1)$.
   Hence $d(a, C) = d(a, c_1) = d(b, c_2) = d(b, C)$.

9. b) 20

11. a) i) $\binom{4}{2}(0.75)^2(0.25)^2$

   ii) $\binom{4}{2}(0.75)^2(0.25)^2 + \binom{4}{3}(0.75)^3(0.25) + \binom{4}{4}(0.75)^4$

   b) At least 3.

Chapter 13

Section 13.1 — p. 394

1. $(10)(11)^2$; $(10)(11)^3$; $(10)(11)^4$; $(10)(11)^n$

5. a) and b) $f(x) = (x^2 + 4)(x - 2)(x + 2)$; the roots are ± 2.

   c) $f(x) = (x + 2i)(x - 2i)(x - 2)(x + 2)$; the roots are ± 2, $\pm 2i$.

   d) a) $f(x) = (x^2 - 5)(x^2 + 5)$; no rational roots.

      b) $f(x) = (x - \sqrt{5})(x + \sqrt{5})(x^2 + 5)$; the roots are $\pm\sqrt{5}$.

      c) $f(x) = (x - \sqrt{5})(x + \sqrt{5})(x - \sqrt{5}i)(x + \sqrt{5}i)$;
         the roots are $\pm\sqrt{5}$, $\pm i\sqrt{5}$.

7. 8060   9. 4; 6; $p - 1$

Section 13.2 — p. 401

1. a) $x^2 + 3x - 1$ is irreducible over $\mathbf{Q}$. Over $\mathbf{R}$, $\mathbf{C}$,
$$x^2 + 3x - 1 = [x - ((-3 + \sqrt{13})/2)][x - ((-3 - \sqrt{13})/2)].$$

   b) $x^4 - 2$ is irreducible over $\mathbf{Q}$.
   Over $\mathbf{R}$, $x^4 - 2 = (x - \sqrt[4]{2})(x + \sqrt[4]{2})(x^2 + \sqrt{2})$;
   $x^4 - 2 = (x - \sqrt[4]{2})(x + \sqrt[4]{2})(x - \sqrt[4]{2}i)(x + \sqrt[4]{2}i)$ over $\mathbf{C}$.

   c) $x^2 + x + 1 = (x + 2)(x + 2)$ over $\mathbf{Z}_3$. Over $\mathbf{Z}_5$ $x^2 + x + 1$ is irreducible;
   $x^2 + x + 1 = (x + 5)(x + 3)$ over $\mathbf{Z}_7$.

   d) $x^4 + x^3 + 1$ is irreducible over $\mathbf{Z}_2$.

3. 7^5

5. a) Yes, since the coefficients of the polynomials are from a field.

   b) $h(x) \mid f(x)$, $g(x) \Rightarrow f(x) = h(x)u(x)$, $g(x) = h(x)v(x)$, for $u(x)$, $v(x) \in F[x]$.
   $m(x) = s(x)f(x) + t(x)g(x)$ for some $s(x)$, $t(x) \in F[x]$, so
   $m(x) = h(x)[s(x)u(x) + t(x)v(x)]$ and $h(x) \mid m(x)$.

   c) If $m(x) \nmid f(x)$, then $f(x) = q(x)m(x) + r(x)$ where $0 \leq \deg r(x) < \deg m(x)$.
   $m(x) = s(x)f(x) + t(x)g(x)$ so $r(x) = f(x) - q(x)[s(x)f(x) + t(x)g(x)]$
   $= (1 - q(x))s(x)f(x) - q(x)t(x)g(x)$, so $r(x) \in S$.
   With $\deg r(x) < \deg m(x)$ we contradict the choice of $m(x)$. Hence $r(x) = 0$ and
   $m(x) \mid f(x)$.

7. a) The g.c.d. is $(x - 1) = (1/17)(x^5 - x^4 + x^3 + x^2 - x - 1)$
$$- (1/17)(x^2 + x - 2)(x^3 - 2x^2 + 5x - 8)$$

b) The g.c.d. is $1 = (x + 1)(x^4 + x^3 + 1) + (x^3 + x^2 + x)(x^2 + x + 1)$

c) The g.c.d. is $x^2 + 2x + 1 = (x^4 + 2x^2 + 2x + 2)$
$$+ (x + 2)(2x^3 + 2x^2 + x + 1)$$

9. a) $f(x) \equiv f_1(x) \pmod{s(x)} \Rightarrow f(x) = f_1(x) + h(x)s(x)$;
$g(x) \equiv g_1(x) \pmod{s(x)} \Rightarrow g(x) = g_1(x) + k(x)s(x)$.
Hence $f(x) + g(x) = f_1(x) + g_1(x) + (h(x) + k(x))s(x)$,
so $f(x) + g(x) \equiv f_1(x) + g_1(x) \pmod{s(x)}$,
and $f(x)g(x) = f_1(x)g_1(x) + (f_1(x)k(x) + g_1(x)h(x) + h(x)k(x)s(x))s(x)$,
so $f(x)g(x) \equiv f_1(x)g_1(x) \pmod{s(x)}$.

b) These properties follow from the corresponding properties for $F[x]$. For example, for the distributive law
$$[f(x)]([g(x)] + [h(x)]) = [f(x)][g(x) + h(x)] = [f(x)(g(x) + h(x))]$$
$$= [f(x)g(x) + f(x)h(x)] = [f(x)g(x)] + [f(x)h(x)]$$
$$= [f(x)][g(x)] + [f(x)][h(x)].$$

c) If not, there exists $g(x) \in F[x]$ where deg $g(x) > 0$ and $g(x) \mid f(x), s(x)$. But then $s(x)$ would be reducible.

d) A nonzero element of $F[x]/(s(x))$ has the form $[f(x)]$ where $f(x) \neq 0$ and deg $f(x) <$ deg $s(x)$. With $f(x), s(x)$ relatively prime, there exist $r(x), t(x)$ with $1 = f(x)r(x) + s(x)t(x)$, so $1 \equiv f(x)r(x) \pmod{s(x)}$ or $[1] = [f(x)][r(x)]$. Hence $[r(x)] = [f(x)]^{-1}$.

11. a) $[2x + 1]$ b) $[x + 1]$
    c) $[2x + 1]$ d) $[2x]$

13. a) p^n b) $\phi(p^n - 1)$

15. 101, 103, 107, 109, 113, 121, 125, 127, 128, 131, 137, 139, 149.

Section 13.3 — p. 406

1. a)
1	2	3	4
2	1	4	3
4	3	2	1
3	4	1	2

b)
1	2	3	4
3	4	1	2
2	1	4	3
4	3	2	1

c)
1	4	2	3
2	3	1	4
4	1	3	2
3	2	4	1

3. $a_{ri}^{(k)} = a_{rj}^{(k)} \Rightarrow f_k f_r + f_i = f_k f_r + f_j \Rightarrow f_i = f_j \Rightarrow i = j$

5. a)
2	3	4	5	1
3	4	5	1	2
4	5	1	2	3
5	1	2	3	4
1	2	3	4	5

b)
3	4	5	1	2
5	1	2	3	4
2	3	4	5	1
4	5	1	2	3
1	2	3	4	5

c)
4	5	1	2	3
2	3	4	5	1
5	1	2	3	4
3	4	5	1	2
1	2	3	4	5

d)
5	1	2	3	4
4	5	1	2	3
3	4	5	1	2
2	3	4	5	1
1	2	3	4	5

In standard form these become

a) 1 2 3 4 5   b) 1 2 3 4 5
    2 3 4 5 1   3 4 5 1 2
    3 4 5 1 2   5 1 2 3 4
    4 5 1 2 3   2 3 4 5 1
    5 1 2 3 4   4 5 1 2 3

c) 1 2 3 4 5   d) 1 2 3 4 5
    4 5 1 2 3   5 1 2 3 4
    2 3 4 5 1   4 5 1 2 3
    5 1 2 3 4   3 4 5 1 2
    3 4 5 1 2   2 3 4 5 1

7. Introduce a third factor such as four types of transmission fluid or four types of tires.

Section 13.4 — p. 412

1.

Field	Number of points	Number of lines	Number of points on a line	Number of lines on a point
$GF(5)$	25	30	5	6
$GF(3^2)$	81	90	9	10
$GF(7)$	49	56	7	8
$GF(2^4)$	256	272	16	17
$GF(31)$	961	992	31	32

3. There are nine points and 12 lines. These lines fall into four parallel classes.

 i) Slope of 0: $y = 0$; $y = 1$; $y = 2$

 ii) Infinite slope: $x = 0$; $x = 1$; $x = 2$

 iii) Slope 1: $y = x$; $y = x + 1$; $y = x + 2$

 iv) Slope 2 (as shown in the figure): (1) $y = 2x$ (2) $y = 2x + 1$ (3) $y = 2x + 2$

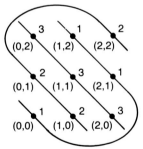

The Latin square corresponding to the fourth parallel class is

$$
\begin{array}{ccc}
3 & 1 & 2 \\
2 & 3 & 1 \\
1 & 2 & 3
\end{array}
$$

5. a) $y = 4x + 1$ b) $y = 3x + 10$ or $2x + 3y + 3 = 0$

   c) $y = 10x$ or $10y = 11x$

7. a) Vertical line: $x = c$. The line $y = mx + b$ intersects this vertical line at the unique point $(c, mc + b)$. As b takes on the values of F there are no two column entries (on the line $x = c$) that are the same.

   Horizontal line: $y = c$. The line $y = mx + b$ intersects this horizontal line at the unique point $(m^{-1}(c - b), c)$. As b takes on the values of F, no two row entries (on the line $y = c$) are the same.

   b) Let L_i be the Latin square for the parallel class of slope m_i, $i = 1, 2$, $m_i \neq 0$, m_i finite. If an ordered pair (j, k) appears more than once when L_1, L_2 are superimposed, then there are two pairs of lines: (1) $y = m_1 x + b_1$, $y = m_2 x + b_2$; and (2) $y = m_1 x + b_1'$, $y = m_2 x + b_2'$ which both intersect at (j, k). But then $b_1 = k - m_1 j = b_1'$ and $b_2 = k - m_2 j = b_2'$.

Section 13.5 — p. 417

1. $v = 9$, $b = 12$, $r = 4$, $k = 3$, $\lambda = 1$.

3. $\lambda = 2$

1	2	3	4		1	3	5	7		2	3	4	7		3	4	5	6
1	2	5	6		1	4	6	7		2	5	6	7					

5. a) No   b) No

7. a) $\lambda(v - 1) = r(k - 1) = 2r \Rightarrow \lambda(v - 1)$ is even.
   $\lambda v(v - 1) = vr(k - 1) = bk(k - 1) = b(3)(2) \Rightarrow 6 \mid \lambda v(v - 1)$.

   b) $(\lambda = 1)$ $6 \mid \lambda v(v - 1) \Rightarrow 6 \mid v(v - 1) \Rightarrow 3 \mid v(v - 1) \Rightarrow 3 \mid v$ or $3 \mid (v - 1)$.
   $\lambda(v - 1)$ even $\Rightarrow (v - 1)$ even $\Rightarrow v$ odd.
   $3 \mid v \Rightarrow v = 3t$, t odd $\Rightarrow v = 3(2s + 1) = 6s + 3$ and $v \equiv 3 \pmod 6$.
   $3 \mid (v - 1) \Rightarrow v - 1 = 3t$, t even $\Rightarrow v - 1 = 6x \Rightarrow v = 6x + 1$
   and $v \equiv 1 \pmod 6$

9. a) pm/n b) $[pm(m - 1)]/[n(n - 1)]$.

11. a) $v = b = 31$; $r = k = 6$; $\lambda = 1$

    b) $v = b = 57$; $r = k = 8$; $\lambda = 1$

    c) $v = b = 73$; $r = k = 9$; $\lambda = 1$

Miscellaneous Exercises — p. 420

1. $n = 9$

3. a) There is no real number r such that $r^2 + 1 = 0$, so $x^2 + 1$ is irreducible in $\mathbf{R}[x]$.

   b) Define $f: \mathbf{R}[x]/s(x) \to \mathbf{C}$ by $f(a + bx) = a + bi$. $a + bx = c + dx \Leftrightarrow a = c$, $b = d \Leftrightarrow a + bi = c + di$, so f is a one-to-one function. For any $a + bi \in \mathbf{C}$, with $a, b \in \mathbf{R}$, $a + bx \in \mathbf{R}[x]/(s(x))$ and $f(a + bx) = a + bi$, so f is onto. $f((a + bx) + (c + dx)) = f((a + c) + (b + d)x) = (a + c) + (b + d)i = (a + bi) + (c + di) = f(a + bx) + f(c + dx)$; $f((a + bx)(c + dx)) = f((ac + bdx^2) + (ad + bc)x) = f((ac - bd) + (ad + bc)x)$ (since $x^2 \equiv -1 \pmod{s(x)}$) $= (ac - bd) + (ad + bc)i = (a + bi)(c + di) = f(a + bx) \cdot f(c + dx)$. Hence f preserves addition and multiplication and is an isomorphism.

5. $\{1, 2, 4\}, \{2, 3, 5\}, \{4, 5, 7\}$

7. a) r 1's in each row; k 1's in each column.

b) $A \cdot J_b$ is a $v \times b$ matrix whose (i, j)-th entry is r, since there are r 1's in each row of A, and every entry in J_b is 1. Hence $A \cdot J_b = rJ_{v \times b}$. Likewise, $J_v \cdot A$ is a $v \times b$ matrix whose (i, j)-th entry is k, since there are k 1's in each column of A and every entry in J_v is 1. Hence $J_v \cdot A = k \cdot J_{v \times b}$.

c)   The (i, j) entry in $A \cdot A^{tr}$ is obtained from the componentwise multiplication of rows i and j of A. If $i = j$, this results in the number of 1's in row i, which is r. For $i \neq j$, the number of 1's is the number of times x_i and x_j appear in the same block—which is given by λ. Hence $A \cdot A^{tr} = (r - \lambda)I_v + \lambda J_v$.

d)

$$\begin{vmatrix} r & \lambda & \lambda & \lambda & \ldots & \lambda \\ \lambda & r & \lambda & \lambda & \ldots & \lambda \\ \lambda & \lambda & r & \lambda & \ldots & \lambda \\ \lambda & \lambda & \lambda & r & \ldots & \lambda \\ \ldots & \ldots & \ldots & \ldots & & \ldots \\ \lambda & \lambda & \lambda & \lambda & \ldots & r \end{vmatrix}$$

$$\overset{(1)}{=} \begin{vmatrix} r & \lambda - r & \lambda - r & \lambda - r & \ldots & \lambda - r \\ \lambda & r - \lambda & 0 & 0 & \ldots & 0 \\ \lambda & 0 & r - \lambda & 0 & \ldots & 0 \\ \lambda & 0 & 0 & r - \lambda & \ldots & 0 \\ \ldots & \ldots & \ldots & \ldots & \ldots & \ldots \\ \lambda & 0 & 0 & 0 & \ldots & r - \lambda \end{vmatrix}$$

$$\overset{(2)}{=} \begin{vmatrix} r + (v - 1)\lambda & 0 & 0 & 0 & \ldots & 0 \\ \lambda & r - \lambda & 0 & 0 & \ldots & 0 \\ \lambda & 0 & r - \lambda & 0 & \ldots & 0 \\ \lambda & 0 & 0 & r - \lambda & \ldots & 0 \\ \ldots & \ldots & \ldots & \ldots & \ldots & \ldots \\ \lambda & 0 & 0 & 0 & \ldots & r - \lambda \end{vmatrix}$$

$$= [r + (v - 1)\lambda](r - \lambda)^{v-1} = (r - \lambda)^{v-1}[r + r(k - 1)] = rk(r - \lambda)^{v-1}$$

((1) Multiply column 1 by -1 and add it to the other $v - 1$ columns.
(2) Add rows 2 through v to row 1.)

Chapter 14

Section 14.1—p. 428

1. a)   To represent the air routes traveled among a certain set of cities by a particular airline.

   b)   To represent an electrical network. Here the vertices can represent switches, transistors, etc., and an edge (x, y) indicates the existence of a wire connecting x to y.

   c)   Let the vertices represent a set of job applicants and a set of open positions in a corporation. Draw an edge (A, b) to denote that applicant A is qualified for position b. Then all open positions can be filled if the resulting graph provides a matching between the applicants and open positions.

3. 6

5. a)

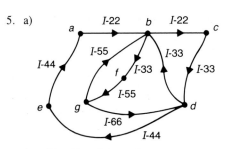

b) $\{(g,d),(d,e),(e,a)\}$;
$\{(g,b),(b,c),(c,d),(d,e),(e,a)\}$.

c) Two: one of $\{(b,c),(c,d)\}$ and one of $\{(b,f),(f,g),(g,d)\}$.

d) No

e) Yes: Travel the simple path
$\{(c,d),(d,e),(e,a),(a,b),(b,f),(f,g)\}$.

f) Yes. Travel the path
$\{(g,b),(b,f),(f,g),(g,d),(d,b),(b,c),(c,d),(d,e),(e,a),(a,b)\}$.

7. If $\{a,b\}$ is not part of a cycle, then its removal disconnects a and b (and G). If not, there is a path P from a to b and P together with $\{a,b\}$ provides a cycle containing $\{a,b\}$. Conversely, if the removal of $\{a,b\}$ from G disconnects G, then there exist $x,y \in V$ such that the only path P from x to y contains $e = \{a,b\}$. If e were part of a cycle C, then the edges in $(P - \{e\}) \cup (C - \{e\})$ would provide a second path connecting x to y.

9. a) Yes   b) No   c) $n-1$

11. The partition of V induced by $\Re$ yields the (connected) components of G.

Section 14.2 — p. 436

1. i)

	R		Y		W		B	
B	1	Y R	2	B Y	3	R W	4	W
	W		B		Y		R	
	W		B		Y		R	

ii) No solution.

iii)

R	1	W W	2	B Y	3	R B	4	Y
	Y		R		B		W	

3. (1) No   (2) No   (3) Yes. Correspond a with u, b with w, c with x, d with y, e with v and f with z.

5. $\binom{v}{2} - e$

7. a) If $G_1 = (V_1, E_1)$ and $G_2 = (V_2, E_2)$ are isomorphic, then there is a function $f: V_1 \to V_2$ that is one-to-one and onto and preserves adjacencies. If $x, y \in V_1$ and $\{x,y\} \notin E_1$, then $\{f(x), f(y)\} \notin E_2$. Hence the same function f preserves adjacencies for $\overline{G_1}$, $\overline{G_2}$ and can be used to define an isomorphism for $\overline{G_1}$, $\overline{G_2}$. The converse follows in a similar way.

b) They are not isomorphic. The complement of the graph containing vertex a is a (simple) cycle of length 8. The complement of the other graph is the disjoint union of two (simple) cycles of length 4.

9.

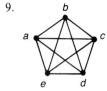

If G is the simple cycle with edges $\{a,b\}$, $\{b,c\}$, $\{c,d\}$, $\{d,e\}$, and $\{e,a\}$, then $\overline{G}$ is the simple cycle with edges $\{a,c\}$, $\{c,e\}$, $\{e,b\}$, $\{b,d\}$, $\{d,a\}$. Hence, G and $\overline{G}$ are isomorphic. Conversely, if G is a simple cycle on n vertices and G, $\overline{G}$ are isomorphic, then $n = \frac{1}{2}\binom{n}{2}$, or $n = \frac{1}{4}(n)(n-1)$, and $n = 5$.

11. a)   Here f must also maintain directions. So $(a, b) \in E_1$ iff $(f(a), f(b)) \in E_2$.

    b)

    c)   They are not isomorphic. Consider vertex a in the first graph. It is incident to one
         vertex and incident from two other vertices. No vertex in the other graph has
         this property.

Section 14.3 — p. 442

1. a) $|V| = 6$;   b) $|V| = 1$ or 2 or 3 or 5 or 6 or 10 or 15 or 30.
    (In the first four cases G is a multigraph; when $|V| = 30$, G is disconnected.)

    c) $|V| = 6$

3. $\delta|V| \le \sum_{v \in V} \deg(v) \le \Delta|V|$. Since $2|E| = \sum_{v \in V} \deg(v)$, $\delta|V| \le 2|E| \le \Delta|V|$ so $\delta \le 2(e/n) \le \Delta$.

5. (Corollary 14.1). Let $V = V_1 \cup V_2$ where $V_1(V_2)$ contains all vertices of odd (even)
   degree. Then $2|E| - \sum_{v \in V_2} \deg(v) = \sum_{v \in V_1} \deg(v)$ is an even integer. For $|V_1|$ odd,
   $\sum_{v \in V_1} \deg(v)$ is odd.
        (Corollary 14.2). For the converse let $G = (V, E)$ have an Euler path with a, b as
   the starting and terminating vertices. Add the edge $\{a, b\}$ to G to form the larger graph
   $G_1 = (V, E_1)$ where G_1 has an Euler cycle. Hence G_1 is connected and each vertex in G_1
   has even degree. Removing edge $\{a, b\}$ from G_1, the vertices in G will have the same
   even degree except for a, b; $\deg_G(a) = \deg_{G_1}(a) - 1$, $\deg_G(b) = \deg_{G_1}(b) - 1$, so the
   vertices a, b have odd degree in G. Also, since the edges in G form an Euler path, G
   is connected.

7. n odd; $n = 2$ 9. Yes

11. a) $\sum_{v \in V} \deg^+(v) = |E| = \sum_{v \in V} \deg^-(v)$

    b)   From (a),

    $$\sum_{v \in V} [\deg^+(v) - \deg^-(v)] = 0.$$

    For each $v \in V$, $\deg^+(v) + \deg^-(v) = n - 1$, so

    $$0 = (n - 1) \cdot 0 = \sum_{v \in V} (n - 1)[\deg^+(v) - \deg^-(v)]$$

    $$= \sum_{v \in V} [\deg^+(v) + \deg^-(v)][\deg^+(v) - \deg^-(v)]$$

    $$= \sum_{v \in V} [(\deg^+(v))^2 - (\deg^-(v))^2],$$

    and the result follows.

13. a) and b)

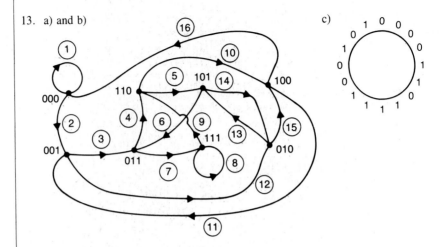

c)

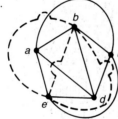

Section 14.4 — p. 455

1.

In this situation vertex b is in the region formed by the edges $\{a, d\}, \{d, c\}, \{c, a\}$ and vertex e is outside of this region. Hence the edge $\{b, e\}$ will cross one of the edges $\{a, d\}$, $\{d, c\}$, or $\{a, c\}$ (as shown).

3. a)

Graph	Number of vertices	Number of edges
$K_{4,7}$	11	28
$K_{7,11}$	18	77
$K_{m,n}$	$m + n$	mn

b) $m = 6$

5. Partition V as $V_1 \cup V_2$ with $|V_1| = m$, $|V_2| = v - m$. Since G is bipartite the maximum number of edges that G can have is $m(v - m) = -[m - (v/2)]^2 + (v/2)^2$, a function of m. For a given value of v, when v is even, $m = v/2$ maximizes $m(v - m) = (v/2)[v - (v/2)] = (v/2)^2$. For v odd, $m = (v - 1)/2$ or $m = (v + 1)/2$ maximizes $m(v - m) = [(v - 1)/2][v - ((v - 1)/2)] = [(v - 1)/2][(v + 1)/2] = [(v + 1)/2][v - ((v + 1)/2)] = (v^2 - 1)/4 = \lfloor (v/2)^2 \rfloor < (v/2)^2$. Hence if $|E| > (v/2)^2$, G cannot be bipartite.

7. 10

9. If not, $\deg(v) \geq 6$ for all $v \in V$. Then $2e = \sum_{v \in V} \deg(v) \geq 6|V|$ so $e \geq 3|V|$ contradicting $e \leq 3|V| - 6$ (Corollary 14.3).

11. a) $2e \geq kr = k(2 + e - v) \Rightarrow (2 - k)e \geq k(2 - v) \Rightarrow e \leq [k/(k - 2)](v - 2)$.

b) 4

c) In $K_{3,3}$, $e = 9$, $v = 6$. $[k/(k - 2)](v - 2) = (4/2)(4) = 8 < 9 = e$. Since $K_{3,3}$ is connected, it must be nonplanar.

d)   Here $k = 5$, $v = 10$, $e = 15$ and $[k/(k - 2)](v - 2) = (5/3)(8) = (40/3) <$
     $15 = e$. Since the Petersen graph is connected, it must be nonplanar.

13. The dual for the tetrahedron (Fig. 14.38(b)) is the graph itself. For the graph (cube) in
     Fig. 14.38(d) the dual is the octahedron, and vice versa. Likewise, the dual of the
     dodecahedron is the icosahedron, and vice versa.

15. i) $\{\{k, m\}, \{m, s\}\}$ ii) $\{\{j, k\}, \{k, s\}, \{m, s\}\}$

17.

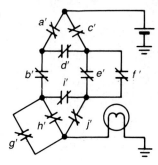

Section 14.5 — p. 463

1. a)   b)   c)   d)

3. a) This graph has no Hamilton cycle but it has a Hamilton path: $a \rightarrow b \rightarrow c \rightarrow d \rightarrow$
       $e \rightarrow f \rightarrow j \rightarrow i \rightarrow h \rightarrow g \rightarrow k$.

   b) Hamilton cycle: $a \rightarrow d \rightarrow b \rightarrow e \rightarrow g \rightarrow j \rightarrow i \rightarrow f \rightarrow h \rightarrow c \rightarrow a$

   c) Hamilton cycle: $a \rightarrow h \rightarrow e \rightarrow f \rightarrow g \rightarrow i \rightarrow d \rightarrow c \rightarrow b \rightarrow a$

5. a) $(1/2)(n - 1)!$ b)   10   c)   9

7. Let $G = (V, E)$ be a simple path with $|V| \geq 4$.

9. For all $x, y \in V$, $\deg(x) + \deg(y) \geq 2[(n - 1)/2] = n - 1$, so the result follows from
   Theorem 14.7.

11. a) i) $\{a, c, f, h\}$, $\{a, g\}$;   ii) $\{z\}$, $\{u, w, y\}$

    b) i) $\beta(G) = 4$;   ii) $\beta(G) = 3$

    c) The complete graph on $|I|$ vertices.

Section 14.6 — p. 471

1. Draw a vertex for each species of fish. If two species x, y must be kept in separate
   aquaria, draw the edge $\{x, y\}$. The smallest number of aquaria needed is then the chromatic number of the resulting graph.

3. a) 2   b)   2 (n even); 3 (n odd)

   c) Figure 14.38(d): 2; Fig. 14.41: 3; Fig. 14.61(a): 2; Fig. 14.61(b): 3.

5. a) 1) $\lambda(\lambda - 1)^2(\lambda - 2)^2$;

       2) $\lambda(\lambda - 1)(\lambda - 2)(\lambda^2 - 2\lambda + 2)$;

       3) $\lambda(\lambda - 1)(\lambda - 2)(\lambda^2 - 5\lambda + 7)$

b) 1) 3;   2) 3;   3) 3
c) 1) 720;   2) 1020;   3) 420

7. a) $\lambda(\lambda - 1)(\lambda - 2)$ b) Follows from Theorem 14.8.

c) Follows by the rule of product.

d)
$$P(C_n, \lambda) = P(L_n, \lambda) - P(C_{n-1}, \lambda) = \lambda(\lambda - 1)^{n-1} - P(C_{n-1}, \lambda)$$
$$= [(\lambda - 1) + 1](\lambda - 1)^{n-1} - P(C_{n-1}, \lambda)$$
$$= (\lambda - 1)^n + (\lambda - 1)^{n-1} - P(C_{n-1}, \lambda) \Rightarrow P(C_n, \lambda) - (\lambda - 1)^n$$
$$= (\lambda - 1)^{n-1} - P(C_{n-1}, \lambda).$$

Replacing n by $n - 1$ yields
$$P(C_{n-1}, \lambda) - (\lambda - 1)^{n-1} = (\lambda - 1)^{n-2} - P(C_{n-2}, \lambda).$$

Hence
$$P(C_n, \lambda) - (\lambda - 1)^n = P(C_{n-2}, \lambda) - (\lambda - 1)^{n-2}.$$

e) Continuing from (d),
$$P(C_n, \lambda) = (\lambda - 1)^n + (-1)^{n-3}[P(C_3, \lambda) - (\lambda - 1)^3]$$
$$= (\lambda - 1)^n + (-1)^{n-1}[\lambda(\lambda - 1)(\lambda - 2) - (\lambda - 1)^3]$$
$$= (\lambda - 1)^n + (-1)^n(\lambda - 1).$$

9. From Theorem 14.11, the expansion for $P(G, \lambda)$ will contain exactly one occurrence of the chromatic polynomial of K_n. Since no larger graph occurs, this term determines the degree as n and the leading coefficient as 1.

11. $\lambda(\lambda - 1)(\lambda - 2)^2(\lambda^2 - 3\lambda + 3)$.

Miscellaneous Exercises — p. 476

1. $n = 17$

3. a) For $v, w \in V$, if v, w differ in k components then there is a path of length k from v to w obtained by starting with v and changing one component at a time until vertex w is reached.

b) $|V| = 2^n$; $|E| = (n)(2^{n-1})$.

5. a) Label the vertices of K_6 with $a, b, \ldots, f$. Of the five edges on a, at least three have the same color, say red. Let these edges be $\{a, b\}, \{a, c\}, \{a, d\}$. If the edges $\{b, c\}$, $\{c, d\}, \{b, d\}$ are all blue, the result follows. If not, one of these edges, say $\{c, d\}$, is red. Then the edges $\{a, c\}, \{a, d\}, \{c, d\}$ yield a red triangle.

b) Consider the six people as vertices. If two people are friends (strangers) draw a red (blue) edge connecting their respective vertices. The result then follows from (a).

7. a) Let I be independent and $\{a, b\} \in E$. If neither a nor b is in $V - I$, then $a, b \in I$, and since they are adjacent, I is not independent. Conversely, if $I \subseteq V$ with $V - I$ a covering of G, then if I is not independent there are vertices $x, y \in I$ with $\{x, y\} \in E$. But $\{x, y\} \in E \Rightarrow$ either x or y is in $V - I$.

b) Let I be a largest maximal independent set in G and K a minimum covering. From (a), $|K| \le |V - I| = |V| - |I|$ and $|I| \ge |V - K| = |V| - |K|$, or $|K| + |I| \ge |V| \ge |K| + |I|$.

9. a) $\alpha(G) = 2$; $\beta(G) = 3$; $\gamma(G) = 4$.

    b) G has neither an Euler path nor an Euler cycle; G does have a Hamiltonian cycle.

    c) G is not bipartite but it is planar.

11. a) $\gamma(G) \geq \text{cl}(G)$.   b) $\text{cl}(G) = \max\{\text{cl}(C_i) \mid 1 \leq i \leq k\}$.

    c) No, if $\text{cl}(G) = 8$ then G has at least $\binom{8}{2} = 28$ edges.

    d) They are equal.

13. a) i) $m = 2$, $n = 8$ ii) $m = n = 4$

    b) i) $K_{m,n}$, for $m \leq n$, has an Euler cycle but not a Hamilton cycle if m and n are both even and $m \neq n$.

        ii) When m, n are both even and $m = n$, then $K_{m,n}$ has both an Euler and a Hamilton cycle.

Chapter 15

Section 15.1 — p. 484

1. a)

    b) 5

3. a) $e = v - \omega$ b) $\omega - 1$

5. If there is a unique path between any pair of vertices in G, then G is connected. If G contains a cycle, then there is a pair of vertices x, y with two distinct paths connecting x and y. Hence G is a tree.

7. a) If the complement of T contains a cut set, then the removal of these edges disconnects G and there are vertices x, y with no path connecting them. Hence T is not a spanning tree for G.

    b) If the complement of C contains a spanning tree, then every pair of vertices in G has a path connecting them and this path includes no edges of C. Hence the removal of the edges in C from G does not disconnect G, so C is not a cut set for G.

9. a) i) 3, 4, 6, 3, 8, 4   ii) 3, 4, 6, 6, 8, 4

    b) No pendant vertex of the given tree appears in the sequence so the result is true for these vertices. When an edge $\{x, y\}$ is removed and y is a pendant vertex (of the tree or one of the resulting subtrees), then the $\deg(x)$ is decreased by 1 and x is placed in the sequence. As the process continues either (i) this vertex x becomes a pendant vertex in a subtree and is removed but not recorded again in the sequence or, (ii) the vertex x is left as one of the last two vertices of an edge. In either case x has been listed in the sequence $(\deg(x) - 1)$ times.

    c)

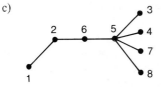

    d) From the given sequence the degree of each vertex in the tree is known.

        Step 1: Set the counter i to 1.

Step 2: From among the vertices of degree 1, select the vertex v with the smallest label. This determines the edge $\{v, x_i\}$. Remove v from the set of labels and reduce the degree of x_i by 1.

Step 3: If $i < n - 2$, increase i by 1 and return to step 2.

Step 4: If $i = n - 2$, the vertices (labels) x_{n-3}, x_{n-2} are connected by an edge if $x_{n-3} \neq x_{n-2}$. (The tree is then complete.)

Section 15.2 — p. 499

1. a) f, h, k, p, q, s, t b) a c) d

   d) e, f, j, q, s, t e) q, t f) 2

   g) k, p, q, s, t h) 4   i) No

3. a) $/ + w - xy * \pi \uparrow z3$ b) 0.4

   c) 1)   2)

5.

7. a) Any depth-first search spanning tree for K_6 is a simple path of length 5.

   b) i) 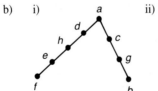   ii)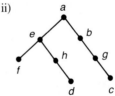

     iii) $f \to e \to h \to d \to a \to c \to g \to b$

   c) i) 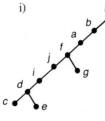   ii) 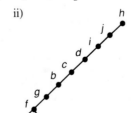   iii)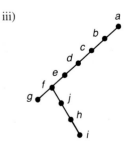

9. *G* is connected.

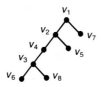

11. Theorem 15.5.

    a) Each internal vertex has *m* sons, so there are *mi* vertices that are the sons of some other vertex. This accounts for all vertices in the tree except the root. Hence $n = mi + 1$.

    b) $\ell + i = n = mi + 1 \Rightarrow \ell = (m - 1)i + 1$

    c) $\ell = (m - 1)i + 1 \Rightarrow i = (\ell - 1)/(m - 1)$
        $n = mi + 1 \Rightarrow i = (n - 1)/m$.
        (Corollary 15.1)

        Since the tree is balanced $m^{h-1} < \ell \le m^h$ by Theorem 15.6.
        $$m^{h-1} < \ell \le m^h \Rightarrow \log_m(m^{h-1}) < \log_m(\ell) \le \log_m(m^h)$$
        $$\Rightarrow (h - 1) < \log_m \ell \le h \Rightarrow h = \lceil \log_m \ell \rceil.$$

13. a)   b) 9   c) $h(m - 1)$

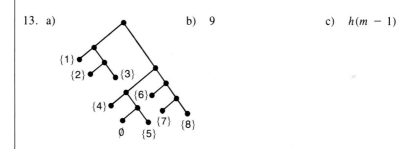

15. 21845; $1 + m + m^2 + \cdots + m^{h-1} = (m^h - 1)/(m - 1)$.

17.

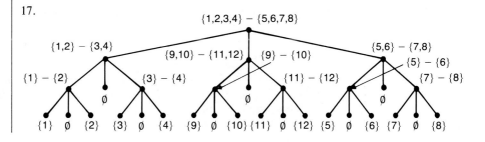

19. a)

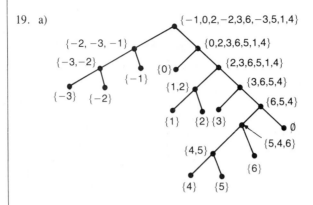

$\{-1,0,2,-2,3,6,-3,5,1,4\}$

$\{-2,-3,-1\}$ $\{0,2,3,6,5,1,4\}$

$\{-3,-2\}$ $\{2,3,6,5,1,4\}$

$\{0\}$

$\{-1\}$ $\{3,6,5,4\}$

$\{1,2\}$

$\{-3\}$ $\{-2\}$ $\{6,5,4\}$

$\{1\}$ $\{2\}\{3\}$ $\emptyset$

$\{4,5\}$ $\{5,4,6\}$

$\{6\}$

$\{4\}$ $\{5\}$

Section 15.3 — p. 508

1. a) tear   b) tatener   c) rant

3. a: 001   e: 01   h: 101
    b: 000001   f: 1001   i: 11
    c: 1000   g: 00001   j: 000000
    d: 0001

Section 15.4 — p. 512

1. The articulation points are b, e, f, h, j, k. The biconnected components are B_1: $\{\{a,b\}\}$; B_2: $\{\{d,e\}\}$; $B_3 = \{\{b,c\},\{c,f\},\{f,e\},\{e,b\}\}$; B_4: $\{\{f,g\},\{g,h\},\{h,f\}\}$; B_5: $\{\{h,i\},\{i,j\}, \{j,h\}\}$; B_6: $\{\{j,k\}\}$; B_7: $\{\{k,p\},\{p,n\},\{n,m\},\{m,k\},\{p,m\}\}$,

3. a) T can have as few as one or as many as $n-2$ articulation points. If T contains a vertex of degree $(n-1)$, then this vertex is the only articulation point. If T is a simple path with n vertices and $n-1$ edges, then the $n-2$ vertices of degree 2 are all articulation points.

    b) In all cases, a tree on n vertices has $n-1$ biconnected components. Each edge is a biconnected component.

5. $\gamma(G) = \max\{\gamma(B_i) \mid 1 \le i \le k\}$.

7. No G: B_1 G': B_1' B_2' B_3'

    B_3 B_2

9. We always have low (y_2) = low (y_1) = 1. (Note: Vertices y_2 and y_1 are always in the same biconnected component.)

Miscellaneous Exercises — p. 515

1. a) If G is a tree, consider G as a rooted tree. Then there are λ choices for coloring the root of G and $(\lambda-1)$ choices for coloring each of its descendants. The result then follows by the rule of product.

    Conversely, if $P(G,\lambda) = \lambda(\lambda-1)^{n-1}$, then since the factor λ only occurs once, the graph G is connected.

$$P(G, \lambda) = \lambda(\lambda - 1)^{n-1}$$

$$= \lambda^n - (n - 1)\lambda^{n-1} + \cdots + (-1)^{n-1}\lambda \Rightarrow G \text{ has } n \text{ vertices}$$

and $(n - 1)$ edges. Hence G is a tree. (By Exercise 8 of Section 15.1)

b) From (a), $P(G, 1) = 0$ while $P(G, 2) = 2 > 0$, so $\gamma(G) = 2$.

c) For any graph $G = (V, E)$, if $e \in E$, then $P(G, \lambda) = P(G_e, \lambda) - P(G'_e, \lambda)$, so $P(G, \lambda) \le P(G_e, \lambda)$. If G is connected but not a tree, delete an edge from each simple cycle of G until the subgraph obtained is a spanning tree T of G. Then $P(G, \lambda) \le P(T, \lambda) = \lambda(\lambda - 1)^{n-1}$. [Recall that $P(G, \lambda)$ is the number of ways to properly color the vertices in G with λ colors. Hence $P(G, \lambda) \le P(T, \lambda)$ is interpreted here as an inequality for two numbers, not two polynomials.]

3. a) 10

   b) $0, a_1, a_1.a_2, a_1.a_2.a_3, \ldots, a_1.a_2.a_3. \ldots .a_8.a_9.a_{10}$

5. If H is a Hamilton path in G, then H is connected, contains every vertex of G, and contains no cycles. Hence H is a spanning tree for G. Since H is a path, $\deg(v)$ (in H) is 2 for all vertices on the path except the endpoints which have (in H) degree 1.

   Conversely, if H is a spanning tree for G, then if each vertex of H has degree (in H) at most 2, H is a simple path. Consequently H is a simple path containing all vertices of G, so H is a Hamilton path for G.

7. a) In going from level h to level $h + 1$ we add on the m sons of exactly one of the leaves at level h.

   b) $x_h = (m/2)(h^2 + h), h \ge 0$.

9. a) 1011001010100

   b)　i) 　　　　　ii)

   c) Since the last two vertices visited in a preorder traversal are leaves, the last two symbols in the characteristic sequence of any binary tree are 00.

11. a) 32; 31; 62

    b) $2^h; 2^h - 1; 2(2^h - 1)$.

Chapter 16

Section 16.1 — p. 522

1. Kruskal's algorithm generates the following sequence (of forests) which terminates in a minimal spanning tree T of weight 18:

   1) $F_1 = \{\{e, h\}\}$,　　　　2) $F_2 = F_1 \cup \{\{a, b\}\}$,　　3) $F_3 = F_2 \cup \{\{b, c\}\}$,

   4) $F_4 = F_3 \cup \{\{d, e\}\}$,　　5) $F_5 = F_4 \cup \{\{e, f\}\}$,　　6) $F_6 = F_5 \cup \{\{a, e\}\}$,

   7) $F_7 = F_6 \cup \{\{d, g\}\}$,　　8) $F_8 = T = F_7 \cup \{\{f, i\}\}$.

3. a) Evansville–Indianapolis (168); Bloomington–Indianapolis (51); South Bend–Gary (58); Terre Haute–Bloomington (58); South Bend–Fort Wayne (79); Indianapolis–Fort Wayne (121).

   b) Fort Wayne–Gary (132); Evansville–Indianapolis (168); Bloomington–Indianapolis (51); Gary–South Bend (58); Terre Haute–Bloomington (58); Indianapolis–Fort Wayne (121).

5. a) To determine an optimal tree of maximal weight, replace the two occurrences of "small" in Kruskal's algorithm by "large."

   b) Use the edges: South Bend–Evansville (303); Fort Wayne–Evansville (290); Gary–Evansville (277); Fort Wayne–Terre Haute (201); Gary–Bloomington (198); Indianapolis–Evansville (168).

7. When the weights of the edges are all distinct, in each step of Kruskal's Algorithm a unique edge is selected.

Section 16.2—p. 533

1. a) $s = 2; t = 4; w = 5; x = 9; y = 4$ b)   18

   c) i) $P = \{a, b, h, d, g, i\}; \overline{P} = \{z\}$

       ii) $P = \{a, b, h, d, g\}; \overline{P} = \{i, z\}$

       iii) $P = \{a, h\}; \overline{P} = \{b, d, g, i, z\}$

3. 1)

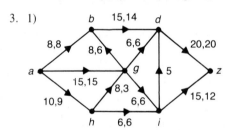

The maximal flow is 32,
which is $c(P, \overline{P})$ for
$P = \{a, b, d, g, h\}$ and $\overline{P} = \{i, z\}$.

2)

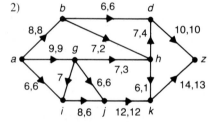

The maximal flow is 23,
which is $c(P, \overline{P})$ for
$P = \{a\}$ and $\overline{P} = \{b, g, i, j, d, h, k, z\}$.

5. Here $c(e)$ is a positive integer for each $e \in E$ and the initial flow is defined as $f(e) = 0$ for all $e \in E$. The result follows because in each application of the labeling procedure, increases in flow come about from the second component of a label, and this component is always a nonnegative integer.

7.

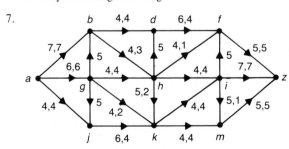

Section 16.3 — p. 542

1. $5/\binom{8}{4} = 1/14$

3. Let the committees be represented as $c_1, c_2, \ldots, c_6$, according to the way they are listed in the exercise.

   a)   Select the members as follows: $c_1 - A$; $c_2 - G$; $c_3 - M$; $c_4 - N$; $c_5 - K$; $c_6 - R$.

   b)   Select the nonmembers as follows: $c_1 - K$; $c_2 - A$; $c_3 - G$; $c_4 - J$; $c_5 - M$; $c_6 - P$.

5. Yes, such an assignment can be made by Brenda. Let X be the set of student applicants and Y the set of part-time jobs. Then for all $x \in X$, $y \in Y$, draw the edge (x, y) if applicant x is qualified for part-time job y. Then $\deg(x) \geq 4 \geq \deg(y)$ for all $x \in X$, $y \in Y$, and the result follows from Corollary 16.3.

7. a)   i) Select i from A_i, $1 \leq i \leq 4$.

          ii) Select $i + 1$ from A_i, $1 \leq i \leq 3$, and 1 from A_4.

   b)   2

9. These graduates cannot all be hired by these six corporations. Let X be the set of 42 graduates and Y the set of 42 open positions allowed under these hiring policies (seven positions per corporation). If set A consists of the ten graduates from the first university, only six edges can be drawn from A to Y. Hence $|A| = 10 > 6 = |R(A)|$, and by Theorem 16.5 there is no complete matching.

11. a) $\delta(G) = 1$. A maximal matching of X into Y is given by $\{\{x_1, y_4\}, \{x_2, y_2\}, \{x_3, y_1\}, \{x_5, y_3\}\}$.

   b)   If $\delta(G) = 0$, there is a complete matching of X into Y, and $\beta(G) = |Y|$, or $|Y| = \beta(G) - \delta(G)$. If $\delta(G) = k > 0$, let $A \subseteq X$ where $|A| - |R(A)| = k$. Then $A \cup (Y - R(A))$ is a largest maximal independent set in G and $\beta(G) = |A| + |Y - R(A)| = |Y| + (|A| - |R(A)|) = |Y| + \delta(G)$, so $|Y| = \beta(G) - \delta(G)$.

   c)   Fig. 16.23(a): $\{x_1, x_2, x_3, y_2, y_4, y_5\}$; Fig. 16.24: $\{x_3, x_4, y_2, y_3, y_4\}$.

Miscellaneous Exercises — p. 546

1. a)   The edge e_1 will always be selected in the first step of Kruskal's algorithm.

   b)   Again using Kruskal's algorithm, edge e_2 will be selected in the first application of step 2 unless each of the edges e_1, e_2 is incident with the same two vertices, i.e., the edges e_1, e_2 form a cycle and G is a multigraph.

3.

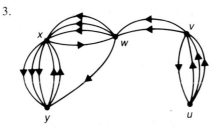

5. There are d_n, the number of derangements of $\{1, 2, 3, \ldots, n\}$.

Index

NOTATION

SPECIAL SETS OF NUMBERS	$\mathbf{Z}$	the set of integers: $\{0, 1, -1, 2, -2, 3, -3, \ldots\}$
	$\mathbf{N}$	the set of nonnegative integers or natural numbers: $\{0, 1, 2, 3, \ldots\}$
	$\mathbf{Z}^+$	the set of positive integers: $\{1, 2, 3, \ldots\} = \{x \in \mathbf{Z} \mid x > 0\}$
	$\mathbf{Q}$	the set of rational numbers: $\{a/b \mid a, b \in \mathbf{Z}, b \neq 0\}$
	$\mathbf{Q}^+$	the set of positive rational numbers
	$\mathbf{Q}^*$	the set of nonzero rational numbers
	$\mathbf{R}$	the set of real numbers
	$\mathbf{R}^+$	the set of positive real numbers
	$\mathbf{R}^*$	the set of nonzero real numbers
	$\mathbf{C}$	the set of complex numbers: $\{x + yi \mid x, y \in \mathbf{R}, i^2 = -1\}$
	$\mathbf{C}^*$	the set of nonzero complex numbers
	$\mathbf{Z}_n$	$\{0, 1, 2, \ldots, n - 1\}$, for $n \in \mathbf{Z}^+$
	$[a, b]$	the closed interval from a to b: $\{x \in \mathbf{R} \mid a \leq x \leq b\}$
	(a, b)	the open interval from a to b: $\{x \in \mathbf{R} \mid a < x < b\}$
	$[a, b)$	a half-open interval from a to b: $\{x \in \mathbf{R} \mid a \leq x < b\}$
	$(a, b]$	a half-open interval from a to b: $\{x \in \mathbf{R} \mid a < x \leq b\}$
ALGEBRAIC STRUCTURES	$(R, +, \cdot)$	R is a ring with binary operations $+$ and $\cdot$
	$R[x]$	the ring of polynomials over ring R
	$(G, \circ)$	G is a group under the binary operation $\circ$
	S_n	the symmetric group on n symbols
	aH	a left coset of subgroup H (in group G): $\{ah \mid h \in H\}$
	$(\mathcal{B}, +, \cdot, {}^-, 0, 1)$	the Boolean algebra $\mathcal{B}$ with binary operations $+$ and $\cdot$, the unary operation ${}^-$, and identity elements 0 (for $+$) and 1 (for $\cdot$)
GRAPH THEORY	$G = (V, E)$	G is a graph with vertex set V and edge set E
	K_n	the complete graph on n vertices
	$\overline{G}$	the complement of graph G
	$\deg(v)$	the degree of vertex v (in an undirected graph G)
	$\deg^+(v)$	the in degree of vertex v (in a directed graph G)
	$\deg^-(v)$	the out degree of vertex v (in a directed graph G)
	$\omega(G)$	the number of connected components of graph G
	$K_{m,n}$	the complete bipartite graph on $V = V_1 \cup V_2$ where $V_1 \cap V_2 = \emptyset$, $|V_1| = m$, $|V_2| = n$
	$\beta(G)$	the independence number of G
	$\gamma(G)$	the chromatic number of G
	$P(G, \lambda)$	the chromatic polynomial of G
	$\alpha(G)$	the domination number of G
	$L(G)$	the line graph of G
	$T = (V, E)$	T is a tree with vertex set V and edge set E
	$N = (V, E)$	N is a (transport) network with vertex set V and edge set E